ENGINEERING
PSYCHOLOGY

THE CENTURY PSYCHOLOGY SERIES
Kenneth MacCorquodale
Gardner Lindzey
Kenneth E. Clark
    Editors

ENGINEERING PSYCHOLOGY

Current Perspectives in Research

WILLIAM C. HOWELL
Rice University

IRWIN L. GOLDSTEIN
University of Maryland

St. Mary's College Library
Winona, Minnesota

APPLETON-CENTURY-CROFTS
EDUCATIONAL DIVISION
New York MEREDITH CORPORATION

Copyright © 1971 by

MEREDITH CORPORATION

All rights reserved

This book, or parts thereof, must not be used or reproduced in any manner without written permission. For information address the publisher, Appleton-Century-Crofts, Educational Division, Meredith Corporation, 440 Park Avenue South, New York, N. Y. 10016.

771-1

Library of Congress Card Number: 79-138829

PRINTED IN THE UNITED STATES OF AMERICA

390-46456-2

TO PATSY AND ARLENE

Preface

The stereotypic "book of readings" has become the convenience food of the education market. Prepackaged and ready to pop into the curriculum, it contains only the finest, most authentic ingredients, blended to suit practically any taste at a cost that all can afford. No longer need the instructor pore over his journals preparing a recipe of original source materials; and gone for the student are the arduous trips to the library. A text, some lectures, and a book of readings; that is all one needs to make every academic meal complete.

While prepackaged "literature" may be better than no original source material at all, it is a pale substitute for a carefully planned program of outside readings blended to the taste of the particular instructor (and suited to the objectives of the particular course). The instructor should be a *chef,* not merely a franchise operator. Consequently, if a "book of readings" offers nothing more than a collection of original papers—irrespective of how significant those papers are—it has no place in the larder of the true academician. To put it even more bluntly, we can see little justification for "convenience" readings except, perhaps, where original material is not generally accessible. If a book composed in this format is to have value, then, we believe that it must serve some higher purpose. But what? Since it could scarcely substitute for the lecture, the only remaining role is that of textbook.

If a textbook is what Webster says it is, ". . . a book containing a presentation of the principles of a subject . . . ,"[1] there is nothing to prevent an author from building such a book around original source material, provided he "presents the principles"—presumably in some coherent fashion. One can, in fact, point to several recent "readings" books which fit this description rather nicely. Our point, then, is simply this: A book of readings can be useful only if the author adopts in its composition the same attitude that he would assume in writing a text. His objective cannot be merely to package suitable "outside material"; it must be to present principles, to develop points of view, to share original thoughts, or to contribute otherwise to the edification and stimulation of interested students. Operating on this premise, we undertook the present project convinced that it should stand or fall on the basis of what *we,* not the other participants, contributed to it. Their work, after all, has already been before the court of scientific opinion, and what happens here can do little to alter the verdict regarding its worth. This is not to minimize the obvious importance of their contribution; it is simply to point out that we bear sole responsibility for the way in which their work is used *here,* and consequently, for the outcome of the entire project. If we have failed in *our* objectives, no articles on earth could salvage the book.

We first became acutely aware of the need for a new text of some sort in engineering psychology when we were given responsibility for teaching courses in

[1] *Webster's New Collegiate Dictionary,* s.v. "textbook."

the area. We simply could find no existing text which reflected to our satisfaction recent developments in the field, particularly with regard to research. And these trends were the very ones we had intended to emphasize in our course offerings! Our only recourse was to rely heavily on selected readings and hope that our lectures were coherent enough to provide the integrative structure normally afforded by a text. Our experience with this approach was just about what one might expect: (*a*) there were some distinct advantages to featuring original source material, but (*b*) without the help of a text, the integrative structure was exceedingly hard to establish. This explains why we decided to embark on this project at all, and also why we preferred the "readings" format. But what about integration—how was it to be achieved in a book of this sort? We attempted to promote this critical quality in two ways: by careful selection and organization of the readings themselves, and by provision of adequate contextual material at appropriate places. Each of these points deserves a bit of elaboration.

First, the matter of selection and organization. This turned out to be by far the most difficult part of the entire project: there are innumerable ways to lay out the subject matter of engineering psychology, most of them justifiable on one basis or another, and for each area so defined there are a host of illustrative articles from which to choose. But on what basis? Should we feature only works of widely acclaimed significance, or present a more representative sample? Should we emphasize method, content or theory—or weight all three equally? Should we include only recent work, or older studies as well? The list of possible considerations goes on and on. The general organization upon which we finally settled looks first at human performance functions in relative isolation, moves on to the man-machine system as a whole, glimpses the important issues in man's *adjustment* to changing system requirements (the matter of training), and finally, touches upon the question of how ambient conditions affect behavior and system performance. Within this framework, particular articles were chosen mainly on the basis of how well they illustrated some point that we wished to make. Overall, however, we attempted to achieve a somewhat *balanced* rather than an *ideal* sample of the research literature and, in keeping with the title, to put the field in *perspective* against the background of history and of related contemporary fields. Thus we included some of the "classical" studies, some newer ones, and some smacking very strongly of sister disciplines.

The integrative "glue" used to fasten the pieces together is embodied in the introductory comments before each Part, and before Chapters Two to Fifteen. These comments are supposed to do more than merely introduce the illustrative research; they are intended as a coherent *context* within which the various articles are interpreted and the entire structure takes on meaning. The main theme is introduced in the third article of Chapter One, and each subsequent discussion is aimed at developing this theme a step further.

If we have been successful in providing a coherent picture of the field as we see it, then it should be possible to build an initial course—at either the advanced undergraduate or graduate level—around this book. This should be true whether or not the instructor agrees with our particular viewpoint. Naturally, we recognize that many courses devoted wholly or partially to man-machine systems affairs are not called *engineering psychology*, and even those so labeled differ widely in content and emphasis (as indeed they should). Hence, even if our objectives are fully met, it is essential that this text, like any other, be supplemented by selected "outside readings" in order that the course assume the desired emphasis. Suggested "leads" toward further study are offered throughout the book in the context of each topic area.

If we have failed in our attempt to compose a suitable text, then we have failed *completely*, and this "readings" book, like most of its predecessors, will prove of little value to anyone—unless, of course, his taste runs to TV dinners.

As every author undoubtedly must, we come to the end of this work realizing that we can never fully acknowledge all the people and organizations to whom we owe a debt of gratitude. Hopefully each of them will recognize his contribution and know that we are appreciative. We must, however, single out a few for special mention.

First, of course, we must thank the direct contributors—the authors and publishers who permitted us to use their material. Second, we are in many ways indebted to the Human Performance Center of the Ohio State University where the idea for this project was born. Many of the attitudes expressed throughout the following pages derive from the tradition of meaningful, quality research maintained by the Center over the years under the guidance of outstanding people such as Arthur Melton, Paul Fitts, and George Briggs. We are both thankful for having been a part of this stimulating group. Third, we are grateful to Kenneth Clark for his review of the entire manuscript, and to Marjorie Kalins of Appleton-Century-Crofts for her unusually helpful assistance. We also appreciate the critical review of parts of the manuscript by David Schum, Darwin Hunt, and Jack Bartlett.

W.C.H.
I.L.G.

Contents

Preface vii

PART ONE INTRODUCTION TO THE FIELD OF ENGINEERING PSYCHOLOGY 1

Chapter One Three Viewpoints 3
1 Psychology and the Design of Machines   F. V. Taylor   3
2 The Emerging Role of Engineering Psychology   J. M. Christensen   16
3 Engineering Psychology Today: Some Observations from the Ivory Tower   W. C. Howell and I. L. Goldstein   27

PART TWO HUMAN PERFORMANCE IN THE MAN-MACHINE SYSTEM 37

Chapter Two Human Factors in Display Design 42
4 Size, Blur, and Contrast as Variables Affecting the Legibility of Alpha-Numeric Symbols on Radar-Type Displays   W. C. Howell and C. L. Kraft   45
5 Development of Design Criteria for Intelligence Display Formats   W. D. Hitt, H. G. Schutz, C. A. Christner, H. W. Ray, and L. J. Coffey   56
6 Relative Motion of Elements in Instrument Displays   W. G. Matheny, D. J. Dougherty, and J. M. Willis   64
7 Operator Decision Performance Using Probabilistic Displays of Object Location   L. M. Herman, G. N. Ornstein, and H. P. Bahrick   69

Chapter Three Human Processing of Information— Receiving and Storing Capabilities 75
8 Decision Processes in Perception   J. A. Swets, W. P. Tanner, Jr., and T. G. Birdsall   78
9 The Magical Number Seven, Plus or Minus Two: Some Limits on Our Capacity for Processing Information   G. A. Miller   87
10 A Model for Some Kinds of Visual Memory Tasks   G. Sperling   102
11 Short-Term Retention of Individual Verbal Items   L. R. Peterson and M. J. Peterson   116
12 Flow of Information Within the Organism   D. E. Broadbent   122

Chapter Four Human Processing of Information— Transmission of Discrete and Continuous Signals 129
13 What is Information Measurement?   G. A. Miller   132

14	On the Rate of Gain of Information	W. E. Hick	143
15	High-Speed Scanning in Human Memory	S. Sternberg	158
16	Searching for Ten Targets Simultaneously U. Neisser, R. Novick, and R. Lazar		163
17	Human Tracking Behavior	J. A. Adams	168
18	The Organization of Skilled Response M. Noble and D. Trumbo		194

Chapter Five Human Processing of Information—Intellectual or Cognitive Processes — 214

19	Information Reduction in the Analysis of Sequential Tasks M. I. Posner		217
20	Man as an Intuitive Statistician C. R. Peterson and L. R. Beach		230
21	The Prediction of Decisions Among Bets	W. Edwards	253
22	Cognitive Aspects of Information Processing	P. M. Fitts	267

Chapter Six Human Factors in Control Design — 275

23	Tactual Coding of Cylindrical Knobs	J. V. Bradley	277
24	An Analysis of Stimulus Variables Influencing the Proprioceptive Control of Movements H. P. Bahrick		288
25	Tests of Ten Control-Display Linkages A. Chapanis and D. A. Mankin		292

PART THREE MULTIMAN-MACHINE SYSTEM PERFORMANCE — 299

Chapter Seven Description and Logical Analysis — 303

26	A New Look at System Research and Analysis	J. S. Kidd	306
27	Dynamic Decision Theory and Probabilistic Information Processing W. Edwards		315
28	Command-and-Control and Management Decision Making R. J. Rhine		332

Chapter Eight Empirical Research — 342

29	Some concepts and Methods for the Conduct of Systems Research in a Laboratory Setting P. M. Fitts, L. Schipper, J. S. Kidd, M. Shelly, and C. Kraft		343
30	Research on a Simulated Bayesian Information-Processing System D. A. Schum, I. L. Goldstein, and J. F. Southard		355
31	Do Large Shared Displays Facilitate Group Effort? S. L. Smith and B. C. Duggar		372

PART FOUR TRAINING IN THE MAN-MACHINE SYSTEM — 381

Chapter Nine Acquisition and Transfer — 385

32	Are We Using What We Know About Training?—Learning Theory and Training W. McGehee		390
33	Military Training and Principles of Learning	R. M. Gagné	397
34	On Transfer and the Abilities of Man	G. A. Ferguson	409

Chapter Ten Evaluation of Training — 418

35	Criterion Measurement and Personnel Judgments R. M. Guion		422

Contents

36	Evaluating the Results of Supervisory Training T. R. Lindbom and W. Osterberg	428
37	Progressive Levels in the Evaluation of Training Programs A. C. MacKinney	432

Chapter Eleven   Training Apparatus — 438

38	Training Devices and Simulators: Some Research Issues R. M. Gagné	441
39	Programed Instruction—Past, Present, Future J. S. Abma	459
40	Computer-Assisted Instruction R. C. Atkinson and H. A. Wilson	467

Chapter Twelve   Directions in Research — 476

41	Job Analysis in the United States Air Force J. E. Morsh	478
42	Development of a Behavior Taxonomy for Describing Human Tasks: A Correlational-Experimental Approach E. A. Fleishman	486
43	Individuality in the Learning Process: Some Issues and Implications G. A. Eckstrand	495
44	Stimulus and Response Fidelity in Team Training G. E. Briggs and W. A. Johnston	506

PART FIVE   ENVIRONMENTAL AND ORGANISMIC CONDITIONS IN MAN-MACHINE SYSTEM PERFORMANCE — 511

Chapter Thirteen   Vigilance — 513

45	The Breakdown of Vigilance During Prolonged Visual Search N. H. Mackworth	516
46	Human Vigilance J. G. Holland	530
47	Effects of Noise and of Signal Rate Upon Vigilance Analysed by Means of Decision Theory D. E. Broadbent and M. Gregory	542
48	Complex Monitoring and Its Relation to the Classical Problem of Vigilance W. C. Howell, W. A. Johnston, and I. L. Goldstein	551

Chapter Fourteen   Other Extreme Input Effects — 565

49	Effect of Sensory Deprivation on Some Perceptual and Motor Skills J. A. Vernon, T. E. McGill, W. L. Gulick, and D. K. Candland	568
50	Adjusting to Overloads of Information J. G. Miller	574
51	On the Concept of Psychological Stress M. H. Appley and R. Trumbull	584

Chapter Fifteen   Environment Constraints — 596

52	An Example of "Engineering Psychology": The Aircraft Noise Problem K. D. Kryter	600
53	Effects of Frequency of Vibration on Human Performance C. S. Harris and R. W. Shoenberger	606
54	Initial Stimulating Effect of Warmth Upon Perceptual Efficiency E. C. Poulton and D. McK. Kerslake	614

| 55 | Lighting for the Forgotten Man | S. K. Guth and A. A. Eastman | 620 |

General References 627

Author Index 633

Subject Index 643

PART ONE

INTRODUCTION TO THE FIELD OF ENGINEERING PSYCHOLOGY

This is a book about one of psychology's newest and most paradoxical specialties. Engineering psychology, as the name implies, differs from traditional areas in that it is concerned with machines as well as people. Yet it did not really have its origin in engineering, or even, as Grether (1968) points out, in industrial or applied psychology; it grew out of the *experimental* branch of psychology. By all objective standards, it is thriving; yet there is considerable debate as to exactly what it is—and whether, in fact, it belongs in psychology at all! It was created for practical, not intellectual reasons; yet it has always had scientific objectives, and these seem to have gained rather than lost strength over the years. To say the least, then, it is in many ways a highly unorthodox discipline.

A book of this sort customarily begins with a definition of the field and, perhaps, a few paragraphs of historical background. Owing to its multifaceted character, however, engineering psychology has been defined in a variety of ways; we should do grave injustice to certain points of view if we were to present but a single description—our own or anyone else's. Therefore, we have elected to introduce the field with three descriptive commentaries, written by different people, at different times, for different reasons. In combination, they furnish the reader an adequate historical perspective as well.

The article by Taylor, which is now considered a classic, appeared about the time an infant engineering psychology was seeking formal recognition within its parent discipline. It was thus one of the earliest "definitive" statements. Christensen, on the other hand, was addressing the question of how a somewhat more mature field should adjust to the rapid technological changes apparent at the start of the 1960s. What must it do to keep pace with the times?

In the final introductory paper we present our own views, extrapolating what we perceive as trends into the next decade (the 1970s). This not altogether unbiased assessment may reflect events as we think they *should* be

References will be found in the General References at the end of the book.

as much as it does events as they actually *are*. It would be difficult, moreover, to guess how widely these opinions are shared by others associated with the field.

The remainder of the book is really an elaboration upon the theme developed in this introductory chapter. We have assembled a collection of papers which we believe illustrate the trends alluded to here, and we have tried to organize them into a coherent representation of the field today. The textual material with which we have interlaced these papers is intended to serve both an interpretive and an amplification function. In a sense, the entire book is an effort to define engineering psychology.

Chapter One

Three Viewpoints

1

Psychology and the Design of Machines

Franklin V. Taylor

Psychologists have been helping engineers design machines for more than fifteen years. It all began during World War II with the rapid development of radars, sonars, aircraft control systems, and other similar devices. Previous to this time, the only role played by psychologists relative to military mechanisms was that of doing research and giving advice on the selection and training of the operators. However, very early in the war, it became apparent that these Procrustean attempts to fit the man to the machine were not enough. Regardless of how much he could be stretched by training or pared down through selection, there were still many military equipments which the man just could not be moulded to fit. They required of him too many hands, too many feet, or in the case of some of the more complex devices, too many heads.

Sometimes they called for the operator to see targets which were close to invisible, or to understand speech in the presence of deafening noise, to track simultaneously in three coordinates with the two hands, to solve in analogue form complex differential equations, or to consider large amounts of information and to reach life-and-death decisions in split seconds and with no hope of another try. Of course the man often failed in one or another of these tasks. As a result, bombs and bullets often missed their mark, planes crashed, friendly ships were fired upon and sunk. Whales were depth-charged.

From *American Psychologist*, **12**, 1957, 249-258. Copyright 1957 by the American Psychological Association, and reproduced by permission.

This paper is substantially the same as that delivered as an address on April 12, 1957 at the 28th Annual Meeting of the Eastern Psychological Association.

Because of these "human errors," as they were called, psychologists were asked to help the engineers produce machines which required less of the man and which, at the same time, exploited his special abilities. The story of what happened is sufficiently well known not to require any lengthy retelling here. In brief, the psychologists went to work, and with the help of anatomists, physiologists, and, of course, engineers they started a new inter-discipline aimed at better machine design and called variously human engineering, biomechanics, psychotechnology, or engineering psychology. The new field has developed rapidly in the seventeen or eighteen years of its existence, and it has now attained sufficient respectability to be accorded divisional status by the American Psychological Association. At the last meeting of the Council of Representatives, authorization was given for the founding of The Society of Engineering Psychologists as Division 21 of the APA.

It seems fitting, now that engineering psychology has been recognized as a viable entity, that we examine this new field to find out just what it is that psychology is doing for the design of machines. It is probably even more necessary that we also inquire into what the participation in the design of machines is doing for, or to, psychology. Many young people are being lured into human engineering by the abundant opportunities provided for advancement and the tantalizing salaries offered by commercial organizations. It has been suggested by an unassailable authority that a major breakthrough in the field of psychology in recent years has been the psychologists' discovery of money. It may be remarked that it was undoubtedly an engineering psychologist who first got wind of the find.

In all seriousness, however, psychologists who might otherwise conduct basic research may be attracted into this new applied area, and it is therefore important to know what it represents professionally and scientifically in order to evaluate its threat, or its promise. To decide what actions to take relative to encouraging the further development of the field, answers are needed to questions such as the following: To what extent is engineering psychology engineering, to what extent is it psychology, and to what extent is it neither? Is it a fruitful scientific area? Is it, indeed, a scientific area at all?

In the attempt to provide answers to these questions, let us look at psychologists caught in the act, so to speak, of doing human engineering. However, before we can meaningfully analyze the behavior of engineering psychologists, the concept of the man-machine system must be described. Human engineers have for some time now looked upon the man and the machine which he operated as interacting parts of one overall system. In Figure 1 is shown a paradigm of the concept. This may be viewed as a radar device, a pilot-aircraft control system, a submachine diving control station, the captain's station on the bridge of his ship, or, in fact, any man-machine system at all.

In essence, it represents the human operator as an organic data transmission and processing link inserted between the mechanical or electronic displays and controls of a machine. An input of some type is transformed by the mechanisms into a signal which appears on a display. Perhaps it is shown as a pointer reading, a pattern of lights, or a pip on a cathode ray tube. However it appears, the presented information is read by the operator, processed mentally, and transformed into control responses. Switches are thrown, buttons are pushed, or force

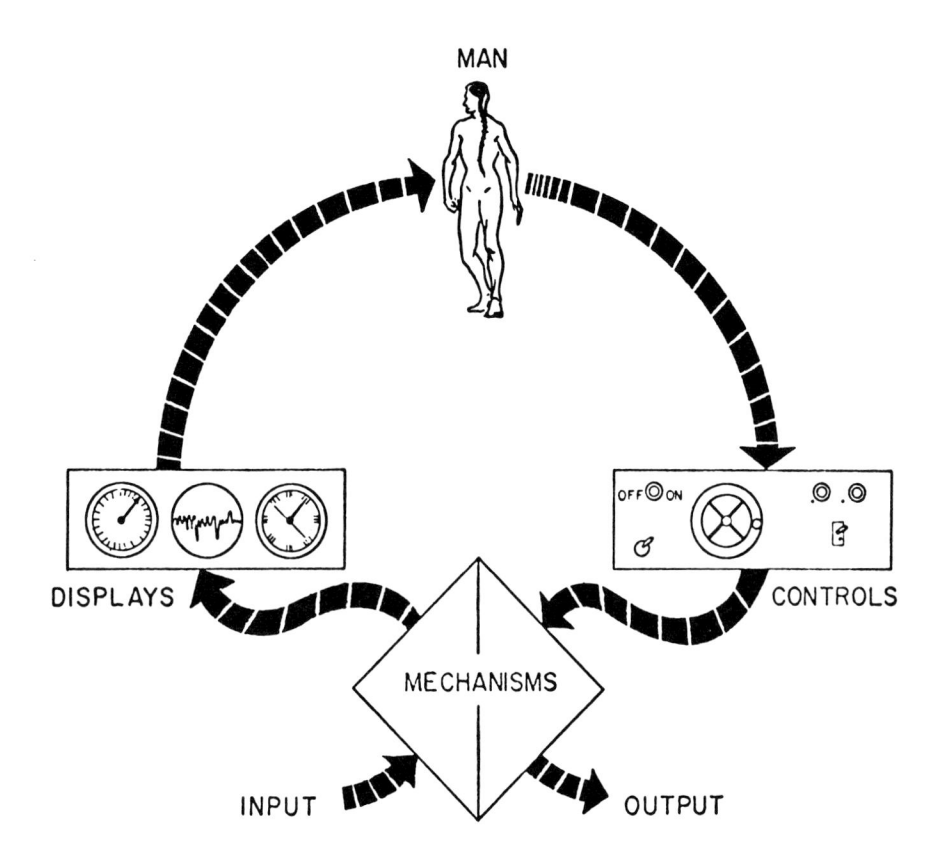

Figure 1. The man-machine system.

is applied to a joy stick. The control signal, after transformation by the mechanisms, becomes the system output, and in some devices it acts upon the displays as well. These latter are called "closed-loop" systems in contrast to "open-loop" systems wherein the displays do not reflect the human's response.

When the man and the machine are considered in this fashion, it immediately becomes obvious that, in order to design properly the mechanical components, the characteristics of the man and his role in the system must be taken into full account. Human engineering seeks to do this and to provide as much assistance to the system designer as possible. Specifically, the psychologist tries to help his engineering colleague in three different ways. First of all, he studies the psychology of the human as a system component. Second, he assists the engineer in experimentally evaluating prototype man-machine systems. Finally, he teams up with engineers to participate actively in the design of machines. Each of these human engineering functions will be described in turn, beginning with the last and the least scientific activity.

HUMAN ENGINEERING TECHNOLOGY

The academic psychologist often forgets, or perhaps never knew, that human engineering is not only a science, it is also a technology; it not only tries to find out things about the interaction of men and machines, it builds the latter. And, surprisingly enough, it is not just the engineers who do the building. There are psychologists also, renegades to be sure but psychologists nevertheless, who are taking an active hand in the design of systems. It is true that with some their apostasy is venial, having progressed only to the stage of writing human engineering handbooks; but with others the defection is more serious, it having developed to the stage where they can spend anything up to full time in systems planning and design with only a twinge or two of longing for the serenity of the research laboratory and the comfort of statistics.

The aim of the human engineering technologist is to apply the knowledge of human behavior, which he and others have gained, to the structuring of machines. He seeks to translate scientific findings into electronic circuits and "black boxes" which in specific situations will compensate for the human's limitations or complement his abilities. Specifically, the practicing engineering psychologist works on an engineering team and participates in the design of man-machine systems. Using procedural analysis techniques, drawing upon his psychological knowledge and attitudes, and employing his common sense and creative ability, the human engineer proceeds to contribute to system development at three levels of complexity.

At the simplest, he designs individual displays, controls, or display-control relationships. At a somewhat more complex level, the human engineering technologist contributes to the design of consoles and instrument panels. At the highest level of complexity, he assists in structuring large systems composed of many mechanical elements and frequently several human beings. In this capacity he helps to determine what information must flow through the system, how it must be processed, how many men are required, what tasks they will perform, and what type of information each one will need. In short, the engineering psychologist helps at this level to determine the configuration of the system.

Human engineering technology is much more extensively practiced by psychologists than is generally recognized by those who are not closely identified with the field. The specific nature of each accomplishment and the difficulty of assigning individual credit for team effort conspire with security and proprietary considerations to keep the lay and psychological public in almost complete ignorance of the technological products of human engineering. However, literally hundreds of devices and systems have been affected to a greater or less extent during the last ten years by the efforts of engineering psychologists. Every major type of military equipment has received some attention, as have also certain nonmilitary products such as aircraft instruments and cabins, flight control towers, artificial limbs, semiautomatic post office sorting equipment, telephone sets, theodolites, experimental equipment for the earth satellite program, control panels for an atomic reactor, and numerous industrial machines.

Although there are no statistics available as to precisely how much time psychologists devote to technology, an informal estimate based on my own experience would suggest that approximately one-third of the engineering psychology effort in government and industry is devoted to the practice of equipment design.

Now, how does this practical activity tie in with psychology? Certainly it cannot be denied that in one sense of the term it is an area of applied psychology. Facts about human behavior are being utilized in the design of machines. Yet in another sense it cannot be regarded as psychology at all, for certainly the design of machines is engineering, regardless of who does it or of the extent of interest on the part of the designer in human behavior. To deny this in favor of the view that system design is applied psychology because of its human reference, converts all engineers to psychologists the moment they take into account the behavior of the human for whom they are designing the machine. Certainly we do not customarily consider all occupations oriented toward human behavior to be psychology. If we did, the APA would undoubtedly be the largest professional society in the world, speaking for actors, school teachers, policemen, politicians, and members of the clergy, in addition to its present membership. So far as I know, such an expansion is not yet contemplated.

A second difficulty which stands in the way of incorporating the design aspects of human engineering under psychology arises from the nature of the goal of systems design. Whereas the primary aim of the practitioners of the more conventional applied psychologies is to control and influence people, the human engineering designer seeks to produce more effective machines. While psychology has, in the past, been applied to improving human performance by selecting, training, and motivating normal men, by curing mentally ill men, and by persuading both to buy toothpaste and television sets, human engineering aims first at building better systems and only secondarily at improving the lot of the operator. Thus, whereas conventional psychology, both basic and applied, is anthropocentric, human engineering is mechanocentric.

Because of these peculiarities of the new field, one is forced to the conclusion that it is of questionable profit to attempt to maintain that human engineering technology is a branch of psychology. Or if one still wishes to do so, it must be admitted that it is psychology most diluted and highly contaminated with physical science and engineering considerations. Although this, of course, does not reflect in any way upon the importance to society of human engineering, it does raise questions concerning the training and professional affiliation of those psychologists who decide to enter this exciting new trans-disciplinary technology.

MAN-MACHINE SYSTEMS EVALUATION

The second way in which the engineering psychologist assists in the design of machines is by taking part in systems evaluations. Like human engineering technology, evaluation studies require a sizeable effort yet receive scarcely any publicity. Evaluations have been performed on headphones, range finders, gunsights, fire control and missile control systems, radar sets, information plotting systems, combat information centers, aircraft control towers, and numerous assorted display and control components. In some instances, the experiments have been carried out in the laboratory with the system inputs being simulated. In other cases, the tests are conducted in the field. But in both situations, the attendant complexities and difficulties of statistical control make this necessary variety of research as trying as any in which psychologists are likely to participate.

The reason that psychologists were called upon in the first place to assist in these evaluations was that they possessed methods for dealing with human variability. In contrast, the engineers generally had worked only with time-stationary components and, therefore, found themselves at somewhat of a loss when they were called upon to assess the performance of devices which were being operated by men.

It is easy to see that psychologists have something definite to offer in regard to this aspect of systems design, and it is not surprising, therefore, that their services are often sought out and accepted. However, the question which we wish to pose, albeit a bit bluntly, is what are the psychologists getting out of it in turn, besides a living? Certainly they are making a contribution to the engineering of better systems, and the consciousness of this may be all that is required for the satisfaction of the individual. Yet, one might wish to know if there were any other returns —to science in general or to psychology in particular.

Admittedly, this is a difficult question which will be more easily answered fifty years from now than at present after only a few years of this activity. But one thing can be said right now about experiments performed on man-machine systems: if one's main object is to learn about human behavior, the use of a complex systems experiment is an uncommonly unwieldy way to go about it.

This may be illustrated with a preposterous, hypothetical systems evaluation. Suppose we wished to compare the performance of a boy on a bicycle with that of a boy hopping on a pogo stick. The main independent variable would be the nature of the boy-machine system; the dependent variable could be, for example, the length of time to travel a quarter mile.

Here we would have a perfectly proper systems test. At least it could be made proper through adequate attention to the training of the boys and to statistical precautions to overcome human variability. Also, the experiment would undoubtedly yield clear and unambiguous results. Assuming that the maintenance of the two conveyances was adequate to the point where neither broke down during the race, the boy-bicycle system would very likely prove to be superior to the boy-pogo stick complex. It is granted that, before positive recommendations could be made concerning the adoption of one vehicle system over the other as a general means of boyish travel, other system criteria such as initial cost, cost of replacing parts, safety, maneuverability, ease of stowage, and consumer acceptance would have to be considered. But this is always true and constitutes no valid criticism of the evaluation. The test itself could have been highly successful; and, if so, we would have learned from it something about systems.

But what have we learned about boys? After all, we are psychologists and are interested in the laws of human behavior. What anthroponomic relationships are revealed by the study? It would seem clear that in our contrived example we have learned next to nothing of interest to psychology. Apart from finding out that boys can learn to operate both bicycles and pogo sticks, the test has disclosed nothing about the characteristics of the human operator. There are several reasons for this.

First of all, the dependent variable in the experiment is a measure of system performance, not human performance. It is perfectly apparent that the fact that one system is better than the other does not mean that the boy in the superior system is doing better in any real sense than the boy who loses the race. Quite the contrary might be true. The boy on the losing pogo stick may actually be doing a better job of pogo stick jumping than the bicycle rider is doing of his bicycle riding. As long as one is dealing with a system performance variable, the behavior of the human in the system can only be inferred, and often the inference is hazardous indeed.

This would not be a serious matter if it were always possible to find some other dependent variable which did directly reflect human behavior. But in many studies no meaningful, uncontaminated, human performance variable exists. Whenever the human responds through some variety of control, his response is inextricably tied up with the physical properties of the control itself. Thus, in our example, it is not possible to measure hopping independently of the physical characteristics of the pogo stick hopped on, nor pedaling in the absence of pedals. Try as one will, one cannot pedal a pogo stick nor effectively hop through the air on a bike. Since it is impossible to separate the manner in which the human applies force from the characteristics of the thing to which the force is applied, there is no way of getting a pure measure of man's behavior in many systems studies and, as a matter of fact, in many laboratory experiments not construed as dealing with systems.

The basic indeterminability of human response has, of course, been recognized for a long time. Philosophers and psychologists have pointed out repeatedly that behavior is an interaction among different kinds of things and that it is arbitrary and misleading to say that one of the things (usually the animate one) is doing the behaving while the others make up the environment. Thus, we say that we are studying the behavior of a man walking (1) and not the behavior of the ground under his feet. Yet, of course, the walking behavior would be impossible without the ground, just as it would without the man. Both, and much else besides, are necessary for the walking to occur, and any measurement of the behavior reflects the characteristics of all of the interacting objects and forces.

Now all this is of very little consequence to psychology so long as the parts of the human's environment which interact with his motor output remain unchanged during an experiment. If, in the walking study, the ground underfoot was always of the same general firmness, levelness, and texture, its contribution to the behavior could be neglected. Similarly, if in a tracking investigation the S always uses the same joy stick working through the same system dynamics, it matters not a whit that the performance measured is that of the man-joy stick system and not of the man alone.

But let the properties of the objects to which the man applies force be varied, unknowingly or deliberately, and it becomes vital to recognize the contaminated nature of the performance measure. Change the ground from hard to muddy, or the tracking control from joy stick to handwheel, and the altered performance resulting is a composite of direct effects and man-environment interactions impossible to untangle without further research. When this is not recognized and the behavorial shift is attributed exclusively to the man, one blunders scientifically.

Conventional psychology has generally avoided this problem of confounded dependent variables by working much more frequently with sensory and state-of-the-organism parameters than with human output variables. Engineering psychology, however, has deliberately undertaken to work with system variables, with the result that the performance measures are almost never pure human response scores. Although this repeated experience has alerted some of the experimenting human engineers to the inferential pitfalls of blindly equating system performance with human performance, there are still those who fall into the trap. One still occasionally hears said, for example, that human tracking performance is improved or degraded by changes in the nature of the control or by alterations in the control dynamics. Such statements may be true but they are certainly not justified, for tracking performance, as measured, is system behavior which can change radically as a result of altered dynamics without reflecting any comparable change on the part of the man.

But it is not only the dependent variable which gives the human engineer trouble in making psychological hay out of systems studies. The independent variables are often even more troublesome because they frequently embody many parameters. Consider the independent variable in our example. When the pogo stick is substituted for the bicycle, actually four sets of dimensions are involved. First of all, the controls are shifted: pedals and handlebars are traded for a spring-mounted step and a pole. Secondly, the system dynamics are changed as they relate to the transformation of human energy into motion along the ground. Third, the sensory inputs to the boys are modified (the displays are altered, so to speak): with the pogo stick the visual world bobs up and down, with the bicycle it glides by; with the pogo stick the boy's weight is all upon his feet, with the bicycle he feels pressure from his seat. Finally, the operator's task is completely transformed: the psychomotor performance of hopping up and down along a Z axis while simultaneously maneuvering along X and Y coordinates, through shifting balance around these axes, is an entirely different stunt from controlling in X through balancing around X, steering around Z, and pedalling around Y—the latter is the task of the bicycle rider.

Now even if one had a measure of human performance as a dependent variable, which one does not, it is clear that next to nothing of psychological interest could be learned by manipulating this multiparameter, independent, system variable.

Since the displays, the controls, the dynamics, and the psychomotor task are all varied simultaneously, the logic of experiment is so completely violated that it is impossible to partial out the individual effects of any of the components upon the performance of the system or of the man. All that one can know in such a systems test is the combined effects of the dimensionally massive, independent variable—in other words, that one system is better than another. This is of value in deciding between systems, but it may be suggested that it is impoverished psychological research.

Of course, our example is a *reductio ad absurdum* intentionally. In many systems experiments, the independent variables are less complex than in the illustration. Yet it is almost always true that system variables comprise more diverse dimensions than do the variables customarily chosen for psychological analysis.

But it is not just the complexity and dimensional confounding typical of system variables which make it hard to derive psychologically relevant facts from man-machine systems tests. A further difficulty stems from the shift in the operator's task (already alluded to) which so often results from the manipulation of the physical parameters of the system. We have pointed out that the psychomotor processes involved in riding a bicycle are entirely different from those underlying hopping about on a pogo stick. Similar radical differences in the operator's task are often to be observed when real systems are compared. One system may require the operator to act analogously to a complex differential equation-solver, while another may require of him nothing more than proportional responding. One radar warning system may require the operator to calculate the threat of each target and to indicate the most threatening; another may compute the threat automatically and place a marker around the target to be signaled.

Clearly, the operator's tasks differ so much from one of these systems to the next that it would never have occurred to a psychologist to compare them. The differences are so gross, so obvious, that they obscure the need for relating the tasks, for placing them on some kind of a useful continuum and for scaling the distances in between. Yet these behaviors must be compared in some way and the knowledge made available to engineers if the human is to be employed effectively in man-machine systems. Changing the operator's task from one of these complex psychomotor processes to another may produce startling improvements in system performance, and the principles determining the substitution of the task must be discovered if systems design is to progress.

But can we consider this to be good psychology? Do those who regard themselves as scientific psychologists wish to spend their time comparing and analyzing vastly dissimilar psychomotor tasks? Is it sophisticated psychology to compare the speed of running with that of walking, or the ability to add in one's head with that of adding on a machine calculator, or the skill of playing a piano with that of operating a phonograph? I think that most psychologists would agree that, although these kinds of comparisons might be relevant to systems design, they are not quite the stuff out of which conventional psychology is made. Likewise, many would no doubt agree that the contaminated variables of systems research are to be avoided whenever possible in psychological investigations. Some might even go so far as to put the two together and suggest that the time spent by psychologists in evaluating systems is a dead loss to the science of psychology. Of such a view we will have more to say shortly.

ENGINEERING PROPERTIES OF THE MAN

The third and final way in which psychologists help in the design of machines is through studying, by conventional means, the behavior of the man as a machine operator. Although, as has just been remarked, psychologists have not yet quite brought themselves to making systematically the gross comparisons required by the system designers, they have undertaken to study selected aspects of the behavior of the man as a system component. The intent here is to provide the engineers or the technologically oriented psychologists with information concerning certain of the characteristics of the man in order that the properties of the machine may be made to harmonize with them.

In contrast to the other two types of assistance furnished, this is precisely what psychologists would be expected to do. Furthermore, with the exception of some experiments on displays and controls which are actually masked systems tests employing confounded variables, the psychology is both satisfying and sanitary. The preponderance of the work is unambiguously directed toward discovering laws of human behavior, and it is as scientific as ever one could wish.

However, although the work in this domain of engineering psychology is every bit as respectable as that of the parent subject, it is far more limited. Whereas psychology in the generic sense embraces all manner of human action, engineering psychology deals with a much more restricted variety of behavior. This class of responses may be characterized in a number of different ways:

1. First off, as was pointed out at the very beginning, the human in a man-machine system can be considered as an information transmission and processing link between the displays and the controls of the machine. When so viewed, his behavior consists of reading off information, transforming it mentally, and emitting it as action on the controls. Thus, the performance may be described as of the type in which the operator's responses image in some way the pattern or sequence of certain of the input events. For example, the S signals when a tone comes on and withholds his response when he hears nothing, or he presses one key when he sees a red light and a different key when he sees a green one, he perceives the range and bearing of a radar target and identifies its location verbally, he moves a cursor to follow the motion of a target image. In all these cases, the essential interest in the behavior focuses upon the correlation in space and time between events in a restricted and predefined stimulus "space" and corresponding events in a preselected response "space."

2. Another way to characterize the behaviors studied in engineering psychology is to indicate that they are voluntary and task-directed or purposive. The operator of a man-machine system is always consciously trying to perform some task. Perhaps it is to follow on a keyboard the successive spatial positions of a signal light, perhaps to see a visual target imbedded in "noise" and to signal its position, possibly it is to watch a bank of displays in order to determine malfunction and to take action where necessary. In all cases, the operator is voluntarily trying to accomplish something specific; he is not just free associating, or living.

3. A third characteristic of the human operator's behavior emerges as a corollary of voluntary control. The class of human responses of interest to the engineering

psychologist involves chiefly the striate muscles. Because it is through the action of this type of effector that men speak and apply force to levers and handwheels, it is these muscles which play the dominant role in the human's control of machines.

4. Finally, practical considerations dictate that vision and audition be the sense modalities most often supplying the input to the human transmission channel. Because of the nature and location of the eyes and ears and because of their high informational capacity, they are ideal noncontact transducers for signal energies emitted by the mechanical or electronic displays of machines.

These four characteristics define the human reactions investigated in engineering psychology as falling within the narrow confines of the classical category of "sensorimotor" or "psychomotor" behavior. But actually, the subject matter is even more limited than this. As was mentioned earlier, the main task of the psychologist-human engineer is to provide system-relevant facts concerning human behavior, and it must be emphasized that not all facts, even though they concern psychomotor performance, can meet this criterion. The hundreds of studies conducted with the pursuit-rotor, for example, have generated very few facts having the remotest relation to the design of systems.

Because it is recognized that not all good sensorimotor psychology is necessarily good engineering psychology, steps are being taken to get at the kinds of behavioral information which the engineers really need. In order to do this, the concepts and models of orthodox psychology are beginning to be replaced by physical and mathematical constructs and engineering models. We have already encountered the notion of the man as an information channel. Systems psychologists also view him as a multipurpose computer and as a feedback control system. The virtue of these engineering models is that they furnish ready-made a mathematics which has already proved itself of value when applied to the inanimate portions of the man-machine system and which may turn out to be useful for the human element as well. In addition, they provide the behavioral scientist with a new set of system-inspired hypothetical constructs and concepts which may redirect his research and stimulate entirely novel lines of inquiry.

Whereas orthodox psychomotor psychology still speaks in a construct language consisting of terms like stimulus, response, sensation, perception, attention, anticipation, and expectancy, the new "hardware" school is rapidly developing a concept argot which, although quite unintelligible to outsiders, is providing considerable inspiration to the initiates. Human behavior for this psychological avantgarde is a matter of inputs, outputs, storage, coding, transfer functions, and bandpass.

And this is far more than a matter of language. The research itself is changing. Questions about human behavior are now being asked experimentally which were literally inconceivable a few years ago. Yet they are the very questions to which engineers desire answers. How stationary and linear is the man? What frequencies can he pass and how many bits per second can he transmit under a variety of different conditions? How does the human's gain change with different system dynamics? How well can he perform as a single integrator, or double integrator, or triple integrator? How effectively can he act as the surrogate for different computer functions? These are some of the experimental questions which engineering psychologists are beginning to ask and which, no doubt, will be asked with increasing frequency as the new field develops.

It is probably not too much to expect that one day soon we will have a completely revised textbook of human engineering, perhaps entitled *The Engineering Properties of the Man,* which will present to engineers in a form which is useful to them the system-relevant facts of psychology as then known. Instead of conventional chapter headings like "Seeing," "Hearing," "Speaking," "Moving," and "Working," it might contain such rubrics as "Mechanical Properties of the Man," "Transduction," "Informational Capacity and Bandwidth," "Linear Properties of the Man" (including analogue addition, integration, differentiation, and multiplication by constants), and "Nonlinear Properties of the Man" (including, it must be confessed, most everything else). Such a treatise, when it is written, will certainly be welcomed by the system designer, and he will waste no time in putting the information to use. Its reception by the orthodox psychologist, however, is somewhat more unpredictable, and it is conceivable that he will consider it, and the research programs which fed into it, more in the nature of an esoteric horror than a blessing. Certainly it must be expected that this "brave new world" of mechanomorphic psychology will, at least at first, be as limited in its appeal as it is in its coverage.

ENGINEERING PSYCHOLOGY AND SCIENCE

We have now had a look at the three ways in which psychologists contribute to the design of machines. We have seen that they act not only as scientists, seeking knowledge for others to use, but also as technologists, actively participating in the planning and design of man-operated mechanisms. In playing the latter role, they have clearly stepped out of their field and entered that of engineering. Even as scientists they seem to have moved away from psychology as classically conceived, for on the one hand they have expanded their subject matter to include the behavior of systems, while on the other they have restricted their interest in human performance to a narrow class of system-relevant psychomotor behaviors.

Must we conclude, therefore, that human engineering, in serving the system designer, will only draw from psychology and not contribute to it? Such might seem to follow from what has gone before, but such a conclusion is almost diametrically opposed to the one which will now be offered. I should like to suggest that the involvement of psychologists in the design of man-machine systems is one of the most important events that has occurred in psychology. I believe that, when psychologists started tinkering with machines and seriously trying to learn how they could better be designed, an opportunity was provided for something to happen of utmost significance to science. And I think it has already begun to happen. This is the destruction of the barrier which has hitherto existed between the psychological sciences and the physical sciences.

Psychologists have conventionally thought and talked in a construct language which is different from that of physics. Traditionally the concepts of psychology tend to be relatively imprecise. At first, this indefiniteness was regarded as almost a necessity; for while physics dealt with physical things, psychology dealt with the mind, and of course the mind was nonobjective. Then later, the mind was abolished, and psychology became anthroponomy, the science of human behavior, and there was less excuse for metaphor. Yet metaphor is with us still, although officially outlawed and in disgrace.

But even when similes are avoided and the concepts are given precise operational definitions, the construct language of psychology is very different from that employed in the physical sciences. First of all, the vocabularies are as dissimilar as are those of English and German. Secondly and more important, the constructs themselves often differ in the nature of their generality, elegance, and fruitfulness, with those of physics far in the lead. Although psychologists have become more scientific in their instrumental procedures, using better and better research tools and employing statistics of ever-increasing power, they are still working with pretty much the same old types of syntactically impoverished concepts. Today we conceptualize the man as doing the kinds of things which he and other creatures with minds have always done: like perceiving, thinking, learning, forgetting, living, and dying. This has tended to result in a perseverative replowing of the same ground, a redoing of the same experiments.

Since psychologists have not conceived of the living organism as an analogue device capable of imitating a wide variety of mathematically describable physical operations, no construct terms have been added to the vocabulary of psychology which overlap directly with those of physics or engineering. Because of this, there has never been any real possibility of describing the behavior of both the man and of the physical objects and events in his environment in the same terms. The language of psychology has had to be used to describe the behavior of the man; the language of physics, the environment. Of course, never before was there a need to develop a scientific notation in which one could express with equal facility the operation of minds and the working of mechanisms.

But that need has now arisen. The advent of human engineering—when psychologists for the first time began to look carefully at mechanical and electronic processes and engineers started to consider seriously the characteristics of human behavior—brought the problem into sudden, clear focus. *One just could not effectively design complexes embodying both men and machines so long as the two components were conceptualized as being entirely different and behaviorally unrelated.* Universal concepts applicable equally to humans or mechanisms were needed. A meta-language of action became a necessity!

The emergence of the systems viewpoint was essential to this important, although simple, intellectual discovery. It made two things obvious for the first time. It drew attention to the fact that in many circumstances the behavior of the man was inseparably confounded with that of the mechanical portions of his environment. This meant that psychologists often could not study human behavior apart from that of the physical and inanimate world—that all along they had been studying the behavior of man-environment systems and not that of the men alone. The inseparability of the behavior of living organisms from that of the physical environment with which they are in dynamic interaction certainly argues against maintaining separate sciences and construct languages: one for the environment, the other for that which is environed.

But the concept of the man-machine system does more than this. Not only does it emphasize the dynamic inextricability of the man and the machine with which he works, it suggests that human and mechanical processes are to some extent interchangeable, although not necessarily equally precise. Thus, the system designer has the choice of having required computations performed by a mechanical computer or by the man—the process is the same in either case, although the accuracy may be vastly different. Again, this recognition that human behavior and mechanical or electronic processes can be surrogates for each other provides an excellent reason for seeking to conceptualize men and machines in terms of the same models.

Engineering psychology has begun to do this as we have already seen. It is beginning to adopt engineering techniques, to ask experimentally how well men can differentiate or integrate or amplify, how their gains change or their frequency response characteristics shift. It is starting to apply to human behavior the trans-science concepts and methods of information theory and feedback servo analysis. It has begun to use cybernetics, not just talk about it.

In short, in starting to contribute to the design of machines, psychologists have begun theoretically and pragmatically to pull together the psychological and physical sciences. Just how far they can be moved toward one another at the concept level has yet to be seen. Certainly today there are no physical or engineering models which are sufficiently complex to be used profitably with any but the most primitive of human behaviors. But, then, there are no models of any type, hardware or software, which are satisfactory.

One can only look at what has already been accomplished, apply his own hunches and prejudices, take a deep breath, and guess. My guess is that psychology, biology, and physics will some day all employ the same physicomathematical meta-language when describing the behavior of those particular system components which fall within their purview. Furthermore, should this ever come about, it will have resulted, at least in part, from the efforts of psychologists to design machines.

REFERENCE

1. Bentley, A. F. The behavioral superfice. *Psychol. Rev.*, 1941, **48**, 39-59.

2

The Emerging Role of Engineering Psychology

Julien M. Christensen

INTRODUCTION

In these days of rapid technological and social changes we look for a maturity in our teenagers that was not expected even a generation ago. Engineering psychology is now in its late "teens" and, in my opinion, is facing problems rather forbidding for a profession so young.

What is the nature of this teenager, engineering psychology? Fitts says it " . . . seeks to understand how human performance is related to task variables [especially engineering design variables] and to formulate theory and principles of human performance that can be applied to the *design* (my underline) of human tasks, human-operated equipment, and man-machine systems (ref. 8)." Later, in the same paper, Fitts reminds us that the engineering psychologist " . . . is also interested in parameters that influence the relations between task variables and performance, such as individual differences, motivation, and level of training." In this dawning of the era of automation, I know that Professor Fitts includes also the social consequences of design as an area in which the engineering psychologist shares responsibility with many other professional people.

I would now like to formulate several points. First, unless research results can be expressed in engineering terms, and thus can affect design, the results do not contribute, at least directly, to engineering psychology. Second, engineering psychology is different from experimental psychology in several critical respects. (Fitts (ref. 8) has covered this topic very well and I will not dwell on it here.) Third, engineering psychologists do research and they serve as staff advisers and consultants to design engineers. Fourth, engineering psychologists do occasionally engage in design work themselves. Years ago Frank Taylor termed such engineering psychologists "human engineering technologists." He said that they contribute to systems development at three levels. To quote him, "At the simplest, he designs individual displays, controls, or display-control relationships. At a somewhat more complex level, the human engineering technologist contributes to the design of consoles and instruments panels. At the highest level of complexity, he assists in

From Aerospace Medical Research Laboratories Technical Report No. 64-88, September 1964.

structuring large systems composed of many mechanical elements and frequently several human beings. In this capacity he helps to determine what information must flow through the system, how it must be processed, how many men are required, what tasks they will perform, and what type of information each one will need. In short, the engineering psychologist helps at this level to determine the configuration of the system (ref. 11)." I really am not too concerned as to what terms are applied as long as we understand that some engineering psychologists have, and others could, make excellent systems planners and designers. I am not willing to accept the design of dials, the design of consoles, and the offering of assistance in structuring larger systems as the ultimate to which a competent engineering psychologist can aspire. I can see no more reason, for example, for keeping an otherwise qualified engineering psychologist out of the business of systems planning, design, and management than I can see in keeping a physician out of the business of designing a steel shaft to replace a broken bone. I do think that sometimes we have become so engrossed in irrelevant theories, unnecessarily fancy experimental designs, filling out forms demanded by certain specifications, etc., that we dilute our effectiveness as engineering psychologists and, consequently, debase our profession. Finally, I feel that generally our universities are failing to prepare engineering psychologists who are capable of providing the kinds of information and services that the design, development, and testing of modern systems require.

SOME POINTS FROM THE HISTORY OF ENGINEERING AND PSYCHOLOGY

I think it not irrelevant to look for a moment at the history of the chief profession that we support and with which we work; namely, engineering. Just as Boring (ref. 2) reminded us that history can be revised ("As time goes on, there come to be second thoughts about the interpretation of it.") so I remind you that this interpretation you are hearing today is a reflection of the distinctive, perhaps idiosyncratic, interaction between certain facts and me. Another psychologist, and almost certainly any engineer, might interpret them differently.

First, we had better define "engineering." Fitts (ref. 8) reported that as early as 1828 the Institute for Civil Engineering of London considered the practice of their profession as more than the strict application of physical sciences. They defined engineering as "The art of directing the great sources of power in nature to the use and convenience of man." It is of more than incidental interest that they too recognized engineering, in part, as an art—the "gut-feeling" of 1828! One hundred and thirty years later the Institute of Industrial Engineering (ref. 12) stated "Philosophically, he [the industrial engineer] is devoted to the ideal of helping the nation to use most effectively its physical facilities and human talents for production of goods and services." This is a recasting of the 1828 definition. *It explicitly includes human talents.* Perhaps we can all agree that no design engineer should assume the title "systems engineer" or "systems manager" unles he is sensitive to and capable of handling those subsystem and system interactions that involve the organic as well as the inorganic. This nation has thousands of splendid design engineers; there is a critical shortage of systems engineers.

The above definition would appear to include the design and utilization of tools as an enterprise that legitimately falls within the purview of engineering. This being so, we can trace engineering back to earliest man, or (depending upon the particular archeological or anthropological authority to whom you subscribe) even to the man-apes. *Australopithecus Prometheus*, for example, used pebble tools, employed thigh bones as weapons, and used scoops made from antelope cannon bones. It is interesting to reflect on the possibility that osteodontokeratic tools probably were concurrent with, or may even have preceded, stone implements. Dart (ref. 6) stated, "I am convinced that long before he knew how to fashion weapons and tools from stone, man had discovered another and livelier material for his primitive skill." Thus, we might conclude that engineering and the human factors components thereof, as reflected in the design and use of the first tools, actually determine the dating of the origin of man. This, incidentally, is the position that is subscribed to by many anthropologists; they have despaired of defining meaningful anatomical distinctions and refer instead to cultural achievements.

Aside from serving as a possible source for dignifying engineering and our profession by anointment with antiquity, what possible lesson can be learned from the events of nearly 1,000,000 years ago? I submit that pebble tools and scoops made from antelope bone were specific, intelligent reactions to *interactions with the environment* and that current principles are no more than that—response to interactions among and between organic and inorganic—responses that have become, as compared with the bone scoop, unbelievably and fascinatingly complex.

I now blithely skip forward a few log units in time (actually almost a million years) in order to consider the impact of the industrial revolution in terms of interactions and man's response to them. (This neglect of hundreds of thousands of years may be justified when one considers the relatively slow rate of development during the "Age of Tools." The causes for this need not concern us here.) I will rely heavily on a previous report of mine for this information (ref. 5).

Phase I—The Age of Machines

The first phase of the industrial revolution, which covered a period of approximately 120 years (1750 to 1870), witnessed the emergence from the Age of Tools to the Age of Machines. This period was characterized by brilliant invention in the textile industry and the application of steam power to the operation of machines. In this period Jacquard used punched card techniques (1801) as a programming aid with weaving equipment and Watt designed a self-regulating governor for his steam engine—the beginning of automation and Cybernetics! Engineering, although crude and unrecognized formally, was attempting to relieve the muscles of mankind. The word "psychology" was not even invented, although this should not engender dismay. There were people, as we have seen, who met the fundamental criteria of "engineer" long before there was such a profession.

Phase II—The Power Revolution

The second phase of the industrial revolution has often been called the "power revolution." This period encompassed the years 1870 to 1945, and was character-

The Emerging Role of Engineering Psychology

ized by fantastic developments and increased efficiency in transportation, communications and agriculture. Human factors pioneers from the disciplines of engineering and psychology made their first formally recognized contributions of a psychological nature and one calls to mind such names as Taylor, the Gilbreths, Muensterberg, Binet, and others.[1] Although the primary emphasis was on the adjustment of man to his work by utilization of the techniques of selection, classification, and training, we would be something less than candid if we did not recognize that Taylor and the Gilbreths made certain contributions that now would be considered applications of engineering psychology.

By 1945 engineering psychology had been formally recognized. Thus, it might be well to pause and briefly to summarize its nature at that time. This was our "knobs and dials" era, and those of us who were part of it occasionally view it with nostalgia. In retrospect it may seem to some an era of simplicity; at the time, however, it seemed to be an era of pioneering that required all the insight and mental resources that could be mustered. The major emphasis of behavioral scientists through World War II had been on selection and classification tests and training procedures, or adapting man to job. The few isolated investigations in engineering psychology in the United Kingdom and the United States contained little evidence that the investigators recognized that they were pioneering a revolutionary enterprise.

As far as the human was concerned, the engineering sciences gave primary attention to the pioneering work of Taylor and the Gilbreths.

Phase III—Machines for "Minds"

Several events that have occurred during and since World War II have so increased the complexity of cultures (and thus increased the number and complexity of individual and group interactions) that it seems to me that this period should be termed Phase III of the Industrial Revolution. We are living during this period and thus do not have the benefit of hindsight to assist us in interpreting the true nature and extent of its impact on the cultures of the world.

The events of Phase III that are of special interest to us are the development of atomic energy, the development of high-speed computers (a powerful tool for use in communication, industrial, military, economic and perhaps even social systems), automation, and space explorations. We are witnessing the wholesale substitution of machines for many of those functions of man that some have considered his "higher" processes.

The Response of Engineering Psychology. Elsewhere (refs. 4, 5) I have covered the response of engineering psychology during this period and will only briefly summarize today. We have witnessed a progression from retrofit of original design to design of elements (knobs and dials) to design of components and subsystems to a clear recognition of the requirement to fit machines to men. If we are to meet the challenge, we must now take a fourth step in our profession and recognize

---

[1] Taylor and the Gilbreths pioneered time-and-motion studies within the field of industrial engineering; Muensterberg was a founder of "applied" psychology; Binet was the father of the intelligence test (Ed.).

that in order to realize maximum efficiency and satisfaction, machines and men can and must be fitted to concepts. We are almost ready to divest ourselves of those preeminently useful lists of the past that explained what men can do best and what machines can do best and, in their stead, these elements of design will be considered only in relation to the conceptual requirements of a particular system as these requirements exist at a particular point in time.

Some will say that this represents a distinction without a difference but I cannot agree. True, the aims of our profession during its first three phases were the same as they are now, i.e., participation in the development of systems that, in terms of specified criteria, are maximally effective. However, even now the nature of complex systems is being considered in some detail at the conceptual level and then, and only then, do the designers determine the components and materials to be used to meet the requirements of the system, and only then do they determine whether these should be organic, inorganic or a combination of both. In the future, systems planners will use simulation more and more as a design tool that will enable them inexpensively and quickly to assess the merits of alternatives without committing substantial resources to any of them. This implies a knowledge and understanding of systems principles (and advancements in simulation) that are not available today.

Why do we need systems research and its companion, research in simulation techniques? Why, in the last decade, have these areas suddenly received increased attention? I hope that the implications from my previous meanderings among anthropology, archeology, engineering, psychology, etc., will support the position I now adopt. Such research is necessary simply because the interactions among our cultural elements (whether organic or inorganic) have become so manifold and so complex that successful and economical design of a system of even modest complexity is impossible without the benefit of sound scientific principles and appropriate supporting tools. Accepting this as a tenet, one then is forced to conclude that professional members of each of the sciences that support the development of systems must assure that their own house is in reasonable order. The principles of each discipline must be so stated (or, by means of transformations be capable of being so stated) that they can be considered appositively with the principles of other contributing disciplines. Do you feel that engineering psychology is ready to meet this challenge?

For some time to come, this—the appositive consideration of data and principles from what in some cases have been unrelated disciplines—will be the thorniest problem confronting the planners and designers of systems. This is why I suggested earlier that we must plan our experimental work in engineering psychology in such a way that the variables are those that planners and designers can manipulate; and present our findings in such a way that planners and designers can make intelligent decisions regarding the consequences of selected values of the independent variables.

Other areas directly in support of systems planning and design that need particular attention are control dynamics, complex visual perception, decision-making, the proper function of computers, design for motivation, and design for work in unusual environments. (This latter, incidentally, must include assessments of the effects of multiple stressors applied simultaneouly or sequentially. Specifically, are the effects on performance additive, logarithmic, or synergistic?)

In a somewhat different vein, we need cross-cultural studies, much like those Hertzberg and his team (ref. 9) have performed in the field of physical anthropology to adequately assess the effects of different cultures, different customs, different educational systems, etc., on systems design and development. Most current engineering psychological data are drawn from studies based on English-speaking people only, and these have been restricted chiefly to Great Britain, Canada, and the United States.

To continue, Chapanis has suggested that we very much need to consider both input and output in the development of better methods of communicating with computers. Someday we will see Kelly's ideas on predictor displays elaborated to include the results of the analysis and synthesis of data from literally hundreds of sources.[2]

In addition, what might be termed "double design" possibilities offer interesting opportunities for engineering psychology. Perhaps I can best illustrate with an example. An engineering psychologist working in the area of remote manipulation should consider the possibilities of redesigning the objects that must be manipulated as well as attempting to improve the remote manipulators.

I suggest also that any branch of science progresses only as its tools and methods improve. While we have one text that might be termed a "methods" book (ref. 2), I think we do far too little work of a methodological nature. We know far too little about the validity of the methods we employ. For example, we frequently employ user opinion as a criterion, but do we really know where and under what conditions it is an effective criterion? We engage in system tests, but we do little to investigate the possibilities of improving our test procedures. We find that traditional laboratory techniques cannot be used, so we seem to despair. Many other examples could be cited.

In systems planning, man is coming more and more to be considered for his cognitive capabilities and his abilities to handle conditions that can't be expressed alpha-numerically. This implies the design of systems in such a manner that man at any moment can effectively exercise his judgment, for this is one of the greatest assets he brings to systems. Bray has termed this the "development of context" in systems design. Finally, let us try, through research, to show systems planners and designers that variability is an outstanding human characteristic that should be used as an asset instead of viewed constantly as a liability. Without variability there would be no learning and no adaptation, two prerequisites for designing into systems one characteristic that distinguishes the mediocre from the outstanding. I speak, of course, of flexibility.

In another publication (ref. 4) I have attempted to trace the history of systems research, to define some of the chief problems in this area, and to forecast future progress. This information is well known to this audience so I will not repeat it here, but will extract some thoughts that are particularly germane to our considerations at present.

The theories of behavior applicable to complex man-machine systems not only must be useful to systems engineers but also should provide a link with existing behavioral knowledge. No new approach can afford to divest itself of the tremendous fund of sound, relevant behavioral knowledge that exists.

[2] Kelly's predictor display is a technique for informing the operator in a manual control system of the projected deviation of that system from its desired course (Ed.).

Research in mathematical model-building with special emphasis on the solution of complex problems involving higher-order interactions is needed. This (mathematics) will prove to be our common communication medium with engineers, physicists, and other scientists. What are we doing to prepare ourselves to speak this language and use this tool effectively?

I hope finally, however, that our approach to systems research in particular is not in danger of becoming stereotyped too soon, that is, we must not insist on a strict operationalism too early in the game. Bateson's expressions regarding the nature of scientific thought may be worth considering in this respect (ref. 1): ". . . whenever we pride ourselves upon finding a newer, stricter way of thought or exposition; whenever we start insisting too hard upon operationalism or symbolic logic or any of these very essential systems of analysis, we lose something of the ability to think new thoughts. And, equally, of course, whenever we rebel against the sterile rigidity of formal thought and exposition and let our ideas run wild, we likewise lose. As I see it, the advances in scientific thought come from a *combination of loose and strict thinking,* and this combination is the most precious tool of science." Engineering psychology, in my opinion, must also support its share of "loose thinkers."

It may be of more than academic interest as to what the exponential increase in interactions referred to earlier means to the individual members of society. Does unit freedom increase or decrease as the number of interactions in which the individual becomes involved increases? While an individual may control more machines, greater computers, more power, perhaps even more people, etc., do not these same elements with which he is interacting also exert greater control over him, essentially reducing individual freedom of choice and action? Such considerations are perhaps beyond the immediate interest or, at least, current responsibilities of the engineering psychologist and the system engineer. Yet to ignore these interactions would patently violate the ultimate intentions of our professions.

The Response of Engineering. The recent period of the era we are considering has witnessed an interesting shift in perspective for certain branches of engineering. To my knowledge, the industrial engineering profession has most clearly elucidated its position with respect to the design of complex systems. Therefore, I would like to restrict my consideration of the engineering profession to them. I do not say that they are typical; I do say that they have something important to say.

Industrial engineering is changing from a discipline concerned primarily with work methods to a system-oriented discipline. I would like to quote from a recent article in the Journal of Industrial Engineering (ref. 12).

> Industrial engineering is concerned with the design, improvement, and installation of integrated systems of men, materials, and equipment. It draws upon specialized knowledge and skill in the mathematical, physical, and social sciences, together with the principles and methods of engineering analysis and design to specify, predict and evaluate the results to be obtained from such systems. . . .
>
> The Industrial Engineering approach is a unique application of engineering design and analysis techniques. The Industrial Engineer's consideration of

people, the design of systems involving people, and the manner by which human performance is analyzed require a fundamental, analytical approach that is essentially different from other engineering disciplines. . . . The Industrial Engineer is distinguished from other engineers in that he:

1. Places increased emphasis on the integration of the human being into the system.
2. Concerns himself with the total problem.
3. Predicts and interprets the economic results.
4. Makes greater utilization of the contributions of the social sciences than do other engineers.

The Industrial Engineer differs from other engineers in the degree that his fundamental concern is with operations that result from both physical and human resources, and are measured in both physical and economic terms. Therefore, he is educated not only in the engineering methods of analysis and design which stem from mathematics and the physical sciences, but also is concerned with psychology, physiology, sociology, economics, costs, and human relations. . . . Philosophically, he is devoted to the ideal of helping the nation to use most effectively its physical facilities and human talents for production of goods and services.

I might observe that we have literally squandered our human talents—even more so than our material resources. The authors point out that industrial engineers must have training and knowledge in virtually every physical, biological, and social science.

A few universities have taken direct action to support this revolutionary movement in industrial engineering. Of these, the program at Northwestern University is, in my opinion, outstanding. These educators had the courage to completely restructure their entire program. They divested their curriculum almost completely of the old "techniques" teaching and have concentrated instead on teaching fundamental engineering with more than cursory attention given to all the professions that must support a modern engineering enterprise. They, to quote Lehrer (ref. 10), ". . . do not rely upon techniques, but rely upon a broad and basic education and upon adequately defining the problem." Further, "Industrial engineering should not be confined just to manufacturing-type activities, but should be truly systems oriented." Again, ". . . creativity is no problem whatsoever when the educational program has a broad orientation, does not rely upon 'techniques,' and when the faculty inspires and motivates the students to learn for themselves." Techniques, including the use of computers, are worked into the curriculum, and emphasized as being what they truly are, tools for the accomplishment of goals—tools, incidentally, any one of which may soon become obsolete. They emphasize that these techinques must never be allowed to have a restrictive influence upon creative thinking. I reemphasize that they do not limit themselves to a consideration of production systems, but include all systems (e.g., management) associated with modern industrial complexes. I think you may agree with me that these views represent a revolution in industrial engineering.

Some may view such actions as threatening to engineering psychology. I do not, although I think the committee of distinguished industrial engineers who wrote the first article from which I quoted (ref. 12) may have somewhat overestimated the capability and knowledge that one man—even an engineer—can achieve! For example, I would have felt better had they disclosed how their systems engineer (for this is essentially what they are defining, not a traditional industrial engineer) planned to work with, and coordinate the work of, the many specialists on whom he must rely if he is to be successful. What requirements will be placed on them? What will be their responsibilities? However, in general, these appear to be the kind of engineers with whom I would be delighted to associate. I recommend that you read Professor Lehrer's paper.

No, rather than a threat, I feel that this "school" of industrial engineering clearly recognizes that any designer of modern systems can enjoy success only if he incorporates into his work human factors considerations as well as data from the physical sciences. We have much in common with these modern industrial engineers, and in those cases where they are serving also as systems engineers I expect that we will find them to be unusually receptive and understanding customers for our services. I do not hesitate to predict, however, that industrial engineers will have to compete with many other professions for the right to serve as systems planners and systems designers. I know of no formal course of training in these areas (although the Northwestern program tends strongly in this direction), and I suspect that for some time to come systems planners, engineers, and managers will be chosen, among other reasons, with respect to how sensitive they are to, and how well they understand the meaning and implications of interaction. Now in some systems, the primary interactions heavily involve the organic. I hope our engineering colleagues won't be chagrined, then, to occasionally find in at least these instances that an outstanding job of systems planning and management is being done by a biological or behavioral scientist. The current interest in bionics,[3] as Hartmann has suggested to me, might tend to hasten this day, because as selected inorganic subsystems become more and more to be modeled after organic subsystems, the compatibility between organic and inorganic will increase, and, I might add, as perhaps will the number of organic scientists who develop an interest and capability in systems planning and development. (You may object to my term "organic scientists" because it implies that there are *inorganic* scientists. I mean, of course, scientists who deal with living organisms. On second thought, this is no worse than the term "life scientist," whose opposite, I suppose, is a "lifeless scientist"—as a matter of fact, I am personally acquainted with a number of the latter!)

I wish I could relate to you instances where departments of psychology are taking action, as Northwestern did in industrial engineering, to completely overhaul those portions of their programs that are intended to meet the needs of engineering psychology. But with two or three possible exceptions, I cannot. All department chairmen in psychology would admit that we live in an industrial and technological as well as a scientific age. All would probably agree that man interacts almost constantly and very intimately with the products and the symbols of this age and, further, that these, in no small part, help mold this "behavior" which all claim is of primary interest to the field of psychology. So how can we ever have a truly comprehensive, satisfactory theory of behavior if we ignore these

---

[3] Bionics, as the subsequent comment implies, is an interdisciplinary effort to use biological principles in the development of "inorganic" or machine subsystems or techniques, thereby furthering understanding in both areas (Ed.).

considerations? The findings of engineering psychology may never provide sufficient substance for the formulation of a comprehensive theory of human behavior, but no theory of human behavior can be considered comprehensive if it cannot account for the findings of engineering psychology.

Are we ready for this age with adequately trained psychologists? Why, for example, shouldn't engineering psychologists direct the planning and designing of at least those systems that have a heavy "organic" component? Why should we be satisfied to sit on the sidelines and simply offer advice? Could it be that we are afraid of such responsibilities? When will we become, to quote Lehrer, "... *truly systems oriented?*" How personally bitter it would be and will be to men like Taylor, Williams, Bartlett, Fitts, Chapanis,[4] and others if we drop the heritage which they bequeathed us.

This is the crisis to which I alluded earlier; it can be solved, however, only with the full participation of our educational institutions. A committee of the Society of Engineering Psychologists has already investigated this problem. Their program should be reevaluated and updated. They should consider the dual functions of gaining knowledge and applying knowledge. Competent, progressive engineering authorities might profitably be asked to sit with the new committee.

I am not recommending that we dispense even partially with training in experimental psychology. As one surveys the programs of, for example, modern industrial engineers it is easy to see that even now we complement each other rather effectively. For example, our training in and knowledge of experimental methods, individual differences, probability, reliability, validity, criterion development, and human capabilities meet specific requirements of engineers, systems planners, and systems designers. However, we must gain a better understanding of the fundamentals of engineering and allied supporting sciences, for how else can we truly understand the needs of one of our best customers? We hope to be of service to him; therefore, it is incumbent on us to acquaint ourselves with his needs and to learn his language.

Program of Action. I submit to you now a program of action. Some of the points are supported by my previous remarks; some are not.

a. The education of engineering psychologists is quite inadequate. The Society of Engineering Psychologists should update the 1961 report on *Training in Engineering Psychology* (ref. 7) and try diligently through our academic members to get the Chairmen of selected departments of psychology to adopt it. This is fundamental to the development, and very survival, of our profession.

b. We have unique contributions to make to the traditional areas with which industrial engineers deal. We should try harder to acquaint them with our services; it would afford a splendid opportunity to show what we can do. Why, for example, would we not expect to make contributions to production systems similar to those we made to pilot display and control systems, air navigation systems, etc.

c. We should reopen the question of whether a Journal of Engineering Psychology is needed. I am concerned, perhaps, not so much with the matter of a proper repository for our research as I am with the availability of a medium for setting down our beliefs, our policies, our aims—in short—the philosophy of en-

[4] All are leaders in the development of the field of engineering psychology (Ed.).

gineering psychology. I would like to see a clear "statement of intent," periodically up-dated for our profession.

d. We should cooperate with other professional organizations, such as the AIIE in exploring the needs for, and possibilities of developing a true systems development curriculum. Skilled systems planners and designers generally recognize immediately the need for information and assistance from such fields as engineering psychology; thus, it would help our profession if there were more of them.

e. Each year the Program Chairman should solicit from the membership their thoughts regarding the critical issues that face our profession and ask leaders in our profession and allied professions to discuss at our annual meeting those deemed most important and timely. If necessary, we should take time for such a symposium from that allotted to the presentation of papers.

f. Because of the shortage of good academic curricula in engineering psychology, we should adopt the practice of the Division of Industrial Psychology of sponsoring a more extensive work shop program at the annual American Psychological Association meeting. This would provide an opportunity for many students and practitioners to learn directly from the outstanding leaders and experts in our field and from leaders in associated disciplines.

In summary, I have suggested that engineering psychology has at least one important choice point confronting it. In addition to researchers, we can remain advisers and designers of little things, or we can, on the applied side, aspire eventually to being systems planners and managers. Dr. Abelson has stated that no psychologist had been asked to comment on the lunar program—probably, I suggest, because none was felt to have sufficient acquaintance with the entire program to judge its implications.

I have given you an example of another profession that took a good hard look at itself, decided it didn't like what it saw, and is doing something about it. I hope we will take such a look. Maybe we will like what this self-examination reveals; maybe we will not. But for Heaven's sake, let's not become complacent at the tender age of 18.

LIST OF REFERENCES

1. Bateson, G., "Experiments in Thinking About Observed Enthonological Materials," *Physiolosophy of Science* 8: 55, 1941.
2. Boring, E. G., *A History of Experimental Psychology*, Second Edition, Appleton-Century-Crofts, New York, 1950.
3. Chapanis, A., *Research Techniques in Human Engineering*, The Johns Hopkins Press, Baltimore, Md., 1959.
4. Christensen, J. M., "Trends in Human Factors," *Human Factors* 1, 1958.
5. Christensen, J. M., "The Evolution of the Systems Approach in Human Factors Engineering," *Human Factors* 4, February 1962.
6. Dart, R. R., *Adventures with the Missing Link*, The Viking Press, New York, 1959.
7. Fitts, P. M., et al., "Training in Engineering Psychology," *American Psychologist* 16(5): 171-177, 1961.
8. Fitts, P. M., *Human Factors Engineering—Concepts and Theory*, The University of Michigan Engineering Summer Conferences, University of Michigan, Ann Arbor, Michigan, 1963.

9. Hertzberg, H. T. E., E. Churchill, C. W. Dupertuis, R. M. White, and A. Damon, *Anthropometric Survey of Turkey, Greece and Italy,* Pergamon Press, New York, 1963.
10. Lehrer, R. N., "Modern Industrial Engineering," *Industrial Engineering in 1975,* AIIE-ASEE Symposium, Detroit, Michigan, 10 May 1961.
11. Taylor, F. V., "Psychology and the Design of Machines," *American Psychologist* **12**(5): 249-258, 1957.
12. Weston, A. (Editor), "The Emerging Role of Industrial Engineering," *The Journal of Indust. Eng.,* 1961.

3

Engineering Psychology Today: Some Observations from the Ivory Tower

William C. Howell and Irwin L. Goldstein

In spite of repeated efforts by its leaders to make explicit its goals, methods, and subject matter (see, for example, Christensen, 1964; Fitts, 1963; Melton & Briggs, 1960; Poulton, 1966; Taylor, 1957), engineering psychology has failed to achieve a widely recognized identity. To most other members of the scientific community, including a distressing number of psychologists, it remains something of an enigma. We believe that there is a good reason for this unfortunate state of affairs, and more important, that the fundamental difficulty extends well beyond the mere issue of recognition. Since we feel that the shape of engineering psychology for years to come may well depend upon how this problem is resolved, it becomes the central concern of the present paper.

We can best begin to understand this problem by surveying the general territory in which the engineering psychologist has established residence. He usually defines it in terms of the proposition that man and machine should be regarded, for both scientific and technological purposes, as functionally integrated parts of a larger goal-oriented unit known as the *man-machine system*. His concern, therefore, is in man-machine system behavior. Although he has been in this area for better than 20 years, the engineering psychologist has by no means had it all to himself. Other branches of psychology (especially experimental, industrial, and social), anthropometry, physiology, and several branches of engineering have shared various parts of it with him and with each other. Looking to the future, newer specialties, such as computer and information sciences, can also be seen

From *Organizational Behavior and Human Performance,* **5**, 1970, 159-169. By permission of Academic Press, Inc.

gaining a foothold. Obviously, the man-machine system concept has had wide appeal.

Now in theory, it is very healthy for different specialties to become interested in the same problems. Science, after all, is supposed to be one big, happy family, working together to solve the mysteries of the universe; the more cohesive its attack, the more complete should be the understanding achieved. It is one thing, however, for recognized disciplines to cooperate in studying some set of phenomena, and quite another for a new, hybrid discipline to set itself up specifically for that purpose. Any hybrid is viewed with some skepticism, particularly by specialties to which it is most closely akin, until it is able to win recognition through the uncontravertable evidence of performance. At the very least, it can expect to be grossly misunderstood; in consequence, it must work doubly hard to make its position clear with regard to goals, methods, and subject matter.

Engineering psychology, of course, is such a hybrid. Moreover, it is a hybrid of a very complex sort as the above sample of related specialties clearly implies. Its language includes terms from the behavioral and physical sciences, engineering, and especially the universal language of mathematics. Its concepts and theories also reflect the influence of both man-oriented and machine-oriented disciplines. What really sets it apart, however, is its desire to accommodate the objectives of science and application under one roof. It is this last facet of the hybridization that seems to have caused the most trouble, within as well as outside the field.

If engineering psychology has yet to achieve full acceptance and recognition, it is not because of its "interdisciplinary" tendencies or its failure to show a satisfactory record of achievement. In support of this contention we need only point out that since 1958 the field has been sufficiently productive to justify independent and regular coverage of its research activities in the *Annual Review of Psychology*. More than a half-dozen journals are devoted exclusively to work in this area, and reports by engineering psychologists appear regularly in many other outlets. Furthermore, a number of organizations—including a division of the American Psychological Association—have been founded to serve the common interests of engineering psychologists.

For all its outward appearance of unity, however, there has never been substantial agreement—even among its foremost spokesmen—as to where engineering psychology belongs on the science-technology spectrum. Do we have a field with two major branches? Two separate but interacting fields? One field with many subfields, each having two branches? Or no field at all? A confusing array of labels has appeared in conjunction with various attempts to resolve this issue, a fact which itself has undoubtedly done a great deal to obscure the image projected to those in other areas.

Without exception, the "founding fathers" singled out the problem of equipment design, a utilitarian concern, as the *reason* for a discipline of engineering psychology (e.g., Chapanis, Garner, & Morgan, 1949; Fitts, 1951, 1958; Geldard, 1953; Taylor, 1957). This seems to have remained the dominant opinion (e.g., see Chapanis, 1965; Christensen, 1964; Grether, 1968; Miller, Malachi, & Farr, 1967; Poulton, 1966). Similarly, leaders down through the years seem agreed that the field should encompass both ". . . a professional and a scientific aspect" (Fitts, 1958, p. 267), the latter being less utilitarian than the former. Where opinions start to differ is with respect to how much is made of the distinction between the two aspects. Some (e.g., Miller et al., 1967) have virtually ignored it, using the terms *engineering psychology, human engineering,* and *human factors engineering* interchangeably. Unquestionably this is also the predominant impression held by "outsiders." Others, however, such as Taylor and Fitts, have been careful to reserve the label engineering psychology for the scientific branch alone.

We feel that now, more than ever before, there is justification for a clear distinction. Our main thesis, which we shall attempt to develop throughout the following pages, is that engineering psychology should be guided directly by scientific objectives and only indirectly by utilitarian ones, using the methods of science to gather basic data which can ultimately be applied by the human factors engineer. In this view engineering psychology—as a field of scientific inquiry—should maintain an identity separate from, but in close touch with, engineering practice. Otherwise it runs the distinct risk of compromising its scientific values in the futile attempt to keep pace with technology, thereby serving neither science nor technology. In a recent review, Poulton (1966) paints a very lucid picture of the field as we would define it here. He says:

> The aim of engineering psychology is not simply to compare two possible designs for a piece of equipment, but to specify the capacities and limitations of the human, from which the choice of the better design should be deducible directly. We are left with a picture of engineering psychology as the experimental psychology of the mid-twentieth century. This contrasts perhaps favorably with the experimental psychology of man and animal in the somewhat artificial environment of the traditional psychological laboratory (p. 178).

This does not necessarily mean, of course, that there is no place for immediately applicable (or "mission oriented") research on human factors. Far from it! At any point in time the engineer or system designer needs all the objective data he can muster to help him make specific design decisions. He need not concern himself unduly about whether a particular specification will be inapplicable five or ten years hence; by that time, the whole design will probably require rethinking. The point is, however, that immediate, specific answers should be the province of the human factors engineer, not the engineering psychologist. There will probably always be plenty of immediate issues that cannot be resolved satisfactorily by direct deduction from general behavior principles; consequently, the need for human factors research will probably continue. Someone, however, must look beyond the problems of the moment, and to our mind no one is better qualified than the engineering psychologist.

In an age in which future demands upon human capabilities are highly problematic, the researcher has little hope of anticipating specific design questions far enough in advance to do anything about them. His only recourse is to solve them after they arise (the approach we just discussed) or to ". . . specify the capacities and limitations of the human . . ." (Poulton, 1966, p. 178), so that whatever the problem, he can make some factual contribution to its solution. The latter is the approach that we have advocated for engineering psychology. This means that the foremost guiding principle for the engineering psychologist should be no different from that for any other scientific psychologist: namely, to seek behavioral principles of widest possible generality. At a minimum he is obliged to consider the extent to which any fact that he reports can be expected to generalize from one situation to another. To summarize, then, we believe the "scientific aspect" of which Fitts spoke should be the primary concern of engineering psychology; moreover, this aspect should be (and, in fact, is) growing increasingly distinct from the "professional aspect" through which scientific data are applied to specific design problems and alternative designs are subjected to direct test (in the true sense of "applied" research). It is now more reasonable than ever to call the former aspect engineering psychology exclusively, and the latter, human factors engineering.

But how does this viewpoint square with our previous comment about the central importance of equipment design? Can a scientist seek general laws of behavior and at the same time concern himself at all with specific devices, systems, or technologies? Does the growing differentiation of science and appliance in man-machine system affairs extend to ultimate goals? These are not easy questions, but they are of the utmost importance at this juncture because their answers hold the key to the entire future of engineering psychology. If equipment design goals are incompatible with those of scientific psychology, then there is no longer a reason for engineering psychology to try to maintain an identity. It should simply dissolve into the other branches of psychology, leaving "human factors" matters exclusively to the engineer. If, on the other hand, there is a basic compatibility of goals, then engineering psychology and human factors engineering can continue to coexist in much the same manner as they do today. How profitable such a coalition might be in future years would depend solely upon the extent to which they are able to maintain effective communications with one another.

In our view, there is no sound logical or empirical basis for the position that scientific and engineering goals are directly incompatible; they are certainly not identical, but neither are they in direct conflict. Essentially the same conclusion has been reached time and again by others who have tried to distinguish "basic" and "applied" research objectives (e.g., Brooks, 1967; Geldard, 1953; Reagan, 1967). Laws of behavior yet to be discovered undoubtedly far outnumber the resources available with which to discover them. It matters little, therefore, which ones we choose to examine first or by what interest we are driven to seek them. Tracing a design problem to its roots in fundamental behavior processes can be just as scientific an endeavor (having just as far-reaching implications) as testing the consequences of some esoteric theory. Likewise, design problems are not immune to facts provided by science. To the extent that engineering psychology can identify those particular aspects of behavior which are most relevant to man-machine system problems and subject them to scientific investigation, it should remain a viable part of scientific psychology. As we have said, however, this all hinges upon the effectiveness of communication between psychologist and engineer and, at the moment, the outlook in this regard is none too bright. Neither design-inspired science nor science-inspired design has much of a future unless means are found to bolster what we perceive as a crumbling liaison. We shall return to this matter presently.

At this point, it might be well to examine a little more closely what it is that constitutes engineering psychology today. We have established that it is a scientific enterprise, but one whose efforts are influenced by system design considerations as well as the usual criteria of science. We have said nothing about the kinds of problems to which this research is addressed, the way in which it is carried out, or the places where one is most likely to find it in progress.

Virtually all sectors of the behavioral research community—academic institutions, military and nonmilitary agencies of government, nonprofit research institutes, industrial firms, and "consulting" firms—are currently making substantial contributions to this work. As might be gathered from our earlier comments, many of the people involved do not call themselves engineering psychologists at all, although they would probably acknowledge the relevance of their work to this area.

Topics of current interest show a corresponding degree of breadth, although this diversity can be a bit misleading. Actually, there is a unifying theme which ties together much of this seemingly unrelated work and gives the content as well as the purpose of engineering psychology an identity. It is a common concern for human performance of skilled tasks; the more a line of research has to say about some form of skilled behavior, the more directly relevant it is to the body of knowledge in engineering psychology. In fact, one need only read Fitts and Posner's (1967) succinct, yet definitive, review of data and theory in this area to appreciate the substantive contribution that "human performance theory" has made to our understanding of human behavior generally. In a very real sense, formulations such as Broadbent's (1958) information flow model and the Tanner-Swets theory of signal detectability (1961) are the molar behavior theories of present-day psychology. Almost without exception, theories concerned with skill recognize and emphasize the integral nature of sensory, perceptual, memory, information-processing, decision, and response processes. Perhaps the most significant contribution of all that engineering psychology has made to the science of behavior is the notion that man can be regarded as a vast communication network: one which receives, processes, and emits vital information in the form of variously coded signals. What, it is asked, are the codes used, the operations performed, and the capacities exhibited by the components in this network? Few branches of experimental psychology have remained untouched by the rather compelling integrative framework offered by the cybernetic or communication model. For example, sensory and physiological psychologists now describe their efforts in terms of "breaking" receptor, neural, and glial "codes"; present-day memory theorists discuss the manner in which events are "tagged" in short- and long-term storage.

Whether the *zeitgeist* of the "information age" predestined that psychology's thinking should be directed toward communication theory, signal detection theory and the like, or whether engineering psychology played the dominant role is perhaps debatable. It is certainly beyond question that engineering psychology, with its focal emphasis upon skilled performance in the man-machine system, has been in the best position to mediate the *zeitgeist* for the rest of psychology. Advances in technology have been of greater concern to the engineering psychologist than to most of his colleagues because of the relevance of such trends for future skill requirements. Historically, the "skills" to which he has devoted the bulk of his research energy have been those which seemed most germane to existing or projected task demands. Thus, we can identify an era, spanning the late 1940's and most of the 1950's, during which continuous perceptual-motor skills (e.g., "tracking") occupied center stage. The reason? During this period, man's immediate future as the principal means of guiding vehicles and other machines seemed secure. Now, of course, that has all changed; man's intellectual capabilities (e.g., information processing, decision making) are receiving greater attention. We can anticipate even further reorientation in this direction as man's physical involvement in system activities grows progressively smaller.

Engineering psychology, then, seeks to discover the laws governing skilled performance, and the skills of greatest interest at any particular time seem to be those for which there is expected to be the greatest need in future systems. Obviously, if this is so, it is essential that the engineering psychologist be aware of general trends in technology, and that the engineer (particularly the "human factors" type) keep abreast of knowledge regarding human skill that he can apply in system design. As we have said, effective communication is required to maintain this coordination. Our feeling is that, at the present time, there is cause for concern on this score. Let us consider how the necessary communication has been implemented in the past; how others have proposed that it be maintained, and how we feel that it might be improved in the future. In tracing these solutions, we shall be forced also to consider some of the principal factors contributing to the widening "communication gap."

In the early days of engineering psychology, communication between research and application interests presented little difficulty because they were so often vested in the same individuals. The only real communication problem was that of convincing traditional engineers that psychology had something to offer in the design of machines. By and large, the founders of the field were trained as experimental psychologists and continued to regard themselves as such (and to be accepted as such by others); they merely had come to focus their research interests upon behavior which had more than passing relevance for machine design. Additionally, either through formal or informal interaction with the engineering disciplines, they had cultivated at least a nodding acquaintance with major concepts and trends in the system design field. In short, these pioneering engineering psychologists could assure that their research would be relevant and that their findings would not go unnoticed in application simply because they wore both hats. And, we might add, they wore them extremely well, for they were indeed men of unusual talent.[1]

In view of their personal success in bridging the gap, it is not surprising that these leaders should advocate a similar approach for future generations. As recently as 1963, Fitts (1963) proposed, as appropriate graduate training, what amounts to a combined program in engineering and psychology, together with liberal helpings of mathematics, physics, and "practical experience" in system design. While admirable in principle, it is unlikely that so rigorous a program would have very wide appeal among prospective graduate students today. It is noteworthy in this regard that at one of the largest Ph.D.-granting institutions in

---

[1] Use of the past tense is not altogether appropriate, for some of these founders are still very active in the field.

the country where such a program is "on the books," not a single student since 1961 has availed himself of this opportunity. This is not to disparage the fortitude of today's graduate student. His reaction is quite natural in view of the information explosion occurring in each of the areas that would comprise his training. Whether we like it or not, the time seems to be upon us when it is no longer feasible either to attract or to train any substantial number of individuals to a high degree of proficiency in both scientific psychology and engineering practice. Even if it were feasible so to train them, it is doubtful that they could keep abreast of developments in both areas for long. Of course, one can still point to a number of living, breathing exceptions to this generalization—unusual people who manage to contribute valuable work to both fields with remarkable consistency. Some, for example, were even trained and employed as engineers before discovering their interest (and pursuing their graduate training) in psychology. Hopefully, we will always be blessed with some individuals who maintain a degree of expertise in more than one field. Our point, however, is that we cannot expect our graduate programs to produce such people as a matter of course.

Scientific engineering psychology and human factors engineering, then, have in our opinion been drifting apart and rightly so. This has occurred at several levels: graduate training, placement of graduates, and activities engaged in by graduates once they are placed. As the gap widens, it becomes more difficult for the research-minded engineering psychologist to assess the salience of potential research projects for system design, and for the human factors engineer to appreciate fully the information available concerning skilled human performance that he could bring to bear on particular design problems. Each must therefore look to the other for help. Unfortunately, however, it is not as easy as it once was to provide such assistance. For one thing, since the two fields no longer speak quite the same language or use altogether the same publication outlets, it has become something of a chore for either to communicate with the other. Then too, the professional rewards to be gained from such interactions may scarcely seem to justify the effort: "Why send my report to an engineering or human factors journal," the engineering psychologist might ask, "when the only people who would fully appreciate its significance are other psychologists?" Finally, of course, each field has plenty of problems of its own to contend with; there remains little time for the problems of others.

One solution to the communication problem has been offered by Christensen (1964). Focusing his attention upon graduate training in psychology, he proposed that these programs be completely revamped so as to offer the student a chance to compete with the engineer for system design and management positions. We have already examined the implications of this approach, and concluded that, in today's world, it is unreasonable to expect any large number of individuals to become proficient in both scientific psychology and engineering. And if we revise our psychology curricula to become predominantly oriented toward the engineering disciplines, then why keep them in psychology at all?

It is our belief that the answer does not lie exclusively in graduate training or, for that matter, at any other single point in the career of the individual psychologist or engineer. Whatever his training, he cannot hope to emulate those of past years who, in and of themselves, formed the vital link between science and application. In the place of such direct personal involvement must come a more formal procedure for exchanging pertinent information. The scientific engineering psychologist should receive his principal training in the appropriate areas of psychology; the human factors engineer should receive his specialized training in areas related to system design. Interdisciplinary training need be extended only so far as to enable each to appreciate the other's point of view, and to grasp the essentials of his "language." In short, we advocate recognition of the fact that there are now two specialties (rather than one with two branches), that specialized training and continuing effort is required for each to keep "current," but that each can still complement the other through formal communication at all levels.

How, then, is effective formal communication to be achieved? First, as we have said, graduate students must be afforded some training in the "language" of both specialties. Ways in which this might be accomplished are as many and varied as are the policies of the institutions offering such programs. At the very least, it would require an increased number of "service" courses in both psychology and engineering; ideally, it might include participation by students in a variety of actual research and design problems.

Second, each area must be willing, upon occasion, to summarize for the other —preferably in the other's journals and language—material of particular significance within his own area. A good example of what we mean is the occasional issue of *IEEE Transactions* (such as Vol. HFE-7, No. 1, 1966) devoted chiefly to the work of scientific engineering psychologists but written for the human factors engineer. To date, such communication has been sporadic and limited principally to the psychologist writing for the engineer. For this approach to be successful, it must be bidirectional, and occur on a far more regular basis. Specific journals do exist, of course, in which such an interchange might and occasionally does take place (*Human Factors* and *Organizational Behavior and Human Performance* being two good examples). At present, however, such outlets are all too often ignored in favor of those having more specialized appeal. One feels considerably more comfortable writing in one's own language to those of like interest, even though his message may be equally as important to those of another tongue.

Finally, and this is really just a corollary to the last point, there must be a change in attitude on the part of both those who *generate* and those who *apply* knowledge unique to engineering psychology. The former must not lose sight of the fact that there is or can be in their work something of value beyond the mere satisfaction of their own curiosity—if only they are willing to help draw it out. The latter must recognize that the scientific psychologist is much more apt to "speak to the issues" if he knows what the issues are; and further, that, just as in years past, he very often *does* have something relevant to say.

If, through the above means or through others which have escaped our notice, the communication link with human factors engineering can be strengthened, engineering psychology may be headed toward fulfillment of the scientific promise which men like Taylor and Fitts envisioned for it many years ago; and it may do so without compromising its involvement in man-machine system affairs. If, instead, the bond continues to weaken, we can conceive of no justification for continuation of such a field within psychology. In our opinion, such a development would make both psychology and engineering the losers.

REFERENCES

Broadbent, D. E. *Perception and communication.* Oxford: Pergamon Press, 1958.
Brooks, H. Applied Science and technological progress. *Science,* 1967, **156,** 1706-1712.
Chapanis, A. *Man-machine engineering.* Belmont, California: Wadsworth, 1965.
Chapanis, A., Garner, W. R., & Morgan, C. T. *Applied experimental psychology: human factors in engineering design.* New York: Wiley, 1949.
Christensen, J. M. The emerging role of engineering psychology. *Aerospace Medical Research Laboratories Technical Report,* No. 64-88, September, 1964.
Fitts, P. M. Engineering psychology, and equipment design. In S. S. Stevens (Ed.), *Handbook of Experimental Psychology.* New York: Wiley, 1951. Pp. 1287-1340.
Fitts, P. M. Engineering psychology. *Annual Review of Psychology,* 1958, **9,** 267-294.
Fitts, P. M. Engineering psychology. In S. Koch (Ed.), *Psychology: A Study of a Science.* New York: McGraw-Hill, 1963. Pp. 908-933.
Fitts, P. A., & Posner, M. I. *Human Performance.* Belmont, California: Wadsworth, 1967.
Geldard, F. A. Military psychology: Science or technology? *American Journal of Psychology,* 1953, **66,** 335-348.
Grether, W. F. Engineering psychology in the United States. *American Psychologist,* 1968, **23,** 743-751.
Melton, A. W., & Briggs, G. E. Engineering psychology. *Annual Review of Psychology,* 1960, **11,** 71-98.
Miller, J. W., Malecki, G. S., & Farr, M. J. ONR's role in human factors engineering. *Naval Research Reviews,* 1967, **20,** 1-10.
Poulton, E. C. Engineering psychology. *Annual Review of Psychology,* 1966, **17,** 177-200.
Reagan, M. D. Basic and applied research: A meaningful distinction? *Science,* 1967, **155,** 1383-1386.
Swets, J. A., Tanner, W. P., Jr., & Birdsall, R. G. Decision processes in perception. *Psychological Review,* 1961, **68,** 301-340.
Taylor, F. V. Psychology and the design of machines. *American Psychologist,* 1957, **21,** 120-125.
Taylor, F. V. Human engineering and psychology. In S. Koch (Ed.), *Psychology: A Study of a Science.* New York: McGraw-Hill, 1963. Pp. 831-907.

PART TWO

HUMAN PERFORMANCE IN THE MAN-MACHINE SYSTEM

If there is any justification for the existence of a field called engineering psychology, it rests, in the final analysis, upon the viability of the man-machine system concept. While this point is implicit in all three of the definitions sampled in Part One, it bears repeating here because of the sort of material that we shall consider in this part. In essence, we shall be examining the *components* of the man-machine system in isolation—or at least those components which are most directly associated with human skill performance.

Whenever the system is disassembled in this manner, content differences between engineering psychology and other disciplines (both within and outside psychology) fade or disappear. All that remains of the distinction is a rather vague matter of orientation; to the extent that research is guided by—or even relevant to—man-machine system objectives, it may be included within the scope of engineering psychology.

To illustrate this point more vividly, consider a problem in color vision: how well can organisms distinguish among various chromatic stimuli? This question has been posed by the sensory physiologist in his effort to understand receptor processes, by the comparative psychologist in his attempt to trace the phylogenetic development of perceptual skills, and by the psychophysicist in his search for laws relating the worlds of physics and experience. Clearly, the interest of each of these specialists is focused upon the organism alone, irrespective of how sophisticated the machinery may be that adorns his laboratory.

The engineering psychologist is also interested in color discrimination, but as a man-machine interface problem. For him the discovery of laws governing this behavior may serve the engineer in his search for ways to improve information displays—in this instance, perhaps, through color coding. In other words, his interest is also in the organism, but only insofar as it functions as a component in a man-machine system.

The point of this example is very simple: in studying a particular aspect

---

References will be found in the General References at the end of the book.

of human performance (such as color discrimination) the engineering psychologist behaves no differently from other psychologists. He may formulate the same kind of problem, use the same techniques, and reach the same conclusions. He may even, upon occasion, find it possible to seek answers in data furnished directly by these more traditional areas. What distinguishes him is principally his concern for man-machine system objectives. He concentrates upon those facets of behavior which, if better understood, would be of greatest value in solving present or future system design problems.

In the present part we shall be sampling research and theory directed more-or-less exclusively toward specific components of the system. The scheme used to organize this material is an adaptation of Taylor's information-flow model presented earlier. Thus, we begin with a machine component—the display—which serves as the point of information input to the organism. Selections included in Chapter Two ask the general question, how should displays be designed so as to enable man to make utmost use of the messages that they attempt to convey?

From the display we follow the course of information, as best we can, through the largely uncharted reaches of man's vast processing network. Recognizing that our ignorance prohibits anything but a tentative classification, we consider *receiving, storing,* and *transmitting* functions as representative, if not of the true sequence of events, at least of the focal points of contemporary research.

Receiving and *storing* operations, of course, encompass the traditional fields of sensory-perception and memory. It is interesting to observe, as an historical footnote, that these venerable areas of psychology are somewhat indebted to the upstart engineering area for their current prominence in the field as a whole. Following the overthrow of introspective psychology by the empirically oriented behaviorists of the 1920s, all areas smacking of cognitive or mental processes went into a serious decline. None were more closely identified with the old order than were sensation and perception; consequently, none felt the wrath of the purge more keenly. It was not until the exigencies of World War II forced a reappraisal of existing psychological knowledge that these areas staged a comeback. It was discovered that classical sensory-perceptual principles could be applied effectively in the design of equipment (notably displays) for use by man, but also that conspicuous gaps in knowledge existed which required immediate attention. Hence the renewed interest in sensory-perceptual *research*.

More recently, a resurgence of interest in memory processes has developed which is traceable, at least in part, to another engineering concern. Our burgeoning technology has put at man's disposal vast quantities of information which, in one way or another, he must selectively process in order to make informed and rapid decisions. Selectivity, however, and other aspects of processing as well, seem to depend heavily upon man's ability to store information—both for brief and for sustained periods of time. How, and how well, can he do it? There are few questions of more fundamental significance for man-machine system design than this.

Transmitting refers to those operations whereby man attempts to reflect in his response as much of the input information as possible. In his excellent review, which appears in Chapter Five, Posner designates this class of functions *information conservation* in contrast to those involving *reduction* or *creation* of information. Owing to the fact that there seem to be two distinguishable emphases within the transmission or conservation research, both *discrete* and *continuous* task requirements are represented in the pages devoted to this topic.

The domain of discrete transmission tasks covers all situations in which man uses a set of distinct, specifiable responses to identify a set of distinct, specifiable signals, often under the pressure of time constraints. It is to such situations that the specialized techniques of *information measurement* are most readily applicable. These techniques, which we introduce with Miller's classic article, were developed by communication engineers in the late 1940s for use in evaluating communication systems. Psychologists quickly saw in them, however, new ways to approach a variety of old psychological problems: ways, for example, to quantify the dependence of any set of responses upon any set of stimuli, irrespective of whether either set is itself quantified.

By the mid-1950s, concepts associated with information measurement had gained a firm foothold on the theoretical as well as the methodological territory of engineering psychology, and for roughly the next half decade the literature was dominated by "information theory" concepts. The years since have seen a gradual decline in this emphasis as new approaches have emerged and the information model has been placed in proper perspective; it continues, however, to exert a prominent influence upon research aimed at discovering how people transmit discrete signals.

In contrast, the information model has had considerably less impact upon either methodology or theory in the domain of continuous transmission tasks. The distinguishing feature of such tasks is not, as the label seems to imply, the smoothness or continuity in what the human processor does; it lies instead in the continuous nature of the stimulus or input signal, and in the fact that both input and response can be regarded as continuous for purposes of analysis. This is perhaps best illustrated by the tracking task, a laboratory adaptation of such real-life problems as driving a car or flying an airplane. A stimulus "target" moves continuously about on the display, describing a course which the tracker is to follow as closely as possible by appropriate movement of a control device (such as a wheel or joystick). Careful analysis of the resulting behavior, however, suggests that man does not respond in a strictly continuous fashion, although his movements may become smoother and more congruent with the input signal as skill develops.

Tracking, then, represents a situation in which behavior is more-or-less discrete and the stimulus or source of information is continuous. Although the information metric can be and has been applied in this situation, other models and techniques have proven far more useful. In particular, notions borrowed from electromechanical servo theory have provided a valuable means for describing human behavior in a variety of control tasks. Several excellent papers by Briggs (1964, 1966) and Poulton (1966b) are recommended as supplementary reading for the reader with a serious interest in this area.

From simple transmission we proceed to what, for want of a better term, we shall call *intellectual* or *cognitive* processes. In this category we lump together all those tasks which seem to require subjects to do more with the input information than simply preserve as much as they can of it in their response. Admittedly, our current understanding of intellectual or cognitive processes is nowhere near sufficient to justify delimiting this category as we have. Our only defense lies again in the criterion of expediency: the papers chosen do reflect the growing emphasis within engineering psychology upon man as a decision maker in a number of seemingly different contexts.

It will be noted, perhaps with some surprise, that no chapter in this part is devoted exclusively to human response characteristics, even though skilled performance has often been viewed naïvely as involving predominantly "motor" processes. Those most intimately concerned with skill research, however, have long warned against this sort of oversimplification, pointing out the central importance of ". . . the spatial-temporal *organization* of receptor-effector-feedback processes, which often are relatively independent of the specific receptor or effector elements initially involved" (Fitts, 1964, p. 244). One does serious injustice, therefore, to the sophistication of both the behavior and the current theoretical accounts of it by accenting the term "motor" in perceptual-motor skills; he might better accent the hyphen! To emphasize this point, we have elected to organize this part so as to include tasks requiring considerable response precision within the broader discussion of transmission and cognitive processing functions.

In Chapter Six, the focus of attention returns from the man back to the machine. What are the factors, it is asked, that should be taken into account in designing controls for man's use? Studies designed to answer this general question have examined system output—usually with regard to positioning or tracking skill—as a function of various physical characteristics of the output device.

Any organizational scheme has its shortcomings, and the one adopted in this part is no exception. Consequently, it is hoped that the reader will heed our repeated advice to attach no great significance to the way in which functions are categorized. Decision is an integral part of detection, for example, and memory, of perception or recognition. Even in the laboratory, the performance of a skilled task involves a constant interplay of many processes. When we spoke earlier of examining man-machine system components "in isolation," we were speaking in relative terms and, perhaps, a bit loosely. What the researcher does, of course, in concentrating upon a particular function is to manipulate systematically those factors which he believes have an important bearing upon the function in question, holding all other factors—and hopefully the functions which they control—as constant as possible. He does not *eliminate* the other functions.

It is also hoped that the reader will not be encouraged by the present organization to ignore chronology. Although it is nowhere explicitly traced, there is a trend in the research illustrated in this part, a trend which seems to reflect man's changing role in the system. The studies with the earliest dates concentrate upon *control* functions and upon the design of specific displays and control devices. Later ones are more concerned with man's performance—especially as an *information processor*—in a far broader range of tasks. Principles of greater generality are being sought in an effort to avoid the obsolescence inherent in task- or system-specific research findings.

Chapter Two

Human Factors in Display Design

That displays should be designed for the convenience of the user seems almost to go without saying. Who with a grain of common sense would argue otherwise? And if our goal is to enhance system performance by making things easier for the user, why can't we simply *ask* him how he would like information to appear? It would scarcely seem necessary to involve human factors specialists, engineering psychology principles, or elaborate research programs in so elementary a problem.

Nevertheless, with all due respect to the integrity of personnel who use displays and engineers who build them, we must take issue with the "common sense" approach. First, it has not proven as easy to implement as one might believe. What is "common sense" to one person may be utter nonsense or irrelevance to another. Adult human beings bring to any situation a host of individual preferences, biases, and expectancies built upon years of experience unique to themselves. Although there are undoubtedly some stereotypic ways of viewing things—at least within a common culture—there is considerable room for differences of opinion in something as complicated as a display design. An astronaut might favor a particular display format for a space vehicle because of its similarity to the ones with which he has had experience as a pilot; to an engineer, on the other hand, a different design might be preferable because of its compactness or ease of maintenance. Whose "common sense" is more valid?

This example serves to point out several other difficulties inherent in trying to implement "common sense." User convenience is not the *only* criterion worthy of consideration; cost, engineering feasibility, retraining requirements, and a host of other factors must also be weighed. Furthermore, a given device may have a number of users—each with a somewhat different set of requirements. Finally, it is not always true that the user is the best judge of what is good for him or for the system, especially if the choice pits something *novel* against something *familiar*. Having no basis upon which to judge how proficient he *could* become with a new display, he tends to prefer a familiar one unless it has proven unusually troublesome for him in the past. This is not to suggest, of course, that user opinion is utterly worthless and should be completely ignored. On the contrary, there are numerous instances—and the manned space program is an excellent case in point—in which potential users have contributed valuable insights to the design of machine components. The point is that subjective evaluation

References will be found in the General References at the end of the book.

alone affords at best a tenuous basis for designing displays "with man in mind."

The second major objection to the "common sense" approach is simply that it has failed so often in the past. The literature in engineering psychology abounds with examples of display designs which have become widely adopted at the expense of system performance or well-being (see, for example, Chapanis, 1965). Apparently, these represent *someone's* "common sense" judgment. Unfortunately, once implemented, such mistakes can be rectified only at great (and often *prohibitively* great) expense. It has been recognized for years that the standard aircraft altimeter is subject to misreading by a pilot under stress, and it is probable that a number of regrettable accidents have occurred as a direct result. Nevertheless, to alter such a display would unquestionably endanger even more lives as pilots, under the stress of a difficult landing, searched in vain for old familiar cues. The prospect of overcoming years of experience gained on a subpar instrument, therefore, has discouraged many attempts at design improvement. Clearly, the moral is to incorporate human factors considerations into the design at the outset, a principle which has been gaining increasingly wide acceptance among those responsible for system planning.

What, then, can the engineering psychologist contribute to display design? The answer, of course, is facts, and methods for collecting facts, regarding human performance as a function of display variables. Without some factual basis upon which to develop "human factors considerations," the human factor specialist would be forced to rely upon his own "common sense" judgments. Admittedly, his point of view might be more detached than that of the user, and his concern for the human might enable him to spot potential difficulties in particular designs proposed by engineers. Nevertheless, any common-sense approach is, as we have seen, severely limited.

Few problems in engineering psychology have spurred a more vigorous or diverse search for empirical data than have those associated with display design. Over the years work in this area has progressed along two very broad but distinguishable fronts. On the one hand, attempts have been made to adapt *general principles* of sensory or perceptual psychology to specific display problems; to extract from known characteristics of one or another sense system guidelines, criteria, or even precise specifications that a given display should meet. This approach has traditionally dealt with stimulus features such as contrast, resolution, color, stroke width, and signal-to-noise ratio. On the other hand, a considerable amount of research has also been applied to specific displays or classes of displays in an effort to establish, through direct comparison, their relative order of merit. Does Design A or Design B yield the better performance under task conditions x, y, and z? Empirical answers to such questions have furnished the basis for several handbooks of specific design principles (see, for example, Baker & Grether, 1954; Woodson, 1954; Morgan, Cook, Chapanis, and Lund, 1963).

It is not within the scope of this chapter to catalog all—or even a significant part—of the display recommendations yielded by these two lines of endeavor. They are indeed voluminous and can easily be found elsewhere. Instead, articles have been selected to illustrate both approaches, and at the same time to show the increasing complexity of problems raised in conjunction with display design.

The first reading in this chapter may be regarded as representative of the general-principle approach. In the Howell and Kraft paper, the issue was: how do the size, contrast, and blur levels of alphanumeric stimuli interact to determine legibility? Since there was no ready answer to this question in facts already available, a new experiment became mandatory. The functions obtained, however, have significance beyond the specific application for which they were intended, i.e., identifying which "alphabets" would prove most useful under various kinds of display degradation. Since they *are* psychophysical functions, they also add to the general body of knowledge regarding sensory processes. Engineering psychology, therefore, both draws upon and contributes to the fundamental understanding of human behavior in seeking answers to questions such as those of display design.

In contrast to the first selection, the last three papers all illustrate the *specific-comparison* approach. Each is an experimental attempt to determine which of several display techniques is most conducive to efficient system performance under a particular set of circumstances. Since the research typically deals with entire *designs* rather than component dimensions or processes, the results tend to be of somewhat limited generality. By the same token, however, they also tend to be more directly applicable to the particular display problem addressed.[1] Recommendations are not *derived from* the results; they *are* the results. This is not to imply that such research is necessarily trivial or unscientific, or that it involves completely arbitrary experimental comparisons (although, of course, all of these shortcomings are in evidence in *some* display work, and probably most prominently in the specific-comparison variety). In fact, the articles included in this chapter were chosen expressly because they minimize these undesirable features. Hitt et al., for example, undertook a format comparison only after a painstaking analysis of relevant task features and feasible display manipulations. Herman, Ornstein, and Bahrick were concerned with the whole issue of whether a decision maker can benefit from *probabilistic* information if it appears in graphic form on a display. Therefore, they compared decision performance using several probabilistic and nonprobabilistic displays.

Only in the Matheny, Dougherty, and Willis article do we encounter a traditional comparison study of highly restricted scope. This investigation was not included because of its system-specificity, but because it illustrates one of the most venerable arguments in the human factors area: inside-out vs. outside-in display of aircraft attitude. Which is better? Generally, it has been found *experimentally* that the outside-in variety, in which the "horizon" is fixed and the "airplane" moves, is superior; pilots, however, probably because of their training, consistently prefer the inside-out variety, in which the "horizon" moves. Here is a classic illustration of the point made earlier about user preference and the problems associated with altering operational designs.

One final point should be made with respect to the general area of display research. With the oft-mentioned evolution in machine technology proceeding so rapidly, engineering psychology has begun to encounter an increasing number

---

[1] A conscious effort is made here and throughout this book to avoid use of the conventional, but grossly misleading, adjectives *basic* and *applied* to distinguish the major classes of research. Our reason is that these terms are heavily loaded with evaluative connotations, even though no one has been able to furnish objective criteria on which to distinguish them.

of display problems against which neither of its traditional modes of attack seems particularly effective. So long as the basic nature of a machine or class of machines (such as automobiles or airplanes) remained unchanged, the engineering psychologist was fairly safe in making choices as to which sensory capabilities to examine and what kinds of displays to compare. He was confident that his recommendations regarding symbol size or inside-out displays would be applicable even if vehicles were made to go faster or to run on a different kind of fuel.

Today, however, entire *concepts* of machine and system design are transient. Of what importance is a solution to the inside-out display question if man no longer controls vehicles? The engineering psychologist must either be prepared to anticipate specific display requirements of the future—necessitating, as Christensen advocates in Chapter One, a complete reorientation and broadening of his formal education—or he must seek more generally applicable principles of human performance which can be adapted to a wider range of possible display problems. The trend, in our opinion, seems to be in the latter direction. We consider it quite likely that within the next ten years there will be no "display" chapter in a book of this sort; that principles of human performance in the processing of information, to which we turn in the next chapter, will stand as the chief contribution of engineering psychology to display design.

4

Size, Blur, and Contrast as Variables Affecting the Legibility of Alpha-Numeric Symbols on Radar-Type Displays

William C. Howell and Conrad L. Kraft

INTRODUCTION

Several investigators (1, 12) have discussed the advantages to be gained in radar air traffic control by displaying certain information in symbols electronically printed on a cathode ray tube (CRT). Only recently, however, with the development of such devices as the Charactron and the Digitron, has it become feasible to use letter and number coding for this purpose. As a consequence of the expected usage of such coding, it becomes important to determine the characteris-

From Wright Air Development Center Technical Report No. 59-536, September 1959.

tics of the symbols which would provide satisfactory legibility under the conditions imposed by the radar-type display.

A number of studies have been reported concerning the influence on eligibility of variables such as type style (4, 9, 13, 16, 17), stroke width (2, 5, 13), size (4, 5, 6), patterning (2, 3, 17), and illumination characteristics (6, 13, 15). Unfortunately, the findings are somewhat equivocal in that general agreement regarding the effects of these variables is lacking. The reasons for this ambiguity are due in part to the following: (a) the discriminations required are very complex and therefore highly susceptible to minor changes in the general method or task environment; (b) the objective criteria used in evaluating legibility have been extremely diverse and, in all probability, not highly comparable; and (c) the questions asked in legibility research have frequently been qualitative and highly specific. For these reasons much of the above data is not necessarily generalizable to the present problem: legibility on a radar-type display.

One series of studies appears in the literature, however, which does bear directly on the current problem. Crook and his associates (5, 6) have investigated the influence of illumination brightness, figure size, and amplitude of apparent vibration, in various combinations, on the speed and accuracy of reading letters or numerals. The materials were black figures on a white ground illuminated by various amounts of dim white or red light. In general, it was found that performance improved very rapidly as size was increased up to a point at which the curve leveled off abruptly. The actual location of this point, as well as the slope of the function itself, varied considerably with the other conditions: under red illumination of .082 ft.-c. it was about 6 pt. type; under white illumination having a luminance of .046 ft.-L. or .94 ft.-L. with .029 in. of vibration, it was about 8 pt. A similar effect was noted for brightness, the break occurring around .05 ft.-L. For apparent vibration, the break approximated 5 min. of visual angle.

Even more marked, however, were the interactions obtained among the above variables. So great were these effects that, within the limited ranges studied, a high level of two variables practically negated any influence of a third variable on performance. Thus, for example, under a good brightness level an apparent 0.04-in. vibration did not affect performance, whereas under a 0.02-ft.-L. level of brightness a 0.02-in. vibration produced a marked proficiency reduction.

In view of these findings, it seemed advisable to approach the CRT display problem with the following considerations: (a) the stimuli and their immediate surroundings should emulate, insofar as possible, those conditions which would obtain in viewing printed characters on a CRT display; (b) those stimulus variables should be selected for investigation which appear most relevant to the CRT display situation; and (c) the interactions among the variables should be given particular attention. The resulting study differed from those of Crook et al. (5, 6) in that the former was aimed at a CRT-reading task rather than a dial-reading task. As a result, the symbols were white-on-black rather than black-on-white, ambient illumination was held constant rather than being varied, the symbol alphabet was enlarged to include both letters and numerals, and all stimuli were presented singly rather than in various combinations.

The dimensions under investigation, although apparently similar to Crook's, were somewhat different operationally. There were three variables: (a) size of the symbols, (b) contrast or brightness of the symbols relative to a constant background brightness, and (c) blurredness of the image which was defined as the rate of transition between the brightness of the symbols and that of the field. The relevance of each of these variables to the problem of displaying symbols on a CRT is readily apparent. Under these conditions size is restricted by the number of targets to be represented in a given display area, contrast by the physical parameters of the electronic beam and the ambient illumination characteristics, and blurredness by the beam focus, tube phosphor graininess, and persistence relative to target movement.

The interactions among these dimensions are especially worthy of investigation. For example, it is important to know the extent to which increased size can *compensate* for increased blur and reduced contrast. Also, it is desirable to know the optimum combination of the three dimensions as well as the loss in legibility incurred by deviations from this optimum.

Another feature of the present study which distinguishes it from earlier research is the emphasis placed upon specific symbols in the alphabet. An attempt was made to determine the specific letter (or numeral) confusions underlying the legibility scores obtained under all conditions. Through this technique it was hoped that data would be made available which could lead to an optimum alphanumeric code selection as well as to an optimum selection of physical dimensions.

The major purpose of the present study, then, was to determine functions relating size, degree of contrast, and blur to legibility. The primary criterion of legibility was rate of information transmission, a metric that reflects both of the customary legibility measures, speed and accuracy of stimulus recognition. In addition, speed and accuracy scores were obtained separately, as were indices regarding the confusability of the individual symbols under the various experimental conditions.

METHOD

Four values were selected from each of the three dimensions (size, blur, and contrast), and these were combined into a 4 x 4 x 4 factorial design. Each of 20 subjects served under all 64 conditions resulting from this combination.

Apparatus. The device used to project the stimuli and to manipulate their size, blur, and contrast levels was loaned to the authors by Drs. Glenn A. Fry and Jay M. Enoch and the Mapping and Charting Research Laboratory of The Ohio State University. This apparatus was intended by its designers for use in photo-interpretation research, and the details of its construction are reported elsewhere (8). Light from a standard projection lamp was passed through a diffuser and a collimating lens which focused on a ground-glass screen. The stimuli, which were introduced just prior to this point on 35-mm. filmstrips, were projected on the ground-glass screen. The filmstrips were housed in a portion of a Soundview 35-mm. projector which included the storage, holding, and stepping mechanisms. The whole filmstrip assembly was mounted on a movable platform which permitted the distance between the diffusing screen and the film to be varied. By manipulating this distance, it was possible to vary by an equal amount the gradient of all light-dark transitions projected on the screen. The relationship between this separation and the light distribution (i.e., blur) has been calculated and found to be directly proportional. Thus, it was possible to manipulate blur systematically by varying the degree of separation between the film and the screen.

The image on the diffusing screen was viewed through a zoomar-type magnification system and a half-silvered mirror illuminated from above by a constant level of diffused light. The former provided a convenient method for manipulating the size variable, and the latter served as the background illuminance for the stimuli on the basis of which contrast specifications could be made.

In order to control accommodation, a collimating lens was introduced between the eye and the rest of the optical system which effected accommodation for infinity. The subject viewed the stimulus through a 2-mm. artificial pupil, the image appearing in the center of a limitless, uniformly illuminated field. This effect was produced by allowing the viewing area to be seen through a rectangular hole cut in the center of a semicircular reflecting surface. This surface was matched in brightness and, insofar as possible, in hue with the viewing area, and it was extended past the limits of peripheral vision permitted by the artificial pupil. The effect was not perfect in that the viewing area could be distinguished from the surround on the basis of hue. The purpose of the surround was served, however, in that it permitted the figures to be seen against a background of homogeneous brightness. The illuminance of this background was such (3.5 ft.-c.) that complete light adaptation by the subject could be expected to occur almost immediately after entering the apparatus.

The filmstrips were advanced by pulses from a relay which closed with the subject's activation of a throat microphone. Subjects were thereby able to bring about the occurrence of a new stimulus merely by responding to the preceding one. This method afforded the subject complete control of stimulus pacing.

Stimuli. All 26 letters of the alphabet and 10 Arabic numerals were used as stimuli. The Mackworth style was chosen on the basis of its demonstrated high legibility over a wide range of illumination levels (13). To avoid the obvious confusions of 1 and I and of 0 and O, the zero and the one were modified: the zero by a diagonal line from the lower left to the upper right and passing through the center, and the one by the addition of a flag and extended base.

The symbols, which appeared to the subject as light on a darker field, were arranged in random orders on high-contrast filmstrips. Each symbol occupied a single frame and had the following dimensions: 1.5-mm. height, 0.2-mm. stroke width, and 0.8-mm. over-all width (except for single stroke letters, such as I). To minimize the necessity for changing filmstrips, the length of each film was made to correspond to the number of stimuli which were to appear in a single trial, i.e., 144 frames. Sixteen such filmstrips were developed, each containing four repetitions of the 36 stimuli in a different random sequence. For a given trial, a random selection was made from among these 16 orders (filmstrips). In this way, all differences in film quality were randomly distributed over all conditions, and the possibility of the subjects memorizing all or part of any sequence was reduced.

Calibration and Selection of Stimulus Values. For the three dimensions investigated, the optimum values were all chosen arbitrarily as the maxima provided by the apparatus. These values, as well as all others measured, are shown in Table 1. The minimum values were arrived at by the following procedure: (a) Using two experienced subjects, absolute (75%) thresholds of recognition were determined separately for each of the three dimensions (the two being held constant at their maximum values). (b) Several j.n.d.'s were marked off along each dimension, and (c) the stimuli were improved in 1-j.n.d. steps on all dimensions concurrently until the 75% recognition threshold (*combined threshold*) was reached. Since this value fell between the 4- and 5-j.n.d. levels, (d) the individual dimensions were reduced systematically until the 75% criterion was met. The result was 4 j.n.d.'s of size, 5 of contrast, and 3 of blur. The two intermediate points along each dimension (steps b and c) were obtained by merely subdividing the physical intervals between the maxima and minima.

Throughout the study, size was expressed in minutes of visual angle for letter height, contrast in units based on the formula [illuminance of figure (B_1) − illuminance of field (B_2)]/illuminance of field (B_2), and blur as the ratio between the width of the transition gradient from figure to ground and the stroke width of the letters. This expression of blur was adopted in preference to the width of the gradient, per se, on the assumption that the former is more applicable to the problem of legibility. It is obvious, for example, that a small letter could be completely obliterated by an absolute gradient that would have little effect on the form of a large letter.

All measures of size were derived from the known magnification characteristics of the system and the measured size of the stimuli on the film. Contrast levels were based on illuminance measures made with a Macbeth illuminometer. The various illuminance values were as follows: (a) surround and limitless field, 3.5 ft.-c.; (b) stimuli prior to blur transition, 46-134 ft.-c.; (c) projection source, 1627 ft.-c.

The characteristics of the individual film strips were determined for the purpose of film comparison. Resolution averaged 22 lines/mm. with extreme values of 17-24 lines/mm. The density of the clear portions (i.e., letters) approximated 15.0% transmission, and that of the dark portions was less than 0.1% transmission on all strips.

Subjects. Twelve male university students served under 18 practice and 64 experimental conditions. The vision of each subject was corrected to 20/20 by inserting the appropriate lenses just behind the artificial pupil. Although a number of the subjects were familiar with the apparatus, the specific task was new to all of them.

Procedure. In order to reduce learning effects and to insure familiarity with the requirements of the task, a total of 18 practice trials preceded the experimental sessions. These trials consisted of identifying 36 stimuli under each of 18 widely differing conditions of blur, contrast, and size.

Each experimental trial required the identification of all 144 stimuli on one of the 16 random filmstrips. Each subject performed one trial under each of the 64 experimental conditions, with the order of performance being randomized. A rest period of approximately 5 min. was introduced after each of the six trials which, together, constituted a session.

Instructions to the subject emphasized the importance of both speed and accuracy. Other investigators have found that under such instructions subjects tend to perform at their own optimum information transmission (H_t) level (11). Only letter and number responses were permitted, although the subject was provided with a means of communicating with the experimenter in the event of equipment failure or similar emergencies.

Scoring. The experimenter checked all responses against a master sheet corresponding to each stimulus sequence. As errors occurred, they were recorded by writing in the incorrect responses. Thus, a running account was kept of the kind as well as the number of errors that occurred under each condition. By combining these sheets it was possible to derive information transmission data, combined error data, and letter confusion data. Magnetic tape records provided verification of these scores.

The average speed at which the subject performed was measured by recording the time required to complete each trial and dividing by the number of stimuli (144). In addition to providing the measure of average speed for number of stimuli per unit time, these scores were combined with the error scores to provide information transmission data that yielded an over-all efficiency (legibility) metric: rate of information transmission.

RESULTS

Information Metric. The error data for each subject under a given condition were combined into separate stimulus-response matrices for each condition, and the amount of information transmitted (bits/sec.) was computed for each matrix.

This computation (14) is based upon the comparison of the uncertainty or entropy in the stimuli, $H(x)$, with that introduced by the observer in giving various responses to the same stimuli, or *response equivocation*, $H_y(x)$. The former entropy is defined as

$$H(x) = - \Sigma_i \, p(i) \, \log_2 p(i) \quad (1)$$

in which $p(i)$ refers to the probability of occurrence of the individual letters or numerals. Response equivocation, on the other hand, depends upon the conditional probability that a specific symbol (i) was presented, knowing the symbol (j) which was given as a response, and is defined as

$$H_y(x) = - \,_i\Sigma_j p(i, j) \, [\log_2 p(i, j) - \log_2 p(j)] \quad (2)$$

Information transmitted is obtained by subtracting Eq. 2 from Eq. 1, giving

$$I_t = H(x) - H_y(x) \quad (3)$$

To this point, all information measures have been expressed in entropy units (known as *bits*), disregarding the time factor. This is introduced simply by di-

TABLE 1
Stimulus Values Used in the Study

	Size (min. visual angle)	Contrast $\frac{B_1-B_2}{B_2}$	Blur Distribution Stroke Width
Apparatus Limit	36.79	37.57	≈ 0
Step 3	26.80	30.14	0.55
Step 2	16.40	19.30	1.50
Approximate 75%* threshold (combined conditions)	6.00	12.14	2.50
Approximate 75%* threshold (individual conditions)**	4.73	1.43	4.67

* Average for two experienced observers
** Not used for stimulus values

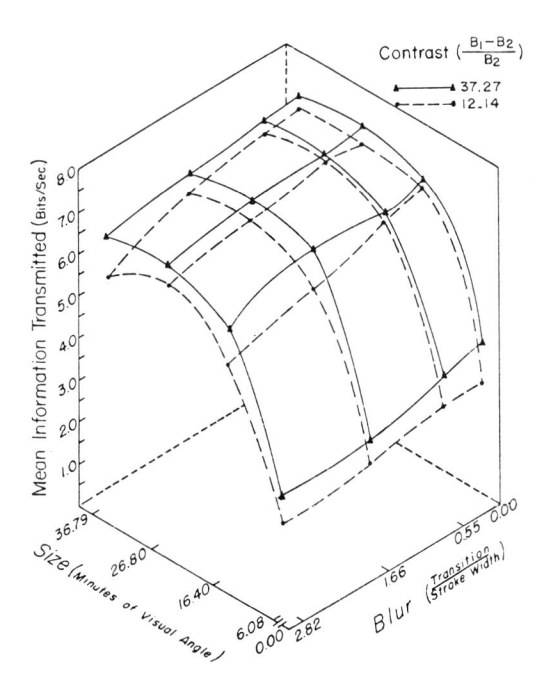

Figure 1. Mean information transmitted (bits/second) as a function of size, blur, and contrast of letters.

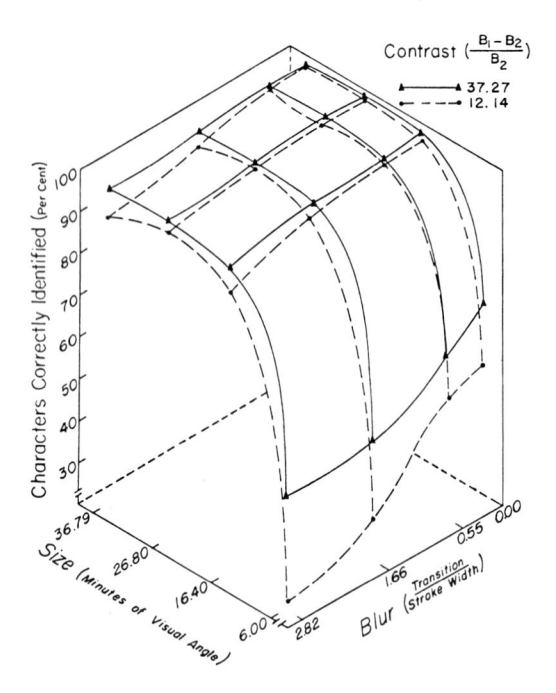

Figure 2. Per cent of characters correctly identified as a function of size, blur, and contrast of letters and numbers.

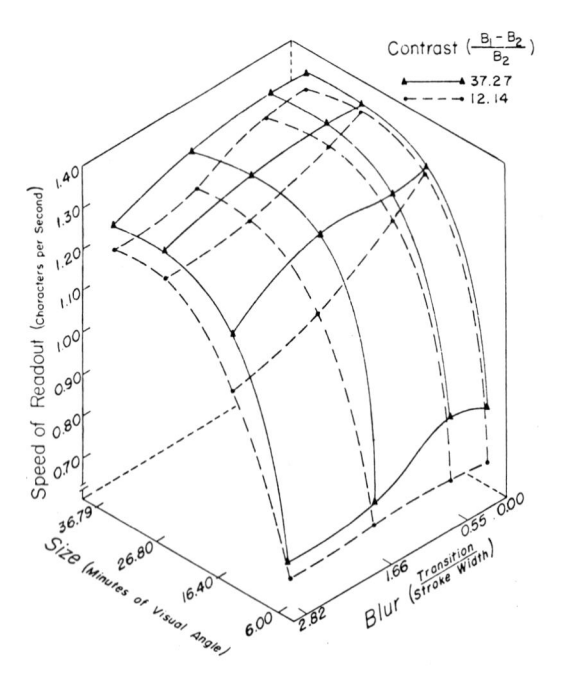

Figure 3. Speed of readout of letters and numbers as a function of their size, blur, and contrast.

TABLE 2
Severity of Specific Confusions as Indicated by Per Cent of Symbol Presentations
(Absolute Frequency of Confusions in Parentheses)

Condition: Blur = 0.00; Contrast = 37.57; Size = 36.79

Symbol	4%-7% (2-3)	8%-11% (4-5)	12%-15% (6-7)	16%-19% (8-9)	20%+ (10+)	% Correct
A						100.0
B						97.9
C						100.0
D						100.0
E						95.8
F		5				95.8
G						100.0
H						100.0
I		Y				93.7
J						95.8
K						95.8
L						100.0
M						97.9
N						100.0
O		Ø				91.7
P						100.0
Q						95.8
R						97.9
S						100.0
T						100.0
U						100.0
V						100.0
W						97.9
X						100.0
Y						97.9
Z						97.9
Ø		G, O, Q				85.4
1						100.0
2						97.9
3						100.0
4						97.9
5						100.0
6						100.0
7						100.0
8						100.0
9						100.0

TABLE 3

Severity of Specific Confusions as Indicated by Per Cent of Symbol Presentations
(Absolute Frequency of Confusions in Parentheses)

Condition: Blur = 0.00; Contrast = 37.57; Size = 6.00

Symbol	4%-7% (2-3)	8%-11% (4-5)	12%-15% (6-7)	16%-19% (8-9)	20%+ (10+)	% Correct
A			4			75.0
B	N, 5, 9	D				62.5
C						93.7
D	3					85.4
E	C					81.2
F		E, P				72.9
G		C		6		64.6
H	F	W	M		N	33.3
I	F, Z					81.2
J	3, I					72.9
K	A, C, R, X, Y	E				62.5
L	E, I, 1					83.3
M	W, Ø	N		H		54.2
N	O, Ø, 8					75.0
O	D					81.2
P		F				83.3
Q		Ø	G		O	35.4
R	C, G, K	E				62.5
S	I, V	3			5	47.9
T	J, Y, 7					75.0
U	O					83.3
V	U	Y				72.9
W	D, H, 4	M			N	27.1
X	I, R, Y	K				60.4
Y	F, I				T	64.6
Z	I, 7	3				81.2
Ø	P, S, U, 8	D			O	35.4
1	I, J	L			Z	68.7
2	7					35.4
3			5	S		62.5
4	A					89.6
5	9	S, 3				64.6
6	A			4		66.7
7		Z				83.3
8	G, R, 9	D, O	B, Ø			20.8
9	Y, 7		P			64.6

TABLE 4
Severity of Specific Confusions as Indicated by Per Cent of Symbol Presentations
(Absolute Frequency of Confusions in Parentheses)

Condition: Blur = 2.50; Contrast = 12.14; Size = 6.00

Symbol	4%-7% (2-3)	8%-11% (4-5)	12%-15% (6-7)	16%-19% (8-9)	20%+ (10+)	% Correct
A	I, J, K, N, R, Y, 1, 3	4				29.2
B	A, E, P, R, X	D, N, Y				20.8
C	E, F, I, O, T, V, 5					41.7
D	B, J, N, O, P, R, 2		5			29.2
E	F, G, K, U, X, 2, 3	C, I				20.8
F	C, J, N, R, X, 6	V, Y		P		14.6
G	A, B, D, E, U, V, Y	C, 6	4			16.7
H	E, S, Ø	O			M, N	2.1
I	E, L, N	J, T	1			39.6
J	D, O, T, Z, 3, 7		I			31.2
K	C, E, M, N, O, V, X, Y, 9		4			18.7
L	C, G, U, Y, 1	A	I			33.3
M	E, K, O, T, V, W	H, R	N			31.2
N	A, G, U, V, 6	M, O				37.5
O	D, E, P, T, Ø, 4, 6	C, N, U				22.9
P	D, N, R, T, V, Y, 4					45.8
Q	D, G, J, U, V, 4				O	4.2
R	D, E, M, P, Ø, 1, 5, 8	K, O	N			10.4
S	C, G, J, M, V, 3, 9	I, P, T	5			12.5
T	A, C, J, P, Z, 1	I	Y			22.9
U	D, F, K, Z, Ø	N, O				33.3
V	C, I, N, 5, 9	O, U	Y			14.6
W	A, D, M, O, U, V, X, Y, Ø	M, N				20.8
X	F, I, J, N, P, R, T, V, 1, 6	A, K				14.6
Y	D, E, J, V, X, 5	T		I		27.1
Z	A, I, J, L, T, 1, 7	C	2			20.8
Ø	A, B, D, G, M, V, 2, 4, 9	O, P				18.7
1	A, D, L, T, V, 4	I, J				25.0
2	A, C, I, J, N, P, S, T, Ø, 7		Z			12.5
3	A, C, J, L, N, O, 1, 8	I, T, 5				12.5
4	A, O, X, 1					54.2
5	K, Y, Ø, 3, 6, 9	S, 1				37.5
6	C, F, G, L, N, O, U, V, 8			4		16.7
7	D, F, O, V, 2, 5, 9		P, T			14.6
8	B, M, N, P, R, X, 4, 5	O, U	Ø			4.2
9	D, N, T, V, 1, 4, 7		P			29.2

TABLE 5
Severity of Specific Confusions as Indicated by Per Cent of Symbol Presentations
(3072 Presentations)

Combination of All Conditions

Symbol	4%-7% (2-3)	8%-11% (4-5)	12%-15% (6-7)	16%-19% (8-9)	20%+ (10+)	% Correct
A						87.5
B						81.3
C						91.4
D						88.8
E						87.0
F						85.9
G	C, 6					82.9
H	M	N				75.8
I		1				79.1
J						87.3
K						81.6
L						88.2
M	H					86.6
N						90.9
O		Ø				81.5
P						89.5
Q		O				73.8
R						83.0
S		5				76.8
T						86.8
U						90.6
V						85.7
W	N					78.4
X						82.5
Y	T					81.4
Z	2					84.7
Ø	O					76.7
1	I					83.1
2		Z				78.4
3	5					81.3
4						93.4
5	S					86.1
6	4					83.7
7						85.8
8	B, Ø					73.4
9	P					83.3

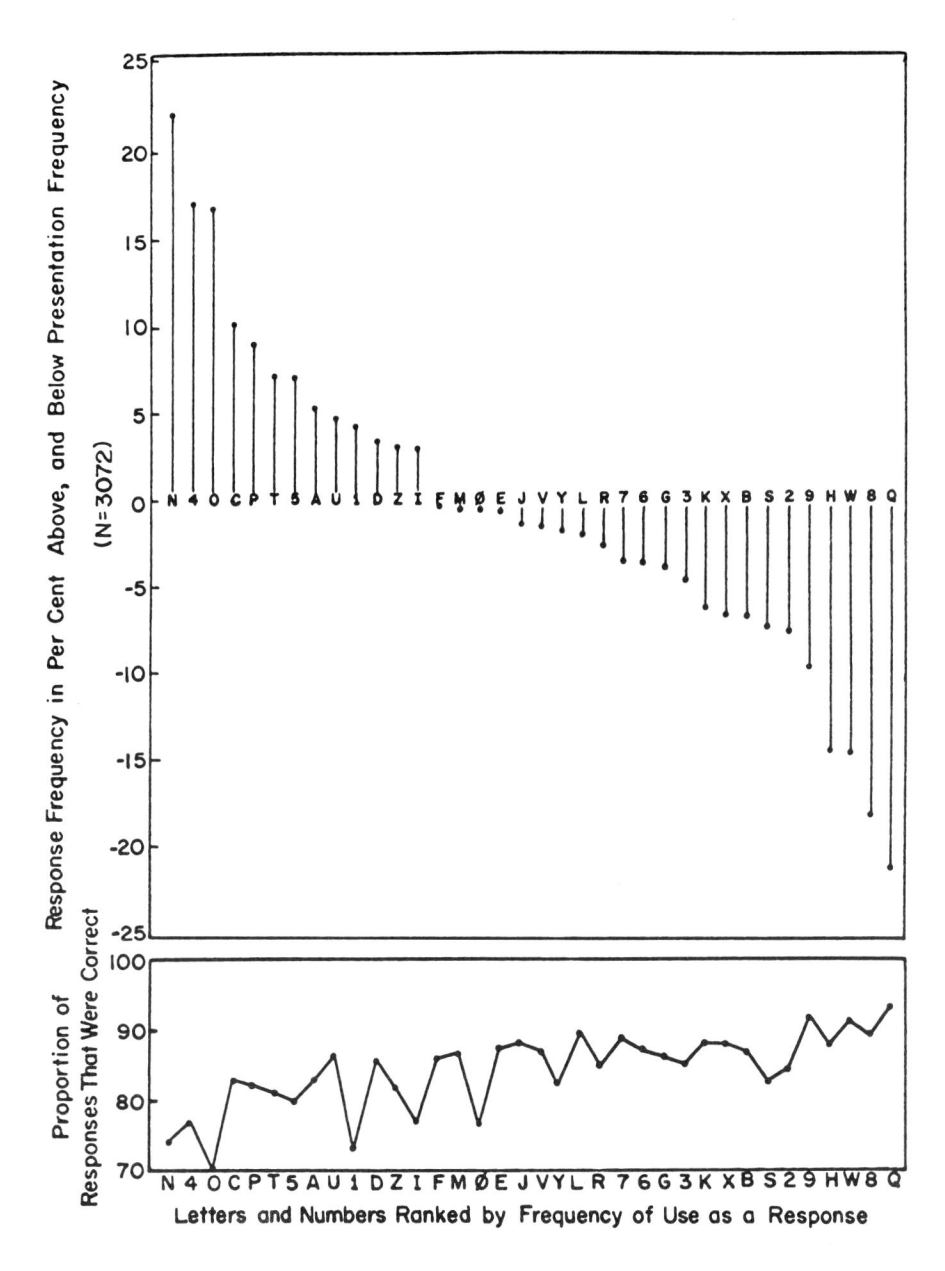

Figure 4. The relative frequency of use of letters and numbers as responses and the proportion of these responses that were correct.

viding the information quantity by the average time required for each response so that

$$H_t \text{ (bits/sec.)} = \frac{H(x) - H_y(x)}{T} \qquad (4)$$

It is obvious from the above discussion that H_t is responsive to the following aspects of performance: time for response, extent to which a stimulus tends to elicit a specific response (in general the correct response), and number of response categories used for a given stimulus. Since each of these factors was considered important in evaluating the legibility of stimuli under the various experimental conditions, the information measure was taken as the primary performance datum.

These information scores were averaged over subjects for each condition to indicate the differences between conditions. Figure 1 is a three-dimensional representation of the results of this analysis. To avoid ambiguity, the contrast variable, which appears as the parameter in this figure, is represented by only the two extreme levels. The surfaces representing the other two contrast levels fall nicely between these extremes.

The differences between conditions were tested statistically using an analysis of variance design. Bartlett's test performed on the original scores indicated that the assumption of homogeneity of variance had been fulfilled. Significant F ratios were obtained for all three of the major variables ($P < .001$) and for two of the interactions, contrast x size ($P < .05$) and contrast x size x blur ($P < .001$).

In order to ascertain the pattern of difference described by these interactions, the Duncan Multiple Comparisons Test (7) was applied to the individual condition means. The major implications of this test may be seen by referring again to Figure 1. First, there is one size (26.80 min.) at which performance was particularly resistant to blur and reduced contrast effects. Increasing size beyond this point resulted in a performance decrement in 13 out of the 16 cases in which such an increase occurred, with nearly half of these decrements being statistically significant ($P < .05$). The only improvements (nonsignificant) in performance with further size increases occurred under the blur-free condition. Second, the performance decrements resulting from either lowered contrast or increased blur alone were accentuated when the two were combined. This effect is particularly noticeable at the largest and the second from smallest sizes.

Accuracy Metric. Error scores for each condition were averaged over subjects and the resulting means were divided by 144 (symbols in a trial) to yield the proportion of incorrect responses per condition. This value was then subtracted from 100 to give the proportion of characters correctly identified as shown in Figure 2. Again, only the extreme contrast levels are shown.

Processing Time Metric. The mean number of characters read per second was computed for all conditions, and the results appear in Figure 3. This speed index includes responding time as well as processing time. However, if it is assumed that responding time remains constant for a given symbol over all conditions, the variation in scores may be taken to reflect differences in processing alone.

One of the major reasons for including the time and error scores as well as the information transmission details is to show the high degree of correlation which exists among these measures. Except for the shape of the three curves at the minimum size, the topography of their surfaces is nearly identical. Correlation coefficients were computed between all three pairs of measures and in all cases exceeded $r = .98$. These values are significant well beyond $P < .01$.

Confusion Data. Confusion matrices for the individual symbols were compiled for every condition. These matrices permitted an analysis of the error scores in terms of the specific letter confusions contributing to them. Since the inclusion of all 64 confusion matrices would have proven unwieldy, only an illustrative sample is presented here. These appear in Tables 2, 3, and 4 as summary tables in which confusions are entered according to their severity. Confusion severity is indicated by the percentage of times a given erroneous response occurred when a given stimulus symbol was presented. Table 2, for example, indicates that no erroneous response was given to any symbol more than 7% of the time under the high contrast, minimal blur, *large size* condition. Under the high contrast, minimal blur, *small size* condition (Table 3), however, N is given as a response to H, O to Q, 5 to S, etc., each more than 20% of the time.

Confusions occurring only once under a given condition (i.e., < 4%) are not recorded in these tables since their occurrence is probably attributable to any of a number of random factors. An indication of the proportion of these errors occurring for a particular symbol may be obtained by adding the confusions given in the table to the per cent correct (last column) and subtracting from 100%, or total presentations.

Table 5 is a summary matrix indicating the confusions occurring under all conditions combined. The information provided by this matrix is the over-all confusability of a given letter under a wide variety of viewing conditions. Here, of course, the total number of presentations for each symbol was 3072, rather than 48 as in the individual tables.

DISCUSSION

It is apparent from the information scores, as well as from the time and error data, that some minimal size (probably around 16 min. of visual angle of letter height) must be exceeded before any practical degree of legibility can be attained. Furthermore, the reduction in legibility below this minimal point is extremely rapid as indicated by the steepness of the curve between the two smallest sizes in Figures 1, 2, and 3. As size is increased above 16 min., however, there is relatively little improvement in legibility except under conditions of reduced contrast and/or increased blur. These findings are in general accordance with those reported by Crook et al. (5, 6), both with respect to the size function and the high degree of interaction obtained among dimensions.

In comparing the present findings with Crook's data, it should be pointed out that one apparent disagreement obtains: the influence of both blur and contrast on performance is relatively small compared with that of size in the present study, whereas similar dimensions of vibration and brightness showed a much greater effect in the Crook study (6). An explanation for this lies in the difference in range of stimulus dimensions used in the two studies. Although size is comparable, Crook appears to have been working at contrast (brightness) and blur (vibration) levels extending considerably below the present ones. Since he found that performance falls off quite suddenly as these values are reduced, it is possible that the present blur and contrast levels were all above the rapid drop-off point. It is difficult, however, to state the exact correspondence between dimensions in these two studies because of the differences in operational specification of the physical dimensions (i.e., front illumination *brightness* vs. transillumination *contrast*, and *vibration* magnitude vs. light *distribution*).

An estimate of the perceptual ranges used in the present study can be determined from data obtained in a subsequent investigation.[1] In the latter study, psychophysical relationships were determined for the dimensions of size, blur, and contrast so that psychological as well as physical stimulus specifications could be used. It appears from these data that the present size range was perceptually more than three times as large as the blur and contrast ranges. Thus, it is not surprising that size should yield the greatest effect.

The functions relating physical to perceived size and blur were both linear, whereas that for contrast was exponential. Therefore, the physical size and blur spacing used in this study was optimal from the perceptual standpoint, whereas that for contrast was slightly biased toward the high contrast values.

Since it is unlikely that sizes smaller than the 16 min. discussed previously would be required in most operational situations (16 min. of letter height, after all, is roughly the equivalent of 5-pt. type at a standard 16-in. reading distance), the most significant data are those obtained for the three largest sizes. Restricting the discussion to these data, several important conclusions can be drawn. First, as noted earlier, legibility does not increase indefinitely as a function of size. In fact, there exists a zone of maximum legibility around 27 min. of visual angle within which blur and contrast effects (within the limits of this study) are relatively insignificant. Both above and below this point legibility decreases generally, the magnitude of this decrease being determined by the amount of contrast and blur present. Second, since the contrast-x-blur-x-size and the contrast-x-size interactions proved significant, it may be assumed that reduced legibility resulting from a change away from the optimum in one dimension may be compensated for by a corresponding change in another. These "trade-off" values may be obtained by comparing directly the desired points on the curves of Figures 1, 2, or 3. Taking Figure 1 as an example, it can be seen that the loss in information transmitted resulting from an addition of 0.55 units of blur to a 16-min., low-contrast target may be completely restored by adding 5 min. to the size, or by increasing contrast some 25%.

The major implications of these findings for operational use are these: (a) in a situation employing white-on-black alpha-numeric symbols, maximum legibility may be attained when no blur exists, contrast is at or above 37%, and size is ap-

---

[1] Howell, W. C., & Kraft, C. L. The judgment of size, contrast, and sharpness of letter forms. *J. exp. Psychol.*, 1961, 61, 30-39.

proximately 27 min. of visual angle (letter height), and (b) if the situation imposes restrictions on any of these values, the loss may be minimized by adjusting the values along the other dimensions in accordance with Figures 1, 2, or 3.

Another method of overcoming restrictions imposed by specific situations is afforded by the confusion matrices. If the type of restriction is known, it is possible to select alphabets on the basis of the appropriate confusion data which will minimize the effect of this restriction. If, for example, an eight-symbol alphabet were needed for a display system in which size rather than resolution was the chief limiting factor, a set composed of A, D, E, I, N, Q, T, and U, would probably lead to the best performance since these are never confused with one another under such conditions (Table 3). Similarly, if all three factors were likely to be degraded at one time, it would be advisable to substitute C, J, and H for E, I, and N in the above list in order to hold confusions to a minimum (Table 4). It should be pointed out that even selected alphabets, such as the above, do not guarantee complete lack of confusions. It is possible, for example, that by restricting the number of alternative symbols in the alphabet, the resulting response and confusion patterns might become changed. The present data, therefore, must be considered only as a first approximation toward optimal selection of restricted alphabets.

In many operational situations the conditions affecting blur, size, and contrast do not remain constant over time. Thus, the optimal alphabet is not one which minimizes confusions for any specific condition, but one which yields the fewest confusions under a wide range of conditions. Such an alphabet can be selected readily from the summary of symbol confusions presented in Table 5.

In regard to the over-all confusion data, it may be noted that the individual symbols differ considerably in the number of times used as a response as well as in the number of times correctly identified. As would be expected, the correlation between these two frequencies is high, rho being .74. On the other hand, the proportion of times a response is correct when given is greater for the low than for the high response frequency symbols (rho = .78). These relationships may be seen clearly in Figure 4, in which the per cent deviation from the expected response frequency is given on the ordinate for each symbol and the ratio of correct responses to frequency of usage appears above each symbol designation.

Viewed in terms of response frequency, neither of these findings is particularly surprising: a symbol given more frequently as a response than the others would be expected by chance to be correct, as well as incorrect, more often. What demands explanation is the fact that some responses occur so much more frequently than do others (nearly half again as often in some cases). After more than 3,000 presentations of each stimulus it would be expected that, on the basis of chance alone, the response frequencies would approach equality.

It is possible that either or both of two processes are operating to bring about this response inequality: (a) certain of the experimental conditions may change the physical *shape* of some images so that they more closely resemble other images, and (b) when ambiguity exists among symbols, the response is undoubtedly weighted according to some subjective probability estimate based on the subject's previous experience with the stimuli. The former explanation is probably most applicable to cases in which spurious resolution occurs as a result of the addition of large amounts of blur to the figure. It is improbable that, within the range of blur used, consistent image changes would have occurred frequently enough to have created the degree of response inequality observed.

The second alternative is not really an explanation; it merely poses the problem of identifying the factors involved in the subjective probability estimates of letter and number occurrence. One of these factors might be the frequency with which the symbols occur in everyday usage. Although some correlation does appear to exist between such usage data and the present frequencies for letters alone, the same does not hold when numbers are added to the alphabet. Numbers are encountered far less frequently than letters in everyday usage, yet three of them (4, 5, and 1) appear among the 10 most frequent responses in the present study and the rest are well scattered throughout the list. For the present, then, a good explanation is still wanting for the observed inequality of response frequencies to letters and numbers.

A final word of caution is in order with regard to the implications of these data for operational use. It should be recognized that the data were obtained using only one style of print, the so-called Mackworth alphabet. Although it is expected that these general results would hold for a number of other alphabets, the specific confusion data may not transfer quite so well. On the other hand, enough similarity exists between the Mackworth and both the AND and AMEL figures that few serious differences would be expected to occur among their confusion patterns.

REFERENCES

1. Alluisi, E. A., and Martin, H. B. *Comparative Information-Handling Performance with Symbolic and Conventional Arabic Numerals: Verbal and Motor Responses.* WADC Technical Report 57-196. Report of the Laboratory of Aviation Psychology under Contract No. AF 33(616)-3612 between The Ohio State University Research Foundation and Wright Air Development Center, Wright-Patterson Air Force Base, Ohio, April 1957.
2. Berger, C. "Stroke-Width, Form, and Horizontal Spacing of Numerals as Determinants of the Threshold Recognition." *Journal of Applied Psychology*, Vol. 28, 1944. Pp. 208-231.
3. Berger, C. "Grouping, Number, and Spacing of Letters as Determinants of Word Recognition." *Journal of General Psychology*, Vol. 55, 1956. Pp. 215-228.
4. Brown, F. R. *Legibility of Uniform Stroke Capital Letters as Determined by Size and Height to Width Ratio and as Compared to Garamond Bold.* NAES Report TED NAMEL-609. Aeronautical Medical Equipment Laboratory, Naval Air Materiel Center, March 1953.
5. Crook, M. N., Hanson, J. A., and Weisz, A. *Legibility of Type as a Function of Stroke Width, Letter Width, and Letter Spacing under Low Illumination.* WADC Technical Report 53-441. Wright Air Development Center, Wright-Patterson Air Force Base, Ohio, March 1954.
6. Crook, M. N., Harker, G. S., Hoffman, A. C., and Kennedy, J. L. *Effects of Amplitude of Apparent Vibration, Brightness, and Type Size on Numeral Reading.* USAF Technical Report 6246. Air Materiel Command, Wright-Patterson Air Force Base, Ohio, September 1950.
7. Duncan, D. B. "Multiple Range and Multiple F Tests." *Biometrica*, Vol. 11, 1955. Pp. 1-42.
8. Fry, G. A., and Enoch, J. M. *Human Aspects of Photographic Interpretation.* Third Interim Technical Report, Contract No. AF 30(602)-1580. Rome Air Development Center, Griffiss Air Force Base, Rome, New York, January 1957.

9. Landsdell, H. "Effects of Form on the Legibility of Numbers." *Canadian Journal of Psychology,* Vol. 8, 1954. Pp. 77-79.
10. Mackworth, N. H. *Legibility of Air Raid Block Letters and Numbers.* Unpublished Report FPRC 423. Applied Psychology Research Unit, British Medical Research Council, Cambridge University, England, 1944.
11. Muller, P. F., Jr., Harter, G. A., and Fitts, P. M. *Information-Processing Capacities of the Human Operator.* Unpublished study of the Laboratory of Aviation Psychology under Contract No. AF 33(616)-43, Wright Air Development Center, Wright-Patterson Air Force Base, Ohio, August 1953.
12. Muller, P. F., Jr., Sidorsky, R. C., Slivinske, A. J., Alluisi, E. A., and Fitts, P. M. *The Symbolic Coding of Information on Cathode Ray Tubes and Similar Displays.* WADC Technical Report 55-375. Report of the Laboratory of Aviation Psychology under Contract No. AF 33(616)-43 between The Ohio State University Research Foundation and Wright Air Development Center, Wright-Patterson Air Force Base, Ohio, October 1955.
13. Schapiro, H. B. *Factors Affecting Legibility of Digits.* WADC Technical Report 52-127. Wright Air Development Center, Wright-Patterson Air Force Base, Ohio, June 1952.
14. Shannon, C., and Weaver, W. *The Mathematical Theory of Communication.* University of Illinois Press, Urbana, Illinois, 1949.
15. Spragg, S. D. S., and Rock, M. L. "Dial-Reading Performance as a Function of Brightness." *Journal of Applied Psychology,* Vol. 36, 1952. Pp. 128-137.
16. Taylor, Cornelia D. "The Relative Legibility of Black and White Print." *Journal of Educational Psychology,* Vol. 25, 1934. Pp. 561-578.
17. Tinker, M. A. "Relative Legibility of Letters and Digits." *Journal of General Psychology,* Vol. 1, 1928. Pp. 472-496.

5

Development of Design Criteria for Intelligence Display Formats

W. D. Hitt, H. G. Schutz, C. A. Christner, H. W. Ray, and L. J. Coffey

INTRODUCTION

In the design of visual displays to be used in an intelligence center, the design engineer needs answers to a number of questions concerning human-factors problems. For instance, what types of formats should be used for the various displays?

From *Human Factors,* 3, 1961, 86-92. By permission of the Human Factors Society.

This work was sponsored by the U.S. Air Force under Contract AF 30(602)-2078 (ref. RADC-TR-60-201).

What types of coding methods should be employed? Is the use of color necessary? What is the best method for displaying trends, or historical data? The purpose of this research program was to attempt to provide answers to questions such as these. The present paper provides introductory and summary information describing the research program. The five following papers describe details of the various experiments.

The function of the intelligence system under study is to collect, process, display, and report pertinent military information about the enemy (or potential enemy). The aspect of this system with which the Battelle research program was primarily concerned is utilization of displays by "Center Control" personnel who exercise management over the data-processing activities. Broadly considered, Center Control exercises a filtering function with respect to directing the processing of system inputs and evaluating the quality of system outputs. An important requirement for visual displays is anticipated in connection with this function. Figure 1 is a general representation of the system under study.

DISCUSSION OF PROBLEM

To provide for greater generality of results and to minimize the possibility of overlooking important parameters, the Battelle research program was developed within a systems framework.

The performance of system managers and analysts, as measured in terms of both speed and accuracy, represents the criterion of system performance to be optimized in the present research program. For the system under study, it is apparent that this kind of performance is a function of a variety of variables. The categories of variables believed to be most important are represented in the following general equation:

$$P = f(M, I, D, C, T, O, E),$$

where

P = Operator performance
M = Methods of presentation
I = Type of information displayed
D = Density of information
C = Complexity of information
T = Operator tasks
O = Operator characteristics
E = Environmental factors

These categories of variables are illustrated in Figure 2, in which a distinction is made between those categories of variables that are under the control of the system designer and those categories that are not. Within the limits of electronic equipment, the system designer has control over methods of presentation. The system designer might also have control over operator characteristics, in so far as he might specify the types of characteristics, such as good visual acuity, high intelligence, etc., necessary for satisfactory performance in the system. Moreover, he controls the environmental factors by specifying the illumination level of the operations room, distance between consoles, etc.

On the other hand, the system designer has little control over the type of information to be displayed and the operator tasks to be performed, both of which are essentially determined by the particular problem situation that is presented.

The first major step in the present research program was to select for study important variables pertaining to methods of presentation. Almost as important, however, were the decisions concerning which of the other categories of variables to include in the experiments and which to exclude. Decisions concerning these secondary categories of variables were as follows:

1. Type of information displayed. No attempt would be made to evaluate the effects of various types of information on operator performance. For the information categories to be used in the experiments, however, efforts would be made to develop items expected to be found in the system under study.

2. Density and complexity of information. Because of the likelihood of significant interactions between display conditions and density-complexity of data it was decided to vary both density and complexity of data so that the degree of interaction between these variables and display conditions could be ascertained.

3. Operator tasks. Because of the likelihood of significant interactions between display conditions and operator tasks, it was decided to select representative tasks and, further, to vary them so that the degree of interaction between tasks and display conditions could be ascertained.

4. Operator characteristics. No attempt would be made to determine the relation between operator characteristics and performance. Subjects selected for the experiments, however, should be fairly representative of the population of intelligence specialists assigned to the system under study.

5. Environmental factors. An effort would be made to optimize major environmental conditions and, in so far as possible, to hold these conditions constant throughout the experiment. (Moreover, the size of the displays would be optimized and then held constant.)

The objective of the Battelle research program was to develop design criteria for intelligence display formats. Three types of visual displays were to be investigated:
1. Alpha numeric.
2. Trend.
3. Cartographic.

FORMULATION AND SELECTION OF RESEARCH PROBLEMS

After an acceptable conceptual framework for the system had been developed, the next step in the research program was to select specific problems for investigation. Problem areas were formulated by considering display parameters in relation to each of the three types of displays. The results of this formulation are presented in Tables I, II, and III.

TABLE I
Problem Areas for Alpha-Numeric Displays

Display parameters	*Variables to consider*
(a) Format:	(1) Arrangement of categories (2) Use of color (3) Black on white versus white on black (4) Size of display (5) Ambient/display lighting ratio
(b) Density:	(1) Number of categories (2) Total number of letters and numbers (3) Amount of data per unit area (4) "Word" size
(c) Number of Categories:	(1) Data are given for how many categories
(d) Coding Dimensions:	(1) Descriptive phrases versus symbols for categories (2) What coding system for symbols (3) Families of symbols (4) Multidimensional coding (5) Use of colored symbols (6) Association characteristics
(e) Rate of Change:	(1) Change of categories (2) Change of all data for categories (3) Change of particular data for categories

TABLE II
Problem Areas for Trend Displays

Display parameters	*Variables to consider*
(a) Format:	(1) Line graph versus bar graph (2) Arrangement of categories (3) Number of trends (4) Black on white versus white on black (5) Size of displays (6) Ambient/display lighting ratio
(b) Density:	(1) Number of trend figures (2) Amount on each figure
(c) Number of Categories:	(1) Number of graphs (2) Number of different lines on graph
(d) Coding Dimensions:	(1) What symbols are used (2) How are symbols organized (3) Multidimensional coding (4) Use of color coding (5) Association characteristics
(e) Rate of Change:	(1) Number of time intervals (2) Change of present time condition (3) Rate of change of new parameter

TABLE III
Problem Areas for Cartographic Displays

Display parameters	Variables to consider
(a) Format:	(1) Degree of pictorial reality
	(2) Area covered
	(3) Use of texture
	(4) Use of color
	(5) Use of elevation
	(6) Mosaic or single map
	(7) Black on white versus white on black
	(8) Size of display
	(9) Ambient/display lighting ratio
	(10) Scale
(b) Density:	(1) Amount of detail of geographic nature
	(2) Number of names of cities, etc.
	(3) Number of symbols
(c) Number of Categories:	(1) Types of information
	(2) How many types of symbols
(d) Coding Dimensions:	(1) How are categories coded
	(2) Use of colored symbols
	(3) What dimensions (e.g. size, shape) for each information category
	(4) Families of symbols
	(5) Degree of reality in code-association characteristics
(e) Rate of Change:	(1) Number of different maps per unit time
	(2) Rate of updating symbols on maps

As an aid in the selection of problems to be investigated, three primary criteria were established:

(1) Importance of problem.
(2) Applicability of results.
(2) Feasibility of conducting study.

By applying these three criteria for problem selection to the problem areas specified in Tables I, II, and III, five specific problems were selected for investigation. The five resulting studies were:

(1) A comparison of vertical and horizontal arrangements of alpha-numeric material.
(2) An evaluation of formats for trend displays.
(3) An evaluation of methods for presentation of graphic multiple trends.
(4) An evaluation of five different abstract coding methods.
(5) An evaluation of the effect of selected combinations of target and background coding on map-reading performance.

EXPERIMENTAL RATIONALE

As mentioned previously, the primary objective of the research program was to develop design criteria for intelligence display formats. This means, of course,

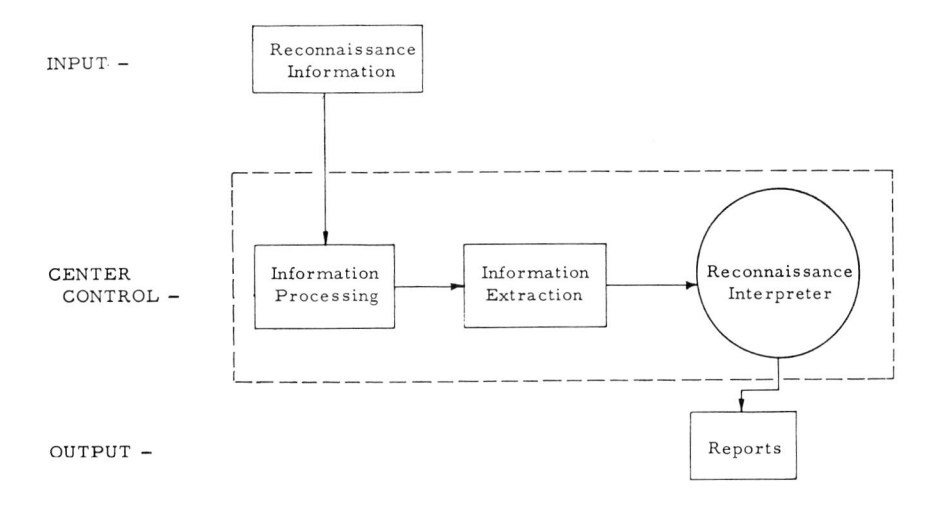

Figure 1. General description of system under study.

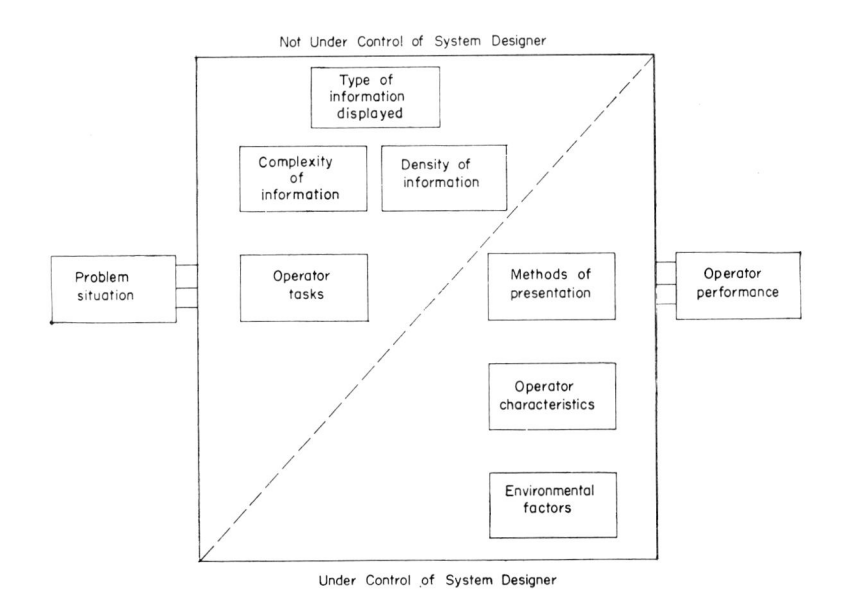

Figure 2. Variables that affect operator performance.

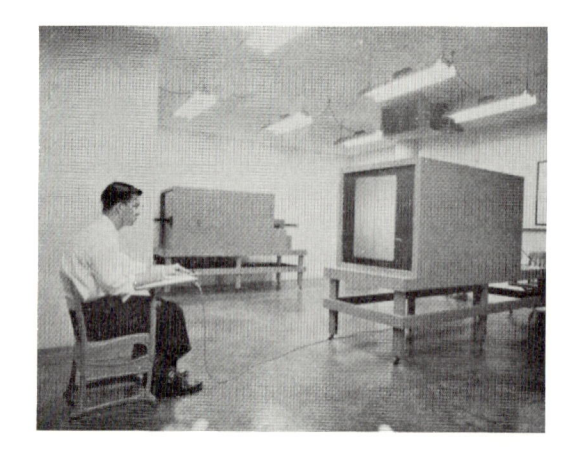

Figure 3. A display unit.

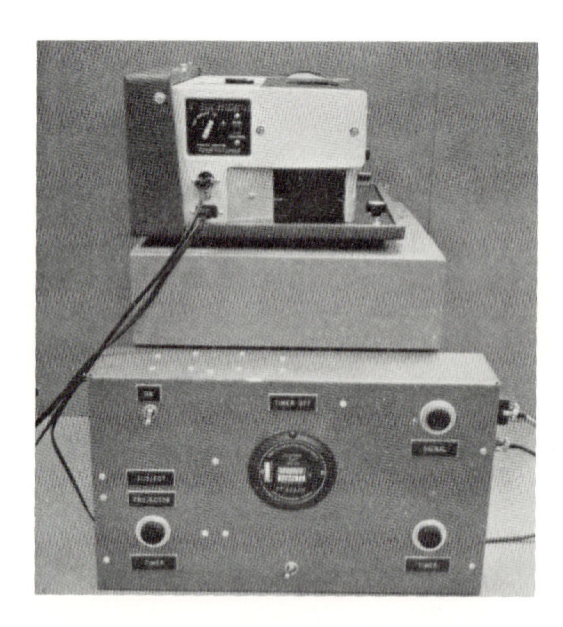

Figure 4. Rear of display unit showing projector and experimenter's control box for response-timing and stimulus-presentation device.

that the experimental results should be formulated in such a manner to be useful to the design engineer.

The major factor that reduces the experimentalist's confidence in generalizing to the operational setting is the obvious gap between the research laboratory and the "real world". To alleviate this problem, the present research program was designed as a partial simulation of the actual operational setting. For example, the major experimental conditions, the secondary variables, the subjects, and the operator tasks were all believed to be fairly representative of what might be found in the present system. No attempt was made, however, to simulate exactly the conditions as found in the actual system.

The present research program thus falls somewhere between fundamental laboratory research at one extreme and operational field testing at the other. Such an approach permits the design of well-controlled experiments in which the variables under study may be varied in a systematic manner; furthermore, the introduction of a certain amount of realism permits a greater degree of generalization. The development of a research program at this level on the "degree-of-reality continuum" means essentially this: If it is concluded that one experimental condition is superior to a second condition, then it is assumed that this same *order* of conditions will prevail in the actual operational system. Because the actual system was not completely simulated, however, we do not have as much confidence that the *magnitude* of the differences between experimental conditions will hold in the actual system.

Experimental Apparatus. Two display units were constructed for rear-projection presentation of 35-mm slides. The projects were fully automatic with respect to slide changing; a tray of 40 cardboard-mounted 35-mm slides could be shown individually simply by pressing a changing switch each time a new slide was desired. The long axis of a 35-mm slide projected at 6 ft. was 30 in.

A display unit consisted of an enclosed housing built on a frame of 2 × 4-in. lumber. The frame was 8½ ft. long, 39 in. wide, and 25 in. high. Legs at the corners and on each side at the center were mounted on casters, which permitted easy movement of the unit. Including the casters, the overall height of the frame was 28 in. A ½-in. sheet of plywood covered the top of the frame.

A 36 × 36-in. framed piece of ¼-in. glass Lenscreen [1] was mounted in a vertical position 6 in. from the front of the frame. An area 6 ft. in length behind the screen was enclosed with ½-in. plywood. The inside of the enclosed area was painted flat black. A small, circular hole at the rear of the enclosed area permitted projection to the center and rear of the Lenscreen. A small wooden stand at the outside rear of the enclosed area, and mounted on top of the frame, held the projector. A display unit is shown in Figure 3.

The response-timing and stimulus-presentation devices consisted of two control boxes, one for the subject's use and the other for the experimenter's use. The subject's control box contained a push-button which could start and stop the timer, as well as introduce and remove a slide. The experimenter's control box contained a time meter that recorded time with $\frac{1}{10}$ of a second accuracy and a push-button

---

[1] Lenscreen is a high-quality display screen produced by a process of coating plate glass with a thin layer of microscopically small lenses.

for signaling the subject. The rear of a display unit showing the projector and the experimenter's control box for the response-timing and stimulus-presentation device is shown in Figure 4.

Regular classroom desk seats were used so that subjects sitting in front of the display unit would have convenient writing space for paper-and-pencil tests. The subject control boxes for the response-timing and stimulus-presentation devices were mounted on the front center, and flush with the desk of each subject's seat.

In all the experiments that employed slides for stimulus material, the above display units were used. In two of the experiments, however, cardboard posters rather than slides were used for stimulus material. The experimental room for these two studies contained a wooden stand with a window blind for the posters and a desk-type chair for the subjects. The illumination was approximately 25 ft.-c. on the screen of the wooden stand, and on the order of 10 to 30 ft.-c. on the Lenscreen.

Selection of Subjects. The population of people expected to be represented by the experimental subjects is indeed a general population, with few specified characteristics. It probably can be assumed, however, that the intelligence specialists working with "second-order" displays in the system under study: (1) are male, (2) are of college caliber in intelligence and technical competence, and (3) possess normal vision.

Thus, to meet these general criteria, 30 male Battelle technicians and engineers were asked to serve as subjects in the experiments. The Snellen Chart and the AO H-R-R Pseudoisochromatic Plates were used to test the subjects for visual acuity and for color vision, respectively.

GENERAL DISCUSSIONS AND CONCLUSIONS

The present research program has added further support to the value of multivariable experiments. The overall efficiency of such experiments in terms of amount of information obtained per unit cost is indeed sufficiently great to offset any difficulties associated with such an approach. Even more important, however, is that multivariable experiments permit the investigation of interactions between (and among) variables. It is apparent that design recommendations based on only main effects can be misleading. In most of the experiments, to follow, the results are qualified by the various interactions. Naturally, if such interactions are so small that they have little practical significance—even though they are statistically significant—then the designer need not be burdened with such findings. When such interactions are important in a practical sense, however, it is essential that recommendations be qualified in accordance with these findings.

A second general conclusion concerns the study of operator tasks. First, it is important to note that significant interactions were found between display conditions and operator tasks. Second, it is equally important to note that independent, or orthogonal, task factors were obtained from the statistical analysis. Inasmuch as task tests—rather than task items—were intercorrelated, the resulting factors were not so "clean" as might be desired. Nevertheless, the statistical results permitted meaningful interpretation of the factors. On the basis of these findings, it seems that many of the future studies on the design of visual displays would benefit if operator task were included as one of the experimental variables. Furthermore, because task factors interact with display parameters, it is important to determine as soon as possible whether or not these task factors are representative of the real-world data-processing operations under study. It would then be possible to make recommendations concerning display parameters according to the relative importance of the various tasks.

A review of the literature emphasizes the need for more standardized (as well as more appropriate) conditions for the study of visual displays. It is a relatively easy task to find studies on visual displays that report contradictory results. By reviewing these studies in detail, however, it might be found that they differed in: subject tasks, actual specification of experimental conditions, experimental procedures, and environmental conditions, as well as characteristics of subjects. If greater standardization could be achieved with respect to the important parameters of visual-display experiments, then it is certain that more consistent and more meaningful reports would be forthcoming.

To assist the individuals responsible for the design of visual displays, an extremely worthwhile goal would be to develop a detailed, up-to-date handbook on "Design Criteria for Visual Displays," which might serve as a "Volume II" to the handbook developed by Baker and Grether (1). Much information is already available for such a handbook, but it is scattered throughout the many Government reports, technical journal articles, and textbooks. If such information were collected and integrated, however, it would serve as an excellent foundation for a visual-display handbook, as well as give direction to future work on visual displays.

REFERENCE

1. Baker, C. A., and Grether, W. F., *Visual Presentation of Information*, WADC Technical Report 54-160, Wright Air Development Center, Wright-Patterson Air Force Base, Ohio, August, 1954.

6

Relative Motion of Elements in Instrument Displays

W. G. Matheny, D. J. Dougherty, and J. M. Willis

Human factors engineers have as one of their responsibilities the making of recommendations for visual displays in various man-machine systems. In fulfilling this responsibility they depend upon data gathered by the research scientists and upon experience with similar systems from which generalizations can be drawn. The Human Factors Research Scientist then has the responsibility of becoming as thoroughly acquainted as possible with the complex of the situation in which his findings are to be applied and to do everything possible to insure that his research investigations include all relevant variables. The conscientious research specialist will always be plagued by the nagging fear that he has not included all relevant variables which might effect performance in the test situation.

Further, it is often not possible to include all of those variables which the researcher strongly suspects would change the outcome or the results of his experiment because of the lag in the state-of-the-art of producing equipment with which he can carry out experiments. This situation breeds the complaint by the Human Factors Engineer who would apply the researcher's findings that his conclusions are too hedged by qualifications and are not firmly decisive and authoritative.

The studies reported here are intended as an illustration of the effect of a relevant variable on the outcome of the evaluation of a display principle and some preliminary investigations of this variable. The display principle in question is the perennial one of the outside-in versus the inside-out display of the information by which the vehicle operator nulls the errors about the longitudinal-lateral axis of the aircraft, i.e., stabilization or attitude control. The variable in whose effect we are interested is that of the motion imparted to the operator by vehicular motion. The study examines the effect upon the interpretation of inside-out and outside-in displays of adding body motion cues provided by vehicular motion.

From *Aerospace Medicine*, 34, 1963, 1041-1046. By permission of Aerospace Medical Association.

This work was performed under the Army-Navy Instrumentation Program, Contract Nonr 1670(00), with the Bell Helicopter Company, Fort Worth, Texas, a Division of Bell Aerospace Corporation, a Textron Company.

In the displays under investigation in this study, the outside-in display may be described as one in which the ground plane frame of reference is fixed with reference to the observer and the symbol representing the aircraft moves relative to this fixed reference. The inside-out display may be described as one in which the frame of reference is such that the ground plane represented within the display, while in actual physical reality being stabilized with respect to the ground plane, may appear to move within the outer frame of reference of the display or, conversely, the outer frame of reference of the display may move around the stabilized apparent ground plane (Figure 1A and Figure 1B).

The relative merits of these two display methods have been under investigation and discussion for some time. A recent report by Kelley, et al. (ref. 1), states that the human engineering research has overwhelmingly supported the outside-in configuration as being superior. The inside-out method of display, however, is the one almost universally used in present-day displays for providing information as to pitch and roll of the aircraft and is the method under investigation for supplying this information within the Army-Navy Instrumentation Program, the latter display being termed the contact analog. Within the Army-Navy Instrumentation Rotary Wing Program evidence has been accumulated which indicates that the inside-out display is more suitable than had been previously supposed. For example, the inside-out type of display mounted in a helicopter was quite successfully flown by pilots with a wide variety of helicopter pilot experience. During these tests the helicopter was flown in a hovering mode —a mode in which the problem of stabilizing the helicopter in pitch and roll is a central and difficult task. It could be safely assumed that the major portion of the relevant variables were present in these tests and the inside-out type of display was found to be quite adequate (ref. 2).

It was, therefore, thought advisable to investigate the adequacy of these two types of displays under conditions in which the relative effect of the additional variable of motion of the vehicle could be determined. A series of experiments was conducted toward this end.

EXPERIMENT NUMBER ONE

Procedure. In this experiment the two major experimental variables investigated were (1) the type of visual display and (2) the presence or absence of correlated vehicular motion. Three types of visual displays were used. Each of the displays incorporated a geometrical projection of a ground plane represented by grid squares with a sharp horizon line and a clear sky. Display A incorporated a small symbol representing an aircraft superimposed upon this ground plane with the aircraft symbol rotating in roll indicative of the aircraft roll (Figure 1A). Display B incorporated an aircraft symbol superimposed upon the grid structured ground plane. However, the aircraft symbol remained stationary with respect to the display frame, while the background grid plane moved in geometric correspondence with motion of the ground plane as it would appear when looking out through a window in the front of the aircraft. Display C incorporated the same ground plane as in A and B with the motion of the ground plane as in B, but with the small symbol of the aircraft removed (Figure 1B). Each of these displays was studied under both conditions of motion present and motion absent. The motion cues were imparted to the observer through the use of a motion simulator constructed under the Army-Navy Instrumentation Program and shown in Figure 2. The Simulator was capable of imparting a roll of 11.3°, either to the right or to the left, in coincidence with the rotation perceived in the visual display. There were, therefore, six experimental conditions under investigation in the Simulator.

Independent groups of 20 subjects each were tested under each of the conditions, the subjects being high school seniors with no visual defects and with no previous flying experience. Subjects were briefed on the nature of the experiment and given instructions and demonstrations of the task required of them. Each subject was required to make judgments of the direction of pitch, roll or pitch and roll combined as given by the visual display or by the visual display and cabin motion combined. The subject responded by calling out the direction of movement. Each subject made 20 judgments of pitch, 20 of roll, and 40 of pitch and roll combined. Errors in judgment of direction of movement and response times were recorded.

A seventh group of 12 subjects was tested in a light two-place helicopter. The subject's view from the helicopter is shown in Figure 3. Under this condition the pilot stabilized the helicopter in the level hovering position. Upon a signal from the experimenter the pilot rolled the helicopter to the right or to the left as required by a pre-determined schedule. The subject responded by calling out the direction of roll. Roll judgments were the only judgments investigated under this condition. Response accuracy and speed of response measures were taken.

Experimental Results—Errors in Judgment of Direction of Roll. The per cent of total judgments, which were in error under each of the test conditions when roll only was presented as the stimulus, is shown in Figure 4. It can be seen that under those conditions in which the cabin motion cues were not present the moving symbol type of presentation was significantly superior. With the addition of the motion cues this superiority was not in evidence and no real differences between the display methods were found. It will be seen that the proportion of errors under the latter condition is about that found when inexperienced subjects made judgments of direction of roll in the helicopter.

When judgments as to the direction of roll were made under conditions in which both pitch and roll combined were presented as the stimulus condition, the findings paralleled those given above (Figure 5). With no cabin motion the symbol movement type of display was interpreted with significantly fewer errors, while the addition of cabin motion erased this difference.

Errors in Judgment of Direction of Pitch. The percentages of total judgments in error in pitch when pitch only was presented as the stimulus are shown in Figure 6. While the difference between conditions A and C is larger, under conditions of no cabin motion than under conditions of cabin motion, these differences do not attain statistical significance.

Figure 7 summarizes the errors in judgment of pitch when these judgments were made to the stimulus condition in which both pitch and roll combined were presented. The difference evident between no cabin motion and cabin motion for Display C is not significant.

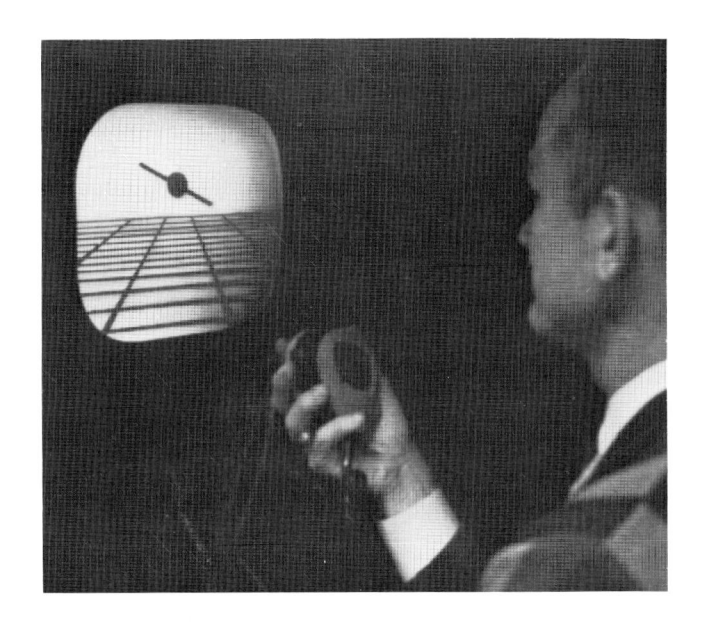

Figure 1A. Outside-in display of aircraft attitude.

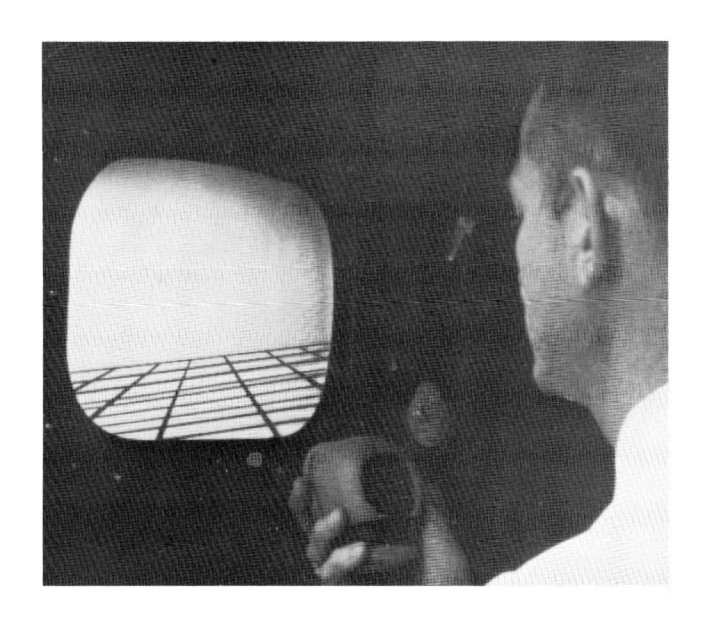

Figure 1B. Inside-out display of aircraft attitude.

Figure 2. Simulator facility cabin capable of movement in six dimensions.

Figure 3. Subject's view from helicopter.

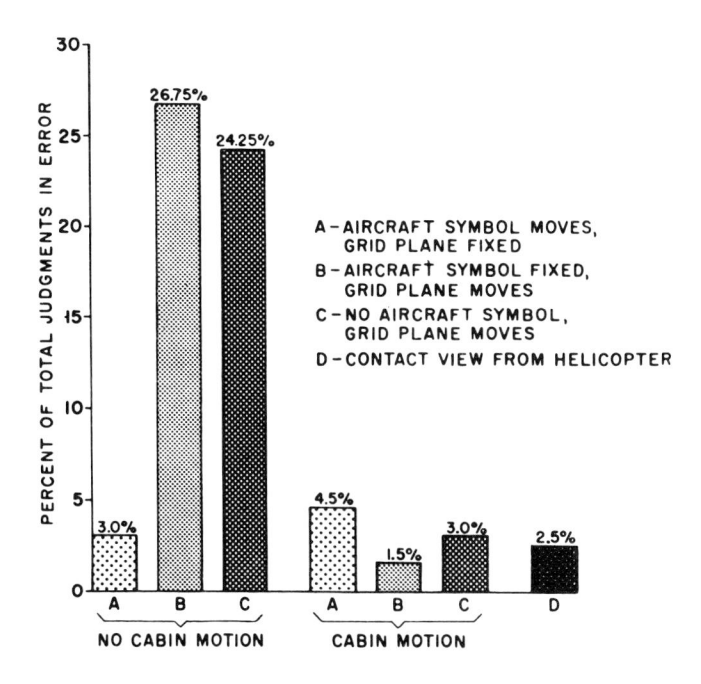

Figure 4. Relative motion of display indices study per cent of roll judgments in error display moving in roll only.

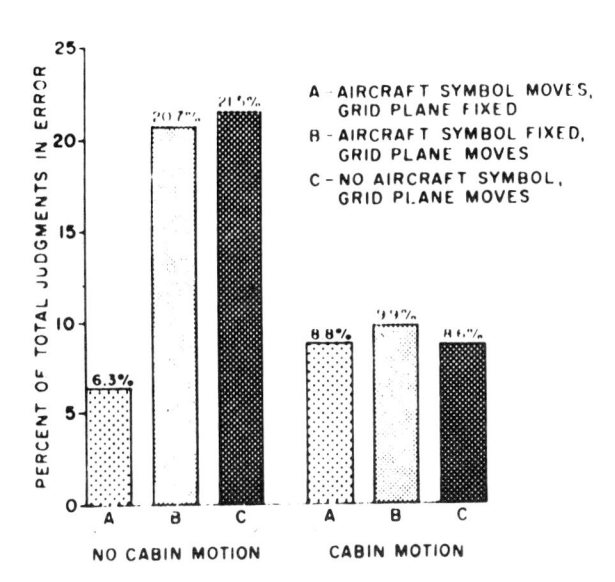

Figure 5. Relative motion of display indices study per cent of roll judgments in error display moving in both pitch and roll.

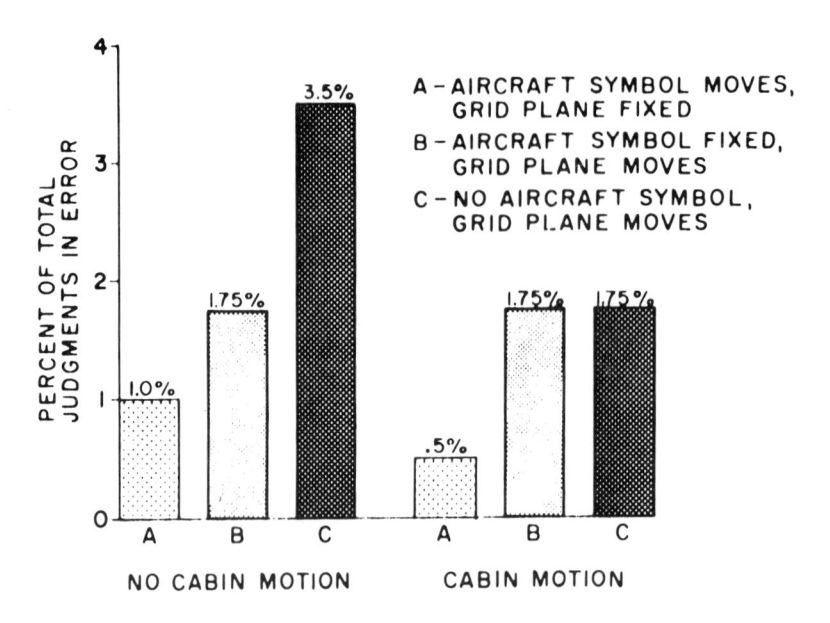

Figure 6. Relative motion of display indices study per cent of pitch judgments in error display moving in pitch only.

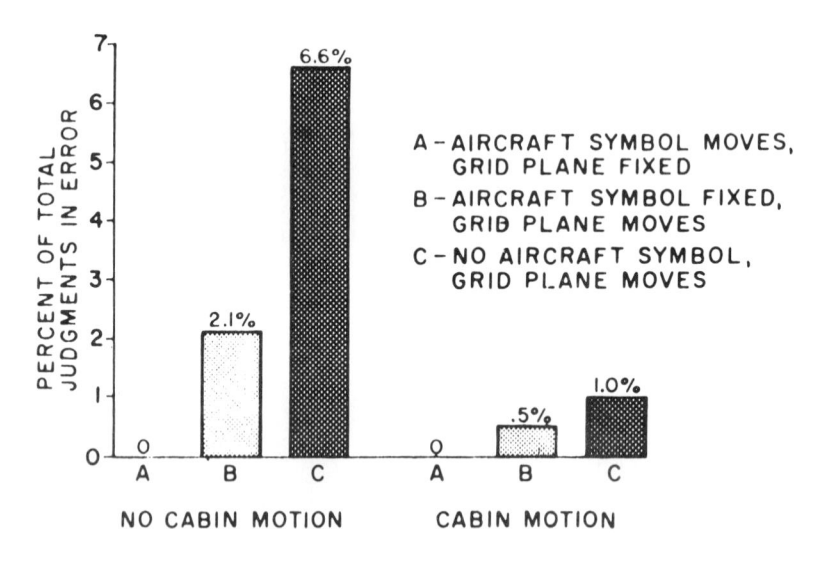

Figure 7. Relative motion of display indices study per cent of pitch judgments in error display moving in both pitch and roll.

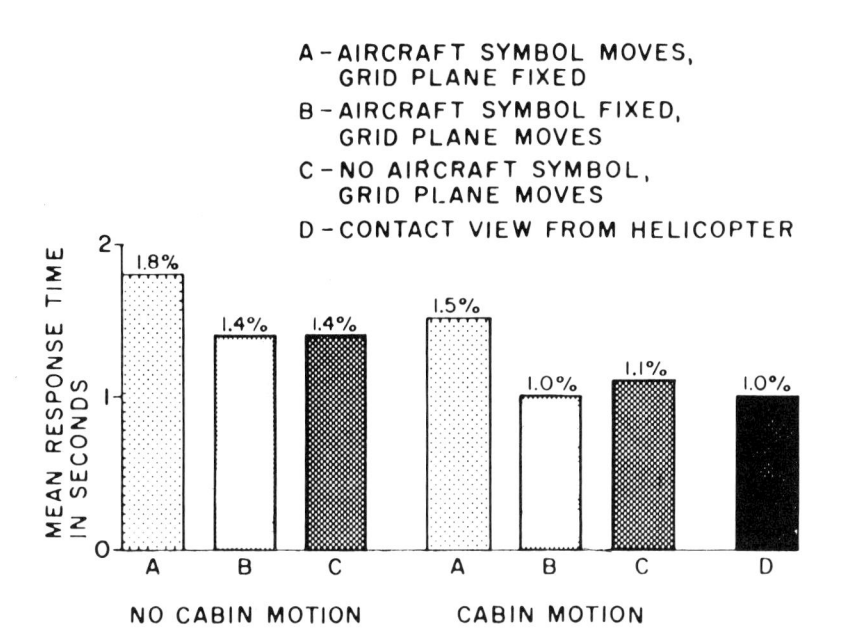

Figure 8. Relative motion of display indices study response time to roll displacements display moving in roll only.

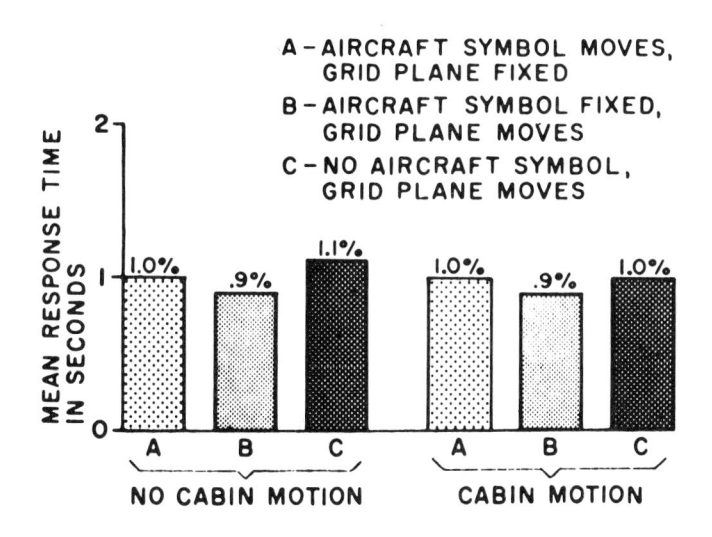

Figure 9. Relative motion of display indices study response time to pitch displacements display moving in pitch only.

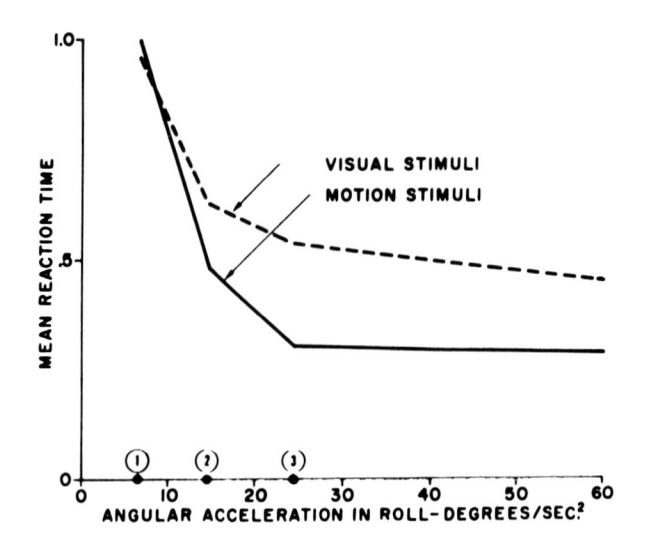

Figure 10. Reaction time for detection of movement as a function of angular acceleration.

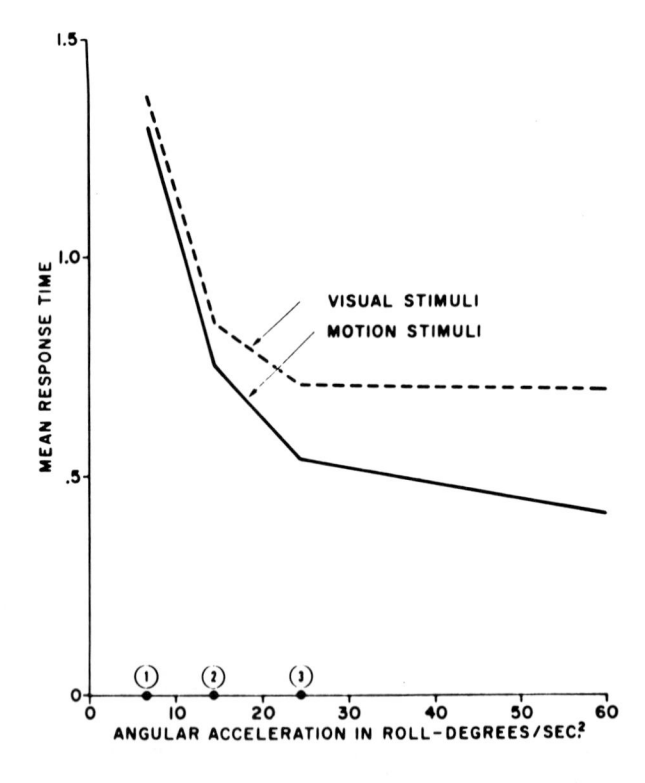

Figure 11. Response time for making judgment of direction of roll as a function of angular acceleration.

Response Time Data. The response time data are summarized in Figures 8 and 9. It can be seen that the mean response time to Display A in which the symbol moves is longer by four-tenths second than that to Displays B and C in which the background grid moves. This difference is statistically significant and understandable in view of the fact that, although the angular displacement of the symbol and of the grid was the same, the linear displacement of the grid at the edge of the screen was greater than was the linear displacement of the tip of the moving symbol. Thus, judgment based on discrimination of linear displacement should be made more quickly when viewing the grid movement display. Similarly, under conditions of cabin motion the mean response time differences between A and B and C are significant.

The addition of cabin motion significantly reduced response time to all displays. The part that motion plays in the interpretation of displays requires a great deal of further investigation. In systems which are capable of responding to relatively high frequency inputs the operator response to motion stimuli may well "lead" his response to the accompanying visual stimuli. Systems which tend to smooth high frequency inputs and are low band pass systems may be first responded to by the operator on the basis of visual stimuli followed by a perception of the motion stimuli. In both types of system varying the "gain" of the display may cause a shift in which stimuli leads—the motion or the visual. Preliminary evidence relative to the part played by motion was obtained in a small experiment conducted at the end of the experiment described above.

EXPERIMENT NUMBER TWO

This experiment was designed to investigate the hypothesis that kinesthetic cues derived from motion of the vehicle may serve as information sources to the operator and that in high frequency response systems such motion cues may precede, in time, the visual cues. The experiment was carried out to obtain preliminary data on two points:

1. The simple reaction time to angular accelerations in roll for the visual sense and for the "kinesthetic" senses, respectively.

2. The relationship between response time in making a judgment as to direction of roll and the angular acceleration in roll for both the visual and kinesthetic senses.

In one phase of this experiment subjects were placed in the motion simulator and asked to respond with either "right" or "left" in accordance with their judgment as to the direction of roll. They were asked to make this judgment under conditions in which they saw only the visual display with no motion of the cabin, and also under the conditions in which they experienced only the cabin motion with no visual display. The subjects made these two types of responses under four levels of angular acceleration of the cabin or display.

In the second phase of this experiment the same conditions of cabin and visual display motion were employed and the same four levels of angular acceleration were used but the subject's response was simply to say "now" as soon as he perceived the stimuli. Time scores were taken and the relationships between time to respond and angular acceleration of the cabin or display are shown in Figures 10 and 11. From Figure 10 it can be seen that for this system when angular accelerations exceed approximately 20 degrees per second squared, the motion cues precede the visual cues by a statistically significant amount. When angular accelerations reach approximately 25 degrees per second squared (point 3), the difference is as much as one reaction time. In slow response systems, such as represented by point 1 in Figure 10 (about 7 degrees per second squared), the reaction times are not significantly different.

A similar situation exists with respect to time to judge direction of movement as shown in Figure 11. Again for fast response systems (points 3 and 4) the kinesthetic cues "lead" the visual cues, while for slow response systems the two sensory modes require more nearly the same time. The differences between kinesthetic and visual response times are significantly different for points 2, 3, and 4 in Figure 11, but not for point 1.

SUMMARY

From Experiment Number One it is evident that motion is an extremely relevant variable in the evaluation of displays in situations in which motion cues are present. The results of Experiment Number Two suggest that in certain systems the operator receives information from his kinesthetic senses in advance of that received through the visual sense. Taken together the results from these experiments suggest not only that motion is a relevant variable, but that the degree to which it duplicates the angular motions of the vehicle being simulated is most important. Lack of the motion cues may lead to erroneous conclusions as to the suitability of displays for systems in which motion cues are present. Motion simulation systems, which exhibit transient accelerations or unrealistic phase differences between motion and visual stimuli, may provide cues to response which are inappropriate to the task of the operator. Such faults in motion simulation may be equally serious for those situations in which the simulator is used as a training device as for those in which it is used as a design and evaluation tool.

Experiment Number One has shown conclusively that motion constitutes a relevant variable in evaluating displays in mobile systems. It further shows that the addition of the motion cues to the test situation changes the conclusions with respect to inside-out versus outside-in displays. Experiment Number Two suggests strongly that the fidelity with which motion cues are simulated is important and bears a good deal of further investigation.

REFERENCES

1. Kelley, Charles R., de Groot, Sybil, and Bowen, Hugh M.: Relative Motion 3: Some Relative Motion Problems in Aviation. Technical Report: NAVTRADEVCEN316-2, U. S. Naval Training Devices Center, Port Washington, N. Y., January 1961.
2. Wilkerson, L. E., and Matheny, W. G.: An Evaluation of a Grid Encodement of the Ground Plane as a Helicopter Hovering Display (RH-1). Bell Helicopter Company ANIP Technical Report D228-421-008, September 1961. ASTIA #AD 268-273.

7

Operator Decision Performance Using Probabilistic Displays of Object Location

Louis M. Herman, George N. Ornstein, and Harry P. Bahrick

In the typical display of object location, the operator views a symbol representing some object and obtains an indication of the object's location by reference to a coordinate system. In this type of display, the operator in effect is provided with a point estimate of object location; no explicit information is displayed to the operator about the error distribution associated with the point estimate. The operator is thus deprived of important information potentially useful in making decisions concerning the object's location.

Studies have demonstrated that man's decision performance in tasks requiring event prediction or event estimation often can be correlated with the parameters of the underlying probability distributions, *e.g.*, with the parameter p of a Bernoulli process,[1] the parameters a and b of a beta distribution,[2] and with apparent local changes in the values of these parameters.[3] It is reasonable to infer, therefore, that man's decision performance could be shaped by the parameters of the error distribution depicted in a particular probabilistic display. In the present research, this inference is tested in an experimental decision situation in which performance of subjects using probabilistic displays of object location is compared with performance of subjects using nonprobabilistic displays (*i.e.*, conventional point-estimate displays).

In the study to be reported here, two methods of presenting error distribution information were investigated: the probability density function and the (cumulative) distribution function. For the probability density function display (termed

---

[1] H. W. Hake and R. Hyman, "Perception of the statistical structure of a random series of binary symbols," *J. Exp. Psychol.*, vol. 45, pp. 64-74; 1953.

[2] E. H. Shuford, *et al.*, "Pre-decisional Processes Related to Psycho-Physical Judgment," Div. Mathematical Psychology, Inst. for Research, State College, Pa., Rept. No. 4; May, 1963.

[3] W. Edwards, "Probability learning in 1000 trials," *J. Exp. Psychol.*, vol. 62, pp. 385-394; 1961.

From *IEEE Transactions on Human Factors in Electronics*, HFE-5, September 1964, pp. 13-19. By permission of the IEEE Editorial Department.

the PDF display), the error distribution about an estimated target position was displayed by use of a bivariate normal density function with the mean of the distribution located at the estimated target position. For the distribution function display, two probability contours centered about the mean were displayed such that the first contour included 50 per cent of the area and the second contour included 90 per cent of the area of the distribution. The display was termed the PC-59 display.

Figure 1 shows examples of the PDF and the PC-59 probabilistic displays and in addition shows a nonprobabilistic display (N display). For the PDF display, gradients of shading are used to simulate the effect of a three-dimensional bivariate normal distribution. No numeric probability information is explicitly provided the operator by this display. In contrast, the PC-59 display gives information that is explicitly numeric, although only a limited amount of such information is provided.

Either probabilistic display will provide more information to the operator than will the N display. It would seem that the PDF display, lacking explicit numeric information, could not provide a firm anchor point (in the sense of classical psychophysics) upon which the operator could base estimates of the parameters of the underlying probability distribution. The PC-59 display, with its limited but unambiguous numeric information, should permit such estimation to be made more readily. Thus, it would appear that the PC-59 display would prove the more useful display to the operator. In the experiments to be discussed, the hypothesis that parameter estimation and decision performance obtained with the PC-59 display will exceed that obtained with the PDF display is tested.

EXPERIMENTAL DECISION PROBLEM

Primary interest in this study was in the degree to which decision-making performance in a realistic decision problem might be improved through use of probabilistic displays of object location. A realistic decision problem based on object location information was generated and operator performance was tested within the framework of this problem. The problem developed was one in which operators chose one of several available time stages during a simulated aircraft search-attack mission to take a particular action against a displayed target. Operators were told that as their "aircraft" drew nearer the target through ten successive, discrete time intervals or stages of the mission, target localization information improved. This improvement was associated with corresponding decreases in the areas of the displayed error distributions for the probabilistic displays. The operators also were told that the target was hostile and therefore, with each succeeding stage an increasing element of danger to the attacker accrued. The operators' assigned task during each mission was to attack the target by firing a weapon of a constant (indicated) kill radius during one and only one of the ten stages of the mission. In Fig. 1, the kill radius of the weapon is indicated by the circle centered above each display. For operators using the N display, no error distribution information is displayed, but they are informed (as are all operators) of the present stage of the mission by the number appearing in the upper left corner of each display.

The decision problem was elaborated to include monetary costs to the operator for exercising decisions at the various stages of the mission and monetary payoffs for favorable outcomes of decisions. Table I shows the basic decision-problem variables: the hit probability at each of the ten stages, the cost of a firing decision at each stage, the constant amount of payoff for hitting the target, and the expected value of each decision. The monetary unit was a "point" worth one half cent. Negative expected values represent expected monetary losses. S was paid 100 points if he hit the target at the selected stage but had to pay the stated cost of firing regardless of firing outcome. Therefore, the expected value of a particular stage is $EV = (p)(r-c) - (1-p(c)) = pr - c = 100p - c$ where

p = hit probability
r = reward for a hit (100 points)
c = firing cost in points.

The operator was provided only the cost and payoff information given in Table I. Two types of operator response measures were then obtained during an experiment: (1) operator estimates of the hit probability at each stage and (2) the expected value of the decisions exercised by the operator. From Table I, it can be seen that the hit probability distribution is a discrete set of 10 Bernoulli processes, each one independent of the others, with parameters respectively of $p_1 < p_2 < \ldots < p_{10}$. The operator thus is asked to estimate p at each stage and an error scope $p - p$ may be obtained. The second response measure is simply the average of the expected values over the set of decisions exercised by an operator during repeated missions. This measure provides an evaluation of the average decision effectiveness of the operator.

EXPERIMENT I

In this experiment, operator probability estimation and decision performance is evaluated as a function of the three basic displays given in Figure 1.

Stimulus Materials

For each probabilistic display, 200 different simulations were prepared. These simulations included all combinations of 10 mission stages, four angles of inclination of the error distribution on the grid, and five eccentricity values for the shape of the distribution. Neither inclination nor eccentricity was treated as an independent variable in the data analysis, but were intended as controls against any possible biasing effects of these variables. For the N display, only 10 different simulations were prepared, corresponding to the 10 different mission stages.

From the hand-drawn simulations, positive transparencies were prepared and projected on to a beaded screen located approximately 11 feet in front of the subjects. The size of the display at that distance was a 3 foot square, permitting good resolution of the display elements. This display mode also permitted testing of several subjects simultaneously.

Procedure

For all display groups, an identical predecision training procedure was initiated. Each group viewed 40 practice missions. During each of the ten stages of each of these missions the experimenter pointed out the true coordinates of the target's position. By reference to the displayed size of the kill circle, the subject was given an indication of whether or not he could have hit the target had he fired at the expected target position during that stage. The subject provided written estimates of hit probability for each stage at the end of training missions 3, 15, and 40.

Following training, each subject "flew" 40 additional decision missions. A stage of a mission lasted 15 seconds, and missions succeeded one another without interruption. During each mission, the subject fired at the target during one and only one of the 10 stages by pressing a button located at his table. He received immediate knowledge of results via a light which flashed when a hit occurred. Hits were programmed randomly over missions, with the probability of a hit at a given stage corresponding to the function shown in Table I. The subject kept a record of his decisions and outcomes, and at the end of the entire experiment he was paid his winnings, if any. The decision situation, therefore, can be classified as a multiple alternative task, with uncertain outcomes, under conditions of risk. Immediately following the 40 decision missions, S was asked to provide an estimate of hit probability for each stage.

The subjects were 18 male undergraduate students at Ohio Wesleyan University, Delaware. Six subjects were assigned randomly to each group.

Results

Probability Estimates. Figure 2 shows for each group the subjective hit probability function (mean estimates for all 6 subjects in a group) at the end of the 40th training and 40th decision mission. Shown also is the true hit probability function. It can be seen that for both the 40th training and decision missions, the PC-59 group produced the most accurate estimates, followed in turn by the PDF and N groups. A Jonckheere ordered test for related groups was applied to the data with the matching variable across groups being stage number.[4] Results of this test at the end of the 40th training mission indicated acceptance at $P = 0.007$ of the ordered hypothesis that the error magnitude would arrange itself $N > PDF > PC-59$. Results at the end of the 40th decision mission indicated acceptance of the same hypothesis at $P = 0.015$.

A test of differences between estimates given at the end of the training missions and those given at the end of the decision missions indicated that all groups showed significant improvement in their estimations at $P \leq 0.025$ (one-tailed).

Decision Performance. Figure 3 shows the firing distribution for the 40 decision missions for each of the three groups. Note the relatively "tight" distribution of

---

[4] A. R. Jonckheere, "A distribution-free k-sample test against ordered alternatives," *Biometrika*, vol. 41, pp. 133-145; 1954.

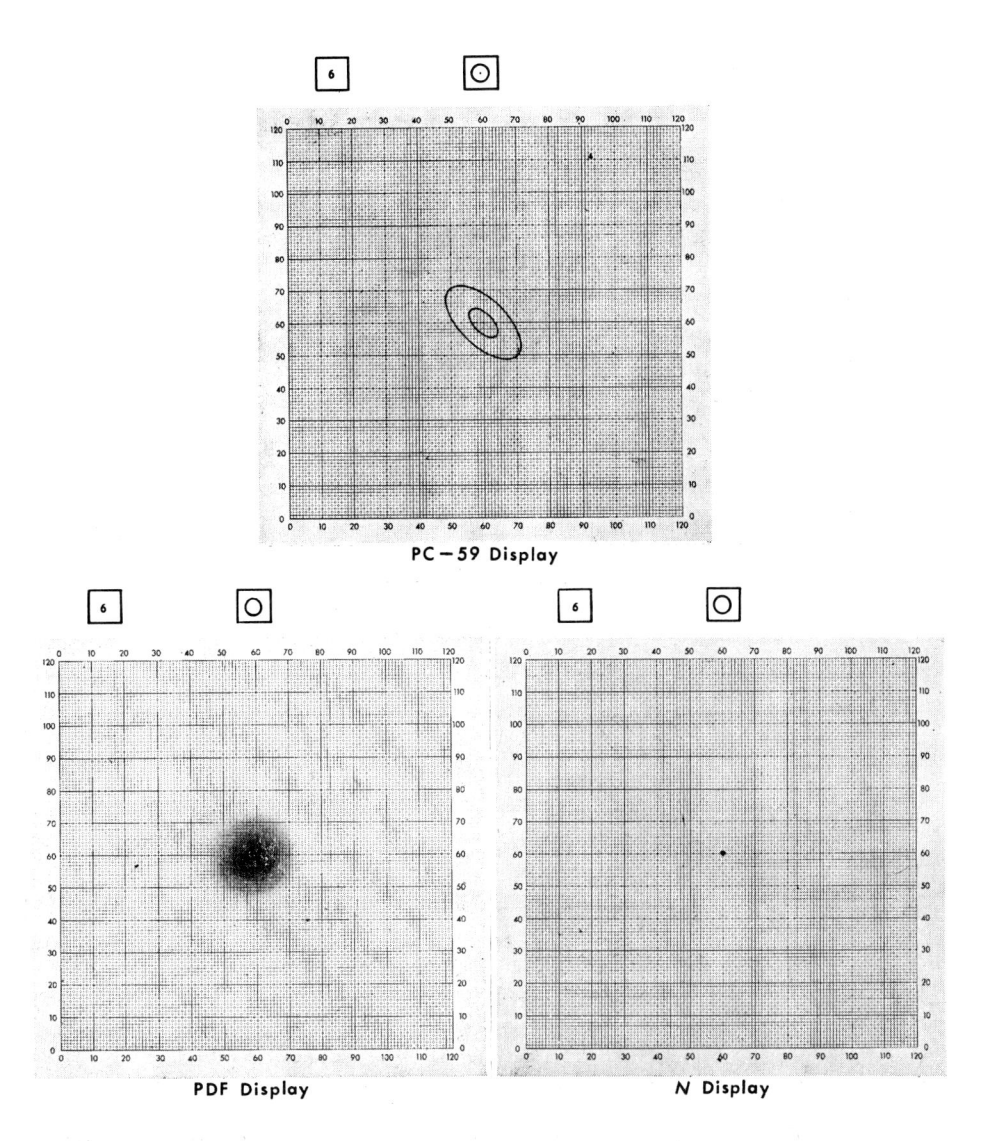

Figure 1. Examples of the three displays used in the experiment. The stage of the mission is shown in the upper left box above each display and the weapon kill-radius in the upper center box. For the two probabilistic displays, note that different angles of inclination and different eccentricity levels are illustrated. For all displays, the coordinate values 60-60 always located the best point-estimate of target location.

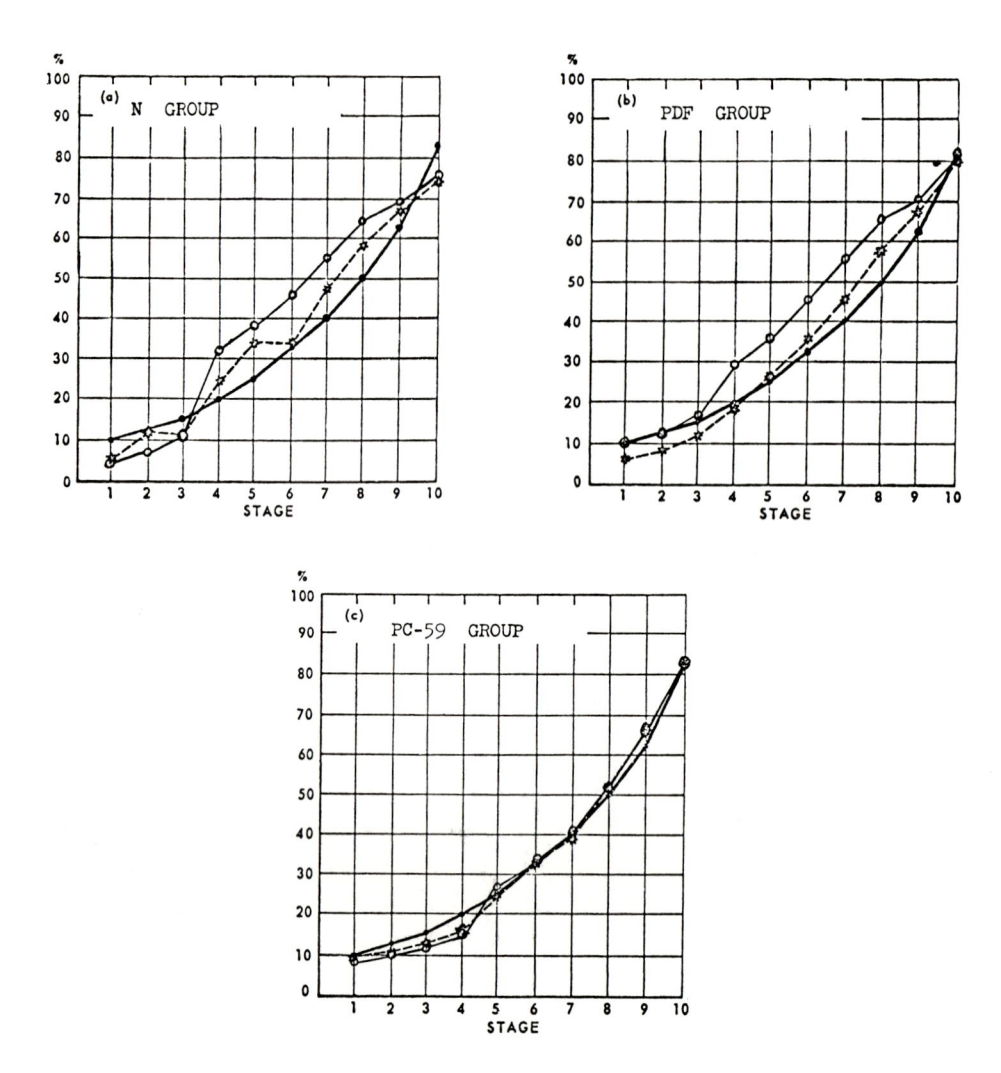

Figure 2. The subjective hit-probability functions for the three groups compared with the true function (solid, thick line). The two subjective functions for each group represent, respectively, the average of the probability estimates at the end of the 40th training mission (open circles) and at the end of the 40th decision mission (asterisks).

TABLE I
Hit Probability, Firing Cost, and Expected Values for Each Mission Stage

Stage	Hit Probability	Firing Cost (points)	Expected Values (points)
1	0.100	15	−5.0
2	0.125	16	−3.5
3	0.150	17	−2.0
4	0.200	19	1.0
5	0.250	21	4.0
6	0.325	25	7.5
7	0.400	36	4.0
8	0.500	49	1.0
9	0.625	64	−1.5
10	0.825	87	−4.5

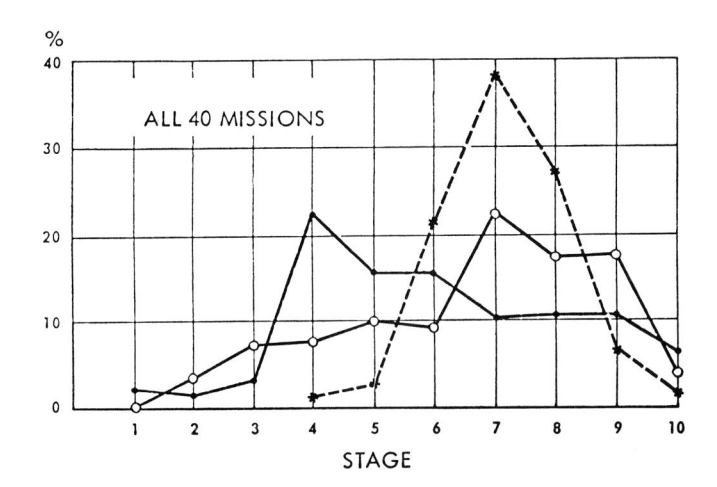

Figure 3. Percentage distribution of firing choices on N (open circles), PDF (filled circles), and PC-59 groups across the ten mission stages.

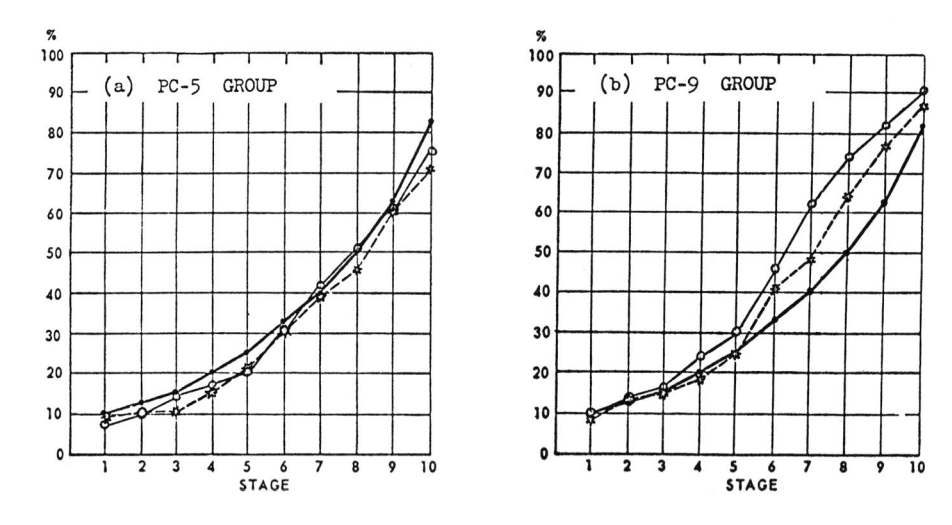

Figure 4. The subjective hit-probability functions at end of the 40th training mission (open circles) and end of the 40th decision mission (asterisks) compared with the true function for the two single-contour groups.

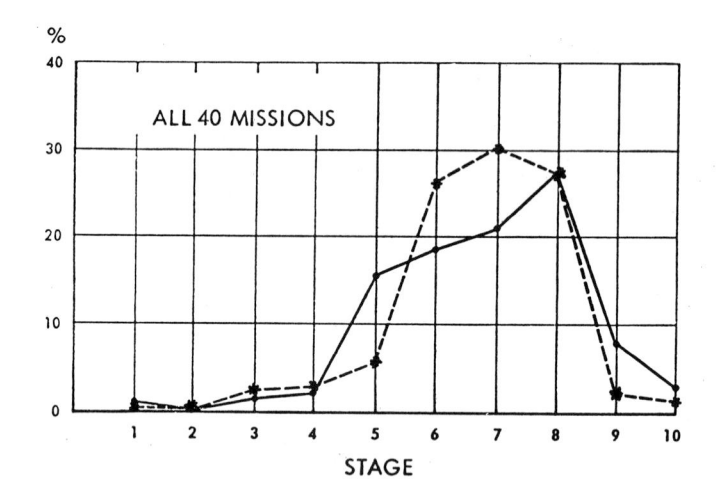

Figure 5. Percentage distribution of firing choices of PC-5 (filled circles) and PC-9 (asterisks) groups across the ten mission stages.

the PC-59 group about stages with positive expected values (4–8). It is seen, however, that no group maximized their firing decisions at the stage of highest expected value (stage six). Group mean expected value scores were: N group 1.49, PDF group 1.92, PC-59 group 3.39 (maximum possible score was 7.5).

A Jonckheere ordered test for independent groups was applied to the data.[5] The ordered hypothesis that the decision performance scores of the three groups would be arranged from lowest to highest as $N < PDF < PC\text{-}59$ was supported at $P = 0.03$.

Clearly then, these results indicate that probabilistic displays can enhance decision performance, and in particular, that the distribution function display (PC-59) results in performance superior to the density function display (PDF).

EXPERIMENT II

This experiment was a study of estimation and decision performance as a function of single contour displays. Two single-contour probabilistic displays were simulated: (1) a PC-5 display showing a contour circumscribing a region with probability of target inclusion of 0.5 and (2) a PC-9 display showing the 0.9 inclusion region. These two displays, it is seen, represent simply a dissociation of the PC-59 display of Figure 1.

There were no differences between the procedures of Experiments I and II, except for the display variables. Subjects were 12 undergraduate male students, six assigned randomly to each display group.

Results

Probability Estimates. Figure 4 shows the subjective hit probability function for each group at the end of the 40th training and 40th decision mission. Note that at the end of the 40th decision mission the PC-5 group underestimates hit probability at all stages whereas the PC-9 group slightly underestimates for the first five stages and overestimates for the remaining stages. Compare these results with that obtained for the PC-59 group at the end of decision mission 40 (Figure 2). It can be seen that the estimates of the PC-59 group, at any given stage, approximate closely the mean of the PC-5 plus PC-9 estimates provided at that same stage. This interesting consistency across all stages appears to indicate that the PC-59 display succeeded in averaging out biasing effects of single contours on probability estimations.

A Walsh test for related samples was applied to the probability estimation data, the matching variable again being stages. Differences between PC-5 and PC-9 groups did not reach the 0.05 level of significance.[6] Each of these groups was subsequently tested against the PC-59 group but again with negative results.

The estimates of the PC-9 group at the end of the 40 decision missions were

---

[5] A. R. Jonckheere, "A test of significance for the relation between m rankings and k ranked categories," *Brit. J. Stat. Psychol.*, vol. 7, pp. 93-100; 1954.

[6] S. Siegel, "Nonparametric Statistics for the Behavioral Sciences," McGraw-Hill Book Company, Inc., New York, N. Y.; 1956.

a significant improvement over their final training estimates, at $P \leq 0.025$. The improvement for the PC-5 group, however, was not significant.

Decision Performance. Figure 5 shows the firing distributions for the PC-5 and PC-9 groups over the 40 missions. These distributions bear a reasonable resemblance to that obtained from the PC-59 group, although exhibiting somewhat less peakedness and greater variability. Both Figures 3 and 5 reveal a general tendency for subjects to fire at later stages in preference to earlier stages (for all groups the mean stage of firing was ≥ 6, with the median $>$ mean). This later firing tendency is in accord with previously noted tendencies of college students (as compared with military personnel) to prefer high probability: low reward to low probability: high reward.[7]

The group mean expected value scores were: PC-5 group 2.83, PC-9 group 3.58. A Mann-Whitney U test indicated that the difference in these scores was not significant. Similar tests carried out on differences between the PC-59 group and each of the single contour groups also were not significant.

It is concluded that biasing effects on probability estimation exist as a function of the particular level of contour employed but these biases do not necessarily have a serious effect on decisions.

DISCUSSION

The two experiments reported have clearly demonstrated that probabilistic displays can enhance certain aspects of probability estimation and decision performance. When using probabilistic displays, subjects can provide accurate estimates of the k parameters of k Bernoulli processes. The displayed probabilistic information, in addition, results in more rapid acquisition of accurate estimates, as well as in continued better accuracy over the longer term, as compared with a group using information from the nonprobabilistic display. Providing probability contour information as contrasted with density function information yields the better estimates. Although significant differences were not found between the PC-59 group and either the PC-5 and PC-9 groups on this estimation metric, there was a tendency for the PC-59 groups to demonstrate less bias in their estimates.

Similarly, probabilistic displays result in better decision performance, as measured by the expected value criterion in this experiment. Again contour displays result in better performance than does a density function display. The exact relation between quality of decision performance and accuracy of probability estimates cannot be logically deduced from this experiment, although there would appear to be strong causal relations. Studies have demonstrated, however, that subjects provided with exact probability information still do not make consistent expected value maximization decisions.[8]

[7] J. Scodel, P. Ratoosh, and J. Minas, "Some personality correlates of decision making under conditions of risk," *Behavioral Sci.*, vol. 4, pp. 19-28; 1959.

[8] H. C. A. Dale, "On choosing between bets," *Quart. J. Exp. Psychol.*, vol. 14, pp. 49-51; 1962.

Chapter Three

Human Processing of Information—Receiving and Storing Capabilities

In view of the attention accorded information processing in the Part Two introduction, little remains to be said concerning the general organization of this chapter. Accordingly, our introductory comments will be addressed chiefly to the question, what do these readings have to say that is of interest to engineering psychology? This question is particularly appropriate here because the topics considered are extremely diverse, and therefore the readings are highly selective. Many are drawn from more "traditional" areas of experimental psychology. Such breadth is achieved, however, only at the expense of depth, making any adequate representation of existing content in such active research areas as human memory or psychophysics completely out of the question.

As a preface to this chapter, therefore, it might be well to review the criteria used in sampling the available literature. First, of course, we considered relevance—do the area and the illustrative research have a bearing upon man-machine system concerns? Second, we gave priority to methodology over content and theory. That is, articles illustrating fruitful research techniques or strategies were chosen in preference to *strictly* theoretical or empirical papers; since all three factors—technique, theory, and observation—are inextricably blended in most research, however, none was completely neglected. Finally, we attempted to include only those articles whose message could be grasped by the relatively nonspecialized reader (this was perhaps the most difficult rule of all to follow, for obvious reasons). Even with so restrictive a filtering process, this chapter and the other two chapters on human information processing emerged as rather long ones, a fact which testifies to the current ebullience of research in this general area. To become fully convinced of this fact, one need only consider the hundreds of studies prompted by our first selection, a theoretical and methodological paper in which Swets, Tanner, and Birdsall introduce psychology to the theory of signal detectability (TSD).

Receiving Capabilities

The impact of TSD upon thinking in psychophysics, the area of psychology whose concern is principally the sensory correlates of physical dimensions, has

References will be found in the General References at the end of the book.

been monumental to say the least. Among other things, TSD has thrown into question the cornerstone of classical psychophysical theory and method, the concept of sensory *threshold*.[1] As explained in the Swets et al. article, the essential TSD position is that it is impossible—regardless of experimental rigor—to measure *pure* sensory or perceptual experience. Included in every psychophysical judgment is a healthy *decision* component, sensitive to such nonsensory "cognitive" factors as the observer's prior expectations regarding the stimulus and his preference for various ways of being "right" and "wrong." Instead of trying to dispose of the decision component in the manner of classical psychophysics, the argument runs, why not try to make it work *for* you? To this end, the psychophysical task is recast into the terms of statistical decision theory and, with certain assumptions, it becomes possible to estimate the relative contribution of both sensory and decision processes to any set of detection or discrimination data.[2] Treated in this fashion, experimental findings fail to produce any evidence of a distinct sensory threshold.

In assessing man's basic receptive capabilities, we cannot afford to stop with detection or discrimination—the tasks of most concern to psychophysics. For the engineering psychologist, it is often of much greater importance to know how well people can judge stimuli on an *absolute* rather than a *relative* basis. Suppose that we wished to color-code the bottles in a hospital dispensary; if we used five colors to denote different drug categories, would a nurse be able to identify at a glance the contents of a *single* bottle? What about ten colors? Twenty? A considerable amount of research devoted to this sort of question is reviewed, together with its theoretical implications, in our second selection. As with TSD, of course, the implications of Miller's article—written from the point of view of "information theory" which we consider a little later on in this chapter—extend well beyond mere *receptive* capacities. Nevertheless, the point of particular interest here is the fact that, compared to the number of *relative* judgments he can make, man is severely limited in his ability to identify stimuli *absolutely*. Therefore, when we attempt to apply psychophysical knowledge to man-machine system problems, it is important to take cognizance of the response demands of the task. It is a big jump from discrimination to identification.

Storage or Memory

Of considerable significance to the field of engineering psychology is a current surge of interest in human memory processes, especially those that seem to be involved in storage of information for relatively brief periods of time. In our view, the particular relevance of *immediate* and *short-term* storage for the man-machine system derives from a consideration of machine vis-à-vis human capabilities. While much remains to be learned about human memory, we do know that it has certain general shortcomings which make it vastly inferior to the

---

[1] Threshold is, briefly, the hypothetical point of transition between one level of sensory experience (including zero) and another.

[2] For TSD there is no basic difference between detection and discrimination tasks; both require the observer to report whether or not he can distinguish between levels of stimulation (or neural activity) coexisting in his field of view.

modern computer for many system applications. For example, the human memory usually takes a long time to enter new items of information into some semblance of "permanent" storage, and the difficulty increases rapidly with the number of items to be stored (among other things). Moreover, even if the items are entered, there is no assurance that these items will be available when needed: other items —prior and subsequent—may interfere with their preservation and/or retrieval. What comes out may be an erroneous or distorted version of what went in. Any system, therefore, that relies on a vast repertoire of stored instructions must gravitate toward automation of long-term storage functions, although, of course, never completely so long as humans function in *any* capacity in the system; everything a person does is dependent to some extent upon long-term memory. In the shorter term, however—minutes, seconds, and even milliseconds—it would often be very cumbersome and quite unnecessary to invoke computer techniques. To be sure, people are severely limited as to how many events they can store even briefly; but so long as these limitations are taken into account in the design of tasks, man can serve as a very effective "buffer" system. In fact, it is now becoming widely recognized that he *must* so function if he is to play any part at all in system affairs. He cannot, for example, even perceive a verbal message without temporarily storing the information in earlier items until later ones arrive. If this last point is granted, then it is equally reasonable to argue that an understanding of the principles governing short-term memory could be put to good use in establishing display, coding, and task characteristics. What, for example, is the best way to code telephone designations—is the all-number code superior to the old letter-number scheme in terms of short-term memory?

Strictly apart from its applied implications, the topic of short-term memory has been found to have important ramifications in selective attention, perception, reaction-time, and of course *long-term* retention (some argue, in fact, that a long-short dichotomy is not justified). It is well beyond the scope of this book, however, to pursue these leads—or, for that matter, to discuss theories of forgetting (see Adams, 1967; Keppel, 1968; and Melton, 1963, for pertinent reviews). The articles selected merely illustrate several of the more stimulating approaches (in terms of research generated) which have appeared in recent years. Sperling's technique seems to attack the very earliest storage process, one which is now commonly designated *immediate, iconic,* or *sensory* storage. He reports a rapid decay in stored information, which levels off at approximately the "immediate memory span" (five items recalled) after about 100 milliseconds.

The classic Peterson and Peterson study illustrates another widely adopted technique for measuring a short-term retention process—in this case, one which seems to operate over a period of up to 18 seconds.

Finally, Broadbent's influential paper represents one of the foremost attempts to integrate the existing facts of memory—both long- and short-term varieties— into a coherent account of human information processing. It might be well to note that Miller's article, which we encounter earlier in this chapter, also contains an important theoretical commentary on short-term memory. In essence, Miller is concerned with the way in which information coding controls the storage process. The Broadbent and Miller articles afford us a convenient entry into the topic of the next chapter, the transmission processes.

8

Decision Processes in Perception

John A. Swets, Wilson P. Tanner, Jr., and Theodore G. Birdsall

THE THEORY

Statistical Decision Theory

Consider the following game of chance. Three dice are thrown. Two of the dice are ordinary dice. The third die is unusual in that on each of three of its sides it has three spots, whereas on its remaining three sides it has no spots at all. You, as the player of the game, do not observe the throws of the dice. You are simply informed, after each throw, of the total number of spots showing on the three dice. You are then asked to state whether the third die, the unusual one, showed a 3 or a 0. If you are correct—that is, if you assert a 3 showed when it did in fact, or if you assert a 0 showed when it did in fact—you win a dollar. If you are incorrect—that is, if you make either of the two possible types of errors—you lose a dollar.

How do you play the game? Certainly you will want a few minutes to make some computations before you begin. You will want to know the probability of occurrence of each of the possible totals 2 through 12 in the event that the third die shows a 0, and you will want to know the probability of occurrence of each of the possible totals 5 through 15 in the event that the third die shows a 3. Let us ignore the exact values of these probabilities, and grant that the two

From *Psychological Review*, 68, 1961, 301-310. Copyright 1961 by the American Psychological Association, and reproduced by permission.

This paper is based upon Technical Report No. 40, issued by the Electronic Defense Group of the University of Michigan in 1955. The research was conducted in the Vision Research Laboratory of the University of Michigan with support from the United States Army Signal Corps and the Naval Bureau of Ships. Our thanks are due H. R. Blackwell and W. M. Kincaid for their assistance in the research, and D. H. Howes for suggestions concerning the presentation of this material. This paper was prepared in the Research Laboratory of Electronics, Massachusetts Institute of Technology, with support from the Signal Corps, Air Force (Operational Applications Laboratory and Office of Scientific Research), and Office of Naval Research. This is Technical Report No. ESD-TR-61-20.

Decision Processes in Perception

probability distributions in question will look much like those sketched in Figure 1.

Realizing that you will play the game many times, you will want to establish a policy which defines the circumstances under which you will make each of the two decisions. We can think of this as a *criterion* or a cutoff point along the axis representing the total number of spots showing on the three dice. That is, you will want to choose a number on this axis such that whenever it is equaled or exceeded you will state that a 3 showed on the third die, and such that whenever the total number of spots showing is less than this number, you will state that a 0 showed on the third die. For the game as described, with the a priori probabilities of a 3 and a 0 equal, and with equal values and costs associated with the four possible decision outcomes, it is intuitively clear that the optimal cutoff point is that point where the two curves cross. You will maximize your winnings if you choose this point as the cutoff point and adhere to it.

Now, what if the game is changed? What, for example, if the third die has three spots on five of its sides, and a 0 on only one? Certainly you will now be more willing to state, following each throw, that the third die showed a 3. You will not, however, simply state more often that a 3 occurred without regard to the total showing on the three dice. Rather, you will lower your cutoff point: you will accept a smaller total than before as representing a throw in which the third die showed a 3. Conversely, if the third die has three spots on only one of its sides and 0's on five sides, you will do well to raise your cutoff point—to require a higher total than before for stating that a 3 occurred.

Similarly, your behavior will change if the values and costs associated with the various decision outcomes are changed. If it costs you 5 dollars every time you state that a 3 showed when in fact it did not, and if you win 5 dollars every time you state that a 0 showed when it fact it did (the other value and the other cost in the game remaining at one dollar), you will raise your cutoff to a point somewhere above the point where the two distributions cross. Or if, instead, the premium is placed on being correct when a 3 occurred, rather than when a 0 occurred as in the immediately preceding example, you will assume a cutoff somewhere below the point where the two distributions cross.

Again, your behavior will change if the amount of overlap of the two distributions is changed. You will assume a different cutoff than you did in the game as first described if the three sides of the third die showing spots now show four spots rather than three.

This game is simply an example of the type of situation for which the theory of statistical decision was developed. It is intended only to recall the frame of reference of this theory. Statistical decision theory—or the special case of it which is relevant here, the theory of testing statistical hypotheses—specifies the optimal behavior in a situation where one must choose between two alternative statistical hypotheses on the basis of an observed event. In particular, it specifies the optimal cutoff, long the continuum on which the observed events are arranged, as a function of (a) the a priori probabilities of the two hypotheses, (b) the values and costs associated with the various decision outcomes, and (c) the amount of overlap of the distributions that constitute the hypotheses.

According to the mathematical theory of signal detectability, the problem of detecting signals that are weak relative to the background of interference is like the one faced by the player of our dice game. In short, the detection problem is a problem in statistical decision; it requires testing statistical hypotheses. In the theory of signal detectability, this analogy is developed in terms of an idealized observer. It is our thesis that this conception of the detection process may apply to the human observer as well. The next several pages present an analysis of the detection process that will make the bases for this reasoning apparent.

Fundamental Detection Problem

In the fundamental detection problem, an observation is made of events occurring in a fixed interval of time, and a decision is made, based on this observation, whether the interval contained only the background interference or a signal as well. The interference, which is random, we shall refer to as *noise* and denote as N; the other alternative we shall term *signal plus noise*, SN. In the fundamental problem, only these two alternatives exist—noise is always present, whereas the signal may or may not be present during a specified observation interval. Actually, the observer, who has advance knowledge of the ensemble of signals to be presented, says either "yes, a signal was present" or "no, no signal was present" following each observation. In the experiments reported below, the signal consisted of a small spot of light flashed briefly in a known location on a uniformly illuminated background. It is important to note that the signal is always observed in a background of noise; some, as in the present case, may be introduced by the experimenter or by the external situation, but some is inherent in the sensory processes.

Representation of Sensory Information

We shall, in the following, use the term *observation* to refer to the sensory datum on which the decision is based. We assume that this observation may be represented as varying continuously along a single dimension. Although there is no need to be concrete, it may be helpful to think of the observation as some measure of neural activity, perhaps as the number of impulses arriving at a given point in the cortex within a given time. We assume further that any observation may arise, with specific probabilities, either from noise alone or from signal plus noise. We may portray these assumptions graphically, for a signal of a given amplitude, as in Figure 2. The observation is labeled x and plotted on the abscissa. The left-hand distribution, labeled $f_N(x)$, represents the probability density that x will result given the occurrence of noise alone. The right-hand distribution, $f_{SN}(x)$, is the probability density function of x given the occurrence of signal plus noise. (Probability density functions are used, rather than probability functions, since x is assumed to be continuous.) Since the observations will tend to be of greater magnitude when a signal is presented, the mean of the SN distribution will be greater than the mean of the N distribution. In general, the greater the amplitude of the signal, the greater will be the separation of these means.

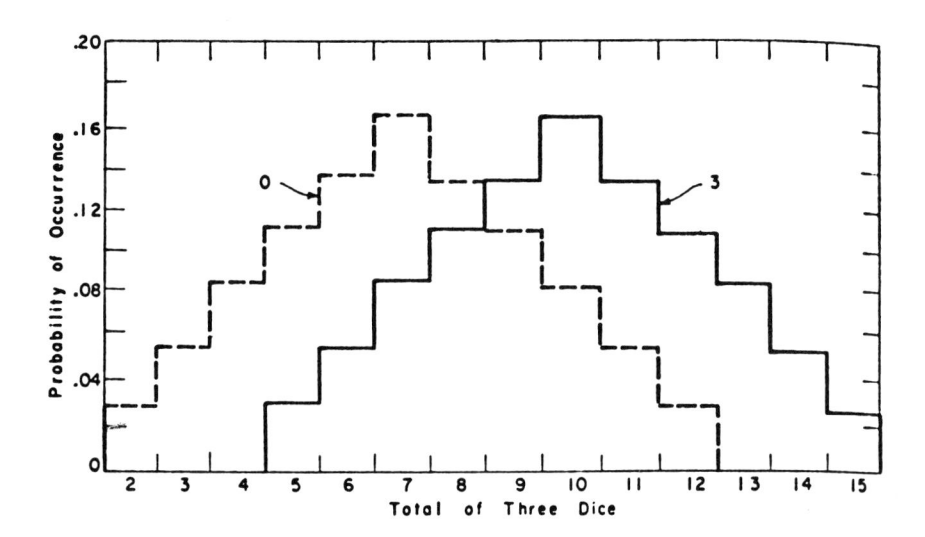

Figure 1. The probability distributions for the dice game.

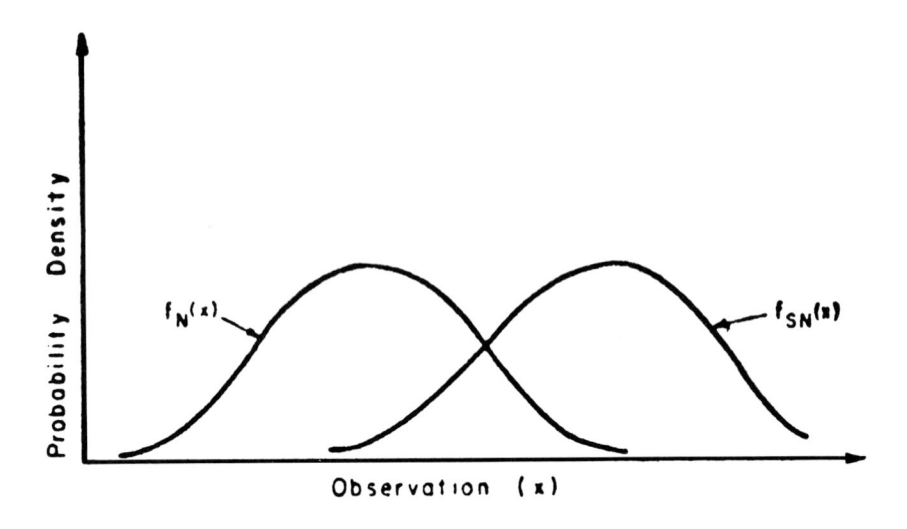

Figure 2. The probability density functions of noise and signal plus noise.

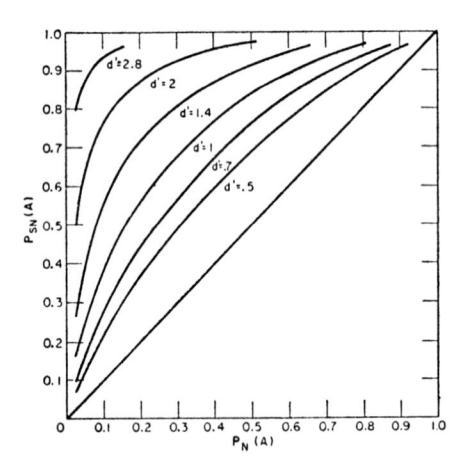

Figure 3. The receiver-operating-characteristic curves. (These curves show $p_{SN}(A)$ vs. $p_N(A)$ with d' as the parameter. They are based on the assumptions that the probability density functions, $f_N(x)$ and $f_{SN}(x)$, are normal and of equal variance.)

Observation as a Value of Likelihood Ratio

It will be well to question at this point our assumption that the observation may be represented along a single axis. Can we, without serious violation, regard the observation as unidimensional, in spite of the fact that the response of the visual system probably has many dimensions? The answer to this question will involve some concepts that are basic to the theory.

One reasonable answer is that when the signal and interference are alike in character, only the magnitude of the total response of the receiving system is available as an indicator of signal existence. Consequently, no matter how complex the sensory information is in fact, the observations may be represented in theory as having a single dimension. Although this answer is quite acceptable when concerned only with the visual case, we prefer to advance a different answer, one that is applicable also to audition experiments, where, for example, the signal may be a segment of a sinusoid presented in a background of white noise.

So let us assume that the response of the sensory system does have several dimensions, and proceed to represent it as a point in an m-dimensional space. Call this point y. For every such point in this space there is some probability density that it resulted from noise alone, $f_N(y)$, and, similarly, some probability density that it was due to signal plus noise, $f_{SN}(y)$. Therefore, there exists a likelihood ratio for each point in the space, $\lambda(y) = f_{SN}(y)/f_N(y)$, expressing the likelihood that the point y arose from SN relative to the likelihood that it arose from N. Since any point in the space, i.e., any sensory datum, may be thus represented as a real, nonzero number, these points may be considered to lie along a single axis. We may then, if we choose, identify the observation x with $\lambda(y)$; the decision axis becomes likelihood ratio.[1]

Having established that we may identify the observation x with $\lambda(y)$, let us note that we may equally well identify x with any monotonic transformation of $\lambda(y)$. It can be shown that we lose nothing by distorting the linear continuum as long as order is maintained. As a matter of fact we may gain if, in particular, we identify x with some transformation of $\lambda(y)$ that results in Gaussian density functions on x. We have assumed the existence of such a transformation in the representation of the density functions, $f_{SN}(x)$ and $f_N(x)$, in Figure 2. We shall see shortly that the assumption of normality simplifies the problem greatly. We shall also see that this assumption is subject to experimental test. A further assumption incorporated into the picture of Figure 2, one made quite tentatively, is that the two density functions are of equal variance. This is equivalent to the assumption that the SN function is a simple translation of the N function, or that adding a signal to the noise merely adds a constant to the N function. The results of a test of this assumption are also described below.

---

[1] Thus the assumption of a unidimensional decision axis is independent of the character of the signal and noise. Rather, it depends upon the fact that just two decision alternatives are considered. More generally, it can be shown that the number of dimensions required to represent the observation is $M - 1$, where M is the number of decision alternatives considered by the observer.

To summarize the last few paragraphs, we have assumed that an observation may be characterized by a value of likelihood ratio, $\lambda(y)$, i.e., the likelihood that the response of the sensory system y arose from SN relative to the likelihood that it arose from N. This permits us to view the observations as lying along a single axis. We then assumed the existence of a particular transformation of $\lambda(y)$ such that on the resulting variable, x, the density functions are normal. We regard the observer as basing his decisions on the variable x.

Definition of the Criterion

If the representation depicted in Figure 2 is realistic, then the problem posed for an observer attempting to detect signals in noise is indeed similar to the one faced by the player of our dice game. On the basis of an observation, one that varies only in magnitude, he must decide between two alternative hypotheses. He must decide from which hypothesis the observation resulted; he must state that the observation is a member of the one distribution or the other. As did the player of the dice game, the observer must establish a policy which defines the circumstances under which the observation will be regarded as resulting from each of the two possible events. He establishes a criterion, a cutoff x_c on the continuum of observations, to which he can relate any given observation x_i. If he finds for the ith observation, x_i, that $x_i > x_c$, he says "yes"; if $x_i < x_c$, he says "no." Since the observer is assumed to be capable of locating a criterion at any point along the continuum of observations, it is of interest to examine the various factors that, according to the theory, will influence his choice of a particular criterion. To do so requires some additional notation.

In the language of statistical decision theory the observer chooses a subject of all of the observations, namely the Critical Region A, such that an observation in this subject leads him to accept the Hypothesis SN, to say that a signal was present. All other observations are in the complementary Subject B; these lead to rejection of the Hypothesis SN, or, equivalently, since the two hypotheses are mutually exclusive and exhaustive, to the acceptance of the Hypothesis N. The Critical Region A, with reference to Figure 2, consists of the values of x to the right of some criterion value x_c.

As in the case of the dice game, a decision will have one of four outcomes: the observer may say "yes" or "no" and may in either case be *correct* or *incorrect*. The decision outcome, in other words, may be a hit (SN · A, the joint occurrence of the Hypothesis SN and an observation in the Region A), a *miss* (SN · B), a *correct rejection* (N · B), or a *false alarm* (N · A). If the a priori probability of signal occurrence and the parameters of the distributions of Figure 2 are fixed, the choice of a criterion value x_c completely determines the probability of each of these outcomes.

Clearly, the four probabilities are interdependent. For example, an increase in the probability of a hit, $p(SN · A)$, can be achieved only by accepting an increase in the probability of a false alarm, $p(N · A)$, and decreases in the other probabilities, $p(SN · B)$ and $p(N · B)$. Thus a given criterion yields a particular balance among the probabilities of the four possible outcomes; conversely, the balance desired by an observer in any instance will determine the optimal location of his criterion. Now the observer may desire the balance that maximizes the

Decision Processes in Perception 83

expected value of a decision in a situation where the four possible outcomes of a decision have individual values, as did the player of the dice game. In this case, the location of the best criterion is determined by the same parameters that determined it in the dice game. The observer, however, may desire a balance that maximizes some other quantity—i.e., a balance that is optimum according to some definition of optimum—in which case a different criterion will be appropriate. He may, for example, want to maximize $p(SN \cdot A)$ while satisfying a restriction on $p(N \cdot A)$, as we typically do when as experimenters we assume an .05 or .01 level of confidence. Alternatively, he may want to maximize the number of correct decisions. Again, he may prefer a criterion that will maximize the reduction in uncertainty in the Shannon (1948) sense.

In statistical decision theory, and in the theory of signal detectability, the optimal criterion under each of these definitions of optimum is specified in terms of the likelihood ratio. That is to say, it can be shown that, if we define the observation in terms of the likelihood ratio, $\lambda(x) = f_{SN}(x)/f_N(x)$, then the optimal criterion can always be specified by some value β of $\lambda(x)$. In other words, the Critical Region A that corresponds to the criterion contains all observations with likelihood ratio greater than or equal to β, and none of those with likelihood ratio less than β.

We shall illustrate this manner of specifying the optimal criterion for just one of the definitions of optimum proposed above, namely, the maximation of the total expected value of a decision in a situation where the four possible outcomes of a decision have individual values associated with them. This is the definition of optimum that we assumed in the dice game. For this purpose we shall need the concept of *conditional probability* as opposed to the *probability of joint occurrence* introduced above. It should be stated that conditional probabilities will have a place in our discussion beyond their use in this illustration; the ones we shall introduce are, as a matter of fact, the fundamental quantities in evaluating the observer's performance.

There are two conditional probabilities of principal interest. These are the conditional probabilities of the observer saying "yes" : $p_{SN}(A)$, the probability of a Yes decision *conditional upon*, or *given*, the occurrence of a signal, and $p_N(A)$, the probability of a Yes decision given the occurrence of noise alone. These two are sufficient, for the other two are simply their complements: $p_{SN}(B) = 1 - p_{SN}(A)$ and $p_N(B) = 1 - p_N(A)$. The conditional and joint probabilities are related as follows:

$$p_{SN}(A) = \frac{p(SN \cdot A)}{p(SN)}$$

$$p_N(A) = \frac{p(N \cdot A)}{p(N)}$$

[1]

where: $p(SN)$ is the a priori probability of signal occurrence and $p(N) = 1 - p(SN)$ is the a priori probability of occurrence of noise alone.

Equation 1 makes apparent the convenience of using conditional rather than joint probabilities—conditional probabilities are independent of the a priori probability of occurrence of the signal and of noise alone. With reference to Figure 2, we may define $p_{SN}(A)$, or the conditional probability of a hit, as the integral of $f_{SN}(x)$ over the Critical Region A, and $p_N(A)$, the conditional probability of a false alarm, as the integral of $f_N(x)$ over A. That is, $p_N(A)$ and $p_{SN}(A)$ represent, respectively, the areas under the two curves of Figure 2 to the right of some criterion value of x.

To pursue our illustration of how an optimal criterion may be specified by a critical value of likelihood ratio β, let us note that the expected value of a decision (denoted EV) is defined in statistical decision theory as the sum, over the potential outcomes of a decision, of the products of probability of outcome and the desirability of outcome. Thus, using the notation V for *positive* individual values and K for costs or *negative* individual values, we have the following equation:

$$EV = V_{SN \cdot A} p(SN \cdot A) + V_{N \cdot B} p(N \cdot B) - K_{SN \cdot B} p(SN \cdot B) - K_{N \cdot A} p(N \cdot A) \quad [2]$$

Now if a priori and conditional probabilities are substituted for the joint probabilities in Equation 2 following Equation 1, for example, $p(SN)p_{SN}(A)$ for $p(SN \cdot A)$, then collecting terms yields the result that maximizing EV is equivalent to maximizing:

$$p_{SN}(A) - \beta p_N(A) \quad [3]$$

$$\text{where } \beta = \frac{p(N)}{p(SN)} \cdot \frac{(V_{N \cdot B} + K_{N \cdot A})}{(V_{SN \cdot A} + K_{SN \cdot B})} \quad [4]$$

It can be shown that this value of β is equal to the value of likelihood ratio, $\lambda(x)$, that corresponds to the optimal criterion. From Equation 3 it may be seen that the value β simply weights the hits and false alarms, and from Equation 4 we see that β is determined by the a priori probabilities of occurrence of signal and of noise alone and by the values associated with the individual decision outcomes. It should be noted that Equation 3 applies to all definitions of optimum. Equation 4 shows the determinants of β in only the special case of the expected-value definition of optimum.

Return for a moment to Figure 2, keeping in mind the result that β is a critical value of $\lambda(x) = f_{SN}(x)/f_N(x)$. It should be clear that the optimal cutoff x_c along the x axis is at the point on this axis where the ratio of the ordinate value of $f_{SN}(x)$ to the ordinate value of $f_N(x)$ is a certain number, namely, β. In the symmetrical case, where the two a priori probabilities are equal and the four individual values are equal, $\beta = 1$ and the optimal value of x_c is the point where $f_{SN}(x) = f_N(x)$, where the two curves cross. If the four values are equal but $p(SN) = 5/6$ and $p(N) = 1/6$, another case described in connection with the dice game, then $\beta = 1/5$ and the optimal value of x_c is shifted a certain distance to the left. This shift may be seen intuitively to be in the proper direction—a higher value of $p(SN)$ should lead to a greater willingness to accept the Hypothesis SN, i.e., a more lenient cutoff. To consider one more example from the dice game, if $p(SN) = p(N) = 0.5$, if $V_{N \cdot B}$ and $K_{N \cdot A}$ are set at 5 dollars and $V_{SN \cdot A}$ and $K_{SN \cdot B}$ are equal to 1 dollar, then $\beta = 5$ and the optimal value of x_c shifts a certain dis-

Decision Processes in Perception

tance to the right. Again intuitively, if it is more important to be correct when the Hypothesis N is true, a high, or strict, criterion should be adopted.

In any case, β specifies the optimal weighting of hits relative to false alarms: x_c should always be located at the point on the x axis corresponding to β. As we pointed out in discussing the dice game, just where this value of x_c will be with reference to the x axis depends not only upon the a priori probabilities and the values but also upon the overlap of the two density functions, in short, upon the signal strength. We shall define a measure of signal strength within the next few pages. For now, it is important to note that for any detection goal to which the observer may subscribe, and for any set of parameters that may characterize a detection situation (such as a priori probabilities and values associated with decision outcomes), the optimal criterion may be specified in terms of a single number, β, a critical value of likelihood ratio.[2]

Receiver-Operating-Characteristic

Whatever criterion the observer actually uses, even if it is not one of the optimal criteria, can also be described by a single number, by some value of likelihood ratio. Let us proceed to a consideration of how the observer's performance may be evaluated with respect to the location of his criterion, and, at the same time we shall see how his performance may be evaluated with respect to his sensory capabilities.

As we have noted, the fundamental quantities in the evaluation of performance are $p_N(A)$ and $p_{SN}(A)$, these quantities representing, respectively, the areas under the two curves of Figure 2 to the right of some criterion value of x. If we set up a graph of $p_{SN}(A)$ versus $p_N(A)$ and trace on it the curve resulting as we move the decision criterion along the decision axis of Figure 2, we sketch one of the arcs shown in Figure 3. Ignore, for a moment, all but one of these arcs. If the decision criterion is set way at the left in Figure 2, we obtain a point in the upper right-hand corner of Figure 3: both $p_{SN}(A)$ and $p_N(A)$ are unity. If the criterion is set at the right end of the decision axis in Figure 2, the point at the other extreme of Figure 3, $p_{SN}(A) = p_N(A) = 0$, is obtained. In between these extremes lie the criterion values of more practical interest. It should be noted that the exact form of the curve shown in Figure 3 is not the only form which might result, but it is the form which will result if the observer chooses a criterion in terms of likelihood ratio, and the probability density functions are normal and of equal variance.

This curve is a form of the *operating characteristic* as it is known in statistics; in the context of the detection problem it is usually referred to as the *receiver-operating-characteristic*, or ROC, curve. The optimal "operating level" may be

---

[2] We have reached a point in the discussion where we can justify the statement made earlier that the decision axis may be equally well regarded as likelihood ratio or as any monotonic transformation of likelihood ratio. Any distortion of the linear continuum of likelihood ratio, that maintains order, is equivalent to likelihood ratio in terms of determining a criterion. The decisions made are the same whether the criterion is set at likelihood ratio equal to β or at the value that corresponds to β of some new variable. To illustrate, if a criterion leads to a Yes response whenever $\lambda(y) > 2$, if $x = [\lambda(y)]^2$ the decisions will be the same if the observer says "yes" whenever $x > 4$.

seen from Equation 3 to be at the point of the ROC curve where its slope is β. That is, the expression $p_{SN}(A) - \beta p_N(A)$ defines a utility line of slope β, and the point of tangency of this line to the ROC curve is the optimal operating level. Thus the theory specifies the appropriate hit probability and false alarm probability for any definition of optimum and any set of parameters characterizing the detection situation.

It is now apparent how the observer's choice of a criterion in a given experiment may be indexed. The proportions obtained in an experiment are used as estimates of the probabilities, $p_N(A)$ and $p_{SN}(A)$; thus, the observer's behavior yields a point on an ROC curve. The slope of the curve at this point corresponds to the value of likelihood ratio at which he has located his criterion. Thus we work backward from the ROC curve to infer the criterion that is employed by the observer.

There is, of course, a family of ROC curves, as shown in Figure 3, a given curve corresponding to a given separation between the means of the density functions $f_N(x)$ and $f_{SN}(x)$. The parameter of these curves has been called d', where d' is defined as the difference between the means of the two density functions expressed in terms of their standard deviation, i.e.:

$$\frac{M_{f_{SN(x)}} - M_{f_{N(x)}}}{\sigma_{f_{N(x)}}} \qquad [5]$$

Since the separation between the means of the two density functions is a function of signal amplitude, d' is an index of the detectability of a given signal for a given observer.

Recalling our assumptions that the density functions $f_N(x)$ and $f_{SN}(x)$ are normal and of equal variance, we may see from Equation 5 that the quantity denoted d' is simply the familiar normal deviate, or x/σ measure. From the pair of values $p_N(A)$ and $p_{SN}(A)$ that are obtained experimentally, one may proceed to a published table of areas under the normal curve to determine a value of d'. A simpler computational procedure is achieved by plotting the points [$p_N(A)$, $p_{SN}(A)$] on graph paper having a probability scale and a normal deviate scale on both axes.

We see now that the four-fold table of the responses that are made to a particular stimulus may be treated as having two independent parameters—the experiment yields measures of two independent aspects of the observer's performance. The variable d' is a measure of the observer's sensory capabilities, or of the effective signal strength. This may be thought of as the object of interest in classical psychophysics. The criterion β that is employed by the observer, which determines the $p_N(A)$ and $p_{SN}(A)$ for some fixed d', reflects the effect of variables which have been variously called the set, attitude, or motives of the observer. It is the ability to distinguish between these two aspects of detection performance that comprises one of the main advantages of the theory proposed here. We have noted that these two aspects of behavior are confounded in an experiment in which the dependent variable is the intensity of the signal that is required for a threshold response.

REFERENCE

1. Shannon, C. E. "A mathematical theory of communication." *Bell Syst. Tech. J.,* 1948, **27,** 379-423, 623-656.

9

The Magical Number Seven, Plus or Minus Two: Some Limits on Our Capacity for Processing Information

George A. Miller

My problem is that I have been persecuted by an integer. For seven years this number has followed me around, has intruded in my most private data, and has assaulted me from the pages of our most public journals. This number assumes a variety of disguises, being sometimes a little larger and sometimes a little smaller than usual, but never changing so much as to be unrecognizable. The persistence with which this number plagues me is far more than a random accident. There is, to quote a famous senator, a design behind it, some pattern governing its appearances. Either there really is something unusual about the number or else I am suffering from delusions of persecution.

I shall begin my case history by telling you about some experiments that tested how accurately people can assign numbers to the magnitudes of various aspects of a stimulus. In the traditional language of psychology these would be called experiments in absolute judgment. Historical accident, however, has decreed that they should have another name. We now call them experiments on the capacity of people to transmit information. Since these experiments would not have been done without the appearance of information theory on the psychological scene,

From *Psychological Review,* 63, 1956, 81-97. Copyright 1956 by the American Psychological Association, and reproduced by permission.

This paper was first read as an Invited Address before the Eastern Psychological Association in Philadelphia on April 15, 1955. Preparation of the paper was supported by the Harvard Psycho-Acoustic Laboratory under Contract N5ori-76 between Harvard University and the Office of Naval Research, U. S. Navy (Project NRI142-201, Report PNR-174). Reproduction for any purpose of the U. S. Government is permitted.

and since the results are analyzed in terms of the concepts of information theory, I shall have to preface my discussion with a few remarks about this theory.

INFORMATION MEASUREMENT

The "amount of information" is exactly the same concept that we have talked about for years under the name of "variance." The equations are different, but if we hold tight to the idea that anything that increases the variance also increases the amount of information we cannot go far astray.

The advantages of this new way of talking about variance are simple enough. Variance is always stated in terms of the unit of measurement—inches, pounds, volts, etc.—whereas the amount of information is a dimensionless quantity. Since the information in a discrete statistical distribution does not depend upon the unit of measurement, we can extend the concept to situations where we have no metric and we would not ordinarily think of using the variance. And it also enables us to compare results obtained in quite different experimental situations where it would be meaningless to compare variances based on different metrics. So there are some good reasons for adopting the newer concept.

The similarity of variance and amount of information might be explained this way: When we have a large variance, we are very ignorant about what is going to happen. If we are very ignorant, then when we make the observation it gives us a lot of information. On the other hand, if the variance is very small, we know in advance how our observation must come out, so we get little information from making the observation.

If you will now imagine a communication system, you will realize that there is a great deal of variability about what goes into the system and also a great deal of variability about what comes out. The input and the output can therefore be described in terms of their variance (or their information). If it is a good communication system, however, there must be some systematic relation between what goes in and what comes out. That is to say, the output will depend upon the input, or will be correlated with the input. If we measure this correlation, then we can say how much of the output variance is attributable to the input and how much is due to random fluctuations or "noise" introduced by the system during transmission. So we see that the measure of transmitted information is simply a measure of the input-output correlation.

There are two simple rules to follow. Whenever I refer to "amount of information," you will understand "variance." And whenever I refer to "amount of transmitted information," you will understand "covariance" or "correlation."

The situation can be described graphically by two partially overlapping circles. Then the left circle can be taken to represent the variance of the input, the right circle the variance of the output, and the overlap the covariance of input and output. I shall speak of the left circle as the amount of input information, the right circle as the amount of output information, and the overlap as the amount of transmitted information.

In the experiments on absolute judgment, the observer is considered to be a communication channel. Then the left circle would represent the amount of information in the stimuli, the right circle the amount of information in his responses, and the overlap the stimulus-response correlation as measured by the amount of transmitted information. The experimental problem is to increase the amount of input information and to measure the amount of transmitted information. If the observer's absolute judgments are quite accurate, then nearly all of the input information will be transmitted and will be recoverable from his responses. If he makes errors, then the transmitted information may be considerably less than the input. We expect that, as we increase the amount of input information, the observer will begin to make more and more errors; we can test the limits of accuracy of his absolute judgments. If the human observer is a reasonable kind of communication system, then when we increase the amount of input information the transmitted information will increase at first and will eventually level off at some asymptotic value. This asymptotic value we take to be the *channel capacity* of the observer: it represents the greatest amount of information that he can give us about the stimulus on the basis of an absolute judgment. The channel capacity is the upper limit on the extent to which the observer can match his responses to the stimuli we give him.

Now just a brief word about the *bit* and we can begin to look at some data. One bit of information is the amount of information that we need to make a decision between two equally likely alternatives. If we must decide whether a man is less than six feet tall or more than six feet tall and if we know that the chances are 50-50, then we need one bit of information. Notice that this unit of information does not refer in any way to the unit of length that we use—feet, inches, centimeters, etc. However you measure the man's height, we still need just one bit of information.

Two bits of information enable us to decide among four equally likely alternatives. Three bits of information enable us to decide among eight equally likely alternatives. Four bits of information decide among 16 alternatives, five among 32, and so on. That is to say, if there are 32 equally likely alternatives, we must make five successive binary decisions, worth one bit each, before we know which alternative is correct. So the general rule is simple: every time the number of alternatives is increased by a factor of two, one bit of information is added.

There are two ways we might increase the amount of input information. We could increase the rate at which we give information to the observer, so that the amount of information per unit time would increase. Or we could ignore the time variable completely and increase the amount of input information by increasing the number of alternative stimuli. In the absolute judgment experiment we are interested in the second alternative. We give the observer as much time as he wants to make his response; we simply increase the number of alternative stimuli among which he must discriminate and look to see where confusions begin to occur. Confusions will appear near the point that we are calling his "channel capacity."

ABSOLUTE JUDGMENTS OF UNIDIMENSIONAL STIMULI

Now let us consider what happens when we make absolute judgments of tones. Pollack (17) asked listeners to identify tones by assigning numerals to them. The tones were different with respect to frequency, and covered the range from 100 to 8000 cps in equal logarithmic steps. A tone was sounded and the listener responded by giving a numeral. After the listener had made his response he was told the correct identification of the tone.

When only two or three tones were used the listeners never confused them. With four different tones confusions were quite rare, but with five or more tones confusions were frequent. With fourteen different tones the listeners made many mistakes.

These data are plotted in Figure 1. Along the bottom is the amount of input information in bits per stimulus. As the number of alternative tones was increased from 2 to 14, the input information increased from 1 to 3.8 bits. On the ordinate is plotted the amount of transmitted information. The amount of transmitted information behaves in much the way we would expect a communication channel to behave; the transmitted information increases linearly up to about 2 bits and then bends off toward an asymptote at about 2.5 bits. This value, 2.5 bits, therefore, is what we are calling the channel capacity of the listener for absolute judgments of pitch.

So now we have the number 2.5 bits. What does it mean? First, note that 2.5 bits corresponds to about six equally likely alternatives. The result means that we cannot pick more than six different pitches that the listener will never confuse. Or, stated slightly differently, no matter how many alternative tones we ask him to judge, the best we can expect him to do is to assign them to about six different classes without error. Or, again, if we know that there were N alternative stimuli, then his judgment enables us to narrow down the particular stimulus to one out of $N/6$.

Most people are surprised that the number is as small as six. Of course, there is evidence that a musically sophisticated person with absolute pitch can identify accurately any one of 50 or 60 different pitches. Fortunately, I do not have time to discuss these remarkable exceptions. I say it is fortunate because I do not know how to explain their superior performance. So I shall stick to the more pedestrian fact that most of us can identify about one out of only five or six pitches before we begin to get confused.

It is interesting to consider that psychologists have been using seven-point rating scales for a long time, on the intuitive basis that trying to rate into finer categories does not really add much to the usefulness of the ratings. Pollack's results indicate that, at least for pitches, this intuition is fairly sound.

Next you can ask how reproducible this result is. Does it depend on the spacing of the tones or the various conditions of judgment? Pollack varied these conditions in a number of ways. The range of frequencies can be changed by a factor of about 20 without changing the amount of information transmitted more than a small percentage. Different groupings of the pitches decreased the transmission, but the loss was small. For example, if you can discriminate five high-pitched tones in one series and five low-pitched tones in another series, it is reasonable to expect that you could combine all ten into a single series and still tell them all apart without error. When you try it, however, it does not work. The channel capacity for pitch seems to be about six and that is the best you can do.

While we are on tones, let us look next at Garner's (7) work on loudness. Garner's data for loudness are summarized in Figure 2. Garner went to some

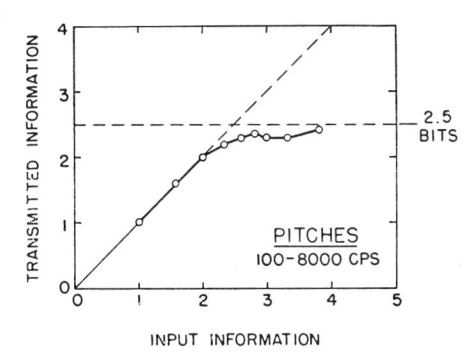

Figure 1. Data from Pollack (17, 18) on the amount of information that is transmitted by listeners who make absolute judgments of auditory pitch. As the amount of input information is increased by increasing from 2 to 14 the number of different pitches to be judged, the amount of transmitted information approaches as its upper limit a channel capacity of about 2.5 bits per judgment.

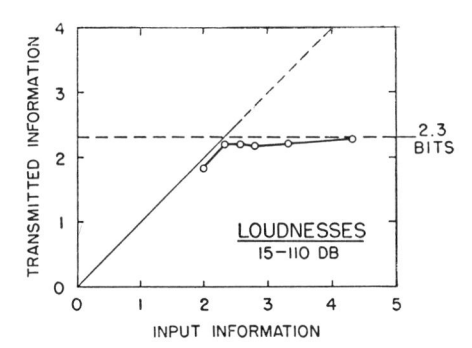

Figure 2. Data from Garner (7) on the channel capacity for absolute judgments of auditory loudness.

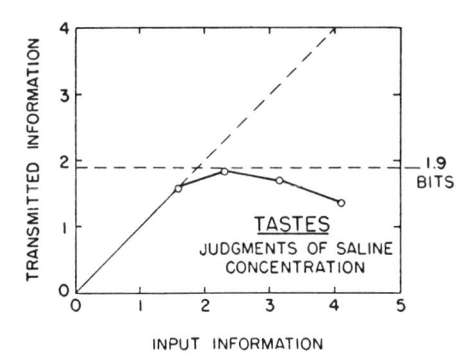

Figure 3. Data from Beebe-Center, Rogers, and O'Connell (1) on the channel capacity for absolute judgments of saltiness.

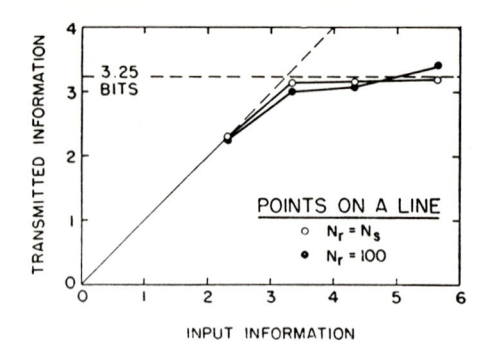

Figure 4. Data from Hake and Garner (8) on the channel capacity for absolute judgments of the position of a pointer in a linear interval.

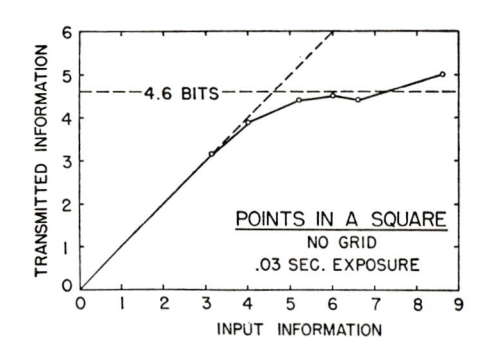

Figure 5. Data from Klemmer and Frick (13) on the channel capacity for absolute judgments of the position of a dot in a square.

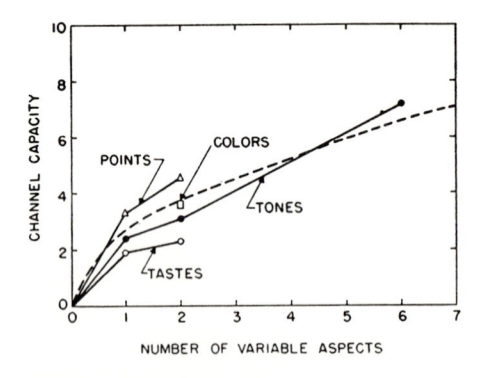

Figure 6. The general form of the relation between channel capacity and the number of independently variable attributes of the stimuli.

trouble to get the best possible spacing of his tones over the intensity range from 15 to 110 db. He used 4, 5, 6, 7, 10, and 20 different stimulus intensities. The results shown in Figure 2 take into account the differences among subjects and the sequential influence of the immediately preceding judgment. Again we find that there seems to be a limit. The channel capacity for absolute judgments of loudness is 2.3 bits, or about five perfectly discriminable alternatives.

Since these two studies were done in different laboratories with slightly different techniques and methods of analysis, we are not in a good position to argue whether five loudnesses is significantly different from six pitches. Probably the difference is in the right direction, and absolute judgments of pitch are slightly more accurate than absolute judgments of loudness. The important point, however, is that the two answers are of the same order of magnitude.

The experiment has also been done for taste intensities. In Figure 3 are the results obtained by Beebe-Center, Rogers, and O'Connell (1) for absolute judgments of the concentration of salt solutions. The concentrations ranged from 0.3 to 34.7 gm. NaCl per 100 cc. tap water in equal subjective steps. They used 3, 5, 9, and 17 different concentrations. The channel capacity is 1.9 bits, which is about four distinct concentrations. Thus taste intensities seem a little less distinctive than auditory stimuli, but again the order of magnitude is not far off.

On the other hand, the channel capacity for judgments of visual position seems to be significantly larger. Hake and Garner (8) asked observers to interpolate visually between two scale markers. Their results are shown in Figure 4. They did the experiment in two ways. In one version they let the observer use any number between zero and 100 to describe the position, although they presented stimuli at only 5, 10, 20, or 50 different positions. The results with this unlimited response technique are shown by the filled circles on the graph. In the other version the observers were limited in their responses to reporting just those stimulus values that were possible. That is to say, in the second version the number of different responses that the observer could make was exactly the same as the number of different stimuli that the experimenter might present. The results with this limited response technique are shown by the open circles on the graph. The two functions are so similar that it seems fair to conclude that the number of responses available to the observer had nothing to do with the channel capacity of 3.25 bits.

The Hake-Garner experiment has been repeated by Coonan and Klemmer. Although they have not yet published their results, they have given me permission to say that they obtained channel capacities ranging from 3.2 bits for very short exposures of the pointer position to 3.9 bits for longer exposures. These values are slightly higher than Hake and Garner's, so we must conclude that there are between 10 and 15 distinct positions along a linear interval. This is the largest channel capacity that has been measured for any unidimensional variable.

At the present time these four experiments on absolute judgments of simple, unidimensional stimuli are all that have appeared in the psychological journals. However, a great deal of work on other stimulus variables has not yet appeared in the journals. For example, Eriksen and Hake (6) have found that the channel capacity for judging the sizes of squares is 2.2 bits, or about five categories, under a wide range of experimental conditions. In a separate experiment Eriksen (5) found 2.8 bits for size, 3.1 bits for hue, and 2.3 bits for brightness. Geldard has measured the channel capacity for the skin by placing vibrators on the chest region. A good observer can identify about four intensities, about five durations, and about seven locations.

One of the most active groups in this area has been the Air Force Operational Applications Laboratory. Pollack has been kind enough to furnish me with the results of their measurements for several aspects of visual displays. They made measurements for area and for the curvature, length, and direction of lines. In one set of experiments they used a very short exposure of the stimulus—$1/40$ second— and then they repeated the measurements with a 5-second exposure. For area they got 2.6 bits with the short exposure and 2.7 bits with the long exposure. For the length of a line they got about 2.6 bits with the short exposure and about 3.0 bits with the long exposure. Direction, or angle of inclination, gave 2.8 bits for the short exposure and 3.3 bits for the long exposure. Curvature was apparently harder to judge. When the length of the arc was constant, the result at the short exposure duration was 2.2 bits, but when the length of the chord was constant, the result was only 1.6 bits. This last value is the lowest that anyone has measured to date. I should add, however, that these values are apt to be slightly too low because the data from all subjects were pooled before the transmitted information was computed.

Now let us see where we are. First, the channel capacity does seem to be a valid notion for describing human observers. Second, the channel capacities measured for these unidimensional variables range from 1.6 bits for curvature to 3.9 bits for positions in an interval. Although there is no question that the differences among the variables are real and meaningful, the more impressive fact to me is their considerable similarity. If I take the best estimates I can get of the channel capacities for all the stimulus variables I have mentioned, the mean is 2.6 bits and the standard deviation is only 0.6 bit. In terms of distinguishable alternatives, this mean corresponds to about 6.5 categories, one standard deviation includes from 4 to 10 categories, and the total range is from 3 to 15 categories. Considering the wide variety of different variables that have been studied, I find this to be a remarkably narrow range.

There seems to be some limitation built into us either by learning or by the design of our nervous systems, a limit that keeps our channel capacities in this general range. On the basis of the present evidence it seems safe to say that we possess a finite and rather small capacity for making such unidimensional judgments and that this capacity does not vary a great deal from one simple sensory attribute to another.

ABSOLUTE JUDGMENTS OF MULTI-DIMENSIONAL STIMULI

You may have noticed that I have been careful to say that this magical number seven applies to one-dimensional judgments. Everyday experience teaches us that we can identify accurately any one of several hundred faces, any one of several thousand words, any one of several thousand objects, etc. The story certainly would not be complete if we stopped at this point. We must have some understanding of why the one-dimensional variables we judge in the laboratory give results so far out of line with what we do constantly in our behavior outside the laboratory. A possible explanation lies in the number of independently variable attributes of the stimuli that are being judged. Objects, faces, words, and the like differ from one another in many ways, whereas the simple stimuli we have considered thus far differ from one another in only one respect.

Fortunately, there are a few data on what happens when we make absolute judgments of stimuli that differ from one another in several ways. Let us look first at the results Klemmer and Frick (13) have reported for the absolute judgment of the position of a dot in a square. In Figure 5 we see their results. Now the channel capacity seems to have increased to 4.6 bits, which means that people can identify accurately any one of 24 positions in the square.

The position of a dot in a square is clearly a two-dimensional proposition. Both its horizontal and its vertical position must be identified. Thus it seems natural to compare the 4.6-bit capacity for a square with the 3.25-bit capacity for the position of a point in an interval. The point in the square requires two judgments of the interval type. If we have a capacity of 3.25 bits for estimating intervals and we do this twice, we should get 6.5 bits as our capacity for locating points in a square. Adding the second independent dimension gives us an increase from 3.25 to 4.6, but it falls short of the perfect addition that would give 6.5 bits.

Another example is provided by Beebe-Center, Rogers, and O'Connell. When they asked people to identify both the saltiness and the sweetness of solutions containing various concentrations of salt and sucrose, they found that the channel capacity was 2.3 bits. Since the capacity for salt alone was 1.9, we might expect about 3.8 bits if the two aspects of the compound stimuli were judged independently. As with spatial locations, the second dimension adds a little to the capacity but not as much as it conceivably might.

A third example is provided by Pollack (18), who asked listeners to judge both the loudness and the pitch of pure tones. Since pitch gives 2.5 bits and loudness gives 2.3 bits, we might hope to get as much as 4.8 bits for pitch and loudness together. Pollack obtained 3.1 bits, which again indicates that the second dimension augments the channel capacity but not so much as it might.

A fourth example can be drawn from the work of Halsey and Chapanis (9) on confusions among colors of equal luminance. Although they did not analyze their results in informational terms, they estimate that there are about 11 to 15 identifiable colors, or, in our terms, about 3.6 bits. Since these colors varied in both hue and saturation, it is probably correct to regard this as a two-dimensional judgment. If we compare this with Eriksen's 3.1 bits for hue (which is a questionable comparison to draw), we again have something less than perfect addition when a second dimension is added.

It is still a long way, however, from these two-dimensional examples to the multidimensional stimuli provided by faces, words, etc. To fill this gap we have only one experiment, an auditory study done by Pollack and Ficks (19). They managed to get six different acoustic variables that they could change: frequency, intensity, rate of interruption, on-time fraction, total duration, and spatial location. Each one of these six variables could assume any one of five different values, so altogether there were 5^6, or 15,625 different tones that they could present. The listeners made a separate rating for each one of these six dimensions. Under these conditions the transmitted information was 7.2 bits, which corresponds to about 150 different categories that could be absolutely identified without error. Now we are beginning to get up into the range that ordinary experience would lead us to expect.

Suppose that we plot these data, fragmentary as they are, and make a guess about how the channel capacity changes with the dimensionality of the stimuli. The result is given in Figure 6. In a moment of considerable daring I sketched the dotted line to indicate roughly the trend that the data seemed to be taking.

Clearly, the addition of independently variable attributes to the stimulus increases the channel capacity, but at a decreasing rate. It is interesting to note that the channel capacity is increased even when the several variables are not independent. Eriksen (5) reports that, when size, brightness, and hue all vary together in perfect correlation, the transmitted information is 4.1 bits as compared with an average of about 2.7 bits when these attributes are varied one at a time. By confounding three attributes, Eriksen increased the dimensionality of the input without increasing the amount of input information; the result was an increase in channel capacity of about the amount that the dotted function in Figure 6 would lead us to expect.

The point seems to be that, as we add more variables to the display, we increase the total capacity, but we decrease the accuracy for any particular variable. In other words, we can make relatively crude judgments of several things simultaneously.

We might argue that in the course of evolution those organisms were most successful that were responsive to the widest range of stimulus energies in their environment. In order to survive in a constantly fluctuating world, it was better to have a little information about a lot of things than to have a lot of information about a small segment of the environment. If a compromise was necessary, the one we seem to have made is clearly the more adaptive.

Pollack and Ficks's results are very strongly suggestive of an argument that linguists and phoneticians have been making for some time (11). According to the linguistic analysis of the sounds of human speech, there are about eight or ten dimensions—the linguists call them *distinctive features*—that distinguish one phoneme from another. These distinctive features are usually binary, or at most ternary, in nature. For example, a binary distinction is made between vowels and consonants, a binary decision is made between oral and nasal consonants, a ternary decision is made among front, middle, and back phonemes, etc. This approach gives us quite a different picture of speech perception than we might otherwise obtain from our studies of the speech spectrum and of the ear's ability to discriminate relative differences among pure tones. I am personally much interested in this new approach (15), and I regret that there is not time to discuss it here.

It was probably with this linguistic theory in mind that Pollack and Ficks conducted a test on a set of tonal stimuli that varied in eight dimensions, but required only a binary decision on each dimension. With these tones they measured the transmitted information at 6.9 bits, or about 120 recognizable kinds of sounds. It is an intriguing question, as yet unexplored, whether one can go on adding dimensions indefinitely in this way.

In human speech there is clearly a limit to the number of dimensions that we use. In this instance, however, it is not known whether the limit is imposed by the nature of the perceptual machinery that must recognize the sounds or by the nature of the speech machinery that must produce them. Somebody will have to

The Magical Number Seven, Plus or Minus Two

do the experiment to find out. There is a limit, however, at about eight or nine distinctive features in every language that has been studied, and so when we talk we must resort to still another trick for increasing our channel capacity. Language uses sequences of phonemes, so we make several judgments successively when we listen to words and sentences. That is to say, we use both simultaneous and successive discriminations in order to expand the rather rigid limits imposed by the inaccuracy of our absolute judgments of simple magnitudes.

These multidimensional judgments are strongly reminiscent of the abstraction experiment of Külpe (14). As you may remember, Külpe showed that observers report more accurately on an attribute for which they are set than on attributes for which they are not set. For example, Chapman (4) used three different attributes and compared the results obtained when the observers were instructed before the tachistoscopic presentation with the results obtained when they were not told until after the presentation which one of the three attributes was to be reported. When the instruction was given in advance, the judgments were more accurate. When the instruction was given afterwards, the subjects presumably had to judge all three attributes in order to report on any one of them and the accuracy was correspondingly lower. This is in complete accord with the results we have just been considering, where the accuracy of judgment on each attribute decreased as more dimensions were added. The point is probably obvious, but I shall make it anyhow, that the abstraction experiments did *not* demonstrate that people can judge only one attribute at a time. They merely showed what seems quite reasonable, that people are less accurate if they must judge more than one attribute simultaneously.

SUBITIZING

I cannot leave this general area without mentioning, however briefly, the experiments conducted at Mount Holyoke College on the discrimination of number (12). In experiments by Kaufman, Lord, Reese, and Volkmann random patterns of dots were flashed on a screen for $1/5$ of a second. Anywhere from 1 to more than 200 dots could appear in the pattern. The subject's task was to report how many dots there were.

The first point to note is that on patterns containing up to five or six dots the subjects simply did not make errors. The performance on these small numbers of dots was so different from the performance with more dots that it was given a special name. Below seven the subjects were said to *subitize;* above seven they were said to *estimate*. This is, as you will recognize, what we once optimistically called "the span of attention."

This discontinuity at seven is, of course, suggestive. Is this the same basic process that limits our unidimensional judgment to about seven categories. The generalization is tempting, but not sound in my opinion. The data on number estimates have not been analyzed in informational terms; but on the basis of the published data I would guess that the subjects transmitted something more than four bits of information about the number of dots. Using the same arguments as before, we would conclude that there are about 20 or 30 distinguishable categories of numerousness. This is considerably more information than we would expect to get from a unidimensional display. It is, as a matter of fact, very much like a two-dimensional display. Although the dimensionality of the random dot patterns is not entirely clear, these results are in the same range as Klemmer and Frick's for their two-dimensional display of dots in a square. Perhaps the two dimensions of numerousness are area and density. When the subject can subitize, area and density may not be the significant variables, but when the subject must estimate perhaps they are significant. In any event, the comparison is not so simple as it might seem at first thought.

This is one of the ways in which the magical number seven has persecuted me. Here we have two closely related kinds of experiments, both of which point to the significance of the number seven as a limit on our capacities. And yet when we examine the matter more closely, there seems to be a reasonable suspicion that it is nothing more than a coincidence.

THE SPAN OF IMMEDIATE MEMORY

Let me summarize the situation in this way. There is a clear and definite limit to the accuracy with which we can identify absolutely the magnitude of a unidimensional stimulus variable. I would propose to call this limit the *span of absolute judgment,* and I maintain that for unidimensional judgments this span is usually somewhere in the neighborhood of seven. We are not completely at the mercy of this limited span, however, because we have a variety of techniques for getting around it and increasing the accuracy of our judgments. The three most important of these devices are (*a*) to make relative rather than absolute judgments; or, if that is not possible, (*b*) to increase the number of dimensions along which the stimuli can differ; or (*c*) to arrange the task in such a way that we make a sequence of several absolute judgments in a row.

The study of relative judgments is one of the oldest topics in experimental psychology, and I will not pause to review it now. The second device, increasing the dimensionality, we have just considered. It seems that by adding more dimensions and requiring crude, binary, yes-no judgments on each attribute we can extend the span of absolute judgment from seven to at least 150. Judging from our everyday behavior, the limit is probably in the thousands, if indeed there is a limit. In my opinion, we cannot go on compounding dimensions indefinitely. I suspect that there is also a *span of perceptual dimensionality* and that this span is somewhere in the neighborhood of ten, but I must add at once that there is no objective evidence to support this suspicion. This is a question sadly needing experimental exploration.

Concerning the third device, the use of successive judgments, I have quite a bit to say because this device introduces memory as the handmaiden of discrimination. And, since mnemonic processes are at least as complex as are perceptual processes, we can anticipate that their interactions will not be easily disentangled.

Suppose that we start by simply extending slightly the experimental procedure that we have been using. Up to this point we have presented a single stimulus and asked the observer to name it immediately thereafter. We can extend this procedure by requiring the observer to withhold his response until we have given him several stimuli in succession. At the end of the sequence of stimuli he then makes his response. We still have the same sort of input-output situation that is required for the measurement of transmitted information. But now we have passed from an experiment on absolute judgment to what is traditionally called an experiment on immediate memory.

Before we look at any data on this topic I feel I must give you a word of warning to help you avoid some obvious associations that can be confusing. Everybody knows that there is a finite span of immediate memory and that for a lot of different kinds of test materials this span is about seven items in length. I have just shown you that there is a span of absolute judgment that can distinguish about seven categories and that there is a span of attention that will encompass about six objects at a glance. What is more natural than to think that all three of these spans are different aspects of a single underlying process? And that is a fundamental mistake, as I shall be at some pains to demonstrate. This mistake is one of the malicious persecutions that the magical number seven has subjected me to.

My mistake went something like this. We have seen that the invariant feature in the span of absolute judgment is the amount of information that the observer can transmit. There is a real operational similarity between the absolute judgment experiment and the immediate memory experiment. If immediate memory is like absolute judgment, then it should follow that the invariant feature in the span of immediate memory is also the amount of information that an observer can retain. If the amount of information in the span of immediate memory is a constant, then the span should be short when the individual items contain a lot of information and the span should be long when the items contain little information. For example, decimal digits are worth 3.3 bits apiece. We can recall about seven of them, for a total of 23 bits of information. Isolated English words are worth about 10 bits apiece. If the total amount of information is to remain constant at 23 bits, then we should be able to remember only two or three words chosen at random. In this way I generated a theory about how the span of immediate memory should vary as a function of the amount of information per item in the test materials.

The measurements of memory span in the literature are suggestive on this question, but not definitive. And so it was necessary to do the experiment to see. Hayes (10) tried it out with five different kinds of test materials: binary digits, decimal digits, letters of the alphabet, letters plus decimal digits, and with 1,000 monosyllabic words. The lists were read aloud at the rate of one item per second and the subjects had as much time as they needed to give their responses. A procedure described by Woodworth (20) was used to score the responses.

The results are shown by the filled circles in Figure 7. Here the dotted line indicates what the span should have been if the amount of information in the span were constant. The solid curves represent the data. Hayes repeated the experiment using test vocabularies of different sizes but all containing only English monosyllables (open circles in Figure 7). This more homogeneous test material did not change the picture significantly. With binary items the span is about nine and, although it drops to about five with monosyllabic English words, the difference is far less than the hypothesis of constant information would require.

There is nothing wrong with Hayes's experiment, because Pollack (16) repeated it much more elaborately and got essentially the same result. Pollack took pains to measure the amount of information transmitted and did not rely on the traditional procedure for scoring the responses. His results are plotted in Figure 8. Here it is clear that the amount of information transmitted is not a constant, but increases almost linearly as the amount of information per item in the input is increased.

And so the outcome is perfectly clear. In spite of the coincidence that the magical number seven appears in both places, the span of absolute judgment and the span of immediate memory are quite different kinds of limitations that are imposed on our ability to process information. Absolute judgment is limited by the amount of information. Immediate memory is limited by the number of items. In order to capture this distinction in somewhat picturesque terms, I have fallen into the custom of distinguishing between *bits* of information and *chunks* of information. Then I can say that the number of bits of information is constant for absolute judgment and the number of chunks of information is constant for immediate memory. The span of immediate memory seems to be almost independent of the number of bits per chunk, at least over the range that has been examined to date.

The contrast of the terms *bit* and *chunk* also serves to highlight the fact that we are not very definite about what constitutes a chunk of information. For example, the memory span of five words that Hayes obtained when each word was drawn at random from a set of 1000 English monosyllables might just as appropriately have been called a memory span of 15 phonemes, since each word had about three phonemes in it. Intuitively, it is clear that the subjects were recalling five words, not 15 phonemes, but the logical distinction is not immediately apparent. We are dealing here with a process of organizing or grouping the input into familiar units or chunks, and a great deal of learning has gone into the formation of these familiar units.

RECODING

In order to speak more precisely, therefore, we must recognize the importance of grouping or organizing the input sequence into units or chunks. Since the memory span is a fixed number of chunks, we can increase the number of bits of information that it contains simply by building larger and larger chunks, each chunk containing more information than before.

A man just beginning to learn radiotelegraphic code hears each *dit* and *dah* as a separate chunk. Soon he is able to organize these sounds into letters and then he can deal with the letters as chunks. Then the letters organize themselves as words, which are still larger chunks, and he begins to hear whole phrases. I do not mean that each step is a discrete process, or that plateaus must appear in his learning curve, for surely the levels of organization are achieved at different rates and overlap each other during the learning process. I am simply pointing to the obvious fact that the dits and dahs are organized by learning into patterns and that as these larger chunks emerge the amount of message that the operator can remember increases correspondingly. In the terms I am proposing to use, the operator learns to increase the bits per chunk.

In the jargon of communication theory, this process would be called *recoding*. The input is given in a code that contains many chunks with few bits per chunk.

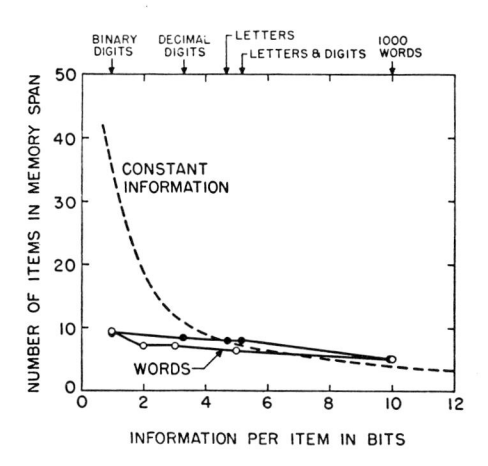

Figure 7. Data from Hayes (10) on the span of immediate memory plotted as a function of the amount of information per item in the test materials.

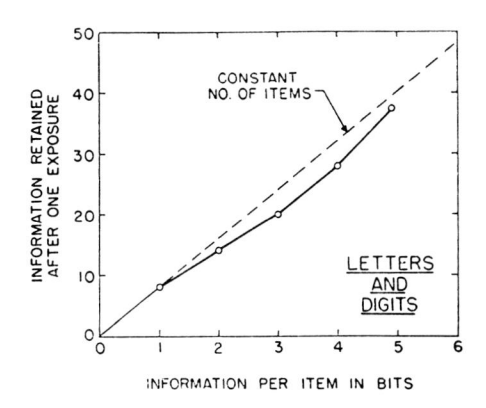

Figure 8. Data from Pollack (16) on the amount of information retained after one presentation plotted as a function of the amount of information per item in the test materials.

TABLE 1
Ways of Recoding Sequences of Binary Digits

	Binary Digits (Bits)	1 0 1 0 0 0 1 0 0 1 1 1 0 0 1 1 1 0							
2:1	Chunks	10	10	00	10	01	11	00	11	10
	Recoding	2	2	0	2	1	3	0	3	2
3:1	Chunks	101	000	100	111	001	110		
	Recoding	5	0	4	7	1	6		
4:1	Chunks	1010	0010	0111	0011	10			
	Recoding	10	2	7	3				
5:1	Chunks	10100	01001	11001	110				
	Recoding	20	9	25					

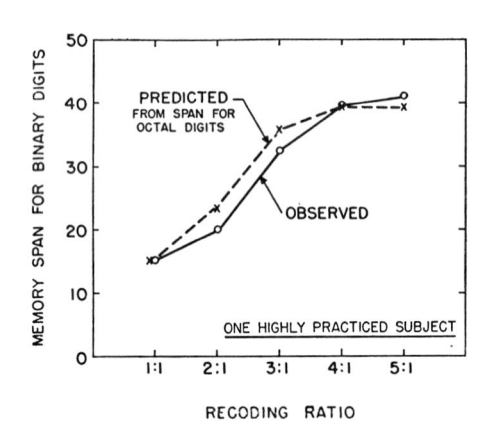

Figure 9. The span of immediate memory for binary digits is plotted as a function of the recoding procedure used. The predicted function is obtained by multiplying the span for octals by 2, 3 and 3.3 for recoding into base 4, base 8, and base 10, respectively.

The operator recodes the input into another code that contains fewer chunks with more bits per chunk. There are many ways to do this recoding, but probably the simplest is to group the input events, apply a new name to the group, and then remember the new name rather than the original input events.

Since I am convinced that this process is a very general and important one for psychology, I want to tell you about a demonstration experiment that should make perfectly explicit what I am talking about. This experiment was conducted by Sidney Smith and was reported by him before the Eastern Psychological Association in 1954.

Begin with the observed fact that people can repeat back eight decimal digits, but only nine binary digits. Since there is a large discrepancy in the amount of information recalled in these two cases, we suspect at once that a recoding procedure could be used to increase the span of immediate memory for binary digits. In Table 1 a method for grouping and renaming is illustrated. Along the top is a sequence of 18 binary digits, far more than any subject was able to recall after a single presentation. In the next line these same binary digits are grouped by pairs. Four possible pairs can occur: 00 is renamed 0, 01 is renamed 1, 10 is renamed 2, and 11 is renamed 3. That is to say, we recode from a base-two arithmetic to a base-four arithmetic. In the recoded sequence there are now just nine digits to remember, and this is almost within the span of immediate memory. In the next line the same sequence of binary digits is regrouped into chunks of three. There are eight possible sequences of three, so we give each sequence a new name between 0 and 7. Now we have recoded from a sequence of 18 binary digits into a sequence of 6 octal digits, and this is well within the span of immediate memory. In the last two lines the binary digits are grouped by fours and by fives and are given decimal-digit names from 0 to 15 and from 0 to 31.

It is reasonably obvious that this kind of recoding increases the bits per chunk, and packages the binary sequence into a form that can be retained within the span of immediate memory. So Smith assembled 20 subjects and measured their spans for binary and octal digits. The spans were 9 for binaries and 7 for octals. Then he gave each recoding scheme to five of the subjects. They studied the recoding until they said they understood it—for about 5 to 10 minutes. Then he tested their span for binary digits again while they tried to use the recoding schemes they had studied.

The recoding schemes increased their span for binary digits in every case. But the increase was not as large as we had expected on the basis of their span for octal digits. Since the discrepancy increased as the recoding ratio increased, we reasoned that the few minutes the subjects had spent learning the recoding schemes had not been sufficient. Apparently the translation from one code to the other must be almost automatic or the subject will lose part of the next group while he is trying to remember the translation of the last group.

Since the 4:1 and 5:1 ratios require considerable study, Smith decided to imitate Ebbinghaus and do the experiment on himself. With Germanic patience he drilled himself on each recoding successively, and obtained the results shown in Figure 9. Here the data follow along rather nicely with the results you would predict on the basis of his span for octal digits. He could remember 12 octal digits. With the 2:1 recoding, these 12 chunks were worth 24 binary digits. With the 3:1 recoding they were worth 36 binary digits. With the 4:1 and 5:1 recodings, they were worth about 40 binary digits.

It is a little dramatic to watch a person get 40 binary digits in a row and then repeat them back without error. However, if you think of this merely as a mnemonic trick for extending the memory span, you will miss the more important point that is implicit in nearly all such mnemonic devices. The point is that recoding is an extremely powerful weapon for increasing the amount of information that we can deal with. In one form or another we use recoding constantly in our daily behavior.

In my opinion the most customary kind of recoding that we do all the time is to translate into a verbal code. When there is a story or an argument or an idea that we want to remember, we usually try to rephrase it "in our own words." When we witness some event we want to remember, we make a verbal description of the event and then remember our verbalization. Upon recall we recreate by secondary elaboration the details that seem consistent with the particular verbal recoding we happen to have made. The well-known experiment by Carmichael, Hogan, and Walter (3) on the influence that names have on the recall of visual figures is one demonstration of the process.

The inaccuracy of the testimony of eyewitnesses is well known in legal psychology, but the distortions of testimony are not random—they follow naturally from the particular recoding that the witness used, and the particular recoding he used depends upon his whole life history. Our language is tremendously useful for repackaging material into a few chunks rich in information. I suspect that imagery is a form of recoding, too, but images seem much harder to get at operationally and to study experimentally than the more symbolic kinds of recoding.

It seems probable that even memorization can be studied in these terms. The process of memorizing may be simply the formation of chunks, or groups of items that go together, until there are few enough chunks so that we can recall all the items. The work by Bousfield and Cohen (2) on the occurrence of clustering in the recall of words is especially interesting in this respect.

SUMMARY

I have come to the end of the data that I wanted to present, so I would like now to make some summarizing remarks.

First, the span of absolute judgment and the span of immediate memory impose severe limitations on the amount of information that we are able to receive, process, and remember. By organizing the stimulus input simultaneously into several dimensions and successively into a sequence of chunks, we manage to break (or at least stretch) this informational bottleneck.

Second, the process of recoding is a very important one in human psychology and deserves much more explicit attention than it has received. In particular, the kind of linguistic recoding that people do seems to me to be the very lifeblood of the thought processes. Recoding procedures are a constant concern to clinicians, social psychologists, linguists, and anthropologists and yet, probably because recoding is less accessible to experimental manipulation than nonsense syllables or T mazes, the traditional experimental psychologist has contributed little or nothing to their analysis. Nevertheless, experimental techniques can be used, methods of recoding can be specified, behavioral indicants can be found. And I anticipate that we will find a very orderly set of relations describing what now seems an uncharted wilderness of individual differences.

Third, the concepts and measures provided by the theory of information provide a quantitative way of getting at some of these questions. The theory provides us with a yardstick for calibrating our stimulus materials and for measuring the performance of our subjects. In the interests of communication I have suppressed the technical details of information measurement and have tried to express the ideas in more familiar terms; I hope this paraphrase will not lead you to think they are not useful in research. Informational concepts have already proved valuable in the study of discrimination and of language; they promise a great deal in the study of learning and memory; and it has been proposed that they can be useful in the study of concept formation. A lot of questions that seemed fruitless twenty or thirty years ago may now be worth another look. In fact, I feel that my story here must stop just as it begins to get really interesting.

And finally, what about the magical number seven? What about the seven wonders of the world, the seven seas, the seven deadly sins, the seven daughters of Atlas in the Pleiades, the seven ages of man, the seven levels of hell, the seven primary colors, the seven notes of the musical scale, and the seven days of the week? What about the seven-point rating scale, the seven categories for absolute judgment, the seven objects in the span of attention, and the seven digits in the span of immediate memory? For the present I propose to withhold judgment. Perhaps there is something deep and profound behind all these sevens, something just calling out for us to discover it. But I suspect that it is only a pernicious, Pythagorean coincidence.

REFERENCES

1. Beebe-Center, J. G., Rogers, M. S., & O'Connell, D. N. Transmission of information about sucrose and saline solutions through the sense of taste. *J. Psychol.*, 1955, **39**, 157-160.
2. Bousfield, W. A., & Cohen, B. H. The occurrence of clustering in the recall of randomly arranged words of different frequencies-of-usage. *J. gen. Psychol.*, 1955, **52**, 83-95.
3. Carmichael, L., Hogan, H. P., & Walter, A. A. An experimental study of the effect of language on the reproduction of visually perceived form. *J. exp. Psychol.*, 1932, **15**, 73-86.
4. Chapman, D. W. Relative effects of determinate and indeterminate *Aufgaben*. *Amer. J. Psychol.*, 1932, **44**, 163-174.
5. Eriksen, C. W. Multidimensional stimulus differences and accuracy of discrimination. *USAF, WADC Tech. Rep.*, 1954, No. 54-165.
6. Eriksen, C. W., & Hake, H. W. Absolute judgments as a function of the stimulus range and the number of stimulus and response categories. *J. exp. Psychol.*, 1955, **49**, 323-332.
7. Garner, W. R. An informational analysis of absolute judgments of loudness. *J. exp. Psychol.*, 1953, **46**, 373-380.
8. Hake, H. W., & Garner, W. R. The effect of presenting various numbers of discrete steps on scale reading accuracy. *J. exp. Psychol.*, 1951, **42**, 358-366.
9. Halsey, R. M., & Chapanis, A. Chromaticity-confusion contours in a complex viewing situation. *J. Opt. Soc. Amer.*, 1954, **44**, 442-454.

10. Hayes, J. R. M. Memory span for several vocabularies as a function of vocabulary size. In *Quarterly Progress Report*, Cambridge, Mass.: Acoustics Laboratory, Massachusetts Institute of Technology, Jan.-June, 1952.
11. Jakobson, R., Fant, C. G. M., & Halle, M. *Preliminaries to speech analysis*. Cambridge, Mass.: Acoustics Laboratory, Massachusetts Institute of Technology, 1952. (Tech. Rep. No. 13.)
12. Kaufman, E. L., Lord, M. W., Reese, T. W., & Volkmann, J. The discrimination of visual number. *Amer. J. Psychol.*, 1949, **62**, 498-525.
13. Klemmer, E. T., & Frick, F. C. Assimilation of information from dot and matrix patterns. *J. exp. Psychol.*, 1953, **45**, 15-19.
14. Külpe, O. Versuche über Abstraktion. *Ber. ü. d. I Kongr. f. exper. Psychol.*, 1904, 56-68.
15. Miller, G. A., & Nicely, P. E. An analysis of perceptual confusions among some English consonants. *J. Acoust. Soc. Amer.*, 1955, **27**, 338-352.
16. Pollack, I. The assimilation of sequentially encoded information. *Amer. J. Psychol.*, 1953, **66**, 421-435.
17. Pollack, I. The information of elementary auditory displays. *J. Acoust. Soc. Amer.*, 1952, **24**, 745-749.
18. Pollack, I. The information of elementary auditory displays. II. *J. Acoust. Soc. Amer.*, 1953, **25**, 765-769.
19. Pollack, I., & Ficks, L. Information of elementary multi-dimensional auditory displays. *J. Acoust. Soc. Amer.*, 1954, **26**, 155-158.
20. Woodworth, R. S. *Experimental psychology*. New York: Holt, 1938.

10

A Model for Some Kinds of Visual Memory Tasks

George Sperling

INTRODUCTION

I shall be concerned with the apparently simple situation in which an observer looks briefly at a complex visual display and then attempts to reproduce part or all of it. Understanding this visual, immediate-memory task is important both in practical problems and in basic psychological problems. From the practical point of view, it is relevant to everyday situations such as a person looking at a number in a telephone book and then attempting to dial it, as well as to the esoteric

From *Human Factors*, **5**, 1963, 19-31. By permission of the Human Factors Society. With annotations by the author.

Invited address, *Symposium on Information Processing in Man: Research Frontiers*, sponsored jointly by the Los Angeles Chapter of the Human Factors Society and the University of Southern California. Held at USC, June 23, 1962.

A Model for Some Kinds of Visual Memory Tasks

problems that arise in matching complex visual displays to human capabilities. I shall not be concerned directly with specific applications; rather, some general principles will be evolved.

The brief visual exposure has a special theoretical significance. Normally, the eye moves in brief, quick motions between its steady fixations upon objects. These movements, called saccads, were first noticed by Javal (1878) and described by Erdmann and Dodge (1898). In such varied tasks as reading, looking at stationary objects and even in visually tracking most moving objects, the eye moves in saccads and takes in information only during the fixation pause between the saccads. As there are several fixations per second, the eye codes information from the environment into a rapid sequence of still pictures. It is natural, therefore, that the problem of what can be seen in a single brief exposure has fascinated researchers for over 100 years.[1] My purpose is to describe some of the components and properties of a preliminary model for the information processing that begins with the observation of a brief visual stimulus and that ends with the observer's response.

Historically, most research in immediate-memory (or span of attention) has been confined to the problem of capacity. That is, experimenters usually have presented subjects with a great variety of stimuli and measured simply the number of items reported correctly. While such experiments reveal something about the capacity of a memory, they do not usually reveal much about its structure. For example, by structure I mean such properties as—to use computer terminology—whether a memory has random access or whether restraints limit the sequence in which items can be remembered or recalled. To ferret out structural details, the experimental technique requires the presentation of one kind of stimulus over and over to determine the various capabilities and limitations of the observer in dealing with this stimulus. In other words, one must determine the limits of the observer's ability to cope with this stimulus depending upon instructions and "irrelevant" stimuli such as various kinds of interference. In principle and in intent, this method has a precedent in the work of Kulpe (1904) and his followers.

The Model

In order to facilitate the exposition, it is desirable at this point to give a brief sketch of the model as follows. (1) The observer sees the stimulus material for a short time. (2) He scans it, selecting certain information to rehearse. (3) He later reports what he remembers of his rehearsal. The experiments seek to clarify and quantify these three main aspects of the model. In the first place it will be established that even in a brief exposure the observer may have available much more information than he can later report. Because the duration for which he "sees" the stimulus normally exceeds the stimulus exposure duration, this component of the model is called visual information storage. The second concern is with the rate at which the observer can select and utilize the visual information.

---

[1] Volkmann (1859) coined the word "tachistoscop" for a device he invented. It utilized falling shutters to replace the electric sparks which had been—up to that time—the primary means of producing brief exposures.

This is called scanning or read-out from visual storage. When it is not limited by the requirement of making muscular responses, the scan rate can be quite high.

The third component of the model deals with the subvocal and/or vocal rehearsal of the selected items and the memory for this rehearsal. The memory is called auditory information storage. It will be suggested that the auditory component may be the limiting factor in a wide range of both visual and auditory reproduction tasks. This single limiting process, common to many tasks, would help account for the item constancy of the so-called span of immediate-memory (cf. Pollack, 1953; Miller, 1956).

I. VISUAL INFORMATION STORAGE

Much of the material in this section has been published elsewhere (Sperling, 1960b; Averbach and Sperling, 1961) but it is presented briefly for completeness and because the results are needed for the model. Subjects are characteristically able to report about six or fewer items from brief stimulus exposures. This finding dates to the last century (Cattell, 1885; Erdmann and Dodge, 1898). In my own experiments involving immediate memory for visual stimuli, this finding occurs with great consistency. For example, in one of my studies using five experienced subjects, it was found that the average number of symbols correctly reported was equal to the number of symbols in the stimulus when stimuli contained four or fewer symbols, and equal to about 4.5 symbols when stimuli contained five or more symbols. This held for various spatial arrangements of the symbols, and for mixtures of letters and numbers as well as for letters alone. There was, moreover, no change in the number of letters correctly reported within the entire range of exposure durations tested, from 15 to 500 msec. In these presentations the stimulus was preceded and followed by a dark field. With another apparatus capable of shorter exposures, I found no change in the number of letters reported correctly even for exposures as short as 5 msec and with light pre- and post-exposure fields.[2] At such short exposures the apparent contrast of the stimulus letters was greatly reduced (black letters appeared light grey) but this did not affect the number of letters reported correctly. These results taken together, define the invariant span of attention, or span of immediate-memory.

The span of immediate-memory, however, is not due to a limit on what the subject can see. This was first proved by using a partial report procedure in the following experiment (Sperling, 1960b). Subjects were presented stimuli consisting of 12 letters and numbers in three rows of 4 symbols each. The exposure duration was 50 msec. The stimulus exposure was immediately followed by a tonal signal. The subjects had been told to report only one row of letters and the signal indicated to the subject the particular row to be reported. Subjects were able to report correctly 76 per cent of the called-for letters even though they did not know in advance which particular row would be called for. This result indicates that after termination of the exposure, subjects still had available

---

[2] Unpublished experiments conducted at the Bell Telephone Laboratories, 1958.

A Model for Some Kinds of Visual Memory Tasks

(somewhere inside them) 76 per cent of the 12 symbols, that is, 9.1 symbols. However, when the tonal signal was delayed for only one second, the accuracy of report dropped precipitously from 76 to 36 per cent. Note that 36 per cent of 12 symbols is 4.3 symbols; the previously measured memory span for this material was also 4.3 symbols.

The explanation for these results is that the visual image of the stimulus persists for a short time after the stimulus has been turned off, and that the subjects can utilize this rapidly fading image. In fact, naïve subjects typically believe that the physical stimulus fades out slowly.

The visual basis of this very short-term information storage can be demonstrated more convincingly by comparing two different kinds of stimulus presentation. In the first type, the pre- and post-exposure fields are dark; in the second, the pre- and post-exposure fields are light. It is well known that the first kind of presentation can produce persisting after-images of the stimulus and the second kind of presentation does not. Stimuli of 18 letters were exposed for 50 msec. After a variable delay following the exposure, any one of six different tonal combinations was presented to indicate the particular letters to be reported. The results of the partial report procedure applied to these stimuli are shown in Figure 1.

The reader will recall the procedure for estimating the number of letters available to the subject. The fraction of letters reported correctly in random samples (partial reports) of the stimulus is multiplied by the number of letters in the stimulus. These values are indicated by the ordinates of Figure 1. The abscissa indicates the time after the exposure at which the instruction calling for the partial report was given. Figure 1 shows that letters in excess of the memory span were available to the subject for one-half second when the pre- and post-exposure fields were light, and for nearly five seconds when the pre- and post-exposure fields were dark. The exposure itself was exactly the same in each case. The ability of noninformational visual fields occurring before and after the stimulus to control the accuracy of partial reports strongly suggests their dependence on a persisting visual image. Conversely, the number of letters given in the usual immediate-memory report is almost independent of the exposure conditions and does not change appreciably if the subject is required to delay his report for five seconds (or longer) instead of being allowed to report immediately. Therefore, the number of letters given in memory reports can be presented by a single bar in Figure 1.

The short-term memory for letters in excess of the immediate-memory span will be called visual information storage (VIS). Some established properties and some hypotheses regarding VIS are listed below.

(1) The effective input for VIS is a local change in retinal light intensity. The contents of VIS depend on visual stimulation. Such factors as stimulus intensity, contrast, duration, pre- and post-exposure fields, etc., are particularly important.

(2) The contents of VIS can become available to subsequent components of the model as a sequence of items through "scanning" or "read-out."

(3) VIS is two-dimensional (for example, contents can be scanned either in a vertical or horizontal sequence).

(4) Maximum information in VIS is at least 17 letters when measured with

stimuli containing 18 letters. Stimuli containing more than 18 letters would probably have yielded higher estimates. However, the resolution is disturbed when items that can be resolved individually are spaced too closely together.

(5) Subsequent stimuli can replace or interfere with the previous VIS contents (see below). This process is not the passive addition of a new stimulus to the fading trace of its predecessor but the active replacement of an earlier stimulus by a later one (Sperling, 1960a).

(6) Contents of VIS normally decay rapidly, decay times varying from a fraction of a second to several seconds.

(7) Long durations of visual storage can occur in the form of after-images which appear to move when the eye moves and therefore are probably localized in the retina. It has not been determined how the central nervous system is involved in the kind of VIS discussed here.

(8) It is tempting to speculate on the purpose of VIS because its properties seem so well suited to the requirements of a system like the eye, which processes information in temporally discrete chunks. The function of the persistence (the storage aspect of VIS) seems to be to maintain a visual image from one fixation of the eye to the next. The function of erasure is to permit the new image following a saccad to overwrite the trace of the previous one without interference to itself and also to "erase" the blur resulting during movement of the eye. The minimum duration of storage that has been recorded (¼ sec) is still long enough to preserve the image between eye movements. The minimum time between saccads is typically too long to allow the image produced by the second saccad to interfere with more than the tail end of the image produced by the first. Thus VIS acts as a buffer which quickly attains and holds much information to permit its relatively slow utilization later. VIS also segregates and isolates from each other successive bursts of visual information (e.g., images).

II. SCANNING

In order to determine the rate at which information can be utilized, it is necessary to gain precise control of actual stimulus availability, that is, the contents of VIS. This cannot be accomplished by controlling exposure duration alone. No matter how brief a *single* stimulus flash may be, if it is of sufficient contrast for easy legibility of letters, then it will be available for about a quarter of a second or longer.

The idea that stimulus duration does not determine stimulus availability is not new. For example, in 1868, Exner published a psychophysical study of the apparent duration of short flashes in which he found that short flashes exert their effect over a considerable time span. Baxt (1871), also working in Helmholtz's laboratory, performed a logical sequel to this work. He followed the stimulus exposure after a delay with a bright second flash which was intended to obliterate the persisting after-image of the stimulus.[3] In one of his experiments he used a stimulus flash of 5.0 msec and a bright second flash of 120.0 msec dura-

---

[3] The method of Baxt was described by Ladd (1887) and James (1890) in their textbooks but it is no longer well known. Consequently, it has been rediscovered, recently by Lindsley and Emmons (1958), Gilbert (1959) and by the author.

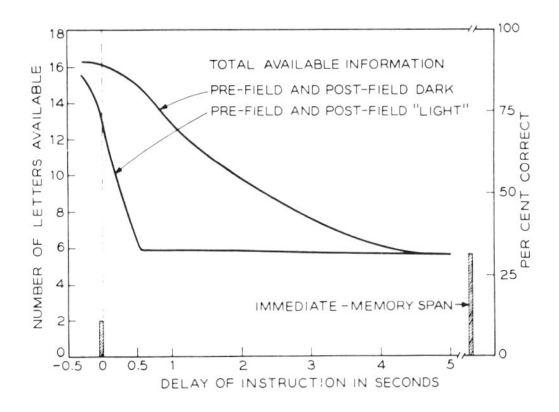

Figure 1. Information available to one observer from two kinds of stimulus presentation. The right ordinate is the average accuracy of partial reports; the left ordinate is the inferred store of available letters. Average immediate-memory span for both presentations and for 0.0 and 5.0 sec delay of report is indicated at right. Stimulus exposure is schematically indicated at lower left. (Redrawn from Averbach and Sperling, 1960)

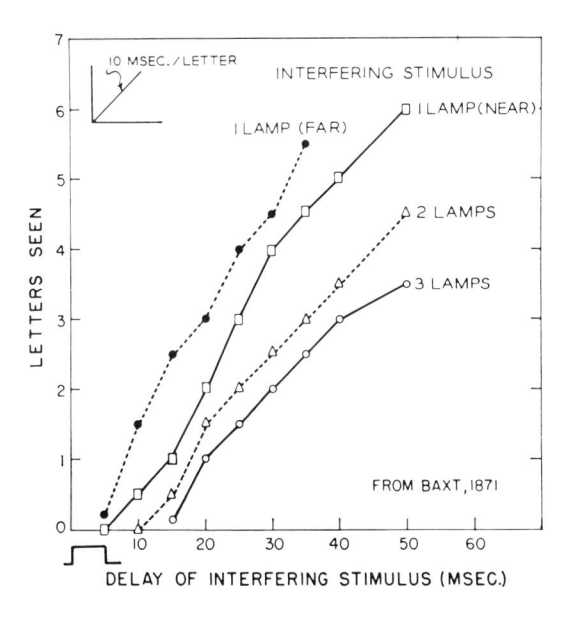

Figure 2. The number of letters seen as a function of the delay between the onset of a 5.0 msec lettered stimulus and a 120 msec blank interfering stimulus. The parameter is the intensity of the interfering stimulus. The time course of the lettered stimulus is indicated at the lower left. (Based on the protocols of Baxt, 1871)

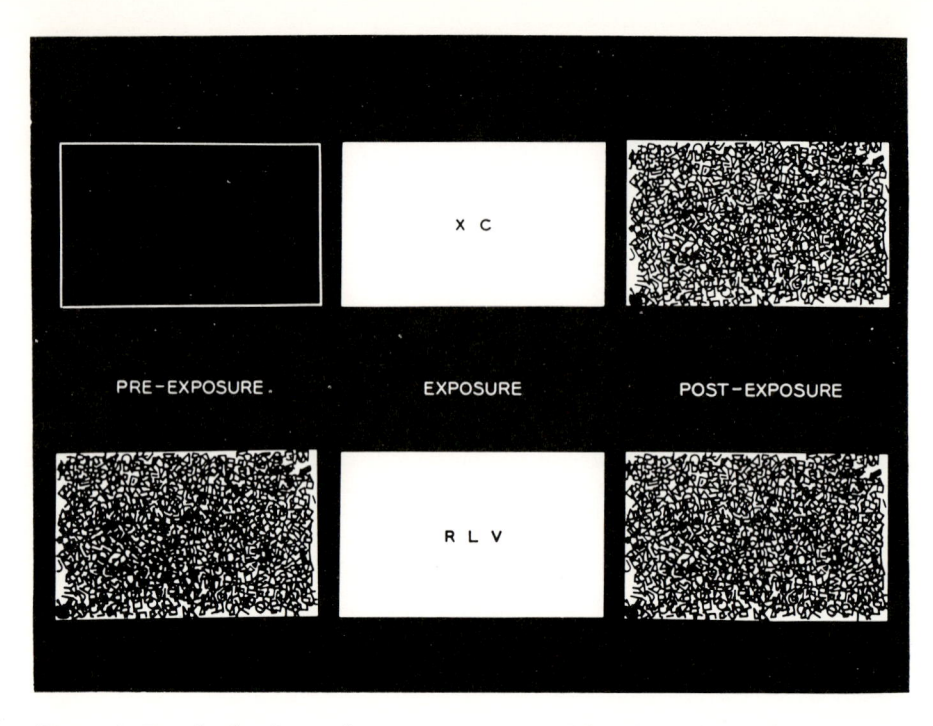

Figure 3. Two kinds of stimulus presentations modified from the method of Baxt. The sequence of stimuli on a trial is indicated from left to right. The luminance of the lettered stimulus was 31 Ft-L.; of the visual noise, 20 Ft-L.

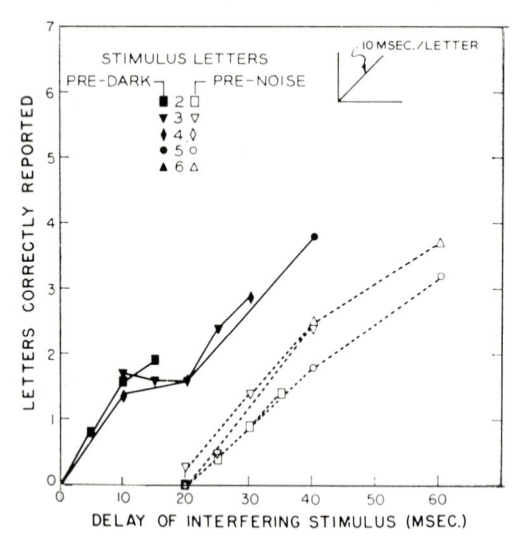

Figure 4. The number of letters correctly reported as a function of the delay between the onset of a lettered stimulus and an interfering visual "noise" stimulus. The number of stimulus letters was also varied. The pre-exposure field was either dark or "noise."

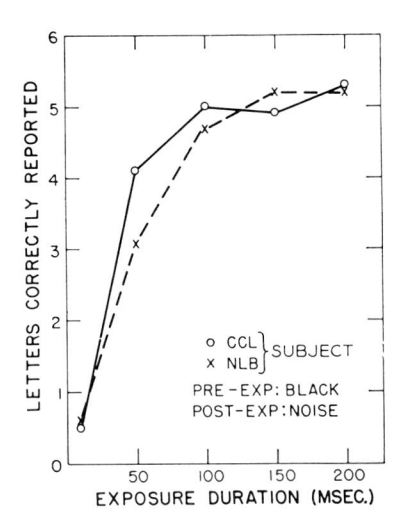

Figure 5. The number of letters reported correctly as a function of the exposure duration or equivalently, the delay of the post-exposure visual "noise" interfering stimulus. The pre-exposure field was dark.

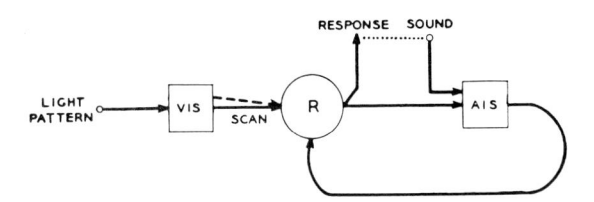

Figure 6. A schematic model for short-term memory tasks. VIS = visual information storage, R = rehearsal component, AIS = auditory information storage. See text for details.

tion. Observers viewed stimuli of 6 or 7 printed letters and reported the number of letters they could see with various time delays between the two flashes. Figure 2 contains a graph that I prepared from Baxt's tabular, introspective data. The abscissa represents the time from the onset of the stimulus to the onset of the "interfering" second flash; essentially, this is the time that the observer had to look at the stimulus. The ordinate represents the number of letters the observer said he could see. The parameter is the intensity of the interfering flash in terms of the arrangement that produced the intensity (three, two, or one lamps, one lamp far). The slope of the data points show that Baxt could see an additional letter for approximately each 10 msec of delay of the interfering flash.

The data of Figure 2 were obtained with dark letters on a white background. Baxt obtained similar results when viewing light letters on a dark surround. He himself did not notice the simple relation of constant slope, perhaps because he did not graph his results.

Baxt's results show that the minimum delay at which detection of a letter is first possible depends on the intensity of the interfering second flash. This flash exerts its interfering effect more rapidly when it is more intense. A similar situation occurs in masking experiments. An intense masking flash can cause marked threshold changes for detection of a test flash even when the test precedes it by as much as forty msec. Had Baxt used hundreds of lamps instead of just three for his brightest interfering stimulus, the minimum delay necessary for detection of a letter might have been increased from 20 msec (with three lamps) to 40 msec or more.

The results of Baxt's 2-flash experiment are more reasonable than those of the single flash experiments. It will be recalled that reducing the duration of a single flash reduced the apparent contrast of the exposed letters but the number reported correctly did not change with exposure duration *per se* except at ridiculously short exposures. It does not make sense to conclude, for example, that if four letters can be read from a single exposure of one msec that the time for reading a letter is ¼ msec. Nevertheless, even the scanning rate of one letter per 10 msec (or 100 letters/sec) in Baxt's experiment seemed so remarkably fast that it seemed worthwhile to attempt to reproduce some of Baxt's conditions. With certain reservations, comparable results were obtained. Some minor changes intended to improve the procedure were also tried. Subjects were required to report letters and were scored for the number correct. (Baxt had asked his observer only to say whether or not he saw a letter.) The blank second field was replaced with a visual "noise" field, consisting of densely scattered bits and pieces of letters. The noise stimulus is much more effective interference than a homogeneous field.

The main objection to Baxt's procedure, however, is a surprising one: I found that the second flash may produce a clearly visible negative after-image of the first stimulus and thereby indirectly fail to stop its persistence. In fact, in certain conditions of presentation, the blank interference produced a negative after-image of the stimulus without a prior positive image being seen (Sperling, 1960c). Ironically, here is a case of simultaneous perfect interference (the positive stimulus is completely invisible) and of perfect persistence (all the stimulus information is available in the after-image). Use of the noise field as the interfering stimulus mitigated this problem somewhat since even an after-image of the stimulus is better hidden by the noise. However, such paradoxical effects as simultaneous interference with and yet persistence of information, certainly argue for caution in interpreting the results of Baxt-type experiments.

Using the noise technique, we sought further clarification of the Baxt procedure. Baxt's results were that the scan rate is the same no matter when the scan is ended. That is, bright flashes presumably ended visual availability of the stimulus sooner than dim flashes. One obvious question that needed to be answered was "is the scanning rate the same no matter when the scan begins." If it were, then this would provide more confidence in the method.

To delay scanning, we used two different pre-exposure fields. One was a dark field which presented no visual recovery problem and the other a noise field. Figure 3 illustrates the sequence in each of the two kinds of presentations. In the first, the pre-exposure field is dark. It is followed by an exposure of variable duration, which is then terminated by a noise post-exposure field. In the second kind of presentation, the pre-exposure field is noise, followed by an exposure of variable duration, which is then also terminated by a noise post-exposure field. The number of letters in the stimuli was varied from two to six in order to check for possible spatial location effects. Data were obtained with two subjects. Figure 4 illustrates the results for the less variable subject.

The data of Figure 4 indicate that the subject gained information from the various stimuli at the rate of one letter per 10 msec and that this rate was independent of the number of letters in the stimulus or of the pre-exposure field. When the pre-exposure field was dark, the subject began to gain information almost immediately upon stimulus exposure. It took this subject about 20 msec to recover from pre-exposure to visual noise. Once she was able to scan, the rate was still about one letter per 10 msec. These results support the hypothesis that pre-exposure could change the time at which a scan begins but not its rate.

Various other pre-exposure fields have also been tried. For example, with a pre-exposure field 100 times brighter than the stimulus field, scanning may be delayed for over 100 msec. Once the stimulus letters become legible, they are scanned at comparable rates. A serious problem in this kind of experiment is that it is very difficult to insure that all letters fall upon the retinal locations which recover sensitivity at the same rate. This particular proceduce therefore has not been pursued in detail. Preliminary data also indicated that rate of scanning was but slightly reduced when the stimulus intensity was reduced to $\frac{1}{100}$ of its prior level. Other factors such as stimulus size, geometry, and contrast have not yet been studied to see how they affect the scan rate.

The point of all these experiments is that, under a variety of conditions, random letters of good contrast are scanned at the same rate; typically, about one letter per 10 msec. However this holds true only for the first three or four letters to be scanned. Figure 5 shows data obtained with the same two subjects viewing a dark pre-exposure field and a noise post-exposure field. One subject reported three, the other four letters in the first 50 msec of exposure. Additional stimulus exposure from 50 msec to 100 msec accounted for about one or two additional letters. Beyond 100 msec the rate of acquiring additional letters is so low as to be virtually indistinguishable from zero on this time scale. Additional data points would have shown the critical break in the curve to occur well before 100 msec.

This kind of experiment perhaps more clearly than any other defines an immediate-memory span for visual materials. Letters up to the immediate-memory span can be scanned at a rate of one letter per 10 or 15 msec. This is so rapid that the rate of acquiring additional letters beyond the immediate-memory span is negligible by comparison.

III. REHEARSAL AND AUDITORY INFORMATION STORAGE

In the last two sections it was shown (1) that letters are visually available to an observer during and after a brief exposure and (2) that they need not be visually available for more than 10 to 15 msec per letter in order to be reported correctly. Yet, the observer usually does not begin his report until several seconds after the exposure, long after his visual store of letters is depleted. In fact, he can delay his report for an additional 30 seconds or more without any loss of accuracy. What are the characteristics of this later, long term memory?

One important clue is obtained from the kinds of errors subjects make. In writing down symbols from a brief stimulus, subjects frequently make auditory confusions; that is, writing D for T, T for 2 and so on. On questioning, subjects assert that they indeed do "say" the letters to themselves prior to report. This kind of behavior is represented in the model by rehearsal (saying the letters to oneself) and by auditory information storage (hearing the rehearsed letters). The properties of auditory information storage (AIS) are in some respects similar to those of VIS. The main functional difference between AIS and VIS in these experiments is the possibility of feeding information already in AIS back into storage again by rehearsal.

The various components of the model and their interactions are represented in Figure 6. The model illustrates that a visual stimulus is first stored in VIS, then scanned and rehearsed. The rehearsal component has two inputs: items scanned from VIS and those heard in AIS. Only one input is rehearsable at any one time and it probably takes a short time to switch from one input modality to the other (cf. Broadbent, 1956, 1958). The rehearsal component produces a verbal response "audible" in AIS or a vocal response audible to another person and indirectly also in AIS.

It will be suggested below that rehearsal cannot proceed faster than about ten syllables per second. Scanning itself may initially be ten times faster, 100 items per second. The difference between the rates of scanning and rehearsal poses a real problem. To account for this difference, it tentatively will be assumed that items may accumulate very briefly in the scan path or rehearsal component until they can be rehearsed.[4] Two scan arrows are drawn from VIS to Rehearsal to represent the assumed temporal overlap in the scan of successive items.

The rehearsed items are maintained in AIS at least for several seconds during

---

[4] Another possible assumption is that there is an error in the interpretation of scan experiments using a Baxt type of stimulus presentation. Two attempts were made to measure directly the stimulus persistence (availability). In the first attempt, the partial report procedure was applied to a Baxt presentation. It was not successful because subjects' performances tended to deteriorate in all respects. In the second procedure, the apparent duration of the visual stimulus was compared to the duration of various simultaneous acoustic signals. For example, a Baxt stimulus presentation from which four letters could usually be read (letter duration = 50 msec) was matched to acoustic stimuli of roughly comparable duration. In such comparisons, it is quite obvious that the apparent duration of the visual letters is a small fraction of a second. Although this is a subjective judgment, it has strong face validity. The result only affirms that the scan rate can be much more rapid than rehearsal.

which time they are usually rehearsed again. Finally, when a response is required, the letters are either rehearsed again as they are being written or, if they are spoken aloud, the feedback loop may be closed externally as well as internally—the spoken sound re-entering AIS.

In summary, there are at least three sources of information for auditory storage: (1) an actual acoustic stimulus, (2) rehearsal of information already in AIS, (3) the scanning process—it is assumed that observers hear themselves make a verbal response as they scan.

Evidence

In this model, the auditory components assume the major burden of short term visual memory. Data relevant to the auditory part of the model are difficult to obtain because audition involves higher order processes. There are some data, however, which come from three kinds of experiments: (1) experiments that measure AIS directly by using auditory stimuli, (2) experiments that seek to measure rehearsal by instructing subjects with regard to rehearsal, and (3) experiments that use interfering stimuli to interfere selectively with some aspects of the process and thereby reveal others. I shall consider these approaches in order.

(1) *Auditory information storage.* Our common experience leaves little doubt of the fact that there is auditory information storage. If, for example, one is spoken to while occupied with another task, one may continue working for a few moments and yet still be able to recover the message later. Two kinds of recent experiments illustrate this type of auditory storage especially well. In the first kind of experiment the subject is presented with several stimuli at once. For example, Broadbent (1957a) produced a spoken sequence of digits at the subjects' ears, with different digits being spoken simultaneously into each ear. When digits were spoken rapidly, subjects could not shift their attention back and forth from one ear to the other with each pair of spoken digits. Rather, they could only report all the digits heard at one ear first and then attempt to report the digits—if any—that were still audible at the other ear. This ingenious procedure excludes the possibility of any simultaneous recoding of the inputs at both ears. It thereby forces the subject to rely directly upon AIS.

The second kind of experiment (Anderson, 1960) was a direct application to audition of the partial report procedure previously used to measure visual storage. A list of 12 letters in three groups of four each was read to subjects. At various times later, an instruction signal was given the subjects to indicate the particular group to be reported. Anderson's results can be interpreted in terms of auditory storage to mean that initially subjects had about 10 letters available. Accuracy of report deteriorated most rapidly during the first five seconds after the presentation and slowly thereafter. Other factors, such as unlimited opportunity for rehearsal and differential recall for certain groups of letters indicate that more than simple AIS was probably involved. All in all, both experiments demonstrate an auditory information storage capable of retaining information before it has been acted upon and whose contents decay during a period of many seconds.

(2) *Rehearsal.* The role of rehearsal in memory is a problem that psychologists

—with a few exceptions—have hoped would solve itself if only they ignored it sufficiently. Rehearsal is not quite so intractable to measurement as many of us were brought up to believe. For example, subjects may be asked to say a particular sequence of letters as fast as they can and to use a stop-watch to time themselves for the task. They may also be asked to say the letters to themselves without producing any sound and to time themselves as before. Subjects understand this second instruction without difficulty. Elapsed times are about three syllables per second for unfamiliar letter sequences spoken aloud. Silent rehearsal is slightly faster. The maximum rate for highly familiar sequences is about 10 syllables per second.

More controlled and somewhat similar studies of rehearsal were conducted by Brown (1958), Peterson and Peterson (1959) and by Murdock (1961). In the Petersons' experiment, subjects were initially presented with a sequence of letters. They were then told to rehearse them out loud in time with a metronome at two letters per second. Other subjects were given an equivalent time duration that could be used for silent rehearsal. The length of time allowed for rehearsal was varied. Following the initial rehearsal period, the subjects were put to a task intended to interfere with further rehearsal, such as counting backwards. Those subjects who had most time for initial rehearsal—aloud or silent—recalled best, accuracy being proportional to rehearsal time. Spoken rehearsal was more effective than silent rehearsal, a result that will be encountered again below.

These experiments indicate that a short period of rehearsal after an auditory stimulus presentation facilitates later recall. An interpolated period of non-rehearsal interferes with recall as does the competing auditory stimulus of hearing oneself saying irrelevant material.

(3) *Interference.* Brown (1958) sought to interfere with rehearsal by requiring his subjects to read digits aloud. Peterson and Peterson (1959) made their subjects count backwards between stimulus presentation and recall. These interpolated tasks produced a decrease in accuracy of recall which was interpreted as due to the interference of counting with rehearsal. The longer the period of counting, the less accurate the recall, presumably because of the fading of the rehearsed letters in auditory storage and the accumulation of interfering auditory stimuli. Brown had previously noted that interpolated silence (which permits rehearsal) does not have a comparable destructive effect on recall.[5]

To a certain degree, even when he is counting backwards, the subject's cooperation is needed for interference with his rehearsal. He is in effect asked to devote himself fully to the task of counting and to neglect completely the task of recall. It would be desirable to find a method of interference whose success did not depend upon the subject's cooperation—a method comparable to Baxt's method of eliminating visual persistence. Towards this end, I constructed two different acoustic stimuli which might interfere with rehearsal by providing a competing input to AIS. The first of these was a loud noise at 90 dB SPL, analogous to Baxt's interfering field. The second interfering stimulus was a recording of the subject's own voice speaking letters at a rate of 6/sec, also at 90 dB. This stimulus is analogous to visual noise. At each trial, either silence or one of these

---

[5] Brown, J. *Immediate Memory.* Doctoral thesis, University of Cambridge, 1955. Cited in Broadbent (1958), p. 225. See also Averbach and Sperling (1960).

extremely loud stimuli was played into the subject's ears, beginning before stimulus exposure and continuing until after report.

The number of letters subjects reported correctly was the same in noise and in silence, and only very slightly lower for the speech interference although the difference was statistically significant (Sperling, 1962). Two supplementary observations were made which elucidate the failure of the interpolated stimuli to interfere with recall. First, the subjects described the task as one of selective listening, that is, of hearing the speech interference as localized at the ears and their rehearsal as localized inside the head. The speech interference accentuates the subjective aspect of listening to rehearsal so that even subjects who are not normally cognizant of the auditory aspect become so.

The second observation was that normally silent subjects, when listening to speech interference, frequently mouthed or even spoke the letters out loud during the period between the stimulus exposure and writing down the response. This overt behavior reflects an effort to accentuate and emphasize the rehearsal. The conclusion therefore is that speech and perhaps also noise do interfere but that subjects can accentuate their rehearsal in order to cope successfully with the interference.

Summary of Hypotheses about Auditory Information Storage (AIS) and Rehearsal

(1) The primary input to AIS is sound. The contents of AIS probably depend on sound intensity, complexity, background, etc.[6] A secondary and possibly less effective input is by rehearsal.

(2) AIS is locally one-dimensional and directional. That is, a string of letters is stored in sequence and it can be recovered only in that sequence. However, one string of letters may be localized at the left ear and another simultaneous string of letters may be localized at the right ear. Thus, although each stored string of letters is individually one-dimensional, the simultaneous imbedding of many strings in storage implies a very complex overall structure for AIS.

(3) At least 10 non-rehearsed items can be maintained momentarily in AIS. They are probably grouped into several strings.

(4) The contents of AIS decay slowly relative to VIS. Rehearsal of items appears to lengthen their subsequent decay times but it also brings into play more complex mechanisms for long term memory via association which are beyond the scope of this paper.

(5) The contents of AIS, VIS, and of any comparable system in other modalities may be rehearsed. That is, a verbal response is made to items in the storage and this verbal response, which may or may not be audible to another person, is an input to AIS. When a stimulus consists of a number of clearly identifiable

---

[6] Wundt (1912) provides an insightful and provocative analysis of auditory memory for patterns of metronome beats. Wundt could maintain in his consciousness 8 groups of two beats each or 5 groups of eight beats each. This material was presumably not rehearsed verbally. Wundt's introspections would suggest that the contents of AIS are limited most directly by subjective grouping (itemizing) and only to a lesser extent by their duration or complexity (information content).

items, then the characteristics of the AIS–Rehearsal loop will probably be the limiting factor in the immediate-memory span for this stimulus.

(6) Rehearsal usually proceeds at a rate of about 3 syllables per second and cannot proceed faster than about 10 syllables per second.

(7) Only one item may be rehearsed at a time. The interval between the rehearsal of successive items is longer if they are not taken from the same sense modality.

(8) Subsequent acoustic stimuli can interfere with the contents of AIS beyond what might be expected from passive decay. So far, only self-produced verbal stimuli have shown the effect. The interference is most effective against items that have not been rehearsed and diminishes with the number of rehearsals. If rehearsal is permitted during the presentation of interfering stimuli (recorded, in this case) then the interference is barely measurable.[7]

DISCUSSION

The model is perhaps more useful for organizing data and experiments than for precise prediction because it consists of complex and only partially specified components. In this respect it reflects the complexity of its human subject matter and the limitations of its human designer.

Transient versus Steady-State Information Flow

The model was derived only from the "transient" response of a human to information; that is, the situation in which he is given a single burst of information and much time to process it. The responses to a "steady-state" or continuous information input undoubtedly involve further complications and new processes. For example, following a brief stimulus exposure, the auditory feedback loop from AIS through Rehearsal was of primary significance in recall. Visual processes in recall may come to be of importance only with longer exposures. Perhaps this is because following a brief stimulus exposure, most people cannot create in themselves a visual feedback loop, but they can in effect do so during a continuous stimulus exposure by looking repeatedly at certain aspects of the

[7] Broadbent (1957b, 1958) has suggested several models for immediate-memory. While a brief discussion cannot do justice to his ingenious and stimulating proposals, it is worthwhile to indicate some of the major similarities and differences. The obvious, basic similarities are the possibility of storage of unrecorded information and the recognition of a rehearsal loop. The differences lie in the nature and subdivision of these functions and are undoubtedly due in part to the different sets of data which each account seeks to explain. In the present model, items (or stimuli) circulate in the rehearsal loop while Broadbent specifically states that information (not items) circulates. The present account assumes one rehearsal loop and one store for rehearsed material, not several of each. Here VIS and AIS are identified in detail from introspective and direct experimental data while in Broadbent's models, storage of unrecoded information is of a more inferential and hypothetical nature—it can even be by-passed entirely. Broadbent is especially concerned with the case of simultaneous or conflicting inputs. He does not consider the special problem of scanning in vision nor make comparable distinctions between the processing of visual and of acoustic stimuli; etc.

stimulus. This example emphasizes another difficulty in the understanding of steady-state observation conditions. Namely, the sequential flow of information into the observer may depend on his own actions so that every complexity and nuance of human judgment and decision making must be considered in a complete steady-state model. On the other hand, some complex tasks have been devised so that a continuous verbal protocol is available—the subject thinks out loud so-to-speak (e.g., Newell and Simon, 1961). In such tasks, the knowledge of how many and for how long the subject's past utterances are available to him in AIS may well be of significance in the understanding of the process of decision making or "thinking."

Keeping in mind the restriction of the model to "transient" responses, it is still possible to make some hypotheses about the mechanisms underlying humans' response to complex displays and some tentative suggestions on how to improve their response.

(1) *Independence of the immediate-memory span.* Because of verbal rehearsal, humans transmit items, not bits. The limit on the number of items depends on such factors as the code (e.g., syllables per item) and upon the individual characteristics of the AIS–Rehearsal loop (e.g., AIS decay time). In so far as these are independent of the input modality of the stimulus, the information limitations in human responding will be independent of sense modality.

(2) *High information requirements.* Even in a brief exposure a human can take in much more information than he can ultimately transmit. By scanning and rehearsal he effectively samples an uncoded form of the input in VIS and AIS. In many practical situations one cannot dictate to the observer what aspects of the stimulus to sample. For example, in order to provide customer satisfaction, a television picture may have to contain information to match the enormous capacity of VIS and not merely the small amount of information transmitted in a sample from it. On the other hand, in other contexts it may be economical to keep the displayed information at the transmission (and not the storage) level.

(3) *Rapid reading.* Observers can extract small numbers of items very quickly from visual stimuli; 10 to 15 msec per letter is a typical rate for the first three or four items. In the absence of interfering stimuli, even a microsecond flash, if it is seen clearly, will usually be visually available for many times the length of time needed to extract the items up to the immediate-memory span. To transmit the maximal information from a brief exposure of a visual display, therefore, the display should be coded into about four symbols (e.g., digits) to take advantage of the rapid scan capability. Of course, other considerations such as the kind of errors may indicate other kinds of displays. In steady exposures, the rate of item utilization is usually not limited by visual factors but by the rehearsal process to a rate of 10 syllables per second or slower.

(4) *Non-susceptibility to auditory interference.* One useful if obvious result is that properly motivated subjects can perform visual memory tasks during extreme auditory interference and show almost no performance decrement. Specifically, subjects can perform without loss in 90 dB noise and almost without loss when barely tolerable levels of speech (recorded letters spoken by themselves) are being "shouted" into their ears. For practical purposes, therefore, it may be assumed that the number of letters reported from a brief exposure will be the same under optimal or adverse conditions.

(5) *Auditory simplification*. In order to facilitate performance and to avoid errors in visual monitoring tasks, telephone dialing, etc., it is necessary to consider the task from the auditory point of view as well as the purely visual. All possible outcomes should be given names that are easy to rehearse and not likely to be confused. The *laissez-faire* practice of allowing the operator to develop his own terminology is not recommended. When only numerals are involved, there is no problem as these are quite distinct. But we have found, for example, that subjects required to memorize sequences of the letters B,P,D,T, etc., do not do as well as when confronted with an equivalent sequence of letters which do not sound so much alike. Frequently, complex tasks which do not appear to involve audition can be simplified and made less susceptible to errors by means of an appropriate code that will ease the auditory memory load (in syllables) and help to avoid auditory confusions.

REFERENCES

Anderson, Nancy S. Poststimulus cuing in immediate-memory. *J. exp. Psychol.*, 1960, **60**, 216-221.
Averbach, E., & Sperling, G. Short term storage of information in vision. In C. Cherry (Ed.), *Information theory*. London: Butterworth, 1961, pp. 196-211.
Baxt, N. Ueber die Zeit welche nötig ist, damit ein Gesichtseindruck zum Bewusstsein kommt und über die Grösse (Extension) der bewussten Wahrnehmung bei einem Gesichtseindrucke von gegenbener Dauer. *Pflüger's Arch. ges. Physiol.*, 1871, **4**, 325-336.
Broadbent, D. E. Successive responses to simultaneous stimuli. *Quart. J. exp. Psychol.*, 1956, **8**, 145-152.
Broadbent, D. E. Immediate memory and simultaneous stimuli. *Quart. J. exp. Psychol.*, 1957a, **9**, 1-11.
Broadbent, D. E. A mechanical model for human attention and immediate memory. *Psychol. Rev.*, 1957b, **64**, 205-215.
Broadbent, D. E. *Perception and communication*. New York: Pergamon Press, 1958.
Brown, J. Some tests of the decay theory of immediate memory. *Quart. J. exp. Psychol.*, 1958, **10**, 12-21.
Cattell, J. McK. Ueber die Zeit der Erkennung und Benennung von Schriftzeichen, Bildern und Farben. *Philos. Stud.*, 1885, **2**, 635-650.
Erdmann, B., & Dodge, R. *Psychologische Untersuchung über das Lesen*. Halle: M. Niemeyer, 1898.
Exner, S. Ueber die zu einer Gesichtswahrnehmung nöthige Zeit. *S.B. Akad. Wiss. Wien.*, 1868, **58**, 601-633.
Gilbert, L. C. Speed of processing visual stimuli and its relation to reading. *J. educ. Psychol.*, 1959, **50**, 8-14.
James, W. *The principles of psychology*. New York: Holt, 1890.
Javal, L. E. Essai sur la physiologie de la lecture. *Ann. Oculist. Paris*, 1878, **82**, 242-253.
Külpe, O. Versuche über Abstraktion. In *Bericht über den I. Kongress für experimentelle Psychologie*, 1904. Leipzig: Barth, 1904, pp. 56-68.
Ladd, G. T. *Elements of physiological psychology—a treatise of the activities and nature of the mind*. New York: Charles Scribner's Sons, 1887.
Lindsley, D. B., & Emmons, W. H. Perception time and evoked potentials. *Science*, 1958, **127**, 1061.
Miller, G. A. The magical number seven, plus or minus two; some limits on our capacity for processing information. *Psychol. Rev.*, 1956, **63**, 81-97.
Murdock, B. B., Jr. The retention of individual items. *J. exp. Psychol.*, 1961, **62**, 618-625.
Newell, A., & Simon, H. A. Computer simulation of human thinking. *Science*, 1961, **134**, 2011-2017.

Peterson, L. L., & Peterson, Margaret J. Short-term retention of individual verbal items. *J. exp. Psychol.*, 1959, **58**, 193-198.

Pollack, I. Assimilation of sequentially encoded information. *Amer. J. Psychol.*, 1953, **66**, 421-435.

Sperling, G. Bistable aspects of monocular vision. *J. opt. Soc. Amer.*, 1960a, **50**, 1140-1141. (Abstract.)

Sperling, G. The information available in brief visual presentations. *Psychol. Monogr.*, 1960b, **74**, No. 11 (Whole No. 498).

Sperling, G. Negative afterimage without prior positive image. *Science*, 1960c, **131**, 1613-1614.

Sperling, G. Auditory interference with a visual memory task. Paper read at Eastern Psychological Association, Atlantic City, New Jersey, April, 1962.

Volkmann, A. W. Das Tachistoscop, ein Instrument, welches bei Untersuchung des monentanen Sehens den Gebrauch des electrischen Funkens ersetzr. *S.B. Kgl. Sächs. Ges. Wiss. Lpz., math-phys.*, 1859, **11**, 90-98.

Wundt, W. *An introduction to psychology.* Tr. from 2nd German ed., 1911, by R. Pintner. New York and London: Macmillan, 1912; reprinted London: Allen & Unwin, 1924.

11

Short-Term Retention of Individual Verbal Items

Lloyd R. Peterson and Margaret Jean Peterson

It is apparent that the acquisition of verbal habits depends on the effects of a given occasion being carried over into later repetitions of the situation. Nevertheless, textbooks separate acquisition and retention into distinct categories. The limitation of discussions of retention to long-term characteristics is necessary in large part by the scarcity of data on the course of retention over intervals of the order of magnitude of the time elapsing between successive repetitions in an acquisition study. The presence of a retentive function within the acquisition process was postulated by Hull (1940) in his use of the stimulus trace to explain serial phenomena. Again, Underwood (1949) has suggested that forgetting occurs during the acquisition process. But these theoretical considerations have not led to empirical investigation. Hull (1952) quantified the stimulus trace on data concerned with the CS-UCS interval in eyelid conditioning and it is not obvious that the construct so quantified can be readily transferred to verbal learning. One objection is that a verbal stimulus produces a strong predictable

From *Journal of Experimental Psychology*, **58**, 1959, 193-198. Copyright 1959 by the American Psychological Association, and reproduced by permission.

The initial stages of this investigation were facilitated by National Science Foundation Grant G-2596.

response prior to the experimental session and this is not true of the originally neutral stimulus in eyelid conditioning.

Two studies have shown that the effects of verbal stimulation can decrease over intervals measured in seconds. Pillsbury and Sylvester (1940) found marked decrement with a list of items tested for recall 10 sec. after a single presentation. However, it seems unlikely that this traditional presentation of a list and later testing for recall of the list will be useful in studying intervals near or shorter than the time necessary to present the list. Of more interest is a recent study by Brown (1958) in which among other conditions a single pair of consonants was tested after a 5-sec. interval. Decrement was found at the one recall interval, but no systematic study of the course of retention over a variety of intervals was attempted.

EXPERIMENT I

The present investigation tests recall for individual items after several short intervals. An item is presented and tested without related items intervening. The initial study examines the course of retention after one brief presentation of the item.

Method

Subjects. The Ss were 24 students from introductory psychology courses at Indiana University. Participation in experiments was a course requirement.

Materials. The verbal items tested for recall were 48 consonant syllables with Witmer association value no greater than 33% (Hilgard, 1951). Other materials were 48 three-digit numbers obtained from a table of random numbers. One of these was given to S after each presentation under instructions to count backward from the number. It was considered that continuous verbal activity during the time between presentation and signal for recall was desirable in order to minimize rehearsal behavior. The materials were selected to be categorically dissimilar and hence involve a minimum of interference.

Procedure. The S was seated at a table with E seated facing in the same direction on S's right. A black plywood screen shielded E from S. On the table in front of S were two small lights mounted on a black box. The general procedure was for E to spell a consonant syllable and immediately speak a three-digit number. The S then counted backward by three or four from this number. On flashing of a signal light S attempted to recall the consonant syllable. The E spoke in rhythm with a metronome clicking twice per second and S was instructed to do likewise. The timing of these events is diagrammed in Figure 1. As E spoke the third digit, he pressed a button activating a Hunter interval timer. At the end of a preset interval the timer activated a red light and an electric clock. The light was the signal for recall. The clock ran until E heard S speak three letters, when E stopped the clock by depressing a key. This time between onset of the light and completion of a response will be referred to as a latency. It is to be distinguished from the interval from completion of the syllable by E to onset of the light, which will be referred to as the recall interval.

The instructions read to S were as follows: "Please sit against the back of your chair so that you are comfortable. You will not be shocked during this experiment. In front of you is a little black box. The top or green light is on now. This green light means that we are ready to begin a trial. I will speak some letters and then a number. You are to repeat the number immediately after I say it and begin counting backwards by 3's (4's) from that number in time with the ticking that you hear. I might say, ABC 309. Then you say, 309, 306, 303, etc., until the bottom or red light comes on. When you see this red light come on, stop counting immediately and say the letters that were given at the beginning of the trial. Remember to keep your eyes on the black box at all times. There will be a short rest period and then the green light will come on again and we will start a new trial." The E summarized what he had already said and then gave S two practice trials. During this practice S was corrected if he hesitated before starting to count, or if he failed to stop counting on signal, or if he in any other way deviated from the instructions.

Each S was tested eight times at each of the recall intervals, 3, 6, 9, 12, 15, and 18 sec. A given consonant syllable was used only once with each S. Each syllable occurred equally often over the group at each recall interval. A specific recall interval was represented once in each successive block of six presentations. The S counted backward by three on half of the trials and by four on the remaining trials. No two successive items contained letters in common. The time between signal for recall and the start of the next presentation was 15 sec.

Results and Discussion

Responses occurring any time during the 15-sec. interval following signal for recall were recorded. In Figure 2 are plotted the proportions of correct recall as cumulative functions of latency for each of the recall intervals. Sign tests were used to evaluate differences among the curves (Walker & Lev, 1953). At each latency differences among the 3-, 6-, 9-, and 18-sec. recall interval curves are significant at the .05 level. For latencies of 6 sec. and longer these differences are all significant at the .01 level. Note that the number correct with latency less than 2 sec. does not constitute a majority of the total correct. These responses would not seem appropriately described as identification of the gradually weakening trace of a stimulus. There is a suggestion of an oscillatory characteristic in the events determining them.

The feasibility of an interpretation by a statistical model was explored by fitting to the data the exponential curve of Figure 3. The empirical points plotted here are proportions of correct responses with latencies shorter than 2.83 sec. Partition of the correct responses on the basis of latency is required by considerations developed in detail by Estes (1950). A given probability of response applies to an interval of time equal in length to the average time required for the response under consideration to occur. The mean latency of correct responses in the present experiment was 2.83 sec. Differences among the proportions of correct responses with latencies shorter than 2.83 sec. were evaluated by sign tests. The difference between the 3- and 18-sec. conditions was found to be significant at

Figure 1. Sequence of events for a recall interval of 3 sec.

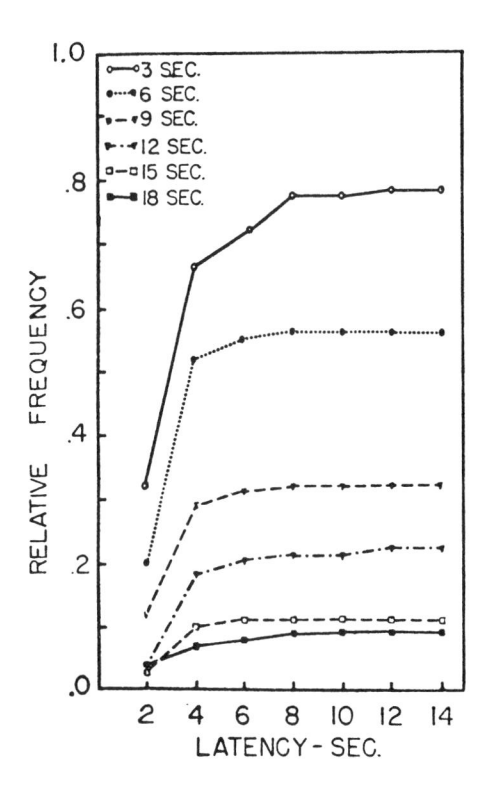

Figure 2. Correct recalls as cumulative functions of latency.

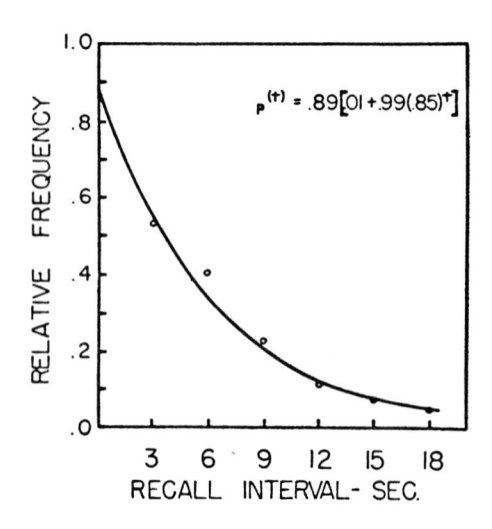

Figure 3. Correct recalls with latencies below 2.83 sec. as a function of recall interval. Equation on graph: $p^{(t)} = .89[.01 + .99(.85)^t]$

TABLE 1

Proportions of items correctly recalled in Experiment II.

Group	Repetition Time (Sec.)	Recall Interval (Sec.) 3	9	18
Vocal	3	.80	.48	.34
	1	.68	.34	.21
	0	.60	.25	.14
Silent	3	.70	.39	.30
	1	.74	.35	.22
	0	.72	.38	.15

TABLE 2

Dependent probabilities of a letter being correctly recalled in the vocal group when the preceding letter was correct.

Repetition Time (Sec.)	Recall Interval (Sec.) 3	9	18
3	.96	.85	.72
1	.90	.72	.57
0	.86	.64	.56

the .01 level. All differences among the 3-, 6-, 9-, 12-, and 18-sec. conditions were significant at the .05 level.

The general equation of which the expression for the curve of Figure 3 is a specific instance is derived from the stimulus fluctuation model developed by Estes (1955). In applying the model to the present experiment it is assumed that the verbal stimulus produces a response in S which is conditioned to a set of elements contiguous with the response. The elements thus conditioned are a sample of a larger population of elements into which the conditioned elements disperse as time passes. The proportion of conditioned elements in the sample determining S's behavior thus decreases and with it the probability of the response. Since the fitted curve appears to do justice to the data, the observed decrement could arise from stimulus fluctuation.

The independence of successive presentations might be questioned in the light of findings that performance deteriorates as a function of previous learning (Underwood, 1957). The presence of proactive interference was tested by noting the correct responses within each successive block of 12 presentations. The short recall intervals were analyzed separately from the long recall intervals in view of the possibility that facilitation might occur with the one and interference with the other. The proportions of correct responses for the combined 3- and 6-sec. recall intervals were in order of occurrence .57, .66, .70, and .74. A sign test showed the difference between the first and last blocks to be significant at the .02 level. The proportions correct for the 15- and 18-sec. recall intervals were .08, .15, .09, and .12. The gain from first to last blocks is not significant in this case. There is no evidence for proactive interference. There is an indication of improvement with practice.

EXPERIMENT II

The findings in Experiment I are compatible with the proposition that the aftereffects of a single, brief, verbal stimulation can be interpreted as those of a trial of learning. It would be predicted from such an interpretation that probability of recall at a given recall interval should increase as a function of repetitions of the stimulation. Forgetting should proceed at differential rates for items with differing numbers of repetitions. Although this seems to be a reasonable prediction, there are those who would predict otherwise. Brown (1958), for instance, questions whether repetitions, as such, strengthen the "memory trace." He suggests that the effect of repetitions of a stimulus, or rehearsal, may be merely to postpone the onset of decay of the trace. If time is measured from the moment that the last stimulation ceased, then the forgetting curves should coincide in all cases, no matter how many occurrences of the stimulation have preceded the final occurrence. The second experiment was designed to obtain empirical evidence relevant to this problem.

Method

The Ss were 48 students from the source previously described. Half of the Ss were instructed to repeat the stimulus aloud in time with the metronome until stopped by E giving them a number from which S counted backward. The remaining Ss were not given instructions concerning use of the interval between E's presentation of the stimulus and his speaking the number from which to count backward. Both the "vocal" group and the "silent" group had equated intervals of time during which rehearsal inevitably occurred in the one case and could occur in the other case. Differences in frequency of recalls between the groups would indicate a failure of the uninstructed Ss to rehearse. The zero point marking the beginning of the recall interval for the silent group was set at the point at which E spoke the number from which S counted backward. This was also true for the vocal group.

The length of the rehearsal period was varied for Ss of both groups over three conditions. On a third of the presentations S was not given time for any repetitions. This condition was thus comparable to Experiment I, save that the only recall intervals used were 3, 9, and 18 sec. On another third of the presentations 1 sec. elapsed during which S could repeat the stimulus. On another third of the presentations 3 sec. elapsed, or sufficient time for three repetitions. Consonant syllables were varied as to the rehearsal interval in which they were used, so that each syllable occurred equally often in each condition over the group. However, a given syllable was never presented more than once to any S. The Ss were assigned in order of appearance to a randomized list of conditions. Six practice presentations were given during which corrections were made of departures from instructions. Other details follow the procedures of Experiment I.

Results and Discussion

Table 1 shows the proportion of items recalled correctly. In the vocal group recall improved with repetition at each of the recall intervals tested. Conditions in the silent group were not consistently ordered. For purposes of statistical analysis the recall intervals were combined within each group. A sign test between numbers correct in the 0- and 3-repetition conditions of the vocal group showed the difference to be significant at the .01 level. The difference between the corresponding conditions of the silent group was not significant at the .05 level. Only under conditions where repetition of the stimulus was controlled by instructions did retention improve.

The obtained differences among the zero conditions of Experiment II and the 3-, 9-, and 18-sec. recall intervals of Experiment I require some comment, since procedures were essentially the same. Since these are between-S comparisons, some differences would be predicted because of sampling variability. But another factor is probably involved. There were 48 presentations in Experiment I and only 36 in Experiment II. Since recall was found to improve over successive blocks of trials, a superiority in recall for Ss of Experiment I is reasonable. In the case of differences between the vocal and silent groups of Experiment II a statistical test is permissible, for Ss were assigned randomly to the two groups. Wilcoxon's (1949) test for unpaired replicates, as well as a t test, was used. Neither showed significance at the .05 level.

The 1- and 3-repetition conditions of the vocal group afforded an opportunity to obtain a measure of what recall would be at the zero interval in time. It was noted whether a syllable had been correctly repeated by S. Proportions correctly repeated were .90 for the 1-repetition condition and .88 for the 3-repetition condition. The chief source of error lay in the confusion of the letters "m" and "n." This source of error is not confounded with the repetition variable, for it is S who repeats and thus perpetuates his error. Further, individual items were balanced over the three conditions. There is no suggestion of any difference in responding among the repetition conditions at the beginning of the recall interval. These differences developed during the time that S was engaged in counting backward. A differential rate of forgetting seems indisputable.

The factors underlying the improvement in retention with repetition were investigated by means of an analysis of the status of elements within the individual items. The individual consonant syllable, like the nonsense syllable, may be regarded as presenting S with a serial learning task. Through repetitions unrelated components may develop serial dependencies until in the manner of familiar words they have become single units. The improved retention might then be attributed to increases in these serial dependencies. The analysis proceeded by ascertaining the dependent probabilities that letters would be correct given the event that the previous letter was correct. These dependent probabilities are listed in Table 2. It is clear that with increasing repetitions the serial dependencies increase. Again combining recall intervals, a sign test between the zero condition and the three repetition condition is significant at the .01 level.

Learning is seen to take place within the items. But this finding does not eliminate the possibility that another kind of learning is proceeding concurrently. If only the correct occurrences of the first letters of syllables are considered, changes in retention apart from the serial dependencies can be assessed. The proportions of first letters recalled correctly for the 0-, 1-, and 3-repetition conditions were .60, .65, and .72, respectively. A sign test between the 0- and 3-repetition conditions was significant at the .05 level. It may tentatively be concluded that learning of a second kind took place.

The course of short-term verbal retention is seen to be related to learning processes. It would not appear to be strictly accurate to refer to retention after a brief presentation as a stimulus trace. Rather, it would seem appropriate to refer to it as the result of a trial of learning. However, in spite of possible objections to Hull's terminology the present investigation supports his general position that a short-term retentive factor is important for the analysis of verbal learning. The details of the role of retention in the acquisition process remain to be worked out.

SUMMARY

The investigation differed from traditional verbal retention studies in concerning itself with individual items instead of lists. Forgetting over intervals measured in seconds was found. The course of retention after a single presentation was related to a statistical model. Forgetting was found to progress at differential rates dependent on the amount of controlled rehearsal of the stimulus. A portion of the improvement in recall with repetitions was assigned to serial learning within the item, but a second kind of learning was also found. It was concluded that short-term retention is an important, though neglected, aspect of the acquisition process.

REFERENCES

Brown, J. Some tests of the decay theory of immediate memory. *Quart. J. exp. Psychol.*, 1958, **10**, 12-21.
Estes, W. K. Toward a statistical theory of learning. *Psychol. Rev.*, 1950, **57**, 94-107.
Estes, W. K. Statistical theory of spontaneous recovery and regression. *Psychol. Rev.*, 1955, **62**, 145-154.
Hilgard, E. R. Methods and procedures in the study of learning. In S. S. Stevens (Ed.), *Handbook of experimental psychology*. New York: Wiley, 1951.
Hull, C. L., Hovland, C. I., Ross, R. T., Hall, M., Perkins, D. T., & Fitch, F. B. *Mathematico-deductive theory of rote learning: A study in scientific methodology*. New Haven: Yale Univer. Press, 1940.
Hull, C. L. *A behavior system*. New Haven: Yale Univer. Press, 1952.
Pillsbury, W. B., & Sylvester, A. Retroactive and proactive inhibition in immediate memory. *J. exp. Psychol.*, 1940, **27**, 532-545.
Underwood, B. J. *Experimental psychology*. New York: Appleton-Century-Crofts, 1949.
Underwood, B. J. Interference and forgetting. *Psychol. Rev.*, 1957, **64**, 49-60.
Walker, H., & Lev, J. *Statistical inference*. New York: Holt, 1953.
Wilcoxon, F. *Some rapid approximate statistical procedures*. New York: Amer. Cyanamid Co., 1949.

12

Flow of Information within the Organism

D. E. Broadbent

The approach which is to be adopted throughout this paper is one which asserts the possibility of making statements about events within an organism (a man or some other kind of animal) purely from the behavior of that organism. This is not altogether a popular view in psychological circles, where the dominant approaches are probably those of pure observation of behavior on the one hand, or of physiological studies and explanations on the other. There is, however, a large gap between the study of behavior and that of neurophysiology, and it can well be argued that the use of simplified diagrams representing the flow of information within the nervous system, without any venturing into physiological detail, will form a most useful bridge between the two studies.

It is, however, essential that one be exceedingly cautious in making any state-

From *Journal of Verbal Learning and Verbal Behavior*, 2, 1963, 34-39. By permission of Academic Press, Inc.

ment about the hidden events within the skin of a man who is trying to remember something. It can only be justified where one can appeal to the principle of causality as a help in drawing conclusions about unobserved events. For example, if we apply stimulus A to a man and he repeatedly produces response A, whereas when we apply stimulus B he produces response B, we may say that in case A there is a complete chain of events of causes and effects, which links the input to the output. The series of events in case A differs in some way at every stage from the other chain of events which links input to output in case B. This cannot really be denied unless we wish to abandon causality.

If on the other hand we have two stimuli A and B, which are different in physical nature but which produce the same response C, then we must conclude that there are two different lines of causation starting at the sense organs in the two cases and running off into the nervous system; but that at some point they come together and produce one single line which issues in the particular response C. For example, it may be that we can show a man a particular shape, or present to his ears a particular pattern of sound, and in both cases he will say the digit 6. The two different stimuli (sight and sound) must then produce different processes inside his brain which at some point converge upon a single output process. For convenient reference later, let us call this process "categorization." In such a case, it is then meaningful to ask whether a learning task which involves one of these categorized stimuli will change subsequent behavior to the other stimulus. If it does, most of us would conclude that the learning in some way involves the common part of the causal chain and not the separate part.

For example, in a recent experiment letters of the alphabet were presented visually to men who were then asked to reproduce the particular series that they had seen. The confusion errors which occurred in recall of this type were similar to those which occur when one is listening to the names of letters of the alphabet spoken through noise (Conrad, 1962). If one has seen the letter V, for instance, one may recall the letter B and not X, even though visually V and X have more in common than V and B. It is likely that all of us will agree in this case that what is happening is that the subjects tended to make a verbal response to the visually presented letters, and that subsequently in recall they responded to their own previous responses: they said the letters over to themselves. Some people then would call this an experiment on mediating responses; another, and perhaps preferable, description is that it was a study on the convergence and divergence of causal lines within the nervous system.

So much then for the general method of approach. The particular kind of diagram we may draw for memory processes is one like this. Let us say that when information comes into the nervous system it passes through a limited capacity channel, which means that it is dificult to take in other information simultaneously. Once through this channel, items of information can be held in a short-term store, but only as long as they can pass repeatedly through the same limited capacity channel as that by which they arrived. So long as they can do this, they can be stored in this way indefinitely; but if the limited part of the system becomes unavailable, perhaps through the admission of some fresh items of information from the outside world, then the storage of the earlier items will fail and they will not be available any longer. In other words, if you look up a telephone number in the directory and then set out to the telephone to dial it, you will be able to remember the number no matter how far away the telephone is, provided that you do not have to deal with any very informative stimuli on your way to the telephone. If somebody asks you a question, you will either have to ignore them or in all probability forget the telephone number.

One must of course also suppose some other kind of storage, in addition to this peculiarly limited variety. One may for instance suppose that whenever the arrival of item A through the limited capacity system is immediately followed by that of item B, this coincidence is counted, and that when the number of sequences A followed by B has become significantly greater than that of A followed by anything else, subsequent stimuli of the type which have produced A will produce responses of the type previously appropriate to B. Thus for example Baddeley (1961) has shown that if one is learning to respond with a nonsense syllable when shown another syllable, it is helpful if the last letter of the stimulus and the first letter of the response form a pair which has often been experienced in ordinary English language. In this case, the stored information about these conditional probabilities does not depend upon the repeated use of the limited capacity channel for its survival.

Why now does one suppose a rather complicated flow diagram of this sort? Basically, one does so because the observed behavior of memories which are old-established is different from those which are recent. In terms of the classical Jost's law, if two learned responses are of equal efficiency at this moment, the one which was more recently established can be disturbed by treatment which will not affect the other. For example, if at this moment I know the telephone number of my Laboratory, and also a number that I have just got from the telephone directory, asking me a question will disrupt the latter but not the former. There are numerous experiments of this kind, all showing that the relationship between present strength and vulnerability to intervening activity is different in the case of recently learned and more long-established memories.

It may still be argued, however, that the effect of interfering activity on memory for recent events is only a more severe form of an interference which will occur even with long-established memories, if one goes in for sufficiently intensive learning of some other kind. For example, I may have trouble even with the number of my own Laboratory (which is 55294) if I take care to learn by heart the number of another Laboratory which is 59254. Even on the theory already given, one must expect such an interference where there is some degree of similarity between the old learning and some new interfering learning; if A has been followed by B on a number of occasions, but more recently A has since been followed by C on an equal number of occasions, A ceases to be of any value in predicting what is going to happen next. So if one learns something and then subsequently learns something else which is in fact similar one may expect some interference between the two memories. This kind of interference would depend upon just how similar the intervening activity is, and indeed has often been shown to do so. On our hypothetical model of the flow of information, however, what matters in short-term memory might well be the time for which interfering activity occupies the limited capacity system, rather than the similarity between the interfering and original activities. Is there then any evidence in the case of short-term memory for the importance of sheer time in deciding how effective a particular interference is?

At this stage one must distinguish two kinds of short-term memory. Firstly, there is the very brief period of a second or so after the actual presentation of a stimulus. During this time, the stimulus has not been "categorized"; or if you prefer, the first response to the stimulus has not appeared. To go back to the original phraseology, the causal line arising from that stimulus does not converge with those of other stimuli which possess the same "significance." During this period of a second or so, there is strong evidence that there is a temporary form of storage which decays with time and which does not depend at all on the nature or amount of any interference. This is shown in such experiments as those of Averbach and Sperling (1960) in which a number of items are flashed on a screen simultaneously, and a marker stimulus is later presented to indicate the particular stimulus which the man is to remember. Since the efficiency of response falls off as the time interval between the original presentation and that of the marker is increased, and since during this period there are no interfering stimuli to or responses from the subject, we must conclude that this short-term storage is impaired simply by passage of time.

Possibly in the same class as this work of Averbach and Sperling is an old experiment in which six numbers were presented to one ear, and two numbers to the other ear, with instructions to reproduce first the six and then the two. This was much more efficiently done if the two digits arrived at the very end of the six, rather than a second earlier (Broadbent, 1957). Here again there was no difference in the number of responses intervening between the presentation of the stimulus and its recall in the two conditions being studied, and those conditions differed rather in the time that elapsed between presentation and recall. It does seem therefore that in this very first second after presentation of the stimulus, before it has been categorized, the stored information decays very rapidly as a function of time and not as a function of intervening activity.

On the other hand, this situation of decay immediately following the presentation of a stimulus is admittedly rather different from that of memory for a stimulus which has been categorized; or, if you will, has received at least one response. We cannot assume, from results such as those of Averbach and Sperling or of Broadbent, that a telephone number which we have once said over to ourselves is now subject to decay if we neglect it and do something else. At this stage of memory, it is extremely difficult to get experimental control of events. Nevertheless, there are some experimental results which make it advisable to keep open the possibility that even this kind of short-term memory suffers from interfering activity simply because of the time which that activity takes rather than its nature or its amount. The first of these are the experiments on controlling rate of recall. If you have to dial a telephone number, you are more likely to make mistakes than if you key the same number on to a keyboard of a calculating machine (Conrad, 1958). This is simply because the process of recall takes longer on a dial telephone, and a very fast rate of recall is a little more efficient than a slightly slower one (Conrad and Hille, 1958). In this case, there is not properly speaking any interfering activity except simply the recall of the material itself which is being remembered; although obviously what is happening is that recall of part of the material is interfering with recall of the rest of it, the difference between fast and slow recall is not one of amount of interfering material but simply of the time which that interference takes up.

Another and particularly interesting variation of the same theme is the use of instructions to learn a particular set of items such as a telephone number, to rehearse it *aloud* (and therefore rather more slowly than one could rehearse it to oneself) and finally after an interval to recall it. It has been shown that under certain conditions saying a telephone number aloud as opposed to saying it to oneself may cause subsequent failures of recall, presumably because the time taken in saying the number is greater and therefore one forgets the beginning of it by the time one reaches the end. Once again the interference is by the information itself which one is trying to recall, and the only difference in conditions is in the time that is taken by the interference. Incidentally, although this point has chiefly been made by Sanders (1961), a Dutch psychologist, there are other experiments which show it (Postman and Phillips, 1961). In this case, the original interpretation was that rehearsal as such may be harmful; but it is rehearsing the material overtly which seems to produce the difficulty, and so far as is known, there is no experiment in which silent rehearsal to oneself has been shown to impair performance. This is therefore another case in which the time taken by interference is all-important.

Nevertheless, it might still be logically possible, even if unlikely, that rehearsal as such is in some way harmful and that there is therefore some interfering effect from the material itself upon itself. It is relevant therefore to report a crude and preliminary experiment by myself and Margaret Gregory, in which we presented from a tape recorder lists of six digits, which were closely followed in each case by the three letters A B C. The subjects were instructed to call out the same three letters in synchrony with the letters on the tape, and then subsequently to write down the six digits that had preceded them. In one condition the three letters were presented rapidly, so that all three together lasted only two-thirds of a second; in another condition the three letters were presented more slowly, for five subjects at intervals of two-thirds of a second and for twelve subjects of one second so that the whole sequence took over twice as long. In either condition the listeners performed repeatedly at the particular speed so that they could synchronize reasonably accurately. They thus had to recall six digits after an interfering activity which was identical in the nature and number of items, but which differed in the time taken by the interference. For each subject, ten lists were performed under one condition, and then ten under the other condition at the same session. To minimize the effects of inter-list differences in difficulty, the same lists were used for both conditions—a point which may be important. Performance was significantly ($P < 0.05$) worse in the case where the letters were presented more slowly. The average number of digits correct for a subject was 43.1 out of 60 for the fast rate, and 38 for the slow rate. Thus here again it seems to be the time taken by the interference rather than its nature or amount which is important.

Nevertheless, this latest experiment is not altogether satisfactory, nor indeed are any of those cited as demonstrating this point in the case of material which has received at least one response. The use of interfering activities which go at various speeds inevitably means a varying degree of control over the other activities of the person who is recalling, and in the case of the latest experiment it is extremely difficult in practice to control how well the synchronization is in fact being carried out. The effect is undoubtedly dependent upon the particular speeds which were chosen. Finally, the differences are extremely slight between the two conditions, even though the time taken by the interference is twice as great in one case as in the other.

Perhaps therefore we ought to emphasize not so much that the interference effects in short-term memory depend upon the length of time taken by the interference, as that they do not seem to depend very much upon the nature of the intervening activity. As already indicated, one would expect long-term memory to show interference from subsequent activity which would be due essentially to an unlearning of the original material, and which would therefore vary very much with the similarity between the original and interpolated material. Empirically indeed it does do so. In the case of short-term memory, however, the intervening activity does not have to be learning but can be almost any kind of performance, provided that it represents a high rate of information transmission through the nervous system. There may be some relationship between the similarity of the original and interpolated material, and the extent of interference, but if so it is extremely slight.

Now, in the information flow diagram with which we started, it is not an essential feature that the short-term store should decay with time when the limited capacity system is occupied elsewhere; but merely that it should break down if there is an interruption in the use of the limited capacity system for memory rather than intake of external information. It might therefore be safer simply to assert that the nature of interference in the intermediate range of short-term memory (greater than a few seconds) is different from that which appears in memories lasting over several days.

To summarize, it is worthwhile drawing inferences about the processes which go on inside a man who is trying to remember something. It is very well established that there is a difference in the internal mechanism in the case of stimuli which have just been presented within the last second, and in the case of stimuli which were presented yesterday but which are still remembered. The point of doubt and of greatest current interest is the intermediate region of memory for events which occurred some minutes ago. In that region the big problem is the categorization of stimuli, when their causal lines within the nervous system converge and they give rise to what other schools of thought would call mediating responses. This is not the place, for instance, to raise the question of the difference in difficulty of recalling the sequence 2 BLUE 7 GREEN 6 RED and the sequence 2 7 6 BLUE GREEN RED, when these sequences are presented rapidly; but that difference is a source of great future interest. So also are the effects of old age; in numerous recent experiments age causes a big increase in the sort of interference that occurs in short-term memory while not making much difference to the sort that happens in long-term memory. These points, however, will no doubt occupy us in succeeding years.

SUMMARY

This paper outlines the general approach of drawing inferences about the flow of information inside the organism and summarizes the evidence for distinguishing short-term from long-term memory. A minor experiment is described, in which short-term memory is compared for items which have been followed by a fixed number of interfering items at fast and at slow speeds. The latter produced inferior performance, which is consistent with a decay theory. The greatest emphasis is placed, however, upon the different role of similarity in short- and in long-term memory.

REFERENCES

Averbach, E., and Sperling, G. Short-term storage of information in vision. In E. C. Cherry (Ed.). *Fourth London Symposium on Information Theory*. London: Butterworths, 1960.

Broadbent, D. E. Immediate memory and simultaneous stimuli. *Quart. J. exp. Psychol.*, 1957, **9**, 1-11.

Baddeley, A. D. Stimulus-response compatibility in the paired-associate learning of nonsense syllables. *Nature*, 1961, **191**, 1327-8.

Conrad, R. Accuracy of recall using keyset and telephone dial and the effect of a prefix digit. *J. appl. Psychol.*, 1958, **42**, 285-288.

Conrad, R. An association between errors and errors due to acoustic masking of speech. *Nature*, 1962, **193**, 1314-15.

Conrad, R., and Hille, B. The decay theory of immediate memory and paced recall. *Canad. J. Psychol*, 1958, **12**, 1-6.

Postman, L., and Phillips, L. W. Studies in incidental learning: IX. A comparison of the methods of successive and single recalls. *J. exp. Psychol.*, 1961, **61**, 236-241.

Sanders, A. F. Rehearsal and recall in immediate memory. *Ergonomics*, 1961, **4**, 29-34.

Chapter Four

Human Processing of Information—
Transmission of Discrete and Continuous Signals

Even the simplest of skill tasks involves a multitude of integrated activities. Consider, for example, the classical choice reaction-time problem in which the operator has merely to indicate by his response (e.g., by pressing the appropriate button) which of a set of stimuli (e.g., light patterns) has appeared. He is encouraged to make his choice as rapidly as possible and is evaluated accordingly. Within the fraction of a second that it takes him to respond, at least the following processes are believed to occur: (a) a raw sensory impression of the stimulus is formed; (b) this image is classified in some way or in several ways on the basis of information called up from memory—e.g., the impression is compared with a list of features or replicas of possible light patterns stored in memory; (c) a choice of motor sequence is made from the individual's response repertoire on the basis of this classification (and, perhaps, other considerations such as how often *in the past* each response has been required); and finally, (d) the response sequence is executed. To complicate things even further, cognitive mechanisms sensitive to such factors as the relative value of speed and accuracy are believed to influence at least the classification and response selection processes. What on the surface seems to be a reasonably straightforward task, then, turns out to implicate a surprising variety of human capabilities. When we extend our gaze beyond the laboratory to such commonplace skills as hitting a baseball or driving an automobile, the processing required of the human reaches astounding proportions.

Our attention in this chapter is upon skills involved in those tasks which can be viewed as having, as their principal objective, *transmission* of information. Signals of one sort or another occur—either as discrete events or as continuous functions over time—and the operator is instructed to respond differentially to each. In other words, he tries to preserve as much of the information in these signals as he can by appropriate responding; if he performed ideally, one could tell from his response exactly which signals had occurred and when. We use the term *transmission*, then, to denote a gross task characteristic, recognizing full well that it encompasses a host of more subtle processing steps within the organism.

Much of the early work on human information-processing skills ignored these intervening steps, concentrating instead upon the gross but objective input-output functions. With the advent of "information theory," as described in Miller's article,

References will be found in the General References at the end of the book.

it became possible to specify how *much* information there was in any set of stimulus events and in any set of response events; even more important, this made it possible to determine precisely how much of the former information was also present in the latter (i.e., to describe how much of the stimulus information was *transmitted* through the responses).

Numerous studies have sought to establish human transmission capacities—both the *amount* that people can transmit per stimulus, and the *rate* at which they can do it—for various stimulus and response characteristics. Several rather broad generalizations have emerged from this work, including the conclusion mentioned in the Miller article in Chapter Three to the effect that people can identify absolutely no more than about nine categories along a single stimulus dimension. Another conclusion, which has since come to be called "Hick's law," is introduced with the second article in this chapter; in essence it holds that people process information at a constant rate. That is, the amount of time that it takes a person to initiate responses to discrete stimuli (i.e., his choice reaction time) increases linearly with the amount of *information* or uncertainty in those stimuli. Considerable evidence has since been gathered in support of this generalization (see, for example, Attneave, 1959; Garner, 1962). As the present selection clearly shows, however, Hick went far beyond this simple empirical statement in attempting to explain processing behavior. In this classic article he discusses a number of possible internal operations whereby a person might go about carrying out such a task, and at least several of these processes would produce the observed constant rate of information transmission. Until recently, though, researchers have shown more interest in verifying the constant rate notion than in pursuing Hick's leads as to underlying mechanisms.

The third and fourth selections illustrate a recent reversal in this trend, one which has seen a definite preference develop toward more molecular explanation of processing phenomena (such as Hick's law). Stimulated by a rediscovery of Hick's speculations, the aforementioned advances in memory research, and several insightful reviews and theoretical papers (e.g., Welford, 1960; Neisser, 1967; Laming, 1968; Smith, 1968), work in the area now centers around a variety of distinct processing models. New research paradigms have been devised to test these models, and a renewed enthusiasm seems to surround the entire transmission enterprise. Sternberg's paper represents one such model and paradigm; it holds that in classifying a presented stimulus so as to choose an appropriate response, a processor searches in a *serial* fashion through all the replicas of potential stimuli stored in his memory. An alternative to the serial search model, one involving *parallel* or simultaneous search, is represented in the Neisser et al. paper. One common conclusion appearing in nearly all of these recent papers is that the information model is inappropriate and perhaps even misleading as a means of explaining reaction-time phenomena (including "Hick's law").

The Hick, Sternberg, and Neisser papers serve to illustrate still another important characteristic of contemporary information-processing research—its integrative nature. Today's theories recognize the fact that people are not *simple communication channels* by any stretch of the imagination. One must, if he wishes to model underlying processes, at least take into account such intimately related functions as memory, decision, and attention. In earlier years, research on these topics was rarely carried out within the same context as "human information transmission." In fact, the different task situations used often served as the basis for classifying research as either *experimental* or *engineering* psychology. That differentiation along these lines is illusory at best is clearly illustrated by integrative schemes such as Broadbent's model which we encountered in Chapter Three.

From the transmission of discrete signals we turn to the problem of processing continuous or time-varying input functions. The chief requisite of tracking and other tasks of this sort is that responses be organized in some fashion so as to preserve both the spatial and temporal characteristics of the input signal; i.e., timing is of the essence. Basic to all vehicular control skills, such processing requirements have long captured the interest of psychologists concerned with selection and training as well as those concerned with human performance principles. In view of the quantity and variety of studies devoted to "tracking" skills and the lack of an adequate theoretical framework within which to organize them, we have elected to present a review article (Adams) and a paper describing a current research program (Noble and Trumbo) as representative of the area. The former summarizes and critically evaluates much of the "classical" research which, like the older work on "discrete" processing, concentrated principally on description of gross input-output functions. In contrast, the Noble and Trumbo approach is aimed at understanding the processes underlying such functions. In particular, Noble and Trumbo have addressed themselves to the task of accounting for the high degree of spatial-temporal organization which characterizes perceptual-motor skill behavior.

Both "discrete" and "continuous" processing research, then, has shifted its emphasis in recent years from gross description to a search for more fundamental understanding of functions underlying skilled performance. The similarity, however, does not end there. In one of the last of his many insightful analyses of perceptual-motor skill, Fitts (1964) expressed the strong belief that so-called "continuous" behavior can be fully explained on the basis of the same processes that control "discrete" behavior. In other words, when reduced to their elements, continuous and discrete tasks draw upon the same fundamental skill processes. To be sure, tasks such as tracking are amenable to a number of unique analytic treatments because they are continuous; as we mentioned earlier, it is possible (and has proven very fruitful) to apply concepts borrowed from adaptive control (servo) theory to the description of certain aspects of human performance (see, for example, Briggs, 1964; Poulton, 1966b; Noble, 1968). Description in these terms has been successful enough, in certain instances, to permit very precise simulation of human control behavior. In fact, computer simulation was so good in one widely-cited study that human trackers were unable to tell, over fairly long periods of time, whether *they* or the machine *simulating* them was actually controlling the system (Goodyear Aircraft Corp., 1955)!

In spite of the fact that amazing fidelity of simulation can, upon occasion, be achieved in continuous terms, there is ample evidence in support of Fitts's view that the underlying behavior is *discontinuous*. This in no way depreciates the value of continuous descriptions. It has long been recognized in other scientific areas that explanation of behavior can be given at several levels of discourse, all of which can be fruitful. That the propagated nerve impulse is a discrete event, for example, does not prevent us from examining records of *combined* neural activity, the continuous appearance of which may enable us better to understand the workings of some more molar unit (e.g., a receptor system).

The Noble and Trumbo article, which closes this chapter, is particularly appropriate as a transition from *transmission* to *intellectual* or *cognitive* processing tasks, which occupy our attention in Chapter Five.

13

What Is Information Measurement?

George A. Miller

In recent years a few psychologists, whose business throws them together with communication engineers, have been making considerable fuss over something called "information theory." They drop words like "noise," "redundancy," or "channel capacity" into surprising contexts and act like they had a new slant on some of the oldest problems in experimental psychology. Little wonder that their colleagues are asking, "What is this 'information' you talk about measuring?" and "What does all this have to do with the general body of psychological theory?"

The reason for the fuss is that information theory provides a yardstick for measuring organization. The argument runs like this. A well-organized system is predictable—you know almost what it is going to do before it happens. When a well-organized system does something, you learn little that you didn't already know—you acquire little information. A perfectly organized system is completely predictable and its behavior provides no information at all. The more disorganized and unpredictable a system is, the more information you can get by watching it. Information, organization, and predictability room together in this theoretical house. The key that unlocks the door to predictability is the theory of probability, but once this door is open we have access to information and organization as well.

The implications of this argument are indeed worth making a fuss about. Information, organization, predictability, and their synonyms are not rare concepts in psychology. Each place they occur now seems to be enriched by the possibility of quantification. One rereads familiar passages with fresh excitement over their experimental possibilities. Well-worn phrases like "perceptual organization," "the disorganizing effects of emotion," "knowledge of results," "stereotyped behavior," "reorganization of the problem materials," etc., begin to leap off the pages.

From *American Psychologist*, 8, 1953, 3-11. Copyright 1953 by the American Psychological Association, and reproduced by permission.

In the first blush of enthusiasm for this new toy it is easy to overstate the case. When Newton's mechanics was flowering, the claim was made that animals are nothing but machines, similar to but more complicated than a good clock. Later, during the development of thermodynamics, it was claimed that animals are nothing but complicated heat engines. With the development of information theory we can expect to hear that animals are nothing but communication systems. If we profit from history, we can mistrust the "nothing but" in this claim. But we will also remember that anatomists learned from mechanics and physiologists profited by thermodynamics. Insofar as living organisms perform the functions of a communication system, they must obey the laws that govern all such systems. How much psychology will profit from this obedience remains for the future to show.

Most of the careless claims for the importance of information theory arise from overly free associations to the word "information." This term occurs in the theory in a careful and particular way. It is not synonymous with "meaning." Only the *amount* of information is measured—the amount does not specify the content, value, truthfulness, exclusiveness, history, or purpose of the information. The definition does not exclude other definitions and certainly does not include all the meanings implied by the colloquial usages of the word. This garland of "nots" covers most of the objectionable exaggerations. In order to demonstrate some properly constrained associations to the word "information," we need to begin with definitions of some basic concepts.

BASIC CONCEPTS

Amount of Information. A certain event is going to occur. You know all the different ways this event can happen. You even know how probable each of these different outcomes is. In fact, you know everything about this event that can be learned by watching innumerable similar events in the past. The only thing you don't know is exactly which one of these outcomes will actually happen.

Imagine a child who is told that a piece of candy is under one of 16 boxes. If he lifts the right box, he can have the candy. The event—lifting one of the boxes—has 16 possible outcomes. In order to pick the right box, the child needs information. Anything we tell him that reduces the number of boxes from which he must choose will provide some of the information he needs. If we say, "The candy is not under the red box," we give him just enough information to reduce the number of alternatives from 16 to 15. If we say, "The candy is under one of the four boxes on the left end," we give more information because we reduce 16 to 4 alternatives. If we say, "The candy is under the white box," we give him all the information he needs—we reduce the 16 alternatives to the one he wants.

The amount of information in such statements is a measure of how much they reduce the number of possible outcomes. Nothing is said about whether the information is true, valuable, understood, or believed—we are talking only about *how much* information there is.

Bit. A perfectly good way to measure the amount of information in such statements (but not the way we will adopt) is merely to count the number of possible outcomes that the information eliminates. Then the rule would be that every time one alternative is eliminated, one unit of information is communicated.

The objection to this unit of measurement is intuitive. Most people feel that to reduce 100 alternatives to 99 is less helpful than to reduce two alternatives to one. It is intuitively more attractive to use ratios. The amount of information depends upon the fraction of the alternatives that are eliminated, not the absolute number. In order to convey the same amount of information, the 100 alternatives should be reduced by the same fraction as the two alternatives, that is to say, from 100 to 50.

Every time the number of alternatives is reduced to half, one unit of information is gained. This unit is called one "bit" of information. If one message reduces k to k/x, it contains one bit less information than does a message that reduces k to $k/2x$. Therefore, the amount of information in a message that reduces k to k/x is $\log_2 x$ bits.

For example, if the child's 16 boxes are reduced to two, then x is 8 and $\log_2 8$ is three bits of information. That is to say, 16 has been halved three times: 16 to 8, 8 to 4, and 4 to 2 alternative outcomes.

Source. The communication engineer is seldom concerned with a particular message. He must provide a channel capable of transmitting any message that a source may generate. The source selects a message out of a set of k alternative messages that it might send. Thus each time the source selects a message, the channel must transmit $\log_2 k$ bits of information in order to tell the receiver what choice was made.

If some messages are more probable than the others, a receiver can anticipate them and less information needs to be transmitted. In other words, the frequent messages should be the short ones. In order to take account of differences in probability, we treat a message whose probability is p as if it was selected from a set of $1/p$ alternative messages. The amount of information that must be transmitted for this message is, therefore, $\log_2 1/p$, or $-\log_2 p$. (Note that if all k messages are equally probable, $p = 1/k$ and $-\log_2 p = \log_2 k$, which is the measure given above.) In other words, some messages that the source selects involves more information than others. If the message probabilities are $p_1, p_2, \ldots, p_k$, then the amounts of information associated with each message are $-\log_2 p_1, -\log_2 p_2, \ldots, -\log_2 p_k$.

Average Amount of Information. Since we want to deal with sources, rather than with particular messages, we need a measure to represent how much information a source generates. If different messages contain different amounts of information, it is reasonable to talk about the average amount of information per message we can expect to get from the source—the average for all the different messages the source may select. This expected value from source x is denoted $H(x)$:

$$H(x) = \text{the mean value of } (-\log_2 p_i)$$
$$= \sum_{i=1}^{k} p_i \, (-\log_2 p_i)$$

What Is Information Measurement?

This is the equation that occurs most often in the psychological applications of information theory. $H(x)$ in bits per message is the mean logarithmic probability for all messages from source x. In all that follows we shall be talking about the average amount of information expected from a source, and not the exact amount in any particular message.

Related Sources. Three gentlemen—call them Ecks, Wye, and Zee—are each making binary choices. That is to say, Ecks chooses either heads or tails and simultaneously Wye also makes a choice and so does Zee. They repeat their synchronous choosing over and over again, varying their choices more or less randomly on successive trials. Our job is to predict what the outcome of this triple-choice event will be.

With no more description than this we know that there are eight ways the triple-choice can come out: HHH, HHT, HTH, HTT, THH, THT, TTH, and TTT. Thus our job is to select one out of these eight possible outcomes. If all eight were equally probable, we would need three bits of information to make the decision.

Now suppose that Ecks tells us each time what his next choice is going to be. With Ecks out of the way we are left with only four combinations of double-choices by Wye and Zee, so we can gain one bit of information about the triple-choice from Ecks. Similarly, if Wye tells us what his choice is going to be, that can also be worth one bit of information. Now the question is this: If Ecks and Wye both tell us what they are going to do, how much information do we get?

Case I: Suppose that it turns out that Ecks and Wye are perfectly correlated. In other words, if we know what Ecks will do, we also know what Wye will do, and vice versa. Given the information from either one of them, the other one has no further information to add. Thus the most we can get from both is exactly the same as what we would get from either one alone. Note that if Ecks and Wye always make the same choice, there are actually only four possible outcomes: HHH, HHT, TTH, and TTT, so we need only two bits to select the outcome.)

Case II: Next, suppose that Ecks and Wye make their choices with complete independence. Then a knowledge of Ecks' choice tells us absolutely nothing about what Wye is going to do, and vice versa. None of the information from one is duplicated by the other. Thus, if we get one bit from Ecks and one bit from Wye, and if there is no common information at all, we must get two whole bits of information from both of them together.

Case III: Finally, suppose that, as will usually be the case when we apply these ideas, Ecks and Wye are partially but not perfectly correlated. If we know what Ecks will do, we can make a fairly reliable guess what Wye will do, and vice versa. Some but not all of the information we get from Ecks duplicates the information we get from Wye. This case falls in between the first two: the total information is greater than either of its parts, but less than their sum.

The situation in Case III is pictured in Figure 1. The left circle is the information we get from Ecks and the right circle is the information from Wye. The symbols $H(x)$ and $H(y)$ denote the average amounts of information in bits per event expected from source Ecks and Wye respectively. The overlap of the two circles represents the common information due to the correlation of Ecks and Wye and its average amount in bits per event is symbolized by T. The left half of the left circle is information from Ecks alone, and the right half of the right circle is information from Wye alone. The symbols $H_y(x)$ should be taken to mean the average amount of information per event that remains to be gotten from source Ecks after Wye is already known. The total area enclosed in both circles together represents all the information that both Ecks and Wye can provide. This total amount in bits per event is symbolized by $H(x,y)$.

$H(x)$ is calculated from the probabilities for Ecks' choices according to the equation given above. The same equation is used to calculate $H(y)$ from the probabilities for Wye's choices. And the same equation is used a third time to calculate $H(x,y)$ from the joint probabilities of the double-choices by Ecks and Wye together. Then all the other quantities involved can be calculated by simple arithmetic in just the way Figure 1 would suggest. For example:

$$H_y(x) = H(x,y) - H(y)$$

or

$$T = H(x) + H(y) - H(x,y).$$

It will be seen that T has the properties of a measure of the correlation (contingency, dependence) between Ecks and Wye. In fact, $1.3863\ nT$ (where n is the number of occurrences of the event that you use to estimate the probabilities involved) is essentially the same as the value of chi square you would compute to test the null hypothesis that Ecks and Wye are independent.

These are the basic ideas behind the general theory. There are many ways to adapt them to specific situations depending on the way the elements of the specific situation are identified with the several variables of the theory. In general, however, most applications of the theory seem to fall into one or the other of two types. I shall refer to these two as the *transmission situation* and the *sequential situation*.

THE TRANSMISSION SITUATION

When information is communicated from one place to another, it is necessary to have a channel over which it can travel. If you put a message in at one end of the channel, another message comes out the other end. So the communication engineer talks about the "input" to the channel and the "output" from the channel. For a good channel, the input and the output are closely related but usually not identical. The input is changed, more or less, in the process of transmission. If the changes are random, the communication engineer talks about "noise" in the channel. Thus the output depends upon both the input and the noise.

Now we want to identify the variables in this transmission situation with the various quantities of information pictured in Figure 1. In order to do this, we let x be the source that generates the input information and let y be the source that generates the output information. That is to say, y is the channel itself. Since x and y are related sources of information, the overlap or common information is what is transmitted. $H(x)$ is the average amount of input information, $H(y)$ is the average amount of output information, and T is the average amount of transmitted in-

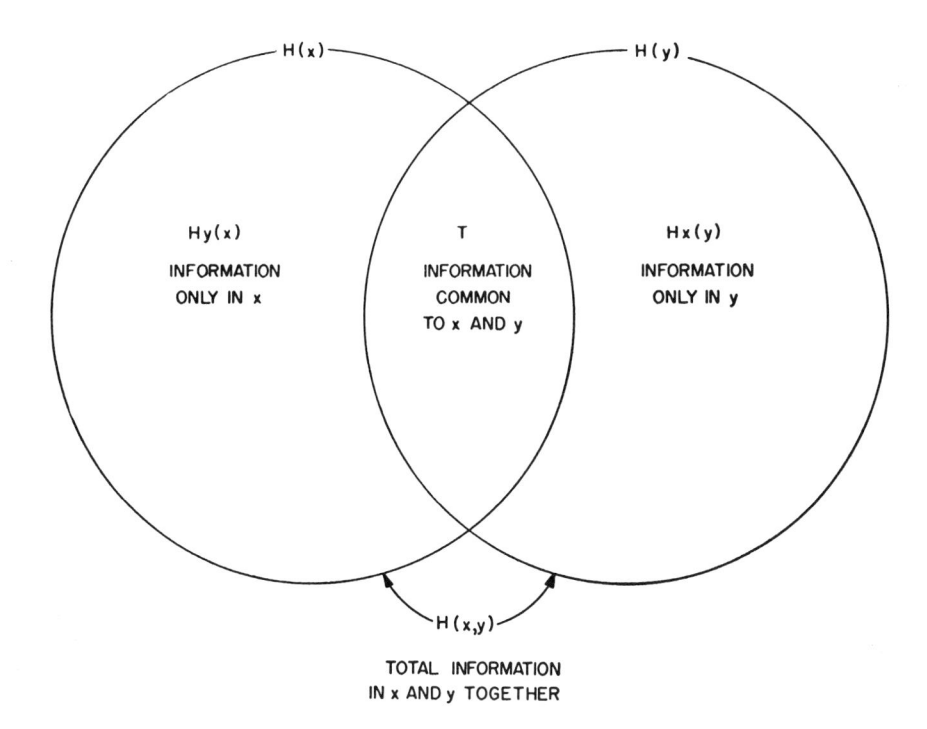

Figure 1. Schematic representation of the several quantities of information that are involved when messages are received from two related sources.

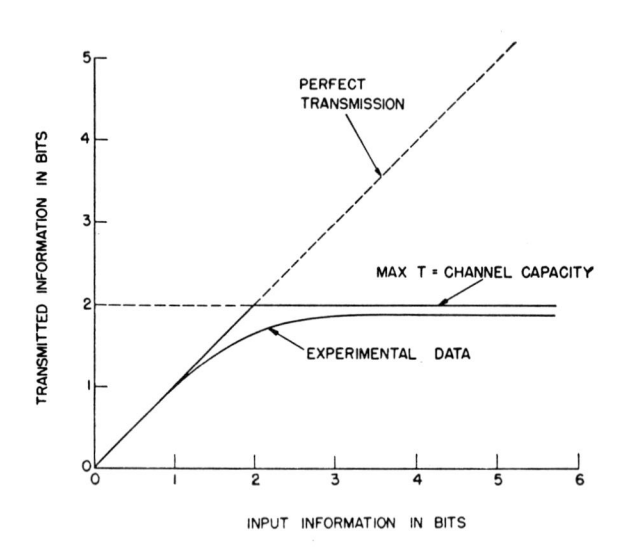

Figure 2. Illustrative graph showing the amount of transmitted information as a function of the amount of input information for a system with a channel capacity of 2 bits.

formation. (To keep terms uniform, we might refer to T as the average amount of "throughput" information.)

What interpretation can we give to $H_y(x)$ and $H_x(y)$? $H_y(x)$ is information that is put in but not gotten out—it is information *lost* in transmission. $H_y(x)$ is often called "equivocation" because a receiver cannot decide whether or not it was sent. Similarly, $H_x(y)$ is information that comes out without being put in—it is information *added* in transmission. $H_x(y)$ is called "noise" with the idea that the irrelevant parts of the output interfere with good communications.

Finally, $H(x,y)$ is the total amount of information you have when you know both the input and the output. Thus $H(x,y)$ includes the lost, the transmitted, and the added information,

$$H(x,y) = H_y(x) + T + H_x(y),$$

equivocation plus transmission plus noise.

This interpretation of the basic concepts of information theory is ordinarily used with the object of computing T, the amount of information transmitted by the channel. A characteristic of most communication channels is that there is an upper limit to the amount of information they can transmit. This upper limit is called the "channel capacity" and is symbolized by C. As the amount of information in the input is increased, there comes a point at which the amount of transmitted information no longer increases. Thus as $H(x)$ increases, T approaches an upper limit, C. This situation is shown graphically in Figure 2, where T is plotted as a function of $H(x)$.

The obvious psychological analogy to the transmission situation is between the subject in an experiment and a communication channel, between stimuli and inputs, and between responses and outputs. Then $H(x)$ is the stimulus information, $H(y)$ is the response information, and T measures the degree of dependence of responses upon stimuli. It turns out that T can be considered as a measure of discrimination, and C is the basic capacity of the subject to discriminate among the given stimuli. That is to say, C can be interpreted as a sort of modern version of the traditional Weber-fraction.

In order to explain how T and C measure the discriminative abilities of the subject, a simple example is useful. Imagine a subject can discriminate perfectly among four classes of stimuli. Any two stimuli in the same class are indistinguishable to him, but two stimuli from different classes are never confused. If we pick the stimuli carefully from different classes, therefore, he can distinguish perfectly which one of two, of three, or of four alternative stimuli we present. However, there is no way we can pick five or more stimuli so that he can discriminate them without any mistakes; at least two must be from the same class and so will be confused. If we select k stimuli to test him with, the best he can do is to reduce k to $k/4$ by saying which of his four classes each stimulus belongs in. He can never reduce the range of possible inputs to less than $k/4$. Thus his channel capacity, C, is $\log_2 4$, or 2 bits, and this is the maximum value of T we can get from him. Since 2^C is the maximum number of discriminably different classes of stimuli for this subject, C is a measure of his basic discriminative capacity.

Another psychological analogy to the transmission situation arises in mental testing. A test is a device for discriminating among people, with respect to some psychological dimension. Each person who takes the test has some true value on this dimension. The result of the test is a score that will, with more or less accuracy, tell us what this value is. So we can think of the test as a communication channel. The true values are the input information, and the test scores are the output information. If it is a good test, if T is large and the noise, $H_x(y)$, is small, then the test may discriminate rather accurately among the people who take it. In other words, 2^T would tell us how many classes of people we can distinguish by using this test.

It requires only a slight extension of this analogy to see the similarity between any process of measurement and the transmission situation. Nature provides the input, the process of measurement is the channel, and the measurements themselves are the output. In this context, the information of the communication engineer is quite similar to the information that R. A. Fisher defined many years ago and used as the foundation for his development of a theory of statistical inference. Considered in this sense, the possible applications of information theory are as broad as scientific measurement itself.

For the psychologist interested in the construction of scales of measurement, information theory will be a valuable tool. He will find that most of the things it tells him he could have learned just as well by more traditional statistical procedures, but the analogy to the transmission situation will undoubtedly stimulate insights and suggest new approaches to old problems.

THE SEQUENTIAL SITUATION

In all that has been said so far it has been implicitly assumed that successive occurrences of the event are independent. When we are dealing with behavioral processes, this assumption is never better than a first approximation. What we are going to do is conditioned by what we have just done, whether we are carrying out the day's work, writing a letter, or producing a random sequence of digits.

Although any behavioral sequence can be analyzed to discover its conditional probabilities, the most interesting example is our own verbal behavior. To take an obvious case, imagine that you are typing a letter and that you have just typed, "I hope we will see you again very." You need at least one more word to complete the sentence. You cannot open the dictionary at random to get this next word. The whole context of the sentence constrains your freedom of choice. The next word depends on the preceding words. Your most probable choice is "soon," although you might choose "often" or "much." You will certainly not choose "bluejay," or "the," or "take," etc. The effect of these constraints built into normal English usage is to reduce the number of alternatives from which successive words are chosen. We have already seen that when the number of possible outcomes of a choice is reduced, some information has been communicated. That is to say, by reducing the range of choice, the context gives us information about what the next item is going to be. Thus when the next word occurs, some of the information it conveys is identical with information we have already received from the context. This repeated information is called "redundancy."

How can the variables in this sequential situation be identified with the various quantities of information pictured in Figure 1? In order to relate them we let x

be the source that generates the context and let y be the source that generates the next word. Since x and y are related sources of information, the overlap or common information from x and y is the redundancy. $H(x)$ is the average amount of information in the first $n-1$ words (the context), $H(y)$ is the average amount of information in the nth word, and T is the average amount of redundant information. $H_y(x)$ is the average amount of information in the context that is unrelated to the next word. $H_x(y)$ is the average amount of information in the next word that cannot be obtained from the context. $H(x,y)$ is the total amount of information we have when all n words, the context plus the next word, are known.

When this interpretation of the basic concepts is used, the quantity of major interest is ordinarily $H_x(y)$, the average amount of information per word when the context is known. $H_x(y)$ can be thought of as the additional information we can expect from each new word in the sequence. Thus $H_x(y)$ is closely related to the *rate* at which information is generated by the source; it measures the average number of bits per unit (per word).

If the successive units in a sequence are chosen independently, then the redundancy T is zero and the context tells us nothing about the next unit. If the next unit is completely determined by the context—for example, in English a "q" is always followed by "u"—then the new information $H_x(y)$ is zero and the occurrence of the next unit adds nothing to what we already know.

Sequences of letters in written English have been studied with this model. It has been estimated that a context of 100 letters will, on the average, reduce the effective number of choices for the next letter to less than three possibilities. That is to say, $H_x(y)$ is about 1.4 bits per letter in standard English. We can compare this result with what would happen if successive letters were chosen independently; then each letter would be chosen from 26 alternatives and would carry $\log_2 26$, or about 4.7 bits of information. In other words, we encode about one-fourth as much information per letter as we might if we used our alphabet more efficiently. Our books seem to be about four times as long as necessary.

It is reasonable to ask why we are so redundant. The answer lies in the fact that redundancy is an insurance against mistakes. The only way to catch an error is to repeat. Redundant information is an automatic mistake-catcher built into all natural languages. Of course, if there is no chance of error, then there is no need for redundancy. The large amount of redundancy that we seem to insist on reflects our basic inefficiency as information-handling systems. Compared with the thousands or millions of bits per second that electronic devices can handle, man's performance figures (always less than 50 bits per second and usually much lower if memory is involved) can charitably be called puny. By making our languages redundant we are able to decrease the rate, $H_x(y)$, to a point where we can cope with what is being said.

Knowledge of the redundancy of English is knowledge about our verbal habits. Since so much of man's behavior is conditioned by these verbal habits, any way to measure them should interest a psychologist. For example, a verbal learning experiment might compare the memorization of ten consonant-vowel-consonant nonsense syllables (30 letters in all) with the memorization of a 30-letter sentence from English text. Since the successive letters in the nonsense syllables are effectively independent, the learner faces many more possible sequences than he does if he knows that the 30 letters are English text. Since he has already learned the redundancies of English, he is required to assimilate less new information from the sentences than from the nonsense syllables. A knowledge of the information in sequences of letters in English text thus gives us an independent, quantitative estimate of previous learning. In short, the sequential application of information concepts enables us to calibrate our verbal learning materials and so to control in a quantitative way factors that we have always discussed before in qualitative terms.

It is not necessary to confine the sequential interpretation to verbal behavior. It can be applied whenever an organism adopts a reasonably stable "course of action" that can be described probabilistically. If the course of action is coherent in such a way that future conduct depends upon past conduct, we say the behavior is predictable or, to some degree, stereotyped. In such cases, the redundancy T can be used to measure the stereotypy. Arguments about the degree of organization in emotional behavior, for example, might be clarified by such a measure.

Taken together, the sequential and the transmission situations suggest a wide range of possible applications in psychology. The idea of reviewing some of the applications already made by psychologists is tempting, but space prevents it here. The reader who wants to follow up these ideas in more concrete terms should find the annotated references given below a good starting point.

SELECTED REFERENCES

The following is not a complete bibliography, but should include most of the papers of direct interest to psychologists.

Aborn, M., & Rubenstein, H. Information theory and immediate recall. *J. exp. Psychol.*, 1952, **44**, 260-266. Artificial languages with varying degrees of contextual constraint were constructed and passages from these languages were memorized. For the less organized languages, the amount of information remembered was a constant.

Cherry, E. C. A history of the theory of information. *Proc. Inst. elect. Engineers*, 1951, **98**, 383-393. This scholarly and interesting paper puts the theory in historical context and contains a valuable bibliography.

Dolansky, L., & Dolansky, M. P. *Table of $\log_2 1/p$, $p \log_2 1/p$, and $p \log_2 1/p + (1-p) \log_2 1/(1-p)$.* Cambridge: Technical Report No. 227, Research Laboratory of Electronics, Massachusetts Institute of Technology, 1952. This table is quite useful in any computation of amounts of information.

Fano, R. M. Information theory point of view in speech communication. *J. acous. Soc. Amer.*, 1950, **22**, 691-696. This paper presents a nonmathematical discussion of the theory and uses it to estimate the rate of transmission of information in speech communication.

Fano, R. M. *The transmission of information.* Cambridge: Technical Reports No. 65 and No. 149, Research Laboratory of Electronics, Massachusetts Institute of Technology, 1949 and 1950. The basic theorems are developed in a different way than Shannon used, and more attention is given to the problem of finding the optimal coding scheme to reduce equivocation.

Frick, F. C., & Miller, G. A. A statistical description of operant conditioning. *Amer. J. Psychol.*, 1951, **64**, 20-36. The sequential interpretation is used to analyze the behavior of rats in a Skinner box during preconditioning, conditioning, and extinction. Here T is used as a measure of stereotypy.

Garner, W. R., & Hake, H. W. The amount of information in absolute judgments. *Psy-*

chol. Rev., 1951, **58**, 446-459. This paper contains an excellent description of the method of calculation of T in the transmission form of the theory, and points out the similarity between T and the contingency coefficient.

Hake, H. W., & Garner, W. R. The effect of presenting various numbers of discrete steps on scale reading accuracy. *J. exp. Psychol.*, 1951, **42**, 358-366. Subjects who estimated the position of a pointer along a linear interval had the same channel capacity C for discriminating positions even though they were given different instructions and made different kinds of errors.

Hartley, R. V. The transmission of information. *Bell Syst. tech. J.*, 1928, **17**, 535-550. Mostly of historical interest today, this paper contains most of the basic ideas so elegantly developed by Shannon twenty years later.

Hick, W. E. Information theory and intelligence tests. *Brit. J. Psychol.*, 1951, **4**, 157-164. Answering test questions is broadly interpreted as a communication problem, and ways of designing tests to maximize the transmission are considered theoretically.

Hick, W. E. On the rate of gain of information. *Quart. J. exper. Psychol.*, 1952, **4**, 11-26. Information measurement is applied to a choice-reaction-time experiment. The amount of information in the choice divided by the reaction time gives a constant—about five bits per second.

Hick, W. E. Why the human operator? *Trans. Soc. Instrument Technology*, 1952, **4**, 67-77. The relative efficiency of men and machines is discussed in terms of their capacities to handle information. Several kinds of evidence are reviewed and an upper limit for men of about 15 bits per second (probably too low) is estimated.

Jacobson, H. Information and the human ear. *J. acous. Soc. Amer.*, 1951, **23**, 464-471. The author calculates that 10,000 bits per second is the maximum capacity of the ear, but argues that the brain can utilize less than one per cent of this information.

Laemmel, A. E. *General theory of communication.* Report R–208-49. Brooklyn: Microwave Research Institute, Polytechnic Institute of Brooklyn, 1949. Reviews the general theory and extends it for engineering applications.

MacKay, D. M. Quantal aspects of scientific information. *Phil. Mag.*, 1950, **41**, 289-311. An attempt is made to distinguish "structural" from "quantitative" information and to relate the work of Fisher, Gabor, and Shannon.

Marks, M. R., & Jack, O. Verbal context and memory span for meaningful material. *Amer. J. Psychol.*, 1952, **65**, 298-300. A repetition of the Miller-Selfridge experiment using the span of immediate memory failed to confirm the results of the previous study. Memory span measurements generally fail to show any constancy in the amount of information retained.

Miller, G. A. *Language and communication.* New York: McGraw-Hill, 1951. In this book an attempt is made to use information theory to integrate and interconnect a review of psycho-linguistics.

Miller, G. A. Language engineering. *J. acous. Soc. Amer.*, 1950, **22**, 720-725. Suggestions are made for the use of information theory in the design of special languages for special uses, e.g. an international language for aviation.

Miller, G. A. Speech and Language. In S. S. Stevens (Ed.), *Handbook of experimental psychology.* New York: Wiley, 1951. The first half of this chapter relates some of the standard studies of verbal behavior to the sequential form of information theory.

Miller, G. A., & Frick, F. C. Statistical behavioristics and sequences of responses. *Psychol. Rev.*, 1949, **56**, 311-324. The sequential form of the theory is developed in an elementary manner and used to define an index of behavioral stereotypy.

Miller, G. A., Heise, G. A., & Lichten, W. The intelligibility of speech as a function of the context of the test materials. *J. exp. Psychol.*, 1951, **41**, 329-335. The transmission form of the theory was used to show the effect of increasing the input information upon the per cent of spoken words correctly perceived.

Miller, G. A., & Selfridge, J. A. Verbal context and the recall of meaningful material. *Amer. J. Psychol.*, 1950, **63**, 176-185. The sequential interpretation of the theory was used to construct learning materials with varying amounts of contextual constraint. Thus the passages differed in average amount of information per word. To a rough approximation, the amount of information recalled was constant for all passages.

Newman, E. B. Computational methods useful in analyzing series of binary data. *Amer.*

J. Psychol., 1951, **64**, 252-262. In addition to a description of an interesting tabulating device for use with binary data, this paper contains a table of $-p \log_2 p$ that is quite useful to computers.

Newman, E. B. The pattern of vowels and consonants in various languages. *Amer. J. Psychol.*, 1951, **64**, 369-379. The measure of redundancy is used to compare the predictability of vowel-consonant sequences in eleven written languages.

Newman, E. B., & Gerstman, L. S. A new method for analyzing printed English. *J. exp. Psychol.*, 1952, **44**, 114-125. A coefficient of constraint is defined and used to obtain results that can be compared to Shannon's estimates of the redundancy in printed English.

Parke, N. G., & Samson, E. W. *Distance and equivalence in sequence space.* Report E4080. Cambridge: Air Force Cambridge Research Laboratories, 1951. Develops the notion suggested by Shannon that the information from two sources x and y be called equivalent if $H_x(y)$ and $H_y(x)$ are zero and that the "distance" between any two nonequivalent sources be defined as $H_x(y) + H_y(x)$.

Ruesch, J., & Bateson, G. *Communication, the social matrix of psychiatry.* New York: Norton, 1951. This book stresses the importance of communication in psychiatric problems, makes some use of information theory and suggests many more.

Samson, E. W. *Fundamental natural concepts of information theory.* Report E5079. Cambridge: Air Force Cambridge Research Station, 1951. The intuitive nature of the concepts of expectation, surprise, and uncertainty is stressed.

Senders, V. L., & Sowards, A. Analysis of response sequences in the setting of a psychophysical experiment. *Amer. J. Psychol.*, 1952, **65**, 358-374. The sequential model is used to analyze response dependencies in a psychophysical experiment.

Shannon, C. E. Communication in the presence of noise. *Proc. Inst. Radio Engineers*, 1949, **37**, 10-21. A geometrical representation is developed for the communication of continuous functions disturbed by noise interference.

Shannon, C. E. A mathematical theory of communication. *Bell Syst. tech. J.*, 1948, **27**, 379-423, 623-656. These two articles comprise the first systematic presentation of the concepts of information measurement. Although the mathematics are difficult for most psychologists, this is still the basic reference in this field. Reprinted in 1949 with an article by Weaver by the University of Illinois Press.

Shannon, C. E. Prediction and entropy of printed English. *Bell Syst. tech. J.*, 1951, **30**, 50-64. A clever device is developed for estimating upper and lower bounds for the redundancy of English and used to estimate the relative efficiency of English text at about 25 per cent.

Weaver, W. Recent contributions to the mathematical theory of communication. In Shannon, C. E., and Weaver, W., *The mathematical theory of communication.* Urbana: Univer. of Illinois Press, 1949. This nonmathematical exposition is an expansion of an article in *Scientific American*, July, 1949, and is intended as an introductory orientation that can be read before one tackles the more mathematical aspects of the theory. Weaver suggests ways to generalize the theory to the broader problems of social communication.

Wiener, N. *Cybernetics.* New York: Wiley, 1948. One chapter of this pioneering volume expounds the author's discoveries in information theory. For most psychologists, this volume is more stimulating than intelligible.

Wiener, N. *The human use of human beings.* Boston: Houghton Mifflin, 1950. In a less mathematical form, the social consequences of the ideas in *Cybernetics* are emphasized and explained in highly readable prose.

Woodbury, M. A. On the standard length of a test. *Psychometrika*, 1951, **16**, 103-106. A parameter called "the standard length" is defined in such a way that a test of standard length gives one unit of information. This paper suggests one way of employing information theory in the analysis of mental tests.

14

On the Rate of Gain of Information

W. E. Hick

I. INTRODUCTION

The work described in this paper was suggested by the observation that the values of choice-reaction times obtained by Merkel (1885), when plotted against the number of alternative stimuli, appeared to lie very close to a smooth uninflected curve. Merkel himself was chiefly interested in the supposed diversibility of reaction time into "cognition time" and "choice time," and does not even give the raw data. However, they are tabulated by Woodworth (1938). Other psychologists of what may be called the "reaction-time era" discussed the increase in reaction time with number of alternatives, attributing it to such causes as the division of attention or a reduction in the effective intensity of the stimulus; but no quantitative theory seems to have emerged. Indeed, as far as the writer is aware, the only reference to a mathematical relation between reaction time and number of alternatives comes later, when Blank (1934) mentions a logarithmic relation, without suggesting any explanation.

As Merkel's data provide important supplementary evidence for the theory put forward here, his method will be briefly described. (The original paper is not very accessible, and the writer is indebted to Mr. A. Leonard for obtaining it and translating the relevant parts.) The display was provided by a kind of tachistoscope in which the numbers 1–5 (Arabic) and I–V (Roman) were printed on a disc. The subject waited with his fingers on ten keys, and, on the illumination of one of these numerals, he released the appropriate key. An interesting piece of experimental technique is the use of a Geissler tube in order to ensure the sudden onset of the illumination. The Geissler tube was an early form of gas-discharge tube, the forerunner of the modern fluorescent lighting. The illumination would be rather weak, but doubtless quite adequate if the subject was moderately dark-adapted. Some of the other archaisms, however, are less pleasing. For example, there is no indication of the order in which the stimuli were given—i.e. whether it was according to the experimenter's whim or determined by some system. The very large practice improvement, even with ten alternatives, suggests that the sequence was easily

From *Quarterly Journal of Experimental Psychology*, 4, 1952, 11-26. By permission of the Experimental Psychology Society.

learnt, since the present writer, using an irregular sequence, found very little improvement with practice. But according to Merkel's point of view, the predictability of the next stimulus in the sequence was very likely irrelevant. One might also criticize his presentation of the different degrees of choice, from one to ten, in ascending order only, although we possibly have that to thank for the remarkable consistency of the results. Figure 1 shows the reaction times he obtained; each point is the mean of about fifty readings from each of nine subjects.

Consideration of these results led the writer to formulate and test a hypothesis of a new kind, and one which would have been impossible in the early days of reaction-time work, because the theoretical framework did not then exist. To put it succinctly, the hypothesis is that the rate of gain of information is, on the average, constant with respect to time, at least within the duration of a single perception. Further qualification may prove to be necessary, but the evidence presented here shows that the hypothesis is true for the conditions employed.

II. THEORETICAL BACKGROUND

More detailed discussion and interpretation will be given later, but some points must be explained here in order to make the experimental approach and results comprehensible.

First of all, the definition of "quantity of information" is that which was originated by communication engineers and has been greatly developed in recent years by Shannon (1949) and Wiener (1948). Briefly, the amount of information given by an event whose probability is p is $-\log p$. It may seem strange that information should depend solely on a probability, but in fact if we introduced any other attribute of the physical—or, for that matter, the psychological—world, the definition would no longer be of information *in general*, but of some particular kind of information. Moreover, it must depend on probability because that is a measure of prior expectation, and information should be that which changes expectation into certainty. By using the negative logarithm, it is ensured that information is always positive (p being a fraction), that it is zero for an event which is absolutely certain to happen and infinite for one which is certain not to happen, and that contributions from independent sources can be added (the probabilities being multiplied together to give the joint probability).

We shall, however, be dealing, not with particular pieces of information, but with average or expected information. We shall average over events of the same kind in order to estimate the probabilities required, and we shall average $-\log p$ over all relevant alternatives to find the expected information. Where the probabilities of the possible alternatives are $p_1, p_2, \ldots p_n$, the expected information (entropy) is the sum of the contributions from each, multiplied by their chances of occurring; thus

$$H = -\sum_{1}^{n} p_i \log p_i$$

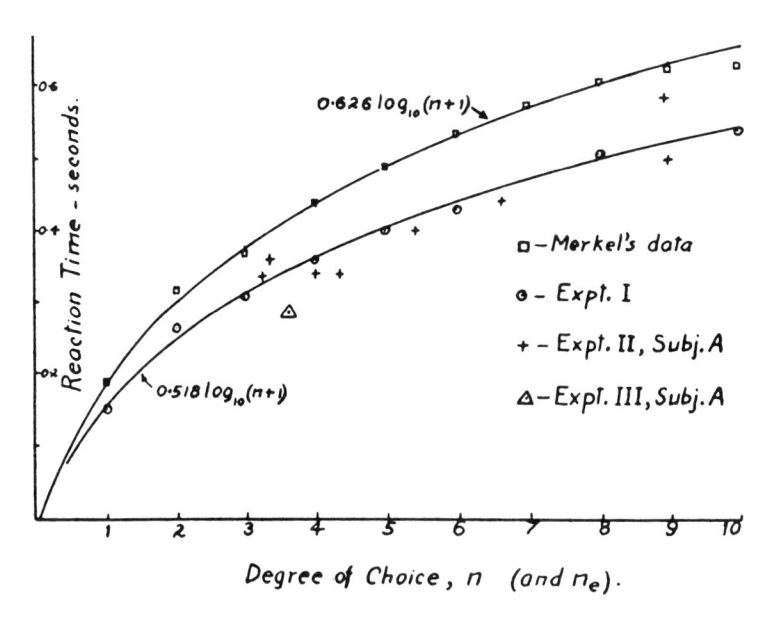

Figure 1. Relations between reaction times and numbers of different stimuli. Some results from fast reactions with errors are also plotted to the same scale, on the basis of the calculated equivalent number of stimuli.

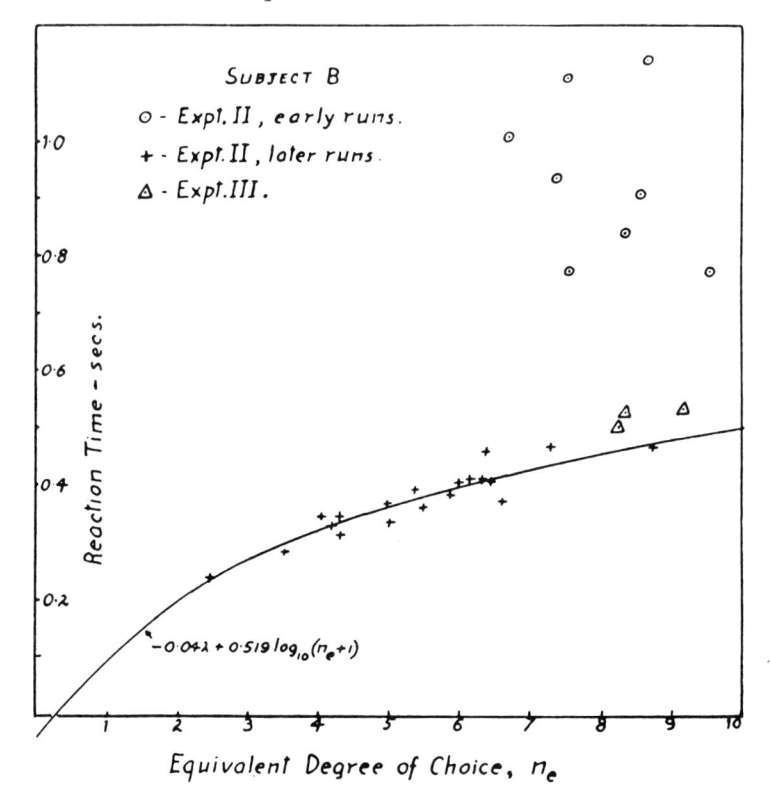

Figure 2. Further results from fast reactions with errors.

TABLE I

Pooled Reaction Times (seconds) and Response Frequencies for Second Subject in Experiment II

Response Categories 11 and 14 accommodate the few cases where more than one key was pressed

Resp. (j)	\multicolumn{10}{c	}{Stimulus (i)}	Totals and Mean RT's
	1	2	3	4	5	6	7	8	9	10	
1	165 / 0·36	2 / 0·35	4 / 0·31	1 / 0·30	0 / –	1 / 0·17	0 / –	0 / –	0 / –	1 / 0·37	174 / 0·36
2	16 / 0·30	167 / 0·35	27 / 0·36	2 / 0·28	0 / –	2 / 0·18	1 / 0·13	0 / –	2 / 0·25	1 / 0·23	218 / 0·34
3	2 / 0·23	24 / 0·32	145 / 0·47	32 / 0·44	5 / 0·32	1 / 0·27	0 / –	1 / 0·20	0 / –	1 / 0·33	211 / 0·44
4	0 / –	1 / 0·27	12 / 0·44	144 / 0·45	38 / 0·45	2 / 0·42	0 / –	0 / –	1 / 0·30	2 / 0·37	200 / 0·45
5	1 / 0·43	0 / –	2 / 0·67	29 / 0·40	139 / 0·45	32 / 0·39	1 / 0·30	1 / 0·43	0 / –	1 / 0·33	206 / 0·43
6	0 / –	0 / –	1 / 0·30	0 / –	6 / 0·39	133 / 0·46	38 / 0·46	0 / –	1 / 0·33	0 / –	179 / 0·46
7	1 / 0·33	0 / –	0 / –	2 / 0·42	6 / 0·37	21 / 0·40	126 / 0·51	48 / 0·45	2 / 0·40	1 / 0·40	217 / 0·48
8	0 / –	0 / –	1 / 0·23	0 / –	0 / –	0 / –	12 / 0·46	95 / 0·53	24 / 0·47	3 / 0·38	135 / 0·51
9	1 / 0·27	0 / –	0 / –	0 / –	0 / –	0 / –	3 / 0·44	40 / 0·43	148 / 0·48	33 / 0·41	225 / 0·46
10	1 / 0·40	0 / –	0 / –	0 / –	0 / –	0 / –	1 / 0·27	2 / 0·35	12 / 0·40	143 / 0·47	159 / 0·46
11	0 / –	0 / –	0 / –	0 / –	0 / –	0 / –	0 / –	0 / –	1 / 0·47	4 / 0·40	5 / 0·41
14	0 / –	0 / –	0 / –	0 / –	0 / –	0 / –	0 / –	1 / 0·33	0 / –	0 / –	1 / 0·33
Totals Mean RT's	187 / 0·35	194 / 0·35	192 / 0·45	210 / 0·44	194 / 0·44	192 / 0·44	192 / 0·49	188 / 0·48	191 / 0·47	190 / 0·45	1930 / 0·436

This is essentially the same entropy as appears in statistical mechanics; it is a measure of our uncertainty as to *what* will happen (as distinct from our doubt as to whether a particular x will happen, which is given by $-\log p_x$, as above). Following Shannon (loc. cit.) we write $H(x)$ for the entropy of the distribution of a variate x.

Shannon shows that if we have a source of messages, signals, or stimuli—whatever we choose to call them—such that their entropy is $H(x)$, and a destination at which signals having the entropy $H(y)$ arrive, the average information actually transmitted is

$$R = H(x) - H_y(x).$$

$H_y(x)$ is the conditional entropy of x when y is known; Shannon calls it the "equivocation." It measures our remaining uncertainty about x even when we know y, on the assumption that the signals are subject to some degree of mutilation in transit by a statistically-definable disturbing influence. If there is no such interference, $H_y(x)$ is, of course, zero, and the information transmitted is the total generated by the source, namely $H(x)$.

In a choice-reaction-time experiment, we have a display which is capable of generating any one of a set of n alternative signals. If they are generated in completely random order, the probability of any particular signal is $1/n$ and

$$H(x) = -n \cdot \frac{1}{n} \log \frac{1}{n} = \log n$$

If the subject makes no mistakes—i.e. if the "equivocation" is zero—his response entropy $H(y)$ is also $\log n$. He is then extracting all the relevant information from the display, because $R = H(x)$. (The fact that $H(y)$ is $\log n$ does not *alone* indicate that any information is being extracted, because the responses might be entirely random.)

Now the hypothesis that the rate of gain of information is constant apparently requires that the reaction time should be proportional to $\log n$. But when $n = 1$ (simple reaction) $\log n = 0$, indicating zero reaction time for this case. Evidently $\log n$ does not include the whole of the display entropy, and obviously the missing portion is due to the possibility of "no stimulus" at any instant during the waiting period. In other words, $\log n$ only measures the uncertainty as to what the stimulus will be, and in the simple reaction there is no uncertainty on this point. But there *is* doubt as to when it will occur; and when it does occur, it must be distinguished from the mass of other, irrelevant, information continually pouring in, so as to be recognized as that which was awaited.

At this stage, the writer made a guess that the possibility of "no stimulus" was treated by the subject as if it had the same probability as any particular stimulus. As it seems to have been a successful guess, as far as the present experimental work can show, discussion of its theoretical significance will be postponed. The result is to make the information gained, assuming no mistakes, equal to $\log(n + 1)$. It can be seen from Figure 1 that the equation

$$\text{RT (seconds)} = 0.626 \log_{10}(n + 1)$$

does fit Merkel's data very well. (As a matter of interest, the function A + B log n was tried by Miss V. R. Cane, but proved to give a slightly worse fit; not so much worse, however, as to put it out of court altogether, if some reason for preferring it should arise.) The slight discrepancy when nine and ten stimuli were used may be due to a higher frequency of mistakes, but Merkel gives no data on this. He excluded wrong responses from his results, but that does not rule out this explanation. It is worth noting that Kraepelin (1894) remarks significantly that the more the choice reaction approaches in character the simple reaction, the more errors tend to occur.

However, it seemed that a prima facie case had now been established for a further extension of this approach to the problem of choice and time.

III. APPARATUS

As the apparatus is rather complicated, only a general description will be given here. The device for delivering a predetermined irregular sequence of stimuli has been described previously (Hick, 1951). It operates on the punched-tape principle and is driven at a constant rate of five seconds per stimulus. Electrical signals in binary code pass to the main part of the apparatus, where they can either activate display elements, such as lamps, in the same code, or be first decoded so that only one out of fifteen elements operates at a time. At the same time, four pens record the stimulus on moving paper, also in the same code.

For the present experiments, ten elements of the decoded display were used. They took the form of ten pea lamps arranged in a somewhat irregular circle. The objects of this arrangement were (a) to place them sufficiently close together to obviate the need for eye movements, yet not so crowded as to form a confusing pattern, and (b) to avoid any very obvious grouping. It is doubtful whether the latter has any relevance at all, since the subject is bound to invent a system of grouping if it is not given to him, but it might have been argued of an externally-imposed system that it really determined the results. The lamps were supplied through a resistance-capacitance network designed to make them light up almost instantaneously. The time between the filament becoming visibly incandescent and the corresponding pens recording was carefully measured by means of a shutter device.

The response was to press the correct one of ten Morse keys on which the subject's fingers rested. But the apparatus can be used with any other form of response which has the effect of selecting one contact out of fifteen (or one combination of not more than four contacts). Thus either the stimulus or the response or both can be given as a pattern or "in clear." The facility for pattern representation has not yet been used.

Since the same pens record the response in the same binary code, we have a fairly complete picture of the events in a run. The chief purposes of using a coding system were to avoid having to keep ten or fifteen pens in working order and to simplify the form of the punched-tape device. There is, however, a disadvantage in that if the subject presses several keys at once, as he is naturally tempted to do, it is not always possible to tell which keys they were. Certain safeguards were incorporated, and these, in conjunction with suitable training, make it safe to say that not more than 0.5 per cent. of the recorded responses can have been wrongly interpreted.

The stimulus sequences are from 100 to 200 stimuli in length, and combine near-equality of frequencies with elimination of the first-order auto-correlation. The sequences were checked by inspection for obvious regularities. Two such sequences were made for each degree of choice from 2 to 10 inclusive, but omitting 7 and 9. Since it was impracticable to make a large number of different sequences, the use of a table of random numbers would not have been a safe expedient.

IV. EXPERIMENT I

The first experiment was carried out mainly to confirm the fitness of the function $\log(n+1)$. It served that purpose, and as the results are involved in later arguments, it will be briefly described.

The experimenter acted as subject, in order to gain an idea of the amount of practice likely to be needed for later variations. The rule adopted was to achieve one errorless run before doing each test run; it seemed necessary to have some such incentive in this excessively tedious task. The total amount of practice given is not exactly known, but cannot be less than 8,000 reactions, since the recorded reactions in this experiment total over 2,400. However, much of this practice was needed to make up for long periods away from the task and minor changes in the display.

The degrees of choice were taken first in ascending order from two to ten, and then in descending order, followed by an irregular order. Provided practice at the same degree of choice was given before a test run, there appeared to be no appreciable carry-over of "set" from the previous degree of choice, as far as could be seen from the reaction times. In other words, it is not enough to know that one is going to do a 5-choice run, say; one must have just done perhaps two or three 5-choice runs, if the effect of having previously done, let us say, a 10-choice run is to be virtually abolished.

Since only two fingers (the left little and ring fingers) were used in the 2-choice task, the data for these two fingers were extracted from the results obtained at other degrees of choice. These are the mean reaction times plotted against the degree of choice (n) in Figure 1. In order to obtain a comparable value for the simple reaction time, the 2-choice punched tape was used, but only one of the stimuli was responded to, the other being ignored. Some such method had to be used with this apparatus, on account of the fixed interval between stimuli.

The curve represents $0.518 \log(n+1)$. It is worth noting that the origin was *not* one of the points to which it was fitted; the fact that it does so—i.e. the fact that the additive constant necessary to give the best fit is negligibly small (less than one millisecond) is more of a coincidence than a sign of precision. The reaction-time measurements themselves may well have a constant error of several milliseconds. Moreover, no allowance has been made for the few milliseconds occupied by peripheral conduction.

Incorrect reactions were omitted from the calculations; as they only amounted to about four per cent., the omission is not thought to be important. However, errors are taken strict account of in the subsequent experiments.

It may be added that the function $A + B \log n$ was again tried, and again gave a slightly worse fit. The difference is too small to mean anything by itself, but as it is the second case in which $\log (n + 1)$ gives the better representation, and as a third will be cited below, it is worth noting.

V. EXPERIMENT II

It was suggested by Mr. J. D. North that if the proposed law has general validity, it ought to apply to the case of partial extraction of information from the stimulus. For example, if the subject can be persuaded to react more quickly, at the cost of a proportion of mistakes, there will be a residual entropy which should vary directly with the reduction in the average reaction time.

Two subjects performed this task. The only special difficulty encountered lay in making enough mistakes to give some points near the origin of the graph, without abandoning altogether the attempt to make the correct responses.

The 10-choice sequences were used, and the results are exhibited as reaction times plotted against the equivalent degree of choice (n_e). For example, if there were no mistakes it would mean that all the information was being extracted, and n_e would be 10. The method of calculation is given in Appendix I; it is enough to say here that n_e is the antilogarithm of the amount of information gained (apart from the component due to the possibility of "no stimulus").

One of the subjects was the experimenter, and his results are shown by the crosses on Figure 1. Each cross represents one run of about 100 reactions, and it will be observed that they are distributed reasonably closely about the theoretical curve (which is also the curve fitted to the data of Experiment I).

The performance of the other subject—a research worker—is shown in Figure 2. This subject (labelled B) was trained practically entirely on the 10-choice sequences, and this may have been partly responsible for the curious separation of the reaction times into two groups. The upper group (circles) represents test runs done while the subject was still learning the code and trying to minimize errors. As there was no sign of improvement, he was then asked to try reacting quickly, with as many errors as he liked. This produced ten of the lower group (crosses). In the hope that he had now acquired the technique, he was again asked to do an accurate run, whereupon he reverted to the middle of the upper group. It was only by going back to the high-speed "set" and then very gradually reducing errors in successive runs that it was possible to extend the readings towards the higher values of n_e. This subject, however, managed to achieve a run containing 70 per cent of errors, without losing control of the situation.

The curve fitted to the lower group represents the function $-0.042 + 0.519 \log (n_e + 1)$. This means, in effect, that extrapolation backwards by the $\log (n + 1)$ formula misses the origin by 42 milliseconds in this case. Since the points are somewhat scattered and none of them is very near the origin, the discrepancy could reasonably be a sampling error. Any curve which purports to represent the information extracted should, of course, pass through the origin, since where there is no choice there is no information. A curve defined by $A + B \log n_e$ was

VI. EXPERIMENT III

Since the same two 10-choice sequences were used a considerable number of times by the same subjects in Experiment II, the question of learning arises. The input entropy $H(x)$ was calculated on the assumption that there was no learning; in other words, that the only thing the subject knew about the sequence presented to him was that the frequencies of the different stimuli were equal.

Of course it was not imagined that the subject would learn the actual frequencies of the stimuli, and at the same time refrain from learning any other properties of the sequence. The supposition was rather the negative one that, provided the sequence had no obvious and striking regularities, it would tend to be treated as if random; that is, at any stage during a test run, the next stimulus would be treated, as far as the recognition process was concerned, as if it had an equal chance of being any one of the ten possibilities.

Neither subject was conscious of any learning of the two sequences, or indeed of any thought that it would be worth attempting, but that is no proof that it did not occur. There were exceptions to this statement in two respects. The first stimulus of both sequences happened to be the same, and no stimulus occurred more than twice in succession. Both subjects were aware of these features. The former resulted in a slightly shorter reaction time to the first stimulus, though still of the order of twice a simple reaction time. The latter should have reduced the reaction times to stimuli immediately following such pairs by at least 0.02 of a second, if the knowledge had been consistently acted upon, since it reduces n by one. In reality, the reduction was found to have the quite negligible value of about 0.0006 of a second. But as this has a standard error of the order of 0.025, we cannot be sure that this piece of knowledge was not being used at all, although full advantage of it was probably not taken.

To find out whether learning of the sequences was introducing a serious error, an entirely new sequence was prepared. It was thought best to retain the same statistical structure with respect to frequencies and first-order serial correlation, although that meant that, again, no stimulus occurred more than twice in succession. One subject (A) did one fast run, and the other (B) did three runs at the other end of the "information scale"—i.e. with few errors. The results are indicated by the triangles in Figures 1 and 2. It will be seen that the reaction times are slightly increased at the higher values of n_c, but that at the low value the reaction time happens to be less than expected.

The only conclusion justified by these results is the very limited one intended —namely that learning of the sequences did not play a large part in determining the previous findings. There is also a suggestion that the effect of learning is more marked, the more information is being extracted. This is, of course, inherently likely; if we consider the extreme case of random response, where the only information gained is to the effect that *some* stimulus has occurred, anything learnt about the sequence is irrelevant.

Merely as a matter of interest, we may consider the estimated corrections to the observed reaction times, to give the times that would be expected in the case of unlearnt sequences having the same statistical structure. Naturally they are not intended to be taken very seriously, in view of the assumptions necessary and the small quantity of data available. Suppose $H'(x) = H(x) + H'$, where $H'(x)$ is the input entropy for the unlearnt type of sequence, and $H(x)$ is the effective input entropy of the present sequences, i.e. $H(x)$ is less than $H'(x)$ by the amount of the learnt information H'. The simplest acceptable hypothesis for H' is that it is proportional to the information gained. In terms of effective degrees of choice, the outcome is that the degree appropriate to random sequences is equal to n_e. The value of k which best fits the unlearnt-sequence data in Figure 2 is about 1.1, indicating that the amount of learnt information utilized was of the order of ten per cent of the total extracted. It may also be remarked that, for this subject and this statistical class of sequence, the rate of gain of information from a stimulus has the average value of 5.6 "bits" per second (one "bit" being the information conveyed by an event whose probability is 0.5).

VII. DIFFERENCES RELATED TO PARTICULAR STIMULI AND RESPONSES

Before discussing more general questions, a particular finding should be mentioned. In Table I will be found the data for the second subject (B) of Experiment II, whose performance appears in Figure 2. The figures for the aberrant early runs have been excluded, as also have those obtained with the unfamiliar sequence. The Table shows the pooled frequency with which each response was evoked by each stimulus, and the lower entry in each cell is the corresponding mean reaction time. The inclusion of response categories numbered 11 and 14 is to accommodate the few cases of more than one key being pressed simultaneously; 11 and 14 are simply the code numbers of the responses as recorded.

Regarding the Table as a matrix, we see that the responses are grouped about the leading diagonal. Had there been no errors, all the responses would, of course, have been on this diagonal. It can also be seen that the reaction times for particular stimuli, averaged over all responses, are far from equal, and the same is true of the times for particular responses, averaged over all stimuli. It is further noticeable that the probability of the correct response is not the same for all stimuli; stimulus No. 8, for example, seems to have been especially difficult. Do these differences reveal, as we might expect them to, a relation between reaction time and some appropriate measure of information?

The basic hypothesis must be, as before, that reaction time is proportional to information extracted. Unfortunately we have no independent measure of the latter, with respect to individual stimulus categories, because it is not now permissible to calculate the stimulus probabilities according to the frequencies with which they were actually given; it is the probability as seen by the subject that determines the input information. All we can do is to estimate the two sets of residual entropies, which may be called $H_i(y)$ and $H_j(x)$. $H_i(y)$ is the remaining uncertainty about the response, given with ith stimulus, and is assessed directly from the frequencies in the ith column and its marginal total. Similarly $H_j(x)$ is the residual uncertainty, given the jth response, about which stimulus evoked it.

These are the two measures of information which we might hope would show some relation to the corresponding reaction times. In fact, the following correlations were obtained:

Between RT_i and $H_i(y)$: $r = 0.798$, $P < 0.01$
" RT_j " " $(i = j)$: $r = 0.947$, $P < 0.001$.

RT_i is the mean reaction time to the ith stimulus and RT_j is that for the jth response. The correlations with $H_j(x)$ were insignificant. The scatter diagrams corresponding to the two significant correlations suggested that two points, which were some distance from the remaining eight, were largely responsible for the correlations. However, even when these two points were neglected, the larger correlation was still significant at better than the 5 per cent level.

It can be said, therefore, that there is almost certainly some relation between the reaction time of a particular response or to a particular stimulus and the corresponding uncertainty (as seen by the outside observer) as to which response will be evoked by the particular stimulus. How the relation comes about cannot be inferred from the present data, which would apparently agree with several plausible hypotheses. The only point it is desired to make is that this is further evidence pointing to the dependence of reaction time upon information, in the technical sense of that term.

VIII. CONCEPTUAL MODELS

With regard to the mechanism responsible for these results, speculation about neural networks is outside the present scope. There is no objection to trying to depict schematically the component operations, but it must be admitted that what analysis of the data has been carried out does little more than draw attention to the difficulties involved in finding any simple scheme.

If we consider the process of recognition or identification operationally, we can liken it to matching a given object to the correct one of m gauges or templets. That is to say, the object or event to be identified must be compared with a standard or set of standards. When the matching standard (or combination of standards) has been found, the object may be said to have been identified, as exactly as is possible with that whole set of standards. There are only four fundamental modes of procedure; any actual mode may be regarded as either one of the four or as having a mixed or intermediate character.

Given that an object can be matched with one, and only one, of m standards provided, the problem is to find that one. One of the four possible methods is to produce m replicas of the object and try them on the standards simultaneously. If we wish to try to picture this replication in neural form, we have only to remember that the nervous system must make a replica of the stimulus in any case, and it only requires a leash of fibres diverging from the original afferents to provide as many replicas as may be needed. The real point, however, is that the time occupied in matching is clearly independent of m. Since Reaction-Time increases with m, according to this scheme it can only be because it takes longer to produce a large number of replicas than a few.

If we examine the process of replication as we know it in other systems, we can distinguish three types. Firstly, there is simultaneous replication, in which any number within the capacity of the system can be produced in the same time as it takes to produce one. This together with inversion of the equation RT = K log (n + 1), leads to a conception of reaction time as a period of continuous accumulation of evidence, which has both advantages and disadvantages over the models discussed here. It is hoped to publish the argument when it has been more fully developed. Secondly, replication may be serial, so that the time taken is proportional to the number produced, in a manner analogous to "manufacture without expansion." This also must be rejected, since reaction time is not linearly related to m. Thirdly, there is self-replication—the "chain-reaction" type of process—in which the number increases in geometrical progression with the passage of time; in other words, the time taken is proportional to the logarithm of the number of replicas required. This at least satisfies one condition. Moreover, self-replication or "increase at compound interest" is very easily provided by a proper arrangement of relays, such as neurones.

In the absence of replication, identification can be effected either by searching or by a system of classification. We can distinguish two extreme types of searching—the purely random and the systematic. In both, the templets—to retain that analogy—are tried one by one, but in the former, any templet may be tried at any stage, and therefore may be tried repeatedly. In systematic searching, no templet is tried more than once, and the correct one must obviously be found in m trials at the most (or $m - 1$ if it is not necessary to try the last one). However, the average number of trials required is, for the random case, m, and for the systematic case, $(m + 1)/2$. That is to say, in both cases the average number is a linear function of the number of alternatives.

Now all the trials are operations of the same kind, and therefore might be expected to take about the same time. Hence if reaction time were due to a search process of this simple sort, it would probably vary linearly with the number of alternatives. It is possible, of course, that a more complicated form of search process would give an approximation to the logarithmic relation required. The complication might take the form of a progressive reduction in the times taken by the elementary operations in the course of a single reaction time; but there seems to be no reason why that should happen, and we can only bear it in mind as a possibility. Alternatively, the probabilities of trying the different templets might be influenced by the choices made earlier in the search. In random searching there is no such influence. In systematic searching, the influence has the simple effect of prohibiting repetitions of the same choice. This, of course, is the best possible method, with the information given; if each trial merely answers the question: "right or wrong?" the average number of trials cannot be less than $(m + 1)/2$. But we wish to manipulate the probabilities so as to make the average number equal to $k \log m$. Unfortunately, this is strictly impossible, because, as m increases, even $(m + 1)/2$ will eventually exceed $k \log m$, no matter how large k is. It is conceivable that a fair approximation to $k \log m$ up to a limited value of m could be obtained, but that has not been attempted.

It may be useful, at this stage, to recapitulate the basic modes of procedure so far considered. They are (a) replication with simultaneous trial, (b) random searching, and (c) systematic searching. Of the three types of (a), self-replication was shown to be the only promising one. Neither (b) nor (c) had anything to recommend them, and we come now to a further reason for regarding them with disfavour—a reason which brings us also to the fourth of the basic modes.

We have seen that to identify one out of m objects requires the extraction of log m units of information. If logarithms to the base 2 are used the units are called "bits." A "bit" is the information we get from applying a test which must turn out in one of two equiprobable ways. Therefore we need only $\log_2 m$ such tests to effect the identification. Now each trial in a search process is certainly a dichotomising test, but the probabilities of "right" and "wrong" are not equal; that is why we need $(m+1)/2$ trials, which is considerably more than $\log_2 m$.

Is it possible to devise a procedure which is ideally efficient, in terms of the average number of dichotomising tests required? The answer is that in principle a near approximation is possible, and that it involves what may be called progressive classification. The process can be represented by the well-known "tree" diagram, shown in Figure 3. The first test places the object—the stimulus, in the present case—in the correct one of two equiprobable classes. According to the result of this first test, the next one is chosen so as to make a similar cut, and so on until the stimulus has been classified with the degree of precision required. The number of tests or stages of classification applied will evidently be $\log_2 m$, where m is the number of terminal sub-classes allowed for, *provided* that m is an integral power of 2. If m is not an integral power of 2, it is necessary to stipulate that the stimulus probabilities be specially adjusted so that each dichotomy has a probability of 0.5 associated with it. Otherwise, if the stimulus probabilities are held equal, the average number of stages is slightly greater than the expected number $\log_2 m$; but it appears that the maximum excess need not be more than about 0.086 of a stage. It can be seen from Figure 3 that the excess is associated with asymmetry of the "tree," and that the asymmetry is least when we merely add terminal twigs to the symmetrical "tree" which is just too small (Figure 3 (b)). The extreme of symmetry would be a single main trunk with a succession of twigs along its length (Figure 3 (c)); this represents the systematic search process referred to previously.

If we can again assume that the component operations—the dichotomising tests —are of like kind and therefore will probably take about the same time, this process of progressive classification agrees closely with the logarithmic relation between reaction time and degree of choice. The only other simple scheme found to satisfy this condition is the self-replication process. Of course there is no real evidence to make us prefer either to the other, but perhaps the classification process may seem a more appropriate model. At any rate, it will not be out of place to consider some of the difficulties in reconciling it with other aspects of the experimental data.

In the first place, although the model accounts for the *average* reaction times lying on a nearly smooth curve, when plotted against the number of alternatives, each individual reaction time is represented as the sum of an exact whole number of stage times. In fact, apart from peripheral delays, each choice reaction time should be an exact multiple of the simple reaction time. Although inevitably blurred by random variations, we might expect some sign of this to show as a periodicity roughly equal to the simple reaction time in the frequency distribution of a large number of choice reaction times. A group of 773 ten-choice times by the same subject were examined for such an effect, but the only periodicity found was much too short, and was eventually traced to a slight tendency to avoid estimating fractions of a scale interval in measuring the times from the paper record. Of course, if the stage-time variations were highly correlated a periodicity of the kind sought would readily be obscured, and might only appear in a very much larger sample. But it is pointed out below that a high positive correlation would make it difficult to explain the observed variances.

The relation between the variability of the reaction time and the degree of choice might have thrown some light on the problem, but unfortunately it could not be ascertained with much confidence owing to the large scatter. The reason for the scatter is simply that, for the purpose of curve-fitting, we are interested in relative rather than absolute deviations. By the usual approximations for large numbers,

$$\sigma_{\bar{x}} = \sigma/\sqrt{N} \text{ and } \sigma_\sigma = \sigma/\sqrt{2N}$$

where $\sigma_{\bar{x}}$ is the standard deviation of the mean and σ_σ is that of the standard deviation (σ) itself. Reduced to comparable scales, they are σ_σ/σ and $\sigma_{\bar{x}}/\bar{x}$ and these, respectively, are in the ratio $\bar{x} : \sigma/\sqrt{2}$. This ratio is of the order of 4 for the present data, so that the standard deviations, when plotted, are bound to—and do—appear that much more scattered than the means.

Nevertheless, it is possible to compare a few hypotheses. According to the classification scheme, $RT_m = k_m RT_3$, where RT_m ($= RT_{n+1}$) is the reaction time for n alternative stimuli and RT_2 is the simple reaction time; k_m is therefore the number of stages of classification. We assume for the moment that k_m is an integer, i.e., that m is an integral power of 2. Hence $k_m = \log m/\log 2$. Now if the stage times vary independently, but with equal variances, the variance of RT_m is

$$V(RT_m) = k_m V(RT_2) = V(RT_2) \log m/\log 2$$

That is to say (granting the hypothesis) that $V(RT_m)$ is calculable from the variance of the simple reaction and the degree of choice; and the converse is equally true. More generally, if s is any power of 2,

$$V(RT_m) = V(RT_s) \log m/\log s$$

Now, the data of Experiment I provide estimates of $V(RT_s)$ for $s = 2$ and $s = 4$, but not, unfortunately, for any higher power of 2, because the case of seven stimuli (corresponding to an effective degree of choice of eight, by the main hypothesis) was not tried.

In order to make use of all the data, we may calculate the average number of stages, k'_m, on the assumption that a known proportion of the m signals are given one more stage of analysis than the remainder. In other words, the "tree" is made asymmetrical by the addition of a sufficient number of terminal twigs. The total variance is now increased by the fact that the number of stages varies about the average number. In the absence of correlations, this effect is additive; in fact,

$$V(RT_m) = k'_m V(RT_2) + \overline{RT_2}^2 V(k)$$

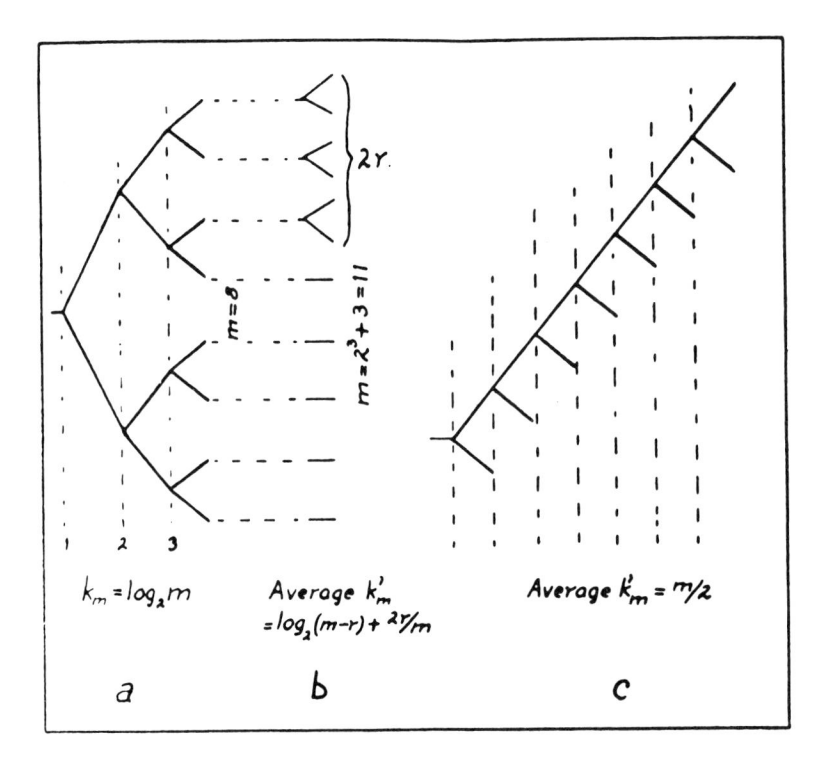

Figure 3. Diagram of the progressive classification process. The symmetrical "tree" (*a*) has the required logarithmic property. The minimum asymmetry (*b*), necessary when *m* is not a power of 2, gives a close approximation. Maximum asymmetry (*c*) depicts the systematic search process.

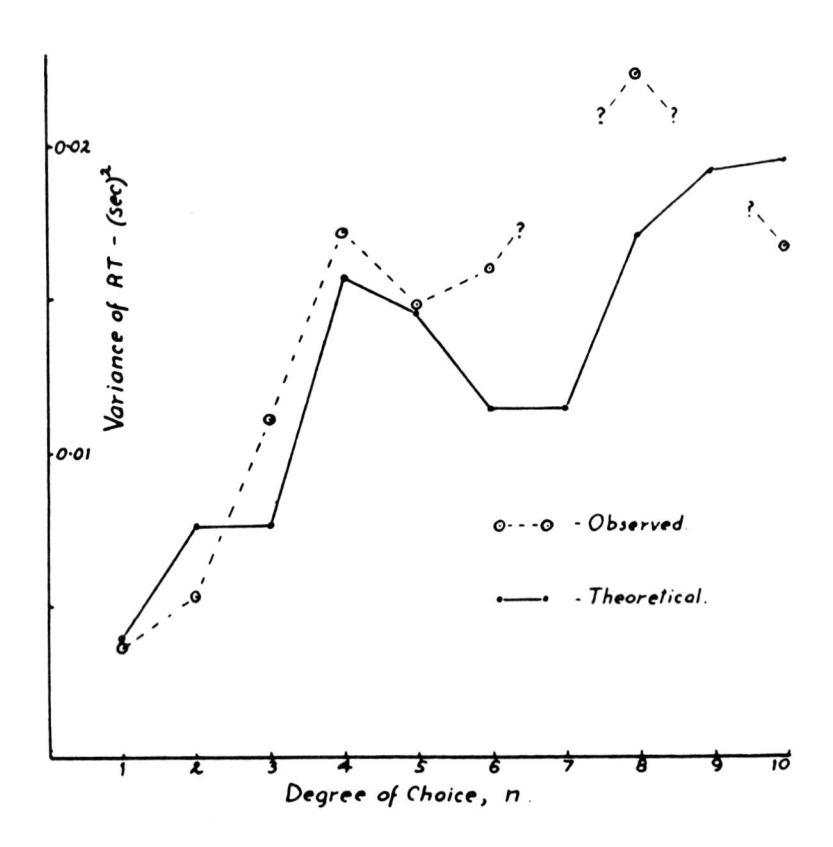

Figure 4. Variances of choice reaction times, compared with theoretical variances.

where $\overline{RT_2}$ is the mean simple reaction time (assumed to be the same as the mean stage time) and $V(k)$ is the variance of the number of stages. We must also allow for the fact that the observed variances are those of reaction times to actual stimuli —not to the supposititious "no stimulus." If the latter is a real signal in the present sense, it seems reasonable to assume that it receives the smaller number of stages of analysis; perhaps the best justification for this is that it slightly improves the fit of the calculated variances.

The outcome is shown in Figure 4. The observed values are for stimuli Nos. 1 and 2 only, as before. The lines joining the points are merely inserted for clarity; they have, of course, no significance. The calculated $V(RT_2)$, which will be seen to differ slightly from the observed value, is the weighted mean derived, according to the hypothesis, from the observed variances. The calculated values do show some tendency to follow the fluctuations of the observed values, but not to any convincing degree. In fact, a rough test suggests that the fit is quite unacceptable, and if the main hypothesis is to be retained, we must postulate some further source of variance not so far accounted for.

However, it is at least possible to say that complete positive correlation between stage-time variations would make the fit very much worse. More generally, if the whole process can be broken down into a sequence of operations of equal average durations, and if their number increases not less rapidly than log m, it seems unlikely that their durations could have large positive correlations. The present data provide no basis for considering combinations of positive and negative correlations.

All that can be added with regard to variances is that those of Experiment II are of similar magnitude to the ones just discussed, and show the same decelerating upward trend with increasing degree of choice (which, it will be remembered, is the theoretical equivalent degree, in this case).

IX. DISCUSSION

Turning now to more general topics, we may consider first some of the implications of the provisional conclusion that information is gained at a constant rate. Perhaps the most important, even though the most obvious, point is that it has proved valid to estimate the stimulus probabilities *ab extra*. Although the differences with respect to individual stimuli and responses suggest that the subjective—or perhaps one should say the psychologically effective—probabilities do not exactly correspond to the objective frequencies, it will be an enormous practical advantage if, for the purpose of estimating average effects, it can be assumed that they do. To discover the limits within which that assumption is justifiable will require a great deal of experimentation. It may be conjectured, for instance, that the effective probabilities are very little affected by increasing inequality in the stimulus frequencies until something like a threshold is reached. Certainly this matter would have to be examined before anything more than the most tentative application of information theory in real-life situations could be made, for it must be seldom that all the relevant possibilities are equiprobable, either subjectively or objectively. However, it may be found both practicable and valid, in some cases, to estimate subjective probabilities by some form of "guessing" technique.

Perhaps the whole matter can be best summed up in the following way. Fairly strong evidence has been obtained that the amount of information extracted is proportional to the time taken to extract it, on the average. Reasons have been adduced which seem to make this proposition inherently likely. But the simplest scheme of operations which fits the general proposition has been found to lead to hypotheses which other aspects of the data largely fail to confirm, although they do not definitely contradict it. At present, therefore, it is impossible to venture beyond the general statement in terms of information theory. This, indeed, may be adequate for practical applications; but it inevitably leaves the details vague; and so they must remain until more evidence or better reasoning is brought to the problem.

Appendix I

The method of estimating the average information per stimulus which is gained in each run was as follows. It has already been mentioned that the information gained (R) is

$$R = H(x) - H_y(x)$$

An alternative and more convenient formula is

$$R = H(x) + H(y) - H(x,y)$$

where $H(x,y)$ is the joint entropy based on the joint probabilities of particular stimulus response pairs. The response frequencies were entered in a table similar to Table I, and R was computed from the above formula, with the appropriate frequency ratios substituted for the probabilities: thus

$$R = \log N - \{\Sigma_i f_i \log f_i + \Sigma_j f_j \log f_j - \Sigma_{ij} f_{ij} \log f_{ij}\}/N$$

where f_i is the marginal total of the ith column, f_j is that of the jth row, f_{ij} is the number in the cell ij, and N is the grand total.

Since R is a quantity of information, it can be formally expressed as the logarithm of a number of equiprobable alternatives; thus

$$R = \log n_e$$

whence the effective degree of choice (n_e) can be obtained. The expression is only formal because, of course, n_e may not happen to be an integer.

Now R is only the information with respect to *which* stimulus occurred; we have still to consider the component due to the occurrence of *some* stimulus. Let $p(s,i)$ be the joint probability of a stimulus occurring and of its being the ith stimulus. Then the total input entropy is

$$H(X) = - \Sigma_i p(s,i) \log p(s,i) - q(s) \log q(s)$$

where $q(s)$ is the probability of no stimulus. After some manipulation this becomes:

$$H(X) = -p(s)\sum_i p_s(i) \log p_s(i) - p(s) \log p(s) - q(s) \log q(s)$$

where $p_s(i)$ is the conditional probability of the ith stimulus, given that some stimulus must occur, and $p(s) = 1 - q(s)$. This can be written briefly as

$$H(X) = p(s)H(x) + H(s)$$

in which $H(x)$ is the same as the $H(x)$ in the formula for R given above. $H(s)$ may be regarded as the uncertainty as to when the stimulus will occur, or the information to be gained from the fact that *some* stimulus has occurred.

Now, if there are no superfluous responses and no failures to respond, we can say that the $H(s)$ component suffers no loss in transmission, the only loss being that sustained by $H(x)$ in its degeneration into R. Therefore the total information transmitted is

$$R_t = p(s)R + H(s).$$

But the immunity of $H(s)$ from depreciation does not necessarily imply that it is independent of R. The temporary capacity of the organism for extracting information from the display is, in a limited sense, indicated by R; in fact, we have expressed this capacity as n_e, the number of equiprobable categories into which the n actual stimuli are, in effect, divided. The further information needed for the selection of one particular response must be drawn from some independent source —an irrational preference or an appeal to chance or something of that kind. In other words, if the stimuli are grouped into n_e categories, the display is being interpreted as if it could generate only n_e different stimuli. It is therefore reasonable to expect that the possibility of no stimulus gives us, altogether, $n_e + 1$ equiprobable signals, by analogy with what we have found to apply to ordinary choice reaction times.

Adopting this assumption, we write $p(s) = n_e/(n_e + 1)$, and the total information transmitted takes the simple form:

$$R_t = \log (n_e + 1)$$

As we have seen (Figures 1 and 2), this function gives a reasonably close fit to the corresponding reaction times, as the main hypothesis requires.

REFERENCES

1. Blank, G. (1934). Brauchbarkeit optischer reactionsmessungen. *Indust. Psychotech.*, **11**, 140-150.
2. Hick, W. E. (1951). A simple stimulus generator. *Quart. J. Exp. Psychol.*, 3, 94-95.
3. Kraepelin, E. (1894). Beobachtungen bei zusammengesetzen reaktionen. *Philos. Stud.*, **10**, 499-506.
4. Merkel, J. (1885). Die zeitlichen verhältnisse der willensthätigkeit. *Philos. Stud.*, **2**, 73-127.

158   Human Processing of Information—Discrete and Continuous Signals

5. Shannon, C. E., and Weaver, W. (1949). *The Mathematical Theory of Communication.* Urbana.
6. Wiener, N. (1948). *Cybernetics.* New York.
7. Woodworth, R. S. (1938). *Experimental Psychology.* New York.

15

High-Speed Scanning in Human Memory

Saul Sternberg

How is symbolic information retrieved from recent memory? The study of short-term memory (1) has revealed some of the determinants of failures to remember, but has provided little insight into error-free performance and the retrieval processes that underlie it. One reason for the neglect of retrieval mechanisms may be the implicit assumption that a short time after several items have been memorized, they can be immediately and simultaneously available for expression in recall or in other responses, rather than having to be retrieved first. In another vocabulary (2), this is to assume the equivalence of the "span of immediate memory" (the number of items that can be recalled without error) and the "momentary capacity of consciousness" (the number of items immediately available). The experiments reported here (3) show that the assumption is unwarranted.

Underlying the paradigm of these experiments is the supposition that if the selection of a response requires the use of information that is in memory, the latency of the response will reveal something about the process by which the information is retrieved. Of particular interest in the study of retrieval is the effect of the number of elements in memory on the response latency. The subject first memorizes a short series of symbols. He is then shown a test stimulus, and is required to decide whether or not it is one of the symbols in memory. If the subject decides affirmatively he pulls one lever, making a positive response; otherwise he makes a negative response by pulling the other lever. In this paradigm it is the identity of the symbols in the series, but not their order, that is relevant to the binary response. The response latency is defined as the time from the onset of the test stimulus to the occurrence of the response.

Because they are well learned and highly discriminable, the ten digits were used as stimuli. On each trial of experiment 1, the subject (4) saw a random series of from one to six different digits displayed singly at a fixed locus for 1.2 seconds each. The length, s, of the series varied at random from trial to trial. There followed

From *Science*, 153, August 5, 1966, 652-654. Copyright 1966 by the American Association for the Advancement of Science.

a 2.0-second delay, a warning signal, and then the test digit. As soon as one of the levers was pulled, a feedback light informed the subject whether his response had been correct. The trial ended with his attempt to recall the series in order. For every value of s, positive and negative responses were required with equal frequency. Each digit in the series occurred as a test stimulus with probability $(2s)^{-1}$, and each of the remaining digits occurred with probability $[2(10-s)]^{-1}$.

Each subject had 24 practice trials and 144 test trials. Feedback and payoffs were designed to encourage subjects to respond as rapidly as possible while maintaining a low error-rate. The eight subjects whose data are presented pulled the wrong lever on 1.3 percent of the test trials (5). Recall was imperfect on 1.4 percent of the trials. The low error-rates justify the assumption that on a typical trial the series of symbols in memory was the same as the series of symbols presented.

Results are shown in Figure 1. Linear regression accounts for 99.4 percent of the variance of the overall mean response-latencies (6). The slope of the fitted line is 37.9 ± 3.8 msec per symbol (7); its zero intercept is 397.2 ± 19.3 msec. Lines fitted separately to the mean latencies of positive and negative responses differ in slope by 9.6 ± 2.3 msec per symbol. The difference is attributable primarily to the fact that for $s = 1$, positive responses were 50.0 ± 20.1 msec faster than negative responses. Lines fitted to the data for $2 \leq s \leq 6$ differ in slope by an insignificant 3.1 ± 3.2 msec per symbol.

The latency of a response depends, in part, on the relative frequency with which it is required (8). For this reason the frequencies of positive and negative responses and, more generally, the response entropy (8), were held constant for all values of s in experiment 1. However, the test-stimulus entropy (predictability) was permitted to co-vary with s.

Both response and test-stimulus entropies were controlled in experiment 2, in which the retrieval process was studied by an alternative method similar to that used in more conventional experiments on choice-reaction time. In experiment 1, the set of symbols associated with the positive response changed from trial to trial. In contrast to this varied-set procedure, a fixed-set procedure was used in experiment 2. In each of three parts of the session, a set of digits for which the positive response was required (the positive set) was announced to the subject (4); there followed 60 practice trials and 120 test trials based on this set. The subject knew that on each trial any of the ten digits could appear as the test stimulus, and that for all the digits not in the positive set (the negative set) the negative response was required. Each subject worked with nonintersecting positive sets of size $s = 1, 2$, and 4, whose composition was varied from subject to subject.

Stimulus and response entropies were both held constant while s was varied, by means of specially constructed populations of test stimuli. Let x_1, y_1, y_2, z_1, ..., z_4 and w_1, ..., w_3 represent the ten digits. Their relative frequencies in the population were x_1, 4/15; each y, 2/15; each z, 1/15; and each w, 1/15. The three sequences of test stimuli presented to a subject were obtained by random permutation of the fixed population and assignment of x_1, the y_1, or the z_1 to the positive response. Thus, the population of test stimuli, their sequential properties, and the relative frequency of positive responses (4/15) were the same in all conditions (9).

A trial consisted of a warning signal, the test digit, the subject's resonse, and a feedback light. Between a response and the next test digit, 3.7 seconds elapsed.

As in experiment 1, feedback and payoffs were designed to encourage speed without sacrifice of accuracy. The six subjects whose data are presented pulled the wrong lever on 1.0 percent of the test trials (5).

The results, shown in Figure 2, closely resemble those of experiment 1. A positive set in experiment 2 apparently played the same role as a series of symbols presented in experiment 1, both corresponding to a set of symbols stored in memory and used in the selection of a response. As in experiment 1, linear regression accounts for 99.4 percent of the variance of the overall mean response-latencies (6). The slope of 38.3 ± 6.1 msec per symbol is indistinguishable from that in experiment 1; the zero intercept is 369.4 ± 10.1 msec. In experiment 2, the relation between latencies of positive and negative responses when $s = 1$ is not exceptional. Lines fitted separately to latencies of the two kinds of response differ in slope by an insignificant 1.6 ± 3.0 msec per symbol.

The linearity of the latency functions suggests that the time between test stimulus and response is occupied, in part, by a serial-comparison (scanning) process. An internal representation of the test stimulus is compared successively to the symbols in memory, each comparison resulting in either a match or a mismatch. The time from the beginning of one comparison to the beginning of the next (the comparison time) has the same mean value for successive comparisons. A positive response is made if there has been a match, and a negative response otherwise.

On trials requiring negative responses, s comparisons must be made. If positive responses were initiated as soon as a match had occurred (as in a self-terminating search), the mean number of comparisons on positive trials would be $(s + 1)/2$ rather than s. The latency function for positive responses would then have half the slope of the function for negative responses. The equality of the observed slopes shows, instead, that the scanning process is exhaustive: even when a match has occurred, scanning continues through the entire series. This may appear surprising, as it suggests nonoptimality. One can, however, conceive of systems in which a self-terminating search would be inefficient. For example, if the determination of whether or not a match had occurred were a slow operation that could not occur concurrently with scanning, self-termination would entail a long interruption in the scan after each comparison.

On the basis of the exhaustive-scanning theory, the zero intercept of the latency function is interpreted as the sum of the times taken by motor response, formation of the test-stimulus representation, and other unknown processes whose durations are independent of the number of symbols in memory. The slope of the latency function represents the mean comparison-time. The two experiments, then, provide a measure of the speed of purely internal events, independent of the times taken by sensory and motor operations. The average rate of between 25 and 30 symbols per second is about four times as high as the maximum rate of "subvocal speech" when the words are the names of digits (11). This difference suggests that the silent rehearsal (12) reported by subjects in both experiments should probably not be indentified with high-speed scanning, but should be thought of as a separate process whose function is to maintain the memory that is to be scanned.

In view of the substantial agreement in results of the two experiments, one difference in procedure merits particular emphasis. A response in experiment 1 was the first and only response based on a particular series, made about three seconds after the series had been presented. In contrast, the positive set on which a re-

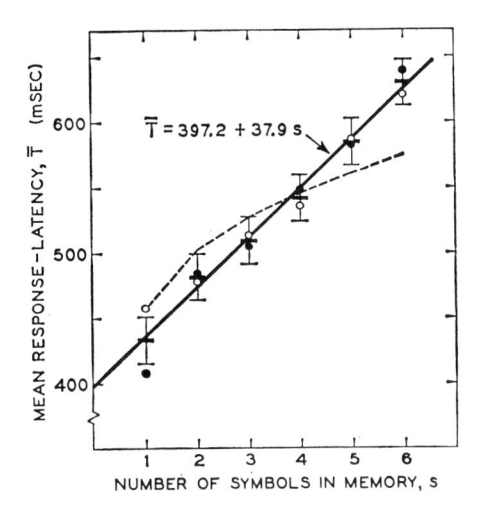

Figure 1. Relation between response latency and the number of symbols in memory, s, in experiment 1. Mean latencies, over eight subjects, of positive responses (filled circles) and negative responses (open circles). About 95 observations per point. For each s, overall mean (heavy bar) and estimates of $\pm\ \sigma$ are indicated (6). Solid line was fitted by least squares to overall means. Upper bound for parallel process (broken curve).

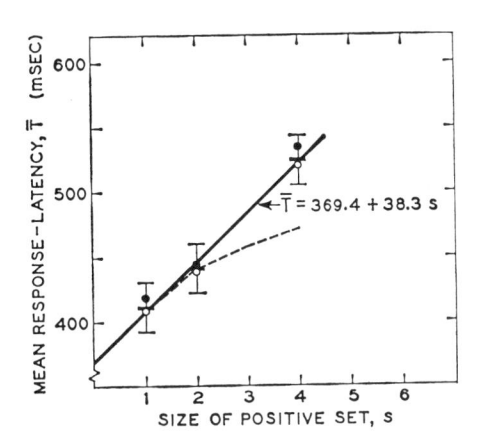

Figure 2. Relation between response latency and the size of the positive set, s, in experiment 2. Mean latencies, over six subjects, of positive responses (filled circles) and negative responses (open circles). About 200 (positive) or 500 (negative) observations per point. For each s, overall mean (heavy bar) and estimates of $\pm\ \sigma$ are indicated (6). Solid line was fitted by least squares to overall means. Upper bound for parallel process (broken curve).

sponse was based in experiment 2 had been used on an average of 120 previous trials. Evidently, neither practice in scanning a particular series nor lengthening of the time it has been stored in memory need increase the rate at which it is scanned.

In accounting for human performance in other tasks that appear to involve multiple comparisons, theorists have occasionally proposed that the comparisons are carried out in parallel rather than serially (13, 14). (This perhaps corresponds to the assumption mentioned earlier that the momentary capacity of consciousness is several items rather than only one. Are the present data inconsistent with such a proposal? Parallel comparisons that begin and also end simultaneously (14) are excluded because the mean latency has been shown to increase with s. A process in which multiple comparisons begin simultaneously is more difficult to exclude if the comparison times are independent, their distribution has nonzero variance, and the response is initiated when the slowest comparison ends. A linear increase in mean latency cannot alone be taken as conclusive evidence against such a process. The magnitude of the latency increase that would result from a parallel process is bounded above, however (15); it is possible to apply the bound to these data (16). This was done for the negative responses in both experiments, with the results shown by the broken curves in Figures 1 and 2. Evidently, the increase in response latency with s is too great to be attributed to a parallel process with independent comparison times (17).

Other experiments provide added support for the scanning theory (16). Two of the findings are noted here: (i) variation in the size, n, of the negative set ($n \geq s$) had no effect on the mean latency, indicating that stimulus confusability (10, 18) cannot account for the results of experiments 1 and 2; (ii) variation in the size of a response-irrelevant memory load had no effect on the latency function, implying that the increase in latency reflects the duration of retrieval and not merely the exigencies of retention.

The generality of the high-speed scanning process has yet to be determined, but there are several features of experiments 1 and 2 that should be taken into account in any comparison with other binary classification tasks (14, 19): (i) at least one of the classes is small; (ii) class members are assigned arbitrarily; (iii) relatively little practice is provided; (iv) high accuracy is required and errors cannot be corrected; and (v) until the response to one stimulus is completed the next stimulus cannot be viewed.

REFERENCES AND NOTES

1. A. W. Melton, *J. Verbal Learning Verbal Behavior*, **2**, 1 (1963).
2. G. A. Miller, *Psychology, the Science of Mental Life* (Harper and Row, New York, 1962), p. 47.
3. These experiments were first reported by S. Sternberg, "Retrieval from recent memory: Some reaction-time experiments and a search theory," paper presented at a meeting of the Psychonomic Society, Bryn Mawr, August 1963.
4. Subjects were undergraduates at the University of Pennsylvania.
5. These trials were excluded from the analysis. Three other subjects in experiment 1 (two in experiment 2) were rejected because they exceeded an error criterion. Their latency data, which are not presented, resembled those of the other subjects.

6. For both experiments the data subjected to analysis of variance were, for each subject, the mean latency for each value of s. So that inferences might be drawn about the population of subjects, individual differences in mean and in linear-regression slope were treated as "random effects." Where quantities are stated in the form $a \pm b$, b is an estimate of the standard error of a. Such estimates were usually calculated by using variance components derived from the analysis of variance.
7. The analyses of variance for both experiments provided a means of testing the significance of differences among individual slopes. Significance levels are .07 (experiment 1) and .09 (experiment 2), suggesting true intersubject differences in slope; the population distribution of slopes has an estimated standard deviation of 8.0 msec per symbol.
8. W. R. Garner, *Uncertainty and Structure as Psychological Concepts* (Wiley, New York, 1962).
9. A result of this procedure is that other factors in choice-reaction time were also controlled: stimulus discriminability (*10*); information transmitted (*8*); and information reduced, M. I. Posner, *Psychol. Rev.* **71**, 491 (1964); P. M. Fitts and I. Biederman, *J. Exp. Psychol.* **69**, 408 (1965).
10. R. N. Shepard and J. J. Chang, *J. Exp. Psychol.* **65**, 94 (1963); M. Stone, *Psychometrika*, **25**, 251 (1960).
11. T. K. Landauer, *Percept. Mot. Skills* **15**, 646 (1962).
12. D. E. Broadbent, *Perception and Communication* (Pergamon, New York, 1958), p. 225.
13. L. S. Christie and R. D. Luce, *Bull. Math. Biophys.* **18**, 89 (1956); A. Rapoport, *Behavioral Sci.* **4**, 299 (1959).
14. U. Neisser, *Amer. J. Psychol.* **76**, 376 (1963); *Sci. Amer.* **210**, 94 (1964).
15. H. O. Hartley and H. A. David, *Ann. Math. Stat.* **25**, 85 (1954).
16. S. Sternberg, in preparation.
17. Exponentially distributed parallel comparisons (*13*) and other interesting theories of multiple comparisons (*18*) lead to a latency function that is approximately linear in $\log s$. Deviations of the overall means from such a function are significant ($P < .03$) in both experiments.
18. A. T. Welford, *Ergonomics* **3**, 189 (1960).
19. I. Pollack, *J. Verbal Learning Verbal Behavior*, **2**, 159 (1963); D. E. Broadbent and M. Gregory, *Nature* **193**, 1315 (1962).
20. Supported in part by NSF grant GB-1172 to the University of Pennsylvania. I thank D. L. Scarborough for assistance, and J. A. Deutsch, R. Gnanadesikan, and C. L. Mallows for helpful discussions.

16

Searching for Ten Targets Simultaneously

Ulric Neisser, Robert Novick, and Robert Lazar

Search is an active process. Consider an S who looks through a list of letters to find (say) the letter "K." As he scans, he must analyze the successive visual configurations to determine whether any of them have K-like properties. There must be some rather definite set of properties which he uses as criteria, although S usually cannot describe them: perhaps indentation at the right, the presence of certain angles, a given ratio of height to width; perhaps contours of particular orientations, or proportions of rectilinear to curvilinear edges. The properties, and the operations that detect them, must vary in different contexts, for different Ss, and at different times. We can be rather sure that no single property is decisive; each will have only a probabilistic relation to the detection of "K." Certainly the stimulus analysis made by S can be rather primitive: in particular, it need not identify the irrelevant letters through which he scans. (Indeed, practiced Ss report that they do not "see" the irrelevant letters in a rapid search.) Nevertheless, some operations of this kind are certainly going on during a scan, and their nature must depend, in part, on the physical characteristics which distinguish "K" from other letters. In another paper (Neisser, 1963) one of us has shown that the rate at which S scans can be consistently interpreted as a measure of the time which these operations require.

A more extensive stimulus analysis must occur when S searches for *either* of two different letters, say "K" or "P." Now he must perform both the operations which detect "K" and those which detect "P"—and there is no reason to suppose that these can be the same. One might prefer to say that he now performs a single, more subtle operation which detects both letters. Such an increase in subtlety might be expected to result in longer processing times, however, and previous re-

---

Reprinted with permission of authors and publisher: Neisser, U., Novick, R., & Lazar, R. Searching for ten targets simultaneously. *Perceptual and Motor Skills*, 1963, **17**, 955-961.

Some of this work was performed while the senior author was at the Lincoln Laboratory of the Massachusetts Institute of Technology, operated with support from the U. S. Army, Navy, and Air Force. In addition, the research was partially supported by the National Science Foundation under Grant No. G-21654 to Brandeis University. The authors also appreciate the helpfulness of the Harvard Center for Cognitive Studies, which made available the laboratory space in which the main experiment was conducted.

sults do not confirm this expectation. It has been shown (Neisser, 1963) that Ss can look for either of two letters as rapidly as for one alone. Indeed, with practice, no more time is needed to examine the input for any of four targets than for two or one.

This finding suggests a particular model of the search process, closely related to the scheme which Oliver Selfridge (1959) first described and which he called "Pandemonium." We assume that a separate set of mechanisms ("analyzers," "demons") is devoted to the detection of each important property, and that all of these mechanisms can function at once, in parallel. Some particular group of analyzers, probabilistically combined, serve to begin activity in a larger, slower system which identifies "K." Another, perhaps overlapping set of analyzers feeds into the system for "P," and other sets are devoted to the other identifiable letters. In such a model, the results obtained with two- and four-target searches are interpreted as evidence that several analyzers can indeed function at once.

The present paper extends this finding to a wider set of simultaneously operating recognition systems: it appears that practiced Ss can keep watch for 10 targets without slowing down their scan.

METHOD

Stimulus Materials. Each list consisted of a single column of 50 "items." Of these, 49 were strings of six letters or numbers (for example NBJ3SG), constructed by randomly permuting a set of 20 context characters and using the first six members of the permutation. The entire set of context characters was B, C, G, I, J, N, O, Q, R, S, T, V, W, X, Y, 1, 3, 5, 6, 7, except that on the last two days of the experiment N, V, and W were replaced by D, L, and 8.

One of the 50 items was randomly selected to contain the target, and one of its six characters (chosen at random) was replaced by an appropriate target letter or number. There were four conditions in the main experiment. In Condition *K*, the target was always the letter "K." In Condition *AFKU9*, the target might, with equal probability, be any of those five characters. Condition *HMPZ4* was similarly defined. In Condition *Ten*, the target might, with equal probability, be any of the 10 letters or numbers used in the other conditions. In addition, a control experiment consisting only of Condition *K* was carried out for reasons that will appear below.

The lists were internally generated by a suitably programmed IBM 7090 computer, and automatically printed on individual pages of standard output paper by an IBM 1401 system. The target letter was reprinted in one corner of the page, together with an indication of its position in the list. Sets of about 20 such pages (lists), all pertaining to the same experimental condition, were kept together as single strips. This facilitated the process of exposing successive lists in the apparatus described below, and also ensured that every S received a given set of lists in the same order.

Apparatus and Procedure. The lists were presented in a device which permits convenient measurement of search times. The entire list was visible to S at once through a glass window in the sloping top of an otherwise light-tight box. The target identification in the corner of each page could not be seen through this window, but was visible to E through a special periscope. A pair of fluorescent lamps was mounted within the box and above the list, so the latter could easily be read through the window when they were on, but was invisible otherwise. An additional fixation light enabled S to focus at the proper depth even when the list could not be seen.

S held a three-position rotary switch. When he was ready to begin scanning, a turn of the switch caused the list to appear and also started an electric timer (calibrated in .01 sec.) visible only to E. S immediately started to search through the list (always from the top down), looking for any of the target characters that were appropriate to the experimental condition in which he was working. (There was always only one such character in the list, and its position was unpredictable from one list to the next.) On finding it, S turned the switch again. This stopped the timer but left the light on, so S could recheck the item at which he stopped. Then another turn put out the light to complete the cycle. After E had recorded target position and search time, he pulled the next list into position under the window and indicated that S might begin again.

In any particular condition, on a given day, S scanned at least 15 lists. The first three were considered practice, while the last 12 were used to calculate the scanning rate as described below. However, S might miss the target on one or more of these 12, i.e., he might scan all the way down to the 50th item in vain. (Errors of commission also occurred, but rarely.) If an error occurred, no time for the list was recorded; an additional list was then presented at the end of the series so that 12 search times would be available for the calculation of rate.

Subjects. Two male and four female students at the Harvard Summer School were employed in the main experiment. Three male and three female Brandeis students were employed in the control experiment. (One of the latter dropped out after 19 days.) All were paid for their services. They knew that the experiment was an attempt to determine the maximum possible scanning speed under the given conditions. Ss were run individually.

Experimental Design. The main experiment lasted 27 experimental "days." Each S underwent all four conditions (*K, AFKU9, HMPZ4, Ten*) in each daily session. The order of conditions varied from day to day. Initial sessions took an hour or more; practiced Ss needed less than half that time. "Daily" sessions were sometimes more than one day apart, and occasionally two were run on the same calendar day at widely separated times.

Ss were to aim for both speed and accuracy. On the first 13 days, instructions and comments by E encouraged high speed unless S had an unusually high error rate. In Day 14 and thereafter, a competition was introduced as additional incentive. The competition involved only a single experimental condition on each day: *Ten* on Day 14, *HMPZ4* on Day 15, etc. Because the same lists were presented to each S, a simple system could be used to reward both speed and accuracy. S received one point for each individual list (of the 12) on which his search time was less than that of any other S. He could not score a point if he made an error. Each day the S with the highest point total received an extra day's pay.

The six Ss of the control experiment were treated in the same way except that (a) they worked only in Condition K, so each daily session took 10 min. or less; (b) their competition began only on Day 15.

RESULTS

Method of Analysis. Each S produced 12 search times in each condition every day. These 12 were combined into a single measure of time-per-item-scanned (the inverse of scanning rate) by the method described previously (Neisser, 1963). A linear function appears when the 12 search times are plotted against the position of the target item in each case. The *slope* of this line represents the time needed to scan over each non-critical item, independently of response factors. For example, if S needs 2 sec. when the target is in the 10th position, 3 sec. when it is 20th, 4 when it is 30th, etc., he is using 0.10 sec. per item.

In every case, lines were fitted to the 12 points by the method of least squares. The slopes of these lines are the principal data of the experiment.[1] Because the results for individual Ss were very similar, they have been combined.

Effect of Experimental Conditions. Figure 1 shows how the mean time-per-item changed over 27 days in each condition of the main experiment. The (mean) results for the control experiment are also shown. It appears that the multi-target conditions are initially slower than K, but after about a dozen days of practice no difference can be detected. It takes no more time to check an item for any of 10 targets (or 5) than for 1 target alone. Space does not permit us to show the results for each one of the individual Ss, but all of the data indicate the same convergence.

The great disparity among conditions in the early part of the experiment can be ascribed to the arbitrary sets of target characters used. Although the set of appropriate targets was always prominently displayed near the apparatus, the display could hardly be referred to during a scan. For rapid scanning, S must be very sure of what he is searching for.

Our theoretical model assumes that S does not identify the contextual characters over which he scans, and this assumption seems to be confirmed by the phenomenal report that these letters are not actually "seen." Nevertheless, there is a theoretical possibility that S simply learns the 20 context characters and searches until he finds some character outside of that set. Such an assumption would easily explain the equivalence of the four experimental conditions. On this assumption, one would expect a marked discontinuity in the search times at Day 26, when three letters of the context were changed. No such discontinuity appears in the figure, nor for any individual S. Hence this hypothesis may be rejected.

---

[1] The raw data, and the parameters of the best-fitting straight lines, are available in the form of easily readable computer output. They can be obtained from the author on request.

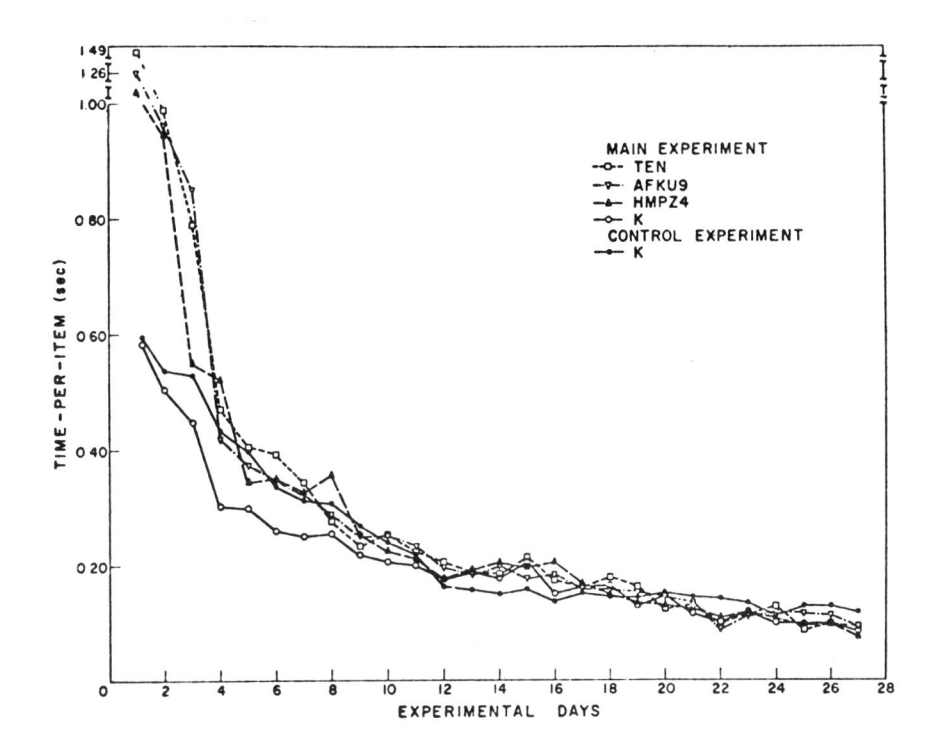

Figure 1. Time-per-item as a function of day of practice. Each point represents the mean time-per-item of 6 Ss except the last 8 points in the control condition, which represent only 5 Ss.

TABLE 1
Mean Errors Per Day in Each Condition

Condition:	K	AFKU9	HMPZ4	Ten
	3.63	3.12	2.57	3.10

Note.—Averaged over the last 14 days and all 6 Ss of the main experiment.

Errors. All Ss made a considerable number of errors of omission (i.e., failures to find the target). Errors became more frequent during the competitive latter portion of the experiment. Over the last 14 days, even the most accurate S averaged over two errors per condition per day; the least accurate S made nearly four. The median S averaged three such errors in 15 lists (the basic 12 plus 3 more to compensate for the errors made), a rate of about 20%. Often the errors seemed to be specific to particular visual configurations: several Ss would miss the same target in the same list.

In terms of our theoretical model, it may be assumed that the average S settles on a set of stimulus analyzers which have only 80% probability of detecting the target in question. More accurate stimulus analysis would presumably require more time than S, trying for speed, is willing to spend. In principle, Ss should be able to trade speed for accuracy. In fact, however, it was very difficult to comply with the instruction "Go more slowly and avoid mistakes." That is, an S who has evolved a particular system of stimulus analysis cannot easily bring new (slower and more accurate) analyzers into play.

Two other, less attractive interpretations of the error rate must be considered. First, it might be suggested that Ss gain in speed simply by skipping over increasingly large portions of each list, making errors if the target happens to fall in the skipped part. From this hypothesis we would predict (a) that faster Ss make more errors, (b) that daily fluctuations in speed and number of errors should be negatively correlated. Neither prediction is confirmed by the data.

We must also consider the possibility that Ss equalize their speed in the four conditions only by varying their error rate. In other words, perhaps *Ten* is really more difficult than *K*, with the difference appearing in accuracy rather than in time-per-item. Table 1, however, shows that this hypothesis is not supported; indeed, the contrary is true. *K* has the most errors, not the fewest. Since all six Ss had a higher error rate in *K* than in *HMPZ4*, the difference between these two conditions can be regarded as significant at the 5% level. It appears that the probability of error is inversely related to the probability that the letter "K" will actually be the target. Perhaps we had the misfortune of selecting a particularly difficult character for our single-target condition. In any case it appears that searching for any of 10 targets is not only as fast but also at least as accurate as searching for only one.

The Control Experiment. These results indicate that Ss *do not* scan faster for one letter than for ten, but fail to show convincingly that this *could not* be done. It is possible that, in spite of the competitive incentive, each S adopted a single speed and used it in all four conditions; that he held himself back on *K*, as it were, rather than change his mode of operation from that of *Ten*.[2] Indeed, casual observation and phenomenal report do suggest that Ss operated with all four conditions in exactly the same way. On occasions when *E* inadvertently used the wrong stimulus list, with a target symbol that S was not expecting, S usually detected it anyway. If we grant, then, that the full stimulus analysis needed for 10 targets was being used even in Condition K, there are two possibilities open.

[2] The possibility that such an "undifferentiated cognitive set" may produce misleading experimental results was pointed out by Fitts and Switzer (1962, p. 328).

First, our theoretical model suggests that such a fixed strategy is entirely plausible. If all the analyzers can act simultaneously, eliminating some of them in Condition K would save no time. But it is also possible, on the contrary, that the tendency to make the full analysis, whether or not it was necessary, forced S to go more slowly on K than he would otherwise have done. To decide between these possibilities we carried out the control experiment, in which new Ss were run for 27 days on K only. If the Ss of the main experiments had been slowed down in K because they were also looking for the other potential targets, our control Ss should have attained higher speeds. Fig. 1 shows that this did not occur.

REFERENCES

Fitts, P. M., & Switzer, G. Cognitive aspects of information processing: I. The familiarity of S-R sets and subsets. *J. exp. Psychol.*, 1962, 63, 321-329.
Neisser, U. Decision time without reaction time: experiments in visual scanning. *Amer. J. Psychol.*, 1963, 76, 376-385.
Selfridge, O. G. Pandemonium: a paradigm for learning. In National Physical Laboratory, *Mechanisation of thought processes.* Vol. 1. London: Her Majesty's Stationery Office, 1959. Pp. 511-526.

17

Human Tracking Behavior

Jack A. Adams

The subject matter of this paper is a critical review and analysis of research, issues, and points of views associated with human behavior in one- and two-dimensional tracking tasks. Tracking tasks have never been given explicit definition and one of the purposes of this paper is to tentatively advance the general

From *Psychological Bulletin*, 58, 1961, 55-79. Copyright 1961 by the American Psychological Association, and reproduced by permission.

This research was supported by the United States Air Force under Contract No. AF 49(638)-371, monitored by the Air Force Office of Scientific Research of the Air Research and Development Command.

A number of psychologists read this manuscript in draft form and contributed to its improvement. The detailed and critical comments of F. C. Bartlett, E. A. Fleishman, C. B. Gibbs, N. B. Gordon, J. A. Leonard, and A. T. Welford were particularly appreciated.

bounds of a tracking situation, but for those who are unfamiliar, a temporary working definition for the moment which enjoys the consensus of most psychologists is as follows:

1. A paced (i.e., time function) externally programmed input or command signal defines a motor response for the operator, which he performs by manipulating a control mechanism.
2. The control mechanism generates an output signal.
3. The input signal minus the output signal is the tracking error quantity and the operator's requirement is to null this error. The mode of presenting the error to the operator depends upon the particular configuration of the tracking task but, whatever the mode, the fundamental requirement of error nulling always prevails. The measure of operator proficiency ordinarily is some function of the time-based error quantity.

The usual tracking task has a visual display but there is no necessity for this. On occasion, auditory tracking tasks have been devised (Forbes, 1946; Humphrey & Thompson, 1952a, 1952b, 1953). The most simple and well-known visual tracking task is the Rotary Pursuit Test (Melton, 1947) which employs a repetitive input signal and, although investigations using the Rotary Pursuit Test are not ordinarily included in that body of research which is considered to study tracking behavior per se, it is nevertheless an unequivocal example of the breed. Tracking studies typically use more elaborate apparatus which allows for controlled manipulation of such variables as the function for the input signal, scale factors, mathematical transformations of the output signal, characteristics of the control mechanism, etc.

While investigations of tracking behavior might legitimately be subsumed under the time-honored rubric "motor skills," this label is misleading in hinting by implication and textbook tradition that motor behavior, such as tracking, is disassociated from so-called "higher processes." British investigators in particular have analyzed the acquired ability to predict input stimulus sequences as a key intervening response class in determining the proficiency of the measured motor responses in tracking tasks, thus emphasizing the interlacing of "higher" and "lower" processes. These British studies will be discussed in detail later, but passing mention of them at the onset seems worthwhile for establishing the archaic connotations of "motor skills." Research by Adams (1957), Fleishman (1954, 1957a, 1957b, 1958), and Fleishman and Hempel (1954, 1955, 1956) on variables influencing individual differences in motor behavior, also documents the inherent complexity of the response totality elicited in motor tasks. Helson (1949), in discussing variables influencing the subject's standard of excellence in a tracking task, includes perceptual and motivational states in addition to motor factors as influential determiners of motor behavior.

BASIC TERMINOLOGY AND FRAME OF REFERENCE

Independent variables influencing tracking behavior will be divided into two classes: task variables and procedural variables. *Task variables* are machine-centered. They are the physical values of the tracking device, and they include such factors as the nature of the input signal, configuration of the display, design of the control system, mathematical transformations relating control displacement and changes in the output signal, etc. *Procedural variables* are man-centered. They are manipulable nontask quantities, and examples of them are instructions, number of practice trials, length of the practice trial, and time between trials. Also, the indicants which are displayed to the operator will be implicitly assumed as simple elements, such as needles or dials, pointers, dots on cathode ray tubes, etc. Special problems that arise when the display is perceptually complex and requires the interpretation of forms, shapes, colors, etc. will be ignored.

THE TRADITION OF ENGINEERING PSYCHOLOGY

A dominant influence in tracking research is the experiments of engineering psychology, with the emphasis being largely on the relations between measures of tracking behavior and *task* variables. The engineering psychologist has as his goal the prediction of the characteristics of man-machine systems, and this goal requires careful attention to the task variables which influence the operator. Representative examples of several hundred task-oriented tracking experiments are studies of control loadings (Bahrick, 1957; Bahrick, Bennett, & Fitts, 1955; Bahrick, Fitts, & Schneider, 1955; Briggs, Bahrick, & Fitts, 1957; Howland & Noble, 1953; Weiss, 1954), input signal characteristics (Hartman, 1957; Hartman & Fitts, 1955; Noble, Fitts, & Warren, 1955), the magnitude of lag between control movement and system output (Conklin, 1957; Warrick, 1949), the effects of visual noise (Briggs & Fitts, 1956; Briggs, Fitts, & Bahrick, 1957), mathematical transformations of the output signal (Briggs, Fitts, & Bahrick, 1958), and compensatory vs. pursuit tracking (Chernikoff & Taylor, 1957; Poulton, 1952b). Task variables, because of their role in determining the behavioral requirements for the operator, are an important class of variables for psychology and engineering psychologists have made a notable contribution in directing attention toward neglected determiners of human behavior. However, this strong task orientation has led to the neglect of procedural variables that influence the operator, and thus the efficiency of the total man-machine system. A recent article (Taylor, 1957) has clearly stated this emphasis:

> . . . human engineering aims first at building better systems and secondarily at improving the lot of the operator. Thus, whereas conventional psychology, both basic and applied, is anthropocentric, human engineering is mechanocentric (p. 252).

This statement succinctly summarized the task-oriented approach of engineering psychology and expresses a downgrading of procedural variables related to training, retention, fatigue, motivation, etc. It is forgotten, or intentionally neglected, that the engineering psychologist must, over the long run, develop the capability to predict the effectiveness of a man-machine system for different states of the operator, and this means a strict scientific accounting of a broad range of variables which influence man. There are few who underestimate the importance of task variables in determining the behavior of a man-machine system, but there seems to be no sound justification for relegating procedural variables to a secondary status. In the beginning, an applied branch of a science might profitably concern itself with rank ordering its variables in terms of their potency in influencing a criterion (Taylor & Garvey, 1959), but this approach does not deserve being elevated to a research philosophy. Sophisticated applied science, just as sophisticated basic science, must work toward a precise accounting of all variables and their interrelations.

Whereas general experimental psychology has often looked to traditional behavioral theory as a basis for its tracking studies, many engineering psychologists, with their mechanocentric views, have turned towards the feedback theory of closed-loop servomechanisms (Bower & Schultheiss, 1958; Brown & Campbell, 1948; Goode & Machol, 1957) as a model for a man-machine tracking system. Basically, a closed-loop servosystem is an electromechanical error-nulling system which compares an input signal with an output signal and works toward reducing the difference between them. Because error nulling is a basic characteristic of systems which include the human operator as a tracking component, some engineering psychologists view physical servotheory as a potential source of descriptive relationships for manual tracking systems. Figure 1 shows the parallel that is ordinarily drawn between a servosystem and a man-machine tracking system. The theory of servomechanisms is a method of mathematical analysis concerned with the description of the output of a complex system as a function of the input, and it allows the system's analyst to state the functional characteristics of his system with some precision. The expression of the input-output relations is by means of a complex ratio called the transfer function which expresses the nature of the transformations that the system imposes on the input signal. In a system comprised of a number of components, a transfer function is determined for each component and these can then be combined to yield an overall transfer function for the system. An important feature of these methods of system analysis is that it is not necessary to painfully trace the signal through each element of a component to compute the input-output transformation represented in the transfer function. Rather, a "black box" approach can be taken where input-output relationships are directly compared without attending to the many intermediate transformations which occur to the signal as it passes through the component.

The servosystem analyst is concerned with input-output relations as they are manifested in two domains: time and frequency. In the time domain the time-varying characteristics of the system are described in terms of overshooting, undershooting, oscillations, steady state errors, etc., in response to a step input. In the frequency domain the output of the system is examined for transformations of a sinusoidal input after the transients have died out. Finally, and perhaps most importantly in this brief exposition on the methods of servosystem analysis, is that the entire mathematical structure is founded on the assumption of linearity. Fundamentally, this assumption means that the system obeys the superposition theorem which states that the system response to the sum of a set of inputs is equal to the sum of the responses made to each input separately. This means that the performance of the system can be predicted for any complex input providing we know the response of the system to each of the constituent inputs comprising the complex input. Another implication for the linear assumption is that it will accurately reproduce input sinusoidal frequencies after transients have died out, although there may be phase shift and amplitude change. Furthermore, it is implicit that the output of the system is solely a function of the input and this functional relationship is described by the transfer function—i.e., for example, it is not a function of such variables as time where the system might perform one class of transformations on the inputs at time t and another class at a later time.

Ellson's paper (1949) best expresses the hope of some engineering psychologists that the transfer function for the human operator might be determined and provide an analytical means of predicting the performance of the total man-machine system, and of optimizing the performance of the system by designing hardware components to complement the response characteristics of man. This goal of mathematically describing the characteristics of man and his machine elements is scientifically admirable but, regrettably, it was doomed from the beginning by the massive barrier of the linearity assumption. Almost self-evident is the fact that the human operator is a nonlinear component of a system with his intricate adaptive propensities toward learning, fatiguing, motivational shifts, etc. and that there is faint possibility of finding *the* transfer function which can be used by system designers to optimize the performance of a system by capitalizing on the transformations that man imposes on a signal as it enters the receptors, makes passage through the organism, and is emitted anew by the responding effector system (Birmingham & Taylor, 1954; Ellson, 1949; Fitts, 1951; Searle & Taylor, 1948). Birmingham and Taylor (1954) have nicely expressed this matter of nonlinearity for the tracking human operator:

> This adaptability on the part of the man is, of course, a great boon to the control designer, since he can rely upon the human to make the most of any control system, no matter how inadequate. It is this which probably constitutes the most important single reason for using men in control loops. Yet, this very adjustability renders any specific mathematical expression describing human behavior in one particular control loop quite invalid for another man-machine arrangement. This suggests strongly that "*the* human transfer function" is a scientific will-o'-the-wisp which can lure the control system designer into a fruitless and interminable quest (p. 1752).

Fitts (1951) has reported on certain limited conditions where human response appears to approximate linearity but, in general, it would seem that the nonlinearities of human behavior negate the usefulness of the servomodel and its mathematical techniques as a serious theoretical instrument for behavior theory or as a tool for the design of man-machine systems. Nonlinearities do, of course, occur in some physical systems but the assumption of linearity is met sufficiently well and often to make the theory of important value for the physical sciences. This could hardly be said for psychology where nonlinearities are an inherent, and indeed the most interesting and challenging, aspect of the human operator. It must be concluded therefore, that present-day servotheory stands in an analogous, not a scientific, relationship to man-machine tracking systems.

Even if analytical methods eventually become available to handle the nonlinearities of closed-loop human behavior, it is unlikely that engineering psychology will be able to make effective use of them if it continues its preoccupation with

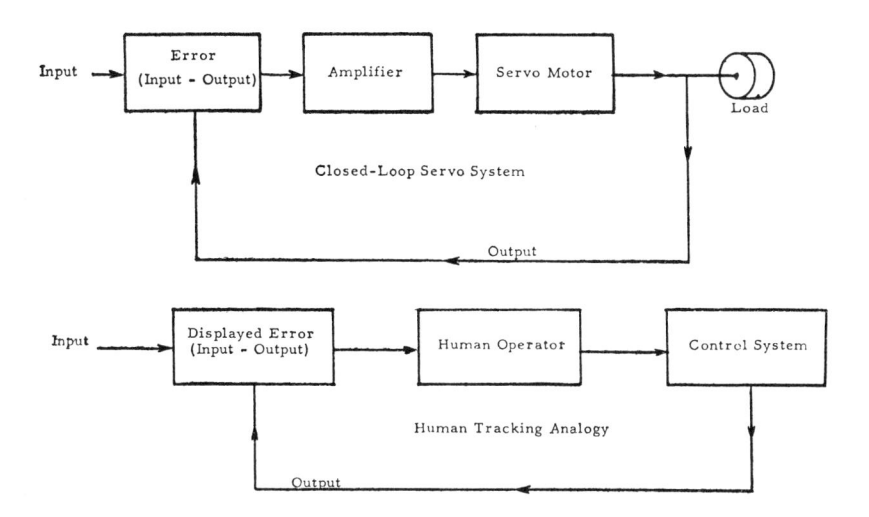

Figure 1. The analogy commonly drawn between a closed-loop electromechanical servosystem and a human operator as an error-nulling agent in a tracking task.

TABLE 1

Matrix Governing the Allowed Transitional States for the Input Signal and the Control System.

	A	B	C	D
A	Yes	Yes	No	No
B	Yes	Yes	Yes	No
C	No	Yes	Yes	Yes
D	No	No	Yes	Yes

(rows indexed by i, columns by j)

Note—The matrix represents a hypothetical four-state one-dimensional tracking task. Cells marked with "Yes" indicate permissible transitions from the ith state at time t to the jth state at time $t+1$. "No" entries are absolute constraints and signify the denial of transition to a jth state from a prior ith state.

task variables (Taylor, 1957; Taylor & Garvey, 1959) and underplays the role of procedural variables which are basic determiners of dispositional states of the operator and contribute substantially to the nonlinearities. Engineering texts on servotheory (Bower & Schultheiss, 1958; Brown & Campbell, 1948; Goode & Machol, 1957) distinguish between *analysis* or the description of a system of existence, and *synthesis* or the prediction of the characteristics of components of the system to achieve certain objectives. Conceivably, we might eventually describe a man-machine tracking system already in existence because the response characteristics of the human operator can be empirically determined for the range of inputs of interest and the operator states that prevail. However, synthesizing is quite different because it requires that we know the laws of human behavior as a function of task and procedural variables and are able to *predict* the characteristics of the human response functions. Questions relating to such operator states as learning and fatigue most certainly will arise and it is evident that these queries will not be answerable if task variables are taken as the primary research domain of engineering psychology. Engineering psychology, it would seem, cannot escape the burden of the same variable and searches for lawfulness which traditionally occupy all psychologists.

In defense of the servotheory approach to tracking, its protagonists have been engaged in proper search for a descriptive mathematical device for man-machine tracking systems which includes provisions for task variables and the properties of response outputs to inputs which are continuous with respect to time. Contemporary behavior theories ordinarily employ measures of behavior, such as frequency and latency, which can be defended as operationally meaningful dependent variables but which are gross summary indices of complex behavior sequences and often do violence to the subtleties of the ongoing behavior. Commonly, psychologists in their laboratory research will elicit elaborate time-based response sequences from an organism and then will ignore completely the time-varying characteristics of the responding in their measurement. In contrast, psychologists studying tracking have recognized, almost from the beginning, that their dependent measures should somehow describe the prominent characteristics of time-based response function. And, because contemporary behavior theories give no attention to time functions, tracking psychologists appear to have suffered disenchantment and have turned to the mathematical schema of closed-loop servotheory, inadequate though it is, because it grapples directly with the measurement and description of time-varying quantities. The fact that servotheory is of little value for quantitative description of man-machine tracking systems should not allow us to forget that the interest in it has reflected a legitimate concern about measurement issues and variables which are important for the response phenomena under investigation.

TRADITION OF GENERAL EXPERIMENTAL PSYCHOLOGY

Basic research on tracking by general experimental psychologists has not had the strong emphasis of task variables. Frequently, in basic research, the experimental task has been a convenient means of eliciting a response class for the purposes of manifesting underlying behavioral processes which are of theoretical rather than practical interest, and consequently tracking tasks have not been studied for their own sake. Examples of this approach are many of the tracking studies on the Rotary Pursuit Test with interest in fatigue-like effects or, more exactly, the implications of Hull's (1943) expressions of reactive and conditioned inhibition for behavior (Adams, 1956; Adams & Reynolds, 1954; Kimble & Horenstein, 1948). Other studies of fatigue processes (Floyd & Welford, 1953; Payne & Hauty, 1954; Siddall & Anderson, 1955) using tracking tasks have had a similar general concern and have shown little interest in the study of tracking for its own sake. The interest in task variables per se which has preoccupied engineering psychology has been largely absent in the research of general experimental psychology. This has been a healthy countertrend to the task emphasis of engineering psychology but the approach of using virtually any convenient task to elicit a response class can be considered a deficiency because it shows a lack of appreciation for the influence of task variables on behavior, and the possible interactions that can be expected to occur between task and procedural variables. These studies seem to have implicitly assumed that behavioral laws will transcend particular characteristics of a task, but this is an unlikely possibility because of the extensive work in engineering psychology showing the potent influence of task variables on performance. There is good reason to expect that many task variables will interact with those variables which have been of interest in testing theoretical deductions. To illustrate, if it were eventually found that a major cause for the depressant effects of massed practice on the tracking response was that work inhibition degraded the quality of proprioceptive feedback, the behavior functions would, as a minimum, have to be expressed in relation to the interaction of intertrial interval and those control system variables which determine proprioceptive feedback. Helson (1949), in a report of the Foxboro investigations which were an early series of systematic tracking studies, points out that both task and procedural variables are pertinent to a complete understanding of human behavior. Lewis (1953) has urged closer attention to the relations between the physical organization of tasks and the complexities of behavior.

An important line of tracking research, which can be subsumed under the rubric of general experimental psychology, has been dominated by British investigators of the Applied Psychology Research Unit, Cambridge University, and mainly concerns efforts to delineate the intrinsic characteristics of the overt motor tracking response, and to identify and assess the response classes which intervene between the displayed stimuli and the measured motor response. Examples of these interests are the question of whether the apparently smooth, continuous tracking response is fundamentally intermittent (Chernikoff & Taylor, 1952; Craik, 1947, 1948; Davis, 1956; Elithorn & Lawrence, 1955; Hick, 1948; Poulton, 1950; Searle & Taylor, 1948; Taylor & Birmingham, 1948; Vince, 1948a, 1948b, 1949; Welford, 1952) and the conditions under which the human operator learns to predict or anticipate changes in the input signal (Bartlett, 1951; Craik, 1947, 1948; Leonard, 1953; Poulton, 1952a, 1957a, 1957b, 1957c; Vince, 1953, 1955). These studies have manipulated both task and procedural variables and have, in many respects, been the most influential of all in improving our scientific understanding of tracking behavior because they have attempted, in a detailed and analytical fashion, to clarify the various response facets of tracking behavior and the variables determining them. It is perhaps safe to say that these studies have stood as a numerical minority in tracking research, and this is unfortunate because such information stands as the foundation of any systematic empirical and theoretical organizations of tracking behavior. Neither the studies of tracking qua tracking which have arisen out of the applied interests of engineering psychology, nor the studies of theoretical psychology where tracking tasks have been used as a matter of convenience, can progress very far until their findings are related to the complex characteristics of tracking behavior. Analytical tracking studies in this vein will be discussed in some detail in later sections of this paper.

AREAS OF NEGLECT

With some exceptions, engineering psychology and general experimental psychology have tended to gloss over three fundamental topics which must be given more attention if we are eventually to have the beginning of a theory of tracking behavior:

1. Tracking tasks have never been defined other than by convention. Early interests in tracking behavior arose out of applied situations where a continuously generated error quantity had to be nulled by continuous operator movements. Laboratory studies of tracking follow this applied tradition of a continuous task, although on occasion discrete displacements of the input signal have been used (Craig, 1949; Ellson, Hill, & Craig, 1949; Rund, Birmingham, Tipton, & Garvey, 1957; Searle & Taylor, 1948; Taylor & Birmingham, 1948; Vince, 1948b, 1949). An attempt must be made, at least in a preliminary way, to define the allowable variations in input, both in type and functional form, as well as the characteristics of the control system used for responding.

2. Not enough attention has been given to the emphasis (largely British) on a more detailed description of behavior in tracking. Recognition must be given to the presence and interaction of several overt and intervening response and stimulus classes, and how these factors act to determine the characteristics of the measured motor response.

3. Relatively little interest has been expressed in multidimensional tracking tasks having two or more stimulus sources in the same or different sense modalities, and corresponding dimensions in the control system for response to each source. Most tracking research has been performed on one-dimensional tasks. The implications of various ways of organizing multiple inputs and the control systems for response to them need more formalization and research.

This paper will, in turn, discuss issues, problems, and research associated with each of these three areas.

Definition of Tracking

A one-dimensional tracking task will be defined by the following conditions:

1. An externally driven input signal defines an index of desired performance and the operator actuates the control system to maintain alignment of the output signal of the control system with the input signal. The discrepancy between the two signals is the error and the operator responds to null the error. Two basic types of tracking tasks are differentiated by how this error quantity is represented:

(a) Pursuit Tracking. The display has two indicants. One is actuated by the input signal and the other is linked to the output signal of the control system. The two indicants are presented directly to the operator and he responds to null the error difference between them. (b) Compensatory Tracking. The error to be nulled is not the difference between two directly observed indicants primarily linked to the input and output signal as in pursuit tracking. Instead, the error observed in pursuit tracking is abstracted and used to actuate a single indicant in relation to a fixed reference. The operator's task, just as in pursuit tracking, is to null this error. The principal difference between pursuit and compensatory tracking is that with the latter the operator never observes the uncontaminated action of the input or output signal directly—only the error difference between them.

2. The input signal is time-based and independent of the operator's response, i.e., the task is *paced*. A paced task is distinguished from a self-paced task where stimulus changes are a function of operator responding (Adams, 1954).

3. The control system has constraints that enforce certain transitional courses of action on the human operator. Instead of being able to move the control from a given position to any other position, the operator must move through defined intervening states of the control system. For example, consider a one-dimensional visual tracking task using a pivoted control lever with hypothetical control Positions A, B, C, and D. If the operator is at Position B at time t, he has a three-choice decision for moving the control at time $t + 1$, each with a probability of being correct: he can repeat the response of time t and leave the control at Position B, or he can move the control to either Position A or C. At the two extreme limiting positions of the control, only two choices are involved: leave the control where it is, or move it to the position adjoining the limiting one. By this definition, any task where the operator has free transitional access to all of the control system states is prohibited from being a tracking task.

4. The states of the input signal have the same transitional constraints as the control system. The input signal, in changing from time t to $t + 1$, must change according to constraints defined by the control system. By imposing the same constraints on the input signal and the control system, the tracking task is given a degree of feasibility for the human operator and means that the input cannot take any action which, in principle, cannot be met by action of the control system. This does not mean that a tracking task must allow near perfect performance by the operator. The input function may be a high frequency sine wave to which the operator can never achieve a high level of proficiency, but this is a behavioral matter and not a function of inherent design features of the task. Table 1 presents the permissible transitional states for the hypothetical four-state tracking task discussed above.

This definition is general and does not specify the characteristics of the input signal or the control system, other than indicating certain transitional restraints for both. The input states and the responses to them can be discrete or continuous, and the input can have any degree of regularity from nearly random (true randomness is denied by conditional restraints of the type shown in Table 1) to completely repetitive. The use of discrete states of the input signal deserves more than the passing attention it has been given in the past because they are particularly amenable to statistical structuring in terms of first and higher order probabilities (with the restraints noted). Another advantage of discrete inputs is that their duration is easily manipulable, making the number of events per unit of time an important dimension for investigation. This time variable has been termed the "speed or pacing factor" (Adams, 1954; Conrad, 1951, 1954; Wagner, Fitts, & Noble, 1954) and is analogous to number of cycles per second when a continuous input is used. One promising measure expressing the statistical coherency of a discrete input signal and the duration of its events is the informational measure of bits per unit of time (Shannon & Weaver, 1949). The rate of change, as well as higher derivatives, can also be a variable for discrete input events but no attempts have ever been made to explore these more complex dimensions.

The Complexity of Behavior in One-Dimensional Visual Tracking

The purpose of this section is to discuss some of the characteristics of the response classes which can be identified in one-dimensional visual tracking, as well as the issues surrounding them. Visual tracking will be analyzed because almost all tracking research has used the visual modality. However, in whatever broad empirical and theoretical conceptualizations of tracking behavior that might eventually mature, it will be necessary to structure the characteristics of tracking in other sense modalities too. But since other modalities such as audition have received only exploratory attention (Humphrey & Thompson, 1952a, 1952b, 1953), it seems unduly speculative at this time to include them.

Rather than the servotheory approach which has been the frame of reference of some investigators, an attempt will be made to demonstrate, on the basis of the available experimental evidence, that tracking behavior involves a linked chain of overt and internal stimuli and responses and is much more complex than implied by the prominent error-nulling characteristics of the servoanalogy. While the servoanalogy is adequate enough for its schematic purposes, the behavioral phenomena cannot be viewed so simply. There are three major areas for discussion: the observing response which orients receptors to sense stimulus events on the display, the prediction responses where the operator learns to anticipate future characteristics of the input signal, and the hypothesis that the measured motor response, even in continuous tracking, is intermittent and not smooth graded movements that might appear to a casual observer.

Most of the phenomena will be discussed in greatest detail under the heading of pursuit tracking, and the presence of the same or similar phenomena in compensatory tracking will, in most cases, be obvious. Behavioral considerations which are uniquely characteristic of compensatory tracking will be treated separately.

PURSUIT TRACKING

Observing Response

The sensing of the displayed indicants driven by the input and output signals, as well as the error difference between them, is by the observing response. These three environmental quantities each play an important role in pursuit tracking and their moment-to-moment state is sampled as the observing response orients the receptors to them. The input indicant is the desired state, the error difference between the input and output signal represents how well the desired state is achieved, and the output indicant gives knowledge of results on how specific sequences of motor movements are represented on the display. Some general attention has been given to the general role of the observing response (Wyckoff, 1952), but within the context of tracking it is considered as having two functions: head and/or eye movements to direct the visual receptors to spatially separated stimuli, and the discrimination of stimulus change. The head and/or eye movements can be considered overt aspects of the observing response and potentially measurable (Mackworth & Mackworth, 1958). However, the discrimination function of the observing response is an inferred phenomena, with its locus unspecified.

Common experience dictates the necessity for an observing response but there is also experimental evidence which documents its importance. Adams (1955), using the Rotary Pursuit Test, found that operations of repeatedly activating the visual observing response independently of the arm-hand goal response, and which presumably served to fatigue the observing response, resulted in a goal response decrement and permitted the inference that the performance level of the goal response is partly determined by the strength of the intervening observing response. Another relevant line of evidence is a study by Poulton (1952b) where it was found that pursuit tracking performance deteriorated when the two pointers on the display were increased in their spatial separation. One interpretation of this finding is that the greater spatial separation required more extensive orienting of the observing response with the result that less time, on the average, was devoted to each pointer. Viewing the observing response as the mechanism by which stimuli are sampled, the wider the spatial separation the less frequently each source of environmental stimuli is sampled and the less likely that an appropriate response will be made. Bearing on this sampling function of the observing response is a vigilance experiment by Jerison and Wallis (1957) where it was found that the scanning of three stimulus sources resulted in a lower rate of detecting aperiodic stimulus change than when only one source had to be watched.

Prediction Responses

The input signal in pursuit tracking actuates an indicant which is directly observed by the operator. To the extent that the operator can predict the regularities inherent in this input signal he will be able to anticipate the correct response movement and initiate at a time to minimize error. In the absence of a predictive capability the operator must wait for the change in the input signal to actually occur on the display, with the result that his response will generate tracking error as a function of a delay of at least one reaction time interval.

Helson (1949) and his associates, in their Foxboro studies of tracking during World War II, were perhaps the first to suggest that prediction behavior is manifest in reaction time values far less than those obtained in classical reaction time experiments. Bartlett (1951) has written an excellent paper on the role of anticipatory behavior which seems to be little known and referenced in the United States. The most extensive research on the prediction of directional course changes in the input signal has been by Poulton (1952a, 1952b, 1957a, 1957b, 1957c), and he distinguishes between two general classes of prediction: (a) receptor anticipation, which is analogous to the foreperiod of the classical simple reaction time experiment where a preparatory signal is presented to the operator in advance and establishes a "set" for response, and (b) perceptual anticipation, where no advance information is intentionally given each time but the operator nevertheless is able to predict the course of future signals on the basis of his past experience. It is this latter type of anticipation which is of greatest interest in tracking in that any knowledge of a future state of the input signal must be an acquired or learned prediction; the definition of a tracking task does not provide for foreknowledge of a state of the input signal. In one study (1952b) Poulton evaluated anticipation in pursuit tracking as a function of practice and two levels of input complexity— a simple harmonic motion and a complex harmonic course. Taking an anticipation of change in the input signal as a response of duration less than the expected reaction time of about .20 seconds, Poulton found that the subjects were predicting the simple harmonic course both early and late in practice, and that the success of prediction was a positive function of practice. Although overall tracking error decreased with practice on the complex input course, there was no evidence for improvement in anticipation and Poulton concluded that the improvement was largely attributable to increased manual dexterity. In this study, Poulton also investigated the smoothness of tracking, defined by the number of unnecessary discrete changes of speed that were made. The fewer the number of such changes, the better the performance. With the simple harmonic course, it was found that smoothness of response increased with practice but no such changes were found for the complex input. Poulton viewed his measure of smoothness as an additional index of anticipation because, when the operator was not anticipating, he would tend to wander off course and his tracking record would show a greater number of corrective movements. He observes that smoothness is a less sensitive measure of beneficial anticipation than response time because the operator may be tracking with a large lag but nevertheless tracking smoothly. Yet, the fact that the subjects tracked most smoothly for the harmonic input course which also produced the greatest degree of anticipation suggested that the covariation of these two measures reflects the same underlying ability to predict stimulus change in direction.

Another study by Poulton (1952a) used the same pursuit tracking apparatus as in his previous study (1952b) and investigated the accuracy with which an operator could predict the position of the input indicant for various amounts of time in the future. At the sound of a hammer blow the operator had to move the output indicant to the position anticipated for the input indicant when a bell sounded .50, 1.5, or 3.5 seconds later. This procedure was regularly repeated and resulted in a series of discrete responses predicting the position of the input indicant. The accuracy of prediction was better than chance for both simple harmonic and complex harmonic inputs, with the accuracy being greater for simple harmonic motion. On the basis of these experiments, Poulton concluded that course anticipation is an important determiner of the overall proficiency level in pursuit tracking. He hypothesized that higher input speeds place a greater premium on prediction because, as the speed of the input signal increases, the failure to anticipate means that a greater segment of the input course span will pass during the subject's reaction time period if he waits for stimulus change to actually occur before responding and a larger error will develop. An excellent review of the role of prediction in tracking and other types of visual-motor tasks has been published by Poulton (1957b).

A series of investigations by Gottsdanker (1952a, 1952b, 1955, 1956) is closely related to those of Poulton. Gottsdanker's studies were concerned with the prediction of velocities and accelerations of input rather than directional changes in the course, and were subsumed under the label prediction motion. The experimental approach required the subject to track a continuous input viewed through a narrow slit. The input was printed on paper in the form of parallel lines 5 millimeters apart, and the subject responded by trying to keep a pencil point between the two lines. He was told that when the input disappeared he was to project its path into the future as if he were attempting to follow an airplane that had gone behind a cloud. Some of the input paths had constant velocities but others had motions that were positively or negatively accelerated. In general, his findings show that constant velocities are accurately predicted, but that the prediction of accelerations tended to be of a constant velocity rather than the required increase or decrease in velocity. Gottsdanker interpreted this to mean that the subject responds on the basis of averages or integrations of preceding velocities. Two studies by Vince (1953, 1955) used a technique very similar to Gottsdanker's in investigations of what she termed "intellectual processes" in skilled performance. Another paper of interest on this topic, but not directly related to tracking, is by Leonard (1953).

The studies by Poulton and by Gottsdanker have involved the learning of prediction during the course of actual practice on a tracking task. A related line of investigation, which has been given some attention in another study by Poulton (1957a), is the effect of training to predict the stimulus source *prior* to actual motor practice in the total tracking task. This can be viewed as a part-whole transfer of training approach, where prediction responses are considered part of the response totality in tracking. Granting this, prediction responses should be trainable prior to whole-task practice and, in being a part of the total response complex, should have their strength reflected in the dependent motor response whose proficiency reflects the strength of all the response classes in the complex. This approach is quite similar to verbal pretraining methods where the operator is required to learn verbal responses to task stimuli prior to motor responses in the whole task (Arnoult, 1957; Goss, 1955). Although verbal pretraining studies have not dealt specifically with the problem of prediction responses, they are concerned with learned mediating responses where response produced stimuli are hypothesized to provide additional discriminative cues for the motor response (Goss, 1955; Osgood, 1953). Conceptually therefore, they appear quite similar to prediction responses and one might hypothesize that an adaptation of these same methods can be used for prior training in the prediction of input events in a tracking task. However, with our impoverished knowledge of the underlying nature of anticipatory mechanisms, it is plausible that prediction has nothing to do with mediating responses but, indeed, may be fundamentally a proprioceptive-oriented phenomenon. Giving proprioception a role in anticipatory behavior needs only the reasonable assumption that motor movements are conditioned to traces of proprioceptive stimuli and that, with practice, the occurrence of a proper configuration of proprioceptive stimuli will tend to elicit the next correct motor sequence. Certainly this is not to deny intellective processes or mediating responses as variables in prediction, but it does suggest that there might be at least two facets that deserve experimental inquiry. "Prediction response" is a commonly used label for anticipation in this paper but eventually it may prove to be a poor term if proprioception proves to be a paramount influence. The verbal pretraining studies throw the balance of the explanatory weight at present in the direction of mediating responses as the basis for anticipation, but definitive research on this topic remains to be done.

Characteristics of the Measured Motor Response

The basic nature of the motor movement activating the control system in a tracking task has been the subject of extensive discussion and controversy. The issue is whether the motor response is a continuous function of time or whether it is discontinuous and intermittent. The intermittency hypothesis stems from arguments that a responding effector has a period of refractoriness or reduced excitability before it can be made to respond in full strength again. Because this evidence stems from molar behavior data, it is called psychological refractory phase to distinguish it from the physiological refractory phase of individual nerve fibers. The similarity of psychological refractory phase and physiological refractory phase is in terms of reduced responsiveness following stimulation and response, but the levels of analysis of the two classes of phenomena are so different that it is perhaps safest to view them as analogous rather than stemming from a common underlying process.

Probably the first statement of psychological refractory phase was by Telford (1931) who found that reaction time to the second of a pair of auditory stimuli was lengthened if the time spacing of the two stimuli was reduced to .50 seconds, and he concluded that the subject becomes refractory in a manner comparable to the refractoriness of isolated nerves. Using Telford's study as a point of departure, Vince (1948a, 1948b) asked whether refractoriness is present in continuous tracking to give the motor response an intermittent, impulsive quality. She concluded that intermittent corrections every .50 seconds is a basic feature of human tracking responses in a manner quite comparable to Telford's finding for discrete stimuli and a reaction time response. If her interpretation is correct, the notion of psychological refractory phase becomes an important general principle. But in criticism of Vince's findings, psychological refractory phase refers to the periodicity of *motor movements* and not tracking *error*. Her conclusions were based on tracking error records and periodicities in them are not a function of motor movements alone but of the *difference* between the output signal generated by the motor movements and the input signal. Periodicities in the error function may be correlated with periodicities in motor movements but they are contaminated by the influence of the input signal and are not an unequivocal index on which to base conclusions about psychological refractory phase as a mechanism for inducing intermittent motor corrections.

With the exception of the foregoing studies by Vince, research on motor intermittence has been with discrete tasks, although many of the investigators have freely implied the generality of the phenomenon to include continuous tracking. Mainly, these studies conclude that reaction time to a second stimulus of a pair will be lengthened if the interstimulus interval is less than .50 seconds. A limit to this generalization is that very brief interstimulus intervals cause the stimuli to be perceived as a single entity, with the result that only a single response occurs. Vince (1948b, 1949) and Hick (1948) have both used discrete tracking tasks and have provided additional corroborative evidence on psychological refractory phase. Craik (1947, 1948) and Welford (1952) use these data for theoretical discussions on the generality of psychological refractory phase as a determinant of intermittency in responding. Poulton (1950) criticized the tendency to regard the refractory interval of .50 seconds as a human constant because the quasirandom presentation of stimuli did not allow the operator to form a proper preparatory set. When allowance is made for the acquisition of a preparatory set by having predictable stimuli, Poulton found that the refractory phase interval reduces to .20–.40 seconds. Davis (1956) and Elithorn and Lawrence (1955) also discuss the role of anticipatory set and the psychological refractory period. A general discussion of research and views on this topic is presented by Fitts (1951).

Another aspect to the intermittency hypothesis is that the duration of patterned movements to discrete stimuli can be less than visual-motor reaction time. This has implied response discontinuity to some investigators because the subject is executing response sequences momentarily independent of the magnitude of the visually perceived error. Because the response is not continuously guided by the primary visual tracking error quantity, it is, for a time, open-loop or intermittent (Searle & Taylor, 1948; Taylor & Birmingham, 1948). These authors conclude that response movements are under a kind of a "cam control" where the visually perceived error triggers a cammed sequence. On the basis of past experience the "cam" runs off the continuously varying force pattern, including starting and stopping, and all without visual or proprioceptive feedback.

Admitting the possibility that the continuous control of movements during an interval less than that required for visual-motor reaction time can be proprioceptive feedback, Chernikoff and Taylor (1952) conducted a study to see if kinesthetic reaction time was sufficient to account for the control of the response. They concluded that continuous tracking behavior is best described by the intermittency hypothesis, analogous to cam control where very brief movement sequences are run off in the absence of visual and proprioceptive guidance. Lashley (1951) in a parallel line of argument, is in agreement that kinesthetic reaction time cannot explain many facts of motor responding such as the finger movements of a skilled pianist moving at about 16 per second. These rates are too fast to allow kinesthetic feedback after each one, and Lashley postulates that some central sensory control is operating, presumably in a fashion similar to the cam hypothesis stated by Taylor and his associates. Craik (1947) holds a similar view. Arguing from piano playing to tracking is tenuous however, if for no other reason than that a musical composition provides foreknowledge of a requirement for movement sequences, and reaction time to each one is known to be greatly shortened under these special conditions (Vince, 1949). Advance notice of stimuli is not a characteristic of tracking tasks. Moreover, Poulton's work (e.g., 1952b) has shown that learning to anticipate stimulus sequences is revealed in greatly shortened reaction time values. It is hardly surprising that a trained musician can sometimes sidestep the restraints of an elementary afferent-efferent loop and receive guidance from learned, internal sources.

Gibbs (1954a, 1954b) in two important papers effectively argues against the hypothesis that continuous motor movements do not have continuous kinesthetic feedback guiding them. He points out that arguments based on kinesthetic reaction time fail to distinguish between the connecting and conducting functions of the central nervous system. Gibbs observes that kinesthetic reaction time to discrete stimuli can be considered the connecting time between kinesthetic stimulation and overt motor response, and this has little bearing on continuous kinesthetic or neural conduction during voluntary movement. Gibbs bases his discussion on physiological data by Matthews (1933) which showed that a muscle had "tension" afferents and "stretch" afferents which, respectively, provide sensing of static position and of movement of a limb. Tension afferents respond primarily when the muscle is at rest and has an electrical discharge approximately proportional to the logarithm of the tension. Stretch afferents, on the other hand, respond when the muscle is stretched in movement and has a rate of electrical discharge proportional to the rate of stretch, and Gibbs holds that this is the source of *continuous* kinesthetic feedback monitoring. The subject must "know" limb position in guiding his movements and Gibbs holds that this is obtained by integrating the rate function. The notion of a finite integration period might suggest that Gibbs' hypothesis is essentially the same as the intermittency hypothesis because successive integrations might be revealed as intermittent movements of .50 seconds as limb position is successively "computed." Actually, the implications are quite different because Gibbs' hypothesis would seem to hold that there are conditions where an integration interval of .50 seconds would apply but that integration intervals of longer duration are equally possible. Gibbs' physiological hypothesis would seem to allow for perfectly smooth tracking movements of relatively long duration and, indeed, this is a common observation in tracking records. Oddly enough, relatively long periods of smooth responding in continuous tracking have not served as grounds for seriously challenging the intermittency hypothesis. Craik (1948) and Noble et al. (1955) remark on these smooth responses and offer the ad hoc explanation that intermittent movements are occurring in accordance with the principle of psychological refractory phase but that the subject's acquired capability to predict input sequences has overlaid a smoothing effect. While prediction responses may well have some sort of smoothing influence, it also may be true that the intermittence hypothesis is false for continuous tracking and that relatively long, smooth responses frequently occur in the absence of prediction behavior. Gibbs' work emphasizes the rather simple fact that the intermittency hypothesis has its validity derived from research on discrete tasks and its generalization to continuous tracking may be inappropriate.

Gibbs' use of Matthews' findings raises the interesting idea that proficiency in making accurate accelerations in tracking is related to the subject's ability to discriminate changes in the rate of kinesthetic impulses. One interpretation of Gottsdanker's findings (1952a, 1952b, 1955, 1956) that the subject poorly predicts velocity changes is that he cannot kinesthetically discriminate with enough accuracy those velocity changes which he visually perceives. However, this interpretation must be approached cautiously because it fails to consider that the inability to discriminate velocity changes conceivably could be on the visual-perceptual side rather than the kinesthetic. To interpret Gottsdanker's data properly we must, by independent operations, determine the relative capabilities of perceptual and kinesthetic discrimination of acceleration. If the operator cannot perceptually discriminate the velocity changes involved, then the motor response system is not receiving adequate information and the overt response cannot be expected to reflect information that has not been received. Or, conversely, the operator may be perfectly able to discriminate the velocity change perceptually but he may be unable to translate it into the proper accelerated movement because he cannot make sufficiently accurate kinesthetic discriminations. Some work on the perceptual discrimination of instantaneous changes in velocity has been done by Hick (1950) and Brandalise and Gottsdanker (1959). Too, this general line of reasoning suggests the hypothesis that the relative effectiveness of position, rate, and acceleration tracking may be related to the compatibility of perceptual and kinesthetic events.

COMPENSATORY TRACKING

The discussion of research and problems under the heading of pursuit tracking applies also to compensatory tracking. Whatever differences exist are resident in the different ways in which the two types of tracking tasks have their data organized on the display. The presentation of only the error quantity in compensatory tracking means that performance usually will be poorer for two reasons:

1. The operator cannot see the input signal directly which means that he is handicapped in the acquisition of prediction responses.

2. The operator cannot see the output signal directly so he is handicapped in receiving knowledge of results. In addition to influencing the acquisition of simple visual-motor learning where prediction behavior is absent, this factor also influences the acquisition of prediction responses because the operator cannot unequivocally verify the results of any particular prediction response.

Depending upon task circumstances, some prediction behavior can be expected to form under compensatory tracking conditions. The error signal is a function of both the input and the output signals, and at times the regularities in the input will be discernible. Poulton (1952b) has shown that prediction behavior does occur with practice in compensatory tracking but that prediction is impressively superior in pursuit tracking. Undoubtedly this is one of the factors which almost always renders pursuit tracking superior to compensatory tracking (Hartman & Fitts, 1955; Poulton, 1952b).

Nor can we assume that the absence of a direct presentation of the output signal means that knowledge of results is completely absent. There is evidence from a study by Chernikoff and Taylor (1957) that when the input signal of a continuous tracking task is a low frequency input the subject receives fairly adequate knowledge of results, probably because the motor movements produce output frequencies which are higher than the input frequency changes. This is deduced from the slightly better performance that was found in this study for compensatory over pursuit tracking when the input was a low frequency signal. At higher frequencies, they found that pursuit tracking maintained its well-known advantage over compensatory tracking.

TWO-DIMENSIONAL TRACKING

A two-dimensional tracking task has two stimulus sources commanding response, with each source having its own separate input signal and a dimension of the control system for response. An example of a two-dimensional visual tracking task would be two voltmeter stimulus sources with a left-hand control lever for response to one and a right-hand lever for response to the other. Or, the two stimulus sources could have a bisensory distribution, with one visual and one auditory. Our ignorance of variables involved in the various ways to organize a two-dimensional tracking task dictates that only a limited examination of some of the issues be made. The discussion will be restricted to two cases: spatially separated visual sources, and bisensory sources where one is auditory and the other is visual. Nothing of importance is known of the effects of control system design as it bears on the distribution of the two response dimensions among the possible effector systems, so it will not be discussed. Nor will the relative advantages of pursuit and compensatory displays be discussed for whatever special implications might be found for two-dimensional tasks.

Because almost all of the research in tracking has employed one-dimensional visual tasks, it is unfortunately necessary to attempt this preliminary discussion of two-dimensional tasks on a rather thin foundation of empirical findings. Perhaps the dearth of analytical data on more complex tracking tasks is because of the implicit view of many psychologists that it is desirable to progress in research from simple to complex systems, and that the laws of complex systems will tend to fall into place once the relationships for simpler tasks are established. On the other hand, it is possible to defend the position that parallel law-seeking at two levels of analysis will result in two bodies of laws, each appropriate for its own domain. As these two bodies of knowledge develop, specific research can then be directed towards finding the empirical composition laws which express the interactions relating the laws of the two strata. If this view is allowed, it does not seem necessary that the study of multidimensional tracking tasks should await the codification of laws governing one-dimensional tracking.

To facilitate exposition, the following terminology has been adopted. Each stimulus source and its dimension of the control system will be called a component task of the total task. Response in the component task will be termed the component response of the total response. As before, the observing response will serve to orient the receptors to the events emitted by the stimulus sources.

Visual Tracking

Observing Response. One of the distinguishing features of a two-dimensional visual tracking task is that there is not only the need for scanning *within* a source but also the more demanding requirement to scan *between* sources. This added response requirement is importantly a product of the task variable called load (Conrad, 1951, 1955). Load is defined as the number of stimulus sources and has an expected interaction with the rate of events emitted from each source. This latter variable has been termed speed (Conrad, 1951, 1954). Performance deteriorates both with increase in speed and load. Moreover, it has been shown that response proficiency is a function of the extent to which events in spatially distributed sources overlap in time and command two simultaneous responses (J. F. Mackworth & N. H. Mackworth, 1956; N. H. Mackworth & J. F. Mackworth, 1956, 1957). Another important task variable which would certainly interact with speed and load in determining the observing response is the amount of the spatial separation of the sources. Tracking proficiency as a function of the amount of spatial separation has not been systematically studied.

Prediction Responses. An important but unverified implication for prediction responses in a two-dimensional visual task is that they might reduce the major requirement for visually scanning the stimulus sources and improve tracking performance. Prediction responses in a one-dimensional task are known to benefit motor performance. We might hypothesize that in a two-dimensional task there is not only prediction within each source but also prediction *between-sources*. If the human operator can learn to predict events within a source, it would seem that he might learn of the covariation between the events of the two sources. Given an event in one source he would have some likelihood of correctly predicting the concurrent event in the other source and consequently would not need to attend visually to his source as often. We know nothing of these matters but between-source prediction is a reasonable expectation.

Component Response Differentiation. Two-dimensional tracking often involves two or more component response effector systems, such as both hands or a hand and a foot, and this raises the issue of motor interaction between the two systems. It is a common observation that initial stages of total response in a multidimensional task are often typified by a level of uncoordinated activity and error far greater than might be expected from low habit strength in each component response separately. But, as practice proceeds, these interactions of component responses tend to drop out completely or show a marked decrease, with each participating component response effector system becoming smoothly proficient. This phenomenon shall be called component response differentiation. Within the framework of his S-R contiguity theory, Guthrie (1952) discusses the acquired differentiation of component responses:

> . . . reduction of habit to essentials makes many habits local responses no longer involving the whole body. When we are practiced we drive and talk, or play the piano and smoke, or skate and greet a friend at the same time. At first this is impossible because driving, playing, skating all include a mass of action that is not essential to the performance but is present because it is part of total associated complex bound together by conditioning. In time, many irrelevant movements are dropped out from the complex and the activity is limited to the muscles and the movements required for the performance. This process is, of course, never complete. Perfect grace, which means the use only of the essential muscles and this use only to the point necessary for the action, is only approximated, never reached (p. 109).

How component response interaction is manifested in a two-dimensional tracking task is not known at this time. However, the extensive literature on experimentally induced muscular tension, which has been organized by Meyer (1953) in terms of physiological hypotheses, leaves little doubt that interaction of simultaneous motor responses occurs. The concern of Meyer's review and analysis was the effects of experimentally induced muscular tension where usually a static, muscular tension-inducing component response accompanies a more central learning activity, such as rotary pursuit or paired-associates learning. The major area of interest for two-dimensional tracking, but where much less is known, concerns total tasks where all component tasks impose a learning requirement on their respective component responses. Perhaps, as Guthrie suggests, the interaction will all but disappear. But until a means of defining and measuring the course of component response differentiation in tracking is uncovered, there is no reason for discussion beyond this passing mention of a potentially important area.

Visual-Auditory Bisensory Tracking

The major issue for two-dimensional tracking with one visual and one auditory source is whether there is interaction which intrinsically prevents the two stimulus event streams from being processed simultaneously. While we might intuitively surmise that a total task organized in this manner will be superior to two-dimensional visual tracking because each stimulus stream, with its own sense modality, gives the operator a higher load carrying capability in that he does not have to time-sample the sources with the observing response as he does in two-dimensional visual tracking, there is no evidence of the conditions for which this can be true, if at all. As a first experimental question it would seem desirable to attack the pure case in two-dimensional bisensory tracking and ask whether it is possible to *simultaneously* process two stimulus streams without impairing interaction effects. Any research program should have a strategy which sets up a hierarchy of research questions whose answers are ordered in terms of their contribution to the delineation of variables and laws, and in bisensory tracking the best strategy is suggested to be one of first determining whether the human operator can process two event streams at once. Having determined the empirical truth or falsity of this hypothesis, we will be in a better position to comparatively examine the relative merits of all-visual and bisensory tasks. Later variables to consider would be the differential capabilities of the visual and auditory senses for different classes of stimulus inputs (Henneman & Long, 1954).

Subjectively we all have the confident feeling that we can handle visual and auditory events simultaneously. It is commonplace to encounter the observation that one can simultaneously read a book and listen to the radio. Yet, as with most anecdotal accounts, they may be true but the absence of experimental controls precludes any proof of the thesis. Thus, an explanation of these experiences of everyday life is just as plausible in terms of rapid sensory shifting from one data stream to another. Experimentally, the issue is a delicate one and will require careful analysis and experimentation to decide it conclusively.

The experimental design necessary to prove or disprove that the human operator is a one-channel system must, as a minimum, show that performance of each component response in a bisensory tracking task will, after practice, be the same as performance when each component task is practiced out of total task context as a separate task. But what interpretation can be given if component response measures in bisensory tracking performance fail to achieve the level attained on part tasks? The hypothesis that the human operator is a single channel data processing system is supported but the investigator is then faced with the new question of the locus of the interaction. There are at least four possibilities which must then be resolved, although it will take some ingenuity and analysis to operationally differentiate them for laboratory testing:

1. The human operator is truly a one-channel system and, when two units of stimuli arise simultaneously, one must be temporarily stored while response occurs to the other. At the completion of the first response, the second stimulus unit is removed from central storage and response is made to it.

2. No storage is required. The operator is capable of simultaneously processing two event streams but there is motor interaction which prevents the two responses from simultaneously occurring with the same effectiveness that would be observed for any one of them separately. In effect, this hypothesis is consistent with Guthrie's position that there is always some interaction between simultaneously functioning response systems, even after very large amounts of practice. Component response differentiation is never complete.

3. No storage is required and there is no interaction of responses at the motor level. However, there is sensory interaction which results in a degradation in performance that would be absent if only one stream of stimuli were being handled. Evidence for sensory interaction is presented in a number of papers (Child & Wendt, 1938; Gilbert, 1941; Gregg & Brogden, 1952; Hartman, 1934; London, 1954; Ryan, 1940).

4. Combinations of the above three possibilities.

The most relevant research on simultaneous bisensory data processing for tracking is by Davis (1957). While he did not study tracking or even strict simultaneity of bisensory events, he did study the effects of very small time intervals between a visual and an auditory stimulus and the experiment makes a significant contribution to the topic. Following the generalizations on psychological refractory phase, Davis asked whether the operator is refractory if the second of two successive stimuli impinges on a different sense modality than the first. Using the reaction time response and stimuli of very brief duration, Davis found that the reaction time to the second signal increased as the interstimulus interval decreased. The data show that the phenomenon which has come to be known as psychological refractory phase operates for two successive stimuli in two sense modalities about as it does for two successive events in a single sense modality. In some fashion, a "queuing of signals," to use the engaging phrase of Davis, occurs whether stimuli arrive over one or two sense channels. Davis finds his data consistent with a model of the human operator as a single channel information system. If we can assume that the processing of simultaneous events is a special zero-interval case of intervals for successive stimuli, the extrapolation of the Davis findings to the zero interstimulus interval suggests a substantial impairment in performance. An empirical study of truly simultaneous events must be done but the Davis experiment is unquestionably provocative on the simultaneity issue.

SUMMARY AND CONCLUSIONS

This paper has reviewed some of the major issues and problems in the study of human tracking behavior. Apart from the complexities that are inherent in the analysis of closed-loop behavior, which is somewhat more complicated than the open-loop situations used by most psychologists in their studies of human behavior, tracking behavior is beset with the added complications of mediating responses and stimuli which are important variables intervening between the display and the measured motor response. Moreover, all of these variables assume further complications when they are cast in the matrix of multidimensional tracking tasks with two or more stimulus sources, each with a corresponding dimension of the control system for response to them. And, not only do multidimensional tasks have complications resulting from a compounding of the effects of variables found in one-dimensional tracking, but they have the added issues of how one or more sense modalities process the incoming data and how the component response systems interact throughout learning to become partly or completely noninteractive (differentiated). We appear to be a long way from understanding these factors and, until we do, we are a long way from the beginnings of any kind of theory of tracking. British research has been most influential in illuminating the characteristics of tracking behavior, with its experimental examination of what is learned (e.g., prediction behavior in tracking), and its study of the intermittency hypothesis. This approach of British investigators would seem to be mandatory for our eventual theoretical description of tracking, and is in some contrast to the approach of engineering psychologists in the United States who tend to emphasize measures of tracking behavior as a function of task variables and often bypass detailed analyses of the learned behavior. Some important exceptions to this emphasis on the domestic scene has been the early work of the Naval Research Laboratory, Gottsdanker, and recent work by Briggs and his associates on learning and transfer as a function of task variables.

If this paper can be said to have a point of view, it is that tracking research is in need of a rapprochement of the interests of the engineering psychologist, with his focus on task variables and the measurement of time-based behavior, and interests of the traditional experimental psychologist who tends to emphasize behavior as a function of variables which determine conceptual states such as habit, work inhibition, motivation, mediating responses, etc. Physical servotheory has been a prominent attempt in engineering psychology to describe tracking behavior, but the absence of variables defining conceptual states long known to influence behavior eliminates it as a psychological theory of any stature, quite apart from its formal shortcomings for the description of nonlinear human behavior. It is unlikely that a theory of tracking behavior will emerge until these conceptual variables are included, along with time series measurement and task variables which traditionally have occupied engineering psychology.

REFERENCES

Adams, J. A. The effect of pacing on the learning of a psychomotor response. *J. exp. Psychol.*, 1954, **47**, 101-105.

Adams, J. A. A source of decrement in psychomotor performance. *J. exp. Psychol.*, 1955, 49, 390-394.

Adams, J. A. Some implications of Hull's theory for human motor performance. *J. gen. Psychol.*, 1956, 55, 189-198.

Adams, J. A. The relationship between certain measures of ability and the acquisition of a psychomotor criterion response. *J. gen. Psychol.*, 1957, 56, 121-134.

Adams, J. A., & Reynolds, B. Effect of shift in distribution of practice conditions following interpolated rest. *J. exp. Psychol.*, 1954, 47, 32-36.

Arnoult, M. D. Stimulus predifferentiation: Some generalizations and hypotheses. *Psychol. Bull.*, 1957, 54, 339-350.

Bahrick, H. P. An analysis of stimulus variables influencing the proprioceptive control of movements. *Psychol. Rev.*, 1957, 64, 324-328.

Bahrick, H. P., Bennett, W. F., & Fitts, P. M. Accuracy of positioning responses as a function of spring loading in a control. *J. exp. Psychol.*, 1955, 49, 437-444.

Bahrick, H. P., Fitts, P. M., & Schneider, R. Reproduction of simple movements as a function of factors influencing proprioceptive feedback. *J. exp. Psychol.*, 1955, 49, 445-454.

Bartlett, F. C. Anticipation in human performance. In G. Ekman, T. Husén, G. Johansson, & C. I. Sandström (Eds.), *Essays in psychology*. Uppsala: Almquist & Wiksells, 1951. Pp. 1-17.

Birmingham, H. P., & Taylor, F. V. A design philosophy for man-machine systems. *Proc. IRE*, 1954, 42, 1748-1758.

Bower, J. L., & Schultheiss, P. M. *Introduction to the design of servomechanisms*. New York: Wiley, 1958.

Brandalise, B. B., & Gottsdanker, R. M. The difference threshold of the magnitude of visual velocity. *J. exp. Psychol.*, 1959, 57, 83-88.

Briggs, G. E., Bahrick, H. P., & Fitts, P. M. The effects of force and amplitude cues on learning and performance in a complex tracking task. *J. exp. Psychol.*, 1957, 54, 262-268.

Briggs, G. E., & Fitts, P. M. Tracking proficiency as a function of visual noise in the feedback loop of a simulated radar fire control system. *USAF Personnel Train. Res. Cent. res. Rep.*, 1956, No. AFPTRC-TN-56-134.

Briggs, G. E., Fitts, P. M., & Bahrick, H. P. Learning and performance in a complex tracking task as a function of visual noise. *J. exp. Psychol.*, 1957, 53, 379-387.

Briggs, G. E., Fitts, P. M., & Bahrick, H. P. Transfer effects from a single to a double integral tracking system. *J. exp. Psychol.*, 1958, 55, 135-142.

Brown, G. S., & Campbell, D. P. *Principles of servomechanisms*. New York: Wiley, 1948.

Chernikoff, R., & Taylor, F. V. Reaction time to kinesthetic stimulation resulting from sudden arm displacement. *J. exp. Psychol.*, 1952, 43, 1-8.

Chernikoff, R., & Taylor, F. V. Effects of course frequency and aided time constant on pursuit and compensatory tracking. *J. exp. Psychol.*, 1957, 53, 285-292.

Child, I. L., & Wendt, G. R. The temporal course of the influence of visual stimulation upon the auditory threshold. *J. exp. Psychol.*, 1938, 23, 109-127.

Conklin, J. E. Effect of control lag on performance in a tracking task. *J. exp. Psychol.*, 1957, 53, 261-268.

Conrad, R. Speed and load stress in a sensori-motor skill. *Brit. J. industr. Med.*, 1951, 8, 1-7.

Conrad, R. Speed stress. In W. F. Floyd & A. T. Welford (Eds.), *Human factors in equipment design*. London: Lewis, 1954. Pp. 95-102.

Conrad, R. Some effects on performance of changes in perceptual load. *J. exp. Psychol.*, 1955, 49, 313-322.

Craig, D. R. Effect of amplitude range on duration of responses to step function displacements. *USAF Air Materiel Command tech. Rep.*, 1949, No. 5913.

Craik, K. J. W. Theory of the human operator in control systems: I. The operator as an engineering system. *Brit. J. Psychol.*, 1947, 38, 56-61.

Craik, K. J. W. Theory of the human operator in control systems: II. Man as an element in a control system. *Brit. J. Psychol.*, 1948, 38, 142-148.

Davis, R. The limits of the "psychological refractory period." *Quart. J. exp. Psychol.*, 1956, **8**, 24-38.
Davis, R. The human operator as a single channel information system. *Quart. J. exp. Psychol.*, 1957, **9**, 119-129.
Elithorn, A., & Lawrence, C. Central inhibition: Some refractory observations. *Quart. J. exp. Psychol.*, 1955, **7**, 116-127.
Ellson, D. G. The application of operational analysis to human behavior. *Psychol. Rev.*, 1949, **56**, 9-17.
Ellson, D. G., Hill, H., & Craig, D. R. The interaction of responses to step function stimuli: II. Equal opposed steps of varying amplitude. *USAF Air Materiel Command tech. Rep.*, 1949, No. 5911.
Fitts, P. M. Engineering psychology and equipment design. In S. S. Stevens (Ed.), *Handbook of experimental psychology*. New York: Wiley, 1951. Pp. 1287-1340.
Fleishman, E. A. Dimensional analysis of psychomotor abilities. *J. exp. Psychol.*, 1954, **48**, 437-454.
Fleishman, E. A. A comparative study of aptitude patterns in unskilled and skilled psychomotor performance. *J. appl. Psychol.*, 1957, **41**, 263-272. (a)
Fleishman, E. A. Factor structure in relation to task difficulty in psychomotor performance. *Educ. psychol. Measmt.*, 1957, **17**, 522-532. (b)
Fleishman, E. A. Dimensional analysis of movement reactions. *J. exp. Psychol.*, 1958, **55**, 438-453.
Fleishman, E. A., & Hempel, W. E. Changes in factor structure of a complex psychomotor test as a function of practice. *Psychometrika*, 1954, **19**, 239-252.
Fleishman, E. A., & Hempel, W. E. The relation between abilities and improvement with practice in a visual discrimination reaction task. *J. exp. Psychol.*, 1955, **49**, 301-312.
Fleishman, E. A., & Hempel, W. E. Factorial analysis of complex psychomotor performance and related skills. *J. appl. Psychol.*, 1956, **40**, 96-104.
Floyd, W. F., & Welford, A. T. *Symposium on fatigue*. London: Lewis, 1953.
Forbes, T. W. Auditory signals for instrument flying. *J. aero. Sci.*, 1946, **13**, 255-258.
Gibbs, C. B. The continuous regulation of skilled response by kinaesthetic feedback. *Brit. J. Psychol.*, 1954, **45**, 24-39. (a)
Gibbs, C. B. Movement and force in sensorimotor skill. In W. F. Floyd & A. T. Welford (Eds.), *Human factors in equipment design*. London: Lewis, 1954. Pp. 103-117. (b)
Gilbert, G. M. Inter-sensory facilitation and inhibition. *J. gen. Psychol.*, 1941, **24**, 381-407.
Goode, H. H., & Machol, R. E. *System engineering*. New York: McGraw-Hill, 1957.
Goss, A. E. A stimulus-response analysis of the interaction of cue-producing and instrumental responses. *Psychol. Rev.*, 1955, **62**, 20-31.
Gottsdanker, R. M. The accuracy of prediction motion. *J. exp. Psychol.*, 1952, **43**, 26-36. (a)
Gottsdanker, R. M. Prediction-motion with and without vision. *Amer. J. Psychol.*, 1952, **65**, 533-543. (b)
Gottsdanker, R. M. A further study of prediction-motion. *Amer. J. Psychol.*, 1955, **68**, 432-437.
Gottsdanker, R. M. The ability of human operators to detect acceleration of target motion. *Psychol. Bull.*, 1956, **53**, 477-487.
Gregg, L. W., & Brogden, W. J. The effect of simultaneous visual stimulation on absolute auditory sensitivity. *J. exp. Psychol.*, 1952, **43**, 179-186.
Guthrie, E. R. *The psychology of learning*. (Rev. ed.) New York: Harper, 1952.
Hartman, B. O. The effect of target frequency on compensatory tracking. *USA Med. Res. Lab. Rep.*, 1957, No. 272.
Hartman, B. O., & Fitts, P. M. Relation of stimulus and response amplitude to tracking performance. *J. exp. Psychol.*, 1955, **49**, 82-92.
Hartmann, G. W. The facilitating effect of strong illumination upon the discrimination of pitch and intensity differences. *J. exp. Psychol.*, 1934, **17**, 813-822.

Helson, H. Design of equipment and optimal human operation. *Amer. J. Psychol.*, 1949, 62, 473-497.

Henneman, R. H., & Long, E. R. A comparison of the visual and auditory senses as channels for data presentation. *USAF WADC tech. Rep.*, 1954, No. 54-363.

Hick, W. E. Discontinuous functioning of the human operator in pursuit tasks. *Quart. J. exp. Psychol.*, 1948, 1, 36-57.

Hick, W. E. The threshold for sudden changes in the velocity of a seen object. *Quart. J. exp. Psychol.*, 1950, 2, 33-41.

Howland, D., & Noble, M. E. The effect of physical constants of a control on tracking performance. *J. exp. Psychol.*, 1953, 46, 353-360.

Hull, C. L. *Principles of behavior.* New York: Appleton-Century, 1943.

Humphrey, C. E., & Thompson, J. E. Auditory Display: II. Comparison of auditory and visual tracking in one dimension: A. Discontinouus signals, simple course. *Johns Hopkins U. Appl. Physics Lab. Rep.*, 1952, No. TC-146. (a)

Humphrey, C. E., & Thompson, J. E. Auditory Display: II. Comparison of auditory tracking with visual tracking in one dimension: B. Discontinuous signals, complex course. *Johns Hopkins U. Appl. Physics Lab. Rep.*, 1952, No. TG-147. (b)

Humphrey, C. E., & Thompson, J. E. Auditory Display: II. Comparison of auditory tracking with visual tracking in one dimension: C. Continuous signals, simple intermediate and complex courses. *Johns Hopkins U. Appl. Physics Lab. Rep.*, 1953, No. TG-194.

Jerison, H. J., & Wallis, R. A. Experiments on vigilance one-clock and three-clock monitoring. *USAF WADC tech. Rep.*, 1957, No. 57-206.

Kimble, G. A., & Horenstein, Betty R. Reminiscence in motor learning as a function of length of interpolated rest. *J. exp. Psychol.*, 1948, 38, 239-244.

Lashley, K. S. The problem of serial order in behavior. In L. A. Jeffress (Ed.), *Cerebral mechanisms in behavior.* New York: Wiley, 1951.

Leonard, J. A. Advance information in sensori-motor skills. *Quart. J. exp. Psychol.*, 1953, 5, 141-149.

Lewis, D. Motor skills learning. Symposium on psychology of learning basic to military training problems. *Res. Develpm. Bd., Comm. Hum. Resour., Rep.*, 1953, No. HR-HDT-201/1.

London, I. D. Research on sensory interaction in the Soviet Union. *Psychol. Bull.*, 1954, 51, 531-568.

Mackworth, Jane F., & Mackworth, N. H. The overlapping of signals for decisions. *Amer. J. Psychol.*, 1956, 69, 26-47.

Mackworth, Jane F., & Mackworth, N. H. Eye fixations recorded on changing visual scenes by the television eye-marker. *J. Opt. Soc. Amer.*, 1958, 48, 439-445.

Mackworth, N. H., & Mackworth, Jane F. Visual search for successive decisions. *Med. Res. Council*, 1956, No. APU 234/56.

Mackworth, N. H., & Mackworth, Jane F. Temporal irregularity in a multi-source task. *Med. Res. Council*, 1957, No. APU 264/57.

Matthews, B. H. C. Nerve endings in mammalian muscle. *J. Physiol.*, 1933, 78, 1-53.

Melton, A. W. (Ed.). *Apparatus tests.* Washington: United States Government Printing Office, 1947. (*AAF Aviat. Psychol. Prog. res. Rep.* No. 4.)

Meyer, D. R. On the interaction of simultaneous responses. *Psychol. Bull.*, 1953, 50, 204-220.

Noble, M. E., Fitts, P. M., & Warren, C. E. The frequency response of skilled subjects in a pursuit tracking task. *J. exp. Psychol.*, 1955, 49, 249-256.

Osgood, C. E. *Method and theory in experimental psychology.* New York: Oxford Univer. Press, 1953.

Payne, R. B., & Hauty, G. T. The effects of experimentally induced attitudes upon task proficiency. *J. exp. Psychol.*, 1954, 47, 267-273.

Poulton, E. C. Perceptual anticipation and reaction time. *Quart. J. exp. Psychol.*, 1950, 2, 99-112.

Poulton, E. C. The basis of perceptual anticipation in tracking. *Brit. J. Psychol.*, 1952, 43, 295-302. (a)

Poulton, E. C. Perceptual anticipation in tracking with two-pointer and one-pointer displays. *Brit. J. Psychol.*, 1952, **43**, 222-229. (b)

Poulton, E. C. Learning the statistical properties of the input in pursuit tracking. *J. exp. Psychol.*, 1957, **54**, 28-32. (a)

Poulton, E. C. On prediction in skilled movements. *Psychol. Bull.*, 1957, **54**, 467-478. (b)

Poulton, E. C. On the stimulus and response in pursuit tracking. *J. exp. Psychol.*, 1957, **53**, 189-194. (c)

Rund, P. A., Birmingham, H. P., Tipton, C. L., & Garvey, W. D. The utility of quickening techniques in improving tracking performance with a binary display. *USN Res. Lab. Rep.*, 1957, No. 5013.

Ryan, T. A. Interrelations of the sensory systems in perception. *Psychol. Bull.*, 1940, **37**, 659-698.

Searle, L. V., & Taylor, F. V. Studies of tracking behavior: I. Rate and time characteristics of simple corrective movements. *J. exp. Psychol.*, 1948, **38**, 615-631.

Shannon, C. E., & Weaver, W. *The mathematical theory of communication.* Urbana: Univer. Illinois Press, 1949.

Siddall, G. J., & Anderson, D. M. Fatigue during prolonged performance on a simple compensatory tracking task. *Quart. J. exp. Psychol.*, 1955, **7**, 159-165.

Taylor, F. V. Psychology and the design of machines. *Amer. Psychologist*, 1957, **12**, 249-258.

Taylor, F. V., & Birmingham, H. P. Studies of tracking behavior: II. The acceleration pattern of quick manual corrective responses. *J. exp. Psychol.*, 1948, **38**, 783-795.

Taylor, F. V., & Garvey, W. D. The limitations of a 'Procrustean' approach to the optimization of man-machine systems. *Ergonomics*, 1959, **2**, 187-194.

Telford, C. W. Refractory phase of voluntary and associative responses. *J. exp. Psychol.*, 1931, **14**, 1-35.

Vince, Margaret A. Corrective movements in a pursuit task. *Quart. J. exp. Psychol.*, 1948, **1**, 85-103. (a)

Vince, Margaret A. The intermittency of control movements and the psychological refractory period. *Brit. J. Psychol.*, 1948, **38**, 149-157. (b)

Vince, Margaret A. Rapid response sequences and the psychological refractory period. *Brit. J. Psychol.*, 1949, **40**, 23-40.

Vince, Margaret A. The part played by intellectual processes in a sensori-motor performance. *Quart. J. exp. Psychol.*, 1953, **5**, 75-86.

Vince, Margaret A. The relation between hand movements and intellectual activity in a skilled task. *Quart. J. exp. Psychol.*, 1955, **7**, 82-90.

Wagner, R. C., Fitts, P. M., & Noble, M. E. Preliminary investigations of speed and load as dimensions of psychomotor tasks. *USAF Personnel Train. Res. Cent. tech. Rep.*, 1954, No. AFPTRC-TR-54-45.

Warrick, M. J. Effect of transmission-type control lags on tracking accuracy. *USAF tech. Rep.*, 1949, No. 5916.

Weiss, B. The role of proprioceptive feedback in positioning responses. *J. exp. Psychol.*, 1954, **47**, 215-224.

Welford, A. T. The "psychological refractory period" and the timing of high-speed performance: A review and a theory. *Brit. J. Psychol.*, 1952, **43**, 2-19.

Wyckoff, L. B. The role of observing responses in discrimination learning. Part I. *Psychol. Rev.*, 1952, **59**, 431-442.

18

The Organization of Skilled Response
Merrill Noble and Don Trumbo

INTRODUCTION

In recent years there has been considerable interest in information processing, decision making, thinking, language behavior, and other complex behavioral processes. There are a number of different approaches to these problems, one of which involves an analytical and detailed assessment of response organization in the context of acquisition and retention of motor skill. Thus, for example, Bartlett (1958) and Woodworth (1958) take the view that an understanding of skill is fundamental to an understanding of higher processes, since these, as well as complex motor skills, often are characterized by a high degree of spatial–temporal organization and patterning of responses.

It long has been recognized that timing as a basis for anticipation is critical to the spatial–temporal organization of response (Adams, 1961, 1966a; Bartlett, 1951, 1958; Broadbent, 1958; Conrad, 1951; Fitts *et al.*, 1959; Helson, 1949; Lashley, 1951; Poulton, 1957a; Woodworth, 1958). Nonetheless, there has been relatively little empirical work directly concerned with anticipation. In the motor-skill area specifically, Poulton (1952) has distinguished between *receptor anticipation* and *perceptual anticipation*. Receptor anticipation refers to situations in which the stimuli to which the subject must respond are available as a preview. Leonard (1953), Poulton (1954, 1964), and Wagner *et al.*, (1954) have found that increased receptor anticipation may benefit performance.

Of more interest for the present report are studies of perceptual anticipation, in which no direct preview is possible, but in which the subject may take advantage of the statistical properties of the stimulus sequence and thus make predictions about the stimulus in space and/or time.

The pioneering work on perceptual anticipation was by Poulton. In one study (Poulton, 1952) performance was compared for pursuit and compensatory track-

---

From *Organizational Behavior and Human Performance*, **2**, 1967, 1-25. By permission of Academic Press, Inc.

The research from our laboratory reviewed in this paper was supported by the Air Force Office of Scientific Research under Grants AFOSR 62-17 and 526-64, and by the National Aeronautics and Space Administration under Grant NsG 606.

ing of both a simple sinusoidal input and a complex input composed of two sine waves. Benefits from perceptual anticipation were substantial for the simple sine-wave input, especially with pursuit tracking. In contrast, little anticipation was apparent with the complex sine-wave input, presumably because performance quickly deteriorates to the level of that for a random input as the complexity of a sinusoidal input is increased (Fitts et al., 1959; Poulton, 1966). Similar results were obtained with the use of input patterns that resembled an irregular triangular waveform (Poulton, 1957a).

Adams and Creamer (1962a, 1962b) found evidence to support the hypothesis that proprioceptive stimuli serve as one basis for temporal anticipation, and Adams and Xhignesse (1960) found that temporal anticipation was important to the efficiency of performance in a task that demanded responding with both hands to two sets of visual stimuli. Stimulus coherence was manipulated by variation in the conditional probabilities involved in the sequence of stimuli. When stimulus coherence [1] was high, the subject could respond as well when the two sets of stimuli were far apart as when they were close together, but when stimulus coherence was low, performance was poorer when the stimulus sources were far apart.

Similarly, it has been found that the addition of a second, auditory, task to be performed concurrently with a visual-motor task which requires relatively precise timing and therefore anticipation, results in degradation of performance on the visual-motor task with the amount of interference inversely related to the degree of redundancy of the visual task (Adams and Creamer, 1962c; Bahrick et al., 1954; Bahrick and Shelly, 1958).

Thus it is clear, on both theoretical and empirical grounds, that anticipation is an important factor in the spatial–temporal organization of responses that characterize skill. Among the factors that affect amount of anticipation in a given task are short- and long-term memory, S–R compatibility, intra- and extra-task interference and cognitive factors (e.g., Fitts, 1966; Posner, 1966). Another is degree of stimulus coherence, about which there is relatively little empirical evidence other than that in which the stimulus sequence was either completely coherent or completely noncoherent. Our research represents an attempt to extend previous work by studying the ways and means by which the subjects organize and pattern their responses as a function of the degree of stimulus coherence.

EXPERIMENTAL TASKS

In initiating our research, we wished to be able to study response organization when stimulus coherence was varied both with respect to the time of occurrence

---

[1] Stimulus coherence refers to the degree to which there is a consistent pattern in a stimulus sequence. A periodic signal, such as a sinusoidal signal, is coherent and so is a repetitious sequence of discrete events, e.g., 1, 2, 3, 1, 2, 3, etc. Such sequences are, of course, redundant, but not all factors that effect amount of redundancy also effect coherence. For example, restrictions upon the range of frequencies and amplitudes and unequal frequencies of occurrences of first-order events all result in increased redundancy, but do not contribute to degree of coherence. Unequal conditional probabilities, on the other hand, contribute both to redundancy and to coherence.

of stimuli and with respect to spatial location of stimuli. To accomplish this, we felt that it would be highly useful to have fine-grained, analytical measures in order to delineate the important characteristics of response organization as skill is acquired and forgotten. This in turn should be of value in inferring and assessing information processing and similar processes that intervene between the reception of overt stimuli and the occurrence of overt responses.

These considerations led us to the choice of motor tasks that require graded responses as a vehicle for our studies of response organization. Of the many classes of motor tasks available, we selected tracking for several reasons. Tracking requires both temporal and spatial patterning of responses, the temporal and spatial aspects of the stimulus input can be varied independently, and with appropriate instrumentation a variety of both overall "outcome" and fine-grained indices are possible. Moreover, these indices are based upon continuous measures which, as Bahrick et al. (1957); Bahrick (1964, 1965) has pointed out, have a number of advantages over those based upon dichotomous scores.

The same considerations led to the decision to use irregular step-function or irregular triangular wave inputs as stimuli, since sinusoidal inputs have the following limitations for the purposes of our research: first, although complex sinusoidal inputs are redundant, apparently the addition of even two harmonically related sine waves to a fundamental results in performance little, if any, better than for a random input having the same mean frequency and amplitude characteristics (Fitts et al., 1959; Poulton, 1966). Second, the addition of visual noise to a sinusoidal input affects performance but not amount of learning (Briggs et al., 1957). Finally, it is somewhat difficult with sinusoidal inputs to separate temporal and spatial errors in responding, although Poulton (1962) has described methods for doing this. In contrast, irregular step-function inputs permit the degree of coherence to be manipulated either by varying the conditional probabilities associated with successive stimulus events, or by varying the proportion of events in a stimulus sequence that are repeated exactly in each repetition of a sequence; e.g., in a sequence of, say, 12 stimulus events, during each repetition of the basic sequence every 2nd, 3rd, 4th, or 6th event may be chosen at random while the remaining events are repeated exactly as before. In either instance, the amount of information or uncertainty (Cross, 1966; Garner, 1962) may be calculated and independent assessment of temporal and spatial characteristics of the response is possible. Irregular triangular wave inputs have the same advantages except that it is more difficult to assess temporal and spatial response factors independently.

Apparatus. The apparatus used has been described previously (Trumbo et al., 1963). Programs are punched on paper tape, read by a tape reader, converted to analog voltages by a digital-to-analog converter, and dislayed on an oscilloscope as a half-inch-long vertical hairline that moves back and forth along the horizontal axis of the CRT. This line is referred to as the target. A similar vertical hairline, the cursor, is controlled by the subject by moving a light-weight lateral arm control pivoted at his elbow. The cursor is displayed below the target on the CRT but overlaps it by ⅛ inch.

With a step-function input, the target jumps at predetermined times from one of as many as 15 possible positions to another along the middle 4 inches of the

5-inch x-axis of the CRT. Similarly, with the irregular triangular wave input, the target moves at a constant rate from one to another of the 15 possible positions.

Performance Measures. The outcome score, integrated absolute error, is obtained by an operational amplifier manifold which obtains the absolute voltage difference between target and cursor and integrates this difference over the duration of the trial. This score is displayed on a voltmeter.

Several fine-grained indices are available either by hand scoring of oscillographic records or by computor methods being developed in collaboration with the Conductron Corporation (Trumbo *et al.*, 1965a).

Only hand-scored analyses are discussed in the present report. Those concerned with temporal factors include amount of time by which the subject initiates his response before the stimulus changes position (leads) and amount of time by which the subject initiates his response after the stimulus has changed position (lags). In some instances, lags of less than a normal reaction time were scored as beneficial anticipations (Adams and Xhignesse, 1960).

Fine-grained indices concerned with spatial characteristics of responses include responses in the wrong direction, the amount by which the primary movement is greater than the amount of stimulus movement (overshoots) and the amount by which the primary movement is less than the stimulus movement (undershoot).

TASK COHERENCE AND SKILL ACQUISITION

Studies from our laboratory can be organized into three classes with respect to task coherency. These are (1) studies concerned with spatial uncertainty wherein the coherency of the direction and/or extent of target displacements is varied, but where temporal uncertainty is zero; (2) studies of temporal uncertainty in which direction and amplitude of target displacements are fixed, but in which the predictability of target durations is varied, and (3) studies in which both spatial uncertainty and temporal uncertainty are varied within the same sequence.

Acquisition and Spatial Uncertainty

Three experiments provide our basic evidence for the effects of spatial uncertainty on skill acquisition. In the first of these (Trumbo *et al.*, 1965) four degrees of coherency were studied at two levels of training and three retention intervals. Coherency was defined in terms of the proportion of fixed and repeating elements in a series of target events. Thus, the Coherent or Predictable Task (P) consisted of a sequence of 12 one-second steps selected from 15 possible positions on the horizontal axis of the display. This sequence was repeated within and throughout all 60-second trials. The Noncoherent, or Random task (R) was generated in the same manner, but was 240 targets, or four trials, long. Two Intermediate tasks (I) were constructed from the basic 12-step (P) sequence. In the I-1 task, every second target was selected anew and randomly for each repetition of the basic sequence, while in the I-2 task, every third target was selected anew on each repetition. Thus, for Tasks P, I-2, I-1, and R, 100%, 67%, 50%, and 0%, respectively, of the events were coherent, that is, were repeated systematically at fixed positions within the sequences. Objectively, these proportions defined the limits of perceptual anticipation in these tasks. Finally a directionally coherent (SP) task was added to the design. This task was constructed in the manner of Task R with the restriction that a simple RLRL directional pattern be maintained. Therefore, while all tasks had temporal certainty (i.e., a fixed one-second step duration), only Task P had zero spatial uncertainty. The random elements of the I tasks had uncertainty both with respect to direction and amplitude of target displacements, the SP Task had uncertainty with respect to amplitude only, and Task R had uncertainty as to direction and amplitude of all target events.

The results were surprising in that after 500 repetitions of the basic sequence, neither intermediate task yielded better performance, in terms of integrated absolute error, than the completely random task. On the other hand, Task SP with the predictable directional pattern resulted in marked improvement with terminal performance about midway between Task R and the completely redundant Task P.

While the integrated error criterion failed to indicate differences between the two partially coherent I Tasks and Task R, differences in response organization became apparent as a result of fine-grained analyses of oscillographic records. Surprisingly, the subjects on Task R showed a somewhat greater tendency to lead the stimulus events than did the subjects on the two I Tasks but considerably less than with Task P. Meanwhile, the subjects in the SP Condition showed leads comparable to those obtained on the completely coherent Task P.

Sequence regularities which permitted perceptual anticipation of direction for all target events thus led to anticipatory responding and lower error scores, whereas sequence regularities with respect to both direction and amplitude of target displacements for as many as 67% of all target events did not result in such anticipation. Subsequent analyses showed that the subjects in the SP Condition led with responses of a restricted range of amplitudes; that is, they appeared to be pooling their experiences with the target displacements to produce a mean amplitude response, resulting in overshooting the short displacements and undershooting the longer ones.

This response strategy, wherein spatial accuracy was sacrificed in order to improve timing, also characterized the best subjects (as determined by integrated error scores) within all tasks. These better subjects developed better timing with practice, but actually increased the frequency of overshoot and undershoot errors. Poorer subjects, on the other hand, frequently improved in spatial accuracy, but showed relatively little change in temporal accuracy. In addition, the better subjects developed a high rate of movement which also tended to reduce error scores, but to decrease spatial accuracy.

In the second study of event uncertainty (Noble *et al.*, 1966) Tasks P, I-2, and R from the first study were repeated together with two new I tasks, I-3 and I-5, with every fourth and every sixth targets, respectively, selected anew on each repetition of the basic 12-target sequence. Training was extended by 64% to 164 minutes of practice over nine days. Under this extended practice, performance on the I-2 task did improve somewhat with respect to Task R. In fact, at the end of training, when fairly stable asymptotic levels under all conditions had been reached, integrated error was perfectly correlated with task uncertainty and the differences in error between adjacent task conditions were about equal. A comparison plotting terminal error scores against the proportion of random events in each task yielded a sigmoidal curve with relatively little increase in error between $p = .33$ (Task I-2) and $p = 1.00$ (Task R). However, a subsequent information analysis of the target sequences yielded a relationship in which integrated error was essentially a linear function of stimulus uncertainty (Cross, 1966).

Data from fine-grained analyses revealed that the temporal accuracy on the I-2 Task was more like that on Task R than on the other I-Tasks. Meanwhile, under the extended practice offered in this study spatial errors (under- and overshooting) first increased, as in the first experiment, but then decreased somewhat toward the end of training, suggesting again that the subjects tended to neglect spatial accuracy early in training while they improved their timing and increased their rate of movement, then later on made some improvement in the accuracy of the amplitudes of their movements.

Intra-sequence analyses indicated that the random events within the I conditions resulted in greater spatial error, but did not disrupt timing. Furthermore, the decrement in spatial accuracy appeared to extend beyond the random target events to subsequent targets, particularly the next target in the sequence. That is, although the random targets were followed by targets which were predictable, the subjects appeared to be unable to prepare an accurate response to these events.

These results indicate that response organization by the subjects in the more predictable tasks involved anticipatory responding to all target events, including the random ones, with rapid and underdamped movements, whereas the subjects on the less predictable tasks (I-2 and R) tended to lag by about one RT, followed by underdamped movements resulting in overshoots and oscillations.

As an alternate means of manipulating task coherency, sequential (digram) probabilities were manipulated in a study by Quigley et al. (1966). Six target positions were used with equal (one-second) durations for all target events. The predictable task (P) consisted of a sequence of six target events repeated faithfully within and throughout all trials. The random task (R) consisted of a 480 target, or 10 trial, random sequence with each of the remaining five alternatives equally probable ($p = .20$) following any given stimulus event. Intermediate tasks were constructed so that the high probability digrams were the same as the fixed digrams in task P, but for each target position there was one additional alternative. Four probability series were constructed with .90/.10, .80/.20, .70/.30, and .50/.50 probabilities for the two alternatives. Thus, the average information per target event was zero, .46, .72, .88, 1.00, and 2.32 bits for the P, 90/10, 80/20, 70/30, 50/50, and R tasks, respectively.

The results showed that with 80 trials of 48 targets per trial, only the P, 90/10, and 80/20 series were different from the random task in integrated error scores. Despite the fact that the 50/50 and 70/30 tasks contained less than one-half the information per target event of Task R, the subjects were unable to use this redundancy to reduce their error scores below those of Task R. Similarly, the subjects on the 80/20 and 90/10 tasks made relatively little improvement as compared with Task P.

Lead-lag analyses were again consistent with error score results; the subjects with the more coherent tasks (P, 90/10, 80/20) led on the average by about 100 msec at the end of training, while the remaining groups lagged by about 50–75 msec. Intra-sequence analyses revealed that the subjects lagged more when alternative targets presented a directional as compared with an amplitude choice. That is, when the high- and low-p alternatives were in different directions from a given target position, greater response latencies were found than when both alternatives were in the same direction.

Construction of the tasks on a probability basis permitted further analysis of response strategies analogous to those evaluated in discrete tasks (e.g., Grant et al., 1951; Humphreys, 1939; Hake and Hyman, 1953). Typically, these latter studies have demonstrated a nonoptimal strategy in which the subjects match the probabilities of alternative events, rather than maximizing by always choosing the most probable events.

Returning to our data, and considering first only those instances where the subjects had a directional choice and where target events were anticipated, the proportions of responses initiated toward high- and low-p events were computed. The results clearly supported a matching strategy: the subjects in the 90/10 condition initiated anticipatory responses in the direction of the more probable target 89.8% of the time by the end of training and the other probability tasks showed comparable matching proportions.

Specification of the response strategy used when alternative target events were in the same direction was less straightforward. It was assumed that maximizing would have resulted in a unimodal distribution of response amplitudes around the high-p target position not unlike that for Task P, whereas matching would result in a bimodal distribution with a second and proportionately smaller mode centered around the low p target. Neither of these strategies fit the data. Instead, the subjects appeared to develop a compromising strategy in which the central tendency of the response distribution was pulled toward the low-p target proportionate to the probability of the low-p event. In addition, the response generalization gradients became flatter as the ratio of the two probabilities decreased.

Acquisition and Temporal Uncertainty

In the two studies described in this section, spatial uncertainty was reduced to zero by using a fixed-amplitude step-function input. Temporal uncertainty was introduced by varying the proportion of fixed and predictable step durations in a sequence of steps in a manner analogous to that described for spatial coherence.

In the first experiment (Trumbo et al., unpublished b), there were seven experimental conditions. Temporal coherence ranged from a predictable condition involving a fixed sequence of six target durations of between .3–1.8 sec in .3-second intervals, to a random condition (R) with a random series of 480 step durations. Intermediate degrees of coherence included I-2, I-3½, and I-5 conditions analogous to those in the spatial uncertainty studies. Sequence length was 6 or 9 targets.

The results showed that, after 40 training trials with the short sequences, performance on Tasks P and I-5 did not differ, but both showed less error than the I-2 condition which again did not differ from Task R. With the longer (9-target) sequences, the range of differences due to predictability was more restricted, and yielded essentially no differences after 40 trials.

Fine-grained analyses demonstrated that all experimental groups began with mean lags equal to about one RT, but, by the end of training, the subjects on the most predictable tasks had average leads of 100–200 msec, while the subjects on the least predictable tasks had mean leads of about zero. When average lead–lag values were computed separately for each target duration there were no differences at Trial 1; all targets, regardless of duration, were lagged by about one RT. However, later in practice, the subjects in all conditions showed definite range effects, that is, they lagged the short-interval targets and led the long-interval targets. For the longest target durations (1500 and 1800 msec) lead time was as great as 700 msec for Task P at the end of training. Thus, despite differential predictability of temporal intervals all groups tended to respond with a somewhat restricted range of temporal values, and to regress toward the mean target duration. Subjects in the more coherent task conditions tended to have shorter lags on the briefest target durations; however, they also tended to overanticipate target displacements from the longest durations, a nonoptimal strategy which increased their error scores.

In a second experiment (Trumbo et al., unpublished b) in which the P, I-3, and R conditions of the previous experiment were used, but with longer training and a wider range of sequence lengths, similar results were obtained. For the shorter sequences (4, 8, and 12 targets), similar differences were obtained between predictable (P) and partially predictable (I-3) temporal sequences, but with a sequence of 16 temporal intervals, neither Tasks P nor I-3 differed from the random task (R) nor from each other.

Acquisition and Temporal-Spatial Uncertainty

As a logical sequel to the studies thus far reveiwed, Cross (1966) designed an experiment in which step-function inputs were varied in direction, amplitude, and time patterns. For different conditions, zero, one, two, or all three of these dimensions were coherent, while the remaining dimensions were randomly programmed. His conditions included a coherent directional pattern (D) wherein the pattern LLRLRR was repeated within and throughout trials but both the amplitude and time of displacements were random; a coherent temporal pattern (T) in which the temporal sequence of 160, 170, 40, 80, 120, and 80 msec was faithfully repeated but both amplitude and direction sequences were random; a pattern (TD) in which both of the above dimensions were coherent and only amplitude (A) was random; a pattern (DA) where both direction and extent patterns were coherent, but not time, and patterns (P) and (R) for which all three dimensions were coherent and random, respectively. The remaining conditions, (A) and (TA), were impossible to obtain within the limits of the display.

Cross's results are of considerable interest on several counts. First, neither a fixed time pattern nor a fixed directional pattern alone resulted in improvement better than that achieved on the completely random task after 104 one-minute trials. However, when two dimensions are coherent—either time plus direction (TD) or direction plus amplitude (DA)—nearly identical error scores were obtained, about one-third of the way between Task R and Task P. Thus, redundancy with respect to one dimension of the task did not improve performance, but redundancy with respect to two dimensions resulted in some, though not a proportionate, reduction in error scores.

Detailed analyses indicated that average lag scores corresponded closely with error data, with Task P showing mean lags well below 100 msec, tasks TD and DA, lags of 200 msec, and tasks T, D, and R, lags of about 300 msec. However, similarity of error and lag data for Conditions DA and TD masked some marked differences in response organization for these two groups. When mean algebraic timing error was compared, Task DA showed the greatest error, increasing to 350 msec by mid-training and remaining constant thereafter. In contrast, Task TD resulted in a decrease in total timing error to about 200 msec at the end of training. A major factor in the discrepancy was the fact that in Task DA the subjects led nearly half of the targets, whereas in Task TD less than 20% of the targets were led by the end of training.

An analysis of spatial errors indicated that, by the end of training, proportions and magnitudes of undershoots and overshoots were about equal for Tasks R and P, despite the fact that the subjects were anticipating nearly 100% of the time on Task P and lagging most of the targets in Task R. This indicates that the responses in anticipation of coherent target events were made with the same accuracy as responses made to the events themselves.

Two additional studies performed in our laboratory provide further evidence on the role of coherence in tasks requiring temporal–spatial patterning of response (Trumbo et al., unpublished a). Both of these studies used constant-rate triangular wave inputs, wherein the target traveled back and forth along the x axis of the display at a constant rate. Task coherence was manipulated by varying the pattern of points at which the target instantaneously reversed directions. These points were the same as the discrete target positions in the step-function experiments. Thus, a coherent sequence might begin at Position 4, travel to Position 8, reverse and travel to Position 1, reverse and travel to Position 5, etc. The coherent sequence consisted of 12 reversal points, the pattern then being repeated within and throughout all trials, while in an intermediate pattern (e.g., I-2), every third reversal point was selected randomly on each repetition of the basic sequence. Finally, the random task (R) was a long sequence of reversal points randomly selected except that each new point required a reversal of target direction.

It should be noted that in these tasks temporal and spatial coherence were completely confounded. That is, information as to *how far* the target will travel in a given direction is at the same time information as to *how long* it will continue in that direction.

In both experiments, a coherent (P), a random (R) and an I task were used. For the I task, in the first study every third reversal point was varied, while in the second study every fourth point was varied. In addition, three rates of target movement were used in each study (.8, 1.6, and 2.4 inches per second, requiring 9°, 18°, and 27°/sec movement of the arm) with one, two, and three repetitions, respectively, of the basic sequence per 54-second trial. Also, in the second experiment, both pursuit and compensatory displays were used. It was assumed that the need for perceptual anticipation would increase with increased rate of target movement and the concomitant increase in the frequency of reversal points. It was reasoned that, at the highest rate, the ability of the subject to keep in phase with the target would depend upon his ability to anticipate the pattern of reversal points, and, possibly, his ability to preprogram several reversals of direction in advance. Similarly, it was assumed that the coherent portions of the input which permit perceptual anticipation would be more readily apparent with the pursuit display, since the compensatory display confounds response and input information.

The results of the first experiment showed that both task coherence and rate had reliable effects on error scores, but not their interaction. At each rate, tasks were ordered from random to coherent, with respect to error scores, with a tendency for the differences among tasks to increase with increased rate, as was predicted when it was assumed that perceptual anticipation was more critical at higher rates. However, the nonsignificant interaction between main effects left these trends as suggestive only. The second study tended to support the assumption that coherence would enhance perceptual anticipation (and, consequently, anticipatory responding) relatively more with the pursuit than with the compensatory display. The effect of coherence was less pronounced in the compensatory task, but greater differences were observed as a function of rate.

Analytical scoring has not been completed for these latter studies; however, visual inspection of continuous response records reveals that the subjects do learn the one component of the task equally coherent to all task conditions, that is, the rate of target movement. Subjects come to match rates comparatively well in all tasks, but their ability to cope with reversal points, particularly at the higher rates, depends on the coherency of the pattern of these points.

LONG-TERM RETENTION AND TASK COHERENCY

A sustained interest within our research program has been in the retention of skill over relatively long periods of disuse. Consequently, many of the studies reviewed in the previous section have included one or more retention intervals ranging from one week to six months in duration.

Spatial uncertainty, in the studies first described in the previous section, did result in some forgetting. Losses were correlated with the length of the retention interval which was varied in the first study of one week, one month, and five months. Absolute losses were positively related to task coherence; however, the more coherent the task, the greater the gains made during training, and, consequently, the greater the potential for loss. Nevertheless, with relative loss calculated as a proportion of gains made in training, Task P again showed the greatest loss followed by the semicoherent (SP) task. In fact, Task R after one week and one month intervals, and Task I-2 after one week showed gains on the recall trial, indicating the benefits of the additional practice offered within this trial, or, possibly, the dissipation of interference accumulated during training.

Except for the I-1 task, all groups showed some relearning during the retention trials. The effects of practice were evident both in recall and relearning performance, especially on Task P, where losses were reduced under extended training and all retention groups regained in relearning trials the level of performance attained at the end of training, but with shorter training only the one-week group recouped their losses.

The greater losses observed for the more coherent tasks suggest that the locus of the loss is in the perceptual anticipation of target events, since the more coherent tasks presumably permit the greatest amount of perceptual anticipation. This interpretation was supported by the analyses of temporal and spatial errors. Both frequency and duration of lags increased and frequency and duration of leads decreased over the retention intervals. Furthermore, these changes were most pronounced for the most coherent tasks (P and SP). Thus, it appeared that retention losses largely reflect losses in timing accuracy, which, in turn, indicate losses in perceptual anticipation of the target events. Those components of performance not dependent upon perceptual anticipation were resistant to loss, as reflected in the failure to find retention loss for Task R. Temporal accuracy appeared to be more crucial than spatial accuracy for skill retention, but this may reflect forgetting of the spatial location of stimuli, forcing the subject to lag.

These findings were replicated in the second study which included a single three-month retention interval. In addition, increases in variability of timing as well as in mean values were observed at retention. An intra-task analysis in this study indicated that retention losses were also related to increased spatial error for the fixed targets immediately following random targets in the I patterns, rather than on the random targets themselves, suggesting that the random events disrupted the perceptual anticipation of subsequent events.

A one-week retention interval was included in the study utilizing sequential dependencies (Quigley et al., 1966). Slight losses were sustained only for the coherent (P) task, although, under an added secondary task load, intermediate task 90/10 also showed considerable loss, but not the less coherent 50/50 or R tasks.

Retention of skill in tasks of temporal uncertainty was evaluated in the second study of this type described earlier (Trumbo et al., unpublished b). Two retention intervals were used: one week and one month. At one week, performance showed no loss, in fact there was evidence of slight improvement on most tasks. However, at one month there were moderate losses which tended to be limited to the coherent tasks with the longer sequences (8, 12, and 16). Thus, as was true of patterns varied in spatial uncertainty, greater retention losses occur for the more coherent temporal patterns. Again, the results suggest a loss in perceptual anticipation as the subjects forget the sequence of, in this case, temporal events.

Finally, long-term retention losses were also studied in the two-rate tracking experiments. In the first study these were one week and one month; in the second study a single interval of six weeks was used. Neither study showed any appreciable loss as a function of retention interval, task coherency, or display mode (pursuit vs. compensatory).

These results are in agreement with Adam's (1966) and with Naylor and

Briggs (1961) who conclude that skill retention may be a function of the continuous nature of the task with the more discrete tasks (i.e., our step-functions) being more subject to retention losses than the more continuous tasks (our triangular, or rate tasks).

OTHER TASK VARIABLES

Task conditions other than task coherence studied in our laboratory may be subdivided into two categories: (a) conditions affecting difficulty of information processing that are intrinsic to the primary task, including sequence length, alphabet size, target rate, and task coding, and (b) increases in information processing load through the addition of a secondary task.

Several of our studies have been concerned with the number of target events in the basic sequence. These include two experiments concerned with spatial uncertainty (Swink et al., 1967; Trumbo et al., 1965c) and two concerned with temporal uncertainty (Trumbo et al., unpublished b). Interest in sequence length stems from two sources. First, sequence length is a basic variable which, as Adams (1964) has indicated, is fundamental to an understanding of this area. Secondly, sequence length has certain interesting analogies to the classical length-difficulty question in verbal learning.

In our first study of sequence length (Trumbo et al., 1965c) fixed sequences of 5, 10, and 15 target steps were employed. The subjects were given equal practice time, i.e., ninety 60-second trials, wherein sequences were repeated 12, 6, and 4 times per trial. Very early in practice, integrated error scores were proportional to sequence length; however, by the second session (Trials 11–30) there was no difference between the two shortest sequences although both differed throughout training from the longest ($N = 15$) sequence. Thus, sequence length did not appear to affect difficulty up to a certain value (between 10 and 15) after which difficulty appeared to be increased.

On the other hand, if difficulty is defined as the ratio of number of repetitions of the sequence needed to reach a performance criterion to the number of items (targets) in the sequence, i.e., repetitions *per item* (Deese, 1958), then this relative difficulty actually decreased with increases in sequence length.

This study also tested the assumption that performance would be inversely related to number of alternatives (i.e., the number of possible target positions) from which the sequences were drawn. Number of alternatives was varied orthogonally with respect to sequence length with values of 5, 10, and 15 alternatives. Contrary to the assumption, number of alternatives had no reliable effect, nor did it interact significantly with sequence length, in affecting error scores.

The second study was designed primarily to define more precisely the function relating sequence length and tracking performance by including a wider range of values. In addition, two training criteria were employed, one in which the subjects were run an equal amount of time, the other in which the subjects received a fixed number of repetitions of the sequences. Finally, both coherent (P) and intermediate (I-3) task conditions were included. Sequence lengths were 8, 12, 16, 24 and 48 targets.

With equal practice time (and, therefore, unequal repetitions) error scores at the end of 100 training trials were significantly related to sequence length for both the coherent (P) and the intermediate (I-3) conditions, although in the P conditions only the 48-target sequence differed reliably from the others, whereas in the I-3 all differences except those between adjacent means were significant.

However, with equal repetitions of the sequences, no significant sequence-length effect or task coherence by sequence-length interaction was found and there were no differences among task means under (P) conditions at either 180 or 360 repetitions. Lead-lag performance for the various sequence lengths was strikingly similar on the coherent tasks; the subjects in all conditions were leading, on the average by 150–250 msec after 180 and 360 repetitions. Meanwhile, under I-3 conditions, only the two shortest sequences resulted in leading after 360 repetitions. Thus, equal repetitions of coherent sequences of unequal length led to equal terminal performance. However, the total numbers of repetitions required to reach two successive error-score criteria were negatively accelerated functions of sequence length, similar to what would be predicted by Thurstone's (1930) general learning equation. This meant, of course, that replotting the data, in the manner of Deese (1958), as the ratio of number of repetitions to number of items, one obtains a negatively accelerated decreasing function with increases in length. In fact, with sequences greater than 8, the number of repetitions per item to reach a criterion level is essentially a constant. Thus, within the range of values employed, there was no evidence to suggest that task difficulty increases disproportionately with increases in the length of completely coherent sequences as has been frequently assumed in verbal learning literature. Except possibly for performance levels characteristic of very early stages of training, it appears that number of repetitions to reach a given performance level may be roughly equal to some constant times the number of items in the sequence, at least for sequences clearly longer than the immediate memory span.

The data from two studies (Trumbo *et al.*, unpublished b) of sequence length with sequences of *temporal* intervals are strikingly similar to those just described for sequences of spatial steps. This was particularly evident in the second experiment where, with sequences of 4, 6, 8, 12 and 16 step durations, total number of repetitions to reach a criterion error score was again a negatively accelerated function of sequence length. Similarly, the ratio of number of repetitions to number of items in the sequence was a decreasing negatively accelerated function of sequence length. Again, for sequences greater than 8, this ratio was essentially a constant. Thus, in both the spatial and the temporal domains, increases in sequence length do not result in disproportionate increase in difficulty, and within the range of values investigated, total number of repetitions to reach a given performance criterion increases approximately as a constant times the number of items in the sequence, at least for values beyond the immediate memory span.

Rate of Target Movement

As previously indicated, the two experiments using triangular wave inputs involved three rates of target movement and, consequently, three rates at which target reversal points occurred.

The results showed pronounced increases in integrated error as rate was increased, with a tendency for task coherency to become more critical at higher target rates. Furthermore, increases in rate had greater effect with a compensatory than with a pursuit display, but task coherence appeared to have more effect on pursuit than on compensatory tracking performance. Although the detrimental effects on tracking performance of increasing input rate (or frequency) do not represent new findings (e.g., Ellson and Gray, 1948; Noble et al., 1955; Poulton, 1957b), the relation between rate and task coherence, particularly using intermediate levels of coherence, do represent new data.

Task Coding

In our one study in this area to date (Trumbo et al., 1965d), two types of cues for coding the task were investigated; verbal pretraining cues, in which a list of numbers, corresponding to the sequence of fixed target positions, was overlearned before tracking, and display cues wherein overlay lines, numbered or unnumbered, were added to the display at each of eight positions the target could assume.

The results showed that both sets of cues facilitated early performance but were less important later in training. However, the unnumbered grid lines resulted in poorer performance than the blank display, suggesting that the subjects may have attempted to use the lines to code the task, but in order to discriminate among them became involved in counting the grid lines. The effects of pretraining plus display specificity were not accumulative, possibly because the two sets of coding information were completely redundant. Either pretraining or the numbered-grid display provided the same information for coding the task as a series of numbered target positions, and, therefore, for the perceptual anticipation of the entire sequence. The fact that the coding cues were affective only early in training was predicted on the grounds that all the subjects would eventually learn the relatively simple input pattern and the coding cues would merely facilitate this cognitive learning process.

Secondary Task Effects

Cognitive processes in skill acquisition, such as short- and long-term memory and perceptual anticipation, appear to be interfered with if the subject is required simultaneously to process information relevant to a second task (Posner, 1966) even though he is able to handle irrelevant visual noise added to a visual stimulus without apparent losses in skill acquisition (Briggs et al., 1957). By the same token, to the extent that one or both tasks can become automatic, the inter-task interference should be reduced (Adams and Chambers, 1962; Bahrick and Shelly, 1958). Thus, one would expect less interference with completely coherent tracking inputs than with the partially coherent tasks, and, similarly, the more redundant the secondary task, the less interference it should produce.

This rationale was the basis for the first of five experiments involving secondary tasks (Trumbo et al., 1967; Noble et al., 1967). Subjects were trained to equal performance levels on the step-function tasks P, I-3 and I-5 described earlier. At a retention session one week after training, the subjects received either the tracking task alone, or together with one of two secondary tasks. Both secondary tasks required the verbal anticipation of a series of the numbers 1–5 presented over the intercom. One series involved differential first-order probabilities of the numbers, the other involved .90/.10 digram probabilities. Thus, both the primary tracking tasks and the secondary tasks involved differences in coherency and information rate.

The results indicated interference from both second tasks on all primary task conditions, but no reliable interaction, i.e., no evidence that the amount of interference was a function of the uncertainty in either primary or secondary task. Even on Task P, which had the greatest potential for becoming automated, tracking performance was degraded to about the same degree as on the less coherent tasks.

The second experiment attempted to localize the source of interference within the secondary task, that is, to determine whether the interference resulted from the added noise, the added response requirement, or the added information processing requirement of the second task. Secondary and no-secondary task conditions were again compared. In addition, a noise condition simply presented the stimulus component of the second task without any response requirement. Finally, a response condition was included wherein the subjects were free to respond with any of the numbers 1–5 to a series of relay clicks substituted for the numbers of the regular secondary task. This response task ostensibly eliminated the information-processing requirement, while maintaining the response requirement of the second task, whereas the stimulus task eliminated both the response and information-processing requirements, but maintained the "noise" of the second task.

The results were clear-cut. The noise condition had no effect; performance did not differ from the no-secondary task control. Meanwhile, the response task resulted in the same interference as the intact secondary task with its information processing and decision-making requirements. Thus, it appeared that interference resulted from the competition of simultaneous responses, and that the interference was essentially at a peripheral, not a central, level.

However, further reflection on the results cast some doubts on the above interpretation. While it was originally assumed that the "free" response task effectively eliminated cognitive processing of the second task it was later reasoned that, in instructing the subject to respond "with any number 1 to 5, but to use all the numbers," we had presented him with a five-choice response-selection task, thus maintaining at least a modicum of decision-making in the second task. Therefore, a third experiment was designed which again included the secondary task, the response task, and the no-secondary task conditions, but also incorporated (1) a condition in which the subjects were instructed to learn the secondary task, but made no overt responses to it during tracking trials, and (2) a no-choice response task wherein the subjects simply repeated the secondary task numbers as they were presented.

The results indicated no interference from either of the new conditions. However, test trials on the secondary task alone verified that the subjects in the no-overt-response condition learned the second task as well as the overt-response group. Thus, information from the two tasks could be processed without any apparent interference when no overt responding was required to the second task, but, since the subjects who repeated the stimulus numbers also showed no interference, it was concluded that the immediate overt response-selection requirement, rather than the requirement of a second response, *per se,* was the locus of interference.

Two further experiments have indicated that, when the subjects were trained with a secondary task throughout and then transferred to the primary task alone, improvement in primary task performance was incremental and gradual, indicating that, unlike simple irrelevant noise, the presence of a secondary task interferes with skill learning as well as skill performance.

RESPONSE STRATEGIES

The purpose of these studies has been both descriptive and analytic. We have sought to describe in detail precisely the sorts of response organizations or patterns which the subjects develop when faced with tasks varied in stimulus organization or coherence, then to interpret these descriptions as strategies used to cope with various kinds and degrees of uncertainty. This, in turn, had led to interferences about information processing, decision-making, and perceptual anticipation, or imaging, of the task.

Several generalizations seem reasonable, though tentative, at this point. First of all, it is clear that the subjects make use of temporal–spatial coherence of the stimulus pattern in organizing patterns of responses that improve their performance. This was anticipated in view of previous studies (Adams and Creamer, 1962d; Bahrick *et al.,* 1954; Poulton, 1952, 1957a, 1964) in which two extreme values of stimulus coherence were investigated, as well as studies of dual-task performance which included intermediate values of stimulus coherence (Adams and Xhignesse, 1960; Bahrick and Shelly, 1958). The extent to which the coherent information is used, however, is not directly a function of the proportion of predictable elements in the stimulus sequence, although, after considerable practice, criterion performance does appear to be nearly linearly related to input uncertainty, given certain assumptions (Cross, 1966).[2]

With spatial uncertainty, but no temporal uncertainty, the subjects sacrifice spatial accuracy to timing, particularly if the uncertainty is with respect to amplitude, but direction of the required response is known and can be anticipated. For low degrees of coherency (e.g., I-1 or I-2) uncertainty regarding both amplitude and direction for some portion of the event leads to a conservative strategy involving greater lagging of all events than that observed when the same uncertainty holds for all events (Task R). Why partial uncertainty in this form should inhibit anticipatory responding is not clear.

Performance under conditions of *temporal* but not spatial uncertainty suggests that the subjects with the same proportions of random events (in the temporal domain) as on the spatial uncertainty tasks do better relative to the completely coherent task than comparable groups in the spatial uncertainty studies. Thus, for example, an I-5 temporal tracking task was handled as well as the completely

[2] In computing stimulus information, Cross assumed that target displacements from randomly selected positions to fixed positions as well as those from fixed to random positions contained information, whereas displacements from fixed to fixed positions were completely redundant.

coherent (P) task of the same basic sequence length, and performance on the I-2 task was about midway between Tasks R and P with the longer sequences. In contrast, under spatial uncertainty conditions, I-5 yielded significantly more error than Task P and I-2 differed from Task R only slightly after extensive practice. Whether this reflects relatively poorer performance on the part of the subjects with the coherent task, or relatively better performance on the I tasks in temporal tracking is a moot question. Differences in sequence length for spatial and temporal tracking studies may also confound these results. However, it may be speculated that the tendency under all temporal tracking conditions to respond with a restricted range of durations reduces the error associated with the randomly selected durations, since at least some portion of these events would be beneficially anticipated by virtue of the subjects' expectancy of an average duration.

Consistent with our findings with uncertainty in one dimension of the task, simultaneously introducing uncertainty into two or more task dimensions disproportionately affects criterion performance. Thus, as Cross (1966) showed, complete coherence with respect to time or directional patterns did not improve error scores as compared with the complete random task, while patterns completely coherent in two dimensions, but not the third, resulted in relatively small gains. Nevertheless, Cross' multivariate information analysis indicated that performance was roughly proportional to stimulus information in bits. Furthermore, his subjects on tasks with the same number of coherent dimensions (i.e., Tasks TD and DA) achieved comparable error scores, but with quite different temporal–spatial patterns of response.

The strategies developed to cope with sequential probability tasks are complex, but at the same time, highly organized and meaningfully related to the decision-making requirements of the stimulus events. The subjects clearly develop a matching strategy in anticipating directional choices and they appear to develop a compromising strategy with amplitude choice situations. In addition, they show greater decision time with directional than with amplitude choices. These results support our earlier contention that the subjects organize their responses in a manner consistent with, though not always proportional to, the kinds and amount of information available in the stimulus.

The complex strategies developed with the sequential probability tasks, wherein the subjects have differential temporal and spatial responses for the two types of choice situations, are somewhat in contrast with those found in the studies involving proportions of fixed and random events. In the latter, timing of *all* target events appeared to be affected equally by the presence of random events in the sequence, and spatial accuracy was affected for at least some fixed events following a random event.

It should be pointed out that, even with completely coherent sequences, not all the subjects adopted the strategy of anticipation as indicated by lead data. Comparison of the subjects who had the best integrated error scores with those who had the poorest error scores revealed that the latter tended to lag rather than to lead (Trumbo *et al.*, 1965b). Whether this reflects inability to learn the coherent sequence or a different strategy remains to be determined.

The disproportional effect of random events on criterion performance is due, in part, to the disrupting effects of these events on subsequent coherent portions of the input. Thus, the intra-sequence analyses suggest that both the coherent-to-random and the random-to-coherent steps are effectively uncertain events, supporting Cross's (1966) information analysis which made a similar assumption.

As one would expect, task coherency interacts with other task variables to determine task difficulty as reflected in the integrated error criterion. Thus, whereas Thurstone's (1930) general learning equation appears to describe the length–difficulty relation with completely coherent sequences, the same function is not obtained with partially coherent tasks. In the latter case there are fairly wide differences in performance and in response organization for sequences of various lengths. Interactions were also noted between task coherence and target rate and display mode in the rate-tracking studies, and between coherence and sequence length in the temporal tracking studies.

Finally, two extra-task conditions, one with facilitative and one with interference effects on criterion performance, provide some further insights into the information processing in the types of tasks we have studied.

The "coding" study demonstrated that both pretraining on a verbal code and optimal display coding can facilitate skill performance early in training, presumably by providing a highly compatible set of labels for designating and differentiating the various target events.

At the same time, the requirement of simultaneously processing information from a second source may interfere with skill performance and, specifically, with the accurate anticipation of target events, if the second task requires both a selection process and an overt response. Thus, it may be, as Adams (1966b) has suggested, that the information-processing limitation of the human operator can be described as a one-channel mechanism, if it is assumed that the processing involves the decision-making, and, our data indicate, the selection of a second overt response as well.

REFERENCES

Adams, J. A. Human tracking behavior. *Psychological Bulletin,* 1961, **58**, 55-79.
Adams, J. A. Motor skills. *Annual Review of Psychology,* 1964, **15**, 181-202.
Adams, J. A. Some mechanisms of motor responding: an examination of attention. *In* Bilodeau, E. A. (Ed.) *Acquisition of skill.* New York: Academic Press, 1966a. Pp. 175-200.
Adams, J. A. Forgetting of motor responses. *In* Marx, M. H. (Ed.) *Learning: processes.* New York: Macmillan, 1966b.
Adams, J. A., and Chambers, R. W. Response to simultaneous stimulation of two sense modalities. *Journal of Experimental Psychology,* 1962, **63**, 198-206.
Adams, J. A., and Creamer, L. R. Anticipatory timing of continuous and discrete responses. *Journal of Experimental Psychology,* 1962a, **63**, 84-90.
Adams, J. A., and Creamer, L. R. Proprioception variables as determiners of anticipatory timing behavior. *Human Factors,* 1962b, **4**, 217-222.
Adams, J. A., and Creamer, L. R. Data processing capabilities of the human operator. *Journal of Engineering Psychology,* 1962c, **1**, 150-158.
Adams, J. A., and Xhignesse, L. V. Some determinants of two-dimensional visual tracking behavior. *Journal of Experimental Psychology,* 1960, **60**, 391-403.
Bahrick, H. P., Fitts, P. M., and Briggs, G. E. Learning curves—facts or artifacts? *Psychological Bulletin,* 1957, **54**, 256-268.
Bahrick, H. P. Retention curves: facts or artifacts. *Psychological Bulletin,* 1964, **61**, 188-194.

Bahrick, H. P. The ebb of retention. *Psychological Review,* 1965, **72,** 60-73.
Bahrick, H. P., Noble, M. E., and Fitts, P. M. Extra-task performance as a measure of learning a primary task. *Journal of Experimental Psychology,* 1954, **48,** 298-302.
Bahrick, H. P., and Shelly, C. H. Time sharing as an index of automatization. *Journal of Experimental Psychology,* 1958, **56,** 288-293.
Bartlett, F. C. Anticipation in human performances. *In* G. Ekman, T. Husen, G. Johansson, and C. I. Sandström (Eds.), *Essays in psychology.* Uppsala, Sweden: Almquist and Wiksells, 1951. Pp. 1-17.
Bartlett, F. C. *Thinking: an experimental and social study.* London: Allen and Unwin, 1958.
Briggs, G. E., Fitts, P. M., and Bahrick, H. P. Learning and performance in a complex tracking task as a function of visual noise. *Journal of Experimental Psychology,* 1957, **53,** 379-387.
Broadbent, D. E. *Perception and communication.* Oxford, England: Pergamon Press, 1958.
Conrad, R. Speed and load stress in a sensori motor skill. *British Journal of Industrial Medicine,* 1951, **8,** 1-7.
Cross, K. Discrete tracking proficiency as a function of temporal, directional, and spatial predictability. Unpublished Doctoral Dissertation, Kansas State University, 1966.
Deese, J. *The psychology of learning,* 2nd ed. New York: McGraw-Hill, 1958.
Ellson, D. G., and Gray, F. E. Frequency responses of human operators following a sine wave input. *United States Air Force Memorandum Report.* No. MCREXD-694-2N, 1948.
Fitts, P. M. Cognitive aspects of information processing: III. Set for speed versus accuracy. *Journal of Experimental Psychology,* 1966, **71,** 849-857.
Fitts, P. M., Bahrick, H. P., Noble, M. E., and Briggs, G. E. Skilled performance: Parts I and II. *United States Air Force Wright Air Development Center Final Report,* No. AF 41 (657)-70, 1959.
Garner, W. R. *Uncertainty and structure as psychological concepts.* New York: Wiley, 1962.
Grant, D. A., Hake, H. W., and Hornseth, J. P. Acquisition and extinction of a verbal conditioned response with differing percentages of reinforcement. *Journal of Experimental Psychology,* 1951, **42,** 1-5.
Hake, H. W., and Hyman, R. Perception of the statistical structure of a random series of binary symbols. *Journal of Experimental Psychology,* 1953, **45,** 64-74.
Helson, H. Design of equipment and optimal human operation. *American Journal of Psychology,* 1949, **62,** 473-497.
Humphreys, L. G. Acquisition and extinction of verbal expectations in a situation analogous to conditioning. *Journal of Experimental Psychology,* 1939, **25,** 294-301.
Lashley, K. S. The problem of serial order in behavior. *In* L. A. Jeffress (Ed.), *Cerebral mechanisms in behavior.* New York: Wiley, 1951. Pp. 112-146.
Leonard, J. A. Advance information in sensori-motor skills. *Quarterly Journal of Experimental Psychology,* 1953, **5,** 141-149.
Naylor, J. C., and Briggs, G. E. Long-term retention of learned skills: a review of the literature. *Aeronautical Systems Division Technical Report,* No. 61-390, 1961.
Noble, M. E., Fitts, P. M., and Warren, C. E. The frequency response of skilled subjects in a pursuit tracking task. *Journal of Experimental Psychology,* 1955, **49,** 249-256.
Noble, M., Trumbo, D., and Fowler, F. Further evidence on secondary task interference in tracking. *Journal of Experimental Psychology,* 1967 (in press).
Noble, M., Trumbo, D., Ulrich, L., and Cross, K. Task predictability and the development of tracking skill under extended practice. *Journal of Experimental Psychology,* 1966, **72,** 85-94.
Posner, M. I. Components of skilled performance. *Science,* 1966, **152** (No. 3730), 1712-1718.
Poulton, E. C. Perceptual anticipation in tracking with one-pointer and two-pointer displays. *British Journal of Psychology,* 1952, **43,** 222-229.
Poulton, E. C. Eye-hand span in simple serial tasks. *Journal of Experimental Psychology,* 1954, **47,** 403-410.

Poulton, E. C. Learning the statistical properties of the input in pursuit tracking. *Journal of Experimental Psychology,* 1957a, **54,** 28-32.
Poulton, E. C. On the stimulus and response in pursuit tracking. *Journal of Experimental Psychology,* 1957b, **53,** 189-194.
Poulton, E. C. On simple methods of scoring tracking error. *Psychological Bulletin,* 1962, **59,** 320-328.
Poulton, E. C. Postview and preview in tracking with complex and simple inputs. *Ergonomics,* 1964, **7,** 257-266.
Poulton, E. C. Tracking behavior. *In* Bilodeau, E. A. (Ed.) *Acquisition of skill.* New York: Academic Press, 1966.
Quigley, J., Trumbo, D., and Noble, M. Sequential probabilities and the learning and retention of tracking skill. Paper presented at the meeting of the Midwestern Psychological Association, Chicago, May, 1966.
Swink, J., Trumbo, D., and Noble, M. On the length-difficulty relation in skill performance. *Journal of Experimental Psychology,* 1967 (in press).
Thurstone, L. L. The relation between learning time and length of task. *Psychological Review,* 1930, **16,** 44-53.
Trumbo, D., Eslinger, R., Noble, M., and Cross, K. A versatile electronic tracking apparatus (VETA). *Perceptual and Motor Skills,* 1963, **16,** 649-656.
Trumbo, D., Noble, M., and Baganoff, F. Analog computer methods for scoring continuous performance records. *Perceptual and Motor Skills,* 1965a, **21,** 707-714.
Trumbo, D., Noble, M., Cross, K., and Ulrich, L. Task predictability in the organization, acquisition, and retention of tracking skill. *Journal of Experimental Psychology,* 1965b, **70,** 252-263.
Trumbo, D., Noble, M., and Fowler, F. Stimulus coherence and response organization in rate tracking (unpublished a).
Trumbo, D., Noble, M., Fowler, F., and Porterfield, J. Temporal coherence and sequence length in tracking performance. (unpublished b).
Trumbo, D., Noble, M., and Swink, P. Secondary task interference in the performance of tracking tasks. *Journal of Experimental Psychology,* 1967 (in press).
Trumbo, D., Noble, M., and Ulrich, L. Number of alternatives and sequence length in acquisition of a step-function tracking task. *Perceptual and Motor Skills,* 1965c, **21,** 563-569.
Trumbo, D., Ulrich, L., and Noble, M. E. Verbal coding and display coding in the acquisition and retention of tracking skill. *Journal of Applied Psychology,* 1965d, **49,** 368-375.
Wagner, R. C., Fitts, P. M., and Noble, M. E. Preliminary investigations of speed and load as dimensions of psychomotor tasks. *United States Air Force Personnel Training Research Center Report,* 1954, No. 54-45.
Woodworth, R. S. *Dynamics of behavior.* New York: Holt, 1958.

Chapter Five

Human Processing of Information— Intellectual or Cognitive Processes

The information presented thus far should have demonstrated clearly that there is considerable "intellectual" involvement in even the most elementary of skills. Simple detection, for example, depends in large measure upon value judgments, and choice reaction time upon memory search strategies. Although motor skills may take on a greater degree of autonomy with continued practice, even tracking performance seems dependent upon course predictions which the tracker makes on the basis of statistical properties of the signal (as we saw in the Noble and Trumbo paper).

Until recently, however, intellectual functions were considered well beyond the scope of any of these topic areas—in fact, well beyond the scope of engineering psychology in general. Typically, research on "thinking" was limited to concept-formation and problem-solving tasks studied within the context of either learning or individual difference theory. Engineering psychology was regarded as chiefly concerned with "doers," not "thinkers." This attitude, of course, has undergone a drastic change, not only within the engineering area, but within psychology generally. We noted this earlier in connection with threshold theory, where it has been formally recognized that *decision* as well as *sensitivity* factors control detection responses. Rare indeed is the theorist or researcher today who completely avoids the use of cognitive terms.

In engineering psychology, the emergence of "intellectual" or "cognitive" considerations has occurred on two broad fronts: one involving methodology and the other, content. The former aspect was brought into clear focus by Edwards (1961), who stressed the fact that every human subject comes into the experiment with his own conception of what is appropriate or desirable behavior; this may or may not conform to what the experimenter has in mind. In a sense, every experimental task is a decision task of sorts: the subject must decide whether he wants to be fast or accurate, to please the experimenter or outwit him, to follow instructions or not, to surpass other subjects or be "average," and so on. Recognizing this fact rather than simply taking the subject's conformity for granted, we can take steps to control decision factors which might otherwise distort our findings. We can make explicit, for example, through monetary costs and payoffs, that for which the subject should strive.

The second way in which engineering psychology has become concerned with

References will be found in the General References at the end of the book.

intellectual processes is as a highly significant subject matter in its own right. Not only have patently "intellectual" functions gained increasing prominence in man-machine system affairs, but, as we said earlier, decision processes have forced themselves into our ways of viewing even the more "mechanical" types of skills (Bartlett, 1958). Clearly, to understand how people can be most effective in system performance, we must know more about how they function intellectually—particularly with respect to making decisions.

As typically used, the concept of "intellectual" or "cognitive" processes seems to encompass a far wider range of hypothetical activities than does that of "decision" processes (see Bourne and Battig, 1966; Neisser, 1967). However, it is difficult to conceive of any meaningful instance of the former which would not at some time implicate the latter, if by decision we mean choice among alternative responses ("response" being defined broadly to include covert as well as overt activities at various levels of organization). Several investigators (e.g., Edwards, 1960; Gagne, 1959; Bourne and Battig, 1966), after wrestling with the difficult problem of codifying such terms, have arrived at the same general conceptualization: thinking can be viewed ". . . as a three-stage process, consisting in (1) a preparatory reception and categorization or organization of information . . . (2) . . . development and formulation of various alternative response sequences . . . ; and (3) the choice or decision as to which of these courses of action is to be followed" (Bourne and Battig, p. 542). These processes are termed, respectively, *conceptual behavior, problem solving,* and *decision making.* Even the proponents of this schema, however, readily admit that its principal virtue lies in distinguishing among three methodological emphases which dominate research in the area; it is doubtful that anyone would argue strenuously that these categories are completely independent and all-inclusive, or even that they truly reflect what goes on "in the head" (see, for example, Taylor, 1963).

Whatever one's theoretical preference, it is safe to conclude that decision behavior is at least an important aspect of intellectual functioning. It also happens to be the aspect which has received the most attention within the context of engineering psychology.[1] For this reason, we have elected to feature so-called "decision research" in this chapter, attempting to show how this emphasis has crept into the study of a variety of problems. As will become clear in the following discussion, however, not all such research is directly concerned with the action-selection facet of decision behavior; in fact, much of the more exciting work of recent vintage concentrates upon how people *process probabilistic information* preparatory to their actual "choice" response (so-called predecision behavior). (The reader is referred to Hunt and Zink (1964), Kleinmuntz (1968), and Shelley and Bryan (1964) for a sampling of work substantiating this statement.) This is also a reason why we have included decision making as a subtopic under *information processing* rather than accord it completely independent status.

The first selection in this chapter is not strictly concerned with decision making at all. It represents, however, a nice transition between the simpler transmission behavior discussed in Chapter Four and the more complex processing require-

[1] There is a scattered and rather task-specific literature in engineering psychology on problem solving, but it is of minor theoretical interest compared to that on decision behavior.

ments involved in decision and other "intellectual" tasks. The taxonomy which Posner develops here is, in a sense, an alternative to the Bourne and Battig schema discussed above, and one which plays down the traditional dichotomy of "simple" and "complex" processes. Within Posner's framework, most decision tasks would involve *reduction* or *creation* of information rather than just *conservation*.

There are many ways of looking at decision behavior. First, since the most interesting kinds of decision problems are those shrouded in uncertainty, we can study the way in which people deal with—evaluate, draw inferences from—equivocal or probabilistic events. This is what the Peterson and Beach article is all about. Second, if we are to have any hope of understanding or dealing with human decision processes, we must assume that outcomes of choices can be ordered in some reasonably consistent fashion on the basis of preference: if so, we can study the laws governing subjective value or utility. There is no article devoted exclusively to this topic in this chapter; the reason is that the most significant work to date on utility is pitched at a level somewhat beyond the intended scope of this book. For those interested in pursuing the matter further, however, a recent review by Becker (1967) and a book of readings by Edwards and Tversky (1967) will serve as a good starting point. Third, we can study the rules or strategies which people seem to adopt in attempting to derive utility from their uncertain environments through various courses of action. Usually this approach incorporates both the motivational and probabilistic facets of decision problems as the classic paper by Edwards clearly illustrates.

While the three aspects of decision behavior just outlined have been the principal focal points for behavioral research, they by no means exhaust the possibilities. It is often useful, for example, to compare actual decision behavior obtained under various experimental conditions with *optimal* behavior—defined in terms of a normative model—in order to evaluate the feasibility of various man-machine combinations (see, for example, Howell, 1967; Shelley & Bryan, 1964). Also, it is meaningful to ask whether individuals differ reliably in their choice of decision rules or their estimates of decision parameters. If this is so, as seems to be the case, it might be feasible to devise selection procedures for tasks requiring specific kinds of decision behavior. Similarly, it might be meaningful to ask whether people can be *trained* to become more nearly optimal in solving decision problems. Evidence to date on this issue is sparse and equivocal.

Finally, we can study decision models in terms of their ability to account for broader aspects of behavior. In the last article of this chapter, Fitts explores the implications of a random-walk decision rule for choice reaction-time performance. Can we, he asks, account for some of the time spent in reacting to discrete signals in terms of the processor's decision as to how *fast* vs. how *accurate* he wants to be? The essential features of the model and the variables chosen to test it stem directly from work on decision processes per se. In other words, since we have a general idea of how people make decisions, the Fitts article seeks to determine whether we can use this knowledge to help explain how they do something else which we believe includes a significant decision component. The results so far are encouraging.

Although the Fitts paper concludes this chapter, it does not conclude our illus-

19

Information Reduction in the Analysis of Sequential Tasks

Michael I. Posner

Most applications of information theory to psychology have been concerned with tasks which require information to be conserved between input and output. In the standard reaction-time task the S is supposed to effect an energy and location change, but must preserve all the information in the input for errorless performance (Bricker, 1955a). Verbal-learning experiments add to these requirements the necessity for S to serve as a hold circuit and thus delay the response a specified period of time. In studies of the limits of discriminability S is required to preserve in so far as possible all stimulus differences in his judgment, the failure to do so is recorded as an error (Miller, 1956). In each of these tasks any change in the informational content during message transmission is clearly error. As useful as these applications have been it is increasingly apparent that this is not the sole nor even the typical information-processing situation for the human. Thus, Bruner (1957) suggests that "the most characteristic thing about mental life is that one constantly goes beyond the information given [p. 41]." With the reverse emphasis, Gerard (1960) suggests that "the real skill of the talented thinker is in discarding irrelevancies. What is omitted in perception, memory and reasoning is of the highest moment [p. 1939]." Attneave (1962) has been somewhat more explicit—"most of the information that goes into the individual never comes out again. The information that is lost is not necessarily wasted, however. The situation is somewhat like that of an executive who considers a mountain of data . . .

From *Psychological Review*, 71, 1964, 491-504. Copyright 1964 by the American Psychological Association, and reproduced by permission.

A portion of this paper is based upon a dissertation submitted to the University of Michigan in partial fulfillment of the requirement of the PhD degree and supported by the Air Force Office of Scientific Research under Contract No. AF 49 (638)-449. The present version was made possible by a grant from the University of Wisconsin Research Committee from funds supplied by the Wisconsin Alumni Research Foundation. The author wishes to thank P. M. Fitts and R. W. Pew for assistance in the presentation of these ideas.

in order to arrive at a one bit decision [p. 634]." It is with these omissions, condensations, and reductions that this paper will be concerned. It is first useful to outline a general taxonomy of information-processing tasks within which the development of empirical relations can proceed.

TAXONOMY OF TASKS

Figure 1 illustrates a general taxonomy for information-processing tasks which includes not only the commonly studied information-conserving tasks but also information reduction and information creation. This classification considers only the input and output information required for perfect performance. A conservation task such as the standard choice reaction time situation, for example, requires S to preserve all the stimulus information for perfect performance. In actual practice he may reduce information in this task by stimulus equivocation, but such a reduction is clearly an error as indicated in Figure 1. In a reduction task, however, S is required to map more than one stimulus point into a single response. The loss of information in this situation is not error, but is required by the nature of the task. For example, the sum of a set of numbers has less information than its components, yet in losing this information, S has clearly not made an error, but rather has accomplished his task. In the final logical category, the creation task, S is required to map a single stimulus point into more than one response. Multiple association is of this character, but the empirical analysis of this type of processing is beyond the scope of this paper. Both the reduction and the creation model seem to capture something of the cognitive aspects of human activity.

Previous Classifications

The necessity for a taxonomy of this sort has become increasingly apparent in the last few years. Studies which place emphasis on many to one (Morin, Forrin, & Archer, 1961) or one to many (Morin & Forrin, 1963) mappings of stimuli into responses have revealed limitations to the generalization so useful in conservation tasks, that task difficulty increases with information transmitted (Bricker, 1955a). Campbell (1958) noted that Ss tend to show systematic errors in information transmission which may make important additions to the message or which may reduce information through classification. Garner (1962), in his discussion of concept-formation tasks, points out that since these require stimulus equivocation the normal relation between total and internal constraint which exists for conservation tasks is changed. Toda (1963) in reviewing Garner's book recognizes the basic nature of this distinction for task analysis and calls the conservation tasks "normal." Hunt (1962) proposes that learning tasks can be divided into those which require complete information transmission (rote learning), information reduction (concept learning), and information production (probabilistic learning). This type of classification system captures something of Bartlett's (1959) verbal distinction between "closed-system thinking" in which the solution is implicit in the problem and "open-system thinking" in which S uses the available evidence to leap beyond

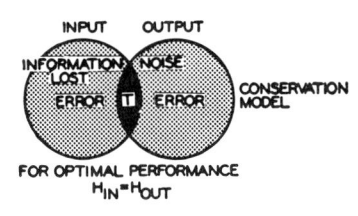

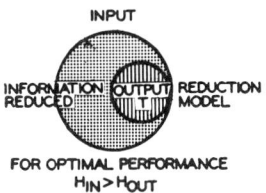

Figure 1. A proposed taxonomy of information-processing tasks.

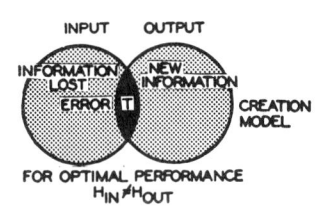

Figure 2. Stimulus and response sets for reaction-time study (after Bricker, 1955b).

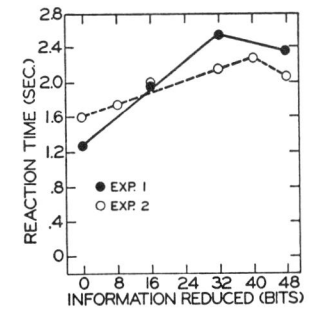

Figure 3. Reaction time as a function of amount of information reduced.

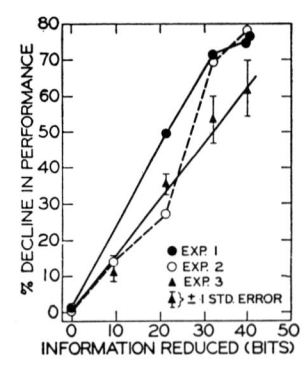

Figure 4. Percentage decline in performance with speeding as a function of the amount of information reduced.

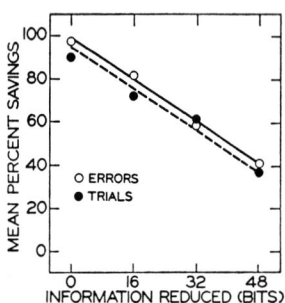

Figure 5. Rate of transfer as a function of amount of information reduction.

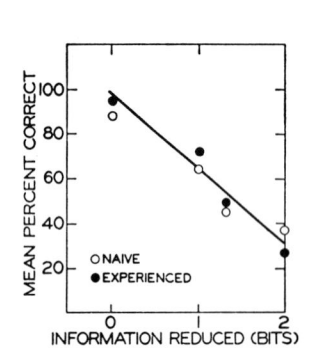

Figure 6. Percentage of correct responses in concept formation as a function of information reduction required (after Shepard, Hovland, & Jenkins, 1962).

TABLE 1

Mean Trials to Criterion (after Metzger, 1958)

	\multicolumn{3}{c}{Information reduced (bits)}	
	0	1	2
Gating		25	38.4
	21.2		
Condensation		40.4	51.2

the input and provide a creative solution. It has a lesser analogy with the classical division of the thought processes into productive and reproductive (Wertheimer, 1959).

Laws of Information Reduction

A classification system is of little value unless it can be shown that tasks falling within a category obey the same general laws. The study of information-conservation tasks has been useful mainly because performance has been shown to have systematic relations to information transmitted. While these relations are by no means perfect, for example memory tasks show relatively little effect of information per symbol (Miller, 1956), they have provided important basic generalizations from which more specific models have been formulated (Fitts, Peterson, & Wolpe, 1963; Sperling, 1963). This paper proposes that for information reduction tasks difficulty is directly related to the difference between stimulus and response information. The author has chosen to call this difference the information reduced. If stimulus uncertainty is held constant this quantity is the inverse of information transmitted, if response uncertainty is held constant and stimulus uncertainty varied it is, of course, independent of information transmitted. The information reduced is equivalent to amount of stimulus equivocation in conservation tasks, but a distinction is made between equivocation introduced by Ss' errors and reduction of information required by the nature of the task.

This simple statement of the relation between information reduction and performance allows a variety of experimental situations to be seen as related through the requirement for Ss to operate on input information to produce a condensed response. Within a restricted task configuration it allows quantitative predictions which will be evaluated in this paper and which should serve to generate new experiments. It is, of course, true that this simple statement does not exhaust our ability to predict, any more than the relation between information transmitted and reaction time completes our knowledge of that task. Within restricted categories more specific models such as the Bourne-Restle model for concept identification may do much better. However, this view provides a basic framework within which more specific formulations can be developed.

Organization of the Paper

This paper will examine first those information-handling tasks which involve the utilization of rules already well learned by S. Under this general heading both tasks which allow S to ignore aspects of the stimulus (gating) and those that require all aspects of the stimulus to be represented in the response but in a condensed form (condensation) will be considered. The second major area of analysis will concern classification tasks which require the S to acquire or learn the use of rules. The paper examines two types of these tasks, first, those who are quite similar to rote learning except that more than one stimulus has the same response (unidimensional) and second, those that meet the full requirements of concept learning by use of multivariate stimulus materials. Within each of these subareas both rules allowing gating and those requiring condensation will be examined.

INFORMATION HANDLING

In attempting to account for compatibility effects in reaction time tasks Fitts (1959) speculated that,

> as the number of intervening information transformations increases, or as the number of hierarchical systems involved in the activity increases, the greater, on the average, is the time required to complete an information-handling task and the greater is the probability of errors [p. 44].

Fitts suggests that such a speculation is not testable by direct observation. This is clearly the case for the information-conserving tasks which he was considering, since the information in the input and output are the same and the experimenter must speculate with respect to intervening transformations. In an information-reduction task, however, the stimulus and response information is different and the experimenter can use the size of transformation between input and output as a directly observable property of the task.

In 1957 Pierce and Karlin studied a task in which Ss were required either to read aloud a name or to classify it into two categories, male or female in one condition and animal or vegetable in another, by pressing a key. When there were only two names the key press was slightly faster than the verbal response. However, as the population of words increased in size the binary categorization required 40% longer than the naming response.

The information transmitted by the classification task is only 1 bit regardless of the size of the population of individual stimuli, whereas the amount of information transmitted under the information conserving conditions increased monotonically with the number of individual stimuli. The authors conclude that the classification must have taken additional processing time, but did not follow up this inversion of the general rule of increasing reaction time with information transmitted.

Crossman (1953) studied card sorting under a number of conditions. One condition involved sorting the cards into two categories, red and black; a second condition required the S to sort into any of four categories, red face, black face, red non-face, black non-face; while the third condition required the S to sort red face and black non-face into one category and the reverse into a second category. Each of these tasks involve a loss of information since they require classification into less categories than the information available in the stimulus would allow. The mean sorting time per card for the one S studied was .7 second for condition one, 1.1 seconds for condition two, while condition three initially required 1.4 seconds and only after some training approached the speed of condition two. Thus whether transmission of 1 bit is faster or slower than 2 bits depends upon whether S is allowed to ignore the face versus non-face distinction of the cards as he is in condition one or whether he is required to process that information in making his classification as he is in condition three.

Gating

We will first look for additional results in reaction time studies which resemble condition one in the Crossman study. Gregg (1954) and Archer (1954) have studied this type of task. In both situations Ss had to classify one or more binary relevant dimensions, while ignoring one or more binary irrelevant dimensions. In both situations Ss knew the rules which allowed them to determine whether or not the information was relevant. Gregg (1954) found a small but significant linear increase in reaction time (RT) with the addition of from 1–3 bits of irrelevant information. The size of the increase depended, in part, on the degree of compatibility between the relevant stimulus and its response. Archer, however, found no increase in RT due to the addition of up to two irrelevant dimensions.

In this type of situation S can ignore large amounts of stimulus information with little or no penalty in RT. What penalty there is appears to be a linear function of the amount of irrelevant information. Both of these studies require S to ignore entire dimensions. When this is the case the way in which the dimensions are separated is probably of importance. For example, studies which present relevant information to the eye and irrelevant to the ear (Broadbent, 1958) and in which S knows to attend to the visual input usually show little decrement due to increased irrelevant information. Obviously this would be even more striking with the opposite coding, when S could merely shut his eyes as a direct physical gate. The studies of Gregg (1954) and Archer (1954) also involve ignoring entire dimensions, but these dimensions overlap spatially with the relevant ones.

Situations which allow Ss to reduce information by ignoring aspects of the stimulus will be called gating tasks. The literature indicates that those tasks which allow gating of entire dimensions show little or no increase in difficulty with increasing irrelevant information, once the gating rules are well known. Tasks which require gating within a dimension may show a more marked increase in difficulty with increasing irrelevant information (Rabbitt, 1963) though the evidence here is hardly complete. In a sense the limiting case is composed of those tasks which, like Pierce and Karlin's and Crossman's condition three, do not allow S to ignore any aspects of the stimulus, but require him to represent it in a reduced or condensed form. These tasks are conceptually different from the gating tasks, although they still involve information reduction and will be called condensation tasks.

Condensation

Bricker (1955b) presented Ss with eight stimulus patterns which had to be classified into eight, four, or two response categories. Thus he considered, in addition to the normal information conserving situation, a 1- and 2-bit reduction task. However, the particular coding in the Bricker study (Fig. 2) makes it difficult to be sure whether or not this can be considered a type of gating task. The eight stimuli could easily be recoded, as suggested by the lines inserted in Figure 2, into four perceptually distinct shapes ignoring differences in location and spacing of the elements: curved outward, curved inward, straight, and wedge, or into two shapes, curved and straight. These codes correspond exactly to the divisions which Bricker made in the response classes. If S can ignore detailed differences between individual patterns this task would be expected to be somewhat like three conservation situations with 1, 2, or 3 bits transmitted (Archer, 1954; Gregg, 1954) and to follow the usual linear relations between RT and information transmitted. Indeed, this is the result. It is interesting to speculate on what would have happened if all possible codes were used. One might expect that most would provide less opportunity for S to ignore stimulus elements in making the classification and the results would become similar to Crossman's condition three and the Pierce and Karlin study. In addition, the Bricker study confounds amount of information reduced with changes in response uncertainty. It would be useful to separate the two effects.

Posner (1962) has used a technique which allows the manipulation of amount of information reduced free from changes in response uncertainty and with systematic variation of the perceptual properties of forms within a response class. Original patterns of eight dots and distortions of these patterns were constructed by means of statistical rules so that a level of uncertainty between an original and its distortions could be stated. The S saw two patterns at a time and was required to state whether they were members of the same class or not. Prior to this procedure the Ss had gone through a rather lengthy learning process where they had learned to give pairs of patterns of varying levels of uncertainty the same names. The results of this learning procedure will be discussed later, but crucial here are the results of the average reaction times when two patterns were exposed which had the same name and when S was correct. These data are shown in Figure 3. The results show reaction time to be linearly related to the uncertainty between patterns at least up to 40 bits. The greater the uncertainty which must be smeared over or reduced in making the classification, the greater the classification time.

Morin, Forrin, and Archer (1961) describe an experiment important in this analysis. The stimuli are slides with either one or two squares or one or two circles. Four conditions are crucial: conservation with two stimuli (1 bit), conservation with four stimuli (2 bit), gating in which the four stimuli are classified into circles or squares, and condensation where the response classes are two circles-one square and one circle-two squares. The initial level of difficulty as shown by RT indicates that the gating task was identical with the 1-bit conservation condition, while the condensation task was much slower than these but slightly faster than the 2-bit conservation task. Fitts and Biederman (1964) first replicated the experiment exactly and then added a new stimulus-response (S-R) code which he had found in prior testing to be more compatible. His results closely replicated the original study in the initial difficulty except that with the new S-R code the reduction task took slightly longer than the 2-bit conservation task. In the Fitts results, unlike the original study, the condensation task continues to be of equal or greater difficulty than the 2-bit conservation task over the full 2 days of training.

The results reviewed so far indicate that a condensation task will be much more difficult in terms of processing time than the conservation task with the same information transmitted (Crossman, 1953; Morin, Forrin, & Archer, 1961; Pierce & Karlin, 1957), and will, after training, approach about the same level of difficulty as a conservation task with the same stimulus uncertainty (Crossman, 1953; Fitts & Biederman, 1964). Gating tasks, however (Archer, 1954; Bricker,

1955b; Crossman, 1953; Morin, Forrin, & Archer, 1961) show little or no increased difficulty over their conservation counterpart with the same information transmitted. Only in the Posner (1962) study, in which stimulus similarity is defined in terms of uncertainty, has a quantitative relationship been discussed between the amount of information reduced and reaction time for a condensation task. In that study information was varied by systematic increases in stimulus uncertainty with response uncertainty constant.

Pollack (1963) has developed the basic paradigm of Pierce and Karlin (1957) into a systematic study of the relationship of amount of information reduction to speed of classification. Pollack used 24 generic word categories each having up to 24 individual instances, such as goat and pig as instances of animal. In the main conditions of the study Ss were required to classify 48 instances into from 2 to 24 categories. When there is only a single instance per category the task is information conserving. The results of these conditions show the expected linear relation between information transmitted and classification time (Bricker, 1955a). However, with the number of response categories fixed there is also a direct and reasonably linear relation between amount of information reduction, from 1 to 4.5 bits per category, and classification time. The slope of this function increases as the number of response categories increases. At low levels of response uncertainty the effect of going from one instance per category (conservation) to two instances (condensation) is much greater than the general straight line trend due to amount of reduction, but the reverse is true at higher levels of uncertainty.

Posner (1962) has performed a study which also allows the production of quantitative relationships between amount of information reduction and performance where stimulus uncertainty is constant and response uncertainty varied. In these experiments the stimulus message was always the same, consisting of a series of eight numbers selected randomly from a population of 1 to 64 and presented aurally to S. The S was required either to record the numbers or to operate upon them by a number of information-reducing operations. The tasks were chosen so that the output information varied from 7.7 to 48 bits and so that the components of the operations involved were familiar. The tasks included an information-conserving recording condition, a task of alternating recording with summing the digits of a number, a partial addition task in which successive pairs of numbers were added together, a 2-bit classification task where the numbers were classified into high and odd, low and odd, high and even, or low and even, and a 1-bit classification task in which high and odd or low and even was an "A" and low and odd and high and even was a "B." It was not possible to compare the tasks directly because errors were so different for each task. Therefore, the rate of change of errors in each task was measured as the interstimulus interval was reduced from 4 to 1 second. The basic notion was that the more difficult the task the greater would be the rate of decline of performance with speeding. The use of an error measure, well-learned responses, and an appropriate correction for chance levels were designed to reduce the effect of changes in response uncertainty per se.

Three separate experiments involved the following designs: (a) each S takes all tasks at one speed, (b) each S takes one task at one speed, (c) each S takes one task at two speeds. The results showed that in every experiment the rate of decline in performance increased monotonically with increasing information reduction. The overall relationship, as shown in Figure 4, is roughly linear with some variation between experimental designs. In general, the designs which successfully controlled for sequential transfer effects between the tasks produced more linear relationships.

These results again demonstrate that condensation tasks, unlike conservation tasks, cannot usefully be thought of as increasing in difficulty with information transmitted. On the contrary, when stimulus information is held constant and S is required to operate upon it to produce a condensed output as in this last study, difficulty declines with increasing information transmitted. If stimulus information is increased with response information constant as in the tasks measuring reaction time to patterns (Posner, 1962) or in the classification of words (Pollack, 1963) then information transmitted is fixed, but task difficulty increases. In all these cases, however, task difficulty is found to be an increasing linear function of amount of information reduced.

One problem with many of the studies cited here is that they do not adequately control the degree of learning. In all cases the operations or classifications were well known by Ss but most studies cannot show that they were equally well known or that learning effects were unimportant during the experimental conditions. It is, therefore, important to look at the effect of information-reducing tasks in a situation where the process of learning rules is observed. This will be discussed in the next section.

CLASSIFICATION LEARNING

Unidimensional Rote Learning

Rote learning, like the standard reaction time task, requires information conservation. The minimum requirement for changing a rote-learning task into one of classification learning is to have more than one stimulus per response. When the stimuli lie along a single dimension the E has minimal control over the relevant and irrelevant aspects of the stimulus, but he can control the number of stimuli to be assigned a single response and the way in which the stimuli within a response category are grouped along the dimension.

Number of Stimuli per Response. One way to vary the amount of information reduction in a classification task is to increase the number of stimuli per response. Most investigators who have attempted this have kept the number of stimuli constant and reduced the number of responses. Hake and Erikson (1955) and Bricker (1955b) studied the learning of patterns of lights when eight, four, and two response categories were used. Richardson (1958) studied the assignment of 16 adjectives to two, four, or eight response categories. In none of these studies were significant differences found in the rate of learning as a function of the number of stimuli per response, once correction had been made for differing chance levels.

There are several problems in studying the information reduction involved in unidimensional rote concepts. First, all of the studies cited confound the amount of information reduction with the number of response terms to be learned and used. Despite this, the results do indicate that an information-reduction task is more difficult to learn than the conservation task with the same information transmitted, but about the same difficulty as a conservation task with the same simulus uncertainty. Thus, when Es look at the classification of a fixed list of stimuli into a variable number of responses little difference is found among conditions. This is in agreement with the results in many reaction-time studies (Crossman, 1953; Fitts & Biederman, 1964; Morin, Forrin, & Archer, 1961). Future studies should run control groups with conservation tasks of the same response uncertainty as the reduction tasks as well as at the same stimulus uncertainty. Second, these studies allow no clear distinction between gating and condensation. As pointed out earlier, Bricker (1955b) used classifications which allow the opportunity for recording rules which could eliminate processing of certain aspects of the stimuli. It is well known that verbal synonyms (Richardson, 1958) are also subject to such mediational rules. However, in these experiments it is not possible to tell if Ss are using such rules or when in the learning they come into play. Finally, the most serious difficulty with these studies is that they do not take into consideration the relationship between the stimuli to be assigned a single response. That is, both the number and the way in which stimuli are grouped should effect the total information reduction involved.

Stimulus Grouping. That the number of stimuli per response can be a potent factor when there is more control over coding within a stimulus grouping is shown in a recent study (Smith, Jones, & Thomas, 1963). This study used eight stimuli varying only along the hue dimension. There were either two, four, or eight response categories. When stimuli were assigned at random to response categories the opportunity for recoding by use of familiar color names was minimal. In this situation performance declined uniformly with increasing information reduction. This is in agreement with the general relation between information reduction and performance, even though as information reduction increased the number of responses to be learned declined. If adjacent colors were assigned the same response the task becomes one of absolute discrimination and performance increases as fewer categories are required (Miller, 1956). This study indicates that both the total information reduction and the way information is distributed within response categories is important to predictions of task difficulty. One can vary the number of stimuli per response, but if the added stimuli are very similar or if they can be coded by a well-learned name, little or no change is made in task difficulty.

Some investigators have varied the informational properties of the task by manipulating the variability of the stimuli within a response category. Richardson (1958), using meaningful adjectives, found that as the similarity of items within a response category increased learning greatly improved; the reverse occurred with increases in similarity between categories. Shepard and Chang (1963), using colors previously graded in similarity by use of paired comparisons, assigned subsets of four stimuli to the same response category. They found that the more similar the subset assigned to a response category the faster the learning and that such differences in within response variability accounted for 78% of the learning variance.

French (1953) systematically studied the effect of stimulus similarity within and between response categories. He varied the similarity between the six visual forms assigned to each of two categories by varying the number of line segments which were identical. In the discrimination situation only half of the set of six similar figures were in the same response category, while in the generalization situation they were all in the same response category. These results show that performance is directly related to the degree of similarity within a response category and inversely related to similarity across different categories.

Posner (1964) has shown that it is possible to express the degree of similarity between patterns in terms of uncertainty. Psychophysical studies showed a linear relation between the amount of uncertainty and perceived distance for patterns of eight dots. Such patterns were used in a paired-associate study in which Ss had to transfer the name learned for one pattern to a new pattern at a given level of uncertainty. It was found that the degree of uncertainty between patterns was linearly related to rate of transfer (Fig. 5).

The result of this study is crucial for three reasons. First, it shows that rate of learning is a decreasing linear function of amount of information within a response category. This is in agreement with studies of stimulus similarity cited above, as one would predict from the linear relation between uncertainty and perceived similarity, and with the results of reaction time studies cited earlier. Second, it allows future studies to combine the number of stimuli per response and uncertainty between stimuli to obtain overall relation between within category uncertainty and rate of learning. Third, it shows how studies of unidimensional classification learning can be systematically compared with multivariate concept-learning tasks which have been explored in informational terms. These relationships will be discussed further in the next section.

Multivariate Concept Learning

The studies which we have been discussing form a bridge between the investigation of rote learning and the investigation of traditional concept formation. They are similar to the concept-formation design in that a single response is made to more than one stimulus. However, they lack one attribute of most concept formation studies (Garner, 1962) namely, they are not multivariate in nature. Garner says,

> this multivariate requirement . . . makes it possible for the experimenter to determine a priori which variables are relevant to changes in the response and which variables are irrelevant to these changes. Then the S is given the task of determining which variables are relevant and once he has learned this fact he can respond correctly to a large group of stimuli with a single response by ignoring stimulus differences which exist with respect to the irrelevant stimulus variables [pp. 311–312].

In these types of studies it becomes possible to distinguish rather clearly between gating and condensation.

Metzger (1958) studied paired-associate learning of patterns which had either three or four distinct elements. Each element was a triangle which could be either large or small. In the three element condition he studied conservation tasks with eight and four stimuli, a gating task with four response categories (one irrelevant element) and a condensation task with random assignment of the eight stimuli to four responses. The results showed that the gating and condensation tasks were not much different from the conservation task with the same stimulus uncertainty though much more difficult than the conservation task with the same response uncertainty. This is in agreement with most of the unidimensional results and agrees with the reaction time results except with respect to the gating task. In the four element conditions it is possible to make some quantitative comparisons of the effects of increasing information reduction on rate of learning for both gating and condensation operations. This is shown in Table 1. The conservation condition is for eight stimuli and eight responses, the 1-bit reduction point represents tasks with eight stimuli and four responses, while the 2-bit reduction point represents 16 stimuli and four responses. The results show that both types of reduction tasks are somewhat more difficult than the conservation task with identical stimulus uncertainty and that both increase in difficulty with increasing information reduction.

Gating. Information measurement has been systematically applied to the investigation of concept identification with multivariate stimuli (Archer, Bourne, & Brown, 1955; Bourne & Haygood, 1959; Walker & Bourne, 1961). These studies are gating tasks since the Ss have to learn to ignore certain irrelevant aspects of the stimulus and then classify all levels of relevant information into distinct categories. The results of these studies show that errors to criterion are an increasing linear function of the amount of irrelevant information which must be ignored in making the response classification. This is quite consistent with the results obtained by Metzger (1959) for paired-associate gating tasks with constant response uncertainty. These results indicate that the clear distinction between gating and condensation tasks is lost in concept-learning studies. Both are increasing linear functions of the amount of information reduced. This is reasonable since the S must learn what aspects of the stimulus are relevant and as the stimuli are made more complex this increases the difficulty. Once the gating rules are learned, however, the gating tasks come to resemble conservation tasks and little effect of irrelevant information on processing time is found (Archer, 1954).

In the concept-identification experiments cited above the number of relevant stimulus levels always equals the number of responses. Battig and Bourne (1961) have discussed this limitation as follows, "one shortcoming of the concept-identification task, and of many other procedures used in experiments purportedly concerned with concept formation, is that in a real sense they do not require concepts to be formed. This is because all of the stimuli to be classified with the same response are identical with respect to the crucial or relevant characteristics [p. 329]."

Condensation. Battig and Bourne (1961) have attempted to remedy this defect by varying both the number of irrelevant dimensions and the variability within a response category. In order to do this they compared, for example, a situation where a single response was made to right, equilateral, and obtuse triangles and another to squares, trapezoids, and parallelograms. In this situation a very simple recoding (four sides versus three sides) could allow S to ignore some of those differences. To determine the actual extent of such recoding the stimulus patterns should be tried in a reaction time task to find out if after learning the RT would approach that for a one bit conservation task. In any case, regardless of the ability of S to ignore such differences, the results so far reviewed in the learning of concepts (Archer, Bourne, & Brown, 1955; Metzger, 1959) would indicate that a linear relation should exist between the amount of information reduced in forming the response class and rate of learning. Hunt (1962) analyzed the Battig and Bourne data and found that the relation between errors to criterion and information within a response category is linear.

The concept-learning study most similar to the condensation tasks discussed in section one is that of Shepard, Hovland, and Jenkins (1962). In these experiments there were always eight stimulus patterns, consisting of two levels on each of three dimensions. There were always two response categories. The problems differed in the way in which the information relevant to the classification was distributed across the dimensions. In one problem all the information involved only one dimension, in another all information involved two dimensions simultaneously, a third type had all three dimensions relevant, and a final type involved some of the information in one, two, and three dimensional terms. The total amount of relevant information within a response category could be calculated by weighting the proportion of information by the number of dimensions which had to be attended in order for it to be extracted. By this method the four types of problems had 1, 2, 2.3, and 3 bits of relevant information, respectively. Figure 6 shows the success of concept learning as a function of the amount of information condensed in obtaining the binary classification. The points of this figure are from Tables 4 and 6 of the original report and represent averages over different sets of a given problem type and over different display conditions. This result indicates that condensation is more important than gating when they are opposed, as in this task. There is some indication that with more extensive training than that shown in Figure 6 the 2-bit reduction tasks begin to become easier than some of the 1.3-bit tasks.

In the learning of concepts the amount of information reduced appears to be a useful summarizing variable. As well as having a consistent relation to task difficulty it also demonstrates the close relation between unidimensional and multivariate concept learning both conceptually and empirically. This measure and the taxonomy based upon type of information processing should be of significant value in an integrated attack upon complex information processing.

REFERENCES

Archer, E. J. Identification of visual patterns as a function of information load. *J. exp. Psychol.*, 1954, 48, 313-317.

Archer, E. J., Bourne, L. E., Jr., & Brown, F. G. Concept formation as a function of irrelevant information and instruction. *J. exp. Psychol.*, 1955, 49, 153-164.

Attneave, F. Perception and related areas. In S. Koch (Ed.), *Psychology: A study of a science.* Vol. 4. New York: McGraw-Hill, 1963.

Bartlett, F. C. *Thinking: An experimental and social study.* New York: Basic Books, 1959.

Battig, W. F., & Bourne, L. E., Jr. Concept identification as a function of intra- and interdimensional variability. *J. exp. Psychol.*, 1961, **61**, 329-333.

Bourne, L. E., Jr., & Haygood, R. C. The role of stimulus redundancy in concept identification. *J. exp. Psychol.*, 1959, **58**, 232-238.

Bricker, P. D. Information measurement and choice time: A review. In H. Quastler (Ed.), *Information theory in psychology*. Glencoe, Ill.: Free Press, 1955. (a)

Bricker, P. D. The identification of redundant stimulus patterns. *J. exp. Psychol.*, 1955, **49**, 73-81. (b)

Broadbent, D. E. *Perception and communication.* New York: Pergamon Press, 1958.

Bruner, J. Going beyond the information given. In, *Contemporary approaches to cognition*. Cambridge, Mass.: Harvard Univer. Press, 1957.

Campbell, D. T. Systematic error on the part of human links in communication systems. *Inform. Cont.*, 1958, **1**, 334-369.

Crossman, E. R. F. W. Entropy and choice time: The effect of frequency imbalance on choice-response. *Quart. J. exp. Psychol.*, 1953, **5**, 41-51.

Fitts, P. M. Human information handling in speeded tasks. IBM Research Report 109, 1959.

Fitts, P. M., & Biederman, I. S-R compatibility and information reduction. *J. exp. Psychol.*, 1965, **69**, in press.

Fitts, P. M., Peterson, J. R., & Wolpe, G. Cognitive aspects of information processing: II. Adjustments to stimulus redundancy. *J. exp. Psychol.*, 1963, **65**, 423-432.

French, R. S. Number of common elements and consistency of reinforcement in a discrimination learning task. *J. exp. Psychol.*, 1953, **45**, 25-33.

Garner, W. R. *Uncertainty and structure as psychological concepts.* New York: Wiley, 1962.

Gerard, R. W. Neuro-physiology: An integration. In, *Handbook of Physiology*, Vol. 3. Washington, D. C.: American Physiological Society, 1960.

Gregg, L. W. The effect of stimulus complexity on discriminative responses. *J. exp. Psychol.*, 1954, **48**, 289-297.

Hake, H. W., & Eriksen, C. W. Effect of number of permissable response categories on learning of a constant number of visual stimuli. *J. exp. Psychol.*, 1955, **50**, 161-167.

Hunt, E. B. *Concept learning: An information processing approach.* New York: Wiley, 1962.

Metzger, R. A comparison between rote learning and concept formation. *J. exp. Psychol.*, 1958, **56**, 226-231.

Miller, G. The magical number seven; plus or minus two: Some limits on our capacity for processing information. *Psychol. Rev.*, 1956, **63**, 81-97.

Morin, R. E., Forrin, B., & Archer, W. Information processing behavior: The role of irrelevant stimulus information. *J. exp. Psychol.*, 1961, **61**, 89-96.

Morin, R. E., & Forrin, B. Response equivocation and reaction time. *J. exp. Psychol.*, 1963, **66**, 30-36.

Pierce, J. R., & Karlin, J. E. Reading rates and the information rate of a human channel. *Bell Sys. tech. J.*, 1957, **36**, 497-516.

Pollack, I. Speed of classification of words into super-ordinate categories. *J. verbal Learn. verbal Behav.*, 1963, **2**, 159-165.

Posner, M. I. An informational approach to thinking. Technical Report, Office of Research Administration Project 02814, Office of Research Administration, Ann Arbor, April 1962.

Posner, M. I. Uncertainty as a predictor of similarity in the study of generalization. *J. exp. Psychol.*, 1964, **68**, 113-118.

Rabbitt, P. M. Ignoring irrelevant information. *Amer. Psychologist*, 1963, **18**, 472. (Abstract)

Richardson, J. The relationship of stimulus similarity and number of responses. *J. exp. Psychol.*, 1958, **56**, 478-484.

Shepard, R. N., & Chang, J. J. Stimulus generalization in the learning of classification. *J. exp. Psychol.*, 1963, **65**, 94-102.

Shepard, R. N., Hovland, C. J., & Jenkins, H. M. Learning and memorization of classifications. *Psychol. Monogr.*, 1961, **75** (13, Whole No. 517).

Smith, T. A., Jones, L. V., & Thomas, S. Effects upon verbal learning of stimulus similarity, number of stimuli per response and concept formation. *J. verbal Learn. verbal Behav.*, 1963, 1, 470-476.

Sperling, G. A model for visual tasks. *Hum. Factors*, 1963, 5, 19-31.

Toda, M. Review of, *Uncertainty and structure as psychological concepts. Psychometrika*, 1963, 28, 293-310.

Walker, C. N., & Bourne, L. E., Jr. The identification of concepts as a function of amounts of relevant and irrelevant information. *Amer. J. Psychol.*, 1961, 74, 410-415.

Wertheimer, M. *Productive thinking.* New York: Harper, 1959.

20

Man as an Intuitive Statistician

Cameron R. Peterson and Lee Roy Beach

"Given . . . an intelligence which could comprehend all the forces of which nature is animated and the respective situation of the beings who compose it—an intelligence sufficiently vast to submit these data to analysis . . . nothing would be uncertain and the future, as the past, would be present to its eyes [Laplace, 1814]." In lieu of such omniscience, man must cope with an environment about which he has only fallible information, "while God may not gamble, animals and humans do, . . . they cannot help but to gamble in an ecology that is of essence only partly accessible to their foresight [Brunswik, 1955]." And man gambles well. He survives and prospers while using the fallible information to infer the states of his uncertain environment and to predict future events.

Man's problems with his uncertain environment are similar to those faced by social enterprises such as science, industry, and agriculture. Satisfactory decisions require sound inferences about prevailing and future states of the environments in which these enterprises operate. Consequently, a great deal of effort has been

From *Psychological Bulletin*, 68, 1967, 29-46. Copyright 1967 by the American Psychological Association, and reproduced by permission.

This research was undertaken in the Engineering Psychology Laboratory, Institute of Science and Technology, University of Michigan, under United States Public Health Service Fellowships MF-12,744 and MH-12,012 from the National Institute of Mental Health and was also supported by the Air Force Office of Scientific Research under Contract AF 49(638)-1731. Many of the ideas expressed here have their origins in discussions and arguments with Ward Edwards, to whom we are deeply indebted. We also would like to thank Kenneth R. Hammond who introduced us to the probabilistic functionalism of the late Egon Brunswik, an approach that has influenced much of the research in this review.

invested in the development of coherent, formal procedures for dealing with fallible information in making inferences. These procedures, complex and sophisticated enough to have become a discipline, are called probability theory and statistics.

Because of the parallels between many of the inference tasks faced by man and by social enterprises, a number of investigators have used formal statistical theory as a point of reference for the study of human inference. For many uncertain situations, statistical theory provides models for making optimal inferences. The psychological research consists of examining the relation between inferences made by man and corresponding optimal inferences as would be made by "statistical man." [1]

The procedure is to use a normative model in order to identify variables relevant to the inference process. In this sense, probability theory and statistics fulfill a role similar to that of optics and acoustics in the study of vision and hearing. Just as optics and acoustics are theories of the environments in which eyes and ears operate, statistics is a theory of the uncertain environment in which man must make inferences. Sense organs do not merely mirror their physical environments, so their behavior cannot be described solely by a description of the environment. Instead, optical and acoustical theories have provided a basis for building descriptive theories that link vision and hearing to the physical dimensions of their environments. In the same manner, the theory of statistical inference can provide a basis for a descriptive theory of imperfect human inference.

A primary reason for selecting the strategy of evolving a theory of human inference from statistics is that the descriptive theory remains couched in the language of, and is structurally related to, the broad framework of the theory of statistical inference. This means that experimental findings from otherwise diverse areas may be logically integrated through reference to that theoretical framework.

The ultimate goal of this research is to develop a theory about human behavior in an uncertain environment, but the scope of this paper is necessarily more limited. First, it includes only behavior interpretable within the framework of statistical decision theory. Within this realm, the complete normative theory includes both statistical inference, as a model about how to gain knowledge of the environment, and decision theory, as a model for selecting courses of action in that environment. The psychological counterparts of these two components are intuitive statistics and psychological decision theory. This review explores only the predecisional process of intuitive statistics; reviews of the psychological decision literature are available elsewhere (Becker & McClintock, 1967; Edwards, 1954, 1961a).

This literature is organized in the familiar outline of an introductory statistics book. First, we examine intuitive descriptive statstics, the process of describing samples of data. We then consider research on intuitive inferential statistics, the process of using samples of data as a basis for making inferences about parent populations. Finally, we review studies of intuitive prediction, the process of

[1] Our use of "statistical man" as a model is analogous to the normative use of the "ideal observer" in signal detectability theory and "economic man" in economics. We mean the statistical logic and procedures appropriate to the task subjects must perform.

using inferences about populations as the basis for predicting future samples to be drawn from those populations.

INTUITIVE DESCRIPTIVE STATISTICS

By and large, psychologists have devoted less attention to studying intuitive descriptive statistics than they have to studying inferences. Perhaps this is because inference is inherently more interesting. Still, inferences about populations require prior summarization of sample data, and it can be argued that intuitive descriptive statistics underlie subsequent inferences.

Typically, experiments on descriptive statistics display a sample of data and ask the subjects for estimates of the proportion, mean, variance, correlation, or some other descriptive statistic. The correspondence between the estimates and the calculated statistics serves as the measure of accuracy.

Judgments of Proportion

Subjects have estimated proportions of both sequential and simultaneous displays of binary events (lights, horizontal and vertical lines, letters, numbers, etc.). The most striking aspect of the results is that the relation between mean estimates and sample proportions is described well by an identity function. The deviations from this function are small; the maximum deviation of the mean estimate from the sample proportion is usually only .03–.05, and the average deviations are very close to zero. Within the constraint of these small discrepancies, experiments have reported two different shapes for the slightly biased function, overestimation of low and underestimation of high proportions (Erlick, 1964; Stevens & Galanter, 1957), and underestimation of low and overestimation of high proportions (Nash, 1964; Pitz, 1965, 1966; Shuford, 1961; Simpson & Voss, 1961). The conflict in these results is particularly difficult to understand because similar procedures were used by Stevens and Galanter (1957) and Shuford (1961) on the one hand, and by Erlick (1964) and Pitz (1965, 1966) on the other.

We view the task of judging a proportion as one of statistical description. The subject never actually counts the elements in a display, however, so the task may also be viewed as one of inference. The displayed stimuli make up the population, and whatever information the subject can glean from observing the display is the sample. Results support this view. Accuracy of estimating proportions increases both with longer presentation times (Erlick, 1961; Robinson, 1964; Shuford & Wiesen, 1959) and with the length of a sequence of elements (Erlick, 1964). Assuming that subjects gather larger samples during longer times or longer sequences, inferences based on those larger samples should have a smaller standard error of estimation and thus greater expected accuracy. Furthermore, with the exception of the .5 position (Nash, 1964), errors are smaller (Robinson, 1964) and fewer (Stevens & Galanter, 1957) and response variance is less (Shuford, 1961) for extreme proportions than for estimates in the middle of the scale. The variance of a sample and therefore the standard error of estimation is theoretically smaller for samples with more extreme proportions, so accuracy should be greater.

Judgments of Means and Variances

The central tendency and variability of samples of binary data are tied to a single statistic, the proportion. By contrast, separate statistics must be used to describe these properties in samples of interval or ratio scaled data. A number of statistics describe central tendency and naïve subjects reflect this variety by giving responses that sometimes correspond to the mean, sometimes to the median, and sometimes to the midrange (Spencer, 1963). When instructions specify the mean as the average to be estimated, the resulting estimates are nearly accurate (Beach & Swensson, 1966). Though there are no apparent biases, the variance among estimates increases with the variance of the sample, the sample size, and the speed of presentation (Beach & Swensson, 1966; Spencer, 1961). Since these variables would influence the standard error of estimate, which would in turn control the variability among estimates from different samples, these results provide further support for the hypothesis that subjects in a descriptive task are actually making inferences.

Just as judgments of means are influenced by the variance of the sample, judgments of variability are influenced by the mean, but in a different way. Hofstatter (1939) obtained judgments of the variability in the lengths of sticks tied in bundles. The judgments increased appropriately as the sample variance increased. However, as the means increased, the judgments decreased, much as though the subjects were estimating the coefficient of variation (standard deviation/mean) rather than the variance. Put another way, it is as though they were judging discrepancies from the mean in relation to the magnitude of the mean, an interpretation related to the Weber fraction, $\Delta I/I$, in psychophysics.

Lathrop (1967) has replicated this aspect of Hofstatter's results. It is as if subjects regard variance as relative to the general magnitude of the stimuli. This is intuitively compelling. Think of the top of a forest. The tree tops seem to form a fairly smooth surface, considering that the tree may be 60 or 70 feet tall. Now, look at your desk top. In all probability it is littered with many objects and if a cloth were thrown over it the surface would seem very bumpy and variable. The forest top is far more variable than the surface of your desk, but not relative to the sizes of the objects being considered. Perhaps this is a place where intuition and typical statistical usage are divergent; statisticians are seldom interested in variances relevant to means, but people may be.

Even when means are taken into consideration, there are still systematic discrepancies between intuitive judgments and objective values of sample variance. These discrepancies can be accounted for in part by the way in which subjects weight deviations of individual data from the sample mean. The mathematical variance is the average of the *squared* deviations. The power to which they are raised dictates the relative weighting of large and small deviations. An increase in the power increases the relative weight of large deviations; a decrease in the power increases the relative weight of small ones. In order to investigate the relative weights assigned by subjects, experimenters have calculated that power that permits the best prediction of intuitive estimates of variability. Hofstatter (1939) found large values, ranging up to 6, when experimental conditions emphasized large deviations. He found small values, ranging down to 0.5, with an emphasis on small deviations. Beach and Scopp (1967) used normally distributed samples and found that a small power, .39, best simulated the judgments of their subjects. In normally distributed samples, most of the data lie relatively near the mean; the resulting prevalence of small deviations may emphasize them. It seems likely that distributions that emphasize extreme scores, such as saddle-shaped distributions, would result in large powers. At any rate, this modification of the normative exponent, and the accompanying psychological interpretation, illustrates a way of modifying a normative statistical model in order to arrive at a model more descriptive of intuitive statistics.

INTUITIVE INFERENTIAL STATISTICS

Although many psychological studies of descriptive statistics may have investigated inference inadvertently, it is the explicit topic of the research discussed next. Experiments on intuitive inference explore how man uses samples of data to reach conclusions about characteristics of his environment. The data provide the basis for his judgments about the covert, underlying statistical structure of events. The theory of statistical inference specifies what kind of inferences should be made from the samples, and the experiments compare inferences made by men with optimal inferences.

Inferences about Population Parameters

Experiments on inference have used the optimal inferences specified by statistics as a basis for evaluating the optimality of human inferences. Note the difference in orientation between this approach and that of studies of intuitive descriptive statistics. The latter use accuracy as the criterion for good performance, that is, they ask "To what degree do estimates agree with the experimenter's measurements of the stimulus being estimated?" Optimality, on the other hand, is the degree to which intuitive inferences agree with optimal inferences given by the statistical model. The distinction is between using God or using statistical man as a criterion for performance. Even in an uncertain and probabilistic environment, an omniscient being would know the actual population parameters. But statistical man must be content to work with only the data in a sample and to make the best possible inference. When a sample is the only information provided to the subject, it is reasonable to use optimality rather than accuracy as the primary criterion for intuitive inference.

Inferences about Proportions. In most investigations of inferences about proportions, subjects observe samples of binary data, and, after each datum in a sequence, they revise their probability estimates of each proportion being the population parameter. These revisions are compared with optimal revisions as calculated by using Bayes' theorem (see Edwards, Lindman, and Savage, 1963, for an extensive discussion of Bayesian statistical inference).

Imagine yourself in the following experiment. Two urns are filled with a large number of poker chips. The first urn contains 70% red chips and 30% blue. The second contains 70% blue chips and 30% red. The experimenter flips a fair coin to select one of the two urns, so the prior probability for each urn is .50. He then draws a succession of chips from the selected urn. Suppose that the sample contains eight red and four blue chips. What is your revised probability that the selected urn is the predominantly red one? If your answer is greater than .50, you favor the same urn that is favored by most subjects and by statistical man. If your probability for the red urn is about .75, your revision agrees with that given by most subjects. However, that revised estimate is very *conservative* when compared to the statistical man's revised probability of .97. That is, when statistical man and subjects start with the same prior probabilities for two population proportions, subjects revise their probabilities in the same direction but not as much as statistical man does (Edwards, Lindman, & Phillips, 1965).

Conservatism is suboptimal, but it is systematic, so research has looked for reasons for it. A number of studies have attempted to find out if conservatism is due merely to procedural variables. Earlier investigations had used probability estimates as the response, and it seemed possible that subjects avoided approaching the bounds of the scale. To check this, probability estimates were compared to unbounded odds estimates (Phillips & Edwards, 1966); odds estimates were only slightly less conservative than the probability estimates. Another hypothesis was that subjects had no incentive to perform well. However, while payoffs decreased response variance, they decreased conservatism only slightly (Phillips & Edwards, 1966). Other variables, such as sample size (Peterson, Schneider, & Miller, 1965) and sequential order of the data (Peterson & DuCharme, 1967; Phillips, Hays, & Edwards, 1966) affect conservatism, but instructions have virtually no influence. In short, while procedural variables influence the degree of conservatism, they do not eliminate it.

The persistence of conservatism in spite of variations in procedure suggests that it has roots in the fundamental aspects of subjects' understanding and use of information. One possibility is that peoples' intuitions about the relation between population and sample differ from the relations specified by statistical theory; or, in more formal terms, subjects have an inaccurate understanding of sampling distributions. In agreement with this hypothesis, when subjects make estimates about sampling distributions, the distributions are too flat (Peterson, DuCharme, & Edwards, in press). Moreover, probability revisions of individual subjects were predicted more accurately by substituting their flat distributions in the appropriate Bayesian equations than by using the theoretical sampling distributions.

In addition to a failure to understand the relation of samples to populations, there is also evidence that subjects have difficulty in aggregating evidence over trials (Edwards, 1966; Phillips, 1966). When they make datum-by-datum revisions throughout a sequence of data, the final subjective probability is far more conservative than when the experimenter optimally combines a series of single estimates made by subjects for each datum in the sequence. The former task requires retention of the previous inference and augmenting it in light of the succeeding datum, while the single estimates require only that subjects assess the meaning of each datum separately. At present, then, conservatism appears to be due in some small part to procedural variables, and in large part both to subjects' misunderstanding of sampling distributions and to their nonoptimal sequential revision of their subjective probabilities.

Inferences about Means and Variances. The experimental paradigm used to study inferences about means and variances is analogous to that used in studies of inferences about proportions. Data that vary along a dimension are sampled from one of two populations, and subjects decide from which of the two populations the data have been drawn. Some experiments using numerical samples had the subject infer which of two hypotheses about the parameter value was correct and state his confidence in the accuracy of that inference.[2] Irwin, Smith, and Mayfield (1956) used populations consisting of decks of cards upon which numbers were written. On the basis of each sample, subjects inferred whether the mean of the population was greater or less than zero. In a second experiment, the cards were sampled from two decks and the task was to infer which of the decks had the larger mean. In both experiments, confidence increased with the size of the sample, with either the difference between the population mean and zero or the difference between the two population means, and as the population variance decreased. Little and Lintz (1965) performed a similar experiment and found that on a trial-by-trial basis, confidence increased with sample size.

These experimenters used the t test as a method of summarizing their independent variables, but they used no normative model in the sense that the term has been used here. That is, they did not use a statistical model to prescribe the optimal confidence statement. The t test would not be the normative model because it yields the probability of the sample of data if the null hypothesis were true. This was not the question the subjects were asked (and it is claimed in some quarters that this is not a question that anybody should be asked; Edwards, Lindman, & Savage, 1963). Rather, they were asked for the probabilities of the alternative hypotheses on the basis of the data, a question answered by Bayesian statistics. Probabilities based on the normative model would be influenced by the three independent variables in the directions found in these experiments, but it is not clear whether subjects were conservative in arriving at their confidence statements.

A Bayesian model has, however, been applied to another experiment that used the paradigm just discussed (Beach & Scopp, 1967). The subjects inferred which of two decks of cards had the larger variance and stated their confidence. Confidence increased as the ratio of the judged sample variances increased, but not as much as prescribed by the model. These results are similar to the conservatism found with population proportions.

When subjects directly infer the central tendency of a population by specifying a value on a continuum of possible values, the inference must in some way rep-

---

[2] We treat the confidence estimates and probability estimates as interchangeable measures of subjective probability when both have been measured on a 0–1.0 scale. For confidence estimates, the subject usually states which event he thinks is most likely to occur and then states his confidence that the choice is correct. For probability estimates, the subject merely states how certain he is that a given event will occur. These estimates are formally equivalent, but it is yet to be demonstrated that they are psychologically so.

resent all the values in the population. Various measures of central tendency represent the population values in different ways; the mode is equal to the most frequently occurring value, the median minimizes the sum of the absolute deviations between itself and the individual values, and the mean minimizes the sum of the squared deviations. For a skewed population distribution, the values of these measures are all different. When subjects base inferences on a sample that is displayed as a | shaped frequency distribution, intuitive inferences of the mode and median are accurate, but inferences of the mean are biased toward the median (Peterson & Miller, 1964). It would be possible to simulate this bias with the approach used to simulate judgments of sample variances (Beach & Scopp, 1967), that is, by modifying the power to which deviations are raised, away from 2 in the direction of 1. This means that subjects were unwilling to weight large deviations heavily. The deviant events were also rare events, so subjects may have regarded them as unrepresentative and thus not more important than the most frequently occurring events.

Much of the research using nonnumerical samples has been conducted within the framework of the theory of signal detectability (Swets, 1964). While we have no intention of reviewing this entire literature, the model of signal detection is a statistical model and several experiments are particularly relevant to intuitive statistics. As in the research discussed above, the formal problem for the subject is one of making an inference about the population from which the observation has been sampled. One population is that of normally distributed random noise. The second population is one of signal plus noise, with the same variance but a different mean than that of the noise population. From the subjects' point of view, the task is one of deciding whether or not a signal was present in the observation.

The majority of signal detection experiments have used auditory, visual, or other sensory stimuli, but the model has also been applied outside the realm of sensory psychophysics. For example, in perceptual defense experiments, the task is to decide whether the observation is a clean word or a taboo word (Dorfman, Grossberg, & Kroeker, 1965); in recognition memory experiments, the task is to decide whether the observation is an old word or a new word (Parks, 1966); in the perception of tilt, the task is to decide whether a line is tipped to the left or the right (Ulehla, 1966); and in one series of experiments, the task was to decide whether a dot was sampled from one spatial distribution or another (Lee, 1963; Lee & Janke, 1964, 1965). The model has even been extended to the judgment of the source of short phrases from a man's magazine or a woman's magazine (Ulehla, Canges, & Dowda, in press) and to reaction time experiments where the subjects' task is to react to a left or a right stimulus light (Edwards, 1965; Stone, 1960). These experiments show that it is possible to interpret a wide range of psychological phenomena within the framework of statistical decision theory. The results are in general accord with the predictions; many deviations from optimal performance are similar to those found in other areas of intuitive statistics. For example, it is possible to manipulate the subjective decision criterion by changing the probability of sampling from a signal distribution or by varying payoffs, but the amount of change in the subjective criterion is less than optimal (Green, 1960; Ulehla, 1966). The subjects also have difficulty in aggregating information across a sequence of trials (Swets & Green, 1961; Swets, Shipley, McKey, & Green, 1959), a result that bears a strong resemblance to the finding of conservatism in the probability-revision experiments discussed above.

Inferences about Correlations. Thus far the tasks discussed have involved populations of events that vary along a single dimension. Nonlaboratory tasks, however, often involve a number of dimensions. Frequently these dimensions are not independent, and therefore it is important to examine intuitive inferences about correlations in multivariate populations.

Experiments using populations that contain two binary dimensions show that subjects do not attend to all cells of the 2 × 2 contingency table when inferring correlation. In some cases, judgments about the relatedness of the two dimensions depend solely upon one cell of the table, the cell in which the two favorable outcomes occur together (Jenkins & Ward, 1965; Smedslund, 1963; Ward & Jenkins, 1965); in other cases, judgments depend upon both cells of the positive diagonal (Inhelder & Piaget, 1958; Ward & Jenkins, 1965). The reason for the conflict is unclear, but even when subjects use the diagonal it appears that they do not fully appreciate the negative evidence represented in the remaining two cells of the 2 × 2 table.

It may be that failure to use all cells of the matrix is restricted to the special case of the 2 × 2 contingency table. Erlick (1966) presented samples from two 5-valued dimensions and had the subjects estimate the degree of positive or negative relatedness. The mean estimates were nearly linear with the objective correlations, except for a tendency to underestimate the magnitude of negative correlations. Beach and Scopp (1966) displayed samples from two 10-valued dimensions; the subjects inferred the sign of the population correlations and stated their confidence in the inferences. For both positive and negative correlations the proportion of optimal inferences and average confidence increased with the magnitude of the sample correlations, although confidence was conservative by comparison with the optimal values. In a more complex multiple regression experiment (Peterson, Hammond, & Summers, 1965b), subjects' estimates of cue weights ranked in the same order as optimal weights, further evidence that subjects do not restrict their attention to only a few cells of a data matrix. "Statistical man" appears to provide a better match to behavior when the stimulus situation becomes more complex, that is, when one moves beyond the special case of a 2 × 2 matrix.

Consistency among Inferences

We have discussed two criteria, accuracy and optimality, for evaluating performance in a statistical task. A third criterion is consistency, the degree to which relations among subjects' inferences correspond to the constraints required of statistical theory.

Optimality implies consistency, and thus optimality is the more stringent of the two criteria. Yet, consistency is an important criterion from a psychological point of view. If one's inferences are suboptimal but they fit together in a consistent manner, then the research problem is to learn why the consistent inferences are suboptimal and to modify the statistical model in order to develop a descriptive psychological theory. If, on the other hand, inferences are also inconsistent, then behavior is far less congruent with statistical theory and the outlook is dim for providing an orderly account of human inference within the framework of statistical theory.

The criterion of consistency requires that relations among sets of inferences be similar to those prescribed by statistical theory, even though the inferences themselves may be inaccurate. Experimenters have obtained inferences about two or more aspects of a population, often two probabilities, and then evaluated how well these inferences fit together when substituted into equations from the appropriate statistical model. Since accurate inferences about probabilities are consistent by definition, investigators usually take steps to insure inaccuracy.

One of the simplest relations to be examined is that the probabilities of an exhaustive set of mutually exclusive events should sum to 1.0. Because most experiments use response devices that automatically normalize, insuring that probability estimates sum of 1.0, few data are available. What data there are come from subsidiary parts of larger studies in which sums were not constrained. The results are conflicting. Phillips et al. (1966) measured the revision of probability estimates for four hypotheses in the light of sequentially presented data. One subject constrained his estimates to equal 1.0, but four other subjects revised their estimates for the most likely hypothesis upward without making corresponding decreases in the probabilities of the less likely hypotheses. In the latter case, of course, the sum of the estimates increased above 1.0 as evidence accumulated over trials. Alberoni (1962) had subjects estimate various binomial sampling distributions for samples of Size 4. The sums of the estimated probabilities for the different outcomes consistently totalled about .85, considerably less than the 1.0 required by probability theory.

When experimenters infer subjective probabilities from choices among bets, the subjective probabilities sometimes sum to approximately 1.0 (Lindman, 1965) and sometimes do not (Leibermann, 1958), and in one case they summed to 1.0 only with certain assumptions about utility for gambling (Tversky, 1964). The unresolved problem of whether or not subjective probabilities inferred from decisions sum to 1.0 has important implications for psychological decision theory, but is too complex to be discussed here. The interested reader is referred to Edwards (1962), Lindman (1965), and Tversky (1964).

Related to the question of whether exhaustive sets of probability estimates sum to 1.0 is the question of whether estimates for unions of events are equal to the sums of estimates for the component events. Beach and Peterson (1966) found that this correspondence held with high reliability when probability distributions were estimated for three different classes of events: a binomial sampling distribution, seven different events of a probability learning task, and the probabilities of each of seven well-known Republicans obtaining the Presidential nomination for the next election.

Experiments have also tested the consistency of probability estimates for the joint occurrence of two independent events. The estimates of the joint event should equal the product of the estimates of the component events. For adult subjects, estimates were roughly similar to the product when they were made for various combinations of skill and chance; but the relation did not hold for children (Cohen, Dearnaley, & Hansel, 1958). Shuford (1959) inferred subjective probabilities from the amount subjects were willing to pay in order to play various bets. Such inferred subjective probabilities for joint events were very nearly equal to the product of the inferred subjective probabilities of the component events.

When events are dependent, it is necessary to deal with conditional probabilities. Subjects perform as consistently as they do in the simpler case of independent events (Peterson, Ulehla, Miller, Bourne, & Stilson, 1965).

So far, we have been discussing structural consistency, the degree to which relations among probability estimates for a specific set of events correspond to the relations demanded by statistical theory. The introduction of change into a static system of probabilities necessitates the evaluation of a second kind of consistency, process consistency. This is the degree to which changes in the system corresponded to the changes demanded by probability theory.

Three experiments investigated consistency among changing probability estimates. In one, subjects observed a sequence of data sampled from one of two populations. After the presentation of each datum in the sequence, they revised probability estimates about which population was being sampled and about which datum would occur on the next trial. The relation between the two revisions was almost identical to that specified by probability theory (Peterson, Ulehla, Miller, Bourne, & Stilson, 1965).

In a more complex situation, subjects were faced with two different tasks. In the first, they revised probability estimates on the basis of a single datum. In the second, they revised probability estimates on the basis of combinations of those data. Consistency demands that revisions based upon the combinations be equal to products of revisions based upon the single datum. The revisions were highly correlated with this demand in all but extremely complicated situations (Beach, 1966).

The third experiment on process consistency was discussed earlier in conjunction with conservatism. Recall that conservative probability revisions were predicted more accurately by using each subject's own conservative estimates of the sampling distribution in the appropriate equations than by using theoretical sampling distributions. Although the subjects had conservative opinions about the sampling distributions, they apparently used revision rules that were nearly the same as those prescribed by probability theory (Peterson, DuCharme, & Edwards, in press).

Consistency need not be restricted to the relation among probability estimates. In the previously discussed investigation of inferences about population variances (Beach & Scopp, 1967), subjects also judged the relative magnitudes of the sample variances. The inferences and the judgments were both inaccurate. Inferences were not systematically related to the ratios of the objective sample variances, as demanded by the normative model, but both the accuracy of the inferences and the subjects' confidence in them increased monotonically with the ratios of the *judged* variances. That is, the subjects' inferences were constrained to be consistent with their inaccurate judgments of the sample variances. If a statistician observed sample variances equal to subjects' judgments, his inferences also would have been monotonically related to the ratios of those variances.

These last experiments, showing consistency between structure and process, illustrate that suboptimal performance may result from appropriate use of erroneous assumptions about the statistical structure of the task. In all of these studies of consistency, incorporation of subjective assumptions into the statistical models leads to improved predictions of performance, a modification that transforms the normative models into descriptive models.

Determining the Size of the Sample

In the experiments discussed so far, the subject has been a passive recipient of the samples of data upon which he based his inferences. In nonlaboratory situations, however, one seldom has such a passive role; an important ingredient of many inference tasks is active control of the amount of data in the sample. Larger samples tend to permit more accurate inferences, but they also cost more in terms of time, effort, and money. The essence of the sampling task, then, is to balance the value of making more accurate inferences against the cost of larger samples.

Formally, there are two ways that the subject can designate the size of the sample. The first is to specify size in advance, observe the data, and then make an inference. The second, called optional stopping, consists of sampling sequentially; after each datum the subject has the option of continuing to sample or of stopping and making his inference. Most research has focused on the latter case. Formal models for optional stopping (Edwards, 1965; Raiffa & Schlaifer, 1961; Schlaifer, 1959, 1961; Wald, 1947) can be summarized by a simple, intuitively appealing rule: Sample another datum if its cost is less than the expected increase in payoff from the information it will provide. In other words, purchase another datum only if it is worth more than it costs. In addition to costs and payoffs, probability variables play roles in determining when sampling should cease. Examples include the probability of each hypothesis prior to a sample, and the expected diagnostic value of the next datum. The models themselves are complex in that all of these variables seem to interact with each other (see, e.g., Edwards, 1965). Our goal here, however, is not to explore these formal models, but to consider the ways in which intuitive sampling processes relate to them.

Experiments have manipulated cost of data, payoff for accurate inferences, or both, and measured the consequent effect upon the number of data purchased. The task in these experiments was to decide from which population data were being sampled. The dependent variable was the number of data purchased prior to making that decision. The manipulations influenced the selected sample sizes, but the magnitude of influence was less than that prescribed by the models (Edwards & Kramer, 1963; Irwin & Smith, 1957; Lanzetta & Kanareff, 1962; Swets & Green, 1961). An exception to this generalization is an experiment in which subjects predetermined the size of the sample they wanted. The size of the payoff had no systematic influence on the number of data purchased (Green, Halbert, & Minas, 1964). Perhaps this is because the optimal stopping procedure more closely resembles nonlaboratory information purchasing tasks than does predetermining the sample size.

Manipulation of the prior probabilities of the alternative hypotheses also influences sample size. When the prior probabilities are reduced by increasing the number of alternative hypotheses, subjects select larger samples before making a decision (Becker, 1958; Messick, 1964). With just two hypotheses, the average amount of data purchased decreases as the prior probabilities become more extreme, that is, depart from .50-.50, but the rate of decrease is somewhat less than that called for by the optimal model (Green et al., 1964). Here, too, there is one exception in which the amount of data purchased was insensitive to the independent variable. Messick (1964) found no effect when he contrasted rectangular with peaked prior distributions.

The story is the same for the expected diagnostic value of data. When diagnostic value is increased by separating proportions for two alternative populations, subjects purchase less data (Becker, 1958; Edwards & Kramer, 1963), but the amount of change is not quite as much as that prescribed by the optimal strategy (Edwards & Kramer, 1963). Once again, there is an exception: Green et al. (1964) found no systematic effect. With normal rather than binomial populations, the diagnostic value is increased by separation of the population means or by decreasing the population variance; fewer data are purchased when they are more diagnostic (Irwin & Smith, 1956, 1957).

The results of these experiments on controlling the size of samples are similar to those obtained in experiments on other inference tasks. Variables that would influence the behavior of statistical man also influence subjects' behavior, but to a smaller degree. This effect may be summarized by the statement that subjects are only partially sensitive to the relevant variables. Recall that the same kind of effect characterizes conservatism (e.g., Peterson, DuCharme, & Edwards, in press; Peterson & Miller, 1965). These two sets of results may be consistent: If subjects are only partially sensitive to variables in probability revision tasks, the hypothesis of consistency requires that they also be only partially sensitive to the same variables in information purchasing tasks.

INTUITIVE PREDICTIONS OF SAMPLES

The first section of this paper considered the intuitive description of statistical characteristics of samples of data; the second section discussed the use of samples as a basis for intuitive inferences about populations. This section examines intuitive predictions about events that are to be sampled from populations.

Samples from Unidimensional Populations

The conceptually simplest prediction task requires trial-by-trial predictions of events that are randomly drawn from a unidimensional population with a stationary probability. When feedback is provided, this is the familiar paradigm of the probability learning experiment. Faced with this task, statistical man would always predict the most frequent event, but subjects do not; over trials the distribution of responses tends to match the distribution of stimuli.

Probability learning experiments constitute the majority of investigations of behavior in the face of uncertainty. Therefore, it is important to interpret this apparently nonrational behavior within the framework of intuitive statistics. One possibility, if we were merely attempting to describe the data, would be to follow the lead of the very successful stochastic learning models and postulate a dice thrower in the subject's head. That is, we would not only assume that the stimuli occur with a degree of randomness, but also that behavior is typified by randomness. This alternative is antagonistic to our point of view that man is an intuitive statistician who seeks to behave optimally. Behavior should be random only when attempting to befuddle a hostile environment and perhaps not even then; otherwise, it should be deterministic. Even in a probabilistic environment one response is usually more profitable than others. That is the response statistical man would choose and that is the response the subject should select.

The behavior to account for in a probability learning task is not matching; demonstration of matching requires that data be summarized across subjects and across blocks of trials. Closer analysis shows that neither the group nor the individual responds randomly with a probability equal to the stimulus probability. Group response proportions change drastically from trial to trial within a block (Overall & Brown, 1959; Toda, 1963), and different subjects yield grossly different response proportions over a block of trials (Peterson & Beach, 1967). Rather than matching, what must be explained is the fact that individual subjects systematically vary their responses from trial to trial instead of always predicting the most frequent event.

The reason that statistical man would always predict the most frequent event is that he understands the implications of drawing events at random from a population with a stationary probability. There is evidence that intuitive theories of randomness do not coincide with the mathematical theory (Brown, 1964; Tune, 1964). When subjects produce "random" sequences of events, they produce too few long runs and too many alternations. Subjective theories about random sequences apparently do not contain the property of independence through trials. The subjective probability of an event on a trial depends upon which events precede it.

Trial-by-trial variations in probability learning experiments also show sequential dependencies; there are too few long runs and too many alternations (Anderson, 1960; Beach & Swensson, 1967; Edwards, 1961b; Jarvik, 1951; Lindman & Edwards, 1961; Tune, 1964). The similarity between the sequential dependencies in these two situations suggests that the subjects' responses in a probability learning task may be determined by their assumptions about random sequences. That is, perhaps each subject has his own theory of randomness, a theory that differs from the mathematical theory in that it admits sequential dependencies. Statistical man using the subjects' theory of randomness in a probability learning experiment might well produce similar response sequences.

Samples from Multidimensional Populations

Though unidimensional sampling is theoretically the simplest case, nonlaboratory tasks are seldom so informationally impoverished. If, for example, you wish to predict the intelligence of a potential employee, you do not rely only on the proportion of previous interviewees who have been intelligent. You rely on the additional information provided by test scores, recommendations, appearance, mannerisms, and so on.

Simulation of such information-rich environments has used multidimensional populations. In relevant experiments, each trial is a random sample from a population with correlated dimensions. The use of cue information is investigated by permitting subjects to observe the outcome of all but one of the dimensions in the sample. These observations, the cues, are used to predict the value of the observation on the remaining dimension, the criterion. Then the sampled value of the criterion is revealed to provide feedback and to permit learning of the relations between the various cue-dimensions and the criterion dimensions.

In the unidimensional experiment, the optimal strategy is to learn which event has the highest probability of occurrence and to predict that event on all trials. The multidimensional case is more complex. Here it is necessary to learn the validities of the different cues, to rely on each cue dimension according to its validity, and to predict the criterion value that has the highest probability on the basis of the evidence provided by all of the cues.

Weighting of Cues. Most empirical research on the problem of cue weighting has used multiple regression as the statistical model. Statistical man, faced with the task of using continuous cue and criterion dimensions, would calculate regression weights for each cue dimension and then use the weights to predict the criterion. The research question is, to what degree are responses the result of appropriate weighting of the cues? (See Hursch, Hammond, and Hursch, 1964, or Peterson, Hammond, and Summers, 1965b.)

Subjective cue-weighting is inferred from a variety of measures: by the correlation between each cue dimension and the responses, by the regression weights of the responses upon the cue dimensions, or by the subjects' direct estimates of the relative importance of each cue dimension in predicting the criterion. Early experimenters were interested in concept formation and the subjects' ability to differentiate relevant cues from complex stimuli (Smedslund, 1955, 1961b; Summers, 1962). They generally obtained poor performance, a result that was probably due to the difficulty in discriminating the cues and criterion rather than to an inability to use the cues correctly after they were discriminated.

More recent studies have used simpler stimuli. The magnitudes of the subjective cue-weights achieve the same rank order as the objective cue-weights and do so in relatively few trials, but the amount of separation among the subjective weights is sometimes less than the separation in statistical man's multiple regression equation. As in the experiments on conservatism and on information purchase, subjects are only partially sensitive to differences in relevant variables; they treat the cues as more equal in predictive value than they actually are (Azuma & Cronbach, 1966; Dudycha & Naylor, 1966; Hammond, Hursch, & Todd, 1964; Peterson, Hammond, & Summers, 1965b; Schenck & Naylor, 1966; Uhl, 1963).

Maximizing versus Distributing Responses. Some experiments report that response distributions approximately match the conditional probability distributions of the criteria (Binder & Feldman, 1960; Estes, 1959). Others find that the response distribution is more peaked than the conditional probability distribution, indicating a deviation from matching in the direction of optimality (Azuma, 1960; Beach, 1964; Goodnow, 1954; Peterson & Ulehla, 1964). Although these results are in conflict about the degree of optimality, they agree that subjects distribute responses rather than maximize. In this respect these results are similar to those obtained in the unidimensional experiments. The explanation in the unidimensional case, misunderstanding of random sequences, is less tenable in the multidimensional case. Until more is known about the microstructure of behavior in this situation, these results remain unexplained within the framework of intuitive statistics.

When the assumptions of regression models are met, the criterion with the highest conditional probability is the value that lies on the regression plane. As in a conditional probability learning experiment, this value of the criterion changes from trial to trial because the sampled cues change. The degree to which responses lie on any linear regression plane is measured by calculating the multiple correlation between cue dimensions and responses. Cue weights reflect the slope of the regression plane; the experiments discussed two paragraphs above show that response regression planes are close to the experimental regression planes, as they should be. The results are conflicting, however, with respect to the degree to which all responses lie on or near that regression plane. When only a single cue dimension is available, all responses do not lie on the plane (line) —they are distributed around it. The variance of the response distributions around the regression line increases as cues become less predictive of the criterion (Gray, Barnes, & Wilkinson, 1965; Schenck & Naylor, 1965).

The nonoptimal behavior found in single-cue experiments does not extend to multiple-cue studies. In the latter, responses are almost completely dependent upon the cues. The very high multiple correlations between responses and cues indicate that almost all responses fall directly on the response regression plane. This result holds not only where criteria are perfectly predictable from cues (Azuma & Cronbach, 1966; McHale & Stolurow, 1962, 1964; Uhl, 1963), but also when they are not (Grebstein, 1963; Peterson, Hammond, & Summers, 1965b; Todd, 1954).

In summary, subjects in conditional probability learning experiments scatter their responses. They do the same thing in single-cue regression experiments. In the seemingly more complex multiple-cue regression experiments, however, almost all responses fall on the response regression plane. It is not clear why the results conflict, but the evidence is abundant. Once again, greater task complexity appears to lead to more nearly optimal performance.

In addition to results on cue weighting and maximizing, there is other evidence that subjects are able to deal with functions relating criteria to cues. They can learn and use functions with both positive and negative slopes (Bjorkman, 1965); they can handle nonlinear as well as linear functions (Summers & Hammond, 1966); and perhaps most impressive of all, when confronted with a cue that they have never seen before, predictions fall on the regression line derived from previous observations (Bjorkman, 1965; Gray, Barnes, & Wilkinson, 1965).

NONSTATIONARY PARAMETER VALUES

We have examined the ability of the intuitive statistician to perform in uncertain but stationary situations. Although the relation between population and sample was a noisy one, the population remained the same over time. The subjects were aware that changes in the sequential sample of data from one time to another were due to random fluctuations.

The nonlaboratory environment, however, is characterized by nonstationary situations as well, situations in which values of parameters change over time. This complicates matters considerably, because temporal fluctuations in the sequential sample can be due either to random perturbations or to real changes in the population. It is therefore necessary to penetrate through random variations, not only to detect population parameters, but also to detect changes in those parameters.

Statistical Models

The statistical procedures used in nonstationary situations are themselves models that assume stationarity. Adapting such a model to a changing situation "consists of finding ways of looking at a changing world so that it seems to be unchanging [Edwards et al., 1965, p. 310]."

Attempts to describe nonstationary situations with stationary statistical models fall into two general classes, which we will call deletion models and detection models. The essence of the deletion model is the analysis of data in small blocks, small enough so that the assumption of stationarity during the block is not too unreasonable, and the deletion of all other data. Another version is to take running averages, a process that slides the blocks through trials by deleting the oldest trial as it adds each new trial. A variation of the running average attributes more weight to recent data than to older data (Dodson, 1961). The deletion models suffer from an arbitrariness in the choice of the number of data included in a given block. The choice requires a compromise between the need for a sample large enough to yield a stable estimate of the population parameter and one small enough to make the assumption of stationarity during the block a reasonable one.

The detection model is less arbitrary. The idea is to compare incoming data with current estimates of the population parameter, until the new data become so divergent that the no-change hypothesis must be rejected in favor of the hypothesis that there has been a change. While these are hypotheses about change, they themselves do not change, thereby permitting the use of conventional statistical models that assume a stationary situation. Thus, detection models yield a hierarchy of inferences; inferences about the population are controlled by inferences about whether or not a change has occurred.

Since conventional statistical models generally assume stationary population parameters, their application to nonstationary situations depends to a considerable degree upon the ingenuity of the user. Few of the experiments have compared the performance of subjects with theoretically optimal performance in nonstationary statistical tasks. The typical procedure is to indicate the trials on which the parameter changed and to display the effect of the change on estimates made by subjects.

Experiments. Experimenters have used nonstationary situations to study the three classes of tasks discussed so far: description, inference, and prediction. In a description task, Robinson (1964) presented sequences of two rapidly flashing lights; the proportion of trials on which each light flashed changed with discrete steps at various points in the sequence. The task was to track the sample proportion by continuously adjusting a pointer on a proportion scale. The behavior could be described better by a detection than by a deletion model. Estimates changed abruptly following changes in the stimulus proportion; deletion models call for more gradual changes in response. Robinson pointed out, however, that the step changes in the proportions being estimated were especially compatible with detection models. It may be that the class of model that will describe behavior more accurately depends upon the characteristic of change in the experimental situation. Whatever the eventual status of these kinds of models, Robinson's results demonstrate that subjects can accurately estimate a time-varying binary probability.

A similar conclusion can be drawn when subjects infer nonstationary values of a population parameter. Rapoport (1964a, 1964b) selected values of the population proportion by a process that changed over time. The subjects used samples of data drawn from these populations to infer the value of the parameter. They made direct estimates and also estimated the interval within which they expected the parameter to fall. Intuitive inferences about the parameter changed in the direction of the shifts in the nonstationary process.

Responses are also sensitive to changes in parameter values when the task is to make sequential predictions about samples to be drawn from nonstationary populations. In probability learning experiments, when the stimulus probability changes over trials, the group response proportion tracks that change (e.g., Estes, 1959; Friedman, Burke, Cole, Keller, Millward, & Estes, 1964). The same thing happens when stimulus probabilities change, not simply as a function of trials, but as a function of the stimulus event of the preceding trial (e.g., Anderson, 1960). Finally, subjective cue-weights track changes in corresponding objective cue-weights when the task is to predict events to be sampled from multidimensional populations (Peterson, Hammond, & Summers, 1965a).

Additional support for the principle that subjects are sensitive to change comes from decision research, particularly research on multistage decisions. Rapoport (1965a, 1965b) developed tasks in which the state of the experiment changed over trials; the change depended on the state of the previous trial, on the decision made by the subject, and on some random process. Costs and payoffs were related both to decisions and to states, and the subject's goal was to make decisions that would maximize his net payoff. Rapoport found that intuitive decisions were remarkably near optimal decisions as prescribed by dynamic programming models (see Rapoport, 1965a, for references on dynamic programming). Although the task was primarily one of decision making and no model of statistical inference was tested, the nearly optimal decisions required sensitive inferences about a complex nonstationary process.

Application of conventional statistical models to changing situations is a complicated process, but the results of these experiments suggest that subjects are very sensitive to change; they are adaptable to nonstationary aspects of probabilistic situations.

SUMMARY AND CONCLUSIONS

The point of view underlying the reseach reviewed in this paper is that man must come to terms with his uncertain environment; he is not aware of all present conditions and he does not always know what will occur in the future. The formal theory of probability with its statistical applications describes the structure of that uncertain environment and the processes governing the occurrence of events within it. In addition, probability theory is normative; it provides optimal models for making inferences under conditions of uncertainty. This normative characteristic is the basis of the concept of "statistical man," a set of procedures for making optimal statistical inferences.

Experiments that have compared human inferences with those of statistical man show that the normative model provides a good first approximation for a psychological theory of inference. Inferences made by subjects are influenced by appropriate variables and in appropriate directions. But there are systematic discrepancies between normative and intuitive inferences. For example, the latter are usually too conservative; subjects apparently fail to extract all the information latent in samples of data. In addition, while intuitive inferences are sensitive to variables relevant to the normative model, the degree of sensitivity is often less than optimal.

A recurrent theme of the research reviewed is that some discrepancies are due to the fact that subjects in an inference task make assumptions different from those of statistical man. If statistical man were to incorporate subjects' assumptions, his inferences would be more descriptive of those made by subjects. Current research integrating subjective assumptions with the concept of statistical man may be a major step toward a psychological theory of intuitive statistical inference.

Such a theory would encompass only a restricted subset of human behavior, but there are some obvious possibilities for expansion. The subset increases considerably when the related normative models of probability theory and decision theory are joined as a basis for a broader psychological model including choice behavior as well as inference processes in an uncertain environment. Research by Piaget and his collaborators suggests another direction for expanding this normative approach to developing psychological models. For example, they have studied children's acquisition of principles such as the conservatism of substance and weight (e.g., see Smedslund, 1961a). Once the principle of conservation has been acquired, the child knows that the amount of substance and the weight of the object must remain unchanged if nothing is added or taken away, even though the form of the object may change. Principles such as the law of conservation are normative in that they lead to correct predictions of future events where alternative notions would lead to error. Thus, research on man as an intuitive statistician and as an intuitive decision maker could be extended to other disciplines offering normative models. The research could consider man as an intuitive scientist, logician, mathematician, and so on, and the resulting psychological theory would indeed apply to a large segment of human behavior.

REFERENCES

Alberoni, F. Contribution to the study of subjective probability. I. *Journal of General Psychology,* 1962, **66,** 241-264.

Anderson, N. H. Effects of first-order conditional probability in a two-choice learning situation. *Journal of Experimental Psychology,* 1960, **59,** 73-93.

Azuma, H. Comparison of a correlation with a probabilistic approach to concept learning. Unpublished doctoral dissertation, University of Illinois, 1960.

Azuma, H., & Cronbach, L. J. Cue-response correlations in the attainment of a scaler concept. *American Journal of Psychology,* 1966, **79,** 38-49.

Beach, L. R. Cue probabilism and inference behavior. *Psychological Monographs,* 1964, **78**(5, Whole No. 582).

Beach, L. R. Accuracy and consistency in the revision of subjective probabilities. *IEEE Transactions on Human Factors in Electronics,* 1966, HFE-7, 29-37.

Beach, L. R., & Peterson, C. R. Subjective probabilities for unions of events. *Psychonomic Science,* 1966, **5,** 307-308.

Beach, L. R., & Scopp, T. S. Inferences about correlations. *Psychonomic Science,* 1966, **6,** 253-254.

Beach, L. R., & Scopp, T. S. Intuitive statistical inferences about variances. Seattle: L. R. Beach, 1967. (Mimeo)

Beach, L. R., & Swensson, R. G. Intuitive estimation of means. *Psychonomic Science,* 1966, **5,** 161-162.

Beach, L. R., & Swensson, R. G. Instructions about randomness and run-dependency in two-choice learning. *Journal of Experimental Psychology,* 1967, in press.

Becker, G. M. Sequential decision-making: Wald's model and estimates of parameters. *Journal of Experimental Psychology,* 1958, **55,** 628-636.

Becker, G. M., & McClintock, C. G. Value: Behavioral decision theory. *Annual Review of Psychology,* 1967, in press.

Binder, A., & Feldman, S. E. The effects of experimentally controlled experience upon recognition responses. *Psychological Monographs,* 1960, **74**(9, Whole No. 496).

Bjorkman, M. Learning of linear functions; comparison between a positive and a negative slope. *Reports from the Psychological Laboratories* (University of Stockholm), 1965, No. 183.

Brown, D. L. Non-independence in subjectively random binary sequences. (Research Bulletin No. 27) Princeton, N. J.: Educational Testing Service, 1964.

Brunswik, E. In defense of probabilistic functionalism: A reply. *Psychological Review,* 1955, **62,** 236-242.

Cohen, J., Dearnaley, E. J., & Hansel, C. E. M. Skill and chance: Variation in estimates of skill with an increasing element of chance. *British Journal of Psychology,* 1958, **49,** 319-323.

Dodson, J. D. Simulation system design for a TEAS simulation research facility. (Report R-194) Los Angeles: Planning Research Corporation, 1961.

Dorfman, D. D., Grossberg, J. M., & Kroeker, L. Recognition of taboo stimuli as a function of exposure time. *Journal of Personality and Social Psychology,* 1965, **2,** 552-562.

Dudycha, L. W., & Naylor, J. C. Characteristics of the human inference process in complex choice behavior situations. *Organizational Behavior and Human Performance,* 1966, **1,** 110-128.

Edwards, W. The theory of decision making. *Psychological Bulletin,* 1954, **51,** 380-417.

Edwards, W. Behavioral decision theory. *Annual Review of Psychology,* 1961, **12,** 473-498. (a)

Edwards, W. Probability learning in 1000 trials. *Journal of Experimental Psychology,* 1961, **62,** 381-390. (b)

Edwards, W. Subjective probabilities inferred from decisions. *Psychological Review,* 1962, **69,** 109-135.

Edwards, W. Optimal strategies for seeking information: Models for statistics, choice reaction time, and human information processing. *Journal of Mathematical Psychology,* 1965, **2**, 312-329.

Edwards, W. Non-conservative probabilistic information processing systems. ESD-TR-66-404, Report 05893-22-F, August 1966, University of Michigan, Institute of Science and Technology, Electronic Systems Division.

Edwards, W., & Kramer, E. F. Boodle from Bayes' at a penny a peek. Ann Arbor: W. Edwards, 1963. (Mimeo)

Edwards, W., Lindman, H., & Phillips, L. D. Emerging technologies for making decisions. In, *New directions in psychology. II.* New York: Holt, Rinehart & Winston, 1965. Pp. 265-325.

Edwards, W., Lindman, H., & Savage, L. J. Bayesian statistical inference for psychological research. *Psychological Review,* 1963, **70**, 193-242.

Erlick, D. E. Judgments of the relative frequency of a sequential series of two events. *Journal of Experimental Psychology,* 1961, **62**, 105-112.

Erlick, D. E. Absolute judgments of discrete quantities randomly distributed over time. *Journal of Experimental Psychology,* 1964, **67**, 475-482.

Erlick, D. E. Human estimates of statistical relatedness. *Psychonomic Science,* 1966, **5**, 365-366.

Estes, W. K. The statistical approach to learning theory. In S. Koch (Ed.), *Psychology: A study of a science.* Vol. 2. New York: McGraw-Hill, 1959. Pp. 380-491.

Friedman, M. P., Burke, C. J., Cole, M., Keller, L., Millward, R. B., & Estes, W. K. Two choice behavior under extended training with shifting probabilities of reinforcement. In R. C. Atkinson (Ed.), *Studies in mathematical psychology.* Palo Alto: Stanford University Press, 1964.

Goodnow, R. Utilization of partially valid cues in perceptual identification. Unpublished doctoral dissertation, Harvard University, 1954.

Gray, C. W., Barnes, C. B., & Wilkinson, E. F. The process of prediction as a function of the correlation between two scaled variables. *Psychonomic Science,* 1965, 3, 231-232.

Grebstein, L. Relative accuracy of actuarial predictions, experienced clinicians, and graduate students in a clinical judgment task. *Journal of Consulting Psychology,* 1963, **37**, 127-132.

Green, D. M. Psychoacoustics and detection theory. *Journal of the Acoustical Society of America,* 1960, **32**, 1189-1203.

Green, P. E., Halbert, M. H., & Minas, J. S. An experiment in information buying. *Journal of Advertising Research,* 1964, **4**, 17-23.

Hammond, K. R., Hursch, C. J., & Todd, F. J. Analyzing of the components of clinical inference. *Psychological Review,* 1964, **71**, 438-456.

Hofstatter, P. R. Über die Schätzung von gruppeneigenschaften. *Zeitschrift für Psychologie,* 1939, **145**, 1-44.

Hursch, C. J., Hammond, K. R., & Hursch, J. L. Some methodological considerations in multiple-cue probability studies. *Psychological Review,* 1964, **71**, 41-60.

Inhelder, B., & Piaget, J. *The growth of logical thinking from childhood to adolescence.* New York: Basic Books, 1958.

Irwin, F. W., & Smith, W. A. S. Further tests of theories of decision in an "expanded judgment" situation. *Journal of Experimental Psychology,* 1956, **52**, 345-348.

Irwin, F. W., & Smith, W. A. S. Value, cost and information as determiners of a decision. *Journal of Experimental Psychology,* 1957, **54**, 229-231.

Irwin, F. W., Smith, W. A. S., & Mayfield, J. F. Tests of two theories of decision in an "expanded judgments" situation. *Journal of Experimental Psychology,* 1956, **51**, 261-268.

Jarvik, M. E. Probability learning in a negative recency effect in a serial anticipation of alternative symbols. *Journal of Experimental Psychology,* 1951, **41**, 291-297.

Jenkins, H. M., & Ward, W. C. The judgment of contingency between responses and outcomes. *Psychological Monographs,* 1965, 79(1, Whole No. 594).

Lanzetta, J. T., & Kanareff, Z. T. Information cost, amount of payoff and level of aspiration as determinants of information seeking in decision making. *Behavioral Science,* 1962, **7**, 459-473.

Laplace, P. S. DE *Essai Philosophique sur les Probabilitiés.* [*Concerning probability.*] (Orig. publ. 1814) In J. R. Neumann (Ed.), *The world of mathematics.* Vol. 2. New York: Simon and Schuster, 1956. P. 1325.

Lathrop, R. G. Perceived variability. *Journal of Experimental Psychology,* 1967, **23**, 498-502.

Lee, W. Choosing among confusably distributed stimuli with specified likelihood ratios. *Perceptual and Motor Skills,* 1963, **16**, 445-467.

Lee, W., Janke, M. Categorizing externally distributed stimulus samples for three continua. *Journal of Experimental Psychology,* 1964, **68**, 376-382.

Lee, W., & Janke, M. Categorizing externally distributed stimulus samples for unequal molar probabilities. *Psychological Reports,* 1965, **17**, 79-90.

Leibermann, B. The auction values of uncertain outcomes in win and loss-type situations. Unpublished doctoral dissertation, Boston University, 1958.

Lindman, H. R. The simultaneous measurement of utilities and subjective probabilities. Unpublished doctoral dissertation, University of Michigan, 1965.

Lindman, H., & Edwards, W. Unlearning the gambler's fallacy. *Journal of Experimental Psychology,* 1961, **62**, 630.

Little, K. B., & Lintz, L. M. Information and certainty. *Journal of Experimental Psychology,* 1965, **70**, 428-432.

McHale, T. J., & Stolurow, L. M. Concept formation with metrically multivalued cues and response categories. Technical Report No. 3, 1962, University of Illinois, Training Research Laboratory.

McHale, T. J., & Stolurow, L. M. More information—Cues or principle? Technical Report No. 5, 1964, University of Illinois, Training Research Laboratory.

Messick, D. N. Sequential information seeking: Effects of the number of terminal acts in prior information. Technical Documentary Report No. ESD-TDR-64-606, 1964, University of North Carolina, Electronic System Division.

Nash, H. The judgement of linear proportions. *American Journal of Psychology,* 1964, **77**, 480-484.

Overall, J. E., & Brown, W. L. The comparison of the decision-behavior of rats and human subjects. *American Journal of Psychology,* 1959, **72**, 258-261.

Parks, T. E. Signal-detectability theory of recognition-memory performance. *Psychological Review,* 1966, **73**, 44-58.

Peterson, C. R., & Beach, L. R. Probability learning? Ann Arbor: C. R. Peterson, 1967. (Mimeo)

Peterson, C. R., & DuCharme, W. M. A primacy effect in subjective probability revision. *Journal of Experimental Psychology,* 1967, **73**, 61-65.

Peterson, C. R., DuCharme, W. M., & Edwards, W. Estimated binomial sampling distributions in the revision of probability estimates. *Journal of Experimental Psychology,* in press.

Peterson, C. R., Hammond, K. R., & Summers, D. A. Multiple probability learning with shifting cue weights. *American Journal of Psychology,* 1965, **78**, 660-663. (a)

Peterson, C. R., Hammond, K. R., & Summers, D. A. Optimal responding in multiple-cue probability learning. *Journal of Experimental Psychology,* 1965, **70**, 270-276. (b)

Peterson, C. R., & Miller, A. Mode, median and mean as optimal strategies. *Journal of Experimental Psychology,* 1964, **68**, 363-367.

Peterson, C. R., & Miller, A. J. Sensitivity of subjective probability revision. *Journal of Experimental Psychology,* 1965, **70**, 117-121.

Peterson, C. R., Schneider, R. J., & Miller, A. J. Sample size in the revision of subjective probabilities. *Journal of Experimental Psychology,* 1965, **69**, 522-527.

Peterson, C. R., & Ulehla, Z. J. Uncertainty, inference difficulty, and probability learning. *Journal of Experimental Psychology,* 1964, **67**, 523-530.

Peterson, C. R., Ulehla, Z. J., Miller, A. J., Bourne, L. E., Jr., & Stilson, D. W. Internal consistency of subjective probabilities. *Journal of Experimental Psychology,* 1965, **70**, 526-533.

Phillips, L. D. Some components of probabilistic inference. Unpublished doctoral dissertation, University of Michigan, 1966.

Phillips, L. D., & Edwards, W. Conservatism in a simple probability inference task. *Journal of Experimental Psychology,* 1966, **72,** 346-354.
Phillips, L. D., Hays, W. L., & Edwards, W. Conservatism in complex probability inference. *IEEE Transactions on Human Factors in Electronics,* 1966, HFE-7, 7-18.
Pitz, G. F. Response variables in the estimation of relative frequency. *Perceptual and Motor Skills,* 1965, **21,** 867-873.
Pitz, G. F. The sequential judgment of proportion. *Psychonomic Science,* 1966, **4,** 397-398.
Raiffa, H., & Schlaiffer, R. *Applied statistical decision theory.* Boston: Harvard Business School, Division of Research, 1961.
Rapoport, A. Sequential decision-making in a computer-controlled task. *Journal of Mathematical Psychology,* 1964, **1,** 351-374. (a)
Rapoport, A. A study of human control in a stochastic multistage decision task. Report No. 43, 1964, University of North Carolina, Psychometric Laboratory. (b)
Rapoport, A. Dynamic programming models for multistage decision making tasks. Reprint 150, 1965, University of Michigan, Mental Health Research Institute. (a)
Rapoport, A. A study of a multistage decision making task with an unknown duration. Reprint 152, 1965, University of Michigan, Mental Health Research Institute. (b)
Robinson, G. H. Continuous estimation of a time-varying probability. *Ergonomics,* 1964, **7,** 7-21.
Schenck, A., & Naylor, J. D. The effect of cue intercorrelation on performance in a multiple-cue choice situation. Paper presented at the meeting of Midwestern Psychological Association, Chicago, May 1965.
Schlaiffer, R. *Probability in statistics for business decisions.* New York: McGraw-Hill, 1959.
Schlaiffer, R. *Introduction to statistics for business decisions.* New York: McGraw-Hill, 1961.
Shuford, E. H. A comparison of subjective probabilities for elementary and compound events. Report No. 20, 1959, University of North Carolina, Psychometric Laboratory.
Shuford, E. H. Percentage estimation of proportion as a function of element type, exposure type, and task. *Journal of Experimental Psychology,* 1961, **61,** 430-436.
Shuford, E. H., & Wiesen, R. A. Bayes' estimation of proportions: The effect of stimulus distribution in exposure time. Report No. 23, 1959, University of North Carolina, Psychometric Laboratory.
Simpson, W., & Voss, J. F. Psychophysical judgments of probabilistic stimulus sequences. *Journal of Experimental Psychology,* 1961, **62,** 416-422.
Smedslund, J. *Multiple-probability learning: An inquiry into the origins of perception.* Oslo: Akademisk, 1955.
Smedslund, J. The acquisition of conservation of substance and weight in children. *Scandinavian Journal of Psychology,* 1961, **2,** 11-20. (a)
Smedslund, J. The utilization of probabilistic cues after 1100 and 4800 stimulus presentations. *Acta Psychologica,* 1961, **18,** 383-386. (b)
Smedslund, J. The concept of correlation in adults. *Scandinavian Journal of Psychology,* 1963, **4,** 165-173.
Spencer, J. Estimating averages. *Ergonomics,* 1961, **4,** 317-328.
Spencer, J. A further study of estimating averages. *Ergonomics,* 1963, **6,** 255-265.
Stevens, S. S., & Galanter, E. H. Ratio scales and category scales for a dozen perceptual continua. *Journal of Experimental Psychology,* 1957, **54,** 377-411.
Stone, M. Models for choice-reaction time. *Psychometrika,* 1960, **25,** 251-260.
Summers, S. A. The learning of responses to multiple weighted cues. *Journal of Experimental Psychology,* 1962, **64,** 29-34.
Summers, D. A., & Hammond, K. R. Inference behavior in multiple-cue tasks involving both linear and nonlinear relations. *Journal of Experimental Psychology,* 1966, **71,** 751-757.
Swets, J. A. *Signal detection and recognition by human observers.* New York: Wiley, 1964.

Swets, J. A., & Green, D. M. Sequential observations by human observers of signals and noise. In C. Cherry (Ed.), *Information theory.* London: Butterworth, 1961. Pp. 177-195.

Swets, J. A., Shipley, E. F., McKey, M. J., & Green, D. M. Multiple observations of signals and noise. *Journal of the Acoustical Society of America,* 1959, **31,** 514-521.

Toda, M. Micro-structure of guess processes: Part C. Report No. 7, June 1963, Pennsylvania State University, Institute of Research, Division of Mathematical Psychology.

Todd, F. J. A methodological study of clinical judgment. Unpublished doctoral dissertation, University of Colorado, 1954.

Tune, G. S. Response preferences: A review of some relevant literature. *Psychological Bulletin,* 1964, **61,** 286-302.

Tversky, A. N. Additive choice structures. Unpublished doctoral dissertation, University of Michigan, 1964.

Uhl, C. N. Learning interval concepts: I. Effects of differences in stimulus weights. *Journal of Experimental Psychology,* 1963, **66,** 264-273.

Ulehla, Z. J. Optimality of perceptual decision criteria. *Journal of Experimental Psychology,* 1966, **71,** 564-569.

Ulehla, Z. J., Canges, L., & Dowda, F. Signal detectability theory applied to conceptual discrimination. *Psychonomic Science,* in press.

Wald, A. *Sequential analysis.* New York: Wiley, 1947.

Ward, W. C., & Jenkins, H. M. The display of information and the judgment of contingency. *Canadian Journal of Psychology,* 1965, **19,** 231-241.

21

The Prediction of Decisions Among Bets

Ward Edwards

This paper presents a very simple mathematical model for predicting choices among bets and an experiment in which that model is tested. The model is based on the concepts of utility and subjective probability and on the theory of games.

From *Journal of Experimental Psychology,* **50,** 1955, 201-214. Copyright 1955 by the American Psychological Association, and reproduced by permission.

This research was done while the author was at The Johns Hopkins University. It was done under Contract N5-ori-166, Task Order 1, between the Office of Naval Research and The Johns Hopkins University. This is Report No. 166-I-191, Project Designation No. NR-145-089, under that contract. I am indebted to Miss Frances Brown, who ran Ss and possessed much of the data, and to Miss Susanne Wollenberg, who also helped with the data processing. A very preliminary report of this experiment, including only the data from Ss 3 and 5, was presented to the Dunlap Symposium on Mathematical Models in the Social Sciences (8).

DEVELOPMENT OF THE MODEL

Four Models for Decision-making

This paper will discuss four similar simple models which have been used to make predictions of choices among bets. They are defined by the following four equations:

$$EV = \sum_i p_i \$_i \quad (1)$$

$$EU = \sum_i p_i u_i \quad (2)$$

$$SEM = \sum_i P^*_i \$_i \quad (3)$$

$$SEU = \sum_i P^*_i u_i \quad (4)$$

Equation 1 defines the concept of *expected value* (*EV*). The symbol p_i stands for the probability of the ith outcome of the bet, the symbol $\$_i$ stands for the value of the ith outcome of the bet in dollars, and $\sum_i p_i = 1$. The *EV* of a bet is the amount which a gambler will receive as a result of making it, on the average. Consequently, the traditional normative theory about choices among bets is that gamblers should choose the bet with the highest *EV* or lowest negative *EV*. (A normative theory is a theory about what people should do, rather than what they do do.)

It is, of course, apparent that people don't do this, and it is even doubtful in some cases that they should. Consequently, Bernoulli (1) suggested instead that gamblers should choose the bet with the highest *expected utility* (*EU*), where *EU* is defined as in Equation 2. The symbol u_i in Equation 2 is the *utility*, or subjective value, of the ith outcome of the bet. This proposal, then, simply is a suggestion that the utility or subjective value of a given amount of money be substituted for its objective value in the calculation of an expected value.

Recently, von Neumann and Morgenstern (13) have revived Bernoulli's suggestion, and have used Equation 2 as a definition of utility. This assumes, of course, that *EU* maximization is a theory about what people do do, rather than a normative theory. Following von Neumann and Morgenstern's suggestion, Mosteller and Nogee (10) assumed the validity of the *EU* maximization model and used it to measure the utility of small amounts of money experimentally.

If it is reasonable to assume that subjective values of money should be substituted for objective values in Equation 1, it is equally reasonable to make the same assumption about probabilities. The notion of subjective probabilities has been discussed for some time; one of the most explicit uses of it (without use of the concept of utility) was made by Preston and Baratta (11). They implicitly suggested that Ss be assumed to make choices among bets so as to maximize what might be called *subjectively expected money* value (*SEM*), which is defined by Equation 3. The symbol P^*_i in Equation 3 stands for the subjective probability corresponding to the objective probability of the ith alternative. No addition theorem applies to the P^* values; however, the sum of the objective probabilities corresponding to the various P^* values in Equation 3 is 1. Having assumed the validity of the *SEM* maximization model, Preston and Baratta used it to calculate a subjective probability scale. Mosteller and Nogee also used Equation 3 to calculate subjective probabilities, but preferred the utility interpretation given by Equation 2. Similarly, it would be possible to analyze Preston and Baratta's data by Equation 2 instead of Equation 3. There is little to choose between these opposed interpretations of either of these two experiments.

In a series of experiments Edwards (3, 4, 5) has shown that Ss prefer to bet at some probabilities rather than others, that these preferences are essentially independent of the amounts of money involved so long as they are not too large and so long as all bets compared with each other have the same *EV*, and that Ss prefer less favorable bets at their preferred probabilities to more favorable bets at disliked probabilities. These experiments show that there are two patterns of probability preferences, one for positive *EV* bets and another for negative *EV* bets. These results, of course, can be easily accounted for by means of the *SEM* model, but are inconsistent with the *EU* model.

All these considerations suggest a more comprehensive model, which includes both utility and subjective probability. This paper is intended to present and test such a model. The model asserts that people choose among bets so as to maximize *subjectively expected utility* (*SEU*), which is defined in Equation 4. The symbols in Equation 4 have the same meaning as before. This model, like the others, does not assert that Ss know the *SEU* values of various bets and deliberately choose the bet with the largest *SEU*; it asserts only that the hypothesis of *SEU* maximization will enable an E to predict the choices that Ss will actually make.

This model is similar to one proposed by Coombs and Beardslee (2). The most important difference is that Coombs and Beardslee assume both P^* and u to be measurable only on an ordered metric, while this model (and the *EU* and *SEM* models also) assume them to be measurable on an interval scale.

For a fuller discussion of the concepts of utility and subjective probability and of the Mosteller and Nogee, Preston and Baratta, and other related experiments, see (6).

THE TEST OF THE MODEL

In order to test the *SEU* maximization model, three things are necessary. Scales of utility and of subjective probability must be determined for Ss whose behavior is to be predicted, and Ss must make choices among bets, so that these choices can be compared with those predicted by the model.

The crux of the problem of testing this model is to devise methods of measuring utility and subjective probability. Classical psychophysical methods seem inappropriate; both probabilities and amounts of money come stated in numerical form, and psychophysical judgments about them are almost certain to be made in terms of their numerical rather than their subjective values.

The Matrix Method

The theoretically ideal method of measuring utility and subjective probability simultaneously in a gambling situation has been suggested by Woodbury (14). The essence of it is that a rectangular matrix with amounts of money on one margin and probabilities on the other defines a series of probability-amount combinations, or bets. An S can be required to inform E of amounts of money such that he is indifferent between them and the bets in the matrix. The SEU maximization model permits a system of simultaneous equations based on these judgments to be set up, enough more equations can be added by linear interpolation between adjacent utility values to make the system soluble, and the solution gives the utility and subjective probability functions needed.

This procedure was tried in this experiment, using the method of adjustment. The data were so variable as to be meaningless, and so will not be further discussed. Less variable data could probably be gotten by the method of constant stimuli, but it is unlikely that they would be good enough to meet the exacting needs of the matrix method.

The N-bets Method of Utility Measurement

If no adequate technique for measuring utility and subjective probability simultaneously can be devised, then it is necessary to measure them separately. The methods used by Mosteller and Nogee (10) and by Preston and Baratta (11) are inappropriate, since they irretrievably confound utility and subjective probability. Some technique must be invented which either eliminates one of these variables or else holds it constant.

The method used in this experiment, which will hereafter be called the N-bets method, is based on a sweeping, convenient assumption. That assumption is that any bet can be considered as an object which has a utility (which can be calculated from Equation 4, once the utility and subjective probability functions are known), and that the utility of N identical bets is N times the utility of one such bet. This is an assumption that what the economists call the *marginal utility* function for bets, but not necessarily for any other object, is a constant. (The marginal utility of a commodity is the first derivative of its total utility curve.)

How can this assumption be used to measure utility? Suppose that S is indifferent between one 50–50 bet in which he wins \$A if he wins and nothing if he loses, and N 50–50 bets in each of which he wins \$0.10 if he wins and nothing if he loses. The assumption discussed above makes it legitimate to write:

$$P^*_{0.5}\, u_{\$A} + P^*_{0.5}\, u_{\$0.00} = N\,(P^*_{0.5}\, u_{\$0.10} + P^*_{0.5}\, u_{\$0.00}) \tag{5}$$

Since both utility and subjective probability are defined as measurable on an interval scale, we may arbitrarily assign two points on each curve. Here and throughout this paper the following arbitrary definitions will be used: $u_{\$0.00} = 0$; $u_{+\$0.10} = 10$; $u_{-\$0.10} = -10$; $P^*_0 = 0$; $P^*_1 = 1$. In cases where \$0.15 small bets are used, $u_{+\$0.15} = 15$ and $u_{-\$0.15} = -15$.

If these definitions and some simple algebra are applied to Equation 5, the terms involving $u_{\$0.00}$ drop out, the $P^*_{0.5}$ factors out, and Equation 5 simplifies to:

$$u_{\$A} = 10 N \tag{6}$$

Since N is known, the utility of $\$A$ can be calculated. The same procedure can be carried out for other amounts of money, and so the whole utility curve can be determined. It should be noted that this utility curve, like those determined by Mosteller and Nogee, is not a total utility curve of the type which is usual in economics. Its zero point is where S is at the moment he makes his choice. Since S wins and loses money during the course of each session, that zero point is not a fixed point on his total utility curve in the classical sense. Consequently, the kind of utility curve which this procedure produces is closely analogous to the kind suggested by Markowitz (9), except that Markowitz takes zero to be S's usual location, while this procedure takes it to be his current location.

The N-bets method offers two advantages. First, it is straightforward and the experiment it suggests is easy to perform. Secondly, it offers a built-in check on some of the assumptions involved. There is no reason why 50–50 bets need be used, and there is no reason why the small bets should involve $0.10. Other probabilities and other sizes of small bets could be used. All probabilities and all sizes of small bets should lead to identical utility curves. If they do not, then it probably means that the N-bets assumption is incorrect, and the marginal utility of bets is not constant. For any given marginal utility function for bets, it is easy to predict the exact nature of the difference between utility functions obtained from different sizes of little bets.

If Ss preferred to bet at some variances rather than others, then the existence of such preferences would invalidate the N-bets method of utility measurement, and also the *EU, SEM,* and *SEU* maximization models. However, Edwards (7) has shown that such preferences, though they do exist, are of minute importance as compared with probability preferences, and so can be neglected.

How to Measure Subjective Probabilities

Once a utility scale is available for a given S, it is very easy to measure subjective probabilities. Suppose that S is indifferent between a certain gain of $0.55 and a ⅜ chance of winning $2.50. The *SEU* maximization model permits us to write:

$$P^*_{3/8} \, u_{\$2.50} + P^*_{5/8} \, u_{\$0.00} = P^*_1 \, u_{\$0.55} \tag{7}$$

This simplifies to:

$$P^*_{3/8} = u_{\$0.55} / u_{\$2.50} \tag{8}$$

Since the two utilities are known, the value of the subjective probability corresponding to an objective probability of ⅜ can be calculated. By determining other indifference points involving other probabilities, other subjective probabilities can be calculated. No assumptions beyond those involved in the SEU maximization model are necessary. The same procedure, carried out with losses instead of gains, gives the subjective probability scale for losing bets.

How to Test the SEU Maximization Model

Once individual utility scales and individual subjective probability scales are available, testing the SEU maximization model is accomplished by having the same Ss make choices among bets. The SEU of each bet for each S can be calculated, and the prediction is that he will always choose the bet with the higher SEU (or smaller negative SEU, if no positive SEU is available). Since utility and subjective probability are measured only on interval scales, they have no true zero points. Consequently, the SEU model cannot be used to predict choices among bets which include both possibilities of winning and possibilities of losing. So simple bets, in which only possibilities of winning or possibilities of losing appear, must be used to test the SEU model.

METHOD

Apparatus. All gambling in this experiment was done on two pieces of gambling apparatus. One was a pinball machine, which has been more fully described elsewhere (3). It has eight cells along its bottom, into any one of which the ball may roll. The Ss are told that the ball is just as likely to roll into any one cell as into any other. Under the surface of the machine eighteen electromagnets are concealed, which completely control which cell the ball ends up in. These electromagnets are operated by a programmer in the next room, and the program is set up so that the instructions to Ss are correct.

The other apparatus was designed because it seemed desirable to have some apparatus which could run off the strings of identical bets necessary for the N-bets method of utility measurement considerably more quickly than they could be run off on the pinball machine. Consequently, a bank of eight counters was set up. The counters were numbered from 1 through 8, and were automatically resettable. In front of them was a button. When S pushed it, one of the eight counters received one electrical impulse. A programmer determined which counter would be operated. By holding the button down, S could operate the apparatus continuously, so that a succession of impulses operated various counters. An auxiliary counter counted the total number of impulses entered into the bank of eight counters since the last resetting. Thus, in order to run off 20 identical bets, all S had to do was to hold down the button until the auxiliary counter read 20, and read off from the eight main counters the numbers of times those counters which paid off for him had been operated. The whole process takes about 20 to 30 sec. This procedure is simply a faster equivalent of playing the pinball machine 20 separate times, each time making an identical bet.

Subjects. The Ss for this experiment were five Johns Hopkins male undergraduates. They were naive about the purposes of this experiment, had not participated

The Prediction of Decisions Among Bets 259

in gambling experiments before, and for the most part were relatively unfamiliar with gambling. The names of those who were invited to be Ss were selected at random, but it was necessary to issue 23 invitations in order to get five acceptances. Therefore this cannot be regarded as a random sample of Johns Hopkins undergraduates. However, so far as the available information goes it is a random sample of those Johns Hopkins undergraduates who are willing to participate in prolonged gambling experiments.

General Procedure. Each S was told that his purpose in the experiment was to make as much money as possible. He was also told that every game he would play in the experiment was mathematically "fair," that is, that in the long run he should break about even. However, he was told that the situation would be biased in his favor by $1.00 per session. This was accomplished by giving him $21 in chips at the beginning of each session and taking back only $20 in chips or cash at the end of each session. Each S knew that he was expected to pay any losses out of his own pocket, and that he would be paid his winnings in cash. However, when an S lost more money than he had available (one S lost $120 in one session), he was allowed to pay with an *IOU* until next time.

After all sessions were finished, each S was run for an extra session, in which the program of the gambling apparatus was prepared so as to cause him to win or lose enough so that he won an average of $1 per hour for the entire experiment. Each S spent a total of about 50 hr. in the experiment.

On all sessions in which the eight-counter apparatus was used, S was permitted to select which counters would be "his" and which would not. The counters which he chose as his would pay off in his favor; the others would be against him. He could not change selections in midsession. This arrangement was necessary because it was obvious that the eight-counter apparatus was programmed, and so it was necessary to make it clear that the program was not prepared unfairly. On the pinball machine, whose electrical connections and mode of operation were not visible, it was not necessary to do this.

Utility Measurement Sessions. The first 12 sessions were devoted to the N-bets method of utility measurement. In Sessions 1–4, the probability of winning was $4/8$ and the amount involved in each little bet was $0.10. In Sessions 5–8, the probability of winning was $4/8$ and the amount involved in each little bet was $0.15. In Sessions 9–12, the probability of winning was $2/8$ and the amount involved in each little bet was $0.10. The amounts of money whose utilities were measured in each session were plus $5.50, $4.50, $3.50, $2.50, $2.00, $1.50, $1.00, $0.90, $0.80, $0.70, $0.60, and $0.50; and minus those same amounts. For each amount of money, seven cards were prepared, each with a different value of N. A sample card read:

A. You have four numbers. If you roll any one of your numbers, you win $5.50. If you do not, you win nothing.
B. You have four numbers. You make 30 identical bets. On each bet, if you roll one of your numbers, you win $0.10. If you do not, you win nothing.

All other cards were similar, except that they included different amounts which could be won in the big bet and different numbers of small bets. To measure negative utilities, the verb was changed from "win" to "lose." In Sessions 5–8, each alternative began "You have two numbers." In Sessions 9–12, each small bet paid off $0.15 instead of $0.10.

In addition to the cards already discussed, other cards were prepared which were identical except that the value of N was left blank. This was necessary because it was possible that the indifference point for a given S might not fall, on one day, within the range of values of N which had been selected. Whenever that happened, E wrote in values of N which seemed more likely to be near the indifference point, and slipped the new cards into the pack. The Ss were, of course, aware that E added cards to the pack from time to time, and they eventually learned that E kept on doing so until S had changed over from A-type alternatives to B-type alternatives, or vice versa. This knowledge would have made it possible for an S who was skilled at gambling to adopt the strategy of changing quickly on losing bets (thus maximizing the number of winning extra cards). To see whether anyone did this, counts were made of the number of extra cards used on winning and losing bets at various stages of the experiment. No indication could be found in these counts that any S used such a strategy. In fact, most Ss tended to resist changing more stubbornly on losing bets than on winning ones.

When the approximate indifference point had been located in the first session, thereafter only cards reasonably near it were used in subsequent sessions. This reduced the total number of cards, and so the length of the sessions. It is probable that the selection of the range of cards presented to S had at least a minor effect on his indifference point. This effect could not be avoided, since some range had to be selected; it could only be hoped that the effect would be minor compared with the utility and subjective probability differences being studied. The effectiveness of the utility and subjective probability scales determined by this method in predicting choices among bets suggests that this hope was fulfilled.

Subjective Probability Measurement Sessions. Sessions 13–18 were subjective probability measurement sessions. The probabilities whose subjective probabilities were measured were ⅛, ⅜, ½, ⅝, ¾, ⅞, and ⅞. Each subjective probability was determined twice in each session, once for positive EV bets and once for negative EV bets. For each probability, cards offering that probability of winning various amounts of money compared with the certainty of winning $0.55 were prepared. A sample card read:

A. You have three numbers. If you roll any one of your numbers, you win $1.50. If you do not, you win nothing.
B. You receive $0.55 for sure.

Other cards used in determining the subjective probability of ½ were identical except for the amount of money which could be won in the A-alternative. Similar sets of cards were prepared for each other probability. As in the utility measurement sessions, extra cards with the amount of money in the A-alternative left blank were available, in case the indifference point fell outside the anticipated range.

Model-Testing Sessions. Sessions 19–22 were the model-testing sessions. For the purpose of testing the model, the bets which have been extensively used in previous probability preference experiments were used again. Four lists of bets were used, with EVs of +$0.75, +$0.52½ and minus those same amounts. Table 1 gives those lists of bets. Each bet was paired with all others at its EV according to Ross's optimal order for pairs in the method of paired comparisons (12). The Ss were required to choose one member of each pair. These bets were run off on the pinball machine. The Ss made each choice four times, once in each session.

RESULTS

Utility Measurement Sessions

Regression lines were fitted to the results of each block of four utility measurement sessions for each S by the method of least squares, or a simpler approximation when little error resulted. The point at which S would presumably choose each alternative 50% of the time (the indifference point) was read from these lines. In three cases (out of 120) in which this procedure gave an obviously unreasonable 50% point, the actually observed 50% point was used instead.

To determine the utility functions, Equation 6 was solved for each amount of money used. The result of these calculations is three utility curves for each S. These curves are shown in Figure 1. It is apparent that there are differences from one curve to another, but these differences do not seem to be systematic; they can probably be attributed to day-to-day variations in judgments. There is no evidence that the curve for Sessions 5–8, in which the $0.15 small bets were used, is consistently smaller for the high values of money than the other two curves, as it would be if the utility of N bets was less than N times the utility of one bet; or consistently larger, as it would be if the utility of N bets was greater than N times the utility of one bet. So it is reasonable to assume that this experiment remained within the sizes of N for which the N-bets assumption is acceptable. Since there does not seem to be any systematic difference between the different utility curves, the three utility values determined for each money value were averaged, and that average is used in all subsequent calculations.

The most outstanding fact about Figure 1 is the relatively high degree of linearity of the utility curves. Most of them adhere fairly closely to the 45° line which represents the equation $u_{\$A} = \$A$ (where $A is measured in pennies). For those cases which deviate substantially from the 45° line, the deviation seems to be greater for the negative branch of the utility curve than for the positive branch. It should not be assumed that the utility of money is equal to its objective value for all possible values even for those Ss who adhere to the 45° line most closely; instead, it seems likely that for large values of $A ($100 and more, for instance), the value of $u_{\$A}$ will be quite different from $A. However, most college students have had extensive experience with receiving and giving up amounts of money such as are used in this experiment. Their subjective impressions of the values of such sums of money are likely to be very close to the true values because of this long learning process. This generalization, if it is true enough often enough, offers a great simplification for future gambling experiments. If it is possible to use the physical value of money instead of its utility in the SEU maximization model, then application of the model will be much easier.

Subjective Probability Measurement Sessions

Graphs similar to those used in determining 50% points for the utility measurement sessions were used for the subjective probability measurement sessions, except that the X-axis contained amount of money rather than value of N. Midpoints were determined by the same methods used for the utility measurement sessions. Then equations like Equation 8 were solved for each positive and negative subjective probability. Graphs of the resulting subjective probability functions are shown in Figure 2. The most outstanding fact about Figure 2 is that in general Ss tend to overestimate their chances of winning money, while retaining a fairly objective estimate of their chances of losing it. This result agrees pretty well with casual observation and common sense. Notice that one subjective probability function goes slightly above 1. This finding, though surprising, is permitted by the *SEU* and *SEM* models.

The Test of the SEU Maximization Hypothesis

The choices made during the model-testing sessions, Sessions 19–22, are not interesting in themselves. They were very similar to the usual results of probability preference experiments (3, 5). The interesting thing about these choices is whether or not they correspond to the predictions made by the *SEU* model. Table 2 shows the number of correct and incorrect predictions for each of the four *EV* levels for each S. It is at once apparent that the model predicts significantly better than chance (i.e., predictions from the model are better than those that could be made by flipping a coin). It is also apparent that the predictions from the model include a number of errors. These errors are of two kinds. First, there are cases in which Ss chose one way on a given pair in some of Sessions 19–22 and the other way on other sessions. Since the model predicts 100% choice of the alternative with the higher *SEU*, any such day-to-day variation in choices results in erroneous predictions. Such errors do little damage to the model. The other and more serious kind of error occurs when S chooses consistently in all four model-testing sessions, but the *SEU* model predicts the opposite of the choices which actually occur. On the average, each S made 18.8 errors of the first kind and only 10.6 errors of the second kind at each *EV* level. (Note that this means that only an average of 2.65 pairs per S per *EV* level were completely wrongly predicted; each completely wrong prediction produces four errors.) An analysis of variance of the errors shows that this difference is very highly significant. Variance due to Ss, of course, is also significant. And the difference between errors on positive *EV* and negative *EV* bets is significant. This last variable interacts with Ss. The nature of the effect, as can be seen from inspecting Table 2, is that errors are in general more frequent on negative *EV* than on positive *EV* bets, but this is not true for some Ss. It is interesting that neither Ss nor positive *EV* bets vs. negative *EV* bets interacts significantly with consistent vs. inconsistent errors.

The next task is to account for the wrong predictions. Some of them, as we have seen, result from inconsistent choices in the model-testing sessions. Others may perhaps be accounted for by the size of the *SEU* differences involved. An obvious hypothesis is that Ss are more likely to choose consistently with predic-

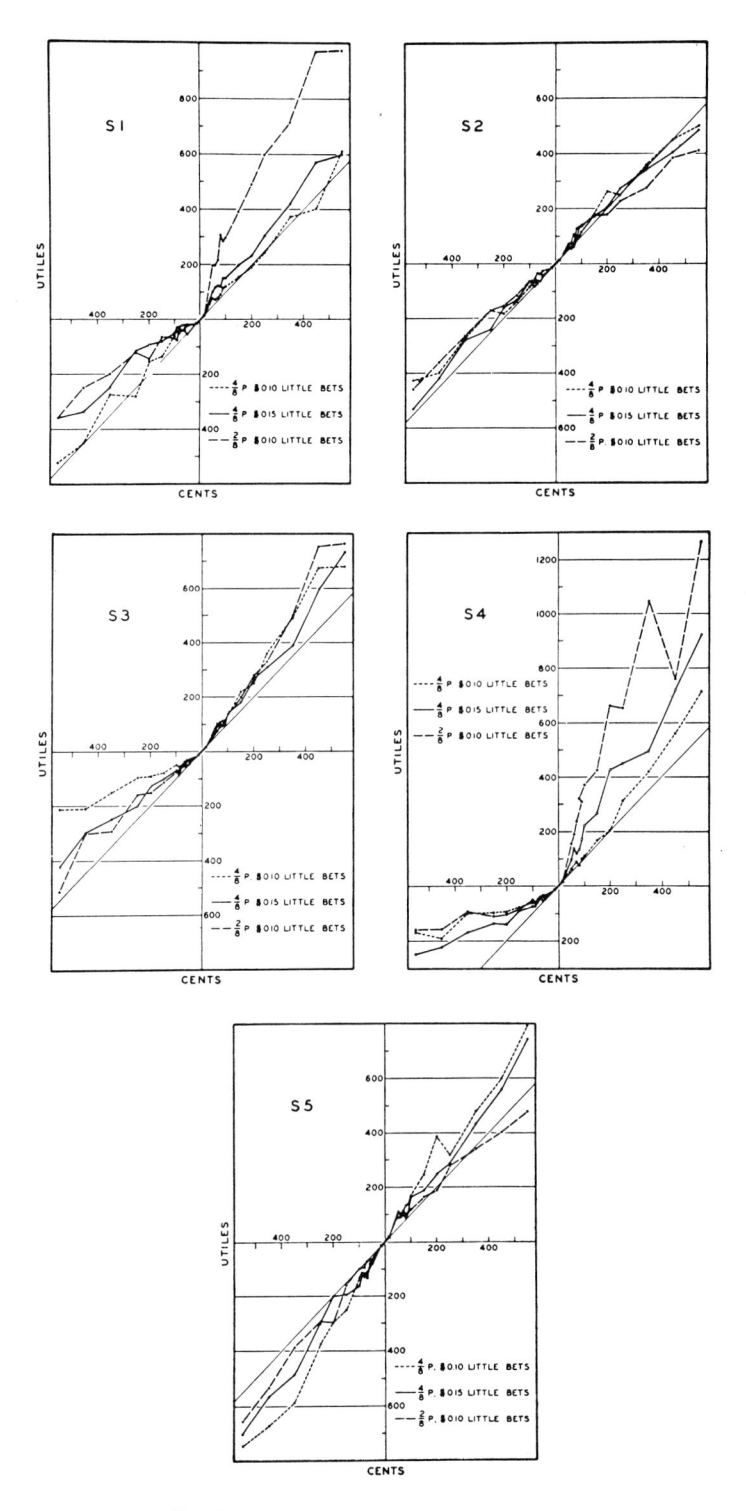

Figure 1. Experimentally determined individual utility curves. The 45° line in each graph is the curve which would be obtained if the subjective value of money were equal to its objective value.

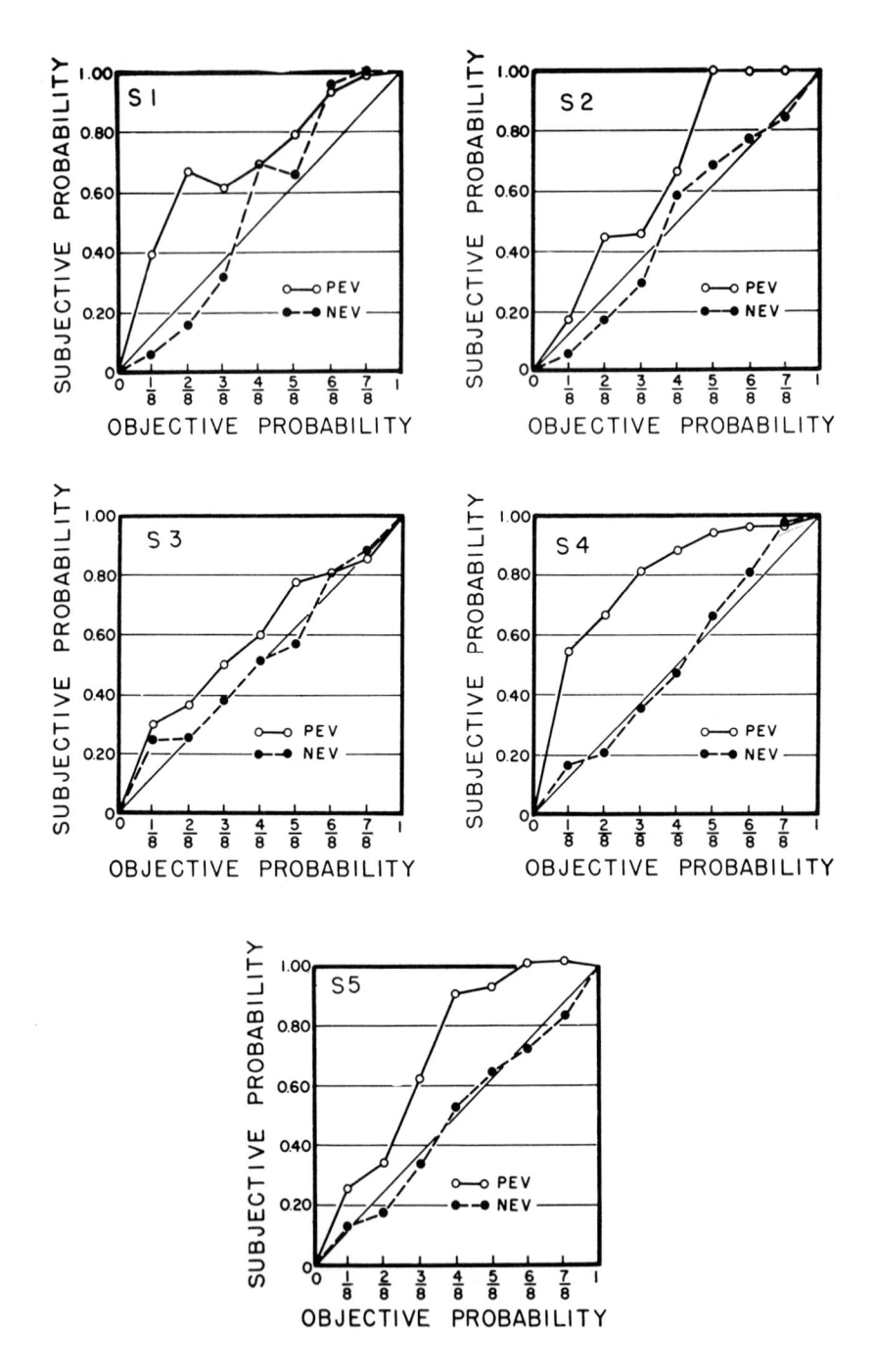

Figure 2. Experimentally determined subjective probability functions. The functions labeled *PEV* are for winning bets. The functions labeled *NEV* are for losing bets. The 45° line is the function which would be obtained if subjective probability were equal to objective probability.

TABLE 1

Bets Used in the Model-Testing Sessions

Positive Expected Value Bets

Probability	Payoff Cells	Amount Paid $0.75 EV	Amount Paid $0.52½ EV
1/8	4	$6.00	$4.20
2/8	1, 7	3.00	2.10
3/8	2, 4, 6	2.00	1.40
4/8	2, 4, 7, 8	1.50	1.05
5/8	2, 3, 5, 7, 8	1.20	0.84
6/8	All but 3 & 6	1.00	0.70
7/8	All but 5	0.86	0.60
8/8	All	0.75	0.53

Negative Expected Value Bets

All negative expected value bets are identical with the corresponding positive expected value bets except that S loses the amount stated if the ball rolls into one of the payoff cells listed, instead of winning it.

Note.—The exact wording of these bets can be found in (5), p. 88.

TABLE 2

Number of Correct and Incorrect Predictions Based on the SEU Maximization Hypothesis

Expected Value Level	1 Right	1 Wrong	2 Right	2 Wrong	3 Right	3 Wrong	4 Right	4 Wrong	5 Right	5 Wrong	Mean Right	Mean Wrong
+$0.52-1/2	100	12	71	41	99	13	107	5	72	40	89.8	22.2
+$0.75	81	31	73	39	100	12	105	7	93	19	90.4	21.6
−$0.52-1/2	78	34	71	41	61	51	95	17	66	46	74.2	37.8
−$0.75	64	48	87	25	45	67	91	21	70	42	71.4	40.6

Note.—By chance 56 correct predictions should be expected in each group. By chi square, 67 predictions is significantly different from 56 at the .05 level; 70 at the .01 level; and 74 at the .001 level. Accordingly, all predictions are significantly better than chance except those for Ss 1 and 3 on the − $0.75 EV level and for Ss 3 and 5 at the − $0.52-1/2 EV level.

TABLE 3

Number of Correct and Incorrect Predictions Based on the *EU* Maximization Hypothesis

Expected Value Level	Subject 1 Right	1 Wrong	2 Right	2 Wrong	3 Right	3 Wrong	4 Right	4 Wrong	5 Right	5 Wrong	Mean Right	Mean Wrong
+$0.52–1/2	12	100	73	39	41	71	37	75	20	92	36.6	75.4
+$0.75	19	93	59	53	40	72	25	87	41	71	36.8	75.2
−$0.52–1/2	38	74	57	55	39	73	103	9	66	46	60.6	51.4
−$0.75	24	88	63	49	47	65	99	13	82	30	63.0	49.0

TABLE 4

Number of Correct and Incorrect Predictions Based on the *SEM* Maximization Hypothesis

Expected Value Level	Subject 1 Right	1 Wrong	2 Right	2 Wrong	3 Right	3 Wrong	4 Right	4 Wrong	5 Right	5 Wrong	Mean Right	Mean Wrong
+$0.52–1/2	100	12	71	41	99	13	107	5	72	40	93.4	18.6
+$0.75	93	19	73	39	100	12	105	7	93	19	92.4	19.6
−$0.52–1/2	78	34	69	43	61	51	93	19	66	46	74.2	37.8
−$0.75	76	36	75	37	45	67	91	21	70	42	72.2	39.8

Note.—By chance 56 correct predictions should be expected in each group. By chi square, 67 predictions is significantly different from 56 at the 0.05 level; 70 at the 0.01 level; and 74 at the 0.001 level. Accordingly, all predictions are significantly better than chance except for S 3 at both negative *EV* levels and S 5 at the − $0.52–1/2 level.

tion when the two bets between which they are choosing are far apart from one another in SEU. This hypothesis is strengthened by the fact that the median SEU difference between pairs of bets is much smaller for the negative EV bets for all Ss than it is for the positive EV bets. This is consistent with the fact that, in general, prediction is less satisfactory on the negative EV bets. In order to test this hypothesis more precisely, product-moment correlation coefficients were calculated between the size of the SEU difference between each pair of bets and the number of choices consistent with prediction on that pair. These correlations are small (median value: + .23), but all of them are positive. The reason they are small is that the number of choices consistent with prediction can have only five values, 0, 1, 2, 3, and 4. When a variable which can assume only a few values is correlated with some other variable, the correlation coefficient is inevitably low. However, a simple sign test shows that the probability of obtaining 20 correlation coefficients with the same sign, if the true correlation is 0, is $2(1/2)^{20}$—very small indeed. This confirms the hypothesis that there is a positive relationship between size of SEU difference and number of choices consistent with prediction.

A Test of the EU Maximization Hypothesis

The next question to ask is: How well can the utilities alone and the subjective probabilities alone predict choices?

To test the predictive value of the utilities considered alone is essentially equivalent to testing the EU maximization model proposed by Bernoulli and von Neumann and Morgenstern. In order to do it, all that is needed is to calculate the EU for each S for each bet, using Equation 2. This was done. The resulting predictions were compared with the choices made in the model-testing sessions. The results of this comparison are shown in Table 3. It is at once evident that the EU maximization hypothesis does not predict better than chance. Indeed, if an analysis of variance of the error scores is made and the error estimate from that analysis is used to estimate the standard error of the mean, then the EU maximization hypothesis predicts significantly ($P = .05$) worse than chance on the positive EV bets. (The results of that analysis of variance, incidentally, show that Ss, the difference between positive EV and negative EV bets, and their interaction significantly affect the number of errors.) None of the various other comparisons which were reported for the SEU model offers any evidence, when made on the EU model, that the EU model is any good. So far as the evidence of this experiment goes, it is necessary to conclude that the EU maximization model fails to predict simple gambling decisions.

The reason for the failure of the EU maximization model is, presumably, the near-linearity of the utility scales used. If the utility scales are actually relatively linear, then any predictions of choices between constant-EV bets must be made on the basis of minor irregularities in the curves, and it is not surprising that such predictions fail to work out. The only successful predictions based on EU values, S 4's NEV bets, are consistent with this argument. This S's negative utility curve deviates substantially from the $u_{\$A} = \$A$ line, and his negative subjective probability curve deviates very little from the $P^* = p$ line.

The relative linearity of the utility curves, if it is a general phenomenon, means that the EU maximization model will never, as a practical matter, be very useful in gambling experiments. This does not, of course, preclude the possibility that it might be useful in other situations, particularly those which use larger sums of money. However, the evidence does strongly suggest that whenever the EU maximization model is useful, the SEU maximization model will be more useful.

A Test of the Subjectively Expected Money (SEM) Maximization Model

We have examined the SEU and EU maximization models. It remains to examine the SEM maximization model, in which prediction of choices is affected only by subjective probabilities. In the original calculation of the subjective probability scales given in Figure 2, utility values were used. If the predictive value of the SEM maximization model is to be tested, it is not permissible to use utilities in the original calculation of the subjective probabilities which go into the model. Consequently, all subjective probabilities were recalculated using money values instead of utility values in equations like Equation 8. The resulting subjective probability functions differ only in minor ways from those given in Figure 2.

For each bet used in the model-testing sessions, an SEM value was calculated from the new subjective probability functions. Then the predictions based on the hypothesis of SEM maximization were compared with the choices in the model-testing session. Table 4 shows the results. It is at once apparent that the SEM maximization model predicts quite well. Comparison of Table 4 with Table 2 shows that the SEM maximization model predicts nonsignificantly better than the SEU maximization model. Analysis of variance of the SEM model errors gives essentially the same picture as that for the SEU model. The same generalizations about errors and about the superiority of positive EV predictions over negative EV predictions which were made for the SEU model apply to the SEM model also. Indeed, the only respect in which the SEM model is inferior to the SEU model is that when the correlations between SEM difference and number of choices consistent with prediction are calculated, only 17 instead of 20 of the correlation coefficients come out positive. The probability of getting so many correlation coefficients of the same sign if the true correlation is 0 is .0026.

It must be concluded that the SEM model predicts choices just as well as the SEU model does. Indeed, examination of the predictions shows that the SEM model and the SEU model usually predict the same or almost the same rank order for the bets. The reason why is that the SEM values correlate highly with the SEU values. The median correlation is .96; the lowest is .82.

DISCUSSION

If the results of this experiment are taken literally, they mean the following things. First, over the range of money used in this experiment (+ \$5.50 to − \$5.50) the utility of money is fairly linearly related to its dollar value for most Ss. Secondly, the SEU maximization model predicts choices among simple bets rather well. Thirdly, there is a positive (though not perfect) correlation between the size of the SEU difference between two bets and the probability that the S will choose the one with the larger SEU. Fourthly, the EU maximization model is no good at predicting choices among simple bets like those used in this experiment. This is presumably because utility is fairly linear with money value over the ranges used in this experiment, and so the sizes of the *EU* differences among the bets used in the model-testing sessions of this experiment are very small. Finally, a model which asserts that Ss choose so as to maximize *SEM* predicts just as well as the *SEU* maximization model, and for the most part makes the same predictions. This fact can be explained, of course, by the fact that the utility curves are approximately linear, so that *SEM* and *SEU* values for any bet are very highly correlated with one another. Most of these conclusions depend for their validity on the assumption that the *N*-bets method is a satisfactory method of utility measurement. The internal consistencies in the results which indicate its acceptability have already been discussed.

The results of this experiment are entirely consistent with the results of the various probability preference experiments. They suggest that the most important determiner of choices among bets is subjective probabilities, and that bets with the same subjective probability ought to be preferred to about the same extent when compared with others at the same *EV* level. That is just what happened in one probability preference experiment (5).

Generalizations from this experiment should be made cautiously. First, neither the *SEU* model nor this experiment offers any way of predicting choices among bets which include possibilities both of winning and of losing. The *SEU* values of the winning possibilities and of the losing possibilities can be algebraically added to get a composite *SEU*, but whether such composite *SEU* values will predict choices is an experimental problem. Secondly, the *SEU* model has no true zero point, and so does not predict whether a given bet will be accepted or refused, if the alternative is simply not gambling. Finally, there is no guarantee that subjective probabilities are independent of the utilities with which they are associated, though the *SEU* model assumes that they are. With very large bets or very small bets, the form of the subjective probability function might be quite different from that found in an experiment like this one for a given S. All of these questions can be settled by appropriate experiments.

More theoretical work needs to be done on the *SEU* model. Ideally, it should be a probabilistic model, predicting the probability of choice of the alternative with the higher *SEU* as a function of the size of the *SEU* difference. The obvious form for a theoretical curve relating probability of choice of alternative *A* over alternative *B* to size of $SEU_A - SEU_B$ is a normal ogive, with the 50% point at a *SEU* difference of 0. But other intuitively plausible curves can also be conceived. At any rate, until better methods of measuring utilities and subjective probabilities are found, there is little reason for going into such theoretical refinements, because there is no hope of testing them experimentally. The improvement of methods of utility and subjective probability measurement is still the most important task in the study of gambling behavior.

SUMMARY

A very simple mathematical model was developed for predicting choices among bets. This model, based on the concepts of subjective value or utility of money and subjective probability, asserts that Ss choose the bet with the maximum subjectively expected utility. An experiment was designed to test this model. The utility of money was measured by assuming that the utility of N identical bets is N times the utility of one such bet—an assumption which the results of the experiment verify. Subjective probabilities were measured by other choices. Finally, still other choices were used to test the model. The model predicted substantially better than chance. Another model, which assumes that Ss choose so as to maximize expected utility, failed to predict choices successfully. A third model, which uses only the concept of subjective probability, predicted as well as the subjectively expected utility maximization model.

It is concluded that the subjectively expected utility maximization model is adequate to account for the results of this experiment, and that subjective probabilities are much more important than utilities in determining choices among bets such as those used in this experiment.

REFERENCES

1. Bernoulli, D. Specimen theoriae novae de mensura sortis. *Comentarii Academiae Scientiarum Imperiales Petropolitanae,* 1738, **5**, 175-192. (Translated by L. Sommer in *Econometrica,* 1954, **22**, 23-36.)
2. Coombs, C. H., & Beardslee, D. C. On decision-making under uncertainty. In R. M. Thrall, C. H. Coombs, & R. L. Davis (Eds.) *Decision processes.* New York: Wiley, 1954.
3. Edwards, W. Probability-preferences in gambling. *Amer. J. Psychol.,* 1953, **66**, 349-364.
4. Edwards, W. Probability-preferences among bets with differing expected values. *Amer. J. Psychol.,* 1954, **67**, 56-67.
5. Edwards, W. The reliability of probability-preferences. *Amer. J. Psychol,* 1954, **67**, 68-95.
6. Edwards, W. The theory of decision making. *Psychol. Bull.,* 1954, **51**, 380-417.
7. Edwards, W. Variance preferences in gambling. *Amer. J. Psychol.,* 1954, **67**, 441-452.
8. Edwards, W. An attempt to predict gambling decisions. In *Mathematical models of human behavior—Proceedings of a symposium.* Stamford, Conn.: Dunlap & Associates, 1955.
9. Markowitz, H. The utility of wealth. *J. polit. Econ.,* 1952, **60**, 151-158.
10. Mosteller, F., & Nogee, P. An experimental measurement of utility. *J. polit. Econ.,* 1951, **59**, 371-404.
11. Preston, M. G., & Baratta, P. An experimental study of the auction-value of an uncertain outcome. *Amer. J. Psychol.,* 1948, **61**, 183-193.
12. Ross, R. T. Optimum orders for the presentation of pairs in the method of paired comparisons. *J. educ. Psychol.,* 1934, **25**, 375-382.
13. von Neumann, J., & Morgenstern, O. *Theory of games and economic behavior.* (2nd Ed.) Princeton: Princeton Univ. Press, 1947.

22

Cognitive Aspects of Information Processing: III. Set for Speed Versus Accuracy

Paul M. Fitts

The present experiment examines the relation between speed and accuracy of responses in a choice reaction-time (RT) task. The approach to this problem is predicated on the conjecture that part of the RT interval is occupied by processes that are analogous to those employed in sequential stimulus sampling and statistical decision making. A model for choice RT incorporating this idea has been proposed earlier by Stone (1960), and has recently been extended by Edwards (1965). It incorporates some aspects of other stochastic models that have been proposed for human latency distributions (McGill, 1962, 1963). The present study and previously reported ones (Fitts, 1964; Fitts, Peterson, & Wolpe, 1963) are intended to provide empirical tests of predictions derived from the sampling and decision model. The specific model and some of its implications are outlined briefly below.[1]

It is assumed that O holds some initial opinion as to the odds, Ω, favoring the occurrence of alternative stimuli, such as Sa vs. $S\bar{a}$. As an illustration, if O assumed equal probabilities then $P(Sa) = P(S\bar{a})$, and $\Omega(Sa)$ would be 1:1, as shown in Figure 1.

Onset of a stimulus is assumed to initiate a sequential sampling process. Each sample is categorized as favoring Sa or $S\bar{a}$, and the likelihood ratio for obtaining this datum (D) under the alternative stimulus conditions is used to obtain a revised odds estimate, $\Omega(Sa|D)$, by a process corresponding to one that would be used by an ideal O employing Bayes' Theorem (see Edwards, Lindman, & Savage, 1963, Formula 9) where

$$\Omega(Sa|D) = \frac{P(D|Sa)}{P(D|S\bar{a})} \Omega(Sa). \qquad [1]$$

From *Journal of Experimental Psychology*, 71, 1966, 849-857. Copyright 1966 by the American Psychological Association, and reproduced by permission.

This research was supported by the Advanced Research Projects Agency, Department of Defense, under Contract No. AF 49(638)-1235 (monitored by the Air Force Office of Scientific Research), with the Human Performance Center. Barbara Radford was responsible for collection of the data.

[1] The author is indebted to W. M. Kincaid and Ward Edwards for many contributions to the formulation of the random-walk model of choice RT as outlined here.

Assuming the presence of noise, the iterative odds revision process should take the form of a random walk (see Figure 1). This random walk is assumed to continue until it reaches some criterion boundary, which is defined in terms of a critical odds value. At this time a response process will be initiated.

The random-walk model has the following qualitative properties relevant to two-alternative choice RT: (a) A symmetrical shift of *both* criterion boundaries toward more nearly equal odds (odds closer to 1:1) should result in faster mean RTs, and increased errors for both responses (Ra and $R\bar{a}$); (b) a unilateral shift of criterion boundary A toward more nearly equal odds should result in Ra being made more often, more rapidly, and with more errors, relative to $R\bar{a}$; (c) an increase in the initial odds favoring Sa should result in faster RTs for Ra, and longer RTs for the alternative response, $R\bar{a}$, provided the criterion boundaries remain fixed (or alternatively should result in an asymmetric readjustment of the boundaries); (d) an increase in the difficulty of discriminating between Sa and $S\bar{a}$, as would result from an increase in similarity or a decrease in signal-to-noise ratio, should prolong the sampling process and result in slower RTs, providing again that the criterion boundaries remain fixed.

An important deduction from the model is the prediction that the mean and shape (but not necessarily the area) of the RT distribution made up of all instances of a given response that are correct should be identical to the RT distribution comprised of all instances in which that response is wrong. This correspondence is implied by the definition of constant odds at all points along the criterion boundary. Stone (1960) has given a proof of this prediction, based on Wald (1947).

It is important to note that the above prediction holds only as long as the criterion boundary remains fixed; if the boundary shifts, speed and errors will become positively correlated and the mean RT for errors will be less than the mean RT for correct responses. Since stationarity is a questionable assumption for a human O, one might postulate a "noisy" criterion, as Swets (1961) has proposed for signal-detection processes. An alternative assumption favored by the present author is that O sets his criterion boundaries as a result of the initial instructions and then continually readjusts them on the basis of the feedback information obtained from his own responses. If this conjecture is correct, then when Os make very few errors they also secure very little information and should find it difficult to maintain stable performance. This assumption leads to the revised prediction that similar RT distributions for correct and wrong responses should most likely be found when Os are making a relatively large proportion of errors.

The preceding discussion has considered only two-alternative choices. Similar arguments apply to the multidimensional case provided it is assumed that the underlying decision process is binary in character. Thus odds at each criterion boundary can be defined as favoring one stimulus against all others, or a binary decision tree can be postulated. The argument also applies to the alternative type of model proposed by Rapoport (1959) and by Christie and Luce (1956), where choice RT is assumed to reflect a number of independent but parallel binary processes with responses being delayed until all parallel processes have been completed.

Finally, all actual RTs of the human O are assumed to include fixed time lags, and thus performance should approach a chance level for some limiting RT value greater than zero.

In summary, the distribution of RTs and the ratio of correct to wrong responses are assumed to be determined by sampling and statistical decision processes, as well as by stimulus properties and by O's level of learning. Criterion boundaries, which determine when response processes will be initiated, are assumed to be influenced by cognitive processes, such as by changes in O's set for speed vs. accuracy and by his use of feedback information.

In order to test predictions from the model, explicit payoff matrices covering various speed-accuracy contingencies were employed as a means for influencing O's initial set, as advocated by Edwards (1961). Immediate augmented feedback was provided, so as to permit Os to monitor and adjust their behavior at a cognitive level. A discrete task was employed and the first half of the 2-sec. interval between a response and the next stimulus was used for the augmented feedback. A relatively difficult task involving four binary stimulus and response elements was used in order to provide opportunity for a considerable range of variation between speed and accuracy of responding.

METHOD

Apparatus. Stimulus sequences were programmed on a punched-tape system that permitted any number of stimulus lights up to 10 to be presented singly or in combination. Responses consisted of the depression of pianolike microswitch keys which were arranged in a pattern to fit the shape of the hands (see Figure 2). Stimuli appeared on a panel 28 in. in front of O, in a spatial pattern corresponding to the finger keys. Each light was assigned to the spatially corresponding key, unused lights and keys being removed, so that the task was high in S-R compatibility. The stimulus, response, and RT in 1-msec. units were automatically recorded on punched tape for subsequent computer analysis. In addition, a maxtrix of four electromagnetic counters provided an immediately available cumulative total of the number of fast and slow, correct and wrong responses.

Subjects. The Os were 18 male college students who were paid for their services.

General Procedure. The first stimulus in a series appeared 2 sec. after a warning signal and remained on for 50 msec. after O pressed any one of the response keys. About 100 msec. after O's first response, two feedback lights came on and remained on for 1 sec. One light of the feedback pair indicated whether the response was correct or wrong; the other indicated whether the response was fast or slow relative to an arbitrary criterion time. Exactly 2 sec. after O's response the next stimulus pattern appeared. This temporal sequence was continued for a block of 100 responses, after which a short rest was given, followed by another block of trials. Eight such blocks, or 800 responses, requiring about 50 min., constituted an experimental session.

When the stimulus pattern required activation of multiple keys, O was allowed

50 msec. within which to complete the correct response pattern after pressing the first key; if any one of the components of a response pattern was delayed longer than 50 msec., or if a wrong key was depressed within this interval, the response was recorded as wrong. This 50-msec. interval is necessary because of O's inability to make exactly simultaneous responses with multiple keys, but is too short to permit O to adopt a deliberate sequential response strategy. When more than one key was depressed, RT was recorded as the time for operating the first key.

Matching Procedure. All Os were pretested in a 10-alternative task for one session of 800 trials. Stimuli were the 10 lights, turned on one at a time in randomized order, and all 10 response keys were used, one at a time (see Figure 2A). The payoff matrix used during this pretest is shown in Table 1. The criterion time for fast vs. slow responses was set arbitrarily at 545 msec., which it was hoped would be at about the +1 SD point in the RT distribution. After all 18 Os had been pretested, they were ranked on the basis of their RT scores and divided into six homogeneous strata. The 3 Os in each stratum were then assigned randomly, 1 to each of three groups, and scheduled for testing so that all Os in a given stratum, beginning with the fastest one, served concurrently, one under each condition, as the experiment progressed.

Experimental Conditions. Following the pretest session, Os were tested for three experimental sessions. Stimuli consisted of the 15 possible patterns provided by four stimulus lights used singly and in all combinations in a randomized sequence. Responses involved the use of the middle finger and forefinger of each hand, singly and in all combinations (see Figure 2B).

The 6 Os assigned to the speed group were tested under conditions which were designed to emphasize speed, and those in the accuracy group under conditions favoring accuracy (see Table 1). These Os were guaranteed a minimum of $1.00 base pay, and were permitted to earn up to $1.00 per session in additional bonus money (at .2¢ per point). Immediate feedback was provided after each response, plus cumulative knowledge of monetary earnings after each block of 100 responses. The Os in the control group were given ambiguous instructions to "work as rapidly as you can without making errors," and received no immediate feedback with respect to either speed or accuracy. They were told their cumulative number of correct and wrong, fast and slow responses after each series of 100 stimuli, and were paid a flat rate of $1.50 per session.

The criterion for fast vs. slow responses was identical for each set of three matched Os and was set at what was hoped would be about the +1 SD point of the RT distribution. The criterion time was 370 msec. for the three fastest Os and it was raised for each successive stratum in an effort to compensate for individual differences and thus enable all Os to earn approximately the same bonus. In preliminary studies efforts had been made to set the criterion time so as to permit Os to achieve about 50% fast responses. However, this led to a marked discrepancy in the frequency of negative feedback for "slow" as contrasted with "wrong" responses and a resulting overemphasis on speed. The +1 SD point was selected in an effort to achieve a better balance between the two types of unsuccessful responses. The use of these criterion times actually provided approximately 84% fast responses for the three groups.

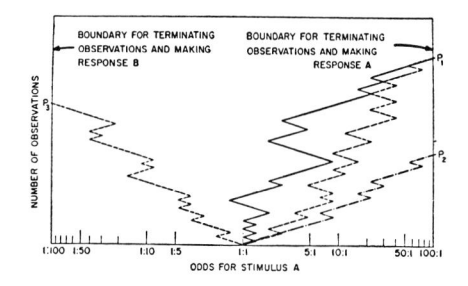

Figure 1. Random-walk model of choice reaction time.

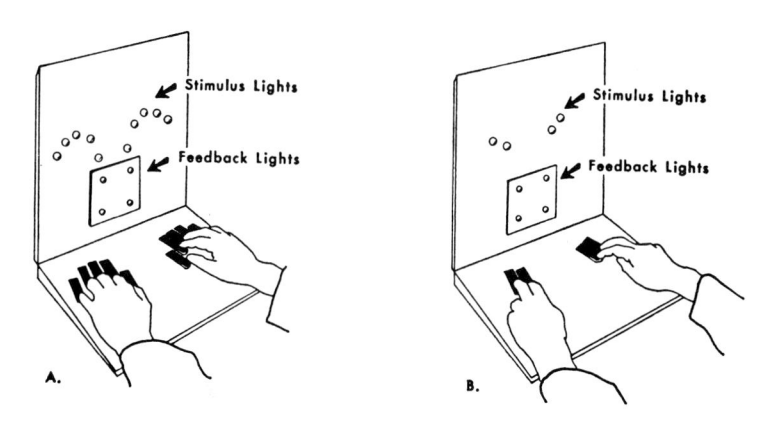

Figure 2. Two stimulus and response console arrangements.

TABLE 1

Payoff Matrices Used for Pretest and Experimental Conditions

Payoff Condition	Correct and Fast	Correct and Slow	Wrong and Fast	Wrong and Slow
Pretest	+1	−0.2	−0.2	−1
Speed	+1	−0.5	−0.1	−1
Accuracy	+1	−0.1	−0.5	−1

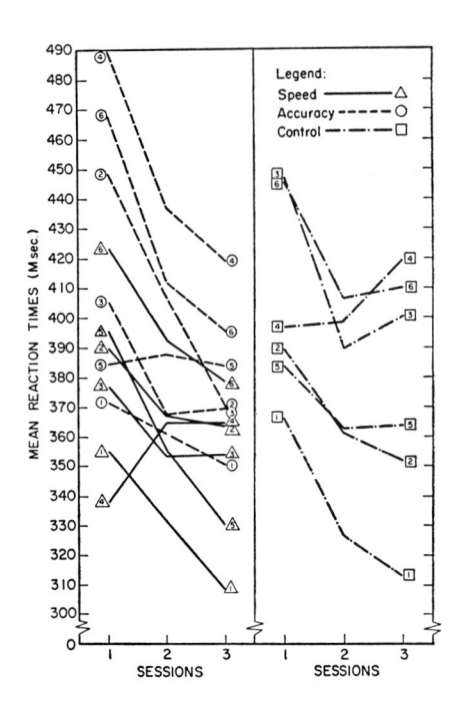

Figure 3. Mean RTs for individual *O*s.

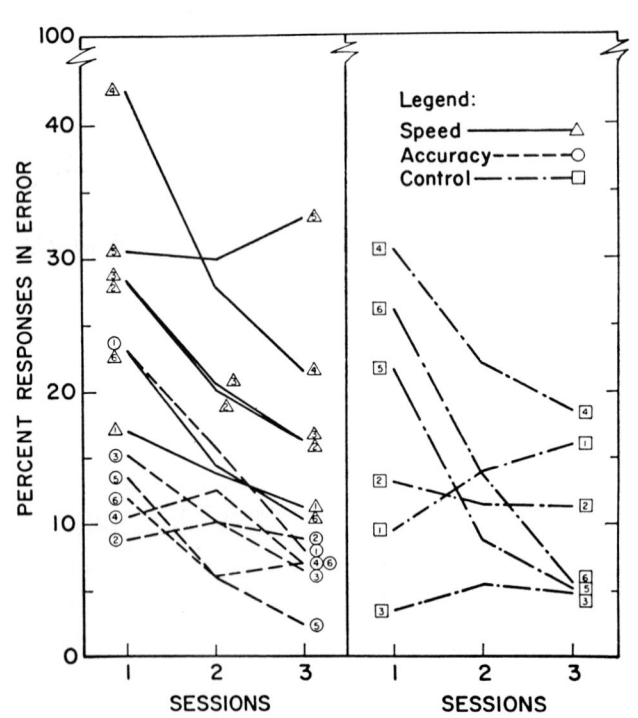

Figure 4. Percentage of responses in error for individual *O*s.

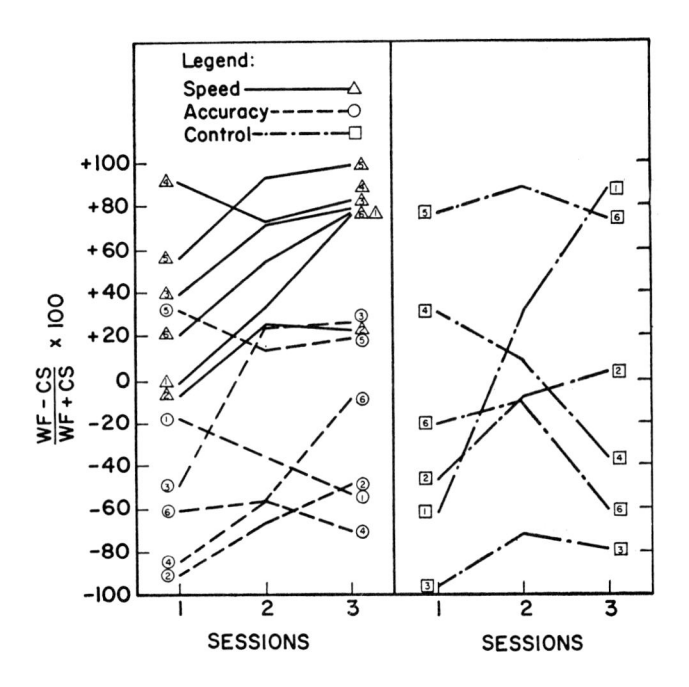

Figure 5. Relative frequency of wrong-fast (WF) vs. correct-slow (CS) responses for individual *O*s.

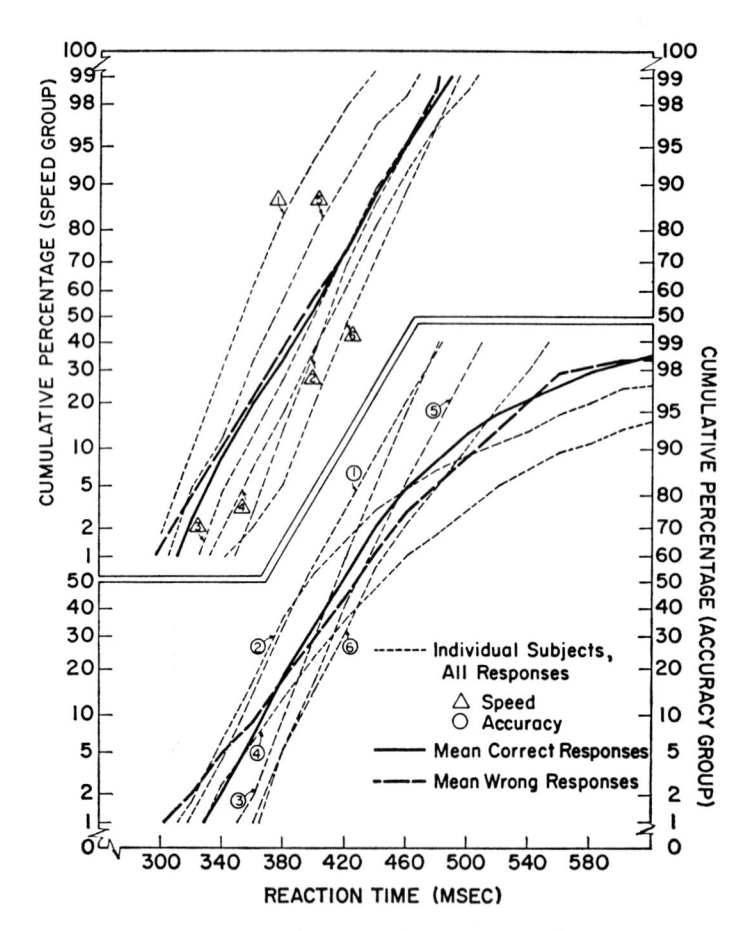

Figure 6. Cumulative RT distributions. (Light dotted lines are for individual Os in the speed and accuracy groups. Heavy lines are cumulative distributions for pooled error and pooled correct responses.)

Figure 7. Information transmission rate vs. response equivocation for each of the 18 Os. (Data are averages for the three experimental sessions.)

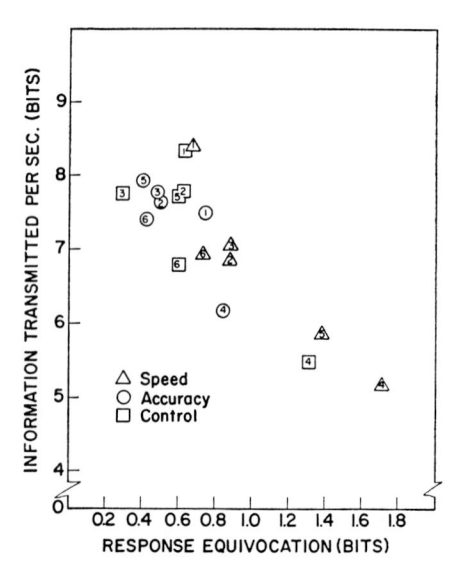

RESULTS

Mean Errors and RTs. Mean RT and percentage error scores for each of the 18 Os are shown in Figures 3 and 4. Payoff conditions had a highly consistent effect, with Os in the speed group responding faster and making more errors, and Os in the accuracy group responding more slowly and making fewer errors. Sets of data for each of the three sessions can be compared for each of the six pairs of matched Os in the two experimental groups; for 17 of these 18 data points, the O in the speed group was faster than the matched O in the accuracy group, and for 16 of 18 pairs, the O in the accuracy group made fewer errors. The mean RTs were 364 and 402 msec., and the mean percentage of errors was 22.6 and 10.3 for the speed vs. accuracy groups, respectively. The comparable scores for the control group were 385 msec. and 13.4% errors, respectively.

Experimental effects are revealed further by comparing the frequencies of correct-slow (CS) responses and wrong-fast (WF) responses for each O relative to the criterion times. A percentage-difference score, based on these two mixed categories, $[(WF - CS)/(WF + CS)] \times 100$, was computed for each O. There was practically no overlap on this score between Os in the two experimental groups (see Figure 5). With respect to this measure, as well as the two previous ones, Os in the control group showed somewhat larger individual differences and somewhat more within-O variability from session to session than did the experimental groups.

The last 300 responses made by each of the 18 Os, including a total of 579 (10.7%) wrong responses, were analyzed in detail. For this set of data 12 of the 210 possible wrong S-R combinations accounted for half of all the errors, the error patterns being similar both across Os and across groups. By far the highest error rates, 32% and 35%, respectively, were found in responses to the 1-2-4 and the 1-3-4 light combinations. Frequent errors were also made in responding to the other sets of three-component stimuli (1-2-3 and 2-3-4). The smallest proportion of errors (1.7%) was made to the pair of middle lights (Lights 2 and 3). Responses to the four single lights also showed good accuracy (between 3.5% and 5.5% errors). Responses were more often wrong because of the use of too few rather than the use of too many finger keys.

Distribution of RTs. The RT distributions for the last session of 800 responses, for each of the 6 Os in the two experimental groups, are shown in Figure 6, plotted as cumulative response frequencies on normal probability paper. It will be seen that the data for individual Os approximate a straight line for each of the 6 Os in the speed group. It therefore appears that under speed conditions response times were very nearly normally distributed. This conclusion is further supported by the finding that, for all three sessions, means and medians for each of the Os in this group were almost identical, the means exceeded the medians by only 2.9 msec. In the accuracy group, 2 Os had very nearly normal distributions of RTs whereas 4 Os showed skewed distributions, 2 being highly skewed. For the latter group, the means exceeded the medians by 17.1 msec.

The RTs for errors were pooled across Os for the last session of 800 trials, providing totals of 911 errors for the speed group and 321 errors for the accuracy group. These cumulative error distributions are shown in Figure 6, along with the distributions for all correct responses. The shapes of these two distributions are quite similar, being more nearly identical for the speed group, as predicted by the assumption that errors provide information for adjusting criterion boundaries. Error- and correct-response distributions for individuals, while more irregular, also showed overall correspondence.

Information Analysis. Response equivocation, $H_s(R)$, and average information transmitted per response, H_t, were computed. Average information rate (H_t/sec) per O was then estimated by dividing average H_t by average RT. It was found that mean group differences in H_t/sec were not statistically significant. However, if treatment effects are ignored, and the data for each of the 18 Os are examined, a surprisingly consistent relation can be observed between response equivocation and information transmission rate (see Figure 7). From Figure 7, it appears that performance deteriorates rapidly once response equivocation exceeds about .6 bits/response in this 3.9-bit task, i.e., as soon as $H_s(R)$ exceeds about 15% of stimulus information. The linear correlation between $H_s(R)$ and H_t/sec is −.86, as compared with correlations of only −.28 between RT and H_t/sec and of only −.24 between RT and $H_s(R)$.

DISCUSSION

We may conclude that sets for speed or accuracy in RT tasks induced by the use of a payoff matrix plus immediate feedback, are relatively effective within and between Os. The use of this combined procedure (payoff + feedback) appears to be a methodological advance over the use of verbal instructions alone (see, e.g., Garrett, 1922; Howell & Kreidler, 1963).

"Errorless" Performance. Many Es, such as Hick (1952) and Hyman (1953), have attempted to induce "errorless" performance on the part of their Os. From the viewpoint of a random-walk model, error rate may become very small in circumstances where the penalty for errors is sufficiently high, but in theory should never become zero as long as speed of response is given any importance at all. Furthermore, the occurrence of errors should not be interpreted as necessarily indicating a qualitative change in performance, since a sufficiently high payoff for speed may induce errors at any level of learning or attention. One is certainly not justified in assuming that data can be rid of spurious effects simply by discarding responses that were in error.

RT Distributions. The finding that the distribution of RTs for errors was very similar to that for correct responses lends support to the conjecture that choice RT is analogous to a sequential sampling and decision process, and that error rate is determined primarily by O's cognitive set for speed vs. accuracy. The actual shape of the distribution appears to be influenced by payoff conditions as well as to be a function of individual differences.

Maximum Information Transmission Rates. The present results suggest that information rate for discrimination and choice behavior may reach a maximum over some limited range of response equivocation greater than zero. Beyond this point the loss in information rate resulting from errors increases relatively much faster than does the gain from increased response speed, producing an information overload effect. The overload effect demonstrated here in the discrete response situation in which speed stress was induced by means of a payoff structure is similar to the finding of information overload induced by speed stress in forced-paced serial RT studies, such as Klemmer and Muller (1953). The presence of such a maximum is contrary to the finding of a relatively constant information capacity of the human motor system over a wide range of movement accuracy (Fitts, 1953, 1954; Fitts & Peterson, 1964; Fitts & Radford, 1966), and does not confirm Hick's (1952) early results on choice RT. However, the present results are entirely consistent with two limiting processes which have been proposed previously. On the one hand, Stone (1960) has pointed out the excessive cost in increased sampling time that is necessary for achieving very low error rates (low equivocation per response); on the other hand, it has been assumed earlier in this article that all measured RTs include some minimum lag time and this added constant would markedly reduce information rate for very short RTs where equivocation per response is excessively high. At most, therefore, it would appear that a constant rate of information transmission should occur for the interchange of speed and accuracy over a rather limited range.

REFERENCES

Christie, L. S., & Luce, R. D. Decision structure and time relations in simple choice behavior. *Bull. math. Biophysics*, 1956, **18**, 89-112.

Edwards, W. Costs and payoffs are instructions. *Psychol. Rev.*, 1961, **68**, 275-284.

Edwards, W. Optimal strategies for seeking information: Models for statistics, choice reaction times, and human information processing. *J. math. Psychol.*, 1965, **2**, 312-329.

Edwards, W., Lindman, H., & Savage, L. J. Bayesian statistical inference for psychological research. *Psychol. Rev.*, 1963, **70**, 193-242.

Fitts, P. M. Influence of response coding on performance in motor tasks. In B. McMillan (Ed.), *Current trends in information theory.* Pittsburgh: Univer. Pittsburgh Press, 1953.

Fitts, P. M. The information capacity of the human motor system in controlling the amplitude of movement. *J. exp. Psychol.*, 1954, **47**, 381-391.

Fitts, P. M. Cognitive factors in information processing. In *Proceedings of the XVIIth international congress of psychology.* Amsterdam: North-Holland Publishing Company, 1964. Pp. 327-329.

Fitts, P. M., & Peterson, J. R. Information capacity of discrete motor responses. *J. exp. Psychol.*, 1964, **67**, 103-112.

Fitts, P. M., Peterson, J. R., & Wolpe, G. Cognitive aspects of information processing: II. Adjustments to stimulus redundancy. *J. exp. Psychol.*, 1963, **65**, 423-432.

Fitts, P. M., & Radford, B. K. Information capacity of discrete motor responses under different cognitive sets. *J. exp. Psychol.*, 1966, **71**, 475-482.

Garrett, H. E. A study of the relation of accuracy to speed. *Arch. Psychol., N. Y.*, 1922, No. 56.

Hick, W. E. On the rate of gain of information. *Quart. J. exp. Psychol.*, 1952, **4**, 11-26.

Howell, W. C., & Kreidler, D. L. Information processing under contradictory instructional sets. *J. exp. Psychol.*, 1963, **65**, 39-46.

Hyman, R. Stimulus information as a determinant of reaction time. *J. exp. Psychol.,* 1953, **45**, 188-196.

Klemmer, E. T., & Muller, P. F. The rate of handling information—key-pressing responses to light patterns. *USAF HFORL Memo. Rep.* 1954, No. 34.

McGill, W. J. Random fluctuations of response rate. *Psychometrika,* 1962, **27**, 3-17.

McGill, W. J. Stochastic latency mechanisms. In D. R. Luce, R. R. Bush, & E. Galanter (Eds.), *Handbook of mathematical psychology.* Vol. 1. New York: Wiley, 1963. Pp. 309-360.

Rapoport, A. A study of disjunctive reaction times. *Behav. Sci.,* 1959, **4**, 299-315.

Stone, M. Models for choice-reaction time. *Psychometrika,* 1960, **25**, 251-260.

Swets, J. A. Is there a sensory threshold? *Science,* 1961, **134**, 168-177.

Wald, A. *Sequential analysis.* New York: Wiley, 1947.

Chapter Six

Human Factors in Control Design

Research oriented directly toward machine design was first introduced in Chapter Two. In that chapter, the components of interest were *displays,* and the human capabilities of greatest relevance for their design were sensory and perceptual processes. Virtually everything that was said in that chapter applies to the present one as well, except that the machine components of interest are *controls.* The human capabilities toward which this work is directed still include receptive characteristics, but the emphasis is shifted toward system output functions—especially those reflecting perceptual-motor skill performance.

Very frequently, display and control design problems are attacked simultaneously because the two are known to be very closely interrelated. For example, we know that the way in which a control can be manipulated most proficiently is largely dependent upon the resulting effect which such movement has upon the element displayed: that is, certain display-control *relationships* yield better performance than others. A general term (*S-R compatibility*) has been coined to denote these relatively permanent interactive biases (Fitts and Seeger, 1953).

We can distinguish at least three general kinds of research falling within the scope of control design (in addition, of course, to the broad distinction made in Chapter Two between general-principle vs. specific-comparison studies). Each of these is illustrated in the present chapter by a representative selection.

The article by Bradley typifies the work directed toward the control identification problem: how can controls be designed so as to discourage people from grabbing the wrong one by mistake? A popular solution has been to make use of our sense of touch, giving each control a distinctive "feel." To do this properly, however, it was first necessary to find out through research just what constitutes "distinctiveness" from the point of view of our tactual sensors. Although this question could obviously be answered most completely by a thorough-going psychophysical attack upon the "dimensions" of tactual experience, the more expeditious approach of specific-comparisons has more often been adopted: e.g., which subset of a fairly arbitrary set of forms is least subject to confusion by experimental subjects? The Bradley study presented here represents something of a compromise between general-principle and specific-comparison research addressed to this problem.

The second kind of "control" study is that concerned with the dynamics of the device. Here the main question is: what effect do various physical charac-

---

References will be found in the General References at the end of the book.

teristics of the control (such as its mass, viscosity, elasticity, etc.) have upon a man's ability to manipulate it skillfully? As Bahrick's paper illustrates, this problem has generally been attacked at a relatively fundamental (general-principle) level, since such parameters constitute a major source of proprioceptive information needed for precise movement.

The third approach illustrated in this chapter is that involving compatibility of display-control relationships. What, it is asked, is the most compatible sort of control arrangement given a specific set of input or display characteristics? This problem almost of necessity demands the specific-comparison variety of research. The Chapanis and Mankin study is a good example of how a direct empirical attack upon such questions ought to be carried out.

For the reader seeking specific answers to control-design problems, we should point out that there exists a wealth of data bearing upon this issue. Again, many of these conclusions are summarized in two handbooks (Morgan et al., 1963; and Woodson, 1954), and also in Bennett, Degan, and Spiegel (1963) and McCormick (1964).

23

Tactual Coding of Cylindrical Knobs

James V. Bradley

INTRODUCTION

Pioneer experiments on tactual coding of knobs were designed to provide a set of tactually identifiable knobs, not to investigate parameters which might be exploited in the design of such knobs. Furthermore these earlier experiments were concerned exclusively with the tactual identifiability of the knobs investigated, not with their manipulability. Therefore, while these experiments have admirably met their limited objectives, they have nevertheless failed to supply certain data relevant to the general design of controls for optimal manipulability as well as tactual discriminability. The present experiment is intended to supplement this earlier research by supplying some of the data required for this purpose.

In the typical knob-coding experiment the knobs constituting the set have simply been an assortment of controls differing widely but unsystematically in shape. Supposing the knobs to be identified by the letters A to Z, the essential

From Wright Air Development Center Technical Report No. 59-182, September 1959.

element of the subject's task has generally been to feel a knob and then state or indicate whether it is some specified knob, say knob X, or "not X," i.e., some unspecified knob other than X in the set of knobs from A to Z. This procedure is open to three major objections which are discussed in the following paragraphs.

By investigating knobs differing unsystematically, one is investigating knobs as such rather than knob parameters. Results are peculiar to the specific shape used but cannot be ascribed to any particular attribute or feature constituting less than the totality of the knob's "shape." Thus the information extracted from the experiment cannot be directly applied to the design of knobs in general.

The response "not X" means "either Y, Z, or one of the knobs from A to W." Therefore, not only are the results for knob X peculiar to the totality of the shape of knob X, but they are also peculiar to the precise set of knob shapes which, in effect, constitute the alternative to the response "knob X."

Finally a high degree of tactual identifiability has generally been secured by resorting to some very bizarre and unknob-like shapes. While there is little doubt that a pyramidal knob is unlikely to be tactually confused with a spherical knob, there is serious question as to the appropriateness of the former for rapidly making precise settings to a fine tolerance. Knobs of noncylindrical shape are unlikely to be very efficient for this purpose. Knobs with sharp angles are likely to be inefficient when considerable torque is required to turn the shaft on which they are mounted; speed of operation is reduced because caution is required to grip a sharp surface tightly without injuring the fingers. Noncylindrical knobs, especially those with odd protrusions, are likely to be inadvertently operated unless they are part of a detented control. Thus tactual identifiability has been dearly bought at the expense of a multitude of the advantages associated with the more conventional control shapes.

The present experiment will attempt to remove these objections by introducing the following modifications of procedure. The investigation will be confined to knobs of an essentially cylindrical shape. Parameters, rather than knobs as such, will be investigated, the parameters being rim surface, diameter and thickness. Instead of being required to choose between the responses "knob X" and "not knob X," the subject will choose one of two specified knobs as his response, e.g., his alternatives might be "knob X" and "knob E."

The research to be reported consisted of a series of separate and self-contained experiments. Each parameter was investigated in a separate experiment in which the other two parameters were held constant. (A major portion of the experiment on rim surface, however, was repeated with a different constant knob diameter.) The effect of wearing gloves and the effect of time limitations on tactual discriminability were also investigated, the latter requiring an additional experiment. The various experiments were quite similar in regard to apparatus and procedure, the greatest deviations occurring in the experiment on time limitations which was more elaborate in these respects. Those elements of apparatus and procedure common to all experiments will therefore be reported once, in a general apparatus and procedure section, any deviations, modifications or additions being described in the section devoted to the experiment in which they occurred. Furthermore, the discussion of results of individual experiments will accompany those results in the section in which they are reported.

APPARATUS AND PROCEDURE

The subject was seated facing a "window" in the apparatus, his right hand resting before a curtained aperture. At the start of a trial the experimenter caused full-sized photographs of two alphabetically-labeled knobs to appear in the subject's window. At this point the subject reached into the curtained aperture and felt a knob which the experimenter had placed there. He had previously been told that the knob would be one of the two whose photographs were displayed in the window. When the subject believed he had felt the knob enough and was ready to respond, he called out the letter corresponding to whichever of the two knobs in his window he believed to be the knob he had felt.

The alphabetical labels were lettered in the white margin of the photographs, either above or below the photograph of the knob, the location of the identifying letter depending on the experiment. The photograph of the knob with the earlier alphabetical designation always appeared in the subject's window to the left of the other photograph, and the space between the two photographs varied with the number of knobs having intervening alphabetical designations. When rim surface or thickness was the parameter under investigation, the photograph showed only the side of the knob; i.e., none of the front or back face was visible. When diameter was the investigated parameter, compass-drawn circles of the appropriate diameter, inked on white paper and alphabetically labeled, were substituted for photographs of the knobs.

The knob to be felt by the subject was mounted in the aperture in the following manner. A long bolt served as the shaft. The bolt was inserted into the knob's shaft hole and the knob was then dropped down the length of the bolt, coming to rest on the bolt's head. The bolt was then inserted into a vertical slot in a block of wood in such a way that the knob was on one side of the wooden block and the experimenter's hand grasping the threaded end of the bolt was on the other. The wooden block was mounted inside the curtained aperture in such a way that the face of the knob was in the subject's frontal plane, i.e., so that the knob and the wooden block simulated a knob mounted on a vertical instrument panel in front of an operator. The experimenter pulled back on the bolt while the knob was being "presented" to the subject. Therefore there was no gap between the back face of the knob and the surface of the wooden block, and there was slight or no rotation of the knob when the subject made twisting movements while grasping or feeling the knob.

EXPERIMENTAL DESIGN

The subject's task in the present experiment represents an idealized choice situation and is subject to certain qualifications. By presenting the subject with *pictures* of the two knobs which constitute his alternatives of response, one is simulating the situation in which an operator is so familiar with his equipment that he retains in his memory a perfect visual image of the controls he must use. Furthermore limiting the response alternatives to two knobs introduces the implicit assumption that an operator using a number of controls may, when he feels one of them, consider the various alternatives a pair at a time without interference. This means that he is, in effect, making a succession of dichotomous choices eventually involving all possible pairings. Neither of these cases is entirely realistic to the practical situation to which inference must ultimately be extended. The same objection may be raised against most laboratory experiments. Nevertheless, it should be borne in mind that the present experiment is investigating a limiting case in which certain ideal conditions obtain and that its results may therefore show less frequent misidentification than would occur in a more practical situation. Against these considerations others must be weighed. The average instrument operator is undoubtedly less nervous and confused, better motivated, and more familiar with his task than is the average subject of a psychological experiment. In any event, using pictures of the knobs was deemed desirable in order to reduce learning effects which might accompany "remembered" alternatives of response. (Learning effects were, in fact, checked and found to be astonishingly slight, presumably due, at least in part, to the procedure in question.) And confining alternatives of response to two knobs permitted extension of results to choices between the knobs of that pair irrespective of the set in which the pair appeared, i.e., irrespective of the nature of other knobs in the tactually coded set.

In each experiment, if N knobs were being compared, the number of pairs of pictures which could be presented to the subject was the number of combinations of N things taken two at a time or N (N-1)/2. For each pair of pictures the felt knob could be either one of the two pictured knobs. Therefore there were N (N-1) possible experimental conditions, or choice situations. In each choice situation the subject chooses his response from two alternatives and would be expected to make the correct choice 50 per cent of the time if the felt knob provided no cues. Therefore statistical significance for frequency of correct choice can be tested by treating it as a binomial variate with parameter $p = \frac{1}{2}$. The binomial test, however, requires independent trials which, in the present case, means that no subject should perform more than once under a given choice situation. This restriction was observed; each subject performed one trial under each of the N (N-1) choice situations. Therefore the significance of the frequency of misidentification under a single choice situation can be obtained by entering binomial tables (7) with n equal to the number of subjects and $p = \frac{1}{2}$. Since complete data will be presented in all cases and since the test requires merely the consultation of probability tables, elaborate sets of significance levels will not be reported for the binomial tests.

The experimental design was a modification of a Latin Square design described elsewhere (2, 18) in which each of n experimental conditions is presented at a different point in sequence to each of n subjects, never being immediately preceded by the same one of the n-1 other experimental conditions. This design was used, unaltered, in those experiments in which the number of subjects equalled the number of experimental conditions. When there were half as many subjects as experimental conditions, the design was modified in such a way that if a pair of pictured knobs be regarded as an experimental condition, sequential effects were balanced. These modifications are described in the following paragraphs.

In the experiments investigating both barehanded and gloved operation, there were N (N-1) choice situations and N (N-1)/2 subjects. An N (N-1) by N (N-1) design was used with subject X running through all N (N-1) conditions barehanded, then becoming "subject" $\frac{N(N-1)}{2}$ + X and re-performing all N (N-1) trials with his hand gloved. For a given hand condition there were twice as many choice situations as subjects so each choice situation could not be presented once in each position in sequence. However, the number of pairs of pictures equalled the number of subjects and under a given hand condition at a given position in the sequence of condition presentation, each of the N (N-1)/2 subjects performed a condition corresponding to a different one of the N (N-1)/2 pairs of pictures.

In the experiment investigating diameter a subject's first $\frac{N(N-1)}{2}$ trials involved a choice situation corresponding to each of the $\frac{N(N-1)}{2}$ pairs of pictures, and his last N (N-1)/2 trials involved the remaining $\frac{N(N-1)}{2}$ choice situations from the N (N-1) choice situations. Again each pair of pictures was represented once in each of the positions in sequence, but not each choice situation.

Experiment I

Parameter Investigated. Rim surface at 2-inch diameter.

Subjects. 45 right-handed male college students. All Ss performed under the 90 choice conditions barehanded first, then performed the 90 conditions in a different order, wearing the MA-1 Double Flying Glove, i.e., a wool insert glove covered by a leather outer shell.

Knobs. Ten ½-inch-thick, 2-inch-diameter, cylindrical, aluminum-alloy knobs of which one was smooth-rimmed, three were fluted, three had a rectangular knurl and three had a diamond knurl. See Figure 1. The alphabetical designations of the various rim surfaces were as follows: A = smooth rim; B = fluted: six troughs (humps and troughs subtending equal chords ½ inch in length, trough depth, i.e., radial distance from bottom of trough to circumference of circle defined by the greatest radius of the knob, 16/128 inch); C = fluted: nine troughs (humps and troughs subtending equal chords 11/32 inch in length, trough depth 11/128 inch); D = fluted: 18 troughs (humps and troughs subtending equal chords 11/64 inch in length, trough depth 5/128 inch); E = full rectangular knurl; F = half rectangular knurl; G = quarter rectangular knurl; H = full diamond knurl; I = half diamond knurl; J = quarter diamond knurl.

Results. For the barehanded and glove condition respectively, Tables I and II give the number of subjects identifying the wrong picture as the knob felt. The row and column coordinates of a cell entry give the pair of knobs from whose pictures the subject must choose the knob felt. Columns give the knob felt and rows the other knob of the pair which, if the subject calls it, is the wrong response.

Figure 1. Knobs used in Experiment I. These photographs were the knob pictures shown to the subject.

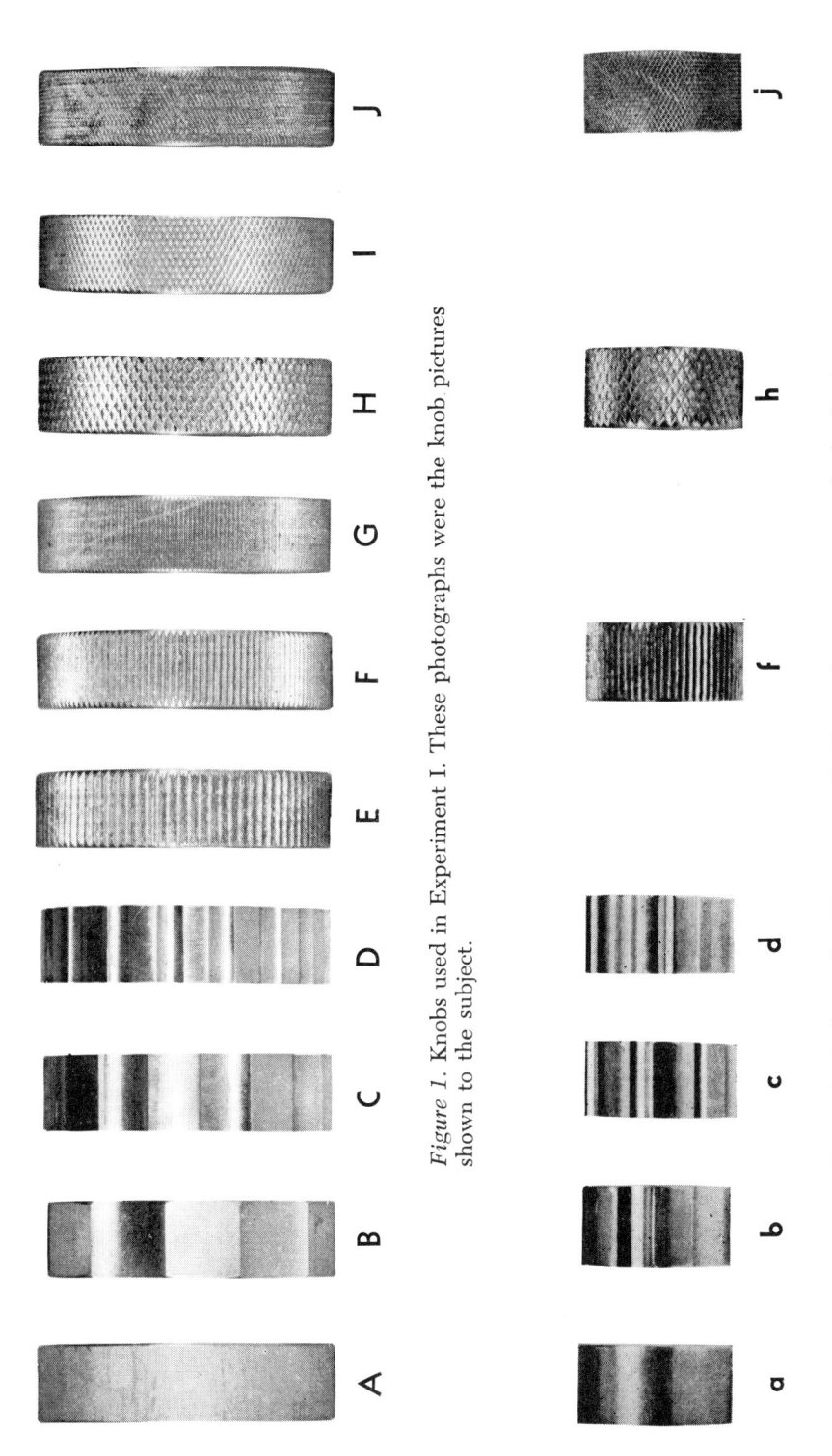

Figure 2. Knobs used in Experiment II. These photographs were the knob pictures shown to the subject.

TABLE I

Number of Times the Row Knob was Called when the Column Knob was Felt Barehanded

		\multicolumn{10}{c}{KNOB FELT}
		A	B	C	D	E	F	G	H	I	J
	A	X									
	B		X	6							
	C		3	X	1						
	D			6	X						
WRONG RESPONSE	E					X	11	1	4	1	
	F					8	X	7	8	5	6
	G					1	7	X		15	9
	H					6	7		X	1	2
	I					1	7	6	5	X	18*
	J						5	4	1	21*	X

* Does *not* differ significantly, at the one-tailed .05 level, from frequency expected by chance. Maximum possible cell entry = 45.

TABLE II

Number of Times the Row Knob was Called when the Column Knob was Felt Gloved

		\multicolumn{10}{c}{KNOB FELT}
		A	B	C	D	E	F	G	H	I	J
	A	X									
	B		X	1							
	C		11	X							
	D		2	7	X						
WRONG RESPONSE	E				1	X	10	8	18*	5	3
	F					14	X	17*	19*	8	13
	G					5	15	X	16	25*	22*
	H					16	14	8	X	2	8
	I					13	12	12	17*	X	20*
	J						8	13	11	26*	X

* Does *not* differ significantly, at the one-tailed .05 level, from frequency expected by chance. Maximum possible cell entry = 45.

TABLE III

Percentage of Trials Resulting in Misidentification when Felt and Alternative Knobs Belonged to Certain Family Types

Hand Condition	Rim Surface of Alternative Knob	Smooth	Fluted	Rectangular Knurl	Diamond Knurl
Bare	Smooth	X	0	0	0
	Fluted	0	5.93	0	0
	Rectangular Knurl	0	0	12.96	11.85
	Diamond Knurl	0	0	8.89	17.78
	Any of Above	0	1.32	5.84	7.90
Gloved	Smooth	X	0	0	0
	Fluted	0	7.78	0	0
	Rectangular Knurl	0	.25	25.56	31.85
	Diamond Knurl	0	0	23.70	31.11
	Any of Above	0	1.81	13.58	17.53

(Column group header: *Rim Surface of Felt Knob*)

TABLE IV

Percentage of Erroneous Responses in all Trials Involving a Given Pair of Pictured Knobs (Each Member of a Pair Serving as Felt Knob for Half of the Trials and as Alternative Knob for the Other Half)

Pair of Knobs from which Subject Must Choose	Bare	Gloved
Full Rectangular & ¼ Diamond	0	3
Full Rectangular & ¼ Rectangular	2	14
Full Diamond & ¼ Rectangular	0	27
Full Diamond & ¼ Diamond	3	21
Full Rectangular & ½ Diamond	2	20
Full Rectangular & ½ Rectangular	21	27
Full Diamond & ½ Rectangular	17	37
Full Diamond & ½ Diamond	7	21
One-half Rectangular & ¼ Diamond	12	23
One-half Rectangular & ¼ Rectangular	16	36
One-half Diamond & ¼ Rectangular	23	41
One-half Diamond & ¼ Diamond	43	51
Full Rectangular & Full Diamond	11	38
One-half Rectangular & ½ Diamond	13	22
One-fourth Rectangular & ¼ Diamond	14	39

(Column header: *Percentage of Choices Which were in Error when Hand was*)

TABLE V
Number of Times Row Knob was Called when Column Knob was Felt Barehanded

		a	b	c	d	f	h	j
				KNOB FELT				
	a	X						
	b		X	4	1			
	c		8*	X				
WRONG RESPONSE	d		2	15*	X		1	
	f				3	X	5	
	h					5	X	
	j					1		X

* Does *not* differ significantly, at one-tailed .05 level, from frequency expected by chance. Maximum possible cell entry = 21.

TABLE VI
Number of Times Row Knob was Called when Column Knob was Felt Gloved

		a	b	c	d	f	h	j
				KNOB FELT				
	a	X						
	b		X	2			1	
	c	1	8*	X				
WRONG RESPONSE	d		2	6	X		2	
	f				8*	X	8*	2
	h				9*	5	X	3
	j			1	5	5	9*	X

* Does *not* differ significantly, at one-tailed .05 level, from frequency expected by chance. Maximum possible cell entry = 21.

TABLE VII
Number of Subjects in each Experiment Making One or More Misidentifications in Comparisons Involving Various Knob Groups

			Number of Subjects Misidentifying	Number of Subjects Not Misidentifying	Chi-Square Sig. Level for Difference Between Two Groups
Comparisons between Nonfluted Knobs (i.e., comparisons within the groups A, F, H, J, or a, f, h, j)	Barehanded	Exp I	21	24	NS
		Exp II	8	13	
	Gloved	Exp I	38	7	NS
		Exp II	16	5	
Comparisons between Fluted Knobs (i.e., Comparisons within the groups B, C, D, or b, c, d)	Barehanded	Exp I	13	32	.001
		Exp II	19	2	
	Gloved	Exp I	19	26	NS (Pr ≅ .10)
		Exp II	14	7	
Comparisons between Fluted and Nonfluted Knobs (i.e., in which B, C or D is Compared with A, F, H, or J, or in which b, c, or d is compared with a, f, h, or j)	Barehanded	Exp I	21	24	NS
		Exp II	10	11	
	Gloved	Exp I	38	7	NS
		Exp II	19	2	

Table III shows the confusability within and between certain families of knobs defined by their general type of rim surface.

Table IV shows relative discriminability between pairs of knurled knobs as a function of depth and type of knurl.

Discussion. The smooth knob was never confused with any other knob, and the fluted knobs were practically never confused with nonfluted knobs. There was, however, considerable confusion between knobs with rectangular and diamond knurls. The amount of confusion between knobs with the same general type of rim surface was least for the fluted knobs, with knobs having rectangular knurl coming a poor second, followed by diamond knurl in third place. And in terms of general insusceptibility to misidentification in comparisons involving all of the knobs tested, the smooth knob was best, followed by the fluted, rectangular knurl, and diamond-knurl knob families.

When one knob had a full, and the other a quarter knurl, tactual discriminability appeared to be good at least when the subject was barehanded. However, confusion increased rapidly with diminishing differences in knurl depth. Discrimination between rectangular and diamond knurls of the same depth was moderately good when the subject was barehanded. But discriminability for knobs differing in both depth and type of knurl was not impressively better than for knobs differing in depth of knurl alone.

All subjects performed the first 90 trials barehanded and the second 90 gloved. The design therefore is not balanced for hand condition. Little, if anything, would have been gained if it had been, since there is scarcely any point in testing statistically the obviously true hypothesis that tactual discriminability of knob rims is impaired by wearing gloves. The data, as they stand, suggest that 90 learning trials do not compensate for the detrimental effect of gloves on tactual discriminability of knob rims.

The rim surfaces of the smooth or knurled knobs could be obtained on knobs of any practical diameter. However, in the case of the fluted knobs, a change in knob diameter necessarily results either in a change in the number of troughs, or in the width of the trough, if humps and troughs are equated as to the chord subtended. And one would expect that either or both of these factors serve as tactual cues in discriminating between knobs. It is therefore desirable to check the tactual identifiability of the fluted knobs, with reference to one another, as well as with nonfluted knobs at a different diameter. This was done in Experiment II, to follow. The diameter used in the present experiment is very nearly the optimal knob diameter (3); however, it is infrequently used in aircraft because of space limitations. In the second experiment a more typically employed diameter was investigated. Another purpose was served by Experiment II. It replicated certain comparisons between nonfluted knobs investigated in Experiment I and therefore provided a test of the assumed irrelevance of the diameter at which these comparisons occur.

Experiment II

Parameter Investigated. Rim surface at 1-inch diameter.

Subjects. 21 right-handed male college students. All Ss performed under the 42 choice conditions barehanded first, then performed the 42 conditions, in a different order, wearing the MA-1 Double Flying glove.

Knobs. Seven ½-inch-thick, 1-inch-diameter, cylindrical, aluminum-alloy knobs were used. The alphabetical designations of the knobs and their various rim surfaces were as follows (see Figure 2): a = smooth rim; b = fluted: six troughs (humps and troughs subtending equal chords which were ¼ inch in length, trough depth $3/128$ inch); c = fluted: nine troughs (humps and troughs subtending equal chords which were $11/64$ inch in length, trough depth $6/128$ inch); d = fluted: 18 troughs (humps and troughs subtending equal chords $11/128$ inch in length, trough depth $3/128$ inch); f = half rectangular knurl; h = full diamond knurl; j = quarter diamond knurl.

Knobs a, f, h, and j had rim surfaces identical to those of knobs A, F, H, and J, respectively, in the preceding experiment investigating rim surface with 2-inch diameter knobs. Knobs b, c, and d differed from knobs B, C, and D, respectively, in that their troughs were not as deep and subtended chords one-half as long as the troughs in the larger knobs; the number of troughs, however, was the same for knobs designated by upper case and lower case of the same letter.

Results. Tables V and VI give results analogous to those presented in Tables I and II for Experiment I.

Table VII compares the pattern of response in Experiment II with that in Experiment I for the analogous choice situation. The response patterns are compared for choices between fluted knobs, between nonfluted knobs, and between fluted and nonfluted knobs. There was a significant difference in response patterns between the two experiments only when fluted knobs were compared with one another. This supports the hypothesis, enunciated in Experiment I, that tactual identifiability of fluted knobs is a function of knob diameter but that this is not the case for nonfluted knobs.

Discussion. Results again tend to break down into the categories found in Experiment I. With a single exception the smooth knob was confused with no other knobs. The fluted knobs tended to be confused mainly with each other, as did the nonfluted knobs. However, this tendency was less clear cut than in Experiment I. Knob d, the fluted knob with 18 troughs, each less than $1/11$ inch wide, was so "finely ribbed" that it resembled a knob with full rectangular knurl. Thus it tended to be confused with the knurled knobs as if it were part of the knurled family of knobs.

Knob D in Experiment I, when compared with the nonfluted knobs, A, F, H, and J, was never misidentified. Knob c in the present experiment was misidentified only once (by a gloved subject) in similar comparisons with nonfluted knobs a, f, h, and j. Since the breadth of trough was the same for the two knobs, this result suggests that breadth of trough may be the critical cue in tactual discriminations involving fluted knobs and that this factor may be independent of diameter.

Experiment III

Parameter Investigated. Rim surface with feeling-and-response time limited.

Subjects. 20 right-handed college students of both sexes.

Knobs. Five ½-inch-thick, 1-inch-diameter knobs, differing in rim surface, were used in this experiment. They were knobs a, c, d, f, and h described in Experiment II and are reported with these alphabetical designations. In the actual running of the experiment these knobs were designated and labeled A, E, I, O, and R, respectively.

Procedural Variations. In the preceding experiments the subject had been given unlimited time in which to feel the knob, decide which of the two pictured knobs was being felt, and call out his conclusion. The present experiment consisted of four parts. In one part, time was unlimited; in the other three parts the subject was required to call out his response within one, two or three seconds of the moment when he first touched the felt knob.

The subject was biassed with 45 volts d.c. by means of a wire from a battery to a bicycle clip worn on the exposed surface of his right arm. The felt knob was the "touchplate" of a thyratron touchplate circuit. (This was accomplished by putting a metal base in the wooden slot which held the knob shaft; the metal base was connected to the grid of the thyratron tube.) When the subject touched the felt knob he began the trial, initiating the following sequence of operations: a red light goes on in front of the subject and remains on for either 0.2, 1.2, or 2.2 seconds, at the end of which time the red light goes off and a green light comes on and remains on for 0.8 seconds; when the green light goes off, a large white light goes on and a time clock starts. The subject was instructed that he should be feeling the knob and deciding upon his response while the red light was on, delaying his response, if necessary, until the onset of the green light, at which point he should respond within 0.8 seconds, i.e., while the green light was on. (Many subjects responded while the red light was on, despite instructions to the contrary.) He was further instructed that if he failed to respond before the green light went off, the white light would light and a time clock would begin to record, both remaining on until a response was given. It was made clear to the subject that failure to respond before the green light went off was regarded by the experimenter as an error. When the subject responded he spoke loudly into a microphone suspended immediately in front of him. The microphone was part of the circuitry for a voice key, and the subject's response brought the trial to an end, extinguishing any signal lights which were lit (and preventing the onset of any others later in the sequence), and stopping the time clock if it was recording. The panel of signal lights in front of the subject was duplicated by a similar panel, wired in parallel, in front of the experimenter. In order to insure operation of the voice key at the moment the subject starts to respond, the five knobs were labeled with the letters A, E, I, O, and R, each of whose sounds start with a vowel and do not require elaborate manipulations of lips or tongue in preparation for the initial sound.

Results. Results are shown in Tables VIII, IX and X. Wilcoxon tests for matched pairs were conducted upon the total number of wrong responses, and upon the total number of responses exceeding the time limit, under each meaningful pair of time conditions. (If there were Z [zero] differences, each was given the midrank, $\frac{Z+1}{2}$, and half of them were regarded as having a plus sign, half a minus sign.) The 31 wrong responses at the one-second time limit significantly exceeded the number recorded under any other time limit.

Likewise the 39 over-time responses at the one-second time limit significantly exceeded the number recorded at either the two- or three-second limits. However, when both time limits compared were two seconds or greater, there were no significant differences in either wrong or over-time responses. Probabilities were such that all of the above statements are valid regardless of whether the one-tailed or the two-tailed .05 level of significance is used.

Discussion. It seems clear that an operator's ability to discriminate tactually between rim-coded knobs is impaired when he is allowed only one second in which to feel the knob and choose between two alternatives. He obviously does better at the two-second than at the one-second limit. However, extending the time limit from two to three seconds does not seem to benefit him appreciably. The nonsignificant increase in discriminability when the three-second time limit is extended to "infinity," therefore, seems explainable, if it is real, as being due to removal of a source of tension, rather than to the actual usefulness, or need for time beyond three seconds.

Experiment IV

Parameter Investigated.   Diameter.

Subjects. The 21 right-handed male college students run in Experiment II were run in the present experiment immediately upon completion of Experiment II. There was an interval of only a minute or so between experiments.

Knobs. Seven ½-inch-thick, smooth rimmed, cylindrical, aluminum-alloy knobs ranging in diameter from ½ inch to 1¼ inches in steps of ⅛ inch. The alphabetical designations of the knobs of various diameters were as follows: a = ⁴⁄₈, b = ⅝, c = ⁶⁄₈, d = ⅞, f = ⁸⁄₈, h = ⁹⁄₈, j = ¹⁰⁄₈.

Results. Results are shown in Tables XI and XII. Knobs differing in diameter by as much as ½ inch were never confused.

Discussion. Since the subjects used in the present experiment had just finished Experiment II, they had had, in effect, 84 learning or practice trials with a one-inch-diameter knob. This concentration of practice upon a single one of the seven diameters tested, however, has not biassed the results in any way which is obvious from the data. The results for the one-inch-diameter knob appear to be in harmony with those for the adjacent, less practiced, knobs.

TABLE VIII

Number of Times Row Knob was Called when Column Knob was Felt under Four Time Limitations

Time Allowed for Feeling Knob and Making Response	Wrong Response	a	c	Felt Knob d	f	h
	a	X		1		
	c	1	X	2		
1 sec.	d		5	X	1	5
	f		1	4	X	5
	h			2	4	X
	a	X				
	c		X	1	1	
2 sec.	d	1		X		
	f			3	X	5
	h			1	3	X
	a	X				
	c		X	1		
3 sec.	d		2	X		
	f	1		2	X	3
	h				5	X
	a	X				
	c		X			
Unlimited	d		3	X		
	f				X	2
	h				1	X

All cell frequencies differ significantly (at two-tailed .05 level) from frequency expected by chance. Maximum possible cell entry = 20.

TABLE IX
Number of Subjects who Failed to Respond within the Time Allowed and (in Parentheses) Number of these Subjects whose Late Response was the Incorrect One

Time Allowed for Feeling Knob and Making Response	Other Knob of Pictured Pair	a	c	Felt Knob d	f	h
1 sec.	a	X		3		1
	c	2	X	3	2	1
	d	1	4(1)	X	2	3
	f	1	3(1)	2	X	5(2)
	h	1	3		2	X
2 sec.	a	X				
	c		X	2	1	1
	d			X		
	f			1(1)	X	
	h			1(1)	2	X
3 sec.	a	X			1	
	c		X			
	d			X		
	f			1	X	
	h	1			3(2)	X

All cell frequencies outside parentheses differ significantly (at two-tailed .05 level) from frequency expected by chance. Maximum possible cell entry = 20.

TABLE X
Summary Data

	Number of Responses Which Were			Percentage of Responses Which Were		
Time Limit	Wrong	Overtime	Both	Wrong	Overtime	Both
1 sec.	31	39	4	7.8	9.8	1.0
2 sec.	15	8	2	3.8	2.0	.5
3 sec.	14	6	2	3.5	1.5	.5
Unlimited	6	—	—	1.5	—	—

	Number of Subjects Making One or More Responses Which Were			Percentage of Subjects Making One or More Responses Which Were		
Time Limit	Wrong	Overtime	Both	Wrong	Overtime	Both
1 sec.	15	16	3	75	80	15
2 sec.	9	6	1	45	30	5
3 sec.	11	4	2	55	20	10
Unlimited	5	—	—	25	—	—

TABLE XI
Number of Times Row Knob was Called when Column Knob was Felt

		\multicolumn{7}{c}{FELT KNOB}	
		a	b	c	d	f	h	j	
	a	X	9*	3	1				4/8
	b		X	3	4	1			5/8
	c		1	X	10*	3			6/8
WRONG RESPONSE	d		1	1	X	3	2		7/8
	f				3	X	4	3	8/8
	h					1	X	4	9/8
	j						4	X	10/8
		4/8	5/8	6/8	7/8	8/8	9/8	10/8	

* Does *not* differ significantly, at one-tailed .05 level, from frequency expected by chance. Maximum possible cell entry = 21.

TABLE XII
Percentage of Trials Resulting in Misidentification for Various Diameter Differences between the Compared Knobs

Size of Alternative Relative to Felt Knob	\multicolumn{6}{c}{Diameter Difference between Alternative and Felt Knobs}
	1/8	2/8	3/8	4/8	5/8	6/8
Smaller	26.2	14.3	2.4	0	0	0
Larger	7.9	1.0	0	0	0	0

TABLE XIII
Number of Times Row Knob was Called when Column Knob was Felt

		FELT KNOB					
		A	B	C	D	E	
WRONG RESPONSE	A	X	2				3/8
	B	2	X		1		4/8
	C	1	7*	X			5/8
	D		4	5	X	1	6/8
	E			1	8*	X	7/8
		3/8	4/8	5/8	6/8	7/8	

* Does *not* differ significantly, at one-tailed .05 level, from frequency expected by chance. Maximum possible cell entry = 20.

TABLE XIV
Percentage of Trials Resulting in Misidentification for Various Differences in Thickness between the Knobs Compared

Size of Alternative Relative to Felt Knob	Thickness Difference between Alternative and Felt Knob			
	1/8	2/8	3/8	4/8
Smaller	3.8	1.7	0	0
Larger	27.5	10.0	0	0

Since the obtained percentage of errors at a given difference in diameter does not appear to vary appreciably with the diameter of the felt knob (at least for the range of diameters investigated), the percentages have been presented, independently of diameter, in Table XII. Had both knobs of a pair been felt, one would expect errors to be correlated negatively with the ratio of diameter difference to the absolute diameter of the first knob felt. However, since this classical procedure was not followed, one need not necessarily expect errors to correlate better with the Weber fraction than with diameter increment.

Some subjects reported that they used the distance between the knob rim and the head of the bolt which acted as knob shaft as a cue to the absolute diameter of the felt knob. The extent to which this influenced results is unknown. However it seems doubtful that discriminability would be radically changed if subjects had to gauge diameter by enclosing the knob with their fingers rather than by feeling the distance-difference between two radii.

Experiment V

Parameter Investigated. Thickness.

Subjects. 20 right-handed college students of both sexes.

Knobs. Five 1-inch-diameter, smooth-rimmed, cylindrical, aluminum-alloy knobs ranging in thickness from ⅜ inch to ⅞ inch in ⅛ inch intervals. The alphabetical designations of the knobs of various thicknesses were as follows: A = ⅜, B = 4/8, C = ⅝, D = 6/8, E = ⅞.

Results. Results are shown in Tables XIII and XIV. Knobs differing in thickness by as much as ⅜ of an inch were never confused.

Discussion. Again the obtained percentage of errors at a given difference in thickness does not appear to vary appreciably with the absolute thickness of the felt knob. This data is therefore presented in Table XIII independently of absolute knob thickness.

GENERAL DISCUSSION

The preceding results have shown that efficient tactual coding can be obtained, without resorting to bizarre shapes, by utilizing certain parameters of cylindrical knobs. Knobs were rarely confused when they belonged to different ones of the rim surface families: smooth, fluted, knurled; nor when they differed in diameter by as much as ½ inch or in thickness by as much as ⅜ inch. Using the above criteria, one could construct 18 tactually identifiable knobs of very reasonable dimensions. Half might be ⅜ inch thick and half ¾ inch thick. For each nine knobs of the same thickness, one-third might have a ¾-inch diameter, one-third a 1¼, and one-third a 1¾-inch diameter. And for each such set of three knobs, three different rim surfaces, one smooth, one fluted and one knurled, might be used. One would conjecture that, unless under extraordinary stress, an operator familiar with the knobs would have little difficulty in mentally scanning the three parameters and the three, or two, gradations of difference along each, as would presumably be necessary to make the correct choice. It is true that the intellectual complexity of the operator's task would be greater than that investigated in the preceding experiments; however, it would scarcely have reached frightening proportions.

The example given in the preceding paragraph intentionally used gradations of sufficient breadth to bring the probability of misidentification close to zero. However, there are situations which would call for a different approach. For example, a design engineer might need only a few tactually identifiable knobs but might have severe limitations as to their maximum dimensions. Suppose that because of space limitations and torque requirements the knobs must range between ¾ inch and 1 inch in diameter, ½ inch and ¾ inch in thickness, and must have a full diamond knurl. If four coded knobs are needed, an obvious solution would be to have two knobs at each extreme of diameter, each knob of the same diameter having a different one of the extremes of tolerable thickness. But difficulty is encountered when one attempts to estimate the relative frequency of misidentification. Table XI shows that knobs ¾ inch and 1 inch in diameter were confused in 6/42 or 14.3% of their comparisons, and Table XIII gives 5/40 or 12.5% as the relative frequency of misidentification for knobs ½ inch and ¾ inch thick. If these figures were the true population percentages of misidentification, one would expect, with justification, that when two knobs differ in both respects, the frequency of misidentification should not exceed 12.5%. However it is not altogether clear how much, if any, the presence of an additional, weaker, tactual cue improves the discriminability of the knobs. What is certain is that one does *not* obtain the correct figure by multiplying the relative frequencies .143 and .125, and design engineers are cautioned against this tempting but entirely fallacious approach. (Skeptics may test its validity by performing some simple calculations on the material presented in Table IV.) Certain mathematical models for prediction of discriminability, in the multiple cue case, on the basis of discriminability of the individual component cues, have been tested empirically. It is intended that the results of these investigations be presented in a subsequent article.

REFERENCES

1. Biel, W. C., Eckstrand, G. A., Swain, A. D., & Chambers, A. N. *Tactual discriminability of two knob shapes as a function of their size.* WADC Technical Report 52-7, January 1952.
2. Bradley, J. V. Complete counterbalancing of immediate sequential effects in a Latin Square design. *J. Amer. Statist. Ass.*, 1958, 53, 525-528.
3. Bradley, J. V., & Arginteanu, J. *Optimum knob diameter.* WADC Technical Report 56-96, ASTIA Document No. 110549, November 1956.
4. Brennan, T. N. N., & Morant, G. M. *Selection of knob shapes for radio and other controls.* Flying Personnel Research Committee Report FPRC 702(a), R.A.F. Institute of Aviation Med., January 1950.
5. Fitts, P. M. *Analysis of factors contributing to 460 "Pilot Error" experiences in oper-*

ating aircraft controls. Army Air Forces, Headquarters, Air Materiel Command, Engineering Division, Memo. Rep. TSEAA-694-12, July 1947.
6. Harter, H. L., & Whiteley, B. R. *Compatibility and identifiability of aircraft control knobs.* Technical Note WCRR-54-3, March 1954.
7. Harvard University, Staff of the Computation Laboratory. *Tables of the cumulative binomial probability distribution.* Cambridge, Mass.: Harvard Univ. Press, 1955.
8. Hunt, D. P. *The coding of aircraft controls.* WADC Technical Report 53-221, August 1953.
9. Hunt, D. P., & Craig, D. R. *The relative discriminability of thirty-one differently shaped knobs.* WADC Technical Report 54-108, December 1954.
10. Jenkins, W. O. *A follow-up investigation of shapes for use in coding aircraft control knobs.* Army Air Forces, Headquarters, Air Materiel Command, Engineering Division, Memo. Rep. No. TSEAA-694-4A, August 1946.
11. Jenkins, W. O. *A further investigation of shapes for use in coding aircraft control knobs.* Army Air Forces, Headquarters, Air Materiel Command, Engineering Division, Memo. Rep. No. TSEAA-694-4B, September 1946.
12. Jenkins, W. O. *Investigation of shapes for use in coding aircraft control knobs.* Army Air Forces, Headquarters, Air Materiel Command, Engineering Division, Memo. Rep. No. TSEAA-694-4, August 1946.
13. Jones, R. E. *A survey of pilot preference regarding knob shapes to be used in coding aircraft controls.* Army Air Forces, Headquarters, Air Materiel Command, Engineering Division, Memo. Rep. TSEAA-694-4E, February 1947.
14. Mainland, D., Herrera, L., & Sutcliffe, Marion. *Tables for use with binomial samples.* New York: New York Univ. College of Med., Dept. of Med. Statist., 1956.
15. Whittingham, D. G. V. *Experimental knob shapes.* Flying Personnel Research Committee Report FPRC 702, R.A.F. Institute of Aviation Med., September 1948.
16. Wilcoxon, F. Individual comparisons by ranking methods. *Biometrics,* 1945, **1**, 80-83.
17. Wilcoxon, F. Probability tables for individual comparisons by ranking methods. *Biometrics,* 1947, 3, 119-122.
18. Williams, E. J. Experimental designs balanced for the estimation of residual effects of treatments. *Australian J. Scientific Research, Series A,* 1949, **2**, 149-168.

24

An Analysis of Stimulus Variables
Influencing the Proprioceptive Control of Movements

Harry P. Bahrick

It is generally known that accurate execution of movements depends upon proprioceptive information reaching the central nervous system. Clinical evidence (7, p. 235) as well as experimental findings (8) indicates that control and perception of movements are very poor when this sensory channel is not functioning.

Little is known, however, about the specific characteristics of proprioceptive stimulation that permit the individual to control changes in position, rate, or acceleration of his limbs. In other words, no detailed theories of proprioception comparable to the specific theories available for some other sense modalities have been developed (7, p. 234). Most of the available knowledge in this area is based upon anatomical investigations of the receptor system, its neural connections, and its central representations. Although several types of receptors have been identified (13, p. 1185; 14), differentiation of their function is as yet not clearly established. It is thought that forces internal to the body act as proprioceptive stimuli, but the processes by which these stimuli are encoded into messages which ultimately form the basis for perception and control of movements are not well understood (7, p. 234).

Behavioral data specifying the relations between stimulus and response characteristics have been difficult to obtain because of problems of controlling proprioceptive stimuli. Investigators have used drugs or faradic currents (8, 9) as means of reducing the effectiveness of proprioceptive stimuli. Recently, an indirect approach to this problem has been attempted. This approach consists of

From *Psychological Review*, 64, 1957, 324-328. Copyright 1957 by the American Psychological Association, and reproduced by permission.

This research was supported in part by the United States Air Force under Contract No. AF 41(657)-70 monitored by the Operator Laboratory, Air Force Personnel and Training Research Center, Randolph Air Force Base, Texas. Permission is granted for reproduction, translation, publication, use, and disposal in whole and in part by or for the United States Government.

Work contributing to this theory was initiated in the Laboratory of Aviation Psychology of the Ohio State University in 1953, and is now being continued there. For the past year the author has continued research contributing to this theory at Ohio Wesleyan University through support from the National Science Foundation.

varying the type and degree of resistance to motion offered by a control which S uses in the execution of movements. The effect of this variation upon S's ability to perceive and control his movements is studied, and an attempt is made to infer characteristics of the proprioceptive system. As a technique of investigating proprioception, this approach has obvious limitations. The forces which S applies to move a control are only indirectly related to the proprioceptive stimulation he receives during the execution of the movement. The cutaneous senses are also stimulated during movement, and unknown transformations are involved between the control force acting upon the limb and the proximal stimuli acting upon receptors in muscles, tendons, or joints.

Despite these substantial limitations, the approach has some theoretical as well as practical advantages. The forces required to move a control, and thus also the control forces acting upon the limb, can be specified as a function of four physical properties of the control, and these properties can all be regulated conveniently by E. They are mass, viscosity, elasticity, and the degree of coulomb friction. In a control such as that used by Howland and Noble (11) these parameters combine according to the following time-varying system equation:

$$L_t = K\theta + B \, d\theta/dt + J \, d^2\theta/dt^2, \tag{1}$$

where the left-hand side of the equation is the force applied by the human arm, the right-hand side represents the component resistive forces offered by the external control. L_t is the torque required to move the control at any instant of time (t), K is the constant of elasticity of the control, B is the viscosity constant, J is the moment of inertia, and θ is the angular displacement of the control with respect to its neutral, or spring-centered, position (6). Coulomb friction has been neglected in this equation, as have the internal resistive forces in the limb itself.

It has already been pointed out that the force which the control exerts upon the limb may not be equated to proprioceptive stimuli. However, one may assume that the forces which do act as proprioceptive stimuli during the movement of limbs are determined by physical properties of our limbs analogous to those specified in the above equation for the control. Previous investigation (4) has already established some of these physical properties of limbs and their significance in relation to the control of movements. These physical properties of limbs are difficult to control, and the present approach attempts to infer their function in proprioception by studying the effects of analogous characteristics of controls where these properties can be manipulated conveniently.

From an applied viewpoint, this approach may be useful in that the data are relevant to the solution of human engineering problems related to the design of controls used in man-machine systems, or to the design of prosthetic devices.

In the present article some general hypotheses are developed about the effect of each of the physical control parameters specified in Equation 1, and data are reported which test these hypotheses.

Inspection of Equation 1 shows that the torque needed to move the control depends upon its position, rate, and angular acceleration, the relative importance of these depending upon the respective values of the elasticity, damping, and inertia constants. Thus, if the elasticity constant K is zero, the torque required to move the control will be independent of its position, but if K is relatively large, the torque will vary largely as a function of position. Analogous relations exist between the damping constant B and angular velocity, and between the moment of inertia and angular acceleration.

It is now hypothesized that a man can use the force cues obtained in moving the control to improve his perception of position, rate, and acceleration of limb motion. Specifically, it is hypothesized that the elasticity constant of the control improves S's ability to perceive and control positions, the damping constant improves perception and control of rate, and the moment of inertia improves the perception and control of acceleration. Thus, an increase in each of these control constants should lead to improvement in the corresponding behavior. At the same time, it is hypothesized that an increase in any of the control constants will affect adversely performance which is aided by the other constants. Thus, increases in K are expected to interfere with the control and perception of rate and acceleration, while increases in B and J will affect adversely the control and perception of position. This hypothesis suggests itself, since the force required to move the control would not be expected to provide useful cues for the control of rate if it changes rapidly with position, and, conversely, it should not offer useful cues for the detection of position if it varied greatly as a function of rate or acceleration.

Several experiments have been conducted in which the accuracy of movement was studied as a function of the physical characteristics of control (1, 2, 3, 10, 11). A few of these (1, 2, 10) were designed specifically to test the above predictions. In one study (2) Ss performed simple circular and triangular control motions with a joystick control which was loaded with various degrees of spring stiffness, or damping, or mass. In each control-loading condition, the movements were first practiced with the help of a visual guide and paced by means of a metronome. The visual and auditory guides were then removed, and Ss were instructed to reproduce the motions as accurately as possible. Photographic records of all motions were obtained and measured for accuracy of temporal and spatial reproduction. It was found that an increase of viscous damping or of inertia of the control resulted in greater uniformity of speed within individual motions, and also in greater uniformity of speed in successive reproductions of the same motion. In the case of the triangular motions, increased mass and increased damping led to greater uniformity of peak velocity on each side of the triangle on successive trials. Spring loading interfered with the control of rate and acceleration, but its effect upon spatial accuracy of the reproduced motion was, in general, not significant. It was suggested that extended practice is needed for effective utilization of cues provided by spring loading.

This hypothesis was checked in a second experiment (1) in which the accuracy of positioning a horizontal arm control was investigated as a function of changes in the torque-displacement relation of the control. Extended practice was given and knowledge of results was provided. It was found that positioning errors are smallest when the ratio of relative torque change to displacement is largest. Under optimum conditions of spring loading, average positioning errors were less than half the amount obtained for a control which was not spring loaded. It was concluded that force cues provided by a spring-loaded control can improve the accuracy of positioning a control, and that the amount of improvement is a function of the relative and absolute torque change per unit of amplitude change.

Further investigation [1] of the usefulness of force cues in regulating the amplitude of motion has supported the above conclusions. It was shown that the transmission of amplitude information can be increased significantly by spring loading the control used by Ss. Optimum results were obtained with a control which provided geometric increments of force as a function of arithmetic changes of amplitude. This condition provides force cues which are equally discriminable over the range of amplitudes employed (12, 15), and yields the largest number of absolutely discriminable categories of amplitude response.

Although the above results support the general hypotheses regarding the effects of K, B, and J constants upon the control of movements, many questions remain unanswered. In order to establish that the observed effects are due to changes in proprioceptive stimulation, it will be necessary to control cutaneous sensitivity. It is hypothesized here that the contribution of cutaneous receptors is most significant in relation to minute manipulatory responses, and least significant for larger movements of the type dealt with here.

Further problems arise because the control parameters under discussion have certain mechanical effects upon the nature of movements, and these must be separated from the effects upon proprioceptive stimulation. Large amounts of damping, for example, make rapid movements difficult and fatiguing, and greater uniformity of movement rate observed under these conditions may reflect mechanical effects rather than improved proprioceptive discrimination. The identification of these mechanical effects becomes more difficult when continuous movements are dealt with, as was shown in the study by Howland and Noble (11). Interactions among the physical parameters of controls may cause complex mechanical effects such as oscillation, and these may obscure or counteract the effects due to augmented proprioceptive stimulation. In general, the analysis of K, B, and J effects is relatively simple for discrete, adjustive movements of the type primarily dealt with so far, but becomes increasingly complicated for complex or continuous motions.

Work now in progress attempts to establish relations between the forces exerted upon a control, and intensity of stimulation at the receptors in the elbow. This analysis is based upon a simplified mechanical model of the arm (16) by means of which forces acting upon the hand are resolved at the elbow joint (5, p. 319). In this manner it may become possible to infer changing intensities of stimulation of receptors at the joint during the course of movements.

Ultimately, the development of proprioceptive theory described here must be supported by a more direct analysis of K, B, and J factors within the body, and their effects upon the perception and control of movements. This, in turn, will require a better understanding of the biophysical principle by which forces internal to the body are brought to bear upon proprioceptive receptors.

REFERENCES

1. Bahrick, H. P., Bennett, W. F., & Fitts, P. M. Accuracy of positioning responses as a function of spring loading in a control. *J. exp. Psychol.*, 1955, **49**, 437-444.
2. Bahrick, H. P., Fitts, P. M., & Schneider, R. The reproduction of simple movements as a function of proprioceptive feedback. *J. exp. Psychol.*, 1955, **49**, 445-454.

[1] Bahrick, H. P. Force cues and the control of movement amplitudes. (In preparation.)

3. Derwort, A. Ueber die Formen unserer Bewegungen gegen verschiedenartige Widerstaende und ihre Bedeutung fuer die Wahrnehmung von Kraefton. *Z. f. Sinnesphysiol.*, 1943, **70**, 135-183.
4. Fenn, W. O. The mechanics of muscular contraction in man. *J. appl. Physics,* 1938, **9**, 165-177.
5. Fick, R. *Handbuch der Anatomie und Mechanik der Gelenke.* Part II. Jena: Verlag von Gustav Fischer, 1910.
6. Fitts, P. M. Engineering psychology and equipment design. In S. S. Stevens (Ed.), *Handbook of experimental psychology.* New York: Wiley, 1951. Pp. 1287-1340.
7. Geldard, F. A. *The human senses.* New York: Wiley, 1953.
8. Goldscheider, A. Untersuchungen ueber den Muskelsinn. I. Ueber die Bewegungsempfindung. In A. Goldscheider, *Gesammelte Abhandlungen,* Vol. II. Leipzig: Barth, 1898.
9. Goldscheider, A. Untersuchungen ueber den Muskelsinn. II. Ueber die Empfindung der Schwere und des Widerstandes. In A. Goldscheider, *Gesammelte Abhandlungen,* Vol. II. Leipzig: Barth, 1898.
10. Helson, H., & Howe, W. H. Inertia, friction, and diameter in handwheel tracking. OSRD Rep. No. 3454, 1943. (PB 406114.)
11. Howland, D., & Noble, M. E. The effect of physical constants of a control on tracking performance. *J. exp. Psychol.,* 1953, **46**, 353-360.
12. Jenkins, W. L. The discrimination and reproduction of motor adjustment with various types of aircraft controls. *Amer. J. Psychol.,* 1947, **60**, 397-406.
13. Jenkins, W. L. Somesthesis. In S. S. Stevens (Ed.), *Handbook of experimental psychology.* New York: Wiley, 1951. Pp. 1172-1190.
14. Mathews, B. H. C. Nerve endings in mammalian muscle. *J. Physiol.,* 1933, **78**, 1-53.
15. Noble, M. E., & Bahrick, H. P. Response generalization as a function of intratask response similarity. *J. exp. Psychol.,* 1956, **51**, 405-412.
16. White, H. E. *Modern college physics.* New York: Van Nostrand, 1948.

25

Tests of Ten Control-Display Linkages

Alphonse Chapanis and Donald A. Mankin

INTRODUCTION

The idea for this study came from installations involving a certain type of new equipment. The control panels for this equipment all contain four controlled ele-

From *Human Factors,* 9, 1967, 119-126. By permission of the Human Factors Society.

The work reported in this article was done under Contract Nonr-4010(03) between the Office of Naval Research and The Johns Hopkins University. This is Report No. 15 under that contract. Reproduction in whole or in part is permitted for any purpose of the United States Government.

ments arranged in a square, and four associated controls arranged in a column to the right of the controlled elements. As they are now being installed, these equipments have different linkages between the controls and their controlled elements, depending on what manufacturer made the equipment. Although we were motivated by this specific kind of equipment, we have abstracted from that particular situation and have asked two more general questions: For vertically-mounted panels of the type illustrated in Figure 1, are some linkages between controls and displays better than others? Further, what are the general design principles that hold for panels of this type?

An arrangement of displays in the form of a square with their associated controls in a vertical column (as illustrated in Figure 1) is undoubtedly not the best that could be designed for installations of this type. Our study makes no attempt to find the *best* arrangement of four controls and four displays. It starts instead with the assumption that, for some reason, the displays and controls must be arranged as shown here. Given this constraint, we have tried only to find out whether some linkages between controls and displays are better than others and what design characteristics distinguish these linkages from each other.

Of the 24 possible linkages between the four controls and four displays we discarded some as being so irregular and unsystematic that we could not conceive of them being used commercially. From the remaining ones we picked ten that either (*a*) might conceivably be used, or (*b*) allowed us to test specific design principles that will be discussed later. It is perhaps worth noting that at least five of the ten linkages we tested are in fact being used at the present time.

METHOD

Apparatus. A single piece of apparatus was constructed for this experiment. It consisted of a vertical panel with the dimensions and form shown in Figure 1. The panel rested on a laboratory table such that the bottom of the panel was 33¼" from the floor. The displays were translucent pieces of glass that could be illuminated from behind to a comfortable level of illumination. The controls were push buttons requiring about 10 oz of pressure to actuate them. In the center of the lower edge of the panel was a little box with a sloping face. Centered on the sloping face was a starting button that will be referred to later.

The ten different linkages between the controls and displays are shown schematically in Figures 2 and 3. The various linkages were made by appropriate electrical connections behind the panel. All subjects saw exactly the same panel.

The panel and subject were inside a soundproof room. The experimenter, and the control and recording apparatus, were outside the room.

Procedure. At the beginning of a test session, the subject read a set of standardized instructions to himself. These instructions also illustrated the linkage he was to use. The subject was allowed to study the instructions and the linkages as long as he liked. When he was ready to start, the instructions were removed so that he could not consult them during the test session. The instructions informed the subject that when he was ready to start he was to place the index finger of his right hand on the starting button. This action illuminated a light on the experimenter's panel outside and informed the experimenter that the subject was ready for a trial. Shortly thereafter, one of the displays would light up. The subject's task was to push the button linked with that light as quickly as possible, using the index finger of his right hand. If he pushed the correct button, the light went out. If he pushed an incorrect button, the light stayed on. The instructions emphasized that if a subject made an error he was to correct it immediately. The subject was then allowed to ask any questions. After these were resolved, the subject was tested for 96 consecutive trials.

Subjects were tested individually. Test sessions varied but were generally between 20 and 30 minutes in length. By means of appropriate timing and recording devices, the experimenter was able to record (a) the time to first response, that is, the time required by the subject to push any key, (b) the time to correct response, that is, the time required by the subject to push the correct key, and (c) all errors made by the subject.

For any one linkage there were eight different testing orders—one for each of the eight subjects assigned to that linkage. Each testing order was a randomized sequence of the display-control pairings with two general restrictions: (1) across the eight testing orders for every linkage each light (and so each button) was tested exactly twice on each of the 96 trials, and (2) for any one testing order, each light (and so each button) was tested exactly four times in each block of 16 trials. The same eight testing orders were used for each of the 10 linkages studied here.

Subjects. The subjects were 80 right-handed, male undergraduate students who participated in this experiment as part of their course requirement for introductory psychology. No other criteria were used in selecting subjects. Subjects were assigned to the ten different linkages in a random order, but with the restriction that an equal number of subjects was tested on each panel.

Experimental Design. The experimental design is illustrated in Table 1.

RESULTS

Response Times

The time to first response is a measure of the simple choice time required for making any response at all. As such, it includes any uncertainties, or hesitancies, the subjects had in making any response. On the other hand, the time to correct response is perhaps the more meaningful measure operationally, although it is the more complex one psychologically since it includes not only simple choice time, but the time to correct all mistakes as well.

Unfortunately, the timing circuit used to record times to correct response did not always operate properly. As a result we ended up with a few missing data points and so were able to do complete statistical analyses only on the times to first response. Despite this loss of a small part of our data, we feel that we are in

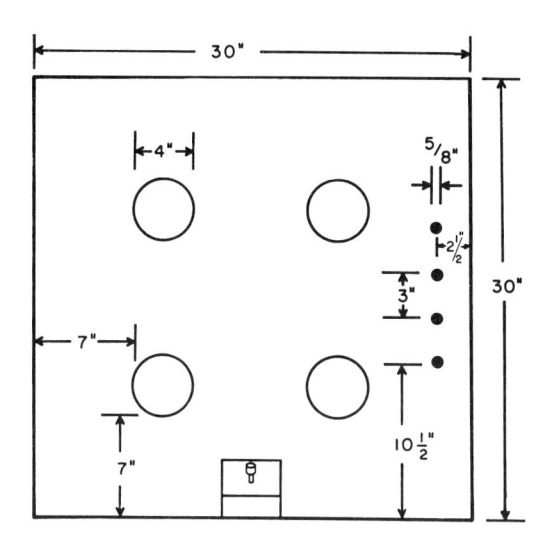

Figure 1. A scale drawing of the control panel used in this study. The displays are represented by the large circular outlines in the center; the buttons by the small circles in the vertical column to the right. The starting button is at the lower center of the panel.

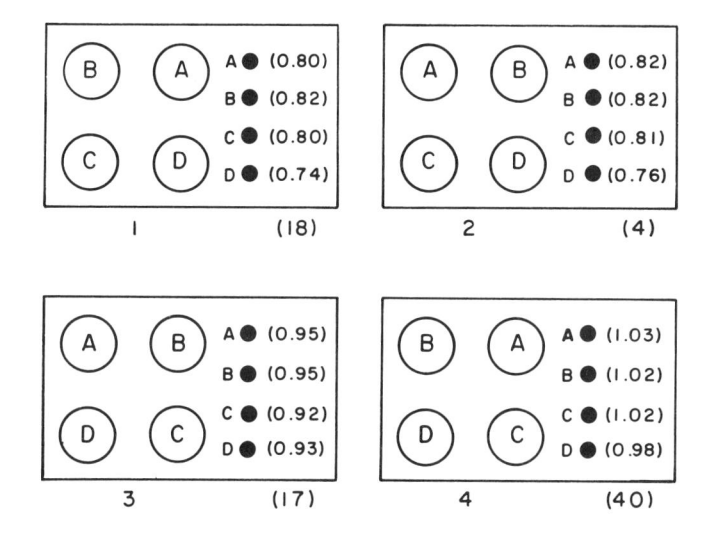

Figure 2. Four of the panels used in this study. Individual linkages are identified by letters which did not, however, appear on the apparatus. Panels are identified by numerals giving the overall ranking from 1 (fastest) to 4 (slowest). The numerals in parentheses underneath each panel show the total number of errors made on that panel. Numerals in parentheses adjacent to each button show the mean response times (in seconds) to individual buttons.

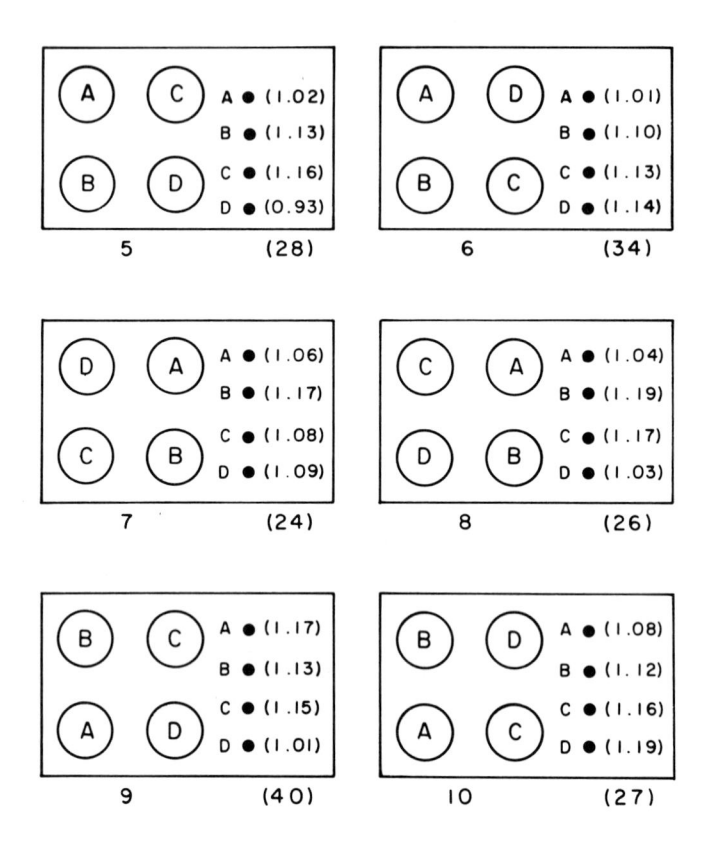

Figure 3. The remaining six panels used in this study. The letters and numerals have the same meanings as those in Figure 2.

TABLE 1
Experimental Design Used in This Study

Linkages (Panels)		*1*	*2*	*3*		*10*
Subjects		$S_1\ldots S_8$	$S_9\ldots S_{16}$	$S_{17}\ldots S_{24}$	…………………………	$S_{73}\ldots S_{80}$
Block I	Trials Tr_1 Tr_2 . . . Tr_{16}					
Block II	Tr_{17} Tr_{18} . . . Tr_{32}					
	. . . .					
Block VI	Tr_{81} Tr_{82} . . . Tr_{96}					

TABLE 2

Analysis of Variance for the Data Shown in Figure 4

Sources of variation	Degrees of freedom	Sums of squares	F ratios
I. Between subjects (S)	79	504.3748	
A. Between linkages (L)	9	115.7095	2.32**
1. Linkages 1, 2, 3, and 4 versus 5, 6, 7, 8, 9 and 10	1	86.4032	15.56*
2. Linkages 1 and 2 versus 3 and 4	1	24.7035	4.45†
3. Linkages 2 and 3 versus 1 and 4	1	0.6285	N.S.
4. All other orthogonal comparisons	6	3.9743	N.S.
B. Between subjects within linkages (S_{wL})	70	388.6653	
II. Between trials (T)	95	32.5496	7.10*
A. Between blocks of trials (B)	5	19.0054	20.70*
B. Between trials within blocks (T_{wB})	90	13.5442	
III. Interaction: S × T	7505	362.2966	
A. Interaction: S × B	395	72.5383	
1. Interaction: L × B	45	11.5509	1.47†
2. Interaction: S_{wL} × B	350	60.9874	
B. Interaction: S × T_{wB}	7110	289.7582	
1. Interaction: L × T_{wB}	810	30.0359	N.S.
2. Interaction: S_{wL} × T_{wL}	6300	259.7223	
IV. Total	7679	899.2210	

*p < 0.001   **0.010 < p < 0.025   † 0.025 < p < 0.050

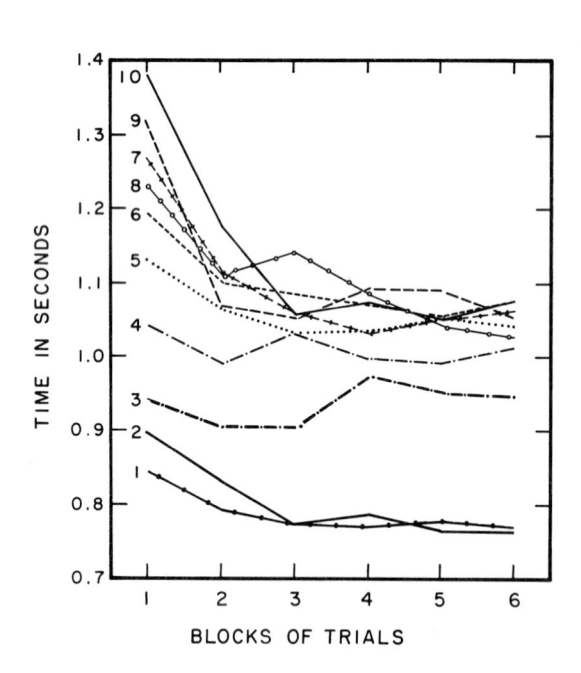

Figure 4. Average response times to each of the ten panels in successive blocks of 16 trials. Each data point is an average of 128 response times (16 trials times 8 subjects).

fact using the more conservative of the two time measures. The evidence for this follows.

First, this study bears certain resemblances to another one by Chapanis and Lockhead (1965). In that experiment, times to first response and times to correct response gave exactly the same *patterns* of findings. The analyses of variance showed, however, that times to correct response gave more significant findings (that is, smaller p's) than did times to first response, this being due in part to the fact that time and errors were positively correlated. Those panels to which subjects responded fastest were, in general, the same ones on which they made the fewest errors. Conversely, those panels over which subjects hesitated the longest were the ones on which they made the greatest number of errors.

The situation in this experiment is exactly comparable to that reported by Chapanis and Lockhead. Using as much of the data as we have for times to correct response, and after making allowance for missing data, it is clear that none of the essential findings to be reported here would be altered in any appreciable way if we had been able to analyze the times to correct response. In addition, the rank-order correlation between time and errors for the ten control-display linkages is +0.52, showing that those linkages over which subjects hesitated the longest were in general the ones on which they made the most errors. Finally, the range of average response times to the ten linkages is greater for times to correct response than it is for times to first response. All of these considerations strongly support the conclusion that our findings hold for times to correct response and that, in using only times to first response, we are in fact using the more conservative of the two time measures.

Figure 4 shows the mean times to first response for each of the ten control-display linkages and Table 2 shows the analysis of variance for this set of data. The data show some large and statistically-significant differences among the ten control-display linkages, and between trials and blocks of trials. There is also a marginally significant interaction between the linkages and blocks of trials.

The first of these findings is, of course, the one of greatest interest to us here. As we said earlier, we picked ten linkages that would allow us to make comparisons of various details of arrangement. With this in mind, the sum of squares attributable to the differences among linkages (with 9 degrees of freedom) was split up into nine orthogonal components, each with one degree of freedom. The comparisons yielding the two highest sums of squares are shown in Table 2 (Lines I.A.1. and I.A.2.). One other interesting comparison (Line I.A.3.) is also given. The pooled sums of squares for the remaining six orthogonal components yielded a variance estimate that did not even equal that of the error term. These six sums of squares are, therefore, merely lumped together in Table 2 (Line I.A.4.).

As we expected, it is clear that certain details of arrangement are highly important and others are not. For example, Line I.A.1. in Table 2 shows that all of the arrangements (the four in Figure 2) in which the displays are arranged in rows, and from top to bottom, are significantly better than all those arrangements in which the displays are arranged vertically in columns (those in Figure 3). In other words, all four arrangements in Figure 2 are significantly better than all those in Figure 3.

The second component of variance (Line I.A.2. in Table 2) shows that the two patterns in the upper part of Figure 2 are significantly better than the two in the lower part of Figure 2. In other words, it appears to make a difference how the two lowest buttons are linked with the two lowest displays.

The third component of variance (Line I.A.3. in Table 2) shows that it does not appear to make any difference how the two uppermost buttons are linked with the upper pair of displays.

The pooled sum of squares attributable to the remaining six degrees of freedom tells us that there are no significant differences among any of the orthogonal comparisons that one might make among the six linkages shown in Figure 3. Incidentally, when we selected our ten linkages originally we were interested in those that have what might be called a *parallel geometry* (2, 4, 5, 8, and 10 in Figures 2 and 3) as compared with those that have what might be called a *circular* or *continuous geometry* (1, 3, 6, 7, and 9 in Figures 2 and 3). This turned out to be of trivial importance. The F-ratio attributable to this comparison is less than 1.00.

Error Data

The total number of errors made on each of the ten linkages is also shown in Figures 2 and 3. Since these error counts are based on 768 trials (96 trials for each of the 8 subjects tested with each linkage), overall error percentages range from 0.5% to 5.2%. These are considerably lower error rates than has been reported in two somewhat similar kinds of experiments (Chapanis and Lindenbaum, 1959, and Chapanis and Lockhead, 1965). On the other hand, the average response times here are somewhat longer than those reported in the two other studies cited. It may be, therefore, that the subjects in this experiment traded accuracy for time (as suggested by Fitts, 1966) by using a somewhat higher accuracy criterion than did those in the two previous ones.

A Kruskal-Wallis analysis of variance by ranks shows that the errors made on the 10 panels differ significantly (p just slightly less than 0.05). Mann-Whitney U tests between the number of errors made for each linkage compared with every other one show, however, that the number of errors made to pattern 2 (Figure 2) is significantly less than the number of errors made to patterns 4, 5, 6, 8, 9, and 10. Differences between the number of errors made to any of the other 39 pairs of comparisons are not statistically significant. We realize, of course, that it is not strictly correct to use the Mann-Whitney U test to make comparisons among all possible pairings of the ten patterns. Unfortunately, there seems to be no non-parametric method of handling data of this kind in post-hoc analyses. In any event, our primary reason for making this kind of analysis is a negative one, namely, to show that the error data do not, in general, reveal very much of significance. A conclusion of somewhat greater generality is that in experiments of this kind response times appear to be more sensitive than errors as an indicator of task difficulty and complexity (see, for example, Chapanis and Lindenbaum, 1959; Chapanis and Lockhead, 1965; Scales and Chapanis, 1954; Shackel, 1959; and Whitfield, 1964).

Although the error data are too few to yield many statistically-significant findings there are some trends so strongly suggestive that we feel they cannot be ignored. For example, of the 258 errors made on all ten linkages 157, or 60.9%, were made because the subject responded as though he expected the pattern to be that of linkage 2 (Figure 2). Further, of the 79 errors made to the first four linkages (those in Figure 2) 67, or 84.8%, were made because the subject responded as though he expected the pattern to be that of linkage 2. Finally, note that 40 errors were made on linkage 4 (Figure 2). Every single one of these errors was a reversal of A with B, or of C with D, that is, as though the subject expected the pattern to be that of linkage 2.

The response time data do not allow us to differentiate between linkages 1 and 2. However, the patterns of error data, even though they are not individually significant by any appropriate statistical test, are all consistently in favor of linkage 2. Linkage 2 appears to be what most subjects expect and most of the errors they make can be attributed to the fact that linkages did not conform to this expectation.

Practice Effects

The data in Figure 4 and Table 2 show that there are some significant changes in performance due to practice. Differences between trials, between blocks of trials, and the interaction of linkages times blocks, are all statistically significant.

In general, there was a marked drop in average response time during the first two blocks of trials (first 32 trials), and a fairly stable level of performance thereafter. Some of the curves are regular; others are not. Those that are irregular seem to reflect idiosyncratic behavior on the part of individual subjects. For example, the rise in average response time for linkage 3 in the 4th, 5th, and 6th blocks of trials is largely due to a single subject. For some unexplained reason, this subject on his 48th trial started giving response times far longer than those he had given in the first three blocks of trials. The rise in the curve for linkage 4 in the third block of trials appears to be due to two other subjects both of whom happened to give their longest response times in that same block. Table 2 shows that there is a significant interaction between linkages and blocks of trials. This interaction was examined by calculating a separate analysis of variance for the data of each linkage. These analyses show that there are no significant differences between trials, or blocks, for linkages 3, 4, 5, and 6. There are significant differences attributable to learning on each of the other six linkages. To sum up, the curves show learning and irregularities of the kind one might expect and are not inherently of great interest.

What is of more interest is to ask whether the differences among the linkages would still persist after subjects had reached a fairly stable level of performance. To answer this question, a separate analysis of variance was made of all the data for blocks 4, 5, and 6 only. These are the data shown in the right half of Figure 4. This analysis shows no significant differences of any kind whatsoever. Although linkages 1 and 2 in Figure 2 still yield lower response times than do the other 8 linkages, one cannot prove that they are *significantly* lower.

Response Times to Individual Lights

Mean response times to each light are shown in Figures 2 and 3. Friedman two-way analyses of variance by ranks show that in all but two cases (patterns 3 and 10) differences in the response times to the individual lights are all highly significant (all p's < 0.01). The two exceptions (patterns 3 and 10) do not even approach significance.

The patterns of response times to individual keys are complex and appear to be the result of two factors: (a) the distance the subject had to reach (greatest for A and shortest for D), and (b) the particular linkage between controls and displays.

DISCUSSION

Of the ten patterns we have tested here one (linkage 2 in Figure 2) appears to be best if we take both time and error data into account. This finding is consistent with, and further extends, design recommendations regarding the layout of displays and controls on panels of this kind (see, for example, Morgan et al., 1963, pp. 302-303). We emphasize, however, that our findings did not attempt to find the *best* way of arranging four displays and four controls on a panel. These data apply only to those situations in which, for some reason, it is necessary to have the displays arranged in a square and the controls in a vertical column to the right of the displays.

REFERENCES

Chapanis, A., & Lindenbaum, L. E. A reaction time study of four control-display linkages. *Hum. Factors,* 1959, **1** (4), 1-7.

Chapanis, A., and Lockhead, G. R. A test of the effectiveness of sensor lines showing linkages between displays and controls. *Hum. Factors,* 1965, **7**, 219-229.

Fitts, P. M. Cognitive aspects of information processing: III. Set for speed versus accuracy. *J. exp. Psychol.,* 1966, **71**, 849-857.

Morgan, C. T., Cook, J. S., III, Chapanis, A., & Lund, M. W. (Eds.) *Human engineering guide to equipment design.* New York: McGraw-Hill Book Company, Inc., 1963.

Scales, E. M., and Chapanis, A. The effect on performance of tilting the toll-operator's keyset. *J. appl. Psychol.,* 1954, **38**, 452-456.

Shackel, B. A note on panel layout for numbers of identical elements. *Ergonomics,* 1959, **2**, 247-253.

Whitfield, D. Validating the application of ergonomics to equipment design: a case study. *Ergonomics,* 1964, **7**, 165-174.

PART THREE

MULTIMAN-MACHINE SYSTEM PERFORMANCE

In Part Two we dissected, as it were, the man-machine system in order to study certain of its vital parts uncomplicated by the confusing pattern of interactions encountered in its normal operation. As we have seen, there are those who believe that a molecular approach of this sort offers the *only* route to a true understanding of system behavior and the only hedge against obsolescence. Naturally, not everyone shares this view. At the opposite pole are some who consider it quite unlikely that molecular research—regardless of its promise for explaining component behavior—can *ever* contribute much to the solution of real system design problems. The reason, they argue, is that the most important characteristics of systems are the very ones stripped away in the simplification process. It is the complex interaction of components which gives a system its characteristic mode of operation; if one wishes to learn something about *system* behavior, then, he must be willing to tolerate complexity and deal with systems intact. He may even find it necessary to forego the convenience of laboratory research, confining his activities to "field" observations and logical analysis of "real" systems.

It is doubtful that many engineering psychologists hold views on the molar-molecular issue as extreme as either of those just presented. Most seem to believe, as we do, that enlightenment is not restricted to any particular level of analysis, and that molecular research, if not a substitute for, is at least a valuable complement to any molar systems program. In any event, there is no disputing the fact that a great deal of molar research has appeared in engineering psychology outlets in recent years. Thus we must assume that a number of psychologists accept systems research as at least *one* fruitful way to understand or resolve system problems. Whether they expect to derive any general laws of system behavior from this approach is another matter entirely; obviously there are great differences of opinion on this question. We shall attempt in the present section, then, to illustrate some of the directions that molar systems research has taken within the context of engineering psychology, and to capture in the process at least several different views on what it all means.

References will be found in the General References at the end of the book.

Existing man-machine systems vary greatly in size, composition, and degree of complexity. An experimental subject and his reaction-time device, for example, could very properly be called a system; so also could an entire corporation or continental defense network. In fact, all that is necessary for something to fall within the *explicit* definition of a system is that it encompass a set of identifiable components interacting in pursuit of a common goal. Normally, however, one *implies* at least a moderate level of complexity when he speaks of "systems research." He usually has in mind some sort of analytic or experimental treatment applied to a recognizable assemblage of men and machines interacting with one another in a relatively "natural" way. A study of operations in an air traffic control center, for example, would qualify as systems research for most people, whereas the rigidly controlled and abstract reaction-time paradigm would not. The initial article in this part comes to grips with this knotty problem of definition, attempting to clarify the basic methodological and philosophical differences between "traditional" and "systems" research.

In keeping with the *implicit* definition of systems research, most of the papers in this part are concerned with organization of more than token complexity. All are of the multiman variety. While this may not be entirely representative of the literature, it serves to underscore the distinctive characteristics—and especially the problems—inherent in research on intact systems. Should the research be conducted in a laboratory setting? If so, how "realistic" should the laboratory vehicle be? What and how much control should there be over interpersonal interactions? How long and in what way should the subjects be trained? What measure or measures are most appropriate as indicants of system performance? What are the most sensible independent variables to manipulate? Although not unique to systems research, all of these problems assume gigantic proportions within the systems context. The reader would do well to orient his study of the present material around these critical issues. He might also apply these questions to relevant material, available elsewhere (e.g., Gagné, 1962b; Kennedy, 1962; Meister & Rabideau, 1965; Pickrell & McDonald, 1964; Walker, 1967).

This part is composed of two chapters: one devoted to theoretical, analytic, and descriptive papers; the other, to empirical (especially laboratory) work. This dichotomy, sad to say, represents more than a mere expository convenience. It pinpoints what many believe to be the principal shortcomings of past systems research: lack of coordination between empirical and theoretical efforts. All too often promising formal constructions stop short of any direct empirical test, and vigorous data-gathering programs run their course in the complete absence of any well-formulated theoretical structure (McGrath & Altman, 1966). The objectives of science are rarely served by such disjointed activity.

We should hasten to add that several notable exceptions to the above pattern have appeared in recent years. In the interest of promoting this trend, we have chosen to feature, in both chapters, at least some articles representing a rapprochement of theoretical and empirical efforts. For example, Edwards' theoretical paper (in Chapter Seven) develops some testable conclusions which are followed up in an empirical report by Schum, Goldstein, and Southard (in Chapter Eight). The latter is but one of a considerable number of recent systems studies prompted, at least in part, by Edwards' proposed application of decision theory to system design.

One very active area of systems research which is not represented here is that of *O.R.*, or *operations research*. Relying heavily upon analytical and mathematical modeling techniques, the O.R. approach is aimed primarily at resolving specific design problems rather than formulating general scientific laws. For this reason its methods vary from problem to problem, and a truly "typical" example becomes difficult, if not impossible, to find. Generally speaking, however, three main steps are involved in most O.R. applications. First, an effort is made to construct a model of the system, or to adapt an existing model, using measured or estimated characteristics of all relevant components. If it is possible to identify a problem as one of a class to which a more-or-less standard kind of model applies (e.g., queuing, inventory control, resource allocation), then all that is necessary is to plug in appropriate parameter values; if this is impossible, the entire *model* must be constructed from scratch. Either way, the next step involves comparing design alternatives (potential solutions) on the basis of the model: this can be done logically or "empirically" (in the sense that the *model* is actually "operated" under the various alternatives [1]) with the objective of predicting which design will maximize the quantity of primary interest in the real system (e.g., profit, efficiency, safety). Finally, a solution, presumably the one identified as "optimal," is implemented in the "real" system.

If human beings play a significant role in the affairs of a system, then the adequacy of any O.R. model as a scientific description of that system must ultimately depend upon how well human performance functions are represented. As we have seen, however, existing data can scarcely be considered definitive in most areas of human performance; hence the model builder must supplement available facts with ample portions of extrapolation, assumption, and "educated" guesswork. Obviously, no one can say how successful his efforts have been unless the model is put to some independent test: how well, for example, do the effects of manipulating the *model* design agree with the effects of manipulating the actual *system* design? Seldom is any such empirical verification possible. In fact, it is precisely because real systems *cannot* be manipulated at will that the O.R. technique of modeling was introduced in the first place. It would be highly impractical, for example, to reorganize a large corporation a number of times just to see which arrangement yielded the best performance. It would also be virtually impossible as

---

[1] "Operation" of models is very frequently carried out on digital computer equipment which, because of its speed, permits exploration of many alternatives in a very short time.

an experimental design problem, because there would be no acceptable way to control such spurious factors as general economic trends. Therefore, models are studied instead, and the results are taken pretty much at face value. From a practical standpoint, this is not as tenuous an approach as it might seem. Even if a model is seriously deficient at various points, it still serves to bring whatever knowledge there is to bear upon the problem at hand in a relatively coherent fashion. Any solution reached as a result should thus be at least as good as one arrived at by less systematic means, and hopefully a lot better.

The logic of the O.R. approach and the obvious impact that it has had upon systems currently in existence, then, argue convincingly for its *practical* worth. What is less clear, perhaps, is the question of how much it adds to our understanding of system behavior *in general,* and whether it will prove superior in the long run, on either scientific or practical criteria, to more fundamental research efforts (i.e., those concerned specifically with system behavior principles). We have chosen in this book to concentrate upon the latter kind of research; it is partly for this reason that we have omitted O.R. papers from the present part. Another reason is that it is simply not possible to do justice to the burgeoning field of O.R. in the space available here. The interested reader will find a wealth of excellent text and reference material devoted exclusively to this approach (e.g., Churchman, Ackoff and Arnoff, 1957; Pierce, 1967; the journal *Management Science*).

In emphasizing the distinction represented by Chapters VII and VIII, we may easily lose sight of an important communality. A recurrent theme in this book concerns the manner in which man's role in system affairs has changed over the years—essentially from that of "doer" to "thinker." Scanning the literature summarized here, we observe directly the evidence of this trend. By all odds the major preoccupation in the research is with the *decision* function; even when the stated objective is improvement of a particular hardware component (e.g., displays, in the Smith and Duggar study), system performance is evaluated in terms of decising-making proficiency. Engineering for the "higher mental" processes is clearly no longer a science-fiction fantasy.

Chapter Seven

Description and Logical Analysis

Three slightly different kinds of non-empirical work are illustrated in this chapter. The first article emphasizes *method;* it explores the philosophical and practical implications of adapting the experimental method to systems research. The second is a *theoretical* paper which brings thinking from several different disciplines into focus upon a particular class of system design problems. The third is an *analytic* description of two important kinds of real-life systems which turn out, upon close inspection, to have a lot in common. Careful analysis of some sort is essential if systems research findings are to have any degree of generality. Let us now proceed to expand upon each of these descriptive comments.

As we said earlier, the paper by Kidd is essentially an overview of the systems research enterprise as of 1962.[1] It begins with a succinct statement of the rationale for experimental systems research, clarifying in the process points of difference and similarity between this and other kinds of experimentation. One notable similarity is the need for a conceptual framework within which to organize research activities and interpret findings. An information-flow or cybernetic model is proposed as a suitable means of achieving this end.

Having laid the formal groundwork for an experimental attack on system problems, Kidd proceeds to discuss one particular form that such an attack might take: a technique known as system simulation. Central to this method is the construction of a laboratory vehicle having certain critical features in common with some class of real systems. The laboratory model is then operated under various experimental conditions using representative *human subjects* in key positions. Generally speaking, the use of real rather than simulated human beings seems to be the main factor that distinguishes this approach from O.R.

Since this Kidd paper appeared, laboratory simulation has indeed achieved a measure of prominence in the formulation of system design principles, although not to the enthusiastic applause of all concerned. One reason for its ambivalent reception is the fact that simulation research represents a compromise between scientific and pragmatic interests, and such compromises rarely satisfy everyone. The scientist finds it difficult to settle for anything short of rigorous experimental control, while the systems designer finds it equally as difficult to take seriously

---

References will be found in the General References at the end of the book.

[1] As one whose professional development has been intimately tied to the development of the systems research field, Kidd can speak from first-hand experience on the entire gamut of system design problems—application as well as research. He is thus eminently qualified to attempt such a broad commentary.

any laboratory research carried out on an unrealistic-looking simulation vehicle. For systems of any appreciable complexity the goals of science and design application are almost always in direct conflict since experimental control can be achieved only at the expense of realism. Amplifying this dilemma is the fact that simulation is at best a costly enterprise; increasing either experimental rigor or fidelity of simulation is done only at great additional expense. All too often elaborately planned simulation programs have ended in failure because the cost of achieving a high level of *fidelity* left nothing with which to finance the *research*. Those programs which seem to have been most successful in generating useful system principles have been those which (*a*) have sought to capture in the laboratory adaptation only the most "essential" features of their real-life counterparts, and (*b*) have not been overly ambitious insofar as experimental design or objectives are concerned. Whether or not the "payoff" for either science or system technology will ultimately justify the expense associated with elaborate simulation research is still open to serious question.

The second article in this chapter illustrates one of the few concerted efforts made so far to link theory, experimentation, and application together in a coherent attack on system problems. In developing PIP, a design principle for combining maximally the unique capabilities of men and machines in decision systems, Edwards does much more than simply provide a clever answer to a nagging problem. He molds several important ideas about human decision behavior into an explicit theoretical account from which a number of testable consequences can readily be deduced.

Among these consequences are some, such as the PIP notion, which have *direct* implication for system design applications and research. Others are more properly tested in "traditional" experimental settings, and therefore bear *indirectly* upon systems affairs. In either case, the theoretical structure greatly enhances the value of any empirical data collected within it: a system experiment not only answers a design question, it advances our knowledge of some general decision process; a "basic" experiment not only contributes to general understanding, it suggests further systems experiments and applications; even the success or failure of a specific *application* can tell us more than it would otherwise. Whether or not it is entirely correct, therefore, the point of view developed by Edwards in this paper has unquestionably gone a long way toward making subsequent research on decision systems count for something.

The final paper in this chapter contributes neither method, theory, nor content to the systems research field. Indeed, one could argue that it has no place in a book devoted to *research*. Our reason for including it lies not so much in what it *says*, but in what can be said *about* it. Specifically it affords us the opportunity to make several observations about the role of research in the development of today's systems.

First of all, this article is typical of the nonempirical, nonquantitative discussions frequently encountered in the systems literature. Were we not to include such an illustration, one would get the erroneous impression that all design decisions today are rooted in some solid base of scientific knowledge afforded by systems research. This, of course, is far from the truth. Both because the scientific "base" is still somewhat limited and because not everyone is willing to abandon "tried and true" methods in favor of those proven only in the laboratory, intuition and opinion still play a significant part in many systems design decisions.

A second point implicit in the Rhine paper is that in spite of all the effort that has been spent analyzing and researching systems, we still have no adequate way to classify them. Consider the two examples discussed by Rhine: military and corporate decision systems. The essence of his argument is that (*a*) these classes of systems have a lot in common, (*b*) significant advances made in the former area (military) could be applied profitably in the latter as well, (*c*) such cross-fertilization has failed to occur because the underlying similarities have escaped notice. If this viewpoint is correct, and we have no reason to doubt its accuracy, it illustrates quite vividly the existing state of affairs in systems classification. Few organizations have been the subject of more intensive study in recent years than these two, yet all of the resulting information has apparently failed to make the simple point that the two belong in the same general category. Without some recognized basis for comparing systems, what is learned in one situation cannot readily be applied in another, as the Rhine paper so aptly illustrates. Research findings, under these circumstances, can have little generality.

A final observation suggested by the Rhine paper is that the key to improved classification may lie in a more satisfactory description of existing systems. What Rhine does is to illustrate a level of description at which he believes meaningful comparisons can be drawn between systems. In contrast, the trend in systems analysis over the years reflects an ever-increasing concern for *detail*. Presumably, it is commonly accepted that masses of analytic data help an organization to understand itself better and thereby enable it to pursue its goals more effectively. Such minute description serves to contribute little, however, to understanding *beyond* that particular system. If anything, it serves to submerge potential similarities among systems in a morass of detail. Perhaps more attention should be given to the *kinds* of data that we seek in systems analysis, and less to the *amount* that we are able to collect. If the Rhine paper has a moral, it is that we have been operating at the wrong level of description for development of general system principles. What constitutes an *appropriate* level is still a moot question, but perhaps one deserving more consideration than it has received heretofore. For further discussion of this issue, we refer the reader to Gagné (1962b) and Meister and Rabideau (1965).

26

A New Look at System Research and Analysis
J. S. Kidd

INTRODUCTION

System research constitutes a rapidly growing adjunct to the main body of human factors technology. While the idea of system is not new and certainly is not the exclusive domain of any particular scientific speciality, its impact on the nature of human factors work seems likely to be profound. Moreover, the combination of the idea of system with the particular problems and goals of human factors work has given rise to many questions in the minds of those who have watched the combination take place. What is the relevance of the system concept for dealing with actual man–machine systems? What is system research? How and why is it done? Where will it lead and is it worth it? It may be worthwhile to attempt some definitions, to develop a set of standardized terms, and to work out a common conceptual framework, as promptly as possible for a start in answering these questions.

No one doubts the reciprocal relationship between theory and practice, particularly when theory is backed up by appropriate and adequate research. However, in a rapidly expanding discipline, one element of the theory–research–practice triad may be excessively retarded or accelerated and a condition of imbalance can result. In human factors work there is an apparent inadequacy to the conceptual support of research and practice. More and more human factors specialists are working on systems or are attempting to study system processes in the laboratory and field. They are seeking limited objectives, usually under severe time constraints and, therefore, are not usually permitted the luxury of conceptual synthesis.

The present document is not intended as a full-fledged theory of man–machine systems. Rather, it is proposed as a categorical framework by means of which theoretical ideas can be organized. Researchers and practitioners have such ideas but often must feel at a loss about what to do with them. Our discipline will be

From *Human Factors*, 4, 1962, 209-216. By permission of the Human Factors Society.

The author acknowledges with appreciation the support and guidance of Dr. Paul M. Fitts and Dr. George E. Briggs.

well served if such ideas can be communicated freely and enter into the normal regime of critical refinement and extension.

APPROACH

One might define the goal of all human factors work as an attempt to enhance the contribution of the human operator to overall system effectiveness. Though some systems operate in certain phases without direct, integral human participation, in a real sense there are no "unmanned" systems. Therefore, the human factors specialist has a legitimate role in the design, development, and evaluation of systems insofar as he can show that different system configurations will influence the degree to which the human operator can perform his assigned function.

Depending on the nature of this assigned function, interest may be focused on such topics as the characteristics of the physical environment within which the man must work, his clothing and personal equipment, the layout of his work place, the readability of displays, and the discriminability of controls. These are topics which treat of the human operator as a more-or-less standard system component; he is one kind of general-purpose, transductive element and must be cared for like a temperature-sensitive transistor.

The general-purpose element is, of course, adjustable and is adjusted to fit a given specific system by means of training; or the differential selection of the already trained. The basic approach considers the system to be a relatively stable setting within which is placed a versatile but essentially pre-programmed unit that must be pampered to an unusual extent.

The system-concept modification of the basic approach tends to add fluidity and uncertainty to this picture. The system itself is seen more as an abstract set of relationships, or information flow channels and control signals. The human operator is regarded as having the power to change himself and to change the system while acting as a part of the system. The system is self-organizing, in part, and achieves this characteristic as a consequence of the presence of the human operator within it.

The goal of system research—as directed by the system idea—is to exploit the changeability of man-machine systems. The search is for conditions or design configurations which not only do not impair the human operator in the performance of his assigned functions but which provide him with the means to create new functions or methods of conducting his affairs as appropriate to the overall conditions of system operations.

Perhaps it goes without saying that system research shares with other investigative disciplines the intention to unearth general principles in the form of quantitative functional relationships which can describe the cause and effect structure for all instances within its zone of discourse. However, because of its avowed connections with the realm of applications, the primary intent is moderated by the need to develop specific guides to the design, the development, and the definition-of-management procedures for particular systems. The reconciliation of the two goals provides an interesting complexity to the creation of useable research methods.

BASIC PARADIGM

Fitts (1959) has defined a man-machine system in the following way: "A system is an assemblage of elements that are engaged in the accomplishment of some common purpose(s) and are tied together by a common information flow network, the output of the system being a function not only of the characteristics of the elements but of their interactions or interrelations."

The terms "purpose," "network," and "interaction" are particularly critical. The conceptual models which are invoked to help understand and describe systems tend to emphasize one or another of these terms, but no single model has been altogether satisfactory. Ideas have been borrowed from several theoretical schemes to give some form to research and its interpretation. Among the most significant sources of theoretical and conceptual material are cybernetics and control system theory. The feedback model seems to provide the best paradigm for dealing with system processes because it helps give substance to the concepts of purposiveness and adaptability (Ashby, 1956). Moreover, the feedback model has proved its fruitfulness through the work of Taylor and his associates (Birmingham & Taylor, 1954; Taylor, 1957) on display-control dynamics and the development of the technique of display quickening.

The present approach to the description and analysis of man–machine systems inevitably centers on the human component. This bias can be justified on the basis of the complexity, versatility and variability of the human element and on the basis of the critical role the human operator is required to fulfill. Placing the human operator in the center of the man-machine system focuses attention on a rather specific action–reaction pattern. The system analyst/human factors specialist is accustomed to look for inputs that terminate at the site represented by the human operator and for outputs that have their origin at the same site. When this is done, it is apparent why the cybernetics model has been so widely adopted as the best overall description of system phenomena; all the elements of the generalized servomechanism can be readily discerned to be present. This fact is significant enough by itself. However, closer scrutiny of the details of the action-reaction processes reveals a further fact that is oft-times overlooked. More than one servo-type feedback mechanism is active in any man-machine system. The main loop connects the operator with his task environment, but secondary loops exist that tie the human operator to other elements within the system proper and to elements in the organizational surround. In his central role, the human operator not only acts to control the environmental elements that define his task but controls and is controlled by agencies within and adjacent to the basic system. The essence of this proposition is illustrated diagrammatically in Figure 1, which attempts to show some of the essential action–reaction flow dynamics of a generalized man–machine system.

The human operator has at least three environments to deal with: the immediate surround composed of fellow components, both man and machine; the operational or task environment; and the organizational environment of presumably allied agencies. There are several obvious implications. For one thing, a time-sharing-of-input requirement is built into every system. Secondly, the system

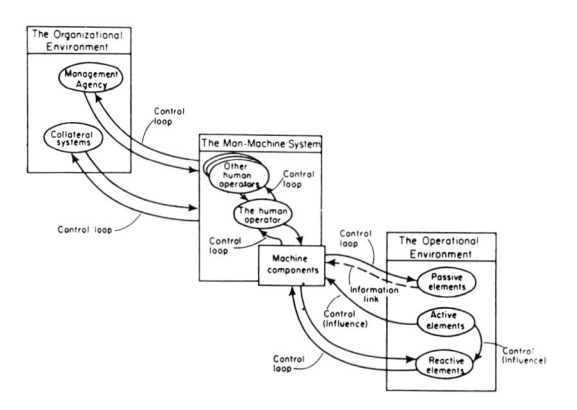

Figure 1. An abstract model of a man–machine system emphasizing the multiplicity of control loops.

proper acts to blend and, hopefully, to integrate the diversity of control signals. Fig. 1 suggests the channels through which control signals flow but indicates nothing of the nature and content of that flow nor its detailed dynamic characteristics. In considering these matters in detail, however, it is necessary to enlarge somewhat the categorical treatment of systems. The dynamic details depend on the *kind* of system under consideration so we need to go from a general model to a taxonomy of types.

SYSTEM TAXONOMY

The problem of taxonomy is to create a set of descriptive categories that have functional relevance. In the present case, the basic model provides a point of departure. By defining the attributes of information-flow along the channels already laid out, it may be possible to provide an analytic tool for the reliable discrimination of various types of man–machine systems. Looking at the model, one basic distinction is clear: inputs from sources external to the system proper fall into two main categories—those arising from the organizational surround and those arising from the task environment. The balance of input volume from these two sources provides one simple discriminant index. The label "degree of system autonomy" might be an appropriate one which could be defined as the ratio of the volume of signals from the operational or task environment to the volume of signals from the organizational environment. The higher this ratio, the greater the system autonomy.

Further distinctions may be made within both primary source categories. For instance, environmental sources come in four types:

(A) background signals, which would include all those events which are detectable by the systems sensors but which are usually irrelevant to system operation (e.g., a school of sonar-reflecting fish encountered by a submarine);

(B) fixed conditions which the system can detect, which impose constraints upon the system, but which do not vary (e.g., geophysical and geopolitical parameters such as the presence of an urban concentration near a jet terminal);

(C) critical non-controllables, which includes factors that are detectable, which vary, which impose constraints upon the system, but which cannot be directly controlled by the system (e.g., weather, enemy technological developments, etc.); and

(D) critical controllables, which include those aspects of the environment that the system is designed to manipulate or control.

These four sub-categories each have two significant attributes; relative signal density or volume of flow and what might be called reliability. Systems should be discriminable on the basis of how much information of each type they are called upon to handle. The system heavily bombarded with type A inputs, for instance, would require a large expenditure of resources on the filtering function since type A inputs are, in effect, noise or a source of distraction from the main task. Systems existing in a preponderantly type B environment would need to rely heavily on mechanisms of accommodation or compensation. Extensive type B inputs can interfere with system performance (Kidd, 1961) by limiting the response options of the system. One would predict that substantial system resources would be diverted to the objective of circumventing or nullifying the type B constraints.

Systems in a predominantly type C environment would depend heavily on rapid adaptability. Since weather affects the flight characteristics of an aircraft, one would expect that all-weather aircraft would have some method for changing aerodynamic configuration built in by design and, indeed, such is the case. Viable systems appear to possess a variety of techniques for facilitating adaptation. As has been suggested, the mere presence of man in the system provides a center of adaptability. In general, adaptation techniques include configurational change, at the component or overall system level, changes in methods or techniques of operation, and even changes in the goal structure of the system.

A configurational change has been described above. In its most sophisticated form, it is based on modular design principles. Procedures changes are illustrated by what occurs during system training (Boguslaw & Porter, 1962). In this instance, the system adjusts under simulated stress to accommodate anticipated problems in the non-simulated work environment. Changes in goal structure are best discussed under the category of organizational inputs.

Systems with predominantly type D inputs are relatively uncomplicated except for the convergent factor of input reliability. All the input types may suffer from a degree of unreliability or residual uncertainty. However, unreliability is least tolerable when associated with type D inputs. The divergencies which are possible are illustrated by a comparison between an air traffic control system and an air defense system. Both systems are responsible for the detection, identification and tracking of flying objects. In the one case, however, all parties to the operation strive to minimize information unreliability; while in the second case, manifold tactics are used (by the elements to be "controlled") to obscure the information available to the system proper. The alert enemy will not be consistent even in his methods of generating uncertainty.

In summary, then, there are four kinds of environmental inputs, each having two facets, volume and residual uncertainty. The relative proportionality of input type and the degree of unreliability associated with each type helps define the class of system.

Inputs from the organizational surround are susceptible to the same analytic procedure. Organizational inputs can be subdivided by source into those deriving from collateral systems and those deriving from a command agency or supravening organization. Those from collateral systems are very comparable to the type C inputs discussed above and need no further elaboration for the present. Inputs from the supravening organization deserve more detailed analysis.

For one thing, the economic support of the system comes, in most instances, from the command agency. (In this important way, the man–machine system is *unlike* a biological organism which extracts its livelihood directly from its task environment.) In the case of the man–machine system, the economic means to acquire and maintain the machine components and the rewards and incentives for the human components come, as it were, from above. This characteristic opens the door for the command agency to engage in performance assessment activities, to interrupt the feedback cycle, and insert value judgments regarding system performance. Moreover, such judgments may be accompanied by sanctions, both positive and negative.

The point is that the command agency functions as a link in what a psychologist would describe as a reinforcement feedback loop. Systems may be differentiated on the basis of the extent to which performance feedback is delayed, the response/reward contingency pattern imposed by or supplemented by the command agency, the extent to which the command agency imposes procedural restrictions, the basis for these procedural restrictions, and the extent to which goals or system objectives are determined by the command agency.

This brief analysis of input characteristics has certain obvious heuristic implications for substantive research as well as for system development efforts. In some degree, it provides a map of the parameters which are external to the system but which can influence system performance. For example, the problems of information flow dynamics and information processing methods—within the system—are accentuated. The application of communications theory has relevance here but there are gaps. Information content analysis and the nature of meaning from both semantic and pragmatic viewpoints are two such gaps that communication theory hardly purports to fill. As far as systems research and analysis are concerned, the only legitimate way to assess the meaning of a symbol or signal input is to observe the response which is elicited (or, in system planning, predict the response which will be elicited). Another example of the need to go beyond the conventional metric of information theory is provided by the concept of redundancy, which is adequate in its original use, but which is somewhat hazy when applied to man–machine system phenomena. For the purpose of man–machine system analysis, the concept might be extended to subsume all processes that reduce uncertainty by repetition, storage, or format techniques. A complete English-sentence message is rather highly redundant because of the high dependency between words in a series *and* because of the grammatical format. When the message itself lacks this kind of intrinsic redundancy, another kind of redundancy can be achieved by message repetition. Moreover, message repetition can occur in several forms: repetition can be serial (the same source or sender emitting the same signal pattern repeatedly over time, which is analogous to the process of determining reliability by the test-retest method); or the repetition may be collateral (the same message is initiated by different sources or transmitted by independent channels, which is analogous to the process of determining validity by comparing results across conditions).

Research may be planned with good prospects, then, when the variables investigated are intended to bear on the issues related to, but extending beyond, communications theory as it is generally conceived. The practical point of all this is that the input parameters described above are likely to be factors which can influence system performance. System research and analysis should be oriented so as to answer the following questions: (*a*) What is relative proportion of inputs from each of the various general sources? (*b*) Can the proportionality be modified to increase the probability of system success? (*c*) Is the content (meaning) of the input messages likely to lead to effective action? (*d*) Can either the substance or the method of presentation be varied so as to lead to more adequate communication? The purport of these questions is not particularly novel. They are ultimately the questions any good system researcher or analyst would ask. The advantage lies in being able to comprehend the system mechanics behind such questions, to understand the causal linkages that determine the answers, and to appreciate more fully how the questions can be attacked and when the attack should commence.

METHODOLOGY—A LIST OF PROBLEMS

More is needed than armchair analytic schemes if prediction and control of system performance are to be achieved. Empirical data, particularly experimentally derived data, are essential. A few years ago, the system scientist was limited to field observations of operational systems or extrapolations from component measures obtained by experimental techniques. Since then, some material progress has been made in applying experimental methods directly to the investigation of full-scale systems or at least major sub-systems. The technique is man–machine system simulation.

Simulation is no cure-all. It is safe to say that simulation does not allow the investigator to test any new variables. Simulation merely provides a research tool that can enhance the meaningfulness and the validity of the results.

One way to describe the method is to make a comparison between simulation and the classical methods of experimental psychology. It would be a mistake, perhaps, to consider man–machine system research as a branch of traditional psychology. There is a shared concern with the vagaries of human behavior and certainly the methodological developments of scientific psychology provide part of the basic foundations for any system inquiry. However, the nature of *what* is measured and the goals of the measurement are significantly different. In the classic procedures of experimental psychology, a prime objective is to make the control of stimuli which impinge upon the subject as nearly complete as possible. The subject is presented with a series of pre-determined stimulus events. The order and character of these events remain unchanged regardless of how the subject responds. It is clear that such a situation is somewhat unreal. Outside the laboratory, an individual's actions are followed by reactions on the part of various elements in his environment. His responses determine, in part, what the reaction will be and the reaction determines, again in part, what his subsequent responses will be. Thus, real-world behavior is highly interactive and may best be represented as a sequential dependency series. The objective of simulation as a research method is to recreate real-world reactivity within the laboratory and at the same time achieve more than a semblance of experimental control.

Another contrast is also appropriate. Simulation for experimental purposes is not the same thing as exercising a mathematical model in the manner of Operations Research. In the OR technique, *all* elements of the situation under study are transmuted into abstract terms. In man–machine system simulation, only some of the environmental factors are made abstract. The flesh-and-blood experimental subject and his immediate physical and social environments are present in concrete form. However, there are points of sharing; for example, an OR model frequently can be used to represent all the non-immediate aspects of the simulated test environment. The use of electronic computers to "activate" the model is another point of commonality. Since the two techniques are often used to investigate similar problems, there is a natural tendency for incomplete differentiation between them.

To complete the description of man–machine system simulation, it may be appropriate to consider the steps in building and using a simulation facility. The first step is the selection of a referent system. If the goal of the research is relatively applied, as was the case in the OSU air traffic control series, this step offers no problems. A system exists in the field which can be copied. At times, however, even in an applied setting, no ready referent system exists. This would be the case for such prospective systems as manned-maneuverable space vehicles. The possibility exists, however, to establish some characteristics by extrapolation from the most comparable system available. (For the present example, it might be a very high performance aircraft such as the X-15.)

If the simulation facility is to be used for "basic" research, it may be advisable to avoid specific representation of *any* existing system. While experience at OSU indicates that basic issues may be attacked in a specific system setting, there is always an implicit limit to the generalizability of the findings. The ideal facility for basic research would consist of some kind of archtypical simulator having those and only those characteristics shared by all or, more reasonably, all of one class of systems.

The next step that causes concern is the selection of performance criteria. Since some definable goal is the reason for the existence of any man–machine system, ultimate criteria are not lacking. However, more often than not, such measures are inconvenient, insensitive, or both. The system researcher must look for component output measures which have a logically valid relationship with overall systems performance. In many instances, multiple criteria are appropriate since each component and sub-system will have its own output, all of which are blended to compose the output of the total system.

The standard requirements for dependent measures—meaningfulness, reliability, and independence—always apply. The difficulties involved in meeting these requirements are universal for all the biologically centered sciences and no pat solutions are to be found here. However, there are some issues that are particularly salient for systems research. One such issue is analogous to the problematic relationship between speed and accuracy of performance in human performance research. Almost all systems have outputs which can be assessed by either term or both. The system researcher has little choice but to account for both facets in spite of the problems involved. A specific example is provided by air traffic control system research. Safety and delay *en route* are both meaningful criteria of system performance. The dual facets appear to be related in reciprocal fashion but the correlation is very far from perfect. The difficulty comes from the fact that any simulated system which yielded enough mid-air collisions to make statistical data processing meaningful would be a very poor system. The comparison of such a system with operational systems which have near perfect safety records is meaningless.

The next step in facilities design would be the choice of means by which the relevant systems characteristics are to be reproduced in the laboratory. The determination of what is relevant in terms of the various environments to be represented has already been defined. One problem is that the probabilistic character of most system input–output-feedback relationships invokes the requirement for some kind of mathematical model which is, in practice, a set of if–then probability statements. The fidelity of simulation depends heavily on the accuracy of this model. A closely related issue is the treatment of the event–time dimension. Some real systems have very long feedback delays. Most productive businesses, for example, require decisions about which the consequences may not be denotable for months or years. Even with extensive support, researchers cannot afford to represent such conditions in the laboratory. Foreshortening of the time intervals between certain processes, then, becomes a highly desirable expedient. Still another facet of this same problem centers on certain "critical-incident" phenomena in system development. For example, under ordinary circumstances, the growth of an organization is a lengthy process. The primary expedient here is usually the introduction of artificially high levels of environmental stress in the hopes of forcing the pace.

The use of probabilistic models and the treatment of the event–time dimension always introduce artificial qualities into the simulation. The effects of these degradations in fidelity on the validity of the findings have yet to be fully assessed. In point of fact, the whole issue of fidelity in simulation is still unsettled. The work at OSU (Kinkade & Kidd, 1959) gives some indication that the essential factor for achieving valid results is the nature of the response requirements imposed on the human operator. Since fidelity is expensive, however, it would be very helpful to have more than an intuitive feeling for the establishment of fidelity values for all the attributes of a simulation facility.

Implicit throughout the preceding discussion of research methods there has been an emphasis on the interdependence of experimental work on the one hand and analytic-theoretical efforts on the other. One way to help insure that both classes of endeavour make headway is to devote some significant proportion of the technological resources available to investigations that have methodological innovations and refinements as a primary target. In the long run, "practical" answers may be achieved more economically by this seemingly devious route.

Acknowledgment

This work was carried out in the Laboratory of Aviation Psychology, and was supported by the United States Air Force under Contract No. AF 33(616)6166, monitored by the Aerospace Medical Laboratory.

REFERENCES

Ashby, W. R. *An introduction to cybernetics.* New York: Wiley, 1956.
Boguslaw, R., & Porter, E. Team functions and training. In Gagne, R. (Ed.) *Psychological Principles in system development.* New York: Holt, Rinehart & Winston, 1962.
Birmingham, H. P., & Taylor, F. V. *A human engineering approach to the design of man-operated continuous control systems.* USN Res. Lab. Rep., April, 1954, No. 4333.
Fitts, P. M. *Notes and selected readings on human engineering concepts and theory.* Ann Arbor, Mich.: Univ. of Michigan, College of Engineering, August, 1959.
Kidd, J. S. *Some sources of load and constraints on operator performance in a simulated radar air traffic control task.* USAF WADD Tech. Rep., March, 1961, No. 60-612.
Kinkade, R. G., & Kidd, J. S. *The use of an operational game as a method of task familiarization,* WADC TR 59-236, August, 1959.

27

Dynamic Decision Theory and Probabilistic Information Processing

Ward Edwards

The most difficult and demanding man–machine systems which men must now design are large military systems which include computers and which are intended for detection and tracking, threat evaluation and diagnosis, or command and control. In such systems, the role of man as decision maker overshadows any other role he may play. The designers and users of such systems now feel frustration about the absence of relevant basic research information, similar in nature to the frustrations which led to upsurges in work on knob-and-dial psychophysics and on tracking in the 1940's. This paper reviews ideas and work done or in progress which seem to me indicative of where basic research on human decision processes is going. These ideas have two closely related foci: dynamic decision theory and Probabilistic Information Processing systems (PIP). I will start by presenting the problem of dynamic decision theory and by proposing a taxonomy of human decision tasks to which such a theory should be applied. Next comes a review of current thought and experimentation in five areas which seem to me crucial to the development of such a theory: information seeking, man as intuitive statistician, sequential prediction, Bayesian information processing, and dynamic programming. Finally, I shall present the idea of a probabilistic information-processing system, both as a kind of system which is urgently needed in several military contexts and as a vehicle for research in dynamic decision theory.

DYNAMIC DECISION THEORY

Description of the Theory

Definition. Decision theory as a topic in psychology seems to come in two distinct and almost noninteracting chunks, which I have named static decision the-

From *Human Factors*, 4, 1962, 59-73. By permission of the Human Factors Society.

Text of a paper presented at the annual meeting of the Human Factors Society of America, Columbus, Ohio, on 14 September, 1961.

ory and dynamic decision theory (Edwards, 1954; 1961a). Static decision theory, of which the subjectively expected utility maximization model in its various versions is the epitome, conceives of a decision maker who is confronted by a well-defined set of possible courses of action. Associated with each such course of action and each possible state of the world is a value; collectively these values form a payoff matrix. Objective probabilities may or may not be associated with the possible states of the world. In any case the decision maker chooses and executes one of his courses of action, receives the value or payoff associated with the intersection of that course of action and the state of the world which actually obtained—and then the world ends. The decision maker (in principle) never gets to make a second decision in which he might apply whatever he may have learned as a consequence of the first.

In dynamic decision theory, decision makers are conceived of as making sequences of decisions. Earlier decisions, in general, produce both payoffs and information; the information may or may not be relevant to the improvement of later decisions. The objective of the decision maker may be taken to be maximization of total profit over the long run. But it is quite likely to be desirable to give up short-run profit in order to increase long-run profit. The most common instance of such a conflict would arise in situations where some courses of action lead to more information and less profit, while others lead to less information and more profit.

In dynamic situations, a new complication not found in the static situations arises. The environment in which the decision is set may be changing, either as a function of the sequence of decisions, or independently of them, or both. It is this possibility of an environment which changes while you collect information about it which makes the task of dynamic decision theory so difficult and so much fun. Most of the manageable cases are those in which the environment is not changing or is changing only in systematic response to the decision maker's decisions. It is not necessary that the environment does not change at all; all that is really necessary is that the environment be what the mathematicians call stationary. This means, roughly speaking, that the environment's behavior is controlled by some variety of statistical process and that the characteristics of that statistical process do not change with time. A roulette wheel or a pair of dice are stationary in this sense. Reasonable mathematical treatment of most decision problems involving stationary environments is possible; reasonable mathematical treatment for decision problems involving nonstationary environments is often unavailable.

A Taxonomy. The considerations of the preceding paragraph indicate that there are six kinds of dynamic decision tasks to study:

1. Stationary environment; fixed information about it; neither environment nor the subject's information is affected by the results of earlier decisions. This is the (unrealizable but perhaps approximatable) static sequential decision task; research on preferences among bets has characteristically been of this kind.

2. Stationary environment; environment unaffected by decisions; but information obtained as a result of earlier decisions is relevant to later decisions. The prototype of this kind of research is the binary prediction or probability learning task with fixed event probabilities. Much research on probability learning is now

going on at Michigan and elsewhere. So far the research at Michigan has indicated that the probability matching hypothesis of Grant, Estes, and many others (including the simpler forms of the Estes and Bush-Mosteller learning models) is not correct; that the gambler's fallacy is an evanescent phenomenon of early trials; that instructions (e.g. about whether it is a gambling or a problem-solving task) have little or no effect on behavior; and that payoff functions have a very vigorous effect. For as much as has yet been reported, see Edwards (1961b).

3. Stationary environment; environment unaffected by decisions; but earlier decisions have mostly to do with whether or not subjects obtain information about environment. The prototype situation is the information-seeking situation, in which subjects must eventually make a major decision, and may, if they wish, purchase information which will (or may) reduce their uncertainty about the decision. Several experiments of this sort are now in progress; the main findings so far are that large and consistent individual differences exist in information seeking, and that in general subjects buy too much information.

4. Nonstationary environment; environment unaffected by decisions; information unaffected by decisions. This is the case of the man affected by forces beyond his control; the best he can hope for is to understand and exploit what is going on. An example is the probability tracking experiment recently completed at Michigan, in which subjects tracked the mean of a binomial distribution displayed by means of two lights, one or the other of which flashed once in each time period. I shall say a little more about this experiment below.

5. Nonstationary environment; environment unaffected by decisions; information controlled by decisions. We are just getting started on a probability learning experiment in which on each trial the subject may choose either to look or to bet. If he chooses to look, he simply observes which of the two predictions would have been correct on that trial. If he chooses to bet, he stakes (say) $0.10 on his prediction, winning that amount if the prediction is correct and losing it if it is wrong; but he is not permitted to observe the result of the trial. Instead, the number of correct and incorrect predictions is accumulated on counters which he cannot see until after the experiment is over. Thus he can either make money or obtain information, but never both at the same time. If the environment were in fact stationary, and he knew this, his optimal strategy would be to look until he had satisfied himself that one side was more frequent than the other, and thereafter never to look again, but instead always to predict the more frequent event. But in a nonstationary environment, no such relatively straightforward optimal strategy exists; the central interest of the experiment lies in seeing how he manages to get a few bets down while still looking often enough to keep up with fluctuating probabilities. Of course we will compare stationary with nonstationary environments in this task, and will manipulate instructions also.

6. Nonstationary environment; environment and therefore information about it affected by decisions. We are now in the middle of a binary prediction experiment in which each prediction of Left makes a Right outcome more probable on the next trial, and vice versa. Variations in the form of these negative feedback functions permit very interesting predictions about asymptotic behavior, predictions independent of any learning model. So far the main finding has been that subjects are very unlikely to reach anything that looks like an asymptote, even after 2000 trials.

The cases which do not appear in the above listing are the cases in which a stationary environment is affected by decisions (it can't be; an environment which can be changed by decisions is necessarily nonstationary), and the case in which an environment is affected by decisions, but information obtained from it is not (it can't happen; an environmental change is meaningless unless it somehow affects what happens to the subject). Cases 2 and 3 are really the same; differences in emphasis lead me to separate them.

Stochastic Learning Models. We usually think of the stochastic learning models as being concerned with the kinds of questions I have lumped together under the heading of dynamic decision theory. At least in these models decision makers are indeed conceived of as making sequences of decisions in which later decisions are influenced (I will not say guided) by the outcomes of earlier decisions. But only the most elementary kinds of states of the world and courses of action are considered, and the hypotheses concerning the nature of the influence exerted on later decisions by earlier outcomes are both unreasonable (no sensible man would obey these models) and unlikely (in real men, the action of reward is complex and very nonautomatic, and path independence is seldom, if ever, realized). The stochastic learning models are difficult to improve and so far attempts to improve them have produced little in the way of better prediction of actual behavior.

The Ideal Dynamic Decision Maker

At the philosophical level, the main difference between the approach of the static model makers and that of the dynamic model makers seems to be that the former attempt to describe optimal, rational courses of action and then ask to what degree and in what ways real men deviate from them, while the latter attempt to construct models from trivial, mathematically simple postulates and hope to find situations in which behavior resembles the resulting predictions. But there is no inevitable relation between the compare-real-with-optimal-behavior approach and the static conception of the problem; optimization models have tended to be static models primarily because the mathematics of optimization in dynamic contexts is quite difficult and relatively new, while the mathematics of optimization in static contexts is relatively simple and relatively older. I believe that this compare-real-with-optimal-behavior approach is indeed available for some dynamic decision problems and that it might be empirically useful to explore it.

From a subject's point of view, what is the difference between decision n and decision $n + 1$ in a sequence of decisions? Aside from whatever intrinsic differences may exist between the two decision problems, the important difference is that when he makes decision $n + 1$, he has the outcome of decision n to guide him, while when he makes decision n, he does not (though he may, of course, have the outcome of $n - 1$, $n - 2$, and so on to guide him in both decision n and decision $n + 1$). How can we represent the advantage which accrues to him on decision $n + 1$ by virtue of knowing the outcome of decision n? Two different approaches to this question seem possible. First, we may talk about the expected value of decision $n + 1$. It has (assuming subjective probabilities, if no objective ones are available) an expected value before decision n is made. It also has a (possibly different) expected value after decision n is made and its result discovered. The difference between these two expected values is a reasonable measure of the value to the decision maker of having the result of decision n available when he makes decision $n + 1$. Unfortunately, the matter is not really that simple. The degree of relevance of decision n to decision $n + 1$ may well be a function of the course of action—and perhaps also of the state of the world—which occurs in decision n. For example, if decision n is a decision whether or not to stop and get gas at gas station n along the road, the decision $n + 1$ is the same for gas station $n + 1$, then the relevance of n to $n + 1$ is a function both of what course of action was taken in n and of the unknown distance between n and $n + 1$—which can be of great importance if you are driving through the desert, for example.

Another way of evaluating decision n is via information theory. There is a well-defined sense in which the information in bits obtained from decision n can be calculated for some, but not all, kinds of decisions; this is most obvious when the decision concerns whether or not to make additional observations in an experiment with optimal stopping. It is likely that the amount of information gained in the course of decision n places some kind of upper bound on the value to decision $n + 1$ of decision n; if no information whatever is gained as a result of n (a most unrealistic hypothesis), n can clearly have no effect on $n + 1$. On the other hand, that is clearly an upper bound rather than an estimate, since it is entirely possible that n may yield lots of information, all of it completely irrelevant to $n + 1$.

What might we as psychologists want to do with this sort of problem? First, we might want to define an ideal dynamic decision maker, comparable in nature, abstractness, and lack of realism to Tanner's ideal psychophysical observer. I am under the impression that the mathematics necessary to do this is already known (though not to psychologists); I will review what little I know about it below. It seems very likely to me that in this notion of the ideal dynamic decision maker, informational as well as expected value and Bayesian considerations will appear. It might even turn out that the notion of information in Shannon's sense can be derived directly from the notion of the upper bound of advantage that knowing the outcome of decision n can provide in making decision $n + 1$. It is, in any case, clear that two sorts of mathematical ideas must necessarily be included in any such ideal dynamic decision maker. One has to do with the amount of information in an observation, and the other has to do with the amount of relevance of information to a decision. We must develop a quantitative theory of relevance. So long as we stick to the ideal decision maker, it seems likely that this can be done, at least for simple cases.

Relevant Research in Progress

After we have an adequate notion of the ideal dynamic decision maker, we will want to compare him in action with real men. Obviously the real men will not shine in such a comparison. But one might hope to find, as Tanner has, that the size and nature of the difference between the ideal and the real dynamic decision maker are lawful, and such laws would clearly be extremely helpful.

What sorts of experiments might one want to perform in order to examine such questions? Four relevant areas in which work already is going on are: information seeking, intuitive statistics, sequential prediction, and Bayesian information processing. I will very briefly review each of these four areas.

Information Seeking. Perhaps the most direct experimental embodiment of the ideas discussed above would be information-seeking experiments, which fall in category 2 above. In such experiments, subjects must make a decision. Before they make it, they may purchase relevant information which will have the effect of modifying their opinions about either probabilities or payoffs. Edith Neimark (1961) and I have each done large experiments on situations of the sort which are simple enough so that the expected value of each possible strategy can be examined separately, thus circumventing mathematical difficulties. John Lanzetta has done a smaller experiment (unpublished draft). Major findings from these three experiments include: large and consistent individual differences; sensitivity of strategy to manipulations of costs, payoffs, and probabilities; general tendency to seek too much information. In all three experiments, the notion of expected value maximization is definitely incorrect in predicting what subjects will do, but changes in strategy in response to changes in payoffs and in probabilities are in the directions, and to some extent in the amounts, which expected value maximization notions would lead you to anticipate (a finding strongly reminiscent of some of Tanner's results).

This situation has been studied both because it is simple to analyse and because of its similarity to many real-life problems. The size and ubiquity of consistent individual differences in these three experiments create discouragement about this approach to the dynamic decision-making problem. On the other hand, the information seeking responses used in these experiments are potentially confusing to subjects because they have explicit costs associated with them. More complex tasks leading up to a final decision without the extra added feature of costs incurred along the way might be better research tools; the surprisingly good results obtained in Bayesian processing experiments add weight to this possibility. Still, one man's poison is another man's potion; these situations, with their large, consistent individual differences, might well be the basis for a test which could discriminate good from bad decision makers.

Intuitive Statistics. I can do no better on this subject than to quote my own review of this literature in the 1961 *Annual Review of Psychology:*

> If the environment is stationary but man's information about it is changing, then a decision task is likely to look very much like a problem in statistics. In fact, most statistical tests can be treated as mathematical models for human behavior, and people can be examined to see if they in fact conform to these models.
>
> Irwin and his collaborators have exploited this possibility in a series of experiments on what they call the "expanded judgment" situation. Subjects are presented with a sequence of cards, on each of which a number is written. Statistical properties of the sequence of numbers are varied, and subjects are required to make judgments about the population which the numbers represent. In the first of these experiments (1956), subjects

judged, after seeing each of 20 cards, whether the mean of a 500-card deck was greater or less than zero, and also indicated their confidence in their judgments. The mean confidence ratings (algebraically added so that they also represent judgments of greater or less than zero) were directly related to the mean of the sample; and the absolute magnitude of the confidence ratings increased with sample size and decreased with sample variability. Another part of the same experiment confirmed these results in a slightly different situation in which cards from two decks were simultaneously exposed, and the subject judged which deck had the larger mean. In another experiment (1956) subjects were required to look at cards until they had seen enough and then stop; they stopped sooner when the means were more different, and later when variability was larger. In yet another experiment (1957), subjects were paid for correct decisions and charged for each card looked at, while lower cost and greater prize increased that number. In addition, the findings of the previous experiment concerning difference between means and amount of variability were confirmed. Unfortunately it is impossible to calculate whether subjects were or were not adopting an expected value maximizing strategy, or something like it. For one thing, a peculiar payoff scheme was used which meant that subjects could not tell the relationship between the amount of imaginary money they had won or lost and the amount of real money they would eventually be paid. Furthermore, subjects had no information about the distribution from which they were sampling, and consequently no optimal strategy can be calculated (except for one unreasonable one based on the peculiarity of the real payoff scheme).

Becker (1958) was also concerned with optional stopping, but he used an experimental design in which the population sampled was binary. Since the population distribution is known except for one parameter, it is easy to calculate optimal strategies in the sense of statistical decision theory. Becker used Wald's model for a two-action decision problem with optional stopping, and compared the behavior of his subjects with the model. He concluded that although people did not perform exactly as the model would require, they came pretty close. He found consistent individual differences which are interpretable as variations in parameters of the model. Unfortunately, Becker used a non-linear and peculiar payoff scheme similar to the one used by Irwin and Smith (1957); it remains unclear what effect variations in real payoff might have on performance.

The upshot of these studies of man (or rather, college student) as statistician is that he makes a fairly good one. In all cases the differences are in the proper directions, though they are not always the proper amounts. (The findings of direct probability estimation experiments discussed earlier are similar.) Of course, the only task studied in these experiments is the estimation of the mean. It is an interesting experimental question whether man can perform other statistical tasks well on an intuitive basis. It seems unlikely, for example, that men can estimate variances as accurately as they can means; in fact, some of the minor findings of the Irwin, Smith, and Mayfield (1956) experiment suggest this. (Edwards, 1961a, pp. 489–490.)

Information-seeking and intuitive statistics share one common characteristic: the subject's information may change, but the environment to which that information is relevant is stationary, and furthermore, the subject knows it. Thus they both belong in category 2 (or 3) of the taxonomy given above. As far as I can see, it would make little sense to study information-seeking or intuitive statistics in any other way. But the other two categories of research on dynamic decision making, sequential prediction and Bayesian information processing, are different; it is possible to study them in a stationary environment, but it is also sensible to study them when the environment is changing, and many of the most interesting problems arise only if you do so.

Sequential Prediction. The entire probability learning literature, with all its variants and offshoots, is appropriate to consider as work on dynamic decision making. Some experiments belong in category 2, others in category 5. I have no intention of reviewing or even referencing that literature here. Nor can I refer to some published review; none exists. But you can get a start into the literature from Edwards (1961b).

Recent variants on this sort of work have taken two new and interesting turns. Kochen and Galanter (1958) have studied sequential prediction with a repeated sequence such that subjects could learn to get all their predictions correct. I find myself unable to form very strong opinions about what, if anything, they found; they collected almost no data. But such experiments might be remarkably interesting especially if it were found that small amounts of noise in the input had very large disruptive effects on performance. A relevant consideration is that, according to unpublished gossip, subjects can without special training become able to predict infallibly sequences of 100 binary events; with training in the use of special mnemonic devices, that length can be increased to 1000.

The other new turn is concerned with the probability tracking task (category 4) which Gordon Robinson and I have been studying; a report of the work, which is Robinson's Ph.D. thesis, will soon be published. A binomial distribution whose mean changes irregularly at irregular intervals is displayed to a subject by using two flashing lights, and he must track that mean continuously with a tracking handle to which a meter reading from 0 to 100 is slaved with no control dynamics. The surprising finding is that men perform this difficult task extraordinarily well—as well as the most sophisticated computer program we have so far been able to think up and far better than any known statistical test of significance. For changes of size 0.12 or larger, subjects will detect that a change has occurred in about 8 flashes, and will come within a 0.05 band around the new mean within about 16 flashes. After coming within that band, subsequent mean error is too small for our equipment to measure, on the order of 0.003. That simply indicates that no consistent bias was found. RMS error, a sort of standard deviation measure, is only 0.07 on the 0-to-1 probability scale; it is appropriate to remember that on that scale the variance is much smaller than the standard deviation.

Bayesian Information Processing. Bayes' Theorem and the rapidly developing field of Bayesian statistics have an especially intimate relationship to dynamic decision theory. They provide the uniquely appropriate tool for combining old information with new, if the information is, or can be made, fundamentally probabilistic in nature. Thus, in conjunction with expected value maximization, they provide a model for how men should use information.

Perhaps the most substantial and successful use of these ideas is in Tanner's notion of the ideal observer, who is definitely Bayesian. I will not review this work here; instead I will simply cite Licklider's excellent review (1958). Tanner has pioneered a new way of looking at the relationship between optimal models and behavior. He, in effect, asks how much worse would the environment have had to be in order that the ideal observer should perform no better than his subjects. By comparing the degraded environment which might have led an ideal observer to do no better than his subjects with the real environment his subjects faced, he gets a measure of their efficiency. That measure is one of his fundamental dependent variables. The fact that it shows orderly and sensible relationships to independent variables does not necessarily argue in favor of the notion that real subjects are ideal observers faced with conditions more difficult than the physical circumstances indicate (perhaps because they come equipped with "noise in the head" which is added to external noise). However, it does argue in favor of using that dependent variable; no one can afford to ignore orderly and sensible experimental findings no matter how plausible, or otherwise, the assumptions used to obtain the measures may be (so long as those assumptions do not compel the sensible relationships found).

Hays, Phillips, and I have been concerned with a more direct examination of human ability to process probabilistic information, in a pseudomilitary task in which explicit prior and conditional probabilities and nonprobabilistic information are provided, and subjects are asked to estimate posterior probabilities. They do it surprisingly well, and training in the Bayesian approach to the task improves their performance. Apparently man can function rather well as a Bayesian information processor. More work along these lines is in progress.

Strictly speaking, Bayesian procedures are applicable only to stationary environments. But there is much less to this limitation than meets the eye. In exactly the same sense, the Bush-Mosteller model is concerned only with stationary processes. A number of devices for using stationary mathematics to fit nonstationary real life are found in engineering and mathematics, and the Bayesian approach lends itself well to some of these devices. Any lawful phenomenon is stationary in the sense that the laws controlling it do not change with time (or that the super-laws controlling these laws do not change).

This discussion has been intended to highlight the currently active psychological research areas which I perceive as most relevant to the dynamic decision problem. These areas have in common mathematical simplicity (to the point of triviality), experimental simplicity, and remoteness from many of the most fascinating real-life problems. It is quite possible that some of the disorderly nature of the findings of such work, especially in information seeking and binary prediction, is a function of the triviality and unrealistic nature of the task from the subject's point of view; the results in Bayesian information processing suggest this. In any event, it is clear that this area of psychological study has been severely handicapped for lack of adequate theoretical underpinnings. Such underpinnings, as they might be embodied in the ideal dynamic decision maker, would probably suggest new and better experimental situations in which to study dynamic decision making, as well as new and better ways of analyzing the data and of making predictions. Although the mathematics of an ideal dynamic decision maker should be applicable to all four of these research areas, I would be willing to bet that that notion would be most appropriately tested in situations more complicated, and perhaps more realistic, than any (except perhaps the Bayesian processing situation) yet studied.

MATHEMATICAL SUBSTRATES

Ideas which might be relevant to the definition of the ideal dynamic decision maker appear in a number of diverse places. I have not seen any previous discussion of the possibility that they might interrelate. They are focussed around three topics: dynamic programming, optional stopping in statistics, and the value of information in Bayesian statistics.

Dynamic Programming

For the general nature of this topic, let me quote Bellman and Kalaba (1958b):

> (Adaptive processes) arise in practically all parts of statistical study, practically engulf the field of operations research, and play a paramount role in the current theory of stochastic control processes of electronic and mechanical origin. All three of these domains merge in the consideration of the problems of communication theory. . . . We propose to illustrate how the theory of dynamic programming can be used to formulate . . . questions that arise in these studies. . . . We wish to study multistage decision processes, and processes which can be construed to be of this nature, for which we do not possess complete information. . . . We may not be in possession of the entire set of admissible decisions; we may not know the effects of these decisions; . . . and we may not even know the over-all purpose of the process. . . . The basic problem is that of making decisions on the basis of the information that we do possess. . . . Particularly processes of this type have been treated in a number of sources, such as the works on sequential analysis . . . , the theory of games . . . , the theory of multistage games . . . , and papers on learning processes.

Bellman and Kalaba invoke, here and in many other publications, a single principle of optimality: "An optimal policy has the property that whatever the initial state and initial decision are, the remaining decisions must constitute an optimal policy with regard to the state resulting from the first decision." This rather meek-sounding notion apparently has rather drastic consequences in making possible the solution of otherwise unmanageable optimum-finding problems.

A particular and interesting instance is a problem first examined by Kelly (1956). A gambler receives advance information about the outcomes of a sequence of horse races over a noisy telephone line which has a probability p of transmitting the correct answer and a probability q of transmitting an incorrect answer. If the gambler has χ dollars and an accommodating bookie, he might be tempted to maximize expected value by betting all the money he has on each race. If he did so, he would surely go broke instead the first time the phone line transmitted a wrong answer. He might, therefore, prefer to maximize the expected value of the logarithm of his capital at the end of N stages of play. If so, he should (assuming even odds, and also assuming that $p > 0.5$) bet $(p - q)\chi$.

If χ is reinterpreted to be his capital at the moment each wager must be made, he should always follow this policy. If he does, his expected capital at each step is $\log \chi + \log 2 + p \log p + q \log q$. Note the emergence of the Shannon information measure from these purely economic considerations.

Bellman and Kalaba (1957a) have greatly generalized and extended Kelly's results. The limited initial capital assumption is, however, retained; such an assumption seems inconvenient and irrelevant to many human decision tasks. Marcus (1958) has carried the problem still further.

An examination of the applicability of these ideas to the study of human decision processes is, I believe, most urgent and is most likely to be a fruitful task for the next two or three years for researchers in this area. But the mathematical content of the literature guarantees that it won't be easy! In case you are interested, the reference list cites a number of other relevant papers by Bellman and collaborators, and a few by others.

Optional Stopping in Statistics

Any model which asserts an optimal information-gathering strategy is an appropriate model for dynamic decision processes. Wald (1947, 1950) has pioneered in working out observational procedures which entertain three alternative courses of action at each choice point: take final action A, take final action B, or take another observation. Although the Wald approach to this problem is based on rather non-Bayesian ideas, his models (or those developed by later workers on the same sort of problem) might well be used as ideal models for experiments of the Irwin type.

The most enthralling new development concerning optional stopping, from my point of view, is what happens to it if you take a Bayesian approach to statistics. Here the question of optional stopping plays no special role, but instead is simply a small part of the much more general discussion of the value of information.

Bayesian Statistics and the Value of Information

Recently, under the leadership of L. J. Savage, Robert Schlaifer, and Howard Raiffa, a new approach to the problem of statistical inference has been gaining wide impetus and acceptance. This approach is called Bayesian mostly because it makes heavy and explicit use of the kind of reasoning which is embodied in Bayes' Theorem.

The heart of Bayesian statistics lies in two notions. One is the familiar idea that the purpose of statistical inference is to guide action wisely. The other is the notion of personal probability. In the version of this notion proposed by L. J. Savage (1954) and explicitly embodied in Bayesian statistical procedures, the prior probability of an event is simply the opinion which a decision maker (whom I will hereafter call you for short) has about the likeliness that the event will occur. I can discover the numerical value of that probability by presenting you with various bets about the event and noting which you will accept and which you will reject; simple calculations will then indicate your subjective probability for the event.

Paradoxically, an important effect of the introduction of personal probabilities into statistics is to make the operations of statistical inference far more objective. In most statistical operations, the actual prior distribution assumed turns out to be completely irrelevant (so long as certain pathological kinds of distributions are not entertained by the decision maker). However, it does not follow that the idea of personal probability plays no real role in Bayesian statistics. Its primary role is philosophical; it permits me to say "The probability that my next child will be a boy is 0.5," "The probability that the Tigers will win the pennant this year is 0.0001," and "The probability that this mean difference would have been obtained if the true mean difference were zero or negative is 0.05," and to mean the same thing by the word "probability" in all three sentences. Note, incidentally, that in the last example I do not use a single-point null hypothesis such as would be necessary in traditional statistical inference. One of the glories of Bayesian statistics is that ridiculous null hypotheses (by ridiculous I refer opprobriously to null hypotheses which have prior probability 0 of being correct) are unnecessary and can be replaced by sensible null (or non-null) hypotheses. Furthermore, the numerical probability that a hypothesis is correct (*not* a numerical probability that the data would have been obtained if the hypothesis were correct) is the normal output of a Bayesian calculation.

Calculation of the value of information pervades Bayesian statistical procedures. Here are several definitions from Raiffa and Schlaifer's splendid book (1961).

> ... An experiment e has always been evaluated by computing the "absolute" utility $u^*(e, z)$ for every possible z (state of the world) and then taking a weighted average of these absolute utilities ... an alternative procedure is available. ... For each z we can compute ... the increase in utility which would result if the decision maker learned that $z = z$ and therefore altered his prior choice of an act a; and we can then take a weighted average of these utility increases. The increase in utility which results or would result from learning that $z = z$ will be called the value of information z ... before z is known the decision maker can compute the expected value of sample information. ... The economic significance of this quantity is due to the fact that the expected terminal utility of any experiment is the expected utility of immediate terminal action augmented by the expected value of the sample information. ... The expected net gain of sampling ... is now naturally defined to be the expected value of the sample information less the expected cost of obtaining it.

The concepts concerning value of information defined above are so heavily used in the Raiffa-Schlaifer book that one or more of them appears on nearly every page. They completely control the decisions about when to stop observing and about optimal sample size, and they are intimately bound up in all phases of Bayesian analysis.

Of particular interest in this connection is a quantity b which plays a central role in Bayesian analysis. It is a measure of the precision of an observation or set of observations; for normal distributions, it is inversely proportional to the variance. It is closely related to another quantity called I, which is the Shannon-Wiener measure of information, familiar in information theory. Good places to start looking at the nature of this relationship and the role of I in Bayesian statistics are pp. 50, 153, and 235 of Savage's epoch-making book (1954).

That is as far as I can carry this discussion of Bayesian statistics. Those who are interested in these ideas might be interested in reading either of two books by Schlaifer (1959, 1961). Both of these books are nonmathematical and nevertheless both are deep. The 1959 book is the more important historically, because it is the first full-dress attempt to apply the Bayesian ideas at the level of practical statistical applications. The 1961 book, however, is probably the more congenial to psychologists, since it is less concerned with dollars-and-cents decisions and more concerned with scientific applications.

An interesting question, to which I do not know the answer, concerns the relationship between Bellman's work on the dynamic decision problem and these Bayesian developments concerning the value of information. They sound as though they are very closely related; quite possibly they complement each other. If so, we may well find that the two sets of ideas may be combined into a very general model for the ideal dynamic decision maker. At any rate, there is a wealth of material here for researchers interested in human rationality, and deviation therefrom, in the acquisition and use of information in the course of sequences of decisions.

A PROBABILISTIC INFORMATION PROCESSING SYSTEM

Recent developments, both experimental and theoretical, in the theory of human and machine information processing have made both necessary and feasible the development of a new class of systems which I have called Probabilistic Information Processing systems, or PIPs. Such PIPs apply, in a practical system context, most of the ideas I have been talking about and at the same time provide motivation and a versatile and powerful research setting for studying human probability estimation, information processing, and decision making.

A great deal of the information which military commanders and other military and business decision makers must use as the basis for their decisions is fallible, incomplete, or both. Sensor returns often contain false positives and missed signals. Other sources of information are similarly imperfect. It is the task of information-processing systems to correct errors in the basic data if possible but, in any case, to base good decisions on this sometimes-bad information.

Customary procedure for coping with missing information is to guess or to seek more information. Customary procedure for coping with fallible information often is to define some kind of criterion of reliability, or of confirmation, or something of the sort, and to instruct parts of the information-processing systems, into which these fallible data enter, not to pass them on unless they exceed the criterion. Such threshold procedures are obviously unsatisfactory; they must inevitably exclude relevant, correct information and pass on incorrect information. Scientists in disciplines which must routinely cope with variable, fallible data have found it necessary to adopt formal procedures based on probability theory to process, interpret, and base decisions on their data. These procedures are collectively called statistics. The basic idea of PIP is that the same techniques which have proved so useful in dealing with fallible scientific data will also prove useful in dealing with fallible military data. In particular, the appropriate techniques are those of Bayesian information processing.

Bayesian procedures take as inputs the prior probability of each conclusion of interest, an observation, and a set of probabilities that the particular observation would have been made given each of the possible diagnoses, and give as output the posterior probability of each conclusion in the light of the evidence. This set of posterior probabilities then serves as the prior probabilities for the next Bayesian calculation, which takes place as soon as a new item of information comes in. Thus, by means of Bayesian calculations a PIP can have at every moment a distribution of opinions about the questions with which it is concerned. If information is good, relevant, and recent, that distribution will have a very high peak, which means that a single conclusion has a very high probability relative to all other possibilities. If information is poor, irrelevant, or old, then the flatness of the distribution will reflect increased uncertainty about what is going on in the outside world.

Two different kinds of questions which a PIP might be designed to answer can be distinguished. One has to do with diagnoses. The simplest kind of diagnostic task is simply the task of identification. Is a particular constellation of objects on which information is available from several sources a missile launching site or not? A more complicated form of the diagnostic task arises when the diagnosis is abstract, as it presumably must be in threat evaluation systems. Is this air activity indicative of impending attack or not?

The other kind of question which a PIP might be designed to answer is concerned with what might be called parameter estimation, and is most relevant to detection and tracking systems. Parameter estimation questions have to do with such variables as speed, altitude, direction, or number; the task of a PIP performing parameter estimations is to define as sharply peaked a distribution function or functions as possible over a very large or perhaps infinite ordered set of possible values. Example: how rapidly is that convoy moving and in what direction?

If the appropriate inputs to PIP can be made, the actual data processing within a PIP should consist primarily of repeated applications of Bayes' Theorem and of related kinds of arithmetic. The prior probabilities, which are the main stumbling block in application of Bayesian procedures to scientific data processing, are no problem for PIP; the output of any calculation is the prior probability distribution for the next calculation. Only the conditional probabilities of the observations given each possible diagnosis must be provided to PIP—in addition, of course, to the observations themselves, which are necessary inputs to any information-processing system.

These conditional probabilities can be calculated directly from knowledge of sensor characteristics in some cases. More often, however, such calculations would be seriously misleading, since it is difficult to take into formal consideration all of the variables relevant to the momentary conditional probability values. However, laboratory research reviewed earlier in this paper indicates that men are extraordinarily good at estimating such probabilities—under some circumstances they are better than optimal linear models. So one fundamental feature of PIP, as I conceive of it, is that men will serve as transducers for probabilities and will thus generate the conditional probabilities required for Bayesian information processing.

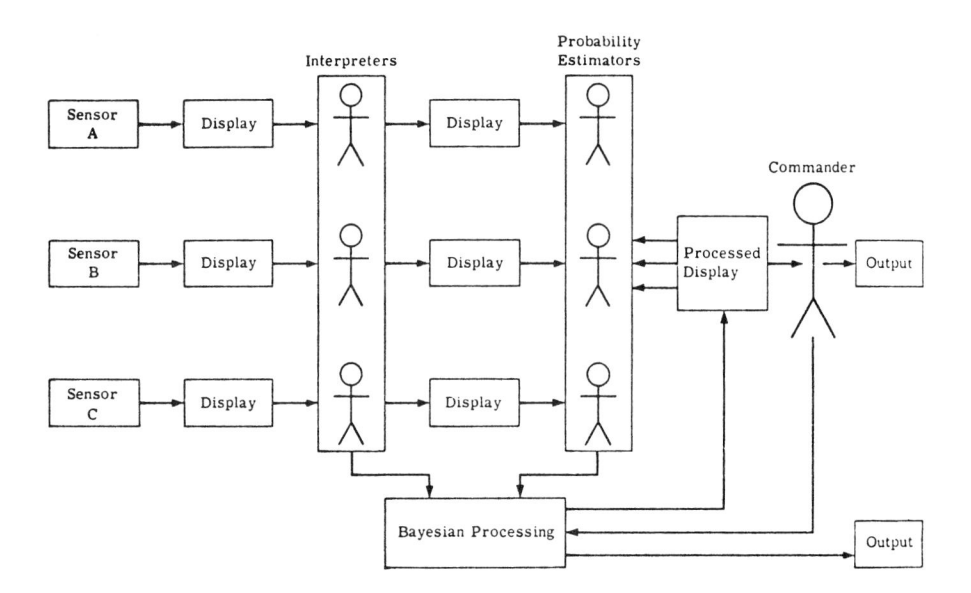

Figure 1. A PIP for threat evaluation.

A PIP for Threat Evaluation. These rather sketchy ideas of how a PIP might work can be given a slightly more concrete form by means of a block diagram. The diagram is a threat evaluation system which attempts to maintain surveillance over the enemy ground area from which a threat might originate and uses a number of different technical sensors. Each sensor return is interpreted, and a display of the interpreted information is presented to a group of probability estimators, one per sensor. These men associate conditional probability estimates (for each possible diagnosis) with each new item of information. These estimates, together with enough information to identify the sensor return and location to which they refer, are passed along to a Bayesian processor. That Bayesian processor can be a man, a group of men, or a large digital computer, depending on the nature and resources of the system. Whatever its nature, the function of the Bayesian processor is to generate a display of the distribution functions obtained by its Bayesian processing. This display, the heart of the system, may take a variety of forms, depending on the nature of the diagnosis being made. If abstract diagnosis of a threat is the issue, a bar graph or a pie diagram over all possible kinds of threats are appropriate displays. Similar displays would be appropriate for identities. If location is the issue, one appropriate display would be an ellipse, with the long axis in the direction of movement, the center on the modal value of the (assumed) bivariate normal distribution of possible positions, the sizes of the two axes as a measure of the variances of the posterior distribution, and the ratio of major to minor axes as a measure of the covariance. Other forms of displays would be appropriate for other kinds of uncertainties. This display of distributions is looked at by at least two kinds of people. One is the probability estimators, who may use the system's uncertainties as well as the interpreted sensor returns in making their estimates; this provides a kind of feedback. Another user of this display is an officer of long experience (here designated commander) who may modify the distributions generated by the processor to take into account matters of enemy strategy and tactics, which are obviously relevant but are not the business of the probability estimators or the sensor interpreters. The output of the system may simply be the processed display. Alternatively, it can be a set of decisions based on that display, or a set of recommendations. In the latter case, the processor, rather than the commander, may originate the recommendations.

A PIP has no advantage over a more traditional deterministic information-processing system unless its special capability, the ability to accept and generate explicit numerical probabilities, is necessary to successful performance of the system mission. It is therefore an improvement over a deterministic system only if the input information is fallible, or the relation of input information to output diagnostic categories is ambiguous or uncertain, or the output is required to be in explicitly probabilistic form. If one or more of these three characteristics obtains, a PIP should be superior to a deterministic system. The extent of that superiority will, of course, depend on specific matters which vary from system to system and from time to time. Under some quite plausible circumstances, a PIP should be able to produce quite usable ouputs, while a deterministic system would be completely baffled.

It is perhaps useful to point out that the strategy of information processing used by PIP differs in an important way from that used by deterministic information-processing systems. Most deterministic information-processing systems begin by performing an operation which might be called "cleaning up the data." In this clean-up operation, information judged irrelevant or likely to be incorrect is excluded, and a tidy, orderly display of relevant information plus first-order deductions from it (e.g. identity) is prepared. Thereafter, an evaluation of the meaning of this cleaned-up information is made.

PIP works differently. It does not achieve order by throwing out information which may or may not be irrelevant or incorrect. Instead it assesses the correctness and relevance of every item of information that comes its way and processes them all by means of an orderly mathematical process which takes formal account of the degree of correctness and of relevance of each item of information. This orderly mathematical process produces an orderly display. But that orderly display already contains an evaluation of the meaning of the information, because that evaluation was applied to each incoming item of information in the course of assessing its relevance to desired system output. So the two stages of operation of deterministic information-processing systems are completely mingled and cannot be separated in PIP.

In conclusion, I would like to point out the research requirements for PIP. Obviously there are a number of system logic problems which must be solved. More interesting to the psychologist, though, is the research on probabilistic displays, on human probabilistic judgment, and on human Bayesian processing which are clearly implied in Figure 1. These problems, of course, are intimately tied up with the dynamic decision theory research areas already reviewed. I believe that the development of dynamic decision theory, and the provision of its research underpinnings, are necessary prerequisites to the translation of PIP from an idea to a real system. I also believe that the merits of PIP as an idea are sufficiently substantial to make its further exploration and eventual embodiment in real systems both necessary and inevitable.

Acknowledgments

The work reported here was for the most part sponsored by the United States Air Force under Contract AF49(638)-769, monitored by the Air Forces Office of Scientific Research of the Air Force Office of Aerospace Research, and carries Document Number AFOSR–1402. Work on the idea of a probabilistic information-processing system was sponsored by the Astrosurveillance Sciences Laboratory, Electronics Research Directorate, Air Force Cambridge Research Laboratories, Office of Aerospace Research (USAF), under contract AF19(604)–7393, monitored by the Operational Applications Laboratory, Deputy for Technology, Electronic Systems Division, Air Force Systems Command.

BIBLIOGRAPHY

Aseltine, J. A., Mancini, A. R., & Sarture, C. W. A survey of adaptive control systems. *IRE Trans. on Automatic Control,* 1958, PGAC-6, 102-108.

Becker, G. M. Sequential decision making: Wald's model and estimates of parameters. *J. exp. Psychol.,* 1958, **55**, 628-636.

Bellman, R. A problem in the sequential design of experiments. *Sankhya,* 1956, **16**, 221-229.

Bellman, R. *Dynamic programming.* Princeton Univer. Press, Princeton, 1957. (a)
Bellman, R. On the application of the theory of dynamic programming to the study of control processes. *Proc. Sympos. on Nonlinear Circuit Analysis.* Polytechnic Inst. Brooklyn, 1957, 199-213. (b)
Bellman, R. Dynamic programming and stochastic control processes. *Information and Control,* 1958, 1, 228-239. (a)
Bellman, R. On the representation of the solution of a class of stochastic differential equations. *Proc. Amer. Math. Soc.,* 1958, 9, 326-327. (b)
Bellman, R., & Kalaba, R. On the role of dynamic programming in statistical communication theory. *IRE Trans. on Inform. Theory,* 1957, IT-3, 197-203. (a)
Bellman, R., & Kalaba, R. Dynamic programming and statistical communication theory. *Proc. Nat. Acad. Sci. Wash.,* 1957, 43, 749-751. (b)
Bellman, R., & Kalaba, R. On communication processes involving learning and random duration. *IRE Natl. Convention Record,* 1958, part 4, 16-21. (a)
Bellman, R., & Kalaba, R. Dynamic programming and adaptive processes—I: Mathematical foundation. *The RAND Corporation P-1416,* 1958. (b)
Bellman, R., & Kalaba, R. Invariant imbedding, wave propagation and the WKB approximation. *Proc. Nat. Acad. Sci. Wash.,* 1959, 44, 317-319. (a)
Bellman, R., & Kalaba, R. Functional equations in adaptive processes and random transmission. *The RAND Corporation P-1573,* 1959. (b)
Bellman, R., & Kalaba, R. On adaptive control processes. *The RAND Corporation P-1610,* 1959. (c)
Busgang, G., & Middleton, D. Optimum sequential detection of signals in noise. *IRE Trans. on Inform. Theory,* 1955, IT-1, 5-18.
Edwards, W. Theory of decision making. *Psychol. Bull.,* 1954, 51, 380-417.
Edwards, W. Behavioral decision theory. *Annu. Rev. Psychol.,* 1961, 12, 473-498. (a)
Edwards, W. Probability learning in 1000 trials. *J. exp. Psychol.,* 1961, 62. (b)
Irwin, F. W., & Smith, W. A. S. Further tests of theories of decision in an "expanded judgment" situation. *J. exp. Psychol.,* 1956, 52, 345-348.
Irwin, F. W., & Smith, W. A. S. Value, cost, and information as determiners of decision. *J. exp. Psychol.,* 1957, 54, 229-232.
Irwin, F. W., Smith, W. A. S., & Mayfield, J. F. Tests of two theories of decision in an "expanded judgment" situation. *J. exp. Psychol,* 1956, 51, 261-268.
Karlin, S., Brandt, R., & Johnson, S. On sequential design for maximizing the sum of n observations. *Ann. Math. Stat.,* 1956, 27, 1061-1074.
Kelly, J. A new interpretation of information rate. *Bell Syst. Tech. J.,* 1956, 35, 917-926.
Kochen, M., & Galanter, E. H. The acquisition and utilization of information in problem solving and thinking. *Information and Control,* 1958, 1, 267-288.
Licklider, J. C. R. Three auditory theories. In S. Koch (Ed.) *Psychology: A study of a science.* I. New York: McGraw-Hill, 1958.
Marcus, M. The utility of a communication channel and applications to suboptimal information handling procedures. *IRE Trans. on Information Theory,* 1958, IT-4, 147-151.
Neimark, Edith. Information-gathering in diagnostic problem-solving: A preliminary report. *Psychol. Record,* 1961, 11, 243-248.
Raiffa, H., & Schlaifer, R. *Applied statistical decision theory.* Boston: Division of Research, Graduate School of Business Administration, Harvard Univ., 1961.
Robbins, H. Some aspects of the sequential design of experiments. *Bull. Amer. Math. Soc.,* 1952, 58, 527-535.
Robbins, J. A sequential decision problem with a finite memory. *Proc. Nat. Acad. Sci. Wash.,* 1956, 42, 920-923.
Savage, L. J. *The foundations of statistics.* New York: Wiley, 1954.
Schlaifer, R. *Probability and statistics for business decisions, an introduction to managerial economics under uncertainty.* New York: McGraw-Hill, 1959.
Schlaifer, R. *Introduction to statistics for business decisions.* New York: McGraw-Hill, 1961.
Wald, A. *Sequential analysis.* New York: Wiley, 1947.
Wald, A. *Statistical decision functions.* New York: Wiley, 1950.

28

Command-and-Control and Management Decision Making

Ramon J. Rhine

INTRODUCTION

We are faced today with the frightening prospect of controlling the movements and actions of a huge force of atomic bombers and intercontinental ballistic missiles. Assuming that an intercontinental ballistic missile takes about a half-hour from a potential launch site to impact and that it takes about 15 min to identify an aggressor's launch and disseminate the information, there remain only 15 min to make the decision and give the command to launch a strategic retaliatory force. Control must be precise to avoid unleashing atomic weapons without a clear-cut decision from the very highest levels. At the same time, the national policy of deterring potential aggressors is based on a capability for instant retaliation. Under these circumstances, minutes, and indeed seconds, are crucial. With problems of this significance and magnitude, it is not surprising to find in recent years the development of automatic command-and-control systems to support human decision making. The need is so great that tremendous energies have been applied to command-and-control problems and important strides forward have been made.

The design, development, and implementation of large-scale military command-and-control systems have been made possible by the recent growth of computer technology. In these military systems, information is computer processed to monitor the status of an operation, to enable key personnel to make proper decisions, and to help ensure prompt implementation of these decisions. The problems and requirements leading to military command-and-control systems are very much the same as those faced by industrial or government management in running a complex operation, despite differences in scope or degree (Skaggs, 1962). Experience gained in the design and engineering of command-and-control systems is directly applicable to management control systems because command-and-control concepts apply to management information processing and high-level decision making.

From *Human Factors*, 6, 1964, 93-100. By permission of the Human Factors Society.

Management-Oriented Command-and-Control

During the time that the military command-and-control systems were being developed, business applications were started in areas where the automation of paperwork increased efficiency. Often computers were put to work as if they were huge filing systems. A computer was found to be a very useful tool in keeping track of financial accounts, and in many companies the accounting systems became dependent for improved effectiveness upon computer applications. Once the computer was in-house, other applications were found: personnel records were readily automated for easy access to details of information; the payroll could be processed automatically; the telephone directory of company personnel could be updated regularly; salary information could be automated; budget reports could be output; etc. Since the computer operation served as a central point for the processing of these kinds of information, automatic data processing began to replace manual processing and filing previously done semi-independently in different departments. Automatic data processing became a tool for "control" in a number of corporate areas, particularly financial areas.

Typically, this type of "control" means keeping track of financial transactions. In contrast, the "control" of command-and-control refers not to keeping track of money, but to the control of a *total operation* (Bramson, 1962). Consequently, the typical automatic data processing set up in business today is less advanced and less complete than would be expected of the type of management system predicated upon command-and-control experience. The command-and-control type of management system is above all the direct tool of top management and top management's staff. As such, it is considerably different from the business system which emphasizes automation of clerical paperwork and/or concentrates upon details of accounting (Daniel, 1961).

Basic Functions of Command-and-Control Systems

All command-and-control systems have certain functions in common. If these functions are the same as those expected of management systems, then command-and-control systems should be reasonable models for management systems. Among the major functions of command-and-control systems are: (1) control, (2) decision making or support of decision making, (3) comparing the results of operations with plans, and (4) providing information for monitoring and as a prelude to decision making. These functions certainly have the ring of management about them; the first, control, is at the very foundation of management theory.

CONTROL

The "control" in command-and-control is illustrated by the SAGE (Semi-automatic Ground Environment) system for air defense. Under this system, when an unknown flying object is detected by air defense radar, its location and other data about it are fed automatically to a computer. The computer processes the information and compares it to what is expected in that particular air space at that particular time. Under specified criteria, the information is displayed to human beings who must decide, among other things, whether a jet fighter should be sent to intercept or visually identify the detected object. If the fighter is sent, its position is also transmitted from radar via computer to a man, who directly controls the actions of the fighter pilot by directing him through radio communication onto an intercepting course with the unknown object.

Control is the exercising of direction over the application of a force in accordance with a set of guidelines. A thermostat is a control mechanism because it exercises direction over the application of an electrical force that turns a heating unit on or off; the temperature setting on the thermostat provides the guideline determining when the electrical force is applied. Similarly, when a baseball pitcher is said to have good control, he is effectively directing a muscular force upon a baseball according to the guidelines laid down by the catcher's signals. He is said to have lost his control when he is no longer able to direct the baseball in accordance with these guidelines. In that case he is removed from his key position in the game, an action which has a parallel in management situations.

Control may be more or less direct. When the baseball pitcher is having a good day, he is able to exercise control over the ball he throws, and the catcher who tells the pitcher what to do has indirect control over the ball and direct control over the pitcher. Similarly, the interceptor pilot who flies an airplane has direct control over its course, but the officer giving directions to the interceptor pilot has direct control over the pilot and indirect control over the course of the aircraft. In industry, the workman has direct control over the number of tons of steel poured in his plant each day, but a manager has indirect control. The clerk who orders a series of items to be placed in a company's inventory is exercising direct control, but indirect control has been exercised by the manager who determines the policies setting boundaries on the amount to which inventories can be accumulated. Management control is control through people who in turn have direct control over other people or over things. Control in command-and-control systems is also exercised via people. The "control" of command-and-control is management control.

The Force of Control

The distinguishing feature of management control, in contrast say, to direct control of a machine or a weapon, is that the force in the definition of management control is not a physical force. In management control the directing force is psychological. The psychological force for control derives from the authoity vested in the manager. For example, psychological force for control derives from the manager's authority to determine salary increases for the individuals he is controlling, or from his authority to say whether they will hold their jobs, or from his authority to define the job to determine the worker's opportunity for advancement. The force of management control is like the notion of power as used in political science. The manager is in a position of power in respect to people reporting to him; by virtue of this power he holds a psychological club over his subordinates. If he is a good manager, the club is kept out of sight.

Instead, he emphasizes wherever possible good leadership practices, rewards, and other positive methods of psychological control. He gets things done without mentioning his weapons, but they are ever present nonetheless.

Not all the power resides with the manager. For example, his subordinates can band together and take actions which compromise the manager in the eyes of his superiors. Since the power is spread between employee and manager and can be tipped in one direction or the other, the good manager exercises his power judiciously. Anybody who has been in an industrial setting very long knows of some manager who had to resign because he could not handle the people working for him without causing widespread unrest and terminations. Perhaps the force of control is more evenly balanced in industrial situations than in military command-and-control operations; nevertheless, in principle, the force of control in both cases derives from the authority structure.

Control and the Man

When dealing with an electrical or mechanical control system it makes sense to say that the system itself exercises control. The thermostat, for example, literally controls temperature. Sometimes, as in the case of the SAGE system, the computer itself is programmed in advance to make many of the lower-level decisions. For example, the SAGE computer will decide that an aircraft is identified when the speed, location, and altitude indicated by radar returns correlate closely with previous information that an aircraft should be in that location at the time of detection. When the computer makes this kind of decision, it is only doing what a computer program directs. Some human has previously specified what decision will be exercised under what precisely defined circumstances. When the situation is relatively simple, well defined, and repetitive, it is economical and technically feasible to give the decision to the computer. Otherwise, decision making is still a job for humans.

The way in which people are included in command-and-control systems is one of the major differences between them and existing business systems. A system is an operating entity which contains many different components, each related to one or more of the others. Therefore, a change to one part of a system affects the functioning of one or more other parts. Simple physical systems like the thermostat consist of a few components related in a relatively straightforward manner; complex systems have many intricately intertwined subsystems and components. Automated command-and-control systems are the most complex systems existing today, for they are man-machine systems containing many interacting components. When people are deliberately included as a central part of the system operation, many complex behavior variables enter the picture and vary according to changes in other aspects of the system. Recognition of this fact is crucial to system design and system operation.

Command-and-control systems are control systems only if conceived in broader terms than physical systems. Besides the equipment for input, processing, and display, command-and-control systems include the methods and procedures for the use of this equipment, computer programs that give it "brains," and the men who operate it. It is only because the total command-and-control system includes men that it is possible to say the system itself exercises *management* control; the men who are part of the system do in fact make decisions and do in fact do things to control the behavior of other men. If the system deals with a complex environment, then the major control functions generally reside with men because no way has yet been found to prepare the machine for all the interacting variables of complex decision situations. The physical and computer program elements of the system provide the people in it (using it) with information enabling them to choose judiciously among alternative commands. Users of the system are the individuals who are operating its machinery and also the high-level decision makers. If a high-level manager uses the system only 15 min a day as a basis for making his decisions and exercising control, then during those 15 min he is part of the man-machine system.

COMMAND

Like control, command can also be illustrated by the SAGE system of air defense. Information about interceptor aircraft being controlled from various locations is continuously being sent to higher headquarters where the total picture is being monitored. Here command decisions will be made if the situation warrants it. For example, it might be decided that the country is in danger, requiring a special state of increased defensive readiness; or it might be decided to move a fighter squadron from one area to another where an attack is more likely. It is this type of decision making which illustrates the command aspect of command-and-control systems. Command-and-control systems are built to ensure the selection of proper decisions and aid in the communication of these decisions. This is also required of a management system.

A command is the communication of a decision expected to guide the actions of its receiver. This definition is deliberately worded in general terms, making no distinction between the actions of men and machines. Not all commands are for men; some commands are communications to machines, sometimes via radio signals, as in the case of space vehicles. For example, the radio commands sent to control the actions of the spacecraft Mariner II during its flight toward Venus are an instance of command consistent with our definition.

Although there is an intimate dependency between control and command, the two concepts are still quite different. The control mechanism of Mariner II activated a physical force that caused the probe to respond. If either the transmitter or receiver failed, the decision to modify the action of the probe could not be properly communicated; then, no matter how well the control mechanism was conceived, it would not perform in accordance with the guiding decision. On the other hand, if the control mechanism failed, no matter how perfectly the decision was sent and received, effective direction of the controlling force is lost. Depending upon the adequacy of transmitted communications, commands can be given that have no effect on control. And, depending upon the adequacy with which a communication is received, control can occur that is diametrically opposed to the command given.

The meaning of command in a management setting is precisely the same as in the spacecraft example, except that management commands are communications sent to and received by people. Just as the communication of a decision may be poorly sent or received by machines, so too, communications may fail between managers. Although communication of decisions between managers may be adequate, again as with machines, the control procedures may still fail to work. In that event effective management direction is lost. Consider, for example, a decision to institute a new product line. A command is given when this decision is communicated from a higher level, such as the president, to a lower level, such as a division head. The president might require that the division head construct an implementation plan and report twice each month on the comparison between the plan and actual results. If this control technique works, remedial action can be taken to keep actual implementation on course. But if the division head is receiving faulty reports, control will break down no matter how well the president's decision was communicated.

Commands are easy to recognize when they are stated in the imperative form: "You will institute this new product line." Since the imperative form communicates the decision rather exactly, managers sometimes state their decisions in just this fashion. However, in the culture of some organizations, it is considered bad form to communicate a decision in the imperative. As often as not a different tone will be taken. For example, an executive might say to a subordinate manager, "Would you take steps to start producing this new product line?" An experienced manager will not miss the meaning of the more softly worded command. Because of the power structure, "Would you do it?" really means "You will do it!" Soft wording does not necessarily rob the communication of a decision of its command nature.

Although commands can be communicated in different ways, the important point is to make sure the message about a decision gets through. Commands cease functioning as commands when they are couched in terms that fail to communicate. When commands are not clearly stated as imperatives or become too sugarcoated, the manager may find that his decisions are not being implemented. "Wouldn't it be nice if we could implement this new product line?" might be taken as a command by some subordinates and as a wish or desire by others. Of course, only a somewhat diffident executive would state such an important decision in these terms. Nevertheless, how many of us have not experienced real-life situations in which a job did not get done because the command intent of a conversation was not properly understood?

THE PLAN

In command-and-control systems, a plan may exist in various forms such as tactical doctrine, the logic of computer programs, or a formal document named a plan. The system is designed to gather data comparing actual occurrences with expected results and to help make and communicate decisions that will narrow any discrepancies between actual and planned occurrences. Precisely the same properties are desirable in a management system.

An organization's plan is its most formal set of decisions about the expected course the organization will follow, together with supporting material. Communication of the decisions in the plan to subordinates is a command, and the set of decisions which this command communicates serves as guidance for control. It has been said that plans are made to be changed, so it might seem on the surface that they should not be called formal commands. But there is nothing in the concept of command that implies finality. When decisions are changed or modified, plans change, which is to say that they are, at once, commands and dynamic.

Some plans are written and some are carried in managers' heads. Writing down a set of decisions tends to make them more organized and integrated. Nevertheless, unwritten plans—call them thoughts, ideas, predilections, attitudes—are no less real than written ones. True, communication of decisions tends to suffer when plans are not written down, but that does not totally rob such communications of their command character. However, top-of-the-head decisions unrelated to any plan, implicit or expressed, face the danger of inconsistencies. Such commands have a real probability of doing as much harm as good. This is especially true if the company is large and complex enough to water down the effectiveness of good intuition.

MONITORING

Command-and-control systems supply data for both status monitoring and the active monitoring, and the same is required of management systems. Status monitoring is the absorbing of information about the current posture of an operation. Status monitoring keeps one informed of the environment of an organization and of its posture at any given time. Although status monitoring may require no immediate action, it *is* important. When the time for action arises, a broad mental picture giving an understanding of the organization's posture is a significant guide to thinking. For example, absorbing information about many facets of the organization's posture and operation might enable a manager to sense a general lack of vigor in operations. The bits of information giving this impression may not themselves lead directly to any single action since no single piece of information dramatizes the situation by itself. But a backdrop is slowly developed that influences, in little ways, actions that occur in the course of day-to-day operations and, in a larger way, reactions to problems requiring major decisions. In a management system, as in a command-and-control system, status monitoring should keep managers informed so that they are not working in a vacuum but can understand what is happening around them.

Systematic status monitoring aids communication because it provides a group of managers with a single consistent set of information, enabling them to have overlapping views of the state of the organization. For example, at a recent meeting of managers in a large company, reference was made to a curve of personnel growth which had been shown at a managers' meeting two weeks previously as background information on the status of the company. The chief executive officer was using this curve to make a point in another context. He simply said, "What I am advocating makes sense in relation to the dotted line for personnel." The 15 or 20 managers in the room indicated in the ensuing discussion that they knew what he meant, although they did not have before them at the moment the particular curve he had in mind.

A major difficulty of status monitoring is the selection of appropriate data. Busy managers cannot be expected to absorb data describing everything about their organization, any more than an admiral or a general can be expected to spend night and day before the displays of a command-and-control system. The data for status monitoring should be carefully screened and selected so that they are of interest to the group which will see them. For example, for a top management group, they should be overviews and summaries relevant to the organization as a whole. The danger of status monitoring is the well-known frustration caused by masses and masses of reports being sent to managers who do not have time to read them. As a result, it is popular to emphasize reporting by exception, i.e., reporting only problems and unusual occurrences. Overemphasis on reporting by exception neglects the need for a broader understanding of the organization's operation and environment and also neglects the advantages of a common basis for communication and understanding. Reporting by exception somewhat simplifies the complex task of the system designer. It is a very inviting path for the lazy to take, for considerable hard thought is necessary to accomplish the difficult job of selecting from a mass of alternatives those sets of information which should be supplied for status monitoring. If healthy progress comes from the solving of difficult problems, status monitoring should not be swept under the rug with a broom labeled "exception reporting," because status monitoring serves an important function.

Active monitoring refers to the receipt of information as a prelude to action. Certain kinds of information are required either regularly or upon request to serve as a basis for systematic decision making. Making a management decision is the process of selecting for implementation one or more alternatives from a set of possible actions. Selection of the proper alternative requires the capability to obtain and evaluate different bits of data, depending upon the alternatives being considered. Command-and-control systems include features enabling the system users to request information that is not ordinarily available on a regularly scheduled basis. Although this feature is not denied to status monitoring, its use is greater for active monitoring. Since active monitoring is a prelude to action, it requires precise information about specific areas of current concern. It is not uncommon, when one set of information is absorbed, to find it stimulates a need for more detailed information on the same topic, or information on some related topic. For example, imagine a company for which an impending contract completion raises the possibility of releasing personnel. Management would want to know the number of personnel on board and the amount of work in hand or expected. If these are badly out of phase, raising the possibility of adjusting manpower resources, more information may be requested such as the breakdown of personnel by skills and seniority, and the types of skills required of jobs in hand and of those expected. Inquiries may also be made about normal attrition rates in comparison with phaseout of the jobs in hand. These data, in turn, might stimulate the need for even more detailed information, depending upon the operating environment and such factors as agreements and relations with a union.

OPERATIONS

Active monitoring implies a rather dynamic concept of operations. It is this which distinguishes the command-and-control type of management system from one which mainly automates clerical, administrative, and accounting functions. Automated command-and-control systems are man-machine systems in which the operational functions center around the men. In many military operations, the system is manned night and day. The activities of the military operation being supported by the system are continuously monitored. Some decision situations are delegated to the routine system operators and others require the guidance of the commanders or their immediate staff. The system itself operates to support the operations of the larger organization of which it is a part.

A management system being developed by a large research and development organization with personnel scattered over the world illustrates the use of a dynamic operation. The system is on a scale less grand than found in military systems, but nevertheless is identical in concept. A Management Operations and Analysis Center is in operation during normal working hours. This office is the hub of all inputs and outputs of the management system. Management Operations and Analysis either distributes computer printouts (reports) unchanged or uses them for further analysis leading to other displays, e.g., 35 mm slides of graphs and charts. Top managers will receive printed reports containing summary level information, and more detailed back-up of these same reports will be supplied to the appropriate middle managers. Through simple display methods, regularly updated visual displays can be projected in a management display room located near the offices of top managers. This information is constantly available to managers, and can be obtained immediately by operating a simple control panel. In addition, the system will have a capability for obtaining special information immediately upon request. Personnel of the Management Operations and Analysis Center are knowledgeable about the company's computer operation and have first priority for computer usage. At the request of specified managers, the interrupt feature of the computer system will be used to obtain certain special data and prepare it for rapid presentation to the requesting manager.

In addition, some oral reporting method is contemplated such as a regular Monday morning briefing for the President and Senior Vice Presidents. This briefing will call to the attention of these top personnel, events of particular management interest which have occurred during the previous week or which are expected to occur in the near future. It will also serve as a vehicle for presentation of special reports of interest which are too numerous to be reported on a cyclical basis. Such briefings will enable the Management Operations and Analysis personnel to receive feedback from top management on special information which should be prepared for them by the next meeting.

The key to the system is an ongoing, day-to-day, dynamic operation. The system does not merely produce reports that arrive each week on a manager's desk to be ignored or elbowed into the wastepaper basket; it is an active system in which operators are continuously working to prepare timely data both on a regular basis and in answer to special needs. The people operating the system have direct access to Top Management. They are kept aware of current company problems so they can anticipate the company's information needs. Of particular importance to the operation, they have control of the company's computing facilities, which makes it possible to assign priorities for use of the computing according to the needs of management decision making.

This concept of a dynamic and ongoing operation is found and taken for granted in military command posts, but it is not common among other organizations. It seems probable with the technology now available that management of more and more large organizations will take advantage of the knowledge gained in command-and-control types of operations to institute ongoing operational systems for the dissemination of management information.

REFERENCES

Bramson, A. D. Unified management operations. *Business Automation,* Jan. 1962, 26-31.
Daniel, D. R. Management information crisis. *Harvard Bus. Rev.,* 1961, 39, 111-121.
Skaggs, B. Computer assistance to top management. *Data Processing,* 1962, 3, 13-19.

Chapter Eight

Empirical Research

Since we have discussed the major points in the part introduction, little remains to be said about the material presented in this chapter. The first article is the empirical embodiment of notions discussed by Kidd in Chapter Seven. Briefly, it describes a particular laboratory vehicle (a simulated air traffic control system) and a program of research designed in connection with it. We should perhaps point out that the *research* summarized here is only a portion of that actually carried out under this project. The work continued for several additional years and is still recognized as one of the most successful efforts of its kind ever attempted. Few large-scale simulation programs have been as fruitful in terms of both specific *applications* and general system *principles* (see, for example, Kidd, 1959).

The second paper describes some of the initial studies in another long-term research program. In this case the vehicle was a somewhat idealized version of a military command-control system adapted to a laboratory setting. It should be quite apparent that this research was concerned with the same basic problem discussed by Edwards in Chapter Seven: processing of probabilistic information by decision systems. For a period of about six years this project studied variations on the PIP theme in order to determine the relative advantages to be gained by automating parts of the inference process (aggregation of prediction data) under various realistic circumstances. For example, how much could a PIP-like design be expected to enhance decisions made on the basis of low- as opposed to high-quality predictive data? What would its advantages be under various speed and load demands? Taken together, the results have been shown to provide substantial support for the notions represented in PIP (Howell, 1967). Thus we again have an example of a systems research program which was reasonably successful in combining practical and scientific objectives.

The Smith and Duggar paper, with which we close the chapter, is perhaps a bit more typical of empirical simulation research than either of the other examples. Its objectives are considerably more limited; its results have more direct and immediate applicability. Essentially, it asks and answers a specific question: do group displays enhance group effort in a decision problem? In terms of speed, at least, the answer seems to be "yes."

As we have emphasized repeatedly in the foregoing pages, research—systems or otherwise—which is designed exclusively to answer a particular question can be applied to other situations only at great risk. Only when a serious effort is made

References will be found in the General References at the end of the book.

to understand the mechanisms *responsible* for a result, such as the facilitating effect of group displays in this case, can the result be generalized with any degree of confidence. One approach to a broader understanding of system data is illustrated in the programs described in the first two papers in this part. Essentially it involves supplementing the systems simulation work with a corollary effort at a more fundamental research level. This enables ideas generated at the simulation level to be subjected to immediate test in more controlled surroundings, while at the same time suggesting new directions for the simulation program to take. Unfortunately, circumstances seem rarely to have fostered this sort of integrated effort.

29

Some Concepts and Methods for the Conduct of System Research in a Laboratory Setting

Paul M. Fitts, Lowell Schipper, Jerry S. Kidd, Maynard Shelly, and Conrad Kraft

General acceptance has recently been gained for the idea that human-factor specialists can make an important contribution to the design of man-machine systems. This contribution lies in helping to decide such important questions as those pertaining to the allocation of functions between men and machines, planning for the work loads on individual operators, and the optimum coupling of men and machines. However, few efforts have been made to bring such system-design problems into the laboratory where they can be subjected to controlled experimental study. This lack of work has been due chiefly to lack of the necessary

From *Air Force Human Engineering, Personnel and Training Research,* Publication 455, Division of Anthropology and Psychology, National Academy of Sciences-National Research Council, Washington, D.C., 1956.

The project was initiated in 1950, work being done for the National Research Council with funds provided by the Air Navigation Development Board. Since 1952 the research program has been supported by contracts (AF 33(616)-43 and AF 33(616)-3612) between the Wright Air Development Center and the OSU Research Foundation, with Drs. Ralph Queal and James McGuire acting as Project Scientists. Dr. Fitts has acted as supervisor of all three projects. The early experimental work on system problems was under the direction of Dr. Schipper; current research on these problems is under the direction of Dr. Kidd. Many other individuals have contributed to this system-research program. Principal among these are Dr. John Versace and Dr. Earl Alluisi. Dr. Alluisi is in charge of the supporting basic research program.

simulation equipment, and of appropriate methods. The present paper reviews the concepts and methods, and some selected results of an experimental program that has been investigating such system-design problems for the past two and a half years. The specific system under study has as its function the safe and efficient control of aircraft in the terminal (approximately 50-mi. radius) area around an airport.

The Planning Phase

The Laboratory of Aviation Psychology is now completing its sixth year of experimental work on human engineering aspects of air traffic control. Earlier, one of the authors together with Dr. George Long conducted a human engineering analysis of air traffic control on the Berlin Airlift (8) and some of the background for the current work goes back to that study.

The first step in initiating the current research program was a one-year study phase which was carried out under the auspices of the NRC Committee on Aviation Psychology. Eleven psychologists from eight different universities and Government laboratories participated in preparing a report of this project (3).

The Research Phase

The research phase of the program began a short while later. A major initial task was to develop suitable simulation equipment and methodology for the study of system-design problems in the laboratory; pending these developments, initial research emphasis was placed on the study of specific display problems. Following an unsuccessful attempt to adapt existing simulation equipment for use in system experiments, an electronic analog simulator was designed specifically to meet our research needs (by Prof. C. E. Warren of the OSU Department of Electrical Engineering and several of his students) and constructed in the laboratory. This device (4) is capable of generating 30 independently-controlled simulated aircraft (or target "blips"), which can be displayed on simulated radar scopes. A photograph of the "pilot" room, and a photograph of a typical simulated radar scope in the "radar control" room, with identity-coded targets, are shown in Figures 1 and 2.

At the present time our system experimentation is planned around two general goals: (a) to solve some of the specific human engineering problems of air-traffic-control system design, and at the same time (b) to provide a contribution to the development of a general theory of the capacities and proficiencies of small groups of people in performing the kinds of perceptual, memory, and decision-making functions required by complex tasks which must be solved continuously in real time.

In this paper we discuss some of the basic concepts which have guided the formulation of the research program, review critical aspects of the methodology we have developed, and present a few illustrations of typical results. For the most part we shall confine the discussion to system research, with only brief reference to related research projects. Although all examples refer to a specific system, many of the methodological problems encountered in studying this system are believed to be quite general.

Figure 1. A view of part of the simulation equipment, showing the panels used to control the positions of the radar blips representing the simulated aircraft in the system.

Figure 2. A view of one of the simulated radar scopes showing radar blips that are identity-coded by means of two electronic arms.

OBJECTIVES AND CONCEPTS

Objectives. The major objectives of the research program are indicated in Table 1. One goal is to provide human engineering principles that can be used by the

TABLE 1

Objectives of the System-Research Program

1. To provide human engineering principles for use in the design and operation of future ATC systems.
2. To provide measures of human capacity and reliability relative to future ATC systems:
   a. When men perform different functions in the system,
   b. Under both emergency and normal conditions,
   c. With different traffic control procedures, and
   d. With various types and rates of aircraft inputs.

engineers who will design future air traffic control systems, and by the operational personnel who will devise the procedures to be employed in operating these systems. We attempt to anticipate future problems and to provide sufficient lead time for our research so that there will be maximum opportunity for findings to be used in the critical planning and development phases of new systems. As an example, we may study the effects of displaying types of information that cannot yet be obtained with current electronic devices, but that may be attainable in five or ten years.

From a psychological viewpoint, another goal of the research is to provide quantitative estimates of human capacity for performing the different types of functions which may characterize future air traffic control and similar complex man-machine systems.

In planning any particular study we first decide what functions must be performed in the general type of system under study. We then allocate certain functions to people and assume that other functions will be performed by automatic or semi-automatic equipment. We next design an experiment which will permit us to measure man's ability to carry out those functions which may have been allocated to him, such as to perceive and handle information, to make decisions, and/or to communicate instructions, as indicated by criteria of system performance.

We are interested in determining human performance capacities under both normal conditions and under unexpected or "emergency" conditions. In some experiments performance is measured under intermediate levels of task difficulty and later human breakdown characteristics are noted when the load is increased to the point where controllers cannot do all that the task demands.

Since human capacity is also a function of the procedures employed by individuals and by groups in solving on-going problems, one of the objectives of our research has been to determine human capacity as a function of procedural or organizational variables. We also have evaluated human capacity as a function of the inputs to the system, i.e., the types of problems which the man-machine system is required to handle. These load variables have to do with the number of stimuli, the similarity of stimuli, and the nature of stimulus changes—for example, aircraft entry rate, heterogeneity of aircraft types, and aircraft speeds.

Concepts. Several concepts or directing tenets have been formulated as guides to our research planning. Several of these concepts are listed in Table 2.

TABLE 2

Concepts Used to Guide System-Research Planning

1. Human Engineering should include the study of
    a. *Existing* systems,
    b. *Simulated future* systems, and
    c. *Subsystems* or specific technical problems.
2. Human operator capabilities should be determined for
    a. *Optimized* as well as *degraded* information, and for
    b. Control *teams* as well as for *individuals*.

The scope of our general research program includes (*a*) field studies of existing systems, (*b*) laboratory experiments on simulated future systems, and (*c*) supporting basic research on specific display- and information-handling problems. Emphasis is placed on the last two topics; we consider here only the system experiments.

One of the major tenets that we have followed is that human capabilities should first be determined under optimized system conditions (e.g., with "idealized" displays and reliable information) and then determined for nonoptimal or degraded systems. Only data obtained under idealized conditions permit an estimate to be made of the upper limits of system performance that could result from future improvements in the machine aspects of the man-machine system. One of the gratifying results of our policy of first studying human performance under idealized conditions is that on several occasions it has been unnecessary to go on to the study of degraded systems. In each case, by the time a series of human factor research studies has been completed, engineering progress had made it possible to eliminate many of the deficiencies of existing systems, and hence had rendered unnecessary the study of the effects of such deficiencies on human performance.

Another of the tenets is that human capacities should be determined for groups of controllers who are collaborating in the solution of a common problem, as well as for individuals working singly. We are interested, for example, in the question of how much two men working together can accomplish in comparison with what one man can do when working alone. By studying group-performance capabilities we hope to discover methods of displaying information, and procedures for the division of responsibilities, that will make possible efficient group effort, and to provide estimates of the number of people required to perform particular functions. We also hope to make a contribution to the theory of small-group proficiency.

A further tenet concerns the importance of maintaining the realism of the total environment in which the research is conducted. Some degree of realism is necessary, and its achievement requires careful attention to details. The environment in our control room is especially conducive to the creation of a realistic attitude on the part of the controllers since many of the characteristics of an actual radar control center are faithfully replicated.

THE EXPERIMENTAL PROGRAM

Experimental Study of System Variables. Most of our system experiments have been conducted through the use of the OSU electronic air traffic control simulator (4). This equipment permits us to investigate three types of system variables. These are shown at the top of Table 3.

TABLE 3

Areas Included in the System-Research Program

1. System Experiments, covering
   a. *Display* variables,
   b. *Load* variables, and
   c. *Procedural* variables.
2. Related Technical Research, including
   a. Visibility and lighting,
   b. Specific display principles,
   c. Information coding, and
   d. Information-handling ability.

Display variables may alternatively be called information variables. They involve the type of information made available to controllers, the degree of precision in the information, and the way in which the information is encoded and displayed. An example of a display variable is the method of indicating identity.

Load variables define the input to the air traffic control system, such as the traffic conditions with which the controller must cope. Traffic variables can be manipulated by predetermining the characteristics of aircraft and the times at which aircraft enter the system.

Procedural variables involve the methods employed by the team that solves the air-traffic-control problem. These variables include communication procedures, procedures by means of which two or more individuals make joint or complementary decisions, and procedures governing the types of instructions that controllers are permitted to issue to aircraft pilots. An important subclass of procedural variables is the way in which two or more men divide responsibilities. As an illustration, a division of responsibility can be based on either geographic or temporal factors, each controller accepting responsibility for different parts of the air space or for different blocks of time; or it can be based on the types of functions to be performed by each man.

Areas of Supporting Basic Research on Specific Display Problems. Supporting research is conducted with two objectives in mind: (*a*) the establishment of basic principles of information transmission, and (*b*) the provision of principles in support of the system-research program. Supporting research is divided into the four phases shown in the lower half of Table 3.

Problems of *visibility* and *lighting* arise because of the unusual visual environment that must be provided in a radar control center. As a part of our work in this area Mr. Kraft has developed the OSU Broad-Band Blue Lighting System (5, 6). This lighting has now been in use at Wright-Patterson AFB for three years, and has recently been adopted with only minor modification by several national agencies.

In the area of *specific displays*, special emphasis has been given to the problem of identity and altitude coding (1, 2, 7, 9). Much of this basic research has made use of information measures as a criterion of performance. In particular, we have been interested in *coding for efficient handling of information*. Usually our first procedure is to scale the stimulus by means of absolute information transmitted. Next, we study the rate at which this information can be transmitted.

METHODOLOGY FOR SYSTEM RESEARCH

During the past two and a half years, eight major system experiments have been completed. The methodology for these studies is constantly evolving so that we shall summarize chiefly the current procedures.

The Types of Control Tasks That Have Been Investigated. The focus of most activity in the simulated system is a plan position (PPI) display showing the ground positions of all aircraft in a 50-mi. radius from the simulated radar site. In some cases additional information is provided on secondary displays. All of the displays have been idealized to the extent that available knowledge and present techniques permit. A good quality air-ground communication system has been simulated. Standard, but abbreviated, voice procedures have been used.

The most typical problem studied in the laboratory is that of a group of military aircraft returning from a combat-type mission. No airborne or ground-navigation equipment other than radar is simulated.

Experiments usually involve a series of 30 to 60 discrete problems, with from 20 to 30 aircraft included in a single problem. These simulated aircraft are given flight characteristics similar to current advanced types such as the B-47 and F-86, or are made to represent possible future types. The mean temporal and spatial separation of aircraft as they enter the control zone has been one of the variables studied.

In most cases the output point of the system has been a hypothetical GCA gate near the center of the PPI display, where aircraft are turned over to a final (GCA) controller. Outgoing aircraft leave the system at designated points around the periphery. The last 10 mi. of the landing approach has not been explicitly studied, but this part of the system has usually been simulated by a GCA controller who had used realistic criteria for determining whether or not aircraft turned over to him could be landed. If any airplane cannot be accepted for landing, a "go-around" is initiated by the GCA controller and the aircraft is returned to the previous controller who must return it to the GCA gate for a second time.

Plans for future studies call for the simulation of other types of terminal operations, such as simultaneous landings on different landing fields, and the delegation of certain responsibilities from controllers to pilots, with controllers assuming a greater degree of monitor activity.

Level of Experience of Controllers and Pilots. In most system experiments, experienced Air Force personnel have been used as controllers. The pilots who operate the other portion of the simulator have been OSU students, selected for previous aviation experience, and thoroughly pretrained before the beginning of an experiment.

The Research Environment. Although we conduct research in a laboratory setting we have devoted a good deal of effort to the creation of a realistic environment, one which produces a high level of motivation and allows individuals to develop sets similar to those that characterize actual field situations. The controllers work under environmental conditions approximating those of an actual radar center. The radar displays are realistic, standard voice procedures are followed precisely, and familiar aircraft types are employed with realistic flight characteristics.

The choice of which variables to treat experimentally and which to treat as fixed parameters is perhaps more of a problem in studying complex man-machine systems than it is in simpler experiments. In each case, the choice has been a compromise between the desire to simulate a realistic situation and the necessity for obtaining unambiguous results. For continuity, we try to carry the same values of parameters through several experiments.

CRITERIA OF SYSTEM PERFORMANCE

If the results of research on man-machine systems are to exercise a significant influence on the design of future systems, the effects of human engineering variables of the type described above must be measured in terms of their effects on over-all man–machine system performance. Provided such meaningful system measures are available, however, then specific measures of the behavior of the people in the system take on added importance because these intervening measures can be related to the performance of the total system. Both types of measures have been employed in our experiments, but we have concentrated initially on the development of system criteria. The most important of these are listed in Table 4.

Listed first in Table 4 are several over-all measures of system efficiency. *Average control time* in the terminal area is the average amount of time required to bring an aircraft from 50 mi. away to the GCA gate, or (for a departing aircraft) to take it to a point 50 mi. from take-off. *Accumulated delay in landing* is the average of the control times for the first few (usually three) aircraft, compared with the average control time for the last few (usually three) aircraft handled during a problem. *Average fuel consumption* is a measure somewhat analogous to average control time, but is computed from altitude and speed as well as time. *Excess fuel consumption* is the amount of fuel used in excess of a theoretical minimum amount. These three are related measures, and have self-evident validity. In addition,

TABLE 4
Criteria of System Performance

1. Measures of efficiency
   a. Average control time
   b. Accumulated delay in landing
   c. Average fuel consumption
   d. Excess fuel consumption
   e. Final heading at the GCA gate
   f. Deviation from runway extension
   g. Average separation at GCA gate
2. Measures of safety
   a. Number of conflicts enroute
   b. Number of conflicts at GCA gate
3. Measures of communication load
   a. Regarding heading
   b. Regarding altitude
   c. Regarding speed

several other efficiency measures are taken. *Final heading at the GCA gate* and *deviation from the runway extension* are measures of the amount of deviation from the heading of the runway when the aircraft arrives at the GCA gate. *Average separation at the GCA gate* is a measure in time of the separation of aircraft as they arrive at the gate. Also listed in Table 4 are several measures of safety and of communications. As most communications in our system are concerned with determinations of or changes in such aircraft characteristics as heading, altitude, and speed, the frequencies of each of these categories are included when a content analysis is made of communications. We use two principal criterion measures of safety. These are *number of midair conflicts enroute,* and *number of conflicts near the GCA gate.* A conflict is defined arbitrarily as a failure to maintain a certain minimal separation measured in units of flying time.

In system research we feel that it is necessary to establish criteria which reflect over-all system performance. Once this is accomplished, it may then be profitable to examine intermediate measures. We are now beginning to use criterion measures which reflect the functioning of parts of the total system. To a certain extent communication measures are of this sort.

THE POWER OF A TYPICAL EXPERIMENT

The development of experimental methods appropriate to system experiments must include the determination of the power of these designs in respect to alternative hypotheses which specify differences of practical significance. As an illustration, in a recent experiment (12) an analysis of the power of the design indicated that the experiment provided better than 90% probability of rejecting the null hypothesis if the independent variables under consideration contributed an actual difference of as much as 10% in such characteristics as control time or fuel consumption. In this experiment there were 64 separate traffic problems of the type studied, and the handling of a total of 1280 simulated aircraft movements. Although a typical experiment is not sufficiently sensitive to detect very small effects reliably, nevertheless sufficient power is available for us to feel relatively confident that we will not reject as unimportant any variables that produce large effects on over-all system performance.

Perhaps the most serious limitation of the experimental designs that we have employed is the use of small numbers of experienced controllers. In many of our studies we have used as few as four controllers. Thus, strictly speaking, we are unable to generalize with much confidence to controllers in general. On the other hand, the use of experienced controllers gives us greater confidence in making statements about the maximum performance that can be expected of human operators.

REVIEW OF RESULTS FROM SEVERAL SYSTEM EXPERIMENTS

As had been pointed out earlier, system variables can be grouped into three classes. These are (a) load or input variables, (b) information or display variables, and (c) procedural or organizational variables. We now will summarize some typical experimental findings for each of these major classes of variables, drawn from the series of studies conducted during the past two years (10, 11, 12, 13, 14, 16).

In Figure 3 are shown some results for a traffic variable, rate of aircraft entry. As rate of entry (number of aircraft entering the terminal area per unit of time) is increased beyond some point, measures of system efficiency have uniformly been found to decline. Thus, in Figure 3, average control time showed a marked upward trend as the average time separation between successive entries decreased beyond one aircraft every 90 sec. for the condition of no identity, and as it decreased beyond one entry every 75 sec. for the condition in which identity was given. The number of conflicts also increased as a function of increased entry rate. These data are for a system in which a single controller was required to handle all aircraft, and entries deviated randomly around the mean temporal separation and appeared anywhere within a 90° sector of the periphery. Each problem consisted of blocks of 20 military-type aircraft. All aircraft were inbound, and had to be moved through a 50-mi. zone in order to reach the GCA gate.

Our most recent experiment (15) was concerned with a more detailed analysis of the load variable, the effect of the degree of spatial and temporal organization upon control efficiency and safety.

A display variable which has been found to have pronounced effect on system efficiency is the addition of an effective means of identifying the blips which appear on the primary display. In Figure 4 is shown the visual code used in most of the system experiments as a means of providing the controllers with knowledge of the identity of incoming aircraft. This coding system is based on the results of several supporting research studies (9). The effect of the use of this identity code on accumulated delay in landing in one of our experiments was shown in Figure 3. Except for the slowest rate, the landing delay was found to be greater for the condition in which there was no identity (12). These data are for single controllers who were asked to handle random-entry problems.

The provision of aircraft identity, however, did not reduce the number of requests for information directed by controllers to pilots, as can be seen in Table 5.

The number of requests for altitude, speed, and heading information were found to be about the same under conditions of identity and of no identity. In this experiment the number of requests for altitude were about three times as great as requests for heading information. However, when a secondary altitude display was introduced in a subsequent experiment (11), a dramatic change in the number of requests for altitude information occurred, as might be expected. Such a reduction in the number of altitude requests directly reduces the load on the communication channel, i.e., reduces voice radio time, and this in itself may be an important gain in certain situations. However, in this particular experiment the availability of an altitude display, and the accompanying reduction in radio time, had no appreciable effect on total system performance as measured by such criteria as control time and fuel consumption.

Organizational variables will be a major focal point of much of our future system research. The largest number of directly cooperating controllers we have used to date, however, is two, supplemented by two additional men who provided contacts at the input and output points. In one investigation the effects of permitting two controllers to work in a face-to-face situation was compared with the effects of a condition in which they were assigned to separate semi-isolated work stations (16). Under conditions of free and direct voice communications between the two controllers (face-to-face situation) not only did the controllers talk more frequently with each other, but the amount of time each spent in communication with pilots also increased. However, measures of system performance, such as fuel economy and control time, showed small changes, and the differences that did occur most often favored the condition of separate work stations.

Another organizational variable that has been investigated concerns the division of control responsibility between two men. Some results from an experiment on this topic (14) are shown in Table 6. The data shown permit an assessment of the effect of the use of two different ways of dividing the work between controllers. Both procedures used a geographic basis for the division of labor. One procedure employed an *in-line* principle in which control zones were determined by the distance from the GCA gate. One controller handled those aircraft farther from the gate, and the second controller handled those nearer to the gate. The other procedure employed a *sector-control* principle. One controller handled all aircraft entering the terminal area north of the GCA gate, and the other controller, all aircraft entering south of the gate. The results indicated that the sector procedure might be slightly superior in efficiency for the faster rates, but the difference was not statistically reliable. In most respects the two controllers used the two procedures about equally well, even though very different modes of cooperation were required.

It is worth noting that negative findings, such as have been obtained in several of the studies we have completed (e.g., little or no effect of a visual altitude display, little difference between sector and in-line division of responsibility, little difference between an omnipresent identity code and use of a light-pencil interrogation device), are often very valuable since they permit system engineers to exercise much greater freedom of choice in designing the over-all system.

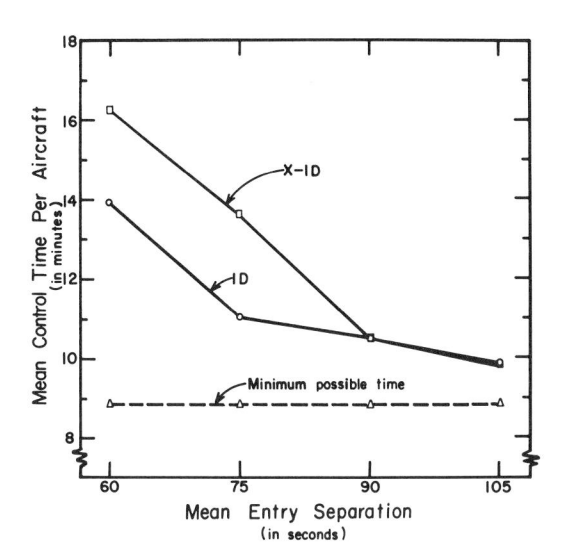

Figure 3. System performance at four entry rates under identity and no-identity conditions. Data are based on the performance of single controllers. ID indicates identity; X-ID, no identity.

Figure 4. The clock-code identification system. The longer arm is read as Alpha, Bravo, Charlie, or Delta; the shorter arm is read as the corresponding clock numeral.

TABLE 5

Summary of the Frequencies of Different Types of Communications Between Controllers and Pilots as a Function of the Presence or Absence of an Identity Code

(Data are frequencies per aircraft movement)

Type of Communication	Commands to Pilots Identity	Commands to Pilots No Identity	Information Requested from Pilots Identity	Information Requested from Pilots No Identity
Heading	3.13	3.55	0.56	0.72
Airspeed	2.38	2.31	0.28	0.30
Altitude	2.20	2.18	1.60	1.59

TABLE 6

Results on System Performance of the Use of Sector vs. In-line Procedures by Two Cooperating Controllers

(Data based on 832 aircraft movements)

Criterion of System Performance	60 Sec. Sector	60 Sec. In-Line	90 Sec. Sector	90 Sec. In-Line
Mean Control Time (in min.)	10.12	10.42	10.16	10.21
Mean Fuel Used per Aircraft (in lbs.)	862	903	876	846

	Sector	In-Line	60 Sec.	90 Sec.
No. of Conflicts	22	19	30	11

LINES OF FUTURE RESEARCH

The content direction of future system research in the Laboratory of Aviation Psychology will emphasize (a) the study of simulated automatic control systems in which people will be asked to monitor the system and to handle emergencies, (b) the study of different kinds of procedures for attaining a high level of effectiveness from a group of men who are working together, (c) the development of displays, work stations, and communication nets suitable for use by groups of three or more controllers, and (d) the study of the optimum numbers of controllers for performing different functions and handling various loads. Continued emphasis will be given to the general problem of the optimum allocation of functions to men and to machines in complex systems, and of optimum allocation of functions among the several men in such systems.

The theoretical and methodological direction of future research will place increasing emphasis on the topic of small-group proficiency, in the setting of a complex man-machine system which must solve externally paced problems in real time. More use will be made of measures of controller interactions and individual behavior in an effort to relate these theoretically to system performance.

Unfortunately, currently there is no effective framework or theoretical model for dealing with system research. The concepts terminology and models of present-day social psychology and personality theory are of little use for our present purposes. Thus, in our further investigations of complex systems we are faced with the development of theory in a relatively new area of psychology, an area lying between traditional experimental psychology, which centers its interest on the study of single subjects under carefully controlled conditions, and social psychology, which studies the behavior of individuals in groups, but only infrequently has been able to bring the environment and the tasks to which these groups are subjected under adequate experimental control. During the past two years, we have succeeded in creating a realistic task in the laboratory, we have developed meaningful and relatively highly reliable measures of group proficiency in problem solving, and we have been able to determine empirically the effects of several important input, display, and procedural variables. Our current major effort is to develop more effective theory for relating the empirical findings to more specific aspects of individual behavior on the one hand, and to a wider type of group problems on the other hand.

REFERENCES

1. Alluisi, E. A., & Muller, P. F. Rate of information transfer with seven symbolic visual codes: motor and verbal responses. *USAF WADC Tech. Rep.*, 1956, No. 56-226.
2. Alluisi, E. A., Muller, P. F., & Fitts, P. M. Rate of handling information and the rate of information presentation. *USAF WADC Tech. Note*, 1955, No. 55-745.
3. Fitts, P. M. (Ed.) *Human engineering for an effective air-navigation and traffic-control system.* Washington, D. C.: National Research Council, 1951.

4. Hixson, W. C., Harter, G. A., Warren, C. E., & Cowan, J. D., Jr. An electronic radar target simulator for air traffic control studies. *USAF WADC Tech. Rep.*, 1954, No. 54-569.
5. Kraft, C. L. A broad-band blue lighting system for radar approach control centers: evaluations and refinements bascsed on three years of operational use. *USAF WADC Tech. Rep.*, 1956, No. 56-71.
6. Kraft, C. L., & Fitts, P. M. A broad-band blue lighting system for radar air traffic control centers. *USAF WADC Tech. Rep.*, 1954, No. 53-416.
7. Learner, D. B., & Alluisi, E. A. Comparison of four methods of encoding elevation information with complex line-inclination symbols. *USAF WADC Tech. Note*, 1956, No. 56-485.
8. Long, G. E., & Fitts, P. M. Human-engineering aspects of the Berlin airlift. *USAF Air Materiel Command, Memo. Rep.*, 1949, No. MCREXD-694-23.
9. Muller, P. F., Jr., Sidorsky, R. C., Slivinske, A. J., Alluisi, E. A., & Fitts, P. M. The symbolic coding of information on cathode ray tubes and similar displays. *USAF WADC Tech. Rep.*, 1955, No. 55-375.
10. Schipper, L. M., & Versace, J. Human engineering aspects of radar air traffic control: I. Performance in sequencing aircraft for landings as a function of control time availability. *USAF WADC Tech. Rep.*, 1956, No. 56-67.
11. Schipper, L. M., Kidd, J. S., Shelly, M., & Smode, A. F. Terminal system effectiveness as a function of the method used by controllers to obtain altitude information: A study in human engineering aspects of radar air traffic control. *USAF WADC Tech. Rep.*, in preparation.
12. Schipper, L. M., Kraft, C. L., Smode, A. F., & Fitts, P. M. The use of displays showing identity versus no-identity: A study in human engineering aspects of radar air traffic control. *USAF WADC Tech. Rep.*, 1957, No. 57-21.
13. Schipper, L. M., Versace, J., Kraft, C. L., & McGuire, J. C. Human engineering aspects of radar air traffic control: II. and III. Experimental evaluations of two improved identification systems under high-density traffic conditions. *USAF WADC Tech. Rep.*, 1956, No. 56-68.
14. Schipper, L. M., Versace, J., Kraft, C. L., & McGuire, J. C. Human engineering aspects of radar air traffic control: IV. A comparison of sector and in-line control procedures. *USAF WADC Tech. Rep.*, 1956, No. 56-69.
15. Shelly, M. W., Kidd, J. S., Jeantheau, G., & Fitts, P. M. Effect of enroute air traffic regulation on radar air traffic control systems employing partially optimized displays: A study in human engineering aspects of radar air traffic control. *USAF WADC Tech. Rep.*, in preparation.
16. Versace, J. The effect of emergencies and communications availability with differing entry rates: A study in human engineering aspects of radar air traffic control. *USAF WADC Tech. Rep.*, 1956, No. 56-70.

30

Research on a Simulated Bayesian Information-Processing System

D. A. Schum, I. L. Goldstein, and J. F. Southard

I. INTRODUCTORY COMMENT

The possibility of automating certain functions within complex threat-diagnosis systems has recently been suggested. Edwards [1] has provided a specific proposal for the design of such systems. The essence of this proposal is that Bayes' theorem suggests a novel and perhaps more efficient allocation of tasks among the men and machine components of diagnostic systems. Diagnoses, in the form of posterior probabilities, could be provided by computer-implemented Bayesian aggregations of expert human estimates of $P(D \mid H)$ [or some other analogous quantity from which $P(D \mid H)$ is recoverable]. The formal justification for these procedures rests upon the optimal information-aggregating properties of Bayes' theorem. Computers, programmed in accordance with Bayes' theorem, could perform these aggregation tasks superbly and almost instantaneously. The empirical justification for these procedures will rest upon a) demonstrations that men can produce $P(D \mid H)$ or some related quantity and b) demonstrations that the diagnoses proceeding from this method are as good as or better than those produced by any other means.

The purpose of the present report is to describe three of five experiments completed to date at the Ohio State University Laboratory of Aviation Psychology on the subject of Bayesian diagnostic systems. Following is a description of the multi-man-machine system simulation facility in which these experiments were performed.

II. SYSTEM SIMULATION FACILITIES

Although the simulation facilities to be described were designed so as to be applicable to a variety of operational situations involving several levels of informa-

From *IEEE Transactions on Human Factors in Electronics*, HFE-7, March 1966, 37-48. By permission of the IEEE Editorial Department.

tion-processing and decision-making functions, the particular context chosen for the present series of investigations is that of an intelligence information-processing and threat-evaluation agency at some unspecified but presumably high level within the Air Force. Final development of this facility coincided with growing interest in the possibility of automated Bayesian diagnosis procedures in threat-diagnosis systems.[1] It seemed logical and necessary to utilize the present system simulation facility in order to evaluate some of the intriguing possibilities suggested in Edwards' paper.

A. Hardware Components

The major hardware components of the simulation facility include an IBM 1401-7094 computer system and associated digital display interface equipment, telephone and closed-circuit TV equipment for intercommunication among members of a subject-operator team, and a large edge-lighted plexiglas display board upon which summarized and updated information can be maintained. The major function of the IBM 1401-7094 system is to generate and present the dynamics of a hostile environment to a subject-operator team that processes and evaluates intelligence data obtained on hypothetical reconnaissance overflights of the hostile territory. The hostile environment consists of a 1000 by 1000-mile area representing the homeland of a fictitious adversary called "Aggressor." The critical events being generated in this hostile environment consist of time-dependent deployments of Aggressor surface and air military forces. There can be as many as 20 different types of deployments represented in the environment. At any given time there can be as many as 25 deployments depicted in various stages of buildup. The specific function of the IBM 7094 is to generate, prior to experimental sessions, specific scenarios depicting these deployments and to provide real-time capabilities for presentation of these scenarios during the experimental sessions. The specific function of the IBM 1401 is to process requests for information which the operator team makes regarding events in the hostile environment. Tabular and map-like presentations of the environmental data are made available, upon request by the subjects, on the digital display consoles (of which there are five) and in the form of printed output from IBM 1401.

Members of the operator team communicate among themselves by means of telephone, closed-circuit TV, and hand-carried messages. By means of standard card inputs to the IBM 1401 or by means of instructions entered into the control register of the digital displays, the subjects can retrieve information from computer storage about environmental events. There are over 259,000 different interrogations the subjects can make with respect to events in the hostile environment. The simulation facilities, therefore, allow generation of complex scenarios depicting deployments of hostile military forces in a circumscribed area. Certain members of the "threat-evaluation" team interrogate computer storage according to

[1] Development of this systems research facility (accomplished over a four-year period prior to February 1963) and the subsequent systems research program have been under the sponsorship of the Behavioral Sciences Laboratory, Aerospace Medical Research Laboratories, Wright-Patterson Air Force Base, Ohio. A more detailed description of this system research facility is to be found in a recent report by Southard, Schum, and Briggs [2].

certain rules and obtain data that describe the characteristics of these deployments as they develop. Other individuals use these data as a basis for diagnoses of the threat posed by each Aggressor deployment.

B. A Simulated Threat-Diagnosis System

The simulated threat-diagnosis system about to be described is not intended as a replica or model of any existing system. Certain critical features common to any threat-diagnostic activity however, formed the basis for the design of the simulated system. The members of the threat-evaluation team perform two basic tasks. First, certain members of the team attempt to uncover and describe the critical events taking place within the hostile environment. These individuals are described in Figure 1 as Intelligence Staff Officers (ISOs). They specify certain areas of Aggressor territory which are to be placed under aerial surveillance. After computer-simulated reconnaissance overflights of these specified areas are performed, the ISOs proceed to interrogate computer storage in which the obtained intelligence data are stored in certain logical categories. In response to these ISO interrogations, verbal and numerical descriptions of Aggressor mobile weapons, vehicles, and aircraft uncovered in the area placed under surveillance are reported to the ISOs by means of the digital displays and computer print-out. From these verbal and numerical descriptions the ISOs infer the existence of units of Aggressor's surface and air forces. In addition, they must establish the spatial and temporal arrangements of these various units in order to identify discrete deployments of Aggressor's forces. In view of the size and complexity of the hostile environment, the ISO task is not an easy one. Each deployment consists of large numbers of military units selected from 33 different types of battalion-size surface units, 14 different types of regiment-size surface units, and 31 different types of squadron-size tactical air units. Each unit is identified on the basis of various combinations of 93 different types of mobile weapons, vehicles, and aircraft. As mentioned previously, there can be as many as 25 deployments, in various stages of development, existing in the environment at any given time. Since the development of these deployments is time-dependent, information about the characteristics of the deployments obsolesces rapidly and new information must be continually requested.

Each deployment can be described in terms of 25 dimensions or attribute data classes. These attributes refer in general to such features as the infantry-armored (main attack), artillery-air (combat support), and logistics constituencies of the deployments and the spatial and temporal arrangement (order of battle) of these forces. Observe in Figure 1 that an individual ISO is responsible for each of these four subsets of the 25 deployment-defining attributes. Each statement about any of the 25 attributes of a certain deployment made by an ISO is probabilistic in nature. Each of the 25 attribute data classes has several possible states or conditions only one of which is applicable to each deployment. Since ISOs often feel considerable uncertainty about the state of each data class observed with respect to a particular deployment, they provide probabilistic estimates about the conditions or states being observed in each data class. Such a probabilistic profile in each of the 25 data classes for every deployment under surveillance forms the basic output of the ISO level of the simulated system.

The second major task performed by the simulated system involves the activities of the individuals labeled "Threat Evaluators" (TEs) in Figure 1. These individuals attempt to determine or diagnose the specific Aggressor intent represented by each of the deployments. In the three experiments to be discussed, performance data were collected from four TEs, each performing independently. As input data the TEs make use of the probabilistic attribute data provided by the ISOs for each deployment. These data are transmitted to the TEs by means of closed-circut television. In terms of the Bayesian paradigm the basic TE task is to estimate the probability of a particular type of deployment (H) given the probabilistic attribute data (D) describing the characteristics of the deployment. These $P(H \mid D)$ or posterior probability estimates represent the basic output of the simulated threat-diagnosis system. The TEs in all experiments were required to normalize their $P(H \mid D)$ estimates, i.e., make them have unit sum across the hypothesis set. In the three experiments being described there were eight possible alternative hypotheses, threats, or deployment types.

The TEs provided another type of probabilistic estimate, namely, the probability that the various attribute data (D) would be observed if H_i (a particular hypothesis) were true. These $P(D \mid H)$ estimates were used by the experimenter to calculate $P(H \mid D)$ by means of a modification of Bayes' theorem (see Section III) for comparison with the human estimates. The "true" $P(D \mid H)$ values, which formed the basis for generation of environmental events, were determined by the experimenters during the development of the simulation facility by reference to certain U. S. Army field manuals which give detailed descriptions of the tactics of the maneuver enemy (Aggressor).

C. Some Methodological Considerations

Several characteristics of the experiments being reported affect the degree to which the results of these experiments will be comparable to the findings of other laboratories. First, the primary human performance data $[P(H \mid D)$ and $P(D \mid H)$ estimations] obtained in all four experiments were collected from subjects who were quite experienced in dealing with contingent events in the simulated hostile environment. Each subject received initial training consisting of from 60 to 114 hours of lectures, problems, and "on-the-job training" as well as further instructions specific to each experiment. The use of such highly trained individuals makes replications difficult but offers some distinct advantages. Presumably, the operation of any real-life military diagnosis or inferential system depends upon highly skilled evaluators. The human behavior of interest in the experiments, therefore, is that of individuals who bring much environmental experience into the various tasks they are assigned. The results of these experiments may not necessarily agree with results obtained from naive subjects who perform only briefly in unfamiliar circumstances.

A second consideration is the fact that the subjects in all three experiments accumulated experience in an environment whose critical events were essentially repetitive in nature. The $P(D \mid H)$ estimates produced by the TEs, for example, were based upon rather careful counts of the number of times each of the states or conditions within the 25 data classes occurred when a particular hypothesis was true. The stimulus environment, therefore, can be described as "frequentistic" in character since relative frequencies of events could be maintained as an aid in judging the probability of occurrence of these events.

A third consideration is that the subjects in all three experiments who estimated $P(H \mid D)$ were always subsequently informed about the "true" hypothesis accounting for the data observation and, consequently, about the quality of their estimates. The assumption was that the true deployment type applicable to a set of previously obtained information could always be subsequently recognized.

Finally, in all three experiments Dodson's [3] modification of Bayes' theorem (MBT) was used for programmed calculations of $P(H \mid D)$. Dodson has extended Bayes' theorem in order to provide for the evaluation of multiple data classes each having several possible states or conditions and to allow for the expression of uncertainty about which of the states within a data class is being observed. Dodson's MBT was chosen because it provided for evaluation of the sort of data which the present simulation facility could generate and not because the experimenters judged that it was intrinsically superior to Baynes' theorem as originally expressed. One assumption of the MBT is that the data classes (of which there were 25 in the present experiments) represent independent event categories. Admittedly, such an assumption may not be valid in many real-world situations. In the contrived experimental situation being described, the states or conditions in each of the 25 data classes for each deployment were generated by the experimenter according to a procedure which assured nearly complete mutual independence among selected levels of the 25 data classes. There is the possibility, however, that this data generation technique is in conflict with intuitive nonindependencies among various data classes which are suggested by the manner in which the data classes were labeled. The effects of data class nonindependence upon $P(H \mid D)$ and $P(D \mid H)$ estimates produced by humans present theoretical and practical problems of some importance. Research on this issue is underway at present. Dodson's MBT is described in (1),

$$P(H_i \mid D_j) = \sum_{k=1}^{\mu_j} P(D_{jk}) \left[\frac{P(H_i)P(D_{jk} \mid H_i)}{\sum_{i=1}^{n} P(H_i)P(D_{jk} \mid H_i)} \right] \quad (1)$$

where k refers to a particular state or condition of D_{jk}, the data class in question; μ_j refers to the total number of states or conditions in the jth data class; $P(D_{jk})$ refers to the probability that the data class is being observed in state k; and i refers to a particular hypothesis among the n possible hypotheses.

Equation (1) describes the evaluation of the probabilistic information in only one data class. In order to evaluate the information in several data classes, the $P(H_i \mid D)$ estimates from one data class become the prior probabilities $P(H_i)$ used in the evaluation of the next data class and so on until the information in all data classes has been incorporated. In the experiments being reported, the TEs produced the estimates of $P(D_{jk} \mid H_i)$ required in (1). Calculations of $P(H_i \mid D)$ on the basis of (1) [incorporating these TE estimates of $P(D_{jk} \mid H_i)$] were compared with the estimates of $P(H \mid D)$ also provided by the TEs. The value of $P(H_i)$ was set at 0.125 for each hypothesis at the beginning of each calculation since there were eight alternative hypotheses. The TEs were similarly told to assume that the a priori probability of occurrence of each of the eight hypotheses (deployment types) was equal. In fact, each of the eight hypotheses was true an equal number of times throughout both experiments. This procedure was instituted out of a desire on the part of the experimenters to simplify an already complex experimental situation. The effects of differing a priori probabilities is an interesting variable in its own right.

Dodson has also described how the MBT could be made to adapt itself to changing environmental events of a frequentistic character. This "self-adapting" vision of the MBT has as its basis arbitrarily set parameters which regulate the rate of information obsolescence and an expression for feedback about the true hypothesis existing in the environment at the time each observation was made. These parameters and the feedback expression were applied by Dodson to the $P(H_i)$ and $P(D_jk \mid H_i)$ terms in (1). They describe how these terms are to be updated on every observational cycle in the sequence. The self-adapting MBT is discussed in detail by Dodson [3] and by Southard, Schum, and Briggs [4]. The self-adapting MBT has limited significance in terms of the present experimental objectives of comparing direct human estimates of $P(H \mid D)$ with those calculated on the basis of MBT aggregations of human $P(D \mid H)$ estimates. For this reason, self-adapting MBT solutions will not be discussed in detail in the present report.

III. RESEARCH USING THE SIMULATED BAYESIAN DIAGNOSTIC SYSTEM

Two introductory studies using the system simulation facilities are reported in detail elsewhere by Southard, Schum, and Briggs [2], [4]. These two experiments produced differing results with respect to a comparison between human and MBT estimates of $P(H \mid D)$. In the first experiment the MBT placed significantly higher estimates of $P(H \mid D)$ in correct hypothesis categories than did a human threat evaluator. In the second experiment human performance was nearly identical to that of the MBT. These differing results can be explained by the fact that there were 20 alternative hypotheses in the first experiment and only four in the second. The TEs in both experiments were of comparable ability and had received identical training. An especially interesting result of the second experiment was the shift, as the experiment progressed, from conservative $P(H \mid D)$ estimates placed by the TE in the correct hypothesis categories to more definite commitment-type estimates. Human conservatism in posterior probability estimation was the typical result in previous studies by Edwards and his colleagues [1] and by Kaplan and Newman [5]. The two introductory studies also documented the reluctance of human threat evaluators to rely upon a computer-implemented MBT diagnostic aid. Following are reports of three full-scale experiments performed using the systems simulator. Primary human performance data [$P(H \mid D)$ and $P(D \mid H)$ estimations] were collected from four TEs each working independently and using identical data.

A. Experiment I

(1) *Purpose, design, and procedures.* The general character of Experiment I was exceedingly simple. Two questions were asked: (a) how does increased experience affect human performance in the task of estimating $P(H \mid D)$, and (b) to what degree will human $P(H \mid D)$ estimates match MBT calculations of $P(H \mid D)$ on the basis of human-estimated $P(D \mid H)$? Not all of the previous studies cited provided sufficient instructions, time, or procedures to allow the subjects to make more correct and more confident estimates of $P(H \mid D)$. There was an indication in the second experiment performed by Southard, Schum, and Briggs [4] that conservatism in estimating $P(H \mid D)$ diminishes with experience. This result, however, needed confirmation since data were collected from only one subject. From one point of view an experiment devoted solely to the effects of experience upon human performance may seem rather trivial and an increase in performance with experience is the sort of thing one would naturally expect. Unfortunately, most studies so far offer little notion about what to expect from persons who are experienced in dealing with events in some environment of concern. The allegation is simply that men are conservative or suboptimal estimators of posterior probabilities and that they do not extract maximum certainty from probabilistic information. The purpose of the present experiment was to determine the extent to which this statement will have to be qualified when men are given the opportunity to become familiar with environmental events and proficient in dealing with probabilistic statements describing these events.

The present experiment consisted of 30 consecutive four-hour sessions. In each session six Aggressor deployments were evaluated by each TE. On the basis of the probabilistic attribute data provided by the ISOs for each deployment, each TE produced his estimates of $P(H \mid D)$. The attribute data were presented to each TE individually and simultaneously by means of closed-circuit television. Since six deployments were evaluated in every session, the TEs produced 180 $P(H \mid D)$ estimates throughout the experiment. In each session they also estimated $P(D_{jk} \mid H_i)$. This involved the generation of an 8 by 103 matrix of $P(D_{jk} \mid H_i)$ values since there were eight hypotheses and 103 possible states within the 25-attribute data classes. TE performance was carefully monitored to assure independence of effort. Aggressor was allowed eight different deployment types or "hypotheses" from the point of view of the TEs.

For each of the 180 Aggressor deployments evaluated during the experiment the experimenter obtained:

(1) Four human estimates of $P(H \mid D)$.
(2) Four MBT solutions of $P(H \mid D)$ calculated individually on the basis of each TE's $P(D_{jk} \mid H_i)$.
(3) One self-adapting MBT calculation of $P(H \mid D)$.

Each TE was provided each session with the six "correct" hypotheses explaining the occurrence of the data observed in the previous session. The two MBT solutions were also made available to the TEs. At the start of each session each TE was able to compare his estimates of $P(H \mid D)$ made in the previous session with the MBT $P(H \mid D)$ calculation based upon his $P(D_{jk} \mid H_i)$ estimates from the previous sessions and with the self-adapting MBT calculations. The TEs did not see one another's estimates.

(2) *Results of Experiment I.* Estimations and calculations of $P(H \mid D)$ in all three experiments were evaluated in terms of two performance measures. The first was called "verified certainty," and it simply indicates the average value or size of $P(H \mid D)$ estimates or calculations placed in the correct hypothesis categories. The second measure, called "dichotomous scores," refers to the number of occasions during the experiment on which the highest or first-choice estimates or calculations were placed in correct hypothesis categories. The major results of Experiment I, in terms of the verified certainty measure, are shown in Figure 2. Performance of the four TEs (as a group), the four MBT solutions incorporating TE estimates of $P(D \mid H)$, and the self-adapting MBT is illustrated for consecutive three-session periods during the experiment. Superiority of the MBT incorporating TE estimates of $P(D \mid H)$ is apparent until about Sessions 15. Beyond this point, however, the three $P(H \mid D)$ curves show remarkable similarity except for one unaccountable irregularity. The top curve in Figure 2 is a gross indication of the accuracy of the attribute data produced by the ISOs. Although measurable in verified certainty terms, this curve does *not* represent $P(H \mid D)$ data. It refers to the average probability value placed by the ISOs in the correct data state levels across all 25 data classes in each deployment observed in the three-session period. The attribute data accuracy curve is included in Figure 2 to illustrate that at least a portion of the $P(H \mid D)$ performance increase for TEs and MBT was due to the slightly increasing input data accuracy as the experiment progressed.

In terms of verified certainty, the overall difference between the TE estimates of $P(H \mid D)$ and the calculations of $P(H \mid D)$ from both types of MBT solution were not statistically significant [TEs vs. MBT incorporating TE estimates of $P(D \mid H)$, $P < 0.22$ (Wilcoxon Test); TEs vs. self-adapting MBT, $P < 0.11$ (Wilcoxon Test)].

The distributions of verified certainty scores for the subjects and the MBT incorporating their $P(D \mid H)$ estimates are shown in Figure 3. The two distributions are quite similar except for the somewhat greater frequency of human estimates in the middle of the distribution. An examination of the individual subjects' distributions revealed that this greater frequency of middle-range estimates was due almost entirely to one subject. The distribution of the verified certainty scores for the self-adapting MBT (not shown) was similar in form to the distributions shown in Figure 3. Bear in mind that these are distributions of $P(H \mid D)$ estimates placed in the *correct* hypothesis categories. Distributions of all $P(H \mid D)$ estimates in the correct or incorrect categories (not shown) for humans and MBT were in fairly close agreement and were also similar in form to those shown in Figure 3.

The verified certainty score distributions do not, of course, indicate the degree of relationship between the human and MBT estimates as these estimates for each deployment are compared. Table 1 shows the rank-order correlation (Rho) between the various human and MBT estimates across all 180 deployments. All correlations are highly significant ($P < 0.001$, df $= 177$).

The other type of performance measure indicates whether or not the first-choice or highest $P(H \mid D)$ estimate (for each deployment) was placed in the correct

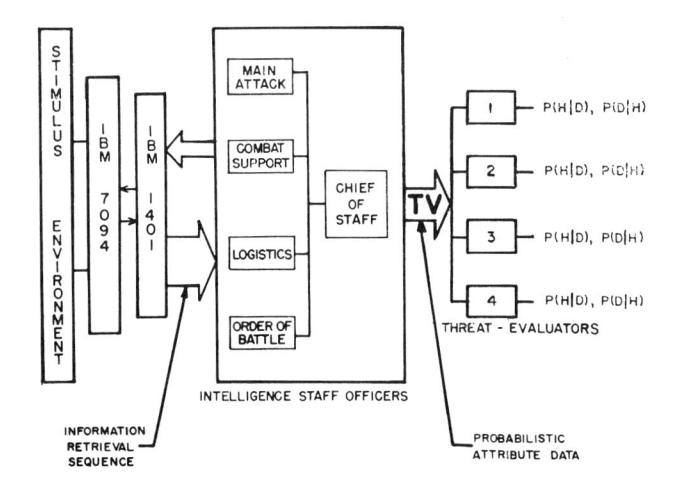

Figure 1. A simulated threat-diagnosis system.

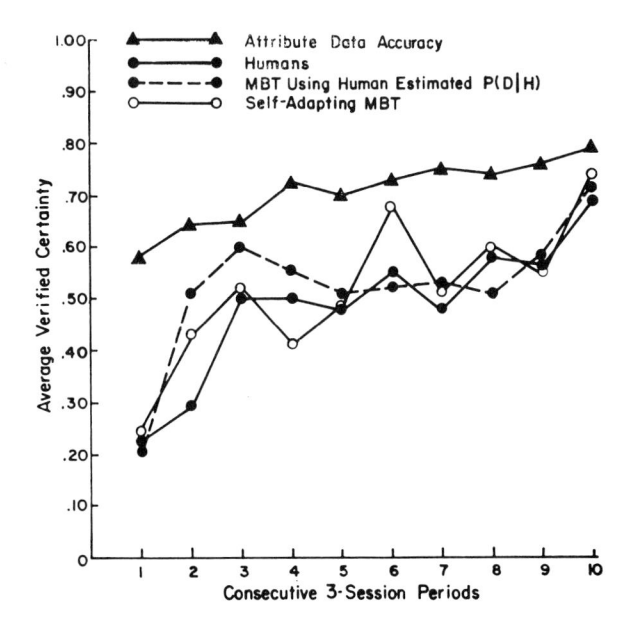

Figure 2. Human and automated posterior probability estimates over 30 experimental sessions. Experiment I.

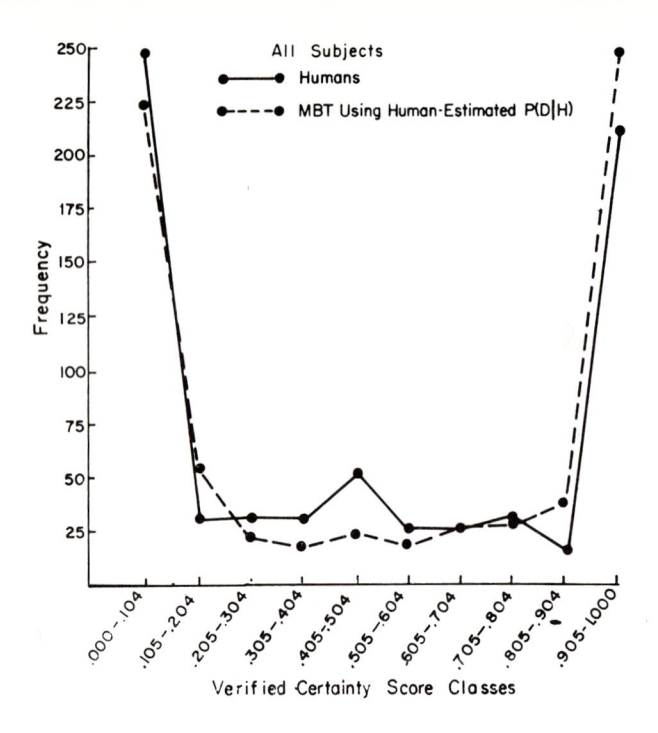

Figure 3. Distributions of verified certainty scores. Experiment I.

TABLE I

Rank Order Correlations Between Human and MBT Verified Certainty Scores in Experiment I

Rank Order Correlation between Verified Certainty Scores for:	Subject	Rho
I. Subjects and the MBT using human-estimated $P(D/H)$	1 2 3 4	0.58 0.53 0.55 0.55
II. Each subject and the self-adapting MBT	1 2 3 4	0.38 0.56 0.55 0.47
III. MBT calculated from each subject's $P(D/H)$ estimates and the self-adapting MBT	1 2 3 4	0.57 0.70 0.59 0.59

TABLE II

Dichotomous Scores in Experiment I

Subject	Subject's $P(H/D)$	MBT Using Human $P(D/H)$	Self-Adapting MBT
1	62	95	
2	111	105	
3	91	96	
4	71	91	
Total	335	387	96

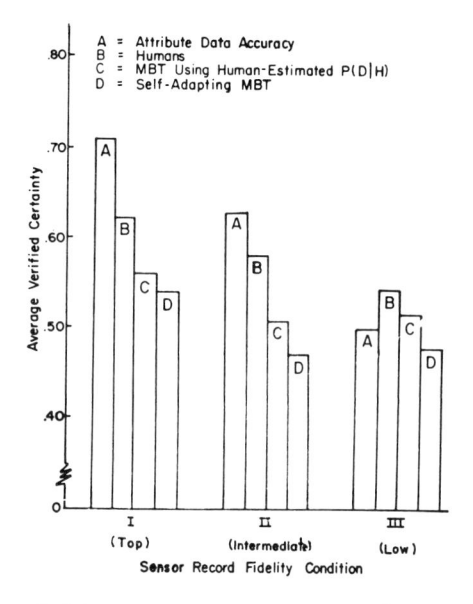

Figure 4. Posterior probability estimation accuracy under several levels of input data fidelity: Experiment II.

TABLE III

Rho Values: Experiment II

Rank Order Correlation between Verified Certainty Scores for:	Subject	Fidelity Condition I	II	III
I. Subjects and MBT Using Human-Estimated $P(D/H)$	1	0.66	0.45	0.50
	2	0.71	0.51	0.30
	3	0.50	0.67	0.51
	4	0.67	0.52	0.32
II. Subjects and Self-Adapting MBT	1	0.52	0.59	0.53
	2	0.64	0.49	0.39
	3	0.60	0.56	0.40
	4	0.63	0.64	0.46
III. Both MBT Solutions	1	0.82	0.75	0.70
	2	0.69	0.74	0.60
	3	0.78	0.81	0.72
	4	0.77	0.69	0.78

TABLE VI
Dichotomous Scores: Experiment III

Mode of $P(H/D)$ Estimation	Subject	Fidelity Condition I	II	III	Total
Subjects	1	30	32	29	91
	2	43	36	37	116
	3	39	34	27	100
	4	47	34	28	99
	Total	149	136	121	406
MBT Using Human-Estimated $P(D \mid H)$	1	34	30	36	100
	2	34	32	37	103
	3	35	33	31	99
	4	36	30	32	98
	Total	139	125	136	400
Self-Adapting MBT		34	35	35	104

TABLE V
Verified Certainty Score Means and Standard Deviations: Experiment III

Mode of $P(H/D)$ Estimation	Fidelity Condition 1 Mean	SD	2 Mean	SD	3 Mean	SD	4 Mean	SD
Subjects	0.397	0.439	0.341	0.348	0.144	0.216	0.131	0.172
MBT Using Human-Estimated $P(D/H)$	0.372	0.248	0.312	0.294	0.186	0.220	0.163	0.222

Mode of $P(H/D)$ Estimation	Time Stress 7 Minutes Mean	SD	4 Minutes Mean	SD	1 Minute Mean	SD
Subjects	0.228	0.226	0.261	0.362	0.271	0.350
MBT Using Human-Estimated $P(D/H)$	0.219	0.249	0.266	0.248	0.290	0.312

TABLE IV
Dichotomous Scores: Experiment II

Mode of $P(H/D)$ Estimation	Fidelity Condition 1	2	3	4
Subjects	82	57	13	6
MBT Using Human-Estimated $P(D/H)$	102	70	39	29

Mode of $P(H/D)$ Estimation	Time Stress 7 Minutes	4 Minutes	1 Minute
Subjects	42	57	57
MBT Using Human-Estimated $P(D/H)$	68	84	88

hypothesis category. During the experiment the MBT solutions calculated on the basis of human-estimated $P(D \mid H)$ placed 387 out of a possible 720 highest estimates in the correct hypothesis categories. The four TEs as a group placed 335 out of 720 in correct categories. This difference, in favor of the MBT, was large enough to be statistically significant ($P < 0.01$, Wilcoxon). These scores are summarized in Table II. (The low total recorded for the self-adapting MBT results from the fact that only one such solution was calculated for each of the 180 deployments.)

(3) *Discussion and interpretation of Experiment I results.* The early MBT superiority illustrated in Figure 2 presumably indicates that more consistency was present in the TE estimates of $P(D \mid H)$ than the TEs actually made use of in their $P(H \mid D)$ estimates. With experience, however, the subjects gained greater appreciation of the environmental consistency indicated by their $P(D \mid H)$ estimates and were better able to reflect this consistency in their $P(H \mid D)$ estimates. There is another explanation for the human performance increase. This explanation was indicated in a questionnaire administered to the four TEs after every four sessions during the experiment. Out of necessity, the subjects when initially confronted with a mass of attribute data, acted to reduce the complexity of the situation and in so doing lost valid or useful predictive information. With experience, however, the better subjects took greater notice of a larger number of data classes in which there existed consistency or predictability. This improved consistency recognition was reflected in improved scores. As Figure 3 illustrates, the TEs were not reluctant to use very high $P(H \mid D)$ estimates. There were no direct costs or pay-offs in these experiments and, therefore, the subjects were never penalized for incorrect diagnoses. High scores in themselves, however, were surely rewarding to the subjects. Under the conditions of the experiment, therefore, the optimal strategy for the humans may not have been basically Bayesian. Beyond a certain threshold of certainty the subject might just as well have estimated 1.0 since there was nothing to lose. Most assuredly, it was an easier task to write 1.0 in one hypothesis category than to give vernier estimates across all eight categories even though such an estimate may have been much higher than the consistency of the data justified.

Throughout the experiment the MBT solutions of $P(H \mid D)$ based upon TE estimates of $P(D \mid H)$ did not place significantly larger average values in true hypothesis categories than the TEs. These same MBT solutions, however, placed a significantly greater number of first-choice $P(H \mid D)$ estimates in true hypothesis categories. The second explanation in the preceding paragraph should help account for this discrepancy. In comparing the relative performance of MBT and subjects, the average verified certainty scores by themselves may be misleading since the subjects are known to maximize $P(H \mid D)$ in hypothesis categories they perceive to be correct (the MBT, of course, performs no such maximization). Both types of measures are needed in order to provide an accurate evaluation of the relative performance of human and MBT.

B. Experiment II

(1) *Purpose, design, and procedures.* Individuals faced with the task of evaluating events in a hostile environment seldom, if ever, have access to information

about these events which is utterly precise or reliable. Indeed, most, if not all, real-life inferences are made on the basis of data whose fidelity has been degraded for some reason or another. In evaluating Bayesian information-processing systems, it is of interest to compare under various levels of input-data fidelity human estimates of $P(H \mid D)$ with those calculated from Bayes' theorem (or from the MBT) on the basis of human-estimated $P(D \mid H)$. In his proposal for the design of a probabilistic information-processing system [in which posterior probabilities are calculated by computers on the basis of Bayes' theorem aggregations of human estimates of $P(D \mid H)$], Edwards [1] suggests that such a system could utilize with profit information so fallible or degraded that it would ordinarily be excluded from consideration. One implication of Edwards' comment is that the value of exploiting the optimal information-aggregation properties of Bayes' theorem ought to be particularly apparent when fidelity of information is low and consistency in large amounts of data becomes difficult for humans to recognize and reflect in their estimates of posterior probabilities. In the present experimental situation, MBT solutions of $P(H \mid D)$ calculated on the basis of human estimates of $P(D \mid H)$ ought to be increasingly superior to human estimates of $P(H \mid D)$ as the fidelity of input data is reduced. The purpose of Experiment II was to compare these two types of posterior probability estimates at several levels of input data fidelity. Input data fidelity, as an experimental variable, was manipulated at the ISO level of the simulated system illustrated in Figure 1.

The input information utilized by the team of ISOs in developing the 25 types of attribute data for each Aggressor deployment consisted of verbal descriptions of the type, number, and activity of Aggressor vehicles, mobile weapons, and aircraft. These descriptions were based upon simulated photo, radar, and infrared sensor records obtained on reconnaissance overflights of the territory of the hypothetical adversary. The overflights and the interpretation of the obtained sensor records were simulated, of course, by the computer facilities. In order to simulate the fact that there might exist several graded quality levels of sensor data, different levels of verbal description were available for each event to be observed by the ISOs. There were three quality levels of photo information and two each of radar and infrared. To illustrate how discriminability among events in the stimulus environment is contingent upon level of sensor description, consider the following example. The verbal description of a 2½-ton cargo truck given to an ISO using a Level I photo (highest level) is "2½-ton cargo truck," an unequivocal description. From a Level II photo, however, the description would read "medium size wheeled vehicle." In this case, the operator could not distinguish a 2½-ton cargo truck from a 5-ton truck, an amphibious armored carrier, or a 140-mm rocket launcher since they are also described as "medium size wheeled vehicles" in reports based upon Level II photos. Using a Level III photo (poorest), the verbal description of a 2½-ton cargo truck would be "self-propelled vehicle." Discrimination is now very poor indeed since many types of mobile weapons and vehicles are described as "self-propelled vehicle" in reports based upon Level III photos. Also, discriminations among critical events, which the 25 attribute data classes reflect, are based upon frequency counts of these vehicles, weapons, and aircraft. One can systematically induce uncertainty into these counts as the quality level of sensor records is reduced. For example, a certain unit in Aggressor's surface forces contains 131 wheeled vehicles as seen using photo

Level I (highest quality). This vehicle count happens to be unique and the unit can always be identified. Before the ISO receives his vehicle total, however, the experimenter can, by means of a programmed algorithm, add or subtract a random number from the original total. This random number is chosen within a certain range which indicates the resolution of the sensor. Suppose, for example, that the degraded sensor was capable of providing an accuracy of $\pm$ 10 percent on vehicle totals. If the *original total* were 131, the experimenter could, on each occasion, add algebraically a random integer between $-$ 13 and 13 to the original total, making the possible range of vehicle totals 118 through 144. This will cause the ISO to confuse this unit with other units whose range of totals falls within the range of totals possible for the unit in question. This procedure for degrading totals can be applied independently of the procedure for degrading verbal descriptions.

In all previous experiments the ISOs were allowed unrestricted access to top quality sensor records. In the present experiment, where fidelity of input is the variable of interest, the experimenter manipulated both the verbal and numerical descriptions provided by sensor records, thus simulating the effects of degraded photo, radar, and infrared images. Following are the chosen levels of the fidelity variable and the method by which these levels were induced:

(a) *Level I–highest fidelity:* No change was induced here from previous experiments. ISOs had unrestricted access to top level photo, radar, and infrared sensor records.

(b) *Level II–intermediate fidelity:* ISOs had access only to the poorest quality photo, radar, and infrared records, *but* only the verbal descriptions showed degradation, i.e., vehicle, weapon, and aircraft totals did not suffer degradation. This condition induced considerable confusion in the estimates provided by the ISOs in 13 of the 25 attribute data classes.

(c) *Level III–poor fidelity:* ISOs had access only to poorest quality photo, radar, and infrared sensor records. In this case, however, the experimenter degraded *both* verbal description and vehicle, weapon, and aircraft totals. This condition induced considerable confusion in the estimates provided by the ISO in 24 of the 25 attribute data classes. Sensor "resolution" was set at $\pm$ 10 percent.

The particular configurations in Levels II and III were chosen because of their even effects upon the 25 attribute data classes, i.e., approximately half were affected by Level II and all but one by Level III. Following is the effect which the experimenter induced by this manipulation. It was mentioned that the ISOs produce probabilistic estimates of the level or state of each attribute data class for every Aggressor deployment. On the basis of top quality sensor records they receive unequivocal verbal descriptions and exact weapon and vehicle totals. In a large number of cases unique identifications of Aggressor surface and air units were possible with this top quality input information and the ISO could estimate the state or level of a certain data class with much confidence. Under the reduced fidelity conditions (for those data classes affected by the degradation level) lack of discriminability forced the ISO to be considerably less certain about the level or state of each data class applicable to each deployment. Increased uncertainty in these attribute data has a direct effect upon $P(D \mid H)$ estimated by the TEs since the establishment of datum-hypothesis relationships becomes more difficult with uncertain data. Since this is so, the $P(H \mid D)$ estimates produced by the TEs and the MBT [incorporating the $P(D \mid H)$ estimates] were ultimately affected.

The experiment consisted of 30 four-hour sessions. In each session, as in the previous experiment, six developmental groupings were evaluated by each TE. The first ten sessions were performed under fidelity Level I, the next ten under fidelity Level II, and the last ten under fidelity Level III. Breaking up the sessions per level in an attempt to balance out possible residual effects was not possible because of the nature of the task. For each mode of estimation each ten-session period was a completely new learning experience. At the beginning of each ten-day session the subjects and the MBT operated upon an entirely different set of environmental contingencies and possible Aggressor deployments (hypotheses) of which there were eight in each condition. Although the 25 types of attribute data remained the same, the relationships between these data and the eight hypotheses (i.e., the true $P(D \mid H)$ values were changed in each ten-session period). Since there were four subjects, six developmental groupings terminating per session, and ten sessions per fidelity condition, 240 estimates of $P(H \mid D)$ were produced for each fidelity level. Only one self-adapting MBT solution was calculated for each grouping making the total under this mode only 60 per fidelity condition (the self-adapting MBT results were not analyzed statistically). The subjects serving as ISOs were the same in this experiment as in the previous one. With one exception the TEs were the same as those who served in Experiment I.

(2) *Results of Experiment II.* The major results of the experiment in terms of verified certainty scores are illustrated in Fig. 4. Along the abscissa are the three levels of sensor record fidelity. The ordinate refers to the average verified certainty scores for the 60 developmental groupings terminating in the ten-session period for each fidelity level. The data points for the subjects and the MBT incorporating the human estimates of $P(D \mid H)$ are averages over 240 $P(H \mid D)$ estimates. Each self-adapting MBT data point is an average over 60 $P(H \mid D)$ estimates. The attribute data points are gross estimates of the average accuracy in all data classes for all developmental groupings terminating in each of the three 10-session periods. The accuracy of the attribute data estimates produced by the ISOs (bar A) decreased rather sharply as the simulated sensor records were degraded. Reduction in accuracy of these data caused a corresponding decrease in $P(H \mid D)$ estimation accuracy by humans and MBT. This $P(H \mid D)$ accuracy decrease was slightly more drastic for the humans (bar B) than that for either the MBT using the human estimates of $P(D \mid H)$ (bar C) or the self-adapting MBT (bar D). Human performance as indicated by these verified certainty scores was superior in all conditions to the performance of either MBT solution. As fidelity decreased, however, this superiority diminished. The distributions of the verified certainty scores were of exactly the same form as those obtained in Experiment I. In every condition there was a distinctly bimodal distribution with extremely high score frequencies at either end of the distribution.

The main effect of *fidelity condition* in terms of verified certainty was statistically significant ($P < 0.04$, Ranks Test). This means that reduced data fidelity caused a significant decrease in overall verified certainty scores for both subjects and MBT. The interaction between fidelity condition and mode of estimation, i.e., human vs. MBT estimations, was not significant ($P < 0.27$, Ranks Test). Throughout the experiment the subjects placed significantly higher average values in correct hypothesis categories than the MBT incorporating their $P(D \mid H)$ estimates ($P < 0.003$, Exact Probabilities Test).

Rank order correlations (Rho) are summarized in Table III. All Rho values were significant ($P < 0.02$, df $= 58$).

The results of the experiment in terms of the dichotomous measure are summarized in Table IV. The scores shown for each individual subject and MBT solution in each of the three conditions refer to the number of occasions (out of a possible 60) on which the first choice or highest $P(H \mid D)$ estimate was placed in the correct category. The first two types of data shown in Table IV were analyzed in the same manner as the verified certainty data. The main effect due to *fidelity condition* and the interaction between *fidelity condition* and *mode of estimation* [human vs. MBT using humans' $P(D \mid H)$] were tested using the Ranks Test for Matched Data. Neither the main effect ($P = 0.273$) nor the interaction ($P = 0.653$) was statistically significant. The main effect due to *mode of estimation* was tested using an exact probability test. The hypothesis of no difference could not be rejected ($P = 0.378$). Fairly interesting, of course, is the fact that the largest difference between the overall human scores and the MBT scores occurred at the lowest fidelity level and favored the MBT.

(3) *Discusson and interpretation of Experiment II results.* In terms of verified certainty scores, the statistical significance of the main effect due to the fidelity variable merely shows the experimental procedure of degrading the simulated sensor records caused an overall decrement in the certainty with which $P(H \mid D)$ estimates were placed in the correct hypothesis categories. The statistical significance of the main effect due to mode of $P(H \mid D)$ estimation is very important but somewhat difficult to interpret since the observed result seems to imply that human performance was greater than "optimal." Again, as in Experiment I, both types of measures must be taken into consideration. By either yardstick, human performance in these complex tasks was fairly amazing. In addition to placing large average values in correct hypothesis categories, the subjects matched the MBT throuhout the experiment in identifying correct hypotheses with highest or first-choice $P(H \mid D)$ estimates. It is true that Condition I of this experiment represented further practice on tasks learned in the first experiment since conditions were identical (and all but one of the subjects served in both experiments). Only in the lowest fidelity level were the MBT solutions superior to the subjects' estimates in terms of the dichotomous measure. As in Experiment I, average verified certainty measures by themselves are apt to be misleading since the subjects could, without penalty, maximize $P(H \mid D)$ in hypothesis categories they considered likely to be correct. The reduction in data fidelity in this experiment was apparently not great enough to override the large amount of experience the subjects had acquired in dealing with events in the simulated environment. The superiority of the MBT in the lowest fidelity level (as indicated by the dichotomous measure) suggested that it might be profitable to observe human performance under even more serious data degradation conditions. The following experiment was performed on the basis of this suggestion.

C. Experiment III

(1) *Purpose, design, and procedures.* The major purpose of Experiment III was to observe human performance in estimating the conditional probabilities $P(D \mid H)$

and $P(H \mid D)$ under even more drastic data-fidelity reduction conditions than those imposed in Experiment II. Again, comparisons of human estimates of $P(H \mid D)$ with those calculated from the MBT incorporating human estimates of $P(D \mid H)$ were of primary interest. Another variable of some operational significance is time stress on threat evaluators. In all previous experiments the TEs were allowed unlimited time to evaluate the 25 items of data for each Aggressor deployment before estimating $P(H \mid D)$. In the present experiment the TEs' estimates of $P(H \mid D)$ were produced under three levels of time stress. Let it be emphasized that the TE estimates of $P(D \mid H)$ were not affected by this time stress variable. The subjects had unlimited time in which to make their $P(D \mid H)$ estimates. The two independent variables of interest in the present experiment were, therefore, *input-data fidelity condition* and $P(H \mid D)$ *estimation-time stress*.

Data fidelity as a variable was manipulated in this experiment in a somewhat different manner from that described in Experiment II. Discussion of the manner in which this manipulation was performed requires further comment about certain features of the simulated hostile environment confronting the subjects. The deployments of Aggressor's military forces, as observed by the subjects, developed or unfolded in a time-dependent fashion. Moreover, events in this simulated environment proceeded in double-time, e.g., in four hours of actual experiment time the subjects observed eight hours of events in Aggressor "world". All of the deployments observed by the subjects developed over a four- to six-day (game time) period. That is, they commenced developing on a certain day and terminated development either four, five, or six days later. As developmental time progressed for each deployment, more and better information about the deployments was produced by the ISO team. The longer a deployment was kept under surveillance the more complete and accurate were the ISO-developed data describing the deployment. Maximum completeness and accuracy of the data set would occur if the ISOs were allowed to observe the deployment until its termination time. Fidelity of the ISO data could be reduced, therefore, by halting the surveillance of a deployment at various lengths of time prior to its termination. The farther back in time from termination that surveillance was halted, the lower would be the fidelity of the ISO data. $P(D \mid H)$ and $P(H \mid D)$ estimations made by the TEs on the basis of these data of reduced fidelity would suffer accordingly. Data fidelity as a variable was manipulated by regulating the amount of time the ISOs were allowed to observe the buildup of the deployments. Four levels of this variable were chosen. In Level 1 the ISOs were allowed to observe deployments until they terminated. The numerical and verbal descriptions of events in the environment were degraded throughout Experiment III as in Condition III of Experiment II. Thus, Level 1 of the fidelity variable in Experiment III was identical to fidelity Condition III (lowest) of Experiment II. In Level 2, surveillance was halted between 3 and 24 hours prior to the termination of deployments; in Level 3 surveillance was halted between 25 and 48 hours prior to termination; and in Level 4 surveillance was halted between 49 and 72 hours prior to termination. This method of data-fidelity manipulation was chosen because surveillance time was easy to regulate systematically in an already difficult experimental situation.

The time-stress variable had three levels: TEs were given either one, four, or seven minutes to evaluate the ISO data before making their $P(H \mid D)$ estimates. The data were presented to the TEs over closed-circuit TV for the controlled length of time specified. Thirty seconds before these data were transmitted, the TEs were informed about the length of time they would have to estimate $P(H \mid D)$ for that deployment. Other than this brief warning, the TEs had no way of knowing beforehand how much time they would have to estimate $P(H \mid D)$ for each deployment. The TEs, however, could keep current records of the status of each developing deployment by processing interim data relayed to them from the ISOs. All TEs had access to identical interim data which was also presented to them over closed-circuit television.

Deployments in each of the 12 possible experimental conditions (three time-stress conditions by four fidelity-level conditions) were presented at random. As in Experiment II, six deployments were evaluated by the TEs in each session. There were 32 four-hour sessions. As in the preceding experiments, there were eight alternative hypotheses (deployment type). The subjects serving as ISOs and TEs were the same as those who had served in Experiment II.

(2) *Results of Experiment III.* Tables V and VI show slight apparent increases in human $P(H \mid D)$ estimation performance as time allowed for response was *decreased.* There are several reasons for believing that the time-stress variable, as manipulated in the present experiment, had no effect at all. These reasons are presented in the discussion which follows. Statistical analysis of the time-stress variable will not be pursued further in this report. The followng results, with respect to the data-fidelity variable, have been accumulated over the three time-stress levels. Table V illustrates the effects of the reduced data-fidelity conditions in terms of the verified certainty measure upon human-estimated and MBT-calculated $P(H \mid D)$. The overall difference between TE performance (as a group) and the MBT calculations was not statistically significant ($P = 0.29$. Exact Probabilities Test). The MBT placed higher average $P(H \mid D)$ values in correct hypothesis categories only in the two lowest fidelity levels. Table VI illustrates the effects of reduced data-fidelity conditions in terms of the dichotomous measure. MBT superiority is evident in all four levels and the overall difference was statistically significant ($P < 0.001$, Ranks Test).

(3) *Discussion and interpretation of Experiment III results.* The manner in which the time-stress variable was implemented made it completely ineffective. This interpretation is based upon subjects' testimony and parallel MBT calculations. The TEs were allowed to keep current on the status of each deployment because it seemed unreasonable to expect that a threat evaluator in any real situation would ever be denied the opportunity of updating his knowledge of affairs prior to a required diagnosis at some time. In the present experiment the TEs did such a good job of following the status of each deployment that when a diagnosis was called for, they merely had to write down their $P(H \mid D)$ estimates. In other words, the data presented to them for diagnosis had already been processed and evaluated during the interim updating periods. The apparent performance increase with time-stress increase is due merely to variation in the difficulty of discrimination of individual deployments seen in the three conditions. The MBT calculations on the basis of the TE estimates of $P(D \mid H)$ substantiate this interpretation. It should be remembered that these calculations were in no way affected by the time-stress variable. Yet, they show increases similar to those of the subjects' estimates of $P(H \mid D)$ as time stress was increased. Because of these considerations, and because graphic plots revealed no apparent interactions

between the two experimental variables, the experimenters felt justified in accumulating the results across the time-stress conditions.

Further reduction in data fidelity over that in Experiment II brought forth predicted MBT superiority at all fidelity levels in terms of the dichotomous measure and at the two lowest fidelity levels in terms of verified certainty. Discrepancy between performance measures is again apparent because of the TEs' tendency to maximize $P(H \mid D)$ in hypothesis categories they perceived to be correct. Although providing very high estimates of $P(H \mid D)$ on occasion, the TEs were definitely inferior to the MBT solutions [incorporating their estimates of $P(D \mid H)$] in identifying correct hypothesis categories with highest or first-choice estimates.

IV. SUMMARY AND CONCLUSIONS

Three experiments have been reported in which posterior probabilities estimated by men have been compared with those calculated using a modification of Bayes' theorem (MBT). The major experimental question has been whether it is better to have men estimate $P(H \mid D)$ directly or whether it is better to have men estimate $P(D \mid H)$ and allow computers programmed according to Bayes' theorem to aggregate these $P(D \mid H)$ estimates. The major results of the experiments were as follows:

(1) Human performance in estimating $P(H \mid D)$ directly from fairly large amounts of data was decidedly superior to what one might have concluded on the basis of previous studies. Though in all studies the subjects were able to place high average $P(H \mid D)$ values in correct hypothesis categories, they were never superior to MBT aggregations of their $P(D \mid H)$ estimates in terms of the number of occasions on which highest $P(H \mid D)$ estimates were placed in correct hypothesis categories.

(2) The role of experience in dealing with environmental events seems to be an especially important determiner of conditional probability estimation performance. The subjects' $P(D \mid H)$ estimates improved with experience as evidenced by increased accuracy of the MBT calculations. In addition, the subjects, either by incorporating more data in their judgments or by perceiving more consistency in the data, were increasingly able to place larger posterior probability estimates in correct hypothesis categories.

(3) Although human estimates of posterior probability may be conservative on occasion, the present research leads one to recognize that there are situations in which men may provide excessively large estimates. Payoff and knowledge of results play as large a role in these probability estimation tasks as they do in any other human task.

(4) Degradation of data fidelity with its consequent obscuration of datum-hypothesis relationships appears to be a significant discriminator between the two methods of $P(H \mid D)$ determination. The MBT aggregations of $P(D \mid H)$ seem to fare better under drastically reduced data fidelity conditions.

(5) It is apparent that more than one type of performance measure is needed if one expects to get an accurate overall view of performance in these probability estimation tasks.

The results of the present experiments permit the conclusion that computer-implemented Bayesian aggregations of expert human estimates of $P(D \mid H)$ [or some related quantity] may well be feasible and of value in systems with a diagnostic mission. Most assuredly, further research in automated Bayesian hypothesis-selection procedures is justified. Hopefully, the present research will give a better idea than previous studies have given about what to expect in real-life systems from experienced people. However, the subjects in the present experiments performed under an essentially unrealistic cost-payoff arrangement. The subjects could, without penalty, attempt to maximize $P(H \mid D)$ in hypothesis categories which were perceived to be correct. The frequentistic nature of the data being processed by the simulated system suggests that the present results may also be relevant to the medical diagnosis situation in which relative frequencies linking symptoms with diseases may be available. Most military diagnosis systems, however, deal with environmental processes not specifiable in terms of long-run frequencies. For this reason, automated Bayesian hypothesis-selection procedures need to be evaluated in nonfrequentistic environments.

REFERENCES

1. W. Edwards, "Probabilistic information processing in command and control systems," Electronics Systems Division, L. G. Hanscom Field, Bedford, Mass., ESD-TDR-62-345, March 1963.
2. J. F. Southard, D. A. Schum, and G. E. Briggs, "An application of Bayes' theorem as a hypothesis-selection aid in a complex information-processing system," Aerospace Medical Research Labs., Wright-Patterson Air Force Base, Ohio, AMRL-TDR-64-71, June 1964.
3. J. D. Dodson, "Simulation system design for a TEAS simulation research facility," Planning Research Corp., Los Angeles, Calif., PRC-R-194, 1961.
4. J. F. Southard, D. A. Schum, and G. E. Briggs, "Subject control over a Bayesian hypothesis-selection aid in a complex information-processing system," Aerospace Medical Research Labs., Wright-Patterson Air Force Base, Ohio, AMRL-TR-64-95, September 1964.
5. R. J. Kaplan and J. R. Newman, "A study in probabilistic information processing," System Development Corp., Santa Monica, Calif., SDC-TM-1150, April 1963.

31

Do Large Shared Displays Facilitate Group Effort?

Sidney L. Smith and Benjamin C. Duggar

In man/machine information processing systems, information must be presented so that it can be perceived and interpreted by the human users. Reliance is generally placed on visual displays of one sort or another, and a decision must be made as to just how large such displays should be.

If a display is small in size, as small as a page in a book, or a sheet of paper in a typewriter, a viewer must be close to it to distinguish the details of the information it presents. It can be used by only one person at a time, or perhaps two. A larger display, such as a briefing chart, or a movie screen, may be viewed by a group of people who use its information simultaneously. Because of these considerations, a design decision concerning the appropriate size of displays, large or small, is often expressed as a choice between group or individual displays.

In terms of the possible value of large group displays, it has been argued that in situations where coordination of group effort is desirable, use of a common, shared display may facilitate system performance (Smith, 1962). Just how this might occur would presumably depend upon the nature of the working group, its hierarchic structure, the nature of its common task, and the way in which the task is shared in terms of functional roles or jobs performed by the individual group members. The value of a group display, therefore, might differ from one setting to another. A military commander and his staff, an industrial manager and his associates, a team of air traffic controllers, a group of people analyzing election returns—in each case, the relative usefulness of common group displays versus separate individual displays will be different.

From *Human Factors,* 7, 1965, 237-244. By permission of the Human Factors Society.

The authors wish to acknowledge the help of Dr. Everett A. Garvin and David E. Moore, who assisted in the data collection for the experimental studies described in this report; of Judith Rubenstein, who assisted in the data analysis; and to Barbara B. Farquhar, who participated both in the final data analysis and in the editing of this report. The research reported in this paper was sponsored by the Air Force Electronic Systems Division, Air Force Systems Command, under Contract AF 19(628)2390. This paper will also be obtainable as ESD Technical Documentary Report No. 64-160. Further reproduction is authorized to satisfy needs of the United States Government. A more detailed account of this research is available from the authors (Smith & Duggar, 1964).

This paper reports the results of an experiment designed to compare directly group and individual displays in the context of a relatively simple group task, that of searching displays for particular items and of counting classes of displayed items.

An earlier study, published by Enoch in 1959, dealt specifically with the effect of display size on visual search. Individual observers were seated at a constant distance from the displays, and thus, the smaller displays subtended a smaller visual angle. Under such conditions, the small displays could be used either to present less data than the larger displays, or to present the same amount of data with decreased legibility.

More recently, Duggar, Autor, and Moore (1964) studied search time as a function of display mode when small displays were placed nearer to the individual observers so that they subtended the same visual angle as the large displays. Here, considerations of visual geometry suggest that the large and small displays were equivalent in the amount of information they could present. And, in fact, no performance differences attributable to display mode were demonstrated.

In the study reported here, the circumstances were similar in that small and large displays of equal visual angle were used. In this case, however, a *group* of viewers had to work together to use displayed information. The question, then, is whether these two logically equivalent display models are in fact equally useful in a group task situation.

PROCEDURE

Twelve groups of four men were run in this study. The subjects were college students, paid volunteers recruited through their school employment office, screened for normal vision. Each group was picked up at their school and driven to the laboratory, which gave the individuals about half an hour to get acquainted with one another if they had not already met.

At the laboratory, the subjects were asked to sit along one side of a table, facing a display screen. The edge of this table was angled so that each subject was able to see the other three. The ambient illumination was about 0.6 footcandles, sufficient for the subjects to converse easily in a direct face-to-face relation. The experimenter explained the experimental procedures and showed the group a sample display to illustrate the task. Half of the groups were shown a single large display, worked with the large displays throughout the first session of the experiment, and shifted to individual displays later. For the other groups this procedure was reversed: they were instructed and began the experiment using small individual displays, identical for each subject, and switched to large displays for the second session.

Each display consisted of a square field on which were presented a number of randomly positioned "tracks," shown as white on a black background. Each track consisted of a single identifying class letter (one of five: F, I, S, H, or U), a three-digit number, and an adjacent vector (arrow) pointing in one of eight cardinal directions. Letter, number, and vector direction were randomly chosen for each track in the display. Letters and numbers were relevant to the group task, but the vectors were not and so can be regarded as "clutter." The displays differed in density, presenting either 20, 60, or 100 track items.[1]

Consider first the large group display. It was presented by rear projection to produce a display field 31 inches square on a screen directly in front of the subjects, centered at eye level, and approximately 60 inches away. The alphanumeric symbols, as projected on the screen, had a height of 12 mm and a stroke width of 3 mm. To the right of the group display was an auxiliary panel which specified the problem the group was to solve in any particular trial. Each subject had an input keyboard on the table before him. This display arrangement is illustrated in Figure 1.

A trial began when the experimenter illuminated the large display. The group task involved searching the display for five particular tracks, one of each letter class, to determine whether or not they were present (on the average, only half of these target tracks were actually present on the display). For each such target track, the class identifying letter and the first two track digits were shown on the auxiliary problem panel. Subjects were asked to indicate when they had found a target by punching in the third track digit, using their individual input facilites. If they failed to find a target track, they were to punch in "NO" as their answer. As responses were made, they appeared on the problem panel.

In addition to this search task, groups were asked to count the number of tracks of each letter class in order to answer the following questions: What is the total number of H and U tracks? By how much does this sum exceed the number of I tracks? Are there more F than S tracks? The problem panel displayed these three questions for every trial, and provided spaces to display the answers punched in by the subjects. The answers to the first two questions were two-digit numbers, and a simple "YES" or "NO" answered the third question.

The input keyboards were so designed that each subject could insert any part of the total answer set. This flexibility permitted the subjects to organize their activities in whatever way seemed desirable to improve group performance. Subjects were instructed that any mode of operation was permissible so long as all parts of the problem were answered for each trial, and that it was not necessary for each member of the group to insert every part of the answer. In another room, the experimenter worked from his own separate display facility to record the inputs made by each member and the time required to complete the trial.

Although each subject did not have to insert an answer to every question, it *was* necessary that he confirm the total group answer by punching a specially designated "confirm" button. When all subjects confirmed in this manner, the display screen went blank. The group answer was complete when an answer of some sort (whether correct or not) was given for every part of the group problem and any discrepant answers from different group members had been resolved. These procedures were made clear to the subjects in two practice trials during the preliminary instructional period. In actual experimental sessions, all group answers met these criteria.

The subjects were instructed to work quickly and accurately. They were told that at the completion of the study a $5 bonus would be awarded to each member of the group with the best overall performance. No information concerning the

[1] An illustrative sample of such displays has already been published in connection with a previous study (Smith, 1963).

Figure 1. Posed photo illustrating use of shared group display.

Figure 2. Group performance time as a function of trial blocks for the two sessions and display modes.

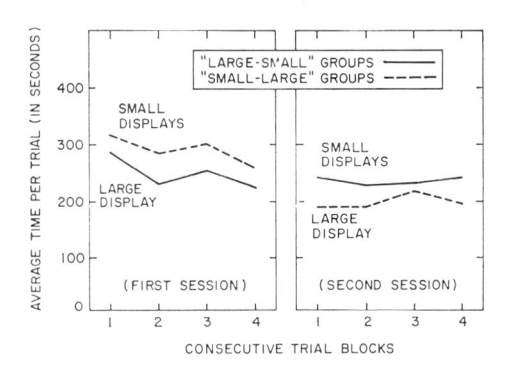

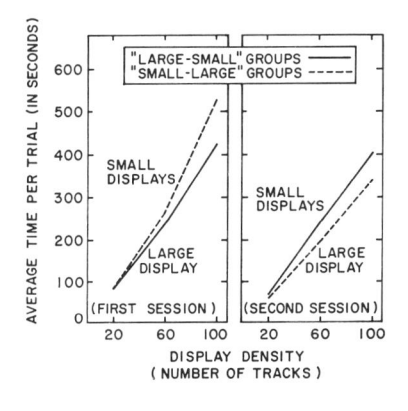

Figure 3. Group performance time as a function of display density for the two sessions and display modes.

TABLE 1

Analysis of Variance for Group Time Scores

Source of Variance	d.f.	Mean Square	F
Display Size (L)	1	115,881	6.44*
Session (S)	1	201,348	11.19**
Trail Block (B)	3	12,637	.70
Display Density (D)	2	2,902,468	161.31**
L x S	1	319	.02
L x B	3	715	.04
L x D	2	36,533	2.03
S x B	3	12,095	.67
S x D	2	47,660	2.65
B x D	6	20,348	1.13
L x S x B	3	2,977	.17
L x S x D	2	4,718	.26
L x B x D	6	4,190	.23
S x B x D	6	2,868	.16
L x S x B x D	6	2,150	.12
Within Replicates	240	17,993	......

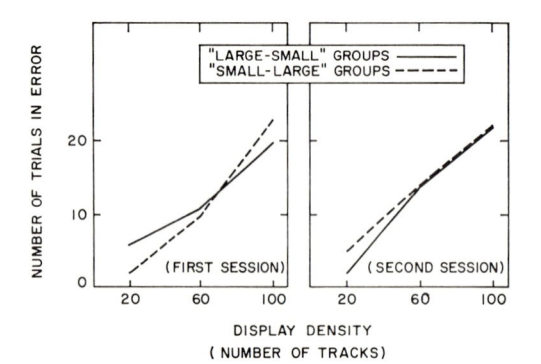

*P < .025
**P < .001

Figure 4. Group performance errors as a function of display density for the two sessions and display modes.

Figure 5. Individual performance time as a function of display density for the two sessions and display modes.

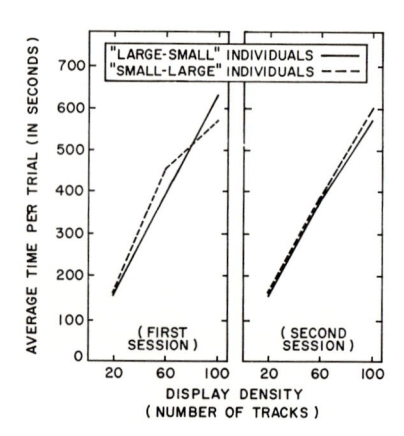

accuracy of their answers was given to the subjects during the course of the experiment.

After the instructional period, the first session of the experiment was begun. There followed a series of 12 trials, with the displays differing from trial to trial in the particular items displayed, the random placement of items on the display field, and in the overall display density (number of items displayed). Displays were presented in a randomly determined order with the restriction that there was one display at each of the three density levels in each "block" of three trials. The session of 12 trials consisted of four such blocks.

When the first series of 12 trials was completed, a recess was called, and each subject was asked to fill out a question sheet. The questions aked him to describe the organization and operation of his group, evaluate the group's performance, and indicate some of his reactions to participating in the study. After a brief period of relaxation, the subjects reassembled, took their same positions at the table, and were given instructions for the second session of the experiment. Their group task remained the same, but the displays they worked from changed in size.

Those groups which had worked first with the large shared display, now used small individual displays. The subjects removed the small displays from a storage rack behind each of the input keyboards and hooked the displays to a viewing rack mounted just above each keyboard, centered at eye level. These displays were identical, in any particular trial, for every subject in the group. The small displays presented a field eight inches square and were viewed from a distance of about 15 inches. They were one-quarter the size of the large group display, with symbology proportionately smaller, and were viewed from one-quarter the distance.

The small displays were prepared on sheet film, and the tracks appeared bright on a dark background just as on the large display. Illumination was provided by projecting a bright lighted field on the lower portion of the large (group display) screen behind the smaller displays. Turning on this light made the items on the small displays visible and defined the start of a trial.

Using their individual displays, and changing them manually from trial to trial, the subjects worked during the second session of the experiment for a series of 12 trials, with four blocks of trials using displays at three density levels as described before. Following this, the subjects were asked to complete a second questionnaire, similar to the first one, but asking some additional questions concerning their preference, if any, for the two display arrangements they had used.

To control for possible differences in legibility between the two display modes, a separate supplementary study was run, using 12 subjects, each of whom worked as an individual to solve the entire problem (all parts) in each trial. The order of the display use was counterbalanced from one subject to another. However, these subjects were required to complete only six trials (two blocks as defined above) in each session of the experiment. Time, error, and questionnaire data were obtained as before.

RESULTS—PERFORMANCE DATA

Trends in average performance time during the two experimental sessions are illustrated in Figure 2. Shown separately are those groups which started with the large display, then shifted to the small individual displays, and those groups for whom the procedure was reversed. The time data are shown by trial blocks, each block containing one trial at each of the three density levels (the average trial time in a block would, in general, correspond to the time required with the intermediate, 60-item displays). These data reflect the joint effect of two variables; groups worked more quickly with the large, shared displays, and also tended to work more quickly during the second experimental session.

Figure 3 summarizes the average time required at different levels of display density for groups using each of the two display mode sequences. Time required increased linearly with display density. It is evident that the performance differences related to display density are much more sizable than the differences between sessions, or between display modes.

Table 1 summarizes results of an analysis of variance for the time data. In this analysis, the scores for the six groups providing data under any particular combination of experimental conditions are treated as replicates. The analysis confirms the statistical reliability of the obvious effect of display density on performance time, and indicates significant (although smaller) effects attributable to display size and experimental session. There was no consistent effect related to trial blocks within each session, nor did any interaction among the various experimental conditions reach statistical significance.

Figure 4 presents the total frequency of error trials, separately for those groups working under different experimental conditions, just as was done for the corresponding time data in Figure 3. In Figure 4, no distinction is made between a trial in which just one part of the group answer was wrong and a trial in which there were multiple errors (when errors in each part of the group answer were counted individually, the same relative differences in error frequency were observed). These data generally match the differences in performance time noted above. That is to say, under circumstances where groups required a considerably longer time to answer the problem, as with displays of high density, more errors were made. However, in comparisons where time differences were relatively small, as between sessions or betwen display modes, there were no significant differences in error frequency.

Performance time data are presented in Figure 5 for individual subjects run in the follow-up study. No consistent difference attributable to display size can be established. There was also no difference in performance time between the two experimental sessions for subjects run individually. However, the strong relation between time required and display density is still manifest. For individual subjects, as for the groups, error frequency increased with display density, but showed no difference between display modes or between experimental sessions.[2]

RESULTS—QUESTIONNAIRE DATA

Questionnaires were completed by the subjects after both the first and second sessions of the experiment. The question sheets were identical in format except

[2] For those readers interested in direct comparison of group and individual data, it may be noted that group performance averaged only 37 percent faster, with 8 percent fewer trials in error.

that at the end of the second questionnaire there was an additional question asking the subjects to compare the two display modes. The subjects were asked to describe their group organization in terms of procedures and checking routines, to report whether their group had a leader, to rate their performance in terms of speed and accuracy, and to estimate their degree of interest in the study and willingness to participate in further experiments. In none of these areas did the answers indicate a clear distinction between groups working with the two different display modes.

Subjects were generally accurate in describing group organization, i.e., their accounts matched the experimenter's observational records of group procedures. The problem of how the group should organize to perform the task, what functional roles should be allocated, what procedures should be instituted for checking answers, tended to engage the attention of the subjects in early trials, including the practice trials during the instructional period. Typically, an organization developed quickly in each group and persisted thereafter, despite change in display mode.

After the first session, 11 of the 24 subjects working with the large shared display named someone in the group as a leader, but only six of the subjects working with small displays did so. However, this difference is not statistically significant. After the second session, these numbers dropped to nine and three, for the same subject groups, now working with small and large displays respectively.

Subjects' estimates of speed and accuracy did not differ between display modes, or between sessions, and there was no consistent relation between estimated and actual performance from one group to another. The subjects consistently underestimated their error rate, a discrepancy averaging as much as 50 percent in some groups.

Subject interest, as rated in the questionnaires, tended to decline somewhat during the experimental sessions, with most subjects mentioning some increase in fatigue in their comments. This decline was approximately the same for all groups, i.e., was unaffected by sequence of display mode. However, even at the end of the experiment, interest ratings were relatively high, averaging somewhere between mild and considerable interest. The proffered performance bonus seemed to act as an effective incentive for the subjects: it was referred to during the experiment in the conversation within groups, suggestions for working more quickly were commonly made, and all groups instituted some sort of checking routine, more or less efficient, to try to ensure the accuracy of their answers.

The subjects, in their questionnaire responses, were divided in their preference for the two modes of display. Fifteen indicated a preference for the large display, some mentioning that it seemed "easier to see." On the other hand, 26 subjects reported that they preferred the small displays, some giving the same argument, "easier to see." One type of comment unique to a stated preference for small displays had to do with the fact that these displays were near enough to be touched, and so the hands (or other objects) could be used to "divide up" the display surface for counting purposes. Twelve of the 26 subjects preferring the small displays mentioned this. A type of comment unique to those subjects preferring the large displays was that they facilitated group effort and cooperation in some undefined way ("more group effort"). Five of the 15 subjects preferring large displays mentioned this.

For the subjects run as individuals in the follow-up study, the results were much the same. None of the questions elicited answers that were significantly different from one display mode to the other. Again, there was no consistent preference for either display mode. Some subjects preferred the large displays, some the small.

DISCUSSION

The general relation of time and error differences to task difficulty, showing an increase with an increasing display density, is not surprising. However, trends in time and error from one experimental session to the next deserve more discussion since they are not necessarily predictable. Presumably these trends reflect a resolution of countervailing effects of learning and fatigue. The observed maintenance and improvement of group performance time during the two experimental sessions suggests that task-oriented motivation remained reasonably high for the subjects. This suggestion is supported by responses to the post-session questionnaires, in which subjects indicated continuing interest in their jobs throughout the experiment.

Still more interesting is the finding that use of the small individual displays resulted in slower group performance. It is possible that some feature associated with the availability of individual displays induced the subjects to institute more careful answer checking procedures, and thus to work somewhat more deliberately. If so, then they should also have been more accurate when working with small displays. However, there were no significant differences in error frequency between the two display modes.

Another possibility is that there was some systematic difference in display legibility favoring the large displays, making them easier to work from, and resulting in faster group performance. The most obvious difference between the display modes was the ability of subjects to enlarge apparent size (subtended visual angle) of displayed items by moving their heads closer to individual displays. However, this difference would seem, if anything, to favor the small displays. The subjects' stated preferences, elicited by the questionnaires, did nothing to clarify this question, since the majority of the subjects who expressed a preference picked the small, individual display mode.

It was as a consequence of these ambiguities that the follow-up study was run, with individual subjects working alone, using both the small and large displays. If the faster group performance with the large displays was a result of some undefined superiority in viewing conditions, then presumably this should also permit individual subjects to work more quickly when using these displays. In fact, no such difference appeared.

Since display size did not affect the performance of individual subjects, it may be supposed that the superiority of the large display in the group study was associated in some way with induced mode of group operation rather than with viewing conditions per se. Various additional attempts were made to discover what features of group operation might account for this result. Questionnaires were examined to determine if there were characteristic differences in leadership recognition in the groups, differences in described functional role allocation, or pattern of operation. The records maintained by the experimenter were analyzed to determine which member of each group inserted each part of the group answer, what the pattern of response was, which members confirmed the group answer first, and so on. All to no avail—these analyses produced ambiguous and/or inconclusive results.

To the experimenters, the most likely explanation seems to derive from casual observations made during the group study that there appeared to be more debate or arguing when the groups were working with the small displays. It was as if each subject had a somewhat greater "vested interest" in an answer he obtained from his "own" display than one from a display shared in common with the other group members.[3] This hypothesized reluctance to accept another's answer, or to modify one's own, might serve to slow group performance without significantly improving accuracy.

One might also assume that improvement in the process of interpersonal interaction accounts for the decrease in average group performance time from the first to the second experimental session, since no such difference appeared for the individual subjects.

To confirm such postulated effects, further experimental investigation would be required. In such a study, the nature of the group task should be changed in ways to emphasize the potential controversial or argumentative aspects of individual member contribution to the group answer. The proper count of a number of displayed items is, after all, in some sense ultimately verifiable. Given a more debatable problem for group discussion and decision, then an explicit attempt should be made to observe, measure, and record the nature of the resulting group interaction. Under such circumstances, one might be able to test this present, tentative conclusion that a large shared display, as compared with separate individual displays, tends to facilitate discussion in a face-to-face group.

REFERENCES

Duggar, B. C., Autor, S. M., & Moore, D. E. *Visual search time as a function of display mode.* MITRE Tech. Ser. Rep. 14, August 1964, The MITRE Corp., Bedford, Mass.

Enoch, J. M. Effect of the size of a complex display upon visual search. *J. Opt. Soc. Amer.*, 1959, 49, 280-286.

Smith, S. L. *Visual displays—large and small.* AFESD Tech. Rep. 62-338, November 1962, Air Force Electronic Systems Divisions, Bedford, Mass. (DDC Document AD-293 826).

Smith, S. L. Color coding and visual separability in information displays. *J. appl. Psychol.*, 1963, 47, 358-364.

Smith, S. L., & Duggar, B. C. *Group versus individual displays for a search and counting task.* MITRE Tech. Ser. Rep. 15, November 1964, The MITRE Corp., Bedford, Mass.

---

[3] It might be argued that this would even be a rational reaction if the subjects suspected that the small displays might be different from one individual to another. However, they were assured that all four individual displays were identical, and this could be confirmed by casual or sometimes explicit comparison by neighboring subjects.

PART FOUR

TRAINING IN THE MAN-MACHINE SYSTEM

The decade following World War II ushered in a period of scientific and technological achievement unparalleled in the history of man. Dramatic advances in the use of atomic energy, the exploration of space, and the application of our ever-expanding wealth of computer techniques have already touched nearly every aspect of our lives. Furthermore, we are just now beginning to appreciate the changes that are in store as we enter what some have termed "the computer age" (Burck, 1965). This general theme is repeated over and over again throughout the present book; it is an inescapable fact and one which has serious implications for virtually every facet of engineering psychology today.

Not the least of these implications is the influence which computer science and automation have had (and will continue to have) upon training requirements in man-machine systems. Each new advance carries with it the potential for changing a vast number of work-skill requirements as well as completely abolishing or creating entire job categories. As virtually everyone must recognize by now, the inevitable trend is toward *skilled* at the expense of *unskilled* classifications. The demand for technical workers is rapidly outstripping the supply, while opportunities at the unskilled level are leveling off and can be expected to diminish. Thus training programs must be developed to equip people just entering the labor force with the skills necessary to fill these more technical jobs. This alone is a formidable undertaking, but by no means the whole story. In the total labor force in 1969, there were 30 million persons who were 45 years of age or older. With job requirements subject to drastic change, many of these workers, trained and experienced in a particular set of skills, must face retraining or unemployment. The economic, technical, and psychological ramifications of the retraining process make this in many ways the most difficult problem of all.

The industrial psychologist has made a continuing contribution to man-machine harmony by devising predictive instruments (e.g., application blanks, tests) aimed at choosing the best man for the job. For the reasons

---

References will be found in the General References at the end of the book.

just described, however, the employer can no longer afford to rely solely upon individual differences and the selection process to fill his available positions no matter how good a test battery he has. Whenever possible, of course, he should select the best man for the job; but even then he will be obliged to train and retrain that man as the job itself changes. In so doing, he may well discover that the man is more capable than he at first suspected. It is now becoming rather clear that in concentrating on *selection,* we have, until recently, neglected the equally important matter of *training* which emphasizes man's *potential* capabilities. Development of knowledge and technology in the training field has been slow in rising to the challenge of industrial technology.

During the time that industry progressed from simple tools to atomic power, there have been few parallel developments in education and training. At the 1873 International Exposition in Vienna, an American school exhibit displayed maps, charts, and textbooks as illustrations of progress in the field of education. Up to the last twenty years, these training devices continued to be the major—if not the only—aids used in training and education. Similarly, the knowledge which we have gained over this period concerning the fundamental processes of *learning* has had relatively little impact upon theories and techniques applied to *training.* It is rather surprising that in spite of the emphasis which psychology has placed upon learning since the 1920s, there has until recently been so little cross-fertilization of basic and applied interests. The facts upon which classical learning theory rests have rarely been extended beyond simple task situations (such as classical conditioning and verbal learning) while the principles sought by training specialists are invariably applied to complex skill situations. In the absence of directly relevant information, the training practitioner is often forced to turn for guidance to very tenuous theoretical notions.

Unfortunately, the practitioner has added to these problems because he has been remiss in evaluating the techniques he is presently using. Management has often operated under the premise that training "works" and that there is no reason to spend time and money on evaluation of the procedures used. A typical survey reveals that fewer than 15 percent of the firms which employ training techniques record objective performance data. Even then, the data collected are often merely achievement scores taken immediately after completion of the training program. Without a follow-up to assess on-the-job performance, such data are of very limited practical significance.

Because of the paucity of critical information, training personnel have found themselves ill-equipped to meet the increased demands for training brought about by the rapidly changing technology of today's world. The articles appearing in the four chapters of this part were chosen to reflect the training psychologist's growing awareness of contemporary problems and past shortcomings.

Most psychologists will probably agree that the purpose of training is to promote, within a training environment, the acquisition of skills, rules, or concepts which will significantly improve an individual's performance in another situation, i.e., that presented by the actual job. Thus, it is necessary to evaluate both the *learning* or acquisition process and the *transfer* process whereby learning is carried over to the criterion task. The opening material (Chapter Nine) in this part is therefore devoted to acquisition and transfer, with particular emphasis upon the critical evaluation of our knowledge of the learning process and its applicability to training techniques. The first two articles present contrasting views on the important variables in the acquisition process; the last article discusses some of the critical problems facing those who wish to understand the transfer process.

While a careful examination of the learning and transfer variables is an important step toward devising improved training procedures, evidence that a program works in *practice* is another question. The evaluation of training techniques is probably our weakest and least clearly understood area. The difficulties are twofold: first, we have the measurement problems common to all performance evaluation, and second, we have deficiencies in the design of studies which purport to provide evaluative data. Few have employed proper controls or have "followed up" by evaluating on-the-job performance. The articles in Chapter Ten explore these problems.

The last two chapters illustrate several areas in which progress is, in fact, being made toward improved training techniques. Hence we close, at least, on a note of optimism rather than despair. Chapter Eleven discusses the sophisticated devices and simulators which have become a critical part of our approach to training in complex systems. The complexity of many jobs and the level of proficiency required for the system to perform well demand the use of such devices in order to save time, money, and in some cases, even lives. Consider, for example, what might happen if commercial jet pilots were trained on the job! The articles review the potential of three current techniques—simulators, teaching machines, and computer-aided instructions—and discuss significant research issues which, hopefully, will enhance the future value of such training aids.

The final chapter, entitled "Directions in Research," represents approaches in research that we feel may dominate the field of training in the near future. The first two articles, which discuss task analysis and task taxonomy, reflect the belief that a dual approach will eventually pay high dividends. Briefly stated, this approach requires that each job be broken down in a systematic manner to determine its behavioral elements, and that the tasks thereby identified be reclassified according to their relationship to particular principles of learning and training. As an adjunct to this approach, the third article discusses the limits which individual variations in behavior impose on training procedures. Here, it is proposed that training techniques should be matched to individual characteristics.

The final selection in this chapter represents the team approach to training. It is now widely recognized that many future (as well as present) jobs will require the effort of several individuals working as a team to operate with maximal efficiency. The entire area of space exploration serves as a good case in point. The same issues associated with task analysis, task taxonomy, and individual differences also apply in the team situation except that the solutions are much less apparent because of the increase in number and level of variables present.

Chapter Nine

Acquisition and Transfer

A large percentage of an individual's time is occupied in training for one or another specific kind of activity. As an infant, his activity is oriented toward such skills as piling up blocks or eating with a spoon; as a school child, it is "the three R's" or various kinds of sports; as an adult, it is the requirements of some job or profession. Whatever the task, the individual is usually afforded some more-or-less systematic instruction at the hands of his parent, teacher, master sergeant, or corporation. While a distinction is sometimes drawn between the terms "education" and "training" (the latter implying more uniform behavior by all recipients), *learning* is the central process in either case. The basic objective of training or education programs must be to foster acquisition of skills, rules, or concepts which will transfer either narrowly or broadly from one situation to another. Obvious though this point may be, however, a wide and deplorable gulf has always separated *learning theory* from *training practice* (Glaser, 1965). It is the nature and consequences of this division that we wish to examine critically in this chapter. The reader should have no difficulty locating summaries of existing knowledge in either of the divergent areas (see, for example, Hilgard and Bower, 1966; Holding, 1965; McGehee and Thayer, 1964).

It is not difficult to find reasons for the lack of integration between fundamental and applied interests in the learning field. In some cases, the trainer is a trainer in name only—he does not know or care anything about the field of learning; hence the program develops as little more than a matter of instructing the trainee to "sit and watch Joe because he knows how to perform the job." Whether Joe really "knows the job" and, more important, whether Joe knows how, or even *wants*, to teach the trainee is certainly questionable. Very often, this type of "program" is excused by the parties responsible for it on the grounds that costs of more elaborate procedures would be prohibitive. Yet, as we shall see later, information on the cost and value of most training programs simply does not exist. Reluctance to change existing practices or even to *look* for relevant data is thus one barrier to a productive interaction of the two interests.

Other people involved in the design of training programs would be willing to use basic data or concepts provided by learning theorists if only they could. Unfortunately, this is not always easy. How, for example, does one generalize from amount of food in a goal box to amount of praise for a job well done? Or worse, how does one justify a training scheme based upon one interpretation of learning data when there are several equally plausible explanations available? Indeed, most

References will be found in the General References at the end of the book.

controversies in learning involve *interpretation* of data, rather than the empirical findings themselves. This confusion causes difficulty for the training specialist. On the one hand, available data do not seem particularly relevant to this needs; and on the other, theoretical interpretations through which he might establish the relevance of such data are vigorously disputed. Consider, for example, the role of *reinforcement* in learning. Most theorists would agree that events such as food or praise administered contingent upon a particular response promote acquisition of that response. There is widespread disagreement, however, over what constitutes a reinforcing event and how it operates to promote learning. Thus even the *simplest* learning situations have produced no consensus as to *what* is learned, *how* it is learned, or *why* it is learned. Little wonder that the training specialist has difficulty in adapting basic learning principles to his immediate needs.

Not all learning research, of course, has been limited to ultra-simple situations. While most theorists have preferred to deal with rats, mazes, eye-blink responses, nonsense syllables, and the like, a few brave souls have attacked the problem at the level of integrated *patterns* of behavior such as perceptual-motor skills. As we saw earlier (Chapters Four and Five), engineering psychology—with its emphasis on skilled performance in realistic tasks—has always been at the forefront of this latter movement (see, for example, Bilodeau, 1966; Fitts, 1964; Noble, 1968). Undoubtedly, research at this level holds much greater appeal for the practitioner. The only difficulty is that integrative theory is also rather fragmentary with respect to these more complex activities. All too often, explanatory concepts have been limited to a particular *kind* of complex behavior (such as pursuit tracking or problem solving), and the higher-order statements which might link them together (and, hopefully, might also tie them more closely to "traditional" learning theory) are still in the earliest stages of development. The training specialist, therefore, finds little more direct satisfaction in the "complex" than he does in the "simple" learning literature, unless he happens to have a task for which there is a ready laboratory analog.

It is easy, of course, to decry our shortcomings in understanding the learning process, and to bemoan the plight of the person who must try to *use* such equivocal knowledge. We must remember, however, that this difficulty is shared to one degree or another by all who would apply basic information to practical problems. As McGehee and Thayer (1964) have noted, the practitioner often expects too much. He wants quick answers to complicated problems, panaceas where none exist. Failing to find them, he feels justified in writing off the entire basic research enterprise as largely irrelevant. We would argue that success in applying any knowledge is dependent in large measure upon what the practitioner *adds* to the situation; rather than discounting incomplete knowledge, he should work doubly hard to extract from it whatever meaning it has for his purposes. To our mind, the training specialist has not always exploited existing knowledge to the full.

There seem, then, to be several factors responsible for the gap between learning research and training practice; the blame can be spread rather evenly among researchers, trainers, and those who sponsor training programs. What is far more important here, however, is establishing the fact that such independent effort can and does work to the detriment of training applications. One need not go far in search of examples. Consider, for instance, the matter of interpreting "the learning curve." By and large, training directors use *average performance* of a group of trainees as their criterion for evaluating progress of the individual and for drawing inferences about the group or the program. Now, psychologists interested in learning have recognized for some time that it is dangerous to infer things about the course of acquisition from averaged performance data (Bahrick, Fitts, and Briggs, 1957; Estes, 1956; Sidman, 1952). For one thing, a mean curve may obscure systematic features of the individual curves comprising it because they occur at different points in time. Or the parameters of individual functions may be entirely different from those describing the mean curve. In any event, there will always be a degree of variability among individual functions, and unless one takes this into account, he cannot properly assess the progress of any *single* individual. Thus, rather than evaluate a trainee on the basis of his performance relative to the *mean* curve, a supervisor should judge him according to his position within the *distribution* of curves. On this basis, a "slow learner" on the *average* criterion might well fall within error tolerance limits for average performance. Failure on the part of trainers to recognize this point has undoubtedly led to special training where none was needed, and to early rejection of suitable workers.

Another notable feature of many learning curves is the *plateau*, which is usually defined as a temporary halt in the improvement of performance. It is usually assumed that when an asymptotic level is reached on the primary performance index, the trainee has mastered the skill and further training is unnecessary. There is always the possibility, however, that this is just a plateau. Fitts (1965), for example, reviewed a number of studies of skill learning in which speed continued to improve long after the primary index (accuracy) had asymptoted. It has also been shown that *time-sharing* efficiency and *resistance to extinction* can be enhanced by such overlearning. Thus, there may be situations in which it is highly desirable to continue training on all or part of the task long after a given criterion is reached. As Fitts pointed out, terminal plateaus may reflect artificial limits of performance rather than limitations of the trainee. Certainly, the training specialist should be aware of these possibilities when he sets out to design a training program.

Fortunately, a growing number of psychologists have become concerned over the lack of integration between research and practice in learning. For example, Hilgard and Bower (1966) devote a chapter in their recent text to learning and the technology of instruction. Another illustration appears in the first article of this chapter. In it, McGehee synthesizes several generalizations from the theories of learning which he feels are useful guidelines for modifying human behavior. Operating on the premise that the learner has a goal and makes responses aimed at attaining that goal, he discusses some of the factors that should *facilitate* and *limit* learning. He ends the article with a warning that we must examine learning variables carefully to make sure they actually apply to the training situation.

Critical examination of the variables that have "worked" in the learning laboratory and their relationship to the skills required in the training situation is the main topic of the second article. Starting with the common premise that "practice makes perfect," and proceeding through a list of the foremost principles derived from learning research, Gagné illustrates the impotency of each of these widely acclaimed concepts when applied to a sample of real-life training problems. He concludes—perhaps a bit hastily in view of the evidence offered—that they are of secondary importance at best for the design of training programs. As we said earlier, however, none of these "principles" has as yet been formulated to everyone's satisfaction; until that happens, we can scarcely afford to dismiss them summarily on the strength of a few negative illustrations. We must test *all* the implications of a concept such as reinforcement before we can safely lay it to rest.

Apart from this one criticism, however, Gagné's position appears well taken. Certainly *at this moment* we are in no position to apply many traditional learning principles directly. Furthermore, we should remain receptive to alternative principles, such as the ones he considers of primary importance, until we are better able to assess their worth. One can scarcely dispute the logic of including careful task analysis among the factors to be considered in designing and evaluating training programs. As an aside, it is also interesting to note that while few psychologists would accept the notion that "practice makes perfect" without considerable qualification, it remains a basic tenet of most training programs.

One final point should be made before we leave Gagné's paper. A very significant implication of this article is that if "traditional" learning principles are ever to be applied successfully, the application will probably come only after we have learned more about the *tasks* on which people are to be trained. Reinforcement, for example, can be applied in many ways, and to subtask as well as terminal response performance. Maybe it is not the *principles* that are trivial so much as it is the way in which they have been *applied*. Perhaps a principle such as *task analysis* must first be applied before we can hope to implement some of the more basic principles such as reinforcement or distribution of practice.

The last article in this chapter discusses *transfer of training*. Acquisition, of course, is just the first step in any training program; for the program to be acclaimed successful, there must be evidence that what is learned will transfer *positively* to the criterion task (i.e., that on-the-job training will be faster, or that performance on the job will be better because of it). All too often this obvious but critical requirement is either assumed or ignored. In view of what we said about the origin of many training tasks, it would be surprising indeed if all of them were to produce a substantial amount of positive transfer. Some may well have little or no impact on criterion performance (*zero* transfer) and, worse yet, some may actually be detrimental (i.e., produce *negative* transfer). This is no purely academic point; numerous instances of both types of consequences have been reported in real-world situations. Dunlap (1947), for example, describes cases of pilot error which were directly attributable to differences in cockpit layout between *training* and *operational* aircraft. Successful acquisition of the wrong habit was, through no fault of their own, the pilots' undoing!

There are no ready formulae by which one can compute the precise transfer effects that will accrue to any training program. Rough guidelines are afforded by classical transfer data, most of which derive from idealized laboratory experiments. As summarized in Holding (1965), the expected *amount* and *direction* of transfer are said to depend upon the *similarity* of stimuli and responses in the original and transfer task as follows:

Task Stimuli	Response Required	Transfer
same	same	high
different	different	none
different	same	positive
same	different	negative

Naturally, this summary greatly oversimplifies the problem, even for the laboratory situation (Ellis, 1965; Holding, 1965). First of all, the words *same* and *different* are not sufficiently precise to allow accurate prediction. It is the *degree* of similarity with which we must reckon, and even then the relationship is far from simple (Bugelski and Cadwallader, 1956; Osgood, 1949). Furthermore, how does one measure similarity independent of the measure of transfer? Ways are being explored, but there is still no universally accepted solution. Finally, what index does one use to measure transfer itself? Ellis (1965) and Murdock (1957) have shown that different percentages of transfer are obtained depending upon the formula used to analyze the same data! Clearly, we are a long way from being able to predict—even in the laboratory—exactly what transfer a unique situation will yield.

Returning to the *training* context, we can see that there are additional problems in the *implementation* of transfer principles. Even if we could describe similarity and transfer unambiguously, it would still be necessary to apply these concepts to the task at hand, and the task at hand is likely to be most complex. For example, what would be the relevant stimulus and response elements in the design of an aircraft simulator? And how would they interact to determine overall transfer? And what good would a high *overall* savings be if one critical element (such as altimeter reading) transferred negatively?

If the experimental laboratory has failed to provide the training specialist with a catalog of useful *facts* on transfer, it has at least suggested some variables and some methodological considerations that he might find useful in seeking his own facts. Undoubtedly, he will be obliged to explore such variables using designs in which *interactions* can make their presence known, for complexity is the hallmark of real-world tasks. Even the theorists recognize that learning, retention, and transfer are intimately related phenomena (Melton, 1963); hence, to study transfer in *any* situation, one must take cognizance of learning and retention variables as well. However, as a review by Naylor and Briggs (1961) suggests, the research to date on these variables has been limited indeed. There is virtually no information, for example, on long-term retention of skills. The final article in the chapter discusses a few of these issues in somewhat greater depth.

32

Are We Using What We Know about Training? — Learning Theory and Training

William McGehee

There is no way to make this subject light. So I am going to start by defining the various concepts in my topic so that we (or at least I) will be reasonably certain about what I am trying to write.

First let us look at the term "theory." This is generally an anathema to the practical man. Yet in spite of its rather disreputable status among the "doers," these same characters make use of theory at least in a crude form. So let us examine the nature of theory.

A theory according to George Kelley is "a way of binding together a multitude of facts so that one may comprehend them all at once" (6, *18*). Kelley maintains that a theory, even if it is not highly scientific, can be useful since it can give meaning to our activities and provide a basis for an active approach to life. So even the theory that "13" is an unlucky number can lighten the life of a bookie. Likewise the theory that role playing, or case study, or visual aids, or the incident method is an answer to a training director's prayer is extremely comforting in his everyday life even if the theory is not highly scientific.

When a theory enables us to make reasonably precise predictions, one may call it scientific. The nearer theories come to precision of predictions, the more useful they are for controlling the phenomena with which they deal. To quote Kelley again "Theories are the thinking of men who seek freedom amid swirling events. The theories concern prior assumptions about certain realms of these events. To the extent that the events may, from these prior assumptions, be construed, predicted, and their relative courses charted, men may exercise control and gain freedom for themselves in the process" (6, *22*).

Learning theory, then, is simply a way of binding together the facts known about the process we call "learning." Learning, itself, is a construct—an abstraction. No one of you has ever seen "learning." You have seen people in the process of learning, you have seen people who behave in a particular way as a result of learning, and some of you (in fact, I guess the majority of you) have "learned" at some time in your life. In other words we infer that learning has taken place

From *Personnel Psychology*, 11, 1958, 1-12. By permission of Personnel Psychology, Inc.

if an individual behaves, reacts, responds as a result of experience in a manner different from the way he formerly behaved. Changes in behavior which result from learning, by definition, exclude those changes which result from maturation or from physiological induced changes arising from fatigue, illness, intoxication and similar pleasant and unpleasant organic experiences. So when we talk about learning theory we are trying to derive from the facts known about the changes in behavior resulting from experience, generalizations which enable us to make predictions concerning changing behavior by experience. The more accurately these generalizations describe the processes by which experience changes behavior, the more useful will learning theory be in controlling the processes.

The central process in industrial training is learning. I am certain that the traditional man from Mars would not deduct this if he listened to training directors talking shop at conventions or weeping in each others' martinis about the lack of top management support for training activities. Very simply, industrial training consists in the organized experiences used to develop or modify knowledges, skills and attitudes of people involved in the production of goods and services.

What training directors are trying to do is to expose people in their companies to those types of experiences which will develop most effectively, or modify, knowledge, skills and attitudes. Whether we are willing to admit it or not, the experiences to which we expose our employees are those which we believe and/or can convince management to believe, will expedite the learning process. In other words, on the basis of "our prior assumptions about a certain realm of events," we establish vestibule training or send a reluctant member of middle management to the Harvard Advanced Management Course. Some of us in doing these things also have some theory as to how experience modifies behavior.

THE PROFESSIONALS AND LEARNING THEORY

Now let us take a look at what the professionals in learning theory have to offer. These are the gents who are not concerned with top management support. They do not have to deal with foremen who have been "learning" men a long time before training directors became epidemic in industry. Certainly they have their problems with the "butcher, the baker, and the candlestick maker" even as you or I but the pressures are different and their goals are different. So I think we should understand a few things about these professionals before we try to see how their efforts at learning theory are applicable to industrial training.

First, these men are trying to develop "scientific" theories i.e. generalizations which closely approximate the phenomena under consideration. Consequently, they have had to adhere to procedures in their experiments which hold all variables constant except the experimental variable. This has resulted in too many reported studies which appear to the man, who is trying to train weavers or mill wrights, as almost puerile. Yet it is only by the careful addition of this bit of knowledge gained from studying the maze learning behavior of the white rats or of that bit from pursuit meter responses of college sophomores, that precise knowledge concerning learning is developed. This, of course, has led to a difficulty in applying these bits of knowledge to everyday problems of training. An experiment which demonstrates conclusively that white rats tend to repeat "rewarded" responses in learning on an elevated maze does not lead easily to the generalization that Rosie the Riveter responds in a similar manner as that of a lower order mammal. So the research, on which many theories of learning are based (since they deal with rats, chimpanzees, pre-school children, and college students) are not easily assimilated into the mores of the individual who must plan a training program to cut learning time for a group of operators assembling electrical relays or sewing seams on towels.

Second, the language used by learning theorists is esoteric and often (what is worse) mathematical. Again this is a penalty of their trade. Not only must generalizations be discovered but they must be stated also in unequivocal terms. Much of the English language is vague and pluralistic. Common words have multiple meanings and each hearer interprets in his own way. For example, Hull states "reaction-evocation potentiality is the product of a function of the habit strength multiplied by a function of the strength of the drive" (5, 242). This can mean different things to different listeners. However, if the same principle can be expressed (as Hull does state it) $_sE_R = f(_sH_R) \times f(D)$ and these symbols are substituted for mathematically, everyone understands precisely what Hull means. Or do they? Actually where common everyday concepts are used, considerable misinterpretations arise. Thorndike, for example, reviewed the old Hedonistic controversy with his law of effect by using the everyday terms like "satisfaction" and "annoyance." It is unfortunate that the necessity for exactness in terminology has obscured (except to the initiated) the general principles of learning postulated by leading theorists like Hull and Tolman.

Finally, these searchers after fundamental truths are human, even as you and I. They bring to their search for underlying principles their own predispositions and they become enmeshed in their own theories. Accordingly not only do they interpret the results of their experiments in terms of their theoretical orientation, they also design experiments which serve to test their theories in terms of their theoretical orientation. Again the hypothetical man from Mars provided he was a non-connectionist, non-reinforcement theorist, and non-Gestalter, could find definite evidence to support each of these theories in the famous study of the learning of simians reported by Kohler. The dispute among theorists concerning the learning process has increased the confusion of the layman without an appreciable gain in controlling the process.

GENERALIZATIONS CONCERNING LEARNING FOR INDUSTRIAL TRAINING

It is not possible, even if I could, in this article to present the subtleties of the major modern learning theories including among others Thorndike's connectionism, Hull's reinforcement theory, Lewin's toplogical psychology or Tolman's sign Gestalt theory. Rather on the basis of what these investigators and many others have learned about learning, I am going to try to make a few simple declarative statements about learning and then try to indicate their implications for industrial training. In trying to write these sentences I must acknowledge my indebtedness to certain individuals who have tried to synthesize the major concepts from various learning theories including J. F. Dashiell (3), D. K. Adams (1), T. R. McConnell

(8), Ernest Hilgard (4), and Lee Cronbach (2). They, however, should not be made to accept any criticism for my synthesis of their syntheses.

From this attempt to synthesize, the following generalizations concerning the learning process have evolved. They may be considered a statement of "what happens" when an individual learns. They can serve as guide posts for ordering experiences planned to modify behavior i.e. train the individual. I shall leave it to you to judge how nearly they describe learning. Later I want to cite their implications for industrial training. These statements are:

The learner has a goal or goals, i.e. he wants something.

The learner makes a response, i.e. he does something to attain what he wants.

The responses, which he makes initially and continues to make in trying to attain what he wants, are limited by:

The sum total of his past responses and his abilities.

His interpretation of the goal situation.

The feedback from his responses, i.e. the consequences of his response.

The learner, having achieved his goal (or goal substitute), can make responses which prior to his goal seeking he could not make. He has learned.

IMPLICATIONS OF GENERALIZATIONS FOR INDUSTRIAL TRAINING

You at this point might easily say "so what." If this is an accurate description of learning, what are its implications for training in industry. As I have indicated, a theory gives the theorist a way to think about his phenomena. It can give him suggestions for predicting and for controlling the phenomena with which he is concerned. Now let us see what thinking about learning as outlined in my declarative sentences leads to in planning and directing industrial training activities.

First, I have postulated that if learning occurs the learner has a goal i.e. he wants something. Now goals can be many kinds—they can be short ranged or long ranged; they can be clearly thought out and verbalized or vague and subverbal. The question I raise here is just how much thought is given in planning any training program to the goals of the individuals who are to be trained. If you are planning training for a production worker, are his goals your goals? I imagine quite different outcomes of training would result if the employee accepted as his goal the production of the maximum units of a given quality in a given time (industry's goal) rather than accepting work group goals of producing enough to keep the foreman off his neck and the time study man from cutting his rate. What also of the training programs designed to improve the human relations behavior of foremen when the foreman's goal is to please a superior whose only concern is costs and units produced.

A recent study (9) gives a clear example of modifying goals which resulted in a significant change in the performance of a group of four operators. These four operators were rewinding bobbins from a spinning frame. In this salvage operation they produced waste at a ratio of 2.50 per week above standard. Methods were used with these operators to change their goal from rewinding so many bobbins a day to rewinding so many bobbins in such a way as to minimize waste. Within a 15 week period of training, these operators learned to handle bobbins in such a way that the waste was reduced to .87 of standard. This reduction with slight variation has persisted for over two years in this operation (10). This represents a considerable savings in material in two years by careful attention to the goals of four operators. There was no decrease in amount of production during this period.

My second postulate is that the learner makes a response, i.e. he does something to attain his goal or goals. In learning rote verbal material or motor skills, we call this "practice"; perhaps in learning facts and developing attitudes and sentiments, we call this "experience."

One of the immediate consequences of accepting this postulate concerning learning is to acknowledge that effective industrial training must make provision for the learner to make a response. Training programs which emphasize elaborate visual aids, demonstrations, clever lectures, and similar devices for inducing learning at the price of learner activity become immediately suspect as to their effectiveness.

The second problem this postulate creates is the question as to whether all responses, "practice or experience," are equally valuable in the goal seeking activity of the learner. Fortunately, the learning theorists have gotten a reasonably clear answer to this question and the answer is "No." Certain valid statements can be made concerning the conditions necessary for effective practice and experience. The following are a few of these statements:

1. The more nearly responses made in a learning situation are those which an individual must make in "playing for keeps" the more quickly he'll learn to play for keeps effectively. For example, if both speed and accuracy are required to perform a task, the learner should be urged to make responses characterized by both speed and accuracy.

2. The order in which various responses are made in learning a task will facilitate or retard effective completion of the task at some subsequent time. If two responses to the similar situations are practiced before either response is firmly established, the situation may call out either one of these responses—one adequate and the other inadequate.

3. Immediate knowledge of the results of a response tends to correct the response most effectively. This was demonstrated in World War I by English's experiment in training rifle shooting and again in several situations in World War II in training radar operators, teaching code, and training director finders (12). Lindahl's study in industry on cutting discs substantiates the same conclusion (7). In spite of these results, very few industrial training programs provide for this feature of immediate knowledge of result.

4. Spacing of practice is extremely important. Again laboratory studies, while differing in details as to the application of this principle, have shown generally that spacing practice sessions results in more efficient learning. We have been able to demonstrate the value of spaced practice in learning several repetitive operations such as spinning and sewing machine operation. The major industrial objection to spaced practice is that the learner is at work for 8 hours a day and should be kept busy. Ingenious trainers, in spite of the 8-hour day, still can arrange for spaced practice. In this connection, I have wondered about the intensive training sessions which characterize many training programs for supervisors and managers. Could spacing these sessions result in more effective learning?

I could cite other points which have emerged from the investigations of the nature of practice and its effectiveness. You are as familiar with them as I am. What I want to stress again is that if we accept "responses" as integral part of learning, we immediately start asking questions about the necessary conditions for these responses to result in the most effective learning.

I would like to repeat my third postulate here since it is more involved verbally than either of the other two: What I have said is that the responses made by a learner in a goal situation are limited both initially and as he continues in the goal situation. I then stated there exist three general classes of limitations. Let us expand about the first limitations. I have said a learner cannot make a response to a goal situation which is not already in his repertoire of responses or is beyond his abilities.

Let me illustrate this by one example in foreman training since I believe the point is overlooked frequently in training designed to modify behavior of supervisors. Many of us, I expect, have sponsored so called "appreciation" courses in time study for foremen. We hope with these courses to secure the wholehearted support of our supervisors in our methods and standard work. We are disappointed if the same buck passing and lack of appreciation of the value of time study continues in our foremen group after completion of the course. Now time study is a relatively complex discipline. Telling people about time study does not equip them with knowledge and skills required to answer employee questions about rates or to discuss intelligently, with time study engineers, proposed rates. Lacking the knowledge and skill required, the foreman is put in an impossible position. He cannot make the responses required because they are not in the sum total of his past responses and abilities.

In designing a course in time study for our supervisors, we give full consideration to the limitation. We designed and exposed a group of our foremen to a course in time study which gave them competence in, rather than appreciation of, time study. We compared their improvement in (1) knowledge of time study (2) handling of time study activities in their department both before and after the course with a matched group of foremen who did not take the course. The trained group showed 26.3% gain in knowledge compared to no change by the untrained group. What is even more important, the trained group showed 42.1% improvement in handling of time study in their department versus a 9% (non-significant) gain for the untrained group. I must add that in both groups each foreman had had at least 10 years experience with time study in his department. Each foreman had had at least one "appreciation" course in time study and several had had two or three such courses (11).

The second limitation to responses are imposed by the learner's interpretation of the goal situation. We found that loopers who were urged to work "carefully" failed to develop speed. We found that battery hands who were allowed to fill batteries using only two bobbins rarely "pick up" a more effective method of handling 4 or 5 bobbins. They saw the situation as fill the batteries with 2 bobbins not fill the battery efficiently. Further in human relations training, we may find the interpretation of the course as a method of manipulating people more easily, not as instruction in how to perceive people differently. The supervisor then learns certain superficial skills but fails to experience basic changes in attitudes toward human relations problems. Operators who interpret methods improvements as threats to job security react in a different manner to those who interpret methods improvements as aids to company and employee progress. Finally, in training in a rather complex job called setting, we found learners develop skills more rapidly when they were urged to visualize the pattern in terms of several strands of yarn rather than reacting to each square on the pattern as an individual strand.

A third limitation placed on responses is the feed-back the learner gets from making the responses. This is clearly illustrated in a skill like playing the piano or driving a golf ball or fly casting. Both visual, auditory, and kinesthetic cues "tell" the performer when the response feels right or wrong. The feed-back which tells the learner that the response is not goal directed encourages him to try some other response; while the response which "feels" good tends to be used again and again.

This limitation points up one practical consideration in industrial training, i.e. the importance of the instructor in the learning situation. It is the instructor who can analyze the adequacy of responses and serve as an effective "feed-back" device for correction of responses. If there is one thing that is known about industrial training, it is that organized training is more effective than unorganized training. I suspect the reason is that organized training provides a more effective feed-back.

My fourth declarative sentence really says that the individual has learned. In other words, he is capable of making responses which he could not make prior to wanting something and responding. It does, however, point up the fact that the process called learning produces responses which can be classified both as adequate and inadequate. Erroneous information, poor skills, and undesirable attitudes are just as much a product of learning as are correct information, effective skills and desirable attitudes. From a practical standpoint, this means that we must continually evaluate the results of placing employees in learning situations. We need to know what facilitates and what retards practice. Many techniques used in modern training such as visual aids and training devices have "face validity," i.e. they look good. Actually such aids to training may not facilitate and may even retard learning. It is entirely possible, for example, that the case study method of training executives may increase the elegance with which a problem is approached but have absolutely no effect upon the quality of the solution.

LEARNING THEORY AND INDUSTRIAL TRAINING

These, then, are some aspects of learning theory in relationship to industrial training as I see them. One more sentence or two. A fully adequate theory of learning will not emerge from the animal laboratories and the classroom. These theories must hold also in the factory and the office. Likewise, learning experience in the factory and office should give cues for laboratory investigations. I would plead therefore for more and more interchange between the learning theorist and the industrial trainer. This cohabitation will aid in giving both "freedom from the swirling events" surrounding the problems of learning and training.

BIBLIOGRAPHY

1. Adams, D. K. A Restatement of the Problem of Learning. *British Journal of Psychology*, 1931, **22**, 150-178.
2. Cronbach, Lee, *Educational Psychology*. New York: Harcourt, Brace, & Co., Inc., 1952. Pp. 617.
3. Dashiell, J. F. A Survey and Synthesis of Learning Theories. *Psychological Bulletin*, 1932, **29**, 261-275.
4. Hilgard, Ernest R. *Theories of Learning*. New York: Appleton-Century-Crofts, Inc., 1948. VI + 408.
5. Hull, Clark L. *Principles of Behavior*. New York: D. Appleton-Century Company, 1943. X + 422.
6. Kelley, George A. *The Psychology of Personal Constructs*. (Vol. I) New York: W. W. Norton & Company, 1955. XVIII + 556.
7. Lindahl, L. G. Movement Analysis as an Industrial Training Method. *Journal of Applied Psychology*, 1945, **29**, 420-436.
8. McConnell, T. R. Reconciliation of Learning Theories. In *National Society for Studies in Education*, 41st Yearbook, Part II. Pp. 243-286.
9. McGehee, W., & Livingstone, D. H. Training Reduces Material Waste. *Personnel Psychology*, 1952, **5**, 115-123.
10. McGehee, W., & Livingstone, D. H. Persistence of the Effects of Training Employees to Reduce Waste. *Personnel Psychology*, 1954, **7**, 33-39.
11. McGehee, W., & Gardner, J. E. Supervisory Training and Attitude Change. *Personnel Psychology*, 1955, **8**, 449-460.
12. Wolfle, Dael. Training. In Stevens, S. S. *Handbook of Experimental Psychology*. New York: John Wiley & Sons, Inc., 1951. Pp. 1267-1286.

33

Military Training and Principles of Learning

Robert M. Gagné

The subject chosen for this address is one which I think I can view with a certain perspective. Stated very briefly, this circumstance arises from the fact of my changes in occupation, from that of an investigator of learning principles in aca-

From *American Psychologist*, 17, 1962, 83-91. Copyright 1962 by the American Psychological Association, and reproduced by permission.

Presidential address delivered at the annual meeting of the Division of Military Psychology, 69th Annual Convention of the American Psychological Association, New York, N. Y., September 5, 1961.

Supported in part by Contract AF 49(638)-975, with the Office of Scientific Research, U. S. Air Force. The opinions expressed are those of the author.

demic laboratories, to a research administrator of programs of military training research in government laboratories, and back again to an academic laboratory. In making the remarks to follow, I claim nothing more than this perspective, which perhaps carries with it a certain detachment, or freedom from involvement with particular learning theory, as well as with particular training problems.

What I should like to talk about are some general impressions concerning the applicability of learning principles to military training. In the time available to me, I cannot really do more than this. It would be satisfying to think that I could review and marshal the evidence from military studies of training in a truly systematic manner. I am not sure that this can be done even with unlimited time. But at any rate, my aim is much more limited than this. Perhaps it can be stated in the following way. Suppose that I were a learning psychologist, fresh out of an academic laboratory, who was to take a new job in charge of a program of research on some type of military training. What principles of learning would I look for to bring to bear on training problems? What kinds of generalizations from laboratory studies of learning would I search for and attempt to make use of in training situations? The answers I shall suggest for these questions require first a consideration of what kinds of principles have been tried, and how they have fared.

SOME REPRESENTATIVE MILITARY TASKS

First, we need to have in mind certain representative military tasks for which training either is or has been given, in order that we can consider in detail the kinds of learning principles that are applicable. Here are three which will serve well as examples: (1) flexible gunnery; (2) putting a radar set into operation; (3) finding malfunctions in an electronic system.

Flexible Gunnery. The gunner of a now obsolete type of bomber aircraft was typically located in the waist or the tail of the plane, and aimed and fired a gun at fighter aircraft attacking on what was called a "pursuit course." To do this he looked at the attacking fighter through a reticle containing a central dot, which he lined up with the target by rotating his gunsight horizontally and vertically. At the same time, he had to "frame" the aircraft within a set of dots arranged in a circle whose circumference could be varied by rotating the round hand-grip by means of which he grasped the gunsight. This is the kind of task the psychologist calls "tracking," on which a great many laboratory studies have been carried out. It was, of course, tracking simultaneously in the three dimensions of azimuth, elevation, and range. To perform this task, the individual had to learn a motor skill.

Putting a Radar Set in Operation. This kind of task is typically what is called a "fixed procedure." That is, the individual is required to push buttons, turn switches, and so on, in a particular sequence. Here, for example, is a set of steps in a procedure used by radar operators to check the transmitter power and frequency of an airborne radar (Briggs & Morrison, 1956):

1. Turn the radar set to "Stand-by" operation
2. Connect power cord of the TS-147
3. Turn power switch on
4. Turn the test switch to transmit position
5. Turn DBM dial fully counter-clockwise
6. Connect an RF cable to the RF jack on the TS-147

There are 14 more steps in this procedure. Notice that each of the steps by itself is easy enough; the individual is quite capable of turning a switch or connecting a cable. What he must learn to do, however, is to perform each step in the proper sequence. The sequence is important, and doing step 5 before step 4 may be not only an error, it may be dangerous. What must be learned, then, is a sequence of acts in the proper order.

Finding Malfunctions in Complex Equipment. This is in many respects a most complex kind of behavior. There are of course some very simple kinds of equipment in which this activity can be reduced to a procedure; and when this is true, the task is one that can be learned from that point of view. But the major job, for complex equipment, is one of troubleshooting, a problem-solving activity that has considerable formal resemblance to medical as well as other kinds of diagnosis. Suppose this is a radar set, again, and that the initial difficulty (symptom) is that no "range sweep" appears on the oscilloscope tube face. Beginning at this point, the troubleshooter must track down a malfunctioning component. He does this first by making a decision as to how he will check the operation of subordinate parts of the system, next by carrying out the check and noting the information it yields, next by making another decision about a next check, and so on through a whole series of stages until he finds the malfunctioning unit. In each of these stages, he presumably must be formulating hypotheses which affect his actions at the next stage, in the typical and classically described manner of problem solving. What does the individual have to learn in order to solve such problems? This is indeed a difficult question to answer, but the best guess seems to be that he must acquire concepts, principles, rules, or something of that nature which he can arouse within himself at the proper moment and which guide his behavior in diagnosing malfunctions.

Here are, then, three types of activities that are not untypical of military jobs, and which are aimed at in military training: a motor skill like flexible gunnery; a procedure like putting a radar set into operation; and troubleshooting, the diagnosing of malfunctions in complex electronic equipment. Each one of these tasks has been examined more or less intensively by military psychologists and learning specialists. Among other things, each of these tasks can be shown to be not entirely unique, but to represent a rather broad class of tasks, in its formal characteristics, which cuts across particular content or occupational areas. For example, flexible gunnery is a tracking skill, which formally resembles many others, like maneuvering an airplane, sewing a seam on a sewing machine, hovering a helicopter, and many others. As for procedures, these are common indeed, and may be found in jobs such as that of a clerk in filling in or filing forms, a cook preparing food, or a pilot preflighting an airplane. Diagnosing difficulties is certainly a widely occurring kind of actvity, whch may be engaged in by the leader of a group who detects the symptom of low morale, as well as by a variety of mechanics who "fix" equipment of all sorts. Accordingly, one should probably not consider these particular examples as peculiar ones; instead, they appear to be representative of a wide variety of human activities.

LEARNING

How are these three kinds of tasks learned? What is it that the learning psychologist can say about them which will enable anyone (the teacher, the curriculum builder, the training manager) to undertake to arrange the external conditions in such a way that the desired performances will be acquired with the minimal expenditure of time, money, and wasted effort?

Suppose that you were, in fact, a psychologist who had studied learning, both animal and human, from the standpoint of experiment and theory, and that you were faced with this problem. How can scientific knowledge of learning be used to improve the process of training? Notice how I have stated this question. I am not asking, how can a scientific approach be applied to the study of training? Nor am I asking, how can experimental methodology be applied to the study of training? There are certainly answers to these questions, which have been provided by several people, notably Crawford (1962). The question is, rather, how can what you know about learning *as an event*, or *as a process*, be put to use in designing training so that it will be maximally effective?

The psychologist who is confronted with this question is likely to appeal, first, to a basic point of view towards learning which is so highly ingrained it may be called an *assumption*. Beyond this, and secondly, he looks for certain *principles* which have been well established by experiment. These are principles which relate certain variables in the learning situation, like time intervals between trials, sequence of trials, kind of feedback after each trial, and so on, to such dependent variables as rate of learning or goodness of performance. Let us try to see what can be done both with the basic assumption and with some of the more important of the principles.

The Assumption. The assumption that many learning psychologists would bring to the problem of designing training is something like this: "The best way to learn a performance is to practice that performance." I should like to show, later on, that this assumption is by no means a good one. But before I do that, I want to consider the question, where does this assumption come from, anyhow? First, it seems to have a cultural basis, by derivation from the writings of John Dewey, preserved in the educational catch-phrase "learning by doing." Second, it appears to come by unwarranted generalization from laboratory prototypes of learning such as the conditioned response. In conditioning, classical or otherwise, one observes learning only *after* the animal has made the first *response*. Thus, performance comes first, and learning is often considered to result from practice of this performance. Third, the assumption comes from theory which deals with conditioning, and which conceives of what is learned as either a response or an association terminating in a response, in either case established by *practicing the response* (with reinforcement). Without going into the matter further at the moment, the basic reason that generalization of this notion to the learning of the human tasks I have mentioned seems questionable is simply that the responses required (turning switches, inserting plugs, moving handles) do not have to be learned at all—they are already there in the human's repertoire.

PRINCIPLES

Beyond this assumption that learning comes about when performances are practiced, what *principles* can the learning psychologist depend on? What kinds of conditions have been found to affect the rate of learning? What findings can he bring to bear on the problem of designing training to be maximally effective?

Let me mention some of the best-known of these principles, not necessarily all of them, using various sources. In part, I shall depend on an excellent article by Underwood (1959). First of all, there is *reinforcement*, the principle that learning will be more rapid the greater the amount of reinforcement given during practice. Other principles include *distribution of practice, meaningfulness*, increasing the *distinctiveness* of the elements of a task, and *response availability*.

These principles would appear to provide the learning psychologist with a fairly adequate bag of tricks with which he can approach the job of designing effective training. There is much evidence in the experimental literature that one can in fact alter the rate of learning by manipulating these variables in the learning situation, whether one is working with single conditioned responses or with verbal material having a somewhat more complex organization. Each of these variables, so far as is known, can be manipulated to make a dependable difference on learning, in the direction of increased as well as decreased effectiveness.

USING THESE ASSUMPTIONS AND PRINCIPLES IN TRAINING DESIGN

How does one fare if he seriously attempts to use this basic assumption and these principles to design effective training situations? *Not particularly well*. The assumption that the most effective learning is provided by practice on the final task leads one astray on many occasions. As for the principles, sometimes they can clearly not be applied, that is, there is no way to manipulate the training situation in the manner suggested by the principle. In other instances, the evidence simply fails to support the principle. When this happens, there may be good theoretical reasons for the event, but this still does not restore one's faith in the usefulness of the principle.

It will be possible here only to give a few examples of military training situations in which these assumptions and principles failed to work, but I have chosen them to be as representative as possible. Let me emphasize again that I do not maintain that these examples demonstrate that the principles are invalid. I simply want to show that they are strikingly inadequate to handle the job of designing effective training situations.

Motor Skill

First let's consider what is perhaps the most difficult case, the learning of a motor skill like gunnery. What happens if we try to employ the assumption that the best way to learn gunnery is to practice gunnery? Using the kind of task required of a flexible gunner, a number of studies were made of the conditions of learning for this performance. One of the earliest ones, during World War II, reported by Melton (1947), showed that different amounts of practice in firing at sleeve targets during one through ten gun-camera missions made no significant difference in the measured proficiency of gunners. A number of other studies of gunnery also indicate the very small and often insignificant effects of practice continued beyond the first three trials or so (Rittenhouse & Goldstein, 1954). Furthermore, several such studies confirm the finding that the major improvement in this performance comes as a result of informing the learners of the correct picture to be achieved in ranging (i.e., so that the dots just touch the wing tips of the target aircraft) (Goldstein & Ellis, 1956). In other words, to summarize the finding very briefly, the evidence is that simple practice on the gunnery task is not a particularly effective training method; instructions about the correct sighting picture for ranging is much more effective in bringing about improved performance. Perhaps there are good theoretical reasons for this. But the fact remains that practicing the performance is *not* the best way to learn.

What about the principles of learning? Well, let's consider the one which a learning psychologist might be inclined to think of first—reinforcement, or the introduction of knowledge of results during practice. Translated into a form applicable to motor skills learning, the principle is that the more adequate the knowledge of results, the more rapid the learning. This variable, too, has been tried out in a number of studies. Typically what was done was to augment the knowledge of results that come to the gunner through his observing his own tracking performance on a screen, by providing an extra cue, such as a buzzer, which sounded whenever the gunner was exactly on target in all three dimensions. The effect of this extra cue, it was found, was to improve the performance during learning. But did this mean that the learning itself was more effective, or simply that the buzzer "propped up" the performance? One seeks the answer to this question by comparing the performance of buzzer-trained and non-buzzer-trained groups on a standard criterion task without the buzzer. When this was done, the findings in several studies were negative (cf. Goldstein & Ellis, 1956), and one (Goldstein & Rittenhouse, 1954) actually showed that learners who had the advantage of augmented knowledge of results (reinforcements) exhibited a lower performance on a second gunnery task.

Other learning principles were unconfirmed in training situations. For example, a carefully executed study could find no evidence for changes in learning as a result of alterations in conditions of practice and rest periods (Rittenhouse & Goldstein, 1954). Still other variables simply cannot be used in the training situation. For example, the meaningfulness of the task is set by the task itself, and cannot be altered by changing the conditions of training. Similarly, the internal similarity of the elements of the task are fixed by the task; one cannot, for example, change the degree of resemblance of the aircraft or of the tracks they follow by simply redesigning the training, without setting about to change the nature of the task itself. (I omit here a discussion of the transfer effects of training with an easy discrimination to performance on a hard discrimination, and vice versa. This is a different principle than the one under discussion, and the evidence about it is not clear-cut.) What about response availability or familiarity? From the evidence on practice previously cited, as well as studies on part-training (cf. Goldstein & Ellis, 1956) it seems fairly clear that the responses in this task (turning knobs, moving the gunsight up and down with a handle) were highly familiar in the first place. No one, so far as I know, ever seriously proposed that they were not.

Perhaps these examples are sufficient to at least raise doubts about the usefulness of the learning psychologist's assumptions and principles, when he attempts to apply them to the practical job of designing training for motor skills. On the whole, it may fairly be said, I think, that the assumption was often wrong and the principles were seldom useful in bringing about training improvement. I caution you again that I am not saying the learning psychologist was unsuccessful in improving training. In many instances, he was very successful. What I am trying to answer is the question, when he was successful, what knowledge or set of principles was he using?

Procedures

There are not many analytical studies of the learning of procedures. Perhaps the reason for this is that learning procedures is relatively such an easy matter, and the methods used to train them seem relatively so obvious, that little work was done on them. Consequently, I shall have to base my arguments primarily on these obvious features, rather than on a great deal of experimental evidence.

Suppose one is faced with the task of training someone to "turn on" a radar set by turning and pushing a series of fifteen switches in a particular sequence. (This is taken to be a simplified version of a representative procedural task.) How does one go about it? If one proposes to conduct training simply by "practicing the task" it becomes obvious almost immediately that this is an inefficient method to use. What is usually done is this: the learner is provided with a *list*, which states, in effect, "First, turn on power switch; second, depress voltage switch; third, set voltage knob to reading 10; etc." (e.g., Briggs & Morrison, 1956). Now the individual may be required to commit the list to memory first, and then proceed to the task; or, he may be allowed to use the list while he practices going through the sequence. The important thing is, however, that it is the *learning of the list* that contributes most to the performance of the task, not the practice of the switch-pressing responses, another example contrary to the principle that the best way to learn is to practice the required performance. I do not say that the performance should never be practiced, simply that something other than direct practice of the final task is more effective for learning procedures, just as is true for motor skills in the example previously described.

Learning principles applied to the training of procedures do not fare very well, either, although again I must note the absence of experimental evidence. One cannot alter meaningfulness, and in most cases the responses required are highly familiar. When they are not, as may be the case when a single step requires the use of an unfamiliar tool, this principle may actually have some limited usefulness. Sometimes the principle of increasing the distinctiveness of the elements of the task can be used, and one would indeed expect it to work. For example, one could put distinctive cues or labels on each of the switches in the 15-switch procedure, and this might be expected to speed up the rate of learning. However, it may be noted that this becomes a matter of changing the task (i.e., the equipment), rather than of changing the conditions of learning. From evidence on the learning of nonsense-syllable lists, one would not expect a variable like distribution of practice to make much difference as a training variable, as Underwood (1959) has noted. Again a review of learning assumptions and principles has indicated limited usefulness.

Diagnosing Malfunctions

When we turn to a consideration of troubleshooting complex equipment, even the most theoretically-dedicated learning psychologist is forced to recognize, almost from the start, that the idea of learning to troubleshoot by simply practicing troubleshooting verges on the ridiculous. The most obvious reason is that one cannot identify a single *task* to be practiced. The troubleshooter is faced with a great variety of initial problem situations, each of which may have a great variety of causes. He cannot possibly practice solving all of them. In fact, it is clear that he must learn not a single task, but a *class of tasks*, or perhaps even several classes of tasks. Yet people do learn to do them, quite successfully, without ever doing anything that can legitimately be called "practicing the final performance."

What they do learn, among other things, is an elaborate set of rules pertaining to the flow of signals through a complex circuit. To a large extent, they learn these rules by looking at and responding to a circuit diagram which is a representation of the equipment rather than the equipment itself. And they use the rules in thinking about the signal flow, that is to say, in making successive decisions leading to a solution of the problem (finding the malfunction).

Since, as I have said, it is impossible to define a single task to be practiced in learning troubleshooting, it is just about equally difficult to apply the principles of reinforcement, meaningfulness, internal differentiation, and so on, to the design of training. If one accepts the task of "learning the rules" as what must be done, it is of course possible to ask the question as to whether such learning variables would apply to that task. This is a job that may some day be done by those interested in research on "learning programing." But it has not been done as yet. The evidence to date (such as it is) has not indicated strong effects, or even significant ones, for the variable of reinforcement in connection with learning programs (Goldbeck & Briggs, 1960). Other variables have not yet been investigated in relation to the learning of rules and principles.

WHAT *IS* APPLICABLE TO THE DESIGN OF TRAINING?

Does this mean that the psychologist has virtually nothing to offer to the problem of designing effective training? Have the results of psychologists' efforts to improve training been entirely negative? Quite to the contrary, it seems to me that efforts can be identified which were quite effective in producing significant improvements in training, and which led to some demonstrably useful designs for training. But the principles which were found to be effective for such purposes were not those that have been mentioned.

Here are the psychological principles that seem to me to be useful in training:

1. Any human task may be analyzed into a set of component tasks which are quite distinct from each other in terms of the experimental operations needed to produce them.

2. These task components are mediators of the final task performance; that is, their presence insures positive transfer to a final performance, and their absence reduces such transfer to near zero.

3. The basic principles of training design consist of: (a) identifying the component tasks of a final performance; (b) insuring that each of these component tasks is fully achieved; and (c) arranging the total learning situation in a sequence which will insure optimal mediational effects from one component to another.

These statements certainly imply a set of principles which would have very different names from those we are now most familiar with. They are concerned with such things as *task analysis, intra-task transfer, component task achievement,* and *sequencing* as important variables in learning, and consequently in training. These principles are not set in opposition to the traditional principles of learning, such as reinforcement, differentiation of task elements, familiarity, and so on, and do not deny their relevance, only their *relative importance*. They are, however, in complete opposition to the previously mentioned assumption "the best way to learn a task is to practice the task."

It should also be pointed out here that I am unable to refer to any well-organized body of experimental evidence for these newly proposed principles. They come instead by inference and generalization from a wide variety of instances of learning and military training. I do not claim more for them than this. But they have to be stated before any systematic experimental work can be done on them.

Let me try now to illustrate a definite meaning for these principles with some examples. Consider first the procedural task described previously. "1. Turn radar set to 'standby' operation; 2. Connect power cord of the TS-147; 3. Turn power switch on; (4) Turn test switch to transmit position; etc." The first step to be undertaken here is to analyze this task; and (with certain minor assumptions on our part), this is seen to be, first, the learning of an order series of responses to things; and second and subordinate to this, the locating of these things. These two *component tasks* have a hierarchical relationship to each other, and immediately suggest the proper *sequencing* for the arrangement of the learning (or training) situation. That is to say, what must first be undertaken is that the learner learn what and where the "things" are (the "standby operation" switch, the "TS-147," the power switch, the test switch, and so forth). This is a matter of identification learning, which has considerable resemblance to the paired-associate learning of the psychological laboratory. Having achieved this subordinate task, it is then possible for the learner to undertake the second, or "serial order of things" task. According to the principle proposed here, maximal positive transfer to this task would be predicted following completely adequate performance on the subordinate task of identifying the "things."

Laboratory experiments which have undertaken to test such a hypothesis seem to be scarce. It is possible, however, to make reference to two studies (Primoff, 1938; Young, 1959) which have some suggestive findings. Generally speaking, when one learns a set of paired associates first, and then undertakes the learning of these units serially, there is high positive transfer; but when one learns units serially first, the amount of transfer to paired-associate learning is very low indeed. These results strongly suggest that there is a *more efficient* and a *less efficient* sequence which can be arranged for the learning of a procedural task, and that this sequence involves learning one subtask before the total task is undertaken. A procedure is a task that can be analyzed into at least two component tasks, one of identification, and the other of serial ordering. The first is subordinate to the second in the sense that it mediates positive transfer to the second, provided it is first completely mastered.

Can this kind of analysis be applied to a more complex task like troubleshooting? Indeed it can, and those psychologists who thought about the problem of training troubleshooting came close to the kind of analysis I have suggested. Generally speaking, they recognized that troubleshooting some particular equipment as a final performance was supported by two broad classes of subordinate tasks. First, there was knowledge of the rules of signal flow in the system, and second, the proper use of test instruments in making checks. The rules of signal flow themselves constitute an elaborate hierarchy of subordinate tasks, if one wants to look at it that way. For example, if the signal with which the mechanic is concerned is the output of an amplifier, then it may be necessary that he know some of the rules about data flow through an amplifier. Thus the task may be progressively analyzed into subordinate components which support each other in the sense that they are predicted to mediate positive transfer.

The task of using test instruments in making checks provides an even clearer example, perhaps. Obviously, one subordinate task is "choosing the proper check to make" (presumably a matter of knowing some "rules"); another is "selecting the proper test instrument" (an identification task); still another is "setting up the test instrument" (a procedural task, which in its turn has components like those previously described); and another is "interpreting the instrument reading" (another task involving a "rule"). Even identifying these component tasks brings to troubleshooting a vast clarification of the requirements for training. If one is able to take another step of arranging the proper sequencing of these tasks in a training program, the difference which results is remarkable. This is the interpretation I should be inclined to make of the studies which have demonstrated significant improvements in troubleshooting training, such as those of Briggs and Besnard (1956); of Highland, Newman and Waller (1956); and of French, Crowder, and Tucker (1956). In providing training which was demonstrably successful, these investigators were giving instruction on a carefully analyzed set of subordinate tasks, arranged in a sequence which, so far as they could tell, would best insure positive transfer to the variety of problem situations encountered in troubleshooting. It was *the identification of these tasks and this sequence* which I believe was the key to training improvement.

A good deal of work also proceeded along these lines, although not always with a terminal phase of measured training effectiveness. For example, a whole series of studies by Miller and Folley, and their associates, were concerned with what was called *task analysis*. They had such titles as these: Line maintenance of the A-3A fire control system: III. Training characteristics (Folley & Miller, 1955); Job anticipation procedures applied to the K-1 system (Miller, Folley, & Smith, 1953); A comparison of job requirements for the line maintenance of two sets of electronic equipment (Miller, Folley, & Smith, 1954). What was all this talk about task analysis? Did it have anything to do with training? My answer is that it had to do with training more than with anything else. These were thoroughgoing and highly successful attempts to identify the variety of tasks contained in a job, and the variety of subtasks which contributed to each task. There was in fact explicit recognition of the idea that successful final performance must be a matter of attaining competence on these subtasks. So here again was the notion that effective training somehow depended on the identification of these subordinate tasks, as well as on their arrangement into a suitable sequence to insure positive transfer to the final performance.

A third source of these ideas in military training research should be mentioned. This was the development of training devices applicable to such jobs as electronic maintenance. It came to be recognized that these devices were in some respects very different from the traditional trainers such as those for developing skill in aircraft maneuvers. They were called "concept trainers," and this, as Briggs' (1959) discussion of them implies, was another name for "teaching machines." As such, they were developed independently of Skinner's ideas, and they were in fact based upon an entirely different set of principles, as is clear from the accounts provided by Briggs (1956), Crowder (1957), and French (1956). Each of these training devices (or teaching machines), aside from its hardware engineering, was developed on the basis of a painstaking task analysis, which identified the subordinate tasks involved in a total task like troubleshooting a particular electronic system. The subordinate tasks thus identified were then incorporated into a sequence designed to insure maximal positive transfer to the final task. There were certainly some programing principles, but they bore little resemblance to those which are most frequently mentioned in recent literature; in my opinion, they were much more important than these.

Still a fourth area of effort in training research was related to these ideas. This was the development of techniques to provide behavioral guides, or "jobs aids" in support of performance in various technical jobs (Hoehn, Newman, Saltz, & Wulff, 1957). In order to do this, it was found necessary to distinguish between those kinds of capabilities which could best be established by thorough training, and those kinds which could be established by minimal training plus the provision of a check list or handbook. Obviously, here again there had to be a detailed task analysis. Subordinate tasks had to be identified which would mediate transfer either to the kind of performance required without a handbook, or the kind required with a handbook. Besides the initial task analysis, it is again evident that this line of work was making use of ideas about component task achievement and intratask transfer.

SUMMARY

Now that I have conveyed the message, my summary can be quite brief. If I were faced with the problem of improving training, I should not look for much help from the well-known learning principles like reinforcement, distribution of practice, response familiarity, and so on. I should look instead at the technique of task analysis, and at the principles of component task achievement, intratask transfer, and the sequencing of subtask learning to find those ideas of greatest usefulness in the design of effective training. Someday, I hope, even the laboratory learning psychologist will know more about these principles.

REFERENCES

Briggs, L. J. A troubleshooting trainer for the E-4 Fire Control System. *USAF Personnel Train. Res. Cent. Tech. Note,* 1956, No. 56-94.

Briggs, L. J. Teaching machines for training of military personnel in maintenance of electronic equipment. In E. Galanter (Ed.), *Automatic teaching: The state of the art.* New York: Wiley, 1959. Ch. 12.

Briggs, L. J., & Besnard, G. G. Experimental procedures for increasing reinforced practice in training Air Force mechanics for an electronic system. In G. Finch & F. Cameron (Eds.), *Research symposium on Air Force human engineering, personnel, and training research.* Washington, D. C.: National Academy of Sciences-National Research Council, 1956. Pp. 48-58.

Briggs, L. J., & Morrison, E. J. An assessment of the performance capabilities of fire control system mechanics. *USAF Personnel Train. Res. Cent. tech. Memo.,* 1956, No. ML-56-19.

Crawford, M. P. Concepts of training. In R. M. Gagné (Ed.), *Psychological principles in system development.* New York: Holt, Rinehart, & Winston, 1962. Ch. 9.

Crowder, N. A. A part-task trainer for troubleshooting. *USAF Personnel Train. Res. Cent. tech. Note,* 1957, No. 57-71.

Folley, J. D., Jr., & Miller, R. B. Line maintenance of the A-3A Fire control system: III. Training characteristics. *USAF Personnel Train. Res. Cent. tech. Memo.,* 1955, No. 55-5.

French, R. S. The K-System MAC-1 troubleshooting trainer: I. Development, design, and use. *USAF Personnel Train. Res. Cent. tech. Note,* 1956, No. 56-119.

French, R. S., Crowder, N. A., & Tucker, J. A., Jr. The K-System MAC-1 troubleshooting trainer: II. Effectiveness in an experimental training course. *USAF Personnel Train. Res. Cent. tech. Note,* 1956, No. 56-120.

Goldbeck, R. A., & Briggs, L. J. An analysis of response mode and feedback factors in automated instruction. Santa Barbara, Calif.: American Institute for Research, 1960. (AIR tech. Rep. No. 2)

Goldstein, M., & Rittenhouse, C. H. Knowledge of results in the acquisition and transfer of a gunnery skill. *J. exp. Psychol.,* 1954, 48, 187-196.

Goldstein, M., & Ellis, D. S. Pedestal sight gunnery skills: A review of research. *USAF Personnel Train. Res. Cent. tech. Note,* 1956, No. 56-31.

Highland, R. W., Newman, S. E., & Waller, H. S. A descriptive study of electronic troubleshooting. In G. Finch & F. Cameron (Eds.), *Research symposium on Air Force human engineering, personnel, and training research.* Washington, D. C.: National Academy of Sciences-National Research Council, 1956. Pp. 48-58.

Hoehn, A. J., Newman, S. E., Saltz, E., & Wulff, J. J. A program for providing maintenance capability. *USAF Personnel Train. Res. Cent. tech. Memo,* 1957, No. ML-57-10.

Melton, A. W. (Ed.) Apparatus tests. *USAAF Aviat. Psychol. Program Res. Rep.,* 1947, No. 4, pp. 917-921.

Miller, R. B., Folley, J. D., Jr., & Smith, P. R. Job anticipation procedures applied to the K-1 system. *USAF Hum. Resources Res. Cent. tech. Rep.,* 1953, No. 53-20.

Miller, R. B., Folley, J. D., Jr., & Smith, P. R. A comparison of job requirements for line maintenance of two sets of electronics equipment. *USAF Personnel Train. Res. Cent. tech. Rep.,* 1954, No. 54-83.

Primoff, E. Backward and forward association as an organizing act in serial and in paired associate learning. *J. Psychol.,* 1938, 5, 375-395.

Rittenhouse, C. H., & Goldstein, M. The role of practice schedule in pedestal sight gunnery performance. *USAF Personnel Train. Res. Cent. tech. Rep.,* 1954, No. 54-97.

Underwood, B. J. Verbal learning in the educative processes. *Harvard Educ. Rev.,* 1959, 29, 107-117.

Young, R. K. A comparison of two methods of learning serial associations. *Amer. J. Psychol.,* 1959, 72, 554-559.

34

On Transfer and the Abilities of Man

George A. Ferguson

In 1954 I published a paper in the Journal of this Association on learning and human ability (5). My intent was to draw together these two fields of enquiry, and present a conceptual framework relating each to the other to mutual advantage. In summary the broad features of that theory are these:

(1) The abilities of man, including the reasoning, number, perceptual, and spatial abilities, and whatever is subsumed under intelligence, are attributes of behaviour, which through learning have attained a crude stability or invariance in the adult, and, as they develop in the child, exhibit considerable stability over limited periods of time at particular age levels.

(2) Biological factors in the formation of ability are not excluded. These fix limiting conditions. The implication is that within these boundaries the range of variation in ability attributable to learning is substantial. Thus emphasis is diverted from biological to environmental determination in the formation of ability.

(3) Cultural factors prescribe what shall be learned and at what age; consequently different cultural environments lead to the development of different patterns of ability. Those abilities which are culturally valid, and correlate with numerous performances demanded by the culture, are those that show a marked increment with age.

(4) Abilities emerge through a process of differential transfer and exert their effects differentially in learning situations. Those that transfer and produce their effects at one stage of learning may differ from those at another.

(5) The concept of a general intellective factor, and the high correlations between many psychological tests, are explained by the process of positive transfer, the distinctive abilities which emerge in the adult in any culture being those that tend to facilitate rather than inhibit each other. Learning itself is viewed as a process whereby the abilities of man become differentiated, this process at any stage being facilitated by the abilities already possessed by the individual.

These are a few of the main points of the theory. Criticism of the 1954 paper

From *Canadian Journal of Psychology,* 10, 1956, 121-131. By permission of the Canadian Psychological Association.

Presidential Address delivered at the Annual Meeting of the Canadian Psychological Association, Ottawa, Ont., June 8, 1956.

suggests a need for some refinement and elaboration. Specific criticism has been directed to the concepts of ability and transfer (1, 4). I propose to elaborate my understanding of these concepts, to review recent experimental evidence on some aspects of the theory, and to comment on a few of its implications and further extensions.

THE MEANING OF ABILITY

The term ability is used in a variety of ways. It may refer to measures of performance in any situation, these measures being subject to error in relation to an underlying latent variable, which is presumed to be a continuous monotonic increasing function of the observed measures of performance. The observed measures locate the individual on the latent variable within some range of error. The concept of a latent variable is analogous to that of a parameter in conventional statistical usage. It is a parametric continuum. The latent variable is a necessary construct in that it permits a concept of error, an error in any situation being a departure from a fixed, true, standard, latent, or parametric value. This meaning of ability is operational, and the introduction of a latent variable, although in itself postulational, and not operational, is not incompatible with this meaning but is a necessary logic adjunct to it. All this is in effect an elaboration of Thurstone's statement that "an ability is a trait defined by what an individual can do" (13). The trait here refers to the latent variable, and the remainder of the statement to measured performance. Scores on psychological tests, measures of performance on learning tasks, and behavioral observations of many kinds are subsumed under this broad meaning of ability. The statement that a learning curve is a description of the change in ability with repetition imples this usage.

Consider now the meaning of ability as a factor in the methodology of factor analysis. This is in effect a derivative of the meaning I have elaborated above. An individual's factor score is a derived measure. It is the weighted additive sum of measures of performance on separate tasks, the weights being obtained by a process of mathematical analysis. Such derived measures are commonly used in physics. For example, the volume of a sphere is given by the formula $V = \frac{4}{3}\pi r^3$, the volume being a function of the radius which is the independent variable. Such functions as this may be called *determined functions,* and the dependent variable is a derived or determined measure. A factor score also implies a latent factor variable, since both the weights and the separate performance measures which determine it are in practice subject to error. The term factor, as distinct from factor score, usually connotes the latent variable. I construe the concept of ability as a factor score to be essentially operational, since it is a determined function of operationally determined variant values.

The term ability is frequently used in a third sense to refer to some attribute of the state of the organism. This state may be vaguely identified with neurophysiological structure and process which is modified by environmental and genetic factors. The state of the organism is presumed to be functionally related, somehow, to observable performance in particular tasks. The persisting notion that differences in intelligence are related to differences in some state of affairs inside people's heads belongs to this class. Likewise the statement that an ability is a learned acquisition, although not entirely learned, implies this usage. The term ability when used in this general way is strictly postulational. No very precise theoretical propositions have as yet been formulated about the nature of ability when construed in this sense, but it may one day be possible to do so. By linking learning and human ability some of the speculations on the neurophysiological bases of learning may have useful implications for ability theory.

No incompatibility exists between the various meanings of the term ability described above. Ability as a latent variable is a necessary logical implication of operationally determined measures of performance. Ability as a factor score is a derived measure. It also implies a latent factor variable. The use of the term ability to refer to some postulated state of the organism is compelled upon us by all those circumstances that require the use of theoretical constructs in science. In the scientific study of ability all these meanings must of necessity co-exist and supplement each other. Of course, different words may be used to distinguish these different meanings, but this is perhaps unnecessary.

In my 1954 paper I used the term ability in all the various ways described above. Likewise I drew a distinction between, on the one hand, performance measures which had attained a crude stability or invariance in relation to learning, and, on the other hand, performance measures which were subject to marked increment or decrement through learning, or its cessation, over relatively short intervals of time. I pointed out that ability was frequently taken to refer in common usage to these more stable measures, and that what was ordinarily meant by intelligence, the perceptual and spatial abilities, reasoning abilities, many of the basic psycho-motor abilities, and the like, seemed to belong to this class. All the above distinctions which were implicit, if not explicit, in my previous work apparently led to certain difficulties in the minds of some of my readers.

THE CONCEPT OF TRANSFER

The concept of transfer occupies a crucial position in any theory attempting to relate learning to human ability. For many years I have been aware of logical difficulties in the commonly accepted concept of transfer. Others, including T. W. Cook (3, 4) have also been aware of this problem. It has long been apparent to Cook, myself, and others that transfer is the more general phenomenon and learning a particular case. In previous theoretical speculation, although pointing to these logical difficulties, I attempted to employ the common interpretation of these terms, but deviated, in the views of some, to an unorthodox position.

I propose now to present a general formulation of the concept of transfer. This formulation generates a number of particular cases. Learning, as conventionally regarded, is one.

Nothing will be said here about the psychological nature of transfer and why or how it occurs. The intent is merely to clarify and restructure the concept, open the way for new forms of experimentation, and provide a conceptual framework within which existing experimental data can be related and reinterpreted.

Transfer, in my view, can best be regarded in terms of the mathematical concept of a function. When two variables are so related that the values of one are dependent on the values of the other, they may be said to be functions of each other. It is customary to distinguish between dependent and independent variables, the value of the dependent variable being dependent on variation in the independent variable. The idea of function is descriptive of change in something with change in something else.

The essence of the idea of transfer, also, is concomitant change, and in the simplest case implies change in performance on one task with change resulting from practice on another.

Consider a simple transfer model comprising measures of performance and amount of practice on two tasks. Denote these by x, y, and t_x, t_y, respectively. Thus we have four variables which may be assumed to be functionally related. The model may, of course, be generalized to include any number of variables, but for the present exposition attention will be directed to the four-variable model only. We may write an expression of the kind $y = \phi(x, t_x, t_y)$, which simply means that performance on one task is some unspecified function of performance on another task, and measures of amount of practice on the two tasks. This expression, and all similar expressions, may be spoken of as transfer functions. Let us now consider particular cases of this general transfer function.

Case I. When the two tasks are considered the same, and y is the same as x for all values of y and x, and t_x is the same as t_y for all values of t_x and t_y, the expression $y = \phi(x, t_x, t_y)$ becomes a two variable function and reduces to $y = \phi(t_y)$ or $x = \phi(t_x)$. These are functions relating measures of performance on a task to measures of amount of practice on that task, and are general expressions for conventional learning curves. In this sense learning is a particular case of transfer and depends on the identity of x with y, and t_x with t_y. Geometrically this means that the model has been reduced from four to two dimensions. All points fall in a plane or the function is a line, instead of a four dimensional surface.

Case II. When t_x and t_y are very large indeed, and additional practice, or its cessation, produces little effect on performance, a state of crude invariance of performance with practice is attained. Here we may measure performance on x and y for a group of individuals and study the relationship between the two sets of measures. I contend that the correlations between many tests, commonly called tests of ability, are in effect correlations between performances which through learning have attained a crude invariance. If this is so it follows in many cases that the correlations between tests of ability can be viewed as particular cases of the more general transfer function. Thus we have a useful linkage between learning and ability.

Case III. When t_y is a constant, that is, where no practice is allowed on the y task, we have a situation involving three variables, y, x, and t_x. We may consider the function $y = \phi(x, t_x)$, or $y = \phi(t_x)$, or $y = \phi(x)$. This latter expression is descriptive of change in y, practice being held constant, with change in x resulting from practice. In effect it implicitly defines the amount of practice in terms of change in performance and not in terms of an independent measure, as, for example, the number of trials. This particular function is of some interest in that it suggests modifications in current thinking on transfer and opens up new lines of experimentation.

The simple transfer function $y = \phi(x)$, when stated in this form, generates, of course, an indefinite number of particular functions. Some of these are deserving of comment. First, consider the case where no change occurs in y with change in x. Here no transfer in fact occurs. Second, consider the case where a constant increment occurs in y with increase in x; the curve describing the relationship is linear and has a positive slope within fixed ranges of the variables. This is a case of linear positive transfer. Similarly the case of linear negative transfer can be recognized. Third, many cases of non-linear transfer functions can be identified. For example, y may show an increase with x over some range of x and no increase or a decrement over some other range. This means that, in the learning of a task, positive transfer may occur at one stage of learning, and no transfer or negative transfer at another. All non-linear cases of this kind may be spoken of as examples of differential transfer, the rate of transfer differing at different stages of learning. Fourth, consider a case where y continues to change after no improvement in x is observed. If examples of this type could be demonstrated experimentally, this would argue for a modification of behaviour through overlearning, despite the absence of any observable change in performance on the learning task itself. This in turn would suggest some continuing modification in the neurophysiological state of the organism during overlearning.

All the above can be greatly elaborated and stated in more precise mathematical form. In general the introduction of the idea of a transfer function argues very simply for the use of the concept of continuous covariation in the study of transfer, and the discarding of discrete concepts. Many of the existing concepts in the field of transfer are in effect, discrete, such as the conventional distinction between positive and negative transfer. The concern of the experimentalist has frequently been with the demonstration of the presence or absence of certain effects, rather than with the mapping of descriptive continuous functional dependencies. The point has been reached in our experimental exploration where concepts of a more general, and consequently more powerful, nature are required. The ideas on transfer I have presented here are a small step in this direction.

The concept of a transfer function has important implications for experimental design. It is possible in practice to design experiments which embody the ideas on transfer presented here. Such designs are in certain cases general extensions of the existing designs used in experiments on transfer. A variety of new problems will of course present themselves. Unfortunately time does not allow a treatment of experimental design here.

The above discussion points to the following very general formulation. At any given point in time the organism may be said to be in a particular state. The concept of the state of a system is of importance in physics. It has a role in psychology also. This state undergoes continuous change because of a large number of circumstances both inside and outside the organism. One set of factors leading to a change in state is the behaviour of the organism in response to specific environmental circumstances, e.g. the performance of a task. Any change of state leads, theoretically, to changes in an indefinitely large number of other possible forms of performance. Any covariation which can be identified between any two or more forms of performance is conceptualized as a transfer function. Certain aspects of the state of the organism attain a crude stability or invariance and are less susceptible than others to modification through continuing behaviour and other factors over limited ranges of time. These, postulationally, are invariants of the state of the organism, which in turn are functionally related to certain invariants in particular observable behaviours. What we conventionally regard as the abilities of man are among these invariants. The whole process of growth and development is directed towards the reduction of uncertainty in the behaviour of the organism, and to the establishment of these invariants. Thus behaviour becomes organized, or structured, and to some extent predictable. The most noteworthy examples of the prediction of human behaviour in psychology are made through a knowledge of these invariants. Although concern here is with the study of ability, what we conventionally regard as personality, as distinct from ability, can be similarly conceptualized. Characteristics of personality, attitudes, and the like can be viewed as attributes of behaviour which have attained some stability through a lengthy learning process. The discovery of these invariants in the behaviour of man is one of the primary objects of psychology endeavour.

RECENT EXPERIMENTAL EVIDENCE

Supporting evidence for the theory relating learning to human ability is substantial, and is scattered broadly through both the psychological and the cultural anthropological literature, and elsewhere. I do not propose to attempt to collate or review this evidence here, but shall confine my remarks to two recent series of studies: those by Edwin A. Fleishman (7, 8, 9) of the Air Force Personnel and Training Research Centre at Lackland Air Force Base, and those by Dr. Alastair Burnett (2) of the Hospital for Mental and Nervous Diseases, St. John's, Newfoundland.

The design of the Fleishman studies involved the administration of learning tasks to groups of subjects and the recording of measures of performance at a number of stages of practice. In one study the learning task was a Complex Coordination Test and in another a Discrimination Reaction Time Test. In both tasks subjects were required to make motor adjustments or manual responses to visual stimulus patterns. In addition to the learning tasks, a variety of tests were administered to the subjects involving spatial, verbal, perceptual, psychomotor, and other abilities. Intercorrelations were obtained between all measures including measures of performance on the learning task at different stages of practice. The correlations were analyzed using standard factorial methods. One study involved 197 and the other 264 subjects.

These studies show conclusively that substantial and systematic changes occur in the factor structure of the learning task as practice continues. The abilities involved at one stage of learning differ from the abilities involved at another stage. Thus conclusive experimental evidence exists to support the hypothesis of differential transfer. Further these studies suggest that certain non-motor abilities, e.g. spatial and verbal, have a greater involvement in the earlier stages of learning of a psychomotor task than in the later stages, whereas the opposite applies to certain motor abilities, e.g. reaction time and rate of movement. This is in line with the notion that cognitive abilities play a more important role in the earlier stages of the learning of a motor task than in the later stages, when performance becomes organized in the form of a habituated psychomotor response pattern. Generalizations on this point are of course precarious because the distinction between cognitive and motor abilities is arbitrary, and because much will depend on the nature of the particular learning task.

Some of Fleishman's results raise important questions on the nature of adult learning. His studies show that factors specific to the tasks themselves increase in importance as practice continues, suggesting the formation of specific or "within-task" abilities, not involving previously established abilities. Acceptance of this finding might require some modificaton of the theoretical position that adult learning is in large measure a process involving the reorganization or integration of that which has been previously learned. It would mean also in practice that our capacity to predict human behaviour would be limited, and would ever remain in some large measure indeterminate. These findings are not acceptable as they stand, and fortunately there are avenues of escape from the experimental data. The most plausible of these is that the tendency for the specific "within-test" variance to increase with practice is a result of the selection of reference variables of ability, and that with the inclusion of new and different reference variables a higher proportion of this variance at the more advanced stages of learning can be brought under control. If adult learning is a reorganizing or integrating process, involving different ability patterns at different stages, it seems reasonable to suggest that individuals may exhibit differing facilities in organizing or integrating their abilities to cope with a new task. Thus we are led to hypothesize certain basic integration abilities, which may have an important role to play in adult learning. Considerable information exists in the literature of factor analysis (6, 10) on the existence of such integration abilities, but this information has not, hitherto, been brought to bear on the problem of adult learning. Fleishman is currently engaged in a comprehensive investigation of the involvement of certain integration abilities in performance at different stages of practice on psychomotor tasks. It is hoped that these studies will show that a higher proportion of the specific within-task variance can be brought under control. If experimentation demonstrates that this is not so, then we may be required to revise existing theories of adult learning and make some radical changes in our current concepts of predicting human behaviour, or evolve another line of attack.

In passing I should mention that the Fleishman experiments provide the empirical buttressing for a number of important hypotheses in my theory which in turn provides a theoretical basis for his experiments. Although I had previously discussed the design of experiments of this type, their conduct, which is very exacting and laborious, lies well beyond the range of the facilities of most university investigators.

Dr. Burnett's work carried on at McGill and at the Hospital for Mental and Nervous Diseases in Newfoundland was directed towards the assessment of the abilities of individuals living in relatively isolated outport communities. Burnett's work shows conclusively that the pattern of ability of children reared in relatively isolated outport communities differs markedly from that of children reared in urban centres. In the isolated Newfoundland environment certain perceptual and motor abilities are developed to a high level, whereas verbal and reasoning abilities are less well developed. A retardation in abstract thinking and concept formation seems to occur. These findings go well beyond the simple observation that tests which may be appropriate in one culture may be inappropriate in another, with its implication for the development of culture-free tests. Everything we know suggests that different environmental demands lead to the development of different ability patterns. The concept of a culture-free test is a misconception because the abilities of man are themselves not culture-free. The extensive body of literature on the abilities of individuals reared in cultures markedly different from our own in general supports this conclusion, and adds substantial evidence for the role of learning in the formation of abilities.

SOME IMPLICATIONS OF THE THEORY

The implications of the theory for contemporary theoretical and experimental work in psychology are numerous. I propose to comment on a few of these implications here.

One implication is that some of the data obtained from the study of human ability using an individual difference approach, may with reinterpretation have relevance for learning theory, and that the methodology of ability studies may be brought to bear on learning problems. The substantial body of data, summarized in my 1954 paper, on the relationship between psychological test performance and age in different cultural groups has, for example, direct relevance for learning. Fleishman's work is an excellent example of the application of a methodology extensively used in the ability field to problems which have, among others, important implications for any theory of learning in the adult. Quite recently I have devoted some thought to the problem of intra-individual variability in ability patterns, or profile analysis, and have formulated hypotheses with implications for learning theory which can be investigated by the study of such profiles. This work bears on the organization or integration of abilities through learning. It is still at a rudimentary stage and further comment on it would be unwarranted.

Learning theorists have frequently preoccupied themselves with somewhat involved theories to explain the nature of the learning process itself. They have, of necessity, tended to restrict themselves to a range of learning tasks of low-order generality. No satisfactory methodology has emerged for describing particular learning tasks, or indicating how one task differs from another, other than by a process of simple inspection. The ability theorists have, however, developed descriptive classificatory systems, which, regardless of their many faults, do have some degree of generality in relation to many forms of human behaviour. Since abilities are clearly involved in the learning process, it follows that particular learning tasks can, at different stages of learning, be described in terms of particular ability patterns. We have, then, a method for describing particular learning tasks and differentiating them one from another. Thus it may be possible to make progress in removing learning theory from the context of particular tasks. Students of human ability have long been aware that, just because two tasks looked on inspection very much the same or different, they were not of necessity the same or different in terms of the behavioural responses which they elicited. In consequence it has been the practice, in the ability field, to describe the task situation or the stimulus in terms of the response. This approach underlies the whole of factor analysis. The description of the stimulus in terms of the response is an approach which has apparently not found favour in the learning field where it undoubtedly has application.

A further implication of the theory resides in its emphasis on environmental factors in the formation of ability. In my early training in psychology I was led to believe that man's abilities were irrevocably fixed and rendered unchangeable by biological endowment. This position is no longer tenable. Although it is conceded that biological factors fix certain boundaries, all the evidence seems to suggest that the range of variation that results from learning is, indeed, very great. If this is so, it immediately raises questions of value and social responsibility. It means that a society, through control of the environment and the educative process, can in some considerable degree determine the patterns of ability which emerge in its members. This has long been known to apply in the formation of attitudes. Its extension to the field of man's abilities broadens the area of social responsibility which the psychologist, through his knowledge of human behaviour, may one day be required to face.

REFERENCES

1. Attridge, B. F., & Sampson, H. A note on Ferguson's learning ability matrix. *Canad. J. Psychol.*, 1955, 9, 84-90.
2. Burnett, A. Assessment of intelligence in a restricted environment. Unpublished Ph.D. thesis, McGill University, Montreal, 1955.
3. Cook, T. W. Repetition and Learning: I, Stimulus and Response. *Psychol. Rev.*, 1944, 51, 25-36.
4. Cook, T. W. Transfer and ability. Unpublished MS, 1955.
5. Ferguson, George A. On learning and human ability. *Canad. J. Psychol.*, 1954, 8, 95-112.
6. Flanagan, J. C. (Ed.) *The aviation psychology programme in the army air forces.* Army Air Forces Aviation Psychol. Program, Res. Rep., 1948, No. 1. Washington, D.C.: U.S. Govt. Printing Office.
7. Fleishman, Edwin A., & Hempel, Walter E., Jr. Changes in factor structure of a complex psychomotor test as a function of practice. *Psychometrika*, 1954, 19, 239-252.
8. Fleishman, Edwin A., & Hempel, Walter E., Jr. The relation between abilities and improvement with practice in a visual discrimination reaction task. *J. exp. Psychol.*, 1955, 49, 301-312.
9. Fleishman, Edwin A. Predicting advanced levels of proficiency in psychomotor skills. Paper read at Joint Air Force–National Research Council Symposium, National Academy of Sciences, Washington, D.C., Nov. 1955.
10. Guilford, J. P., & Lacey, J. I. (Eds.). *Printed Classification Tests.* Army Air Forces Aviation Psychol. Program, Res. Rep., 1947, No. 5. Washington, D.C.: U.S. Govt. Printing Office.
11. Hebb, D. O. *The organization of behavior.* New York: Wiley, 1949.
12. Thurstone, L. L. *Primary Mental abilities.* Chicago: Univer. of Chicago Press, 1938.
13. Thurstone, L. L. *Multiple factor analysis.* Chicago: Univer. of Chicago Press, 1947.

Chapter Ten

Evaluation of Training

As jobs become more complicated, training procedures increase in complexity and cost. A report by H. Oliver Holt, Director of Training Research for the American Telephone and Telegraph Company, indicated that approximately 75 million dollars is being spent annually by this one corporation on salaries of non-management employees who are undergoing formal classroom training (Holt, 1963). This figure does not include any of the costs of training per se (e.g., supplies, salaries of training personnel, and equipment). Other "dollars and cents" figures are just as astounding. It is not unusual, for example, to find advertisements in business magazines promoting short courses for executives with fees running well into five figures. Present-day estimates of the budget for the total education enterprise exceed 50 billion dollars per year. In the next decade, an increasing portion of the budget is expected to be used in developing new modes of training, such as computer-assisted instruction. There can be no doubt that training is expensive, or that more expensive techniques are on the horizon.

Unless we have some basis for evaluating the *success* of these programs, we cannot even guess—much less quantify—their actual cost to society or, on a smaller scale, to the company or municipality through which they are funded. If the methods are effective, then society is merely exchanging a resource that it has (dollars) for one that it needs (skills) in order to fulfill some of its ultimate objectives (such as increased productivity, reduced unemployment, or lowered illiteracy). If, on the other hand, the methods are ineffective, then society is receiving little or nothing in exchange for its dollars, and a net loss (a *true* cost) should be declared. The point is, we cannot tell how much we are really "paying" unless we can gauge the effectiveness of training.

All too frequently, the simple logic of this argument is lost in practice. A personal experience of one of the authors may serve to underscore the practical seriousness of this problem. Recently he and a colleague were invited to visit a neighboring community to discuss a job training program. The stated objective of this program was to place personnel and keep them on the job. An inquiry into the success of the program brought only the reply that the community was most fortunate to have obtained the "famous curriculum" used by a neighboring state. Discussing the matter further, community officials pointed again with obvious pride to the excellence of the curriculum, and noted the high rate of class attendence that it had produced. Further attempts to draw out information pertaining to the *criterion*—i.e., how successful they had been in actually keeping

References will be found in the General References at the end of the book.

people on the job—led only to the discovery that no such data had ever been collected! The possibility that so exceptional a curriculum could possibly fail to yield the desired results in their situation was never even considered. Neither was the question of *how much* improvement would be necessary in order to justify the expense involved. An attitude such as this, of course, precludes any search for more effective training methods.

That the above is by no means an isolated case is borne out in many of the surveys that have appeared on training evaluation in recent years. A study by French (1953), for example, showed that despite huge sums spent on supervisory training, only one company in forty made any scientific evaluation of training. Another report reviewed 476 training studies and failed to find *any* that measured both pre-training and on-the-job performance (Castle, 1952). Moreover, there were no control groups in any of these latter programs.

The serious problems raised by this paucity of evaluative research are vividly illustrated by the difficulties that older workers encounter in finding employment. The phrase "too old to work" does not apply so much to a specific *age* as it does to an *attitude* on the part of employers with regard to specific positions. The employers seldom have any solid evidence to back up their opinions about older workers. Sometimes their attitude represents an "image" adopted by an entire industry. One of the authors, for example, has heard advertising executives proudly proclaim theirs to be "a young man's field." While this is probably an apt description, it is certainly not the result of evaluative studies showing the ineptitude of older workers. One of the prime reasons for such biases is the preconceived idea that older people cannot be trained as readily as younger ones; as the old adage goes, "you can't teach an old dog new tricks." There are, of course, virtually no adequate data to support—or refute—this contention. Weinberg (1963), for example, canvassed some 100 firms where new technologies were being introduced and found only four in which suitable records of performance, education level, and age were even *kept*. Naturally, little can be said on the basis of such scattered data, especially since scientific rigor (e.g., proper control) is rarely observed.

Proposed reasons for the sad state of affairs in training evaluation are, of course, diverse and colored by opinion and vested interest. Some suggest that training supervisors are not sufficiently schooled in proper evaluative techniques. Others blame the researcher, claiming that he has neglected important issues or failed to present his findings effectively. Still others maintain that management is largely at fault. It is claimed that managers are often unwilling to evaluate programs to which they are already committed (and in which they may already have complete confidence); or that they object to the whole idea of allocating funds in support of training research (perceiving this as somebody else's responsibility). Of course, management usually becomes "committed" to a program because it has been "sold" on the idea by someone involved in training. If, then, that program were to be found wanting, both the decision maker and the trainer who "sold" it would be placed in an awkward position; moreover, any subsequent training proposal would undoubtedly meet with even stiffer resistance. Is it better to protect an existing program by *assuming* its worth, or to *measure* its worth and risk having no program at all? When management is sufficiently unenlightened, this indeed becomes a valid concern.

Taken individually, each of the above "explanations" probably tells only part of the story. Together, however, they present a rather vivid illustration of a deep-rooted problem that besets the entire field of engineering psychology—a problem that we first encountered in Chapter One. Very simply, it is that communication between those who generate knowledge and those who apply or stand to benefit from the application of that knowledge is not all that it should be. There is a definite need for the training researcher, the training practitioner, and the manager who must ultimately justify training expenditures to work toward achieving a better rapport in matters such as the evaluation of training programs.

Of course, the situation is not entirely as bleak as we have pictured it to this point. Well-conceived training programs do exist—ones for which there is ample evaluative evidence. A program described by McGehee and Livingstone (1952), for example, was shown to produce a 61.6 percent reduction in material waste. Even more important, a subsequent study by the same investigators showed that waste reduction *continued* for 80 weeks without any follow-up training program (1954). Here, then, was unequivocal evidence that the program was worthwhile in at least one respect; and even had it proved otherwise, the investigators would have been in a position to render the valuable—if often unappreciated—service of recommending revision or abandonment of the program.

The actual *implementation* of training evaluation is by no means a simple task, even if all parties are convinced of its desirability. Probably the most difficult step is the first: specification of appropriate criteria. Obviously, one cannot measure success until he has established what success *is*, and for many of today's jobs this represents a major undertaking. There are usually not one, but *several* relevant considerations; some of these are fairly objective and easily quantified, while others are vague, qualitative, and subjective. A list prepared by Lindahl some years ago (1949) illustrates this point rather nicely. Among factors listed as criteria for training evaluation were: quality of production, time to perform the job, absenteeism, job tenure, waste and breakage, accidents, performance tests, and rating scales. It is perhaps worth noting that the most subjective of these indices —ratings of one sort or another—is by far the most popular today. Guion (1965), for example, cites the figure 81 percent in discussing the use of such measures in validation studies of various kinds over a five-year period.

Because of the magnitude and importance of the criterion problem, we have elected to feature various aspects of it in our selections for this chapter. In the first of these, Guion develops more fully several of the points that we just introduced. He discusses general problems associated with *selection, measurement,* and *use* of criterion judgments, and elaborates on the multidimensional character of success in jobs moulded by our advanced technology. Although his comments are restricted to criteria for *selection* and *placement,* everything he says applies to the field of *training* as well. People are selected, placed, and trained for basically the same reason—to promote job success; therefore, whatever one uses to measure job success applies equally to all three procedures.

Success in some jobs, such as those involving supervision, is doubly hard to evaluate because it can be expressed at several *levels*. The second article in this chapter examines these levels, discussing the relevance of each for training evaluation. If one assumes that successful supervision is ultimately reflected in the behavior of those supervised, then on-the-job records for *employees* should furnish the best index of supervisory training effectiveness. Unfortunately, this level is rarely tapped in selecting a criterion. More often, the easily accessible—yet far less meaningful—level of *classroom performance* is elected as the source of evaluative data.

The implication to be drawn from the first two selections, then, is that evaluation using a questionable criterion is little better than none at all, and perhaps a lot worse. It could, for example, lend credence to a useless program, or support to detractors of a good one. Furthermore, the evidence seems to suggest that in those instances where evaluation has actually been attempted, the criterion often *has* been of the questionable variety. The third article in this chapter continues the discussion of criterion levels and introduces still another way in which evaluation can go wrong. MacKinney discusses three designs that could be used to evaluate a training program, only one of which is capable of providing an unambiguous appraisal. This is the controlled experimental design, in which one of two comparable groups receives training and the other serves as a control. Proficiency measures before and after training serve as the basis for statistical comparison. In other words, a criterion—even a good one—is not enough to establish the consequences of training. Unless a proper control is included, any effect could be the result of spurious factors (as, for example, the classical "Hawthorne effect").

Once again, of course, it would appear that the most *suitable* design is not necessarily the most *popular;* control groups are still more often the exception than the rule in this area. We would add but one further point to what MacKinney has to say—a word of emphasis regarding *when* performance of the two groups is compared. Significant differences obtained immediately after training may not persist on the job. Thus, while immediate post-testing can provide useful data, the "proof of the pudding" should ultimately lie in some comparison between the trained and control groups *on the job*. A series of such comparisons at various points in time is particularly advantageous in that it permits an assessment of the entire function produced by training.

A study by Fleishman, Harris and Burtt (1955) illustrates the importance of examining on-the-job behavior. These researchers designed a training program to increase the amount of consideration (friendship, mutual trust) in foremen's behavior. At the end of the training program, they found a significant increase in consideration scores. However, the increase in the consideration factor did not persist on the job. In this case, the authors discovered that the day-to-day social climate was a stronger variable than the training program. The foremen were not able to continue their use of strong human relations values when the environment (including their supervisors) was not sympathetic to that view.

The reader is referred to a recent book by Campbell and Stanley (1966) for a more detailed discussion of various design considerations.

35

Criterion Measurement and Personnel Judgments
Robert M. Guion

In their review of the year's literature, Wallace and Weitz (1955) suggested that more talking was being done about "the criterion problem" than research. Despite all of the verbal attacks on this problem, they concluded, most workers still accept the most convenient criterion with the hope that it will turn out to be all right. The present note is in the tradition observed by the Wallace and Weitz survey; it is a discussion of an approach to the problem, not a report of concrete research!

A first point of view to be made explicit in this discussion is that the solution to the criterion problem does *not* lie in the typical efforts at statistical refinement. Much of this has been blind numerical manipulation. If we are honest, we will describe many of the very complex criterion refinements with the words, "We don't know what we are doing, but we are doing it very carefully and hope you are pleased with our unintelligent diligence" (Wherry, 1957. p. 1).

THE PRESENT STATUS OF CRITERION MEASUREMENT

Most efforts at personnel prediction start (at least implicitly) from a judgment about the nature of some "ultimate criterion" (Thorndike, 1949; Fiske, 1951) of what may be called the general satisfactoriness of personnel performing some specified function within an organization. This ultimate criterion is a complex array of skills, interests, motives, attitudes, and other imponderables along with the whole gamut of specific behaviors appropriate to the job. Conceptually, the ultimate criterion can apparently be defined as "the total worth of a man to the company—in the final analysis." Such an abstraction is not measurable—at least not until the man's total career with an organization has finally been terminated!

Therefore, an acceptable substitute is usually deemed necessary. The search for such a substitute may begin by examining job descriptions and by making some kind of subjective judgment or analysis of the needs of the organization. At best, the present status of techniques of job and need analysis leave much to be desired; typical techniques of job and need analysis are seldom the best when one

From *Personnel Psychology*, 14, 1961, 141-149. By permission of Personnel Psychology, Inc.

is merely seeking some clues about possible criteria! Typically, in fact, the whole sequence of criterion selection goes something like this:

1. The psychologist has a hunch (or insight!) that a problem exists and that he can help solve it.
2. He reads a vague, ambiguous description of the job.
3. From these faint stimuli, he formulates a fuzzy concept of an ultimate criterion.
4. Being a practical psychologist, he may then formulate a combination of several variables which will give him—as nearly as he can guess—a single, composite measure of "satisfactoriness."
5. He judges the relevance of this measure: the extent to which it is neither deficient or contaminated (Cf. Nagle, 1953).
6. He may judge the relative importance of each of the elements in his composite and assign some varying amount of weight to each.
7. He then finds that the data required for his carefully built composite are not available in the company files, nor is there any immediate prospect of having such records reliably kept.
8. Therefore, he will then select "the best available criterion." Typically, this will be a rating, and the criterion problem, if not solved, can at least be overlooked for the rest of the research.

In short, the sequence of events involved in the usual selection of a criterion involves a series of judgments on the part of the personnel research worker, culminating in the decision to use someone else's judgment as the criterion. This means, then, that the whole superstructure of personnel research—with its multiple correlations and confidence levels and other trappings of a quantitative, scientific methodology—is built upon the weakest of foundations: a residual judgment. (Parenthetically, it must be recognized that ratings may provide excellent criteria when properly applied and carefully worked out. But when ratings become the residual technique—the technique to use after everything else has been rejected because of *obvious* inadequacy—then one may be permitted a raised eyebrow when one reads in the journal, "The best available criterion was the instructor's rating." If a rating is indeed the best available criterion, then of course a personnel researcher must do the best he can under the situational limitations. But he should not delude himself that he has solved the criterion problem.)

Even in more desirable circumstances, where the finally selected criterion is not residual but does indeed conform to the judgments made in the first half-dozen steps of the sequence, the superstructure of research results may rest on a very shaky foundation. Ignore, if possible, the ever present possibility of poor judgment. There still remains the problem of changing circumstances. Changes in the general economy, in the company's competitive position, in the labor market, or (often overlooked) in the distribution of skills or traits within the organization itself are very likely to occur. Such changes may in turn influence one's "expert judgment" about the relative importance of the various elements that make up one's composite criterion measure. Consider, as one example, a company beset by customer complaints about quality of workmanship. If it initiates a personnel research program to help solve its problem, it will need a criterion—and the ultimate criterion must include not only the quality element but a quantity measure as well. In judging the relative importance of these two elements, at such a time, it is quite possible that a sound judgment might consider the quality measure to be perhaps two or three times as important as quantity. Selection research (and perhaps training or administrative changes) are then validated against this criterion, the variance of which is perhaps 70 per cent attributable to quality measures.

If the research is successful, and if the results are incorporated into the employment procedure (and other personnel practices), then in due time the organization should have a highly "quality-conscious" work force. Customer complaints should become markedly reduced and, as in other fairy tales, everyone should live happily ever after.

In most companies, however, there comes a time for a second chapter, and in this chapter the problem that has been solved is quickly replaced by a new one to solve. In the example company, for instance, the emphasis on quality may have resulted, in time, in a deliberate, careful work that moves slowly enough to endanger the company's competitiveness in pricing. While quality is still an important element, its relative importance may therefore shrink enough to reverse the earlier ratio. If this happens, then all of the validation work done with the original weighting needs to be done again with the revised criterion.

The argument of this note is simply that the judgment of relative importance of criterion elements is typically made too soon.

MULTI-DIMENSIONALITY OF CRITERIA

The variety of criterion selection outlined in the foregoing is based upon the idea of a single, over-all ultimate criterion which must somehow be duplicated in the criterion measure actually employed. This idea assumes a generality in criteria analogous to a general factor in intelligence.

Criteria, however, exist along many dimensions, not just one. A major landmark in our collective awareness of this point is the article by Ghiselli (1956), "Dimensional Problems of Criteria." Although there has been much recognition of criterion dimensionality at any given time, Ghiselli has gone further in introducing the concept of "dynamic dimensions" of criteria. The concept here is that the dimensions along which performance occurs may change as a result of experience. This idea is supported by findings in one organization known to the writer. In the organization it was found that tests which can predict sales performance during the first year of employment are different tests from those which predict sales performance after five years. Clearly, first year sales and five year sales are not equivalent criteria. The reason is not as clear. What are the single, identifiable dimensions of sales performance that have changed in this period of time?

Such dimensions must be clearly defined and isolated if marked improvement in personnel selection techniques are to be made. This implies that personnel research must be multi-dimensional in its dependent, criterion measurement as well as in its independent, predictor measurement. The old insistence of the laboratory for a single dependent and a single independent variable may have worked well with rats and mazes, but it is not appropriate to the study of the functioning of the human organism in his many-faceted world of making a living.

Successful multivariate research demands clear knowledge of each of the component parts. Each must be not merely definable, but reliably measurable as well.

There is a profound difference between sound multivariate research, depending upon systematic handling of the variables included, and the verbal, conceptually over-cooked hash that is today's typical "composite criterion."

Too long and too frequently the doctrine is accepted that a single, over-all criterion is indispensable (Nagle, 1953; Toops, 1928). The second argument of this note is that this doctrine should be rejected; in many situations it is doubtful if the doctrine of a single criterion is even sensible, let alone indispensable.

A broad and useful definition of a criterion is behavior, or consequences of behavior, that one wishes to predict. The fallacy of the single criterion lies in its assumption that everything that is to be predicted is related to everything else to be predicted—that there is a general factor in all criteria accounting for virtually all of the important variance in behavior at work and its various consequences of value. Considering that there are two broad classes of criteria, satisfaction and performance (Katzell, 1957), this assumption is obviously silly in view of the often-noted failure to find significant correlations between these two classes of variables (Brayfield and Crockett, 1955). Even within the job performance domain, the assumption of a general factor is frequently not tenable. Within the domain of job satisfaction, evidence does suggest the existence of a general factor (Wherry, 1958). The same evidence, however, also points to a number of other reasonably invariant factors which also account for appreciable portions of the total satisfaction variance.

Where criterion elements or dimensions are shown to be related, then there may be some point in combining them into a general composite. Where they are clearly independent, however, then prediction should also be independent.

JOB AND NEED ANALYSIS

The identification of criterion elements and the determination of their independence levels can be accomplished through some form of factor analysis. Factor analysis, however, requires data. The data from which clear identification of criterion elements must come are collected in job analysis and in the analysis of organizational needs.

The basic question of job analysis is, "What is the nature of the behavior called for by the job or the employment situation?" A corollary question in criterion selection is, "What is the degree of correspondence between the behavior called for and the behavior actually exhibited?" In analyzing needs, the basic question is, in a value-judgment sense, "What consequences should on-the-job behavior produce?" The corollary question for criteria is, "To what degree does each individual's behavior actually lead to these desired consequences?"

These questions indicate that criteria can be classified either as behavior data or as result-of-behavior data. Such a classification is related to, although not identical with, the job analysis classification suggested by McCormick (1959) with his job-oriented and worker-oriented job elements. It seems quite plausible that future research may find that job-oriented job analysis may be preferable for identifying the majority of the result-of-behavior criterion dimensions, and that these will be what Ghiselli (1956) termed the static dimensions, while the worker-oriented job analysis may be more effective in identifying the behavior dimensions, and that these may prove to be the "dynamic dimensions." This is, of course, mere speculation. What can be said now with conviction is that much work must be done to improve techniques of job analysis, and to develop and improve techniques of situational need analysis, before such dimensions can be clearly identified, isolated, and objectively and reliably measured.

THE ROLE OF JUDGMENT

The development of better techniques of job analysis with resulting increased emphasis upon actual worker behavior does not remove judgment from the process. The complete process of criterion development calls for two varieties of value judgment: (a) the judgment that a particular form of behavior or result of behavior is good or desirable [1] and (b) that one behavior or result is more or less equally desirable compared with another.

A major problem in criterion development, therefore, is to find means by which these judgments can be improved. A whole program of personnel research will stand or fall upon the adequacy of the initial judgments which are made.

The first type of judgment can be made more effectively by persons who are well informed. The fund of information should include a clear formulation of the objectives of the organization (Fiske, 1951; Bass, 1952)—a matter for top level policy determination. It should also include definite facts about the relationships between specific behaviors on any given job and those organizational objectives—a matter for objective, descriptive research. Fiske (1951), as a matter of fact, suggests that clear policy plus competent research can eliminate the need for any more value judgments in criterion development.

This continues the fallacy of the single criterion by overlooking the second type of judgment, the designation of certain behaviors or results (criterion elements) as more or less valuable than others. It is this judgment which is most easily improved by the very simple and convenient expedient of postponing it as long as possible.

The sequence of events suggested here as typical of the present state of affairs places this judgment quite early in the personnel research process. This judgment is usually made prior to any research activity save that which may be associated with job analysis. The reliability of this judgment can be improved, however, by postponing it until validation data are in and until current needs of the organization, *at the time a prediction must be made,* are known.

This proposal can be made more explicit by suggesting a new sequence of events in criterion development and subsequent personnel research:

1. Analyze the job and/or the organizational needs by new, yet-to-be-developed techniques.

2. Develop measures of actual behavior relative to the behavior expected, as identified in job and need analysis. These measures are to supplement measures of the consequences of work—the so-called objective criteria commonly tried at present.

[1] Typically, this is a linear judgment; that is, it is judged that if more of something is good, then still more of that something must be better. The wisdom of this assumption needs to be questioned, but that, happily, is beyond the scope of this paper.

3. Identify the criterion dimensions underlying such measures by factor analysis or cluster analysis or pattern analysis.

4. Develop reliable measures, each with high construct validity, of the elements so identified.

5. For each independent variable (predictor), determine its predictive validity for *each one* of the foregoing criterion measures, taking them one at a time.

The first kind of judgment—that a given behavior or its result has some degree of value—is still implied in step 2. Any judgment of the relative importance of those considered to *be* important does not appear in this sequence of research events. This judgment would not be made, in fact, until administrative use is to be made of the research results. If we are concerned with selection research, for example, the judgment would be postponed until the time an employment decision must be made. The employment manager could, for each applicant, make a series of predictions about the kind of behavior to be expected or about the organizational consequences of hiring the applicant. It is likely that, for any given applicant, some predictions would be favorable and others less so. Other applicants might show different patterns, being predicted to be more successful where a first applicant seems risky. Only now, in this proposed sequence, is the judgment of the relative importance of the things to be predicted actually made. At this point, that judgment has the benefit of knowing the relative validities of the various predictions that are possible and the further benefit of knowing the immediate situational needs of the organization.

CONCLUSION

Improvement in personnel research of all kinds—including selection and placement but not these exclusively—will not be dramatic until the criterion problem is solved. This note has made no real, empirical progress in reaching that solution, but it has attempted to point out a serious obstacle and the path for avoiding it.

The basic argument of this paper is that (a) there are in many personnel situations dimensions of job performance and of performance consequences that are quite independent of each other, and that (b) the relative importance of these independent criteria ought not be judged prior to validation research—as is so commonly done in the development of "composite" criteria—but ought to instead be judged *after* the empirical data are in, at the time these data are to be used. The suggestion is that this will result in clearer criterion definition, more reliable criterion measurement, and greatly improved criterion prediction.

REFERENCES

Bass, B. M. "Ultimate Criteria of Organizational Worth." *Personnel Psychology*, V (1952), 157-174.

Brayfield, A. H., & Crockett, W. H. "Employee Attitudes and Employee Performance." *Psychological Bulletin*, LII (1955), 396-424.

Fiske, D. W. "Values, Theory, and the Criterion Problem." *Personnel Psychology*, IV (1951), 93-98.

Ghiselli, E. E. "Dimensional Problems of Criteria." *Journal of Applied Psychology*, XL (1956), 1-4.
Katzell, R. A. "Industrial Psychology." *Annual Review of Psychology*, VIII (1957), 237-268.
McCormick, E. J. "Application of Job Analysis to Indirect Validity." *Personnel Psychology*, XII (1959), 402-413.
Nagle, B. F. "Criterion Development." *Personnel Psychology*, VI (1953), 271-289.
Thorndike, R. L. *Personnel Selection*. New York: John Wiley & Sons, 1949.
Toops, H. A. "Selection of Graduate Assistants." *Personnel Journal*, VI (1928), 457-472.
Wallace, S. R., & Weitz, J. "Industrial Psychology." *Annual Review of Psychology*, VI (1955), 217-250.
Wherry, R. J. "The Past and Future of Criterion Evaluation." *Personnel Psychology*, X (1957), 1-5.
Wherry, R. J. "Factor Analysis of Morale Data: Reliability and Validity." *Personnel Psychology*, XI (1958), 78-89.

36

Evaluating the Results of Supervisory Training

Theodore R. Lindbom and Wesley Osterberg

American industry today is offering more training in supervisory principles and methods than in any other single area. This training is offered with confidence and received with understandable approval, but if you ask questions about results, you are likely to be met with either silence or a sales talk.

Though a great deal has been written about supervisory training, very little has been published about research in this field—chiefly because little has been done. Looking at the literature, you will find descriptions of the various training methods and types of programs that companies have used. You will also find opinions on such subjects as what should be included in training programs, how these should be installed, who should be trained, and what methods should be utilized. Some of these descriptions and opinions, particularly when they come from competent authorities, certainly are not without value. However, the personnel executive seeking objective research findings to help answer some of the many questions that arise in planning supervisory training will find the literature disappointingly meager.

This lack of research evidence is probably most striking if one seeks to answer

Reprinted by permission of the publisher from "Evaluating the Results of Supervisory Training," by T. R. Lindbom and W. Osterberg, *Personnel*, November 1954. © 1954 by the American Management Association, Inc.

Evaluating the Results of Supervisory Training 429

the basic question. "Do formal supervisory training programs have any validity?" —or, in terms more often used in business, "Do they pay off?"

Vital as this question is, there have been few organized attempts made to answer it. If we are to get adequate answers, therefore, we must seek more precise evidence than has thus far been available. The relative effectiveness of various techniques cannot be determined without some form of evaluation. Moreover, existing techniques of no validity are likely to be perpetuated unless careful study shows them to be ineffective. Without evaluation, company managements may continue to pour money into training programs that actually do them, the supervisors, and the employees no good at all.

THREE LEVELS OF EVALUATION

There are, roughly, three alternative levels at which efforts to train supervisors can be evaluated. First, at the classroom or training room level; second, by the supervisor's actual behavior on the job; third, by the behavior of the supervised employees. (There is also the alternative of "no evaluation at all," which, sadly enough, is most widely accepted. New methods and techniques are uusally reported without any objective appraisal. Training a group of foremen in "better" methods of supervision is apparently such a "logical" method of getting better performance that no effort is made to assess the results.)

1. The Supervisor's Classroom Behavior

The first and most popular level at which to evaluate is in the training room or classroom. Two sub-levels can be distinguished. One is simply to give the trainee a pencil-and-paper achievement test *after* he completes the course and arbitrarily set a per cent of correct answers as a passing grade. The other expands the first procedure to include both a "before" and an "after" measurement, with the same or equivalent forms of a pencil-and-paper test. In this second method, some arbitrary increase or a statistically significant increase in score is usually taken as an indication that the program has succeeded.

A further modification of this method requires the conference leader or trainer to rate the trainee before and after the training program. Such factors as sociability, participation in the conferences, and ability to get along with others in the group are rated. Another modification, not often used, has the trainees rate one another. Trainees, too, are often asked to rate the course itself.

An example of evaluation at this first level has been reported by Katzell.[1] He worked with a group of supervisors above the rank of foreman, but below the executive level. The course, consisting of eight weekly meetings of four hours each, concentrated on the application of psychological principles to leadership. Using the before-and-after technique of measurement with File's "How Supervise?" test, he found a significant gain in score after training. He also reported that trainees rated the course favorably.

[1] Katzell, R. A., "Testing a Training Program in Human Relations," *Personnel Psychology*, 1948, V. 1, pp. 319-329.

The disturbing weakness in the methods described above is that an assumption must be made that performance in the training room is related to performance on the job. Otherwise, of course, the evaluation at the training room level has no real meaning.

There are some kinds of supervisory training where classroom examination might be the most feasible measure of results. For example, a course of training in basic economics, not intended to be passed on to employees or to cause any change in the supervisor's on-the-job behavior, might be so classified. However, such training efforts are small in number compared to those in which the supervisor's relationship to his employees is the focus of the training.

2. The Supervisor's Behavior on the Job

A deeper level of evaluation, going a step beyond the classroom, is appraisal of the supervisor's on-the-job behavior. This can be carried out in a variety of ways. One method is to have a trained observer spend time in the department. This has the advantage of thoroughness, but usually raises the question: To what degree was the supervisor's behavior influenced by the training, and to what degree by the presence of the observer?

Reports of superiors provide another basis for checking behavior on the job. The use of such reports, however, often introduces problems of bias, differing standards, and other inaccuracies commonly found in ratings. On the other hand, one advantage of this procedure is that it involves supervisors in the training program through their participation in the evaluation.

Still another means of measuring on-the-job conduct of the supervisor is to analyze the opinions of his subordinates. This is ordinarily done by interviewing the workers or by having them prepare some form of written rating of the supervisor.

A final method of evaluation is to have the foreman report on his own behavior, either through a questionnaire or through interview. Osterberg and Lindbom [2] have reported a study using this method. Sixty-four oil company supervisors who had received five half-days of training in human relations were later surveyed by questionnaire. Among other things, they were asked about changes in their behavior that they felt resulted from the training program. Ninety percent of the respondents reported such changes.

Implicit in self-evaluation, of course, is the problem of unreliability caused by both conscious and unconscious distortion. From the training standpoint, however, this method is most likely to highlight strong and weak points of the program, because the person reporting on behavior changes is the supervisor himself, and he speaks in terms of training he has just experienced.

It must be recognized that evaluation at this level, no matter how it is made, still does not get directly at the final objective of most supervisory training—better *employee* performance. We still must assume that improved supervisor behavior will result in improved employee behavior—and there is no guarantee that this assumption is valid. On the other hand, evaluation of supervisors' behavior on the

---

[2] Osterberg, W. H., & Lindbom, T. R., "Evaluating Human Relations Training for Supervisors," *Advanced Management*, 1953, V. 18, pp. 26-28.

job would be the ultimate for training which is intended merely to change supervisors' but not subordinates' work habits—for example, a course in record keeping or report writing.

3. Employees' Behavior on the Job

Most supervisory training programs, which are concerned with the supervisor's relationship to workers, can best be assessed at the third level of evaluation. Here, behavior of the supervised employees is studied directly and major assumptions are unnecessary in determining the effect of supervisor training on employee performance.

Several aspects of employee behavior can be used to check the results of supervisory training. In most situations, employee productivity—quantity or quality of output, or a combination of both—is probably the most meaningful measure. Productivity ratios, however, are often difficult and costly to calculate, and are affected by many factors other than supervision. The aspect of employee behavior to be observed is determined largely by the nature and objectives of the supervisory training program. Success of a course in developing safety-consciousness among workers, for example, might be evaluated by comparing accident frequency and severity records before and after the training program. Human relations instruction might be appraised by measuring morale of employees before and after training. An example of this method of evaluation has been reported by Lindbom.[3] A standardized employee attitude scale was administered before and three months after a 16-week training program in human relations which was attended by all 50 members of management in a small insurance company. Attitudes of the 126 employees supervised by these trainees were found to be significantly more favorable following the training program.

Direct measurement is not the only way to gauge employee behavior. The supervisor can be asked, by personal interview or by questionnaire, to report on workers' behavior changes. His superior can also be questioned, or an outside observer may be used to obtain the needed information.

CONCLUSIONS

Though the authors believe the third general level of evaluation—getting at the behavior of those supervised—is probably the most meaningful, each of the three levels has its place in different kinds of programs. And any one type of evaluation is better than none at all.

When appraisal at all three levels can be made, it may indicate where a program is breaking down, if desired results are not forthcoming. If no change were found at the classroom measurement level, for example it might indicate that the subject matter simply had not been absorbed. Good classroom results, but no change in the supervisor's on-the-job behavior might mean that the supervisor had not been "sold" on putting into practice what he had learned, or that he was

---

[3] Lindbom, T. R., "Supervisory Training and Employee Attitudes," *Journal of Industrial Training*, 1953, V. 7, pp. 17-20.

not able to do so. Good results at both the classroom and supervisor levels, with no subsequent change in employee behavior, might raise serious doubts about the original choice of training subject matter. Whatever the level of evaluation, something can usually be learned about the effectiveness of the training program.

In any event, if this relatively young area of training is to come of age, more and better techniques of evaluation must be devised for use in uncovering the facts needed for modification and improvement of training. Unless evaluation is given far more serious attention in the years ahead than it has thus far received, the future of supervisory training in American business and industry is anything but bright.

37

Progressive Levels in the Evaluation of Training Programs

A. C. MacKinney

It is an indication of the scientific orientation of personnel management today that the problem of evaluation is receiving increasing attention. This is evident in many areas of personnel practice but nowhere more conspicuously than in management training, where the evaluation of methods and results is a matter of urgent concern.

But evaluation is only the first step toward applying the scientific method to personnel problems. In appraising management training, for example, the question is not merely whether training is effective, but *how* we determine whether it is or not: Is the method of evaluation itself valid?

This question has been discussed in a number of recent articles.[1] It is my purpose here to carry the discussion further and to suggest criteria for judging the validity of training evaluations. The point at issue is not the actual content or results of training programs but the validity of the measures used to evaluate them. Once it is recognized that training can be evaluated at different levels, it should be possible to improve our evaluative procedures and to get more accurate information about the effectiveness of training programs.

Reprinted by permission of the publisher from *Personnel,* November/December, 1957. © 1957 by the American Management Association, Inc.

[1] See, for example, T. R. Lindbom and W. Osterberg, "Evaluating the Results of Supervisory Training," *Personnel,* November, 1954; D. M. Goodacre, "The Experimental Evaluation of Management Training: Principles and Practices," *Personnel,* May, 1957; and P. C. Buchanan, "Testing the Validity of an Evaluation Program," p. 78 of this issue.

In speaking of "levels" of evaluation, it is implied that different procedures can be used to evaluate training, that these procedures are variable in quality, and that they can be arranged in a hierarchy from best to worst. It is this hierarchical arrangement that yields the various levels to be referred to here. The justification for making such a distinction is that the quality of the information provided at each level is different, the so-called higher levels giving better evaluative information. I shall advocate in this paper that we raise the level at which training activities are evaluated.

To begin with, let us consider various systems of levels worked out in previous studies.

LEVELS OF EVALUATION

In the Lindbom and Osterberg system,[2] for example, levels are classified according to the kinds of behavior being evaluated: (1) the trainee's classroom behavior, (2) the trainee's on-the-job behavior, and (3) his subordinates' on-the-job behavior. This classification seems to be a hierarchy of levels, though it is not explicitly called such by its authors. No. 1 is low, for example, and No. 3 is high; that is, the on-the-job behavior of the trainee's subordinates gives much better information than the trainee's classroom behavior. I think that most people who are involved in personnel research would have little quarrel with this opinion.

It is important to note, however, that this method is merely one way of classifying levels of training evaluation. Another possibility is the use of the two-point scale of objective *versus* subjective, the objective evaluation being regarded as separate from the interpretation, biases, and feelings of the person doing the evaluation, and the subjective as reflecting these various elements.

Or we can talk about formal *versus* informal evaluation as another major classification. In formal evaluation, there is a specific plan by which the training is evaluated; informal evaluation is casual, based perhaps on comment or conversation. It is evident that the objective and formal levels are very much alike, as are the subjective and informal levels.

These two simplified classifications of levels of evaluation may be meaningfully contrasted with the classification system discussed by Goodacre.[3] This is based upon a three-unit scale. At the lowest level, training is evaluated in terms of the attitudes of trainees as measured by an attitude scale. The assumption is that effective training should be reflected in more favorable trainee attitudes. At the second level, evaluation is made in terms of the knowledge acquired by the trainees. Effective training should impart skills or knowledge, and these can be measured by achievement tests. The highest level, according to Goodacre, is actual job performance. Effective training should increase the trainee's job proficiency. This level can best be reached, he suggests, through the rating of job performance by the trainee's immediate superior. Here, then, is a considerably more complex system of levels of evaluation.

At this point, let me introduce an alternative classification which is related

[2] Lindbom and Osterberg, *op. cit.*
[3] Goodacre, *op. cit.*

primarily to the design of the evaluation used. In brief, this means that there are various ways of designing the evaluation procedure and that these determine the quality of the information we get.

At the top of our hypothetical scale is the *controlled* experimental study. This is a design in which two groups are used, the one to receive training, and the other to act as a control group. The procedure is as follows: (1) A "before" proficiency measure is taken for both groups; (2) one group is trained while the other is left on the job; (3) an "after" proficiency measure is taken for both groups; (4) the "before" is subtracted from the "after" for both groups to measure the gain in proficiency. If the training did any good, the trained group should have gained significantly more in job proficiency than the untrained group.

THE CONTROLLED EXPERIMENTAL STUDY

It may be asked why a control group is necessary. This can be answered by pointing out that certain changes may take place in an employee's job proficiency simply as a result of remaining on the job and that such changes have to be taken into account in any experimental evaluation of a training program. If it should turn out that the group remaining on the job gained as much in job proficiency or more than the group that was trained, there would be no justification for the additional expense of the training program. Odd as it may seem, it is not at all uncommon for the control group to gain as much as the trained group in job proficiency.

The controlled experimental study uses acceptable criteria of performance. A criterion is a measure of proficiency on the job. An *acceptable* criterion is one that meets the various standards set up for the evaluation of criterion measures.[4] Most people in personnel research would agree, I believe, that the most difficult problem in this field is that of finding acceptable criteria. We shall return to this question later in this article.

In addition, the controlled experimental study utilizes acceptable statistical methods to aid in the interpretation of the results of the evaluation. Since certain changes can take place in job performance as a result of chance alone, statistical procedures are necessary to determine whether or not the changes observed after training might reasonably be considered as other than chance effects.

There is a further consideration in this experimental design. We need some sort of assurance that the changes taking place in the trained group are not merely a result of the amount of attention the trainees receive. As the studies at Western Electric showed, changes in productivity may occur not only as a result of changed working conditions but also in consequence of the amount of attention given to the workers. To eliminate this factor, the control group should receive, so far as is possible, as much attention (exclusive of training) as the trained group.

It is important to note that there is a sizable range within the level of the controlled experimental study because the criteria used for evaluation can actually vary in quality and still be acceptable. This point will be discussed later in considering levels of training evaluation as related to the relevance of the criterion.

---

[4] See, for example, R. L. Thorndike, *Personnel Selection*, Wiley, 1949. The whole of Chapter 5 is excellent, especially pages 124-132.

LOWER LEVELS OF EVALUATION

At the second level in this classification system is the evaluation of training by means of the trained group only. This is not an acceptable experimental design. It must be admitted, however, that this type of evaluation is widely used and is perhaps better than no evaluation at all, assuming that adequate criterion measures and analysis procedures are employed. The evaluation design is a simple one. "Before" and "after" measures are compared and any gain is attributed to the training.[5] Statistics are used to test the significance of the gain from zero.

At the lowest level, we have the evaluation of the trained group only, as above, with the criterion measure taken *after* training, but not before. In this procedure, no statistical analysis is possible and the interpretation of the end result is simply a matter of intuitive judgment. All that can be done is to look at the information gathered after training and make some sort of guess as to whether it is "good enough." Obviously, this is an extremely weak method.

Of these two latter evaluation procedures, neither can be recommended. *In terms of experimental design considerations, the only way to evaluate training properly is by the controlled experimental study.* It is my urgent recommendation that training people should not waste their time and the company's money on anything short of a controlled study utilizing acceptable criteria. Even if such a study requires outside assistance, it is the only way of doing the kind of job that needs to be done.[6]

THE RELEVANCE OF CRITERIA

We noted earlier that the criterion is a measure of proficiency on the job, a measure that tells us whether the performance of an individual or group is effective. In this section, we shall discuss levels of evaluation as they are influenced by the relevance of the criteria used in evaluating training.[7] "Relevance" is roughly equivalent to quality. A relevant criterion is a good criterion, one that accurately reflects the contribution of the group or individual to the organization and does not contain any extraneous factors.

The system of levels based on criteria, like that based on experimental design, represents a hierarchy. The criteria system can be thought of as being superimposed on the design system. In other words, a highly relevant criterion may appear in a good experimental design or in a poor experimental design. A good

---

[5] This may not be valid, as noted earlier.

[6] Selecting a consultant to provide the help needed is not so simple as it sounds. Many of them, though thoroughly honest and reputable, operate with a single theoretical bias that renders them helpless when faced with a problem requiring some experimental work. There is no single pat answer to all personnel problems. Each problem is unique and requires a unique answer that is ground out experimentally. Beware of the consultant who has *the* system!

[7] For a full discussion of this question, see Thorndike, *op. cit.*, pp. 125-127.

design usually has good criteria, but a less relevant criterion may accompany a good design simply because there is no other criterion that might reasonably serve. Hence both extremes of the experimental design system can be used at any level of the criterion system.

The highest level in this system comes about with the use of *objective performance scores* as criteria. Such measures are generally the most relevant ones we can use. The best objective performance score would be a measure of output automatically recorded by machine. Also acceptable are measures of scrap, defects, material or tool usage, manpower cost, etc. It is obvious that there is considerable variability within this level inasmuch as a direct measure of output is better than a less direct measure such as tool control.

Objective performance scores may be determined either for the trainee himself or for his subordinates. It is generally considered to be stronger evidence of the effectiveness of a training program if it is followed by improved performance not only on the part of the trainee but of his subordinates as well. In addition, if the trainee is in a supervisory capacity, evaluation may have to be based upon the performance of his subordinates for the simple reason that no objective criteria are available for supervisory personnel.

There are several dangers in the use of objective performance scores, however. Perhaps the most important of these is bias in the measure, that is, the influence to some unknown degree of factors other than the trainee's contribution to the organization.[8] Some bias in the measure is probably inevitable. Consider, for example, the interdependence of defects, particularly in an integrated assembly-line operation. What one man does early in the process may affect what another man does later on down the line. If defects are used as the criterion, the second man is being evaluated in part by the first man's performance. This is only one of many examples of bias in objective performance scores.

SUBJECTIVE ESTIMATES OF PERFORMANCE

The second level in this system is provided by subjective judgments or estimates of job performance. Here we are not talking about trait rating [9] but about the trainee's actual on-the-job behavior.[10] Trait ratings would be on some other level in this system and will not be treated here.

With subjective performance estimates, as with objective performance scores, there is considerable variability. The effectiveness of a man's performance can be judged by a formalized procedure or by casual comment. The latter extreme is characterized by "Joe's a good man." The former is represented by a well con-

---

[8] For a full discussion of this point, see Thorndike, *op. cit.*, pp. 130-131.

[9] Traits are inferred personal characteristics presumably existing within the individual and hence very well hidden. They have no objective reality of their own but are inferred from certain observable behaviors. "Motivation," "morale," and "maturity" are three common examples. As a general rule, it is wise to avoid rating such abstractions.

[10] Performance estimates may be made by anyone familiar with the trainee's job performance—subordinates, peers, or, most commonly, the supervisor.

structed appraisal program.[11] Subjective performance estimates, like objective scores, may refer either to the trainee or to his subordinates.

The third level comprises evaluations of the trainee's knowledge and understanding of the content of the training course. Obviously such evaluation can be made only of trainees themselves, not of their subordinates, for example. The procedure is to use some kind of achievement measure similar to those in college courses to find out how much the trainee has learned as a result of his exposure to training. As most of us know from our own academic experience, achievement measures are variable in quality. A good test is not easy to construct; certainly it cannot be thrown together as an afterthought but requires considerable study and revision.

At the final level in our system, the criteria are opinions and attitudes. Here again there is a hierarchy within the level: opinions of the training course may come from the trainee's subordinates, from the supervisor, from the trainee, or from the trainer himself. The last source is the least dependable, of course, since it is only natural for the trainer to be favorably impressed with his own work.

Though opinions and attitudes may appear to be similar to performance judgments as criteria for evaluating training, they are in fact quite different. Performance estimates refer to actual changes in job performance that come about as a result of exposure to a training program, whereas opinions refer to the quality of the training program.

Attitudes and opinions may represent sound and thoughtful evaluations or casual snap judgments. In fact, they can be very effective criteria if used together with a well constructed attitude scale and a controlled experimental study. Certainly this combination would rate a higher place in our hierarchy of levels.

COMPROMISE MAY BE NECESSARY

The purpose of this paper is to point out that training may be evaluated on different levels and that the quality of the evaluation is better at the higher of these levels. It recognizes, however, that, as those familiar with applied research in industry know, compromise procedures are often necessary and in such instances, all that can be expected is the closest possible approximation to the best study design.

[11] A great deal of the confusion about appraisal would not exist if we were more careful to differentiate between *program, method,* and *form.* The *program* is the total appraisal effort including method, form, control, development, follow-up, and so on. The *method* is the system used to make the judgments, such as forced choice, rank order, and so on. The *form* is just a piece of paper to write the judgments on. Too many people, I fear, think the piece of paper is the whole program. It isn't.

Chapter Eleven

Training Apparatus

The word *"apparatus"* is used here to signify the general class of devices, aids, and materials that are used in the training process. Technically speaking, it includes such common materials as books and slates—for centures the bulwark of classroom instruction—although we shall limit our attention in this chapter to complex simulators, "teaching machines," and computer-based devices, chiefly products of the last quarter century. Underlying all of these devices are educational philosophies, theories, and methods; one cannot discuss "hardware" without giving some attention to these related matters as well.

We now face a technological revolution in the training center and classroom, just as we do in other facets of our existence. Computerized classrooms, "talking" typewriters, and the like are not future possibilities; they are present realities. The question is, are we ready for them? Can they and *will* they be used to advantage in furthering our educational goals?

Many people in the field of education have expressed serious doubts on this score. For one thing, most of the innovations have not been around long enough to permit an adequate evaluation. And for another, it is doubtful that evaluative data—even if available—would be the chief factor in development and promotion of these devices. The commercial potential is a far more dominant influence. Ready or not, therefore, society can expect growing pressure from manufacturers to promote all manner of new devices and programs. Manufacturers, rather than educators, would thus become our chief educational policymakers—a very disturbing thought for some people.

While such concern is far from groundless and should not be taken lightly, one must put the entire matter in perspective. Certainly, inadequate evaluation is not unique to these *newer* techniques. We saw in Chapter Ten how little progress has been made in the assessment of industrial training programs. Likewise, one would be hard pressed to build a case for "traditional" classroom methods on the strength of existing objective data. It is somewhat ironic that those who are most critical of the evidence supporting new techniques are willing to take their *own* methods completely on faith! In any event, we can only hope that recent developments will provide the impetus for re-examination of *all* methods, old and new alike. Only in this way can we expect to make the best use of our growing repertoire of training alternatives.

The articles in this chapter provide both a description and a critique of three prominent instructional "devices" which have particular relevance for industrial

References will be found in the General References at the end of the book.

training. In addition, we hope through these papers to illustrate some of the principal research issues that have been—and should be—explored in connection with these techniques. The engineering psychologist's interest in these issues stems mainly from two considerations: first, the design of a training device is itself a man-machine system problem; and second, the skills promoted in such devices are hopefully the ones of greatest relevance to real-world systems behavior.

Gagné opens the chapter with a discourse on training devices in general and *simulators* in particular. By far the oldest of the "newer" techniques, simulation training has as its basic philosophy the replication of essential features of the real-world task in a more convenient setting, i.e., one in which better "control" can be exerted and performance can be more easily measured. We discussed this concept briefly in Chapter Eight in connection with multiman-machine system performance. There, the simulator was introduced as a device for *studying* complex system behavior; here it is viewed as a *training* device. In either case, the essential point is that the behavior on the simulator must *generalize* to the class of tasks simulated. This, as Gagné points out, is not necessarily a matter of reproducing *physical* details of the real-world situation. Determining which features are essential for the transfer of skills is by far the most difficult—as well as the most important—aspect of this approach.

Gagné discusses the topic of simulation from the standpoint of research issues surrounding its development and use. We offer here a somewhat amended version of Gagné's list of issues which we feel encompasses the most relevant considerations associated with *any* training device.

(a) *Adherence to training objectives.* A training device should be designed to meet a specific set of training objectives. The device is but one part of a training program, and should be supported by both job analysis and criterion development. Only then can inferences be made concerning its relevance for objectives specified. For example, Gagné maintains that the objectives of performance *improvement* are distinct from those of *measurement* and could very well require different emphases in the simulation design. The point, then, is that the effectiveness of any device depends heavily upon research aimed at establishing its relevance for system goals; the design cannot be judged solely on its own merits.

(b) *Fidelity of simulation and transfer of training.* We discussed this point in Chapter Nine. The idea is that when transfer of training is the chief objective —as is the case in most training programs—fidelity should be in terms of *functional* rather than *structural* considerations. This characteristic can only be established empirically. Adams (1961), for example, cites several investigations on response precision in which a substantial amount of positive transfer occurred without a particularly high level of physical "fidelity" in the training device (the underlying similarity was not immediately apparent). Briggs and his co-workers have reported similar findings with respect to vehicular control (Briggs, Fitts, & Bahrick, 1958; Briggs & Waters, 1958).

(c) *Control.* One of the major advantages of simulation is that it gives the training specialist control over the *training* environment when control over the *real* environment is difficult or impossible to achieve. For example, it might be desirable to arrange for certain weather conditions to occur repeatedly, or to prepare the trainee for potentially dangerous environments, or to operate in "fast" and "slow" time. The only satisfactory means of doing this may be through simulation. Of course, as we discussed earlier, control often comes only at the expense of fidelity. The significance of such trade-offs is itself a worthwhile topic for research.

(d) *Performance measurement.* The value of a simulator is usually no better than the measurements it affords. As Gagné points out, it is important to establish the *reliability* as well as the validity of these measures, especially if a primary objective is performance evaluation. Although this need for reliability seems perfectly obvious, it is often overlooked in practice.

(e) *Cost.* We hesitate to include the factor of cost in the same context as control and precision. We must include it somewhere, however, because it represents a very real and important consideration in simulation and training device decisions. It has been argued that cost effectiveness is the most logical basis on which to determine the usefulness, for example, of simulation vs. on-the-job training. Even under the best of conditions, however, it is difficult to arrive at valid cost estimates, and the conclusions reached as a result may be quite misleading. Actually, this brings us right back to the old criterion problem. Dollars and cents are certainly *one* criterion, but by no means the *only* one. Suppose, for example, one device increases production, but another increases job satisfaction. Are all the factors worth weighing reducible to an ultimate cost index (e.g., production, accidents, absenteeism)? Or is there some inherent value in a worker's sense of well-being?

In the second paper, Abma discusses a number of issues associated with research and practice in programmed instruction (the so-called "teaching-machine"). While most of the considerations fall within the scope of the outline just presented, a few are unique. For example, one of the most active research interests has been the direct comparison of programmed learning with more traditional methods of instruction. As Abma explains, such comparisons, fraught as they usually are with prior biases and experimental hazards, have not been particularly illuminating. More complete accounts of this research may be found in Galanter (1959), Lumsdaine and Glaser (1960), and Smith and Moore (1962). Fortunately, the trend today seems to be toward less emotionally-charged and more promising issues, such as how best to incorporate the strengths of programmed instruction into the overall curriculum, and how to prepare the most effective programs. There is certainly no dearth of important questions which *are* suitable topics for experimental research.

One interesting aspect of these discussions on simulators and teaching machines as training devices is the emphasis that is placed on *individual differences*. In a sense, the field of training has made a full turn back to the concept of individualized instruction. We find ourselves concerned with individual differences in acquisition and transfer as well as with the effects of learning variables on group performance. This trend has been an integral part of the programmed instruction approach from the outset, and much of today's literature centers around concepts related to individual ability, such as *pacing, branching,* and the *needs* of the individual trainee or student. In our third article, this trend is carried to the ultimate with Atkinson and Wilson's description of the astounding promise of computer-assisted instructional techniques. Visionaries feel that capabilities such as free interaction between the individual student and his computer-linked console will revolutionize education for children three to six years of age as well as for technicians learning new concepts in rapidly expanding fields such as aerospace engineering. Further material on computerized instruction can be found in another excellent article by Atkinson (1968). There is, however, one catch in all this discussion of training aids: even though the potential of programmed instruction is undeniable, we must not become so overwhelmed by the "bells and whistles" that we lose sight of the main purpose of these innovations, which is the acquisition and transfer of knowledge and skills. We must constantly review our criteria, and continue in our effort to gather evaluative evidence on *all* our devices and techniques.

38

Training Devices and Simulators: Some Research Issues

Robert M. Gagné

Research on training devices is an area peculiarly fitted to the talents of the experimental psychologist, first, because he knows the field of learning, and second, because he usually likes to tinker with gadgets. The combination of these interests and skills has not only made training devices research attractive to many psychologists, but has also created a demand for their services. This has been true for a number of years, in fact, since World War II when training devices were used widely in the armed forces. And this interest and demand have led to the establishment of a number of agencies and programs in which the psychologist participates in the design, development, and evaluation of training devices. The Army, Navy, and Air Force all have research programs in these areas.

Many kinds of research activity are evident from published reports. Probably a great many changes, presumably desirable ones, have been made in the design of training devices as a result of psychologists' participation in decisions about design characteristics. There are also a number of studies on the "effectiveness" of training devices (e.g., 13, 24, 30) which are generally characterized by sound but unstartling conclusions. The discussion of research problems by Wolfle (31) has remained the major source of systematic thinking for several years. Never-

From *American Psychologist,* **9,** 1954, 95-107. Copyright 1954 by the American Psychological Association, and reproduced by permission.

The opinions and conclusions contained in this report are those of the author, and are not to be construed as reflecting the views or indorsement of the Department of the Air Force.

theless, all these things do not constitute the kind of effort, nor the kind of scientific product which might be expected from research having such a high degree of intrinsic interest and so many ramifications into nearly all fields of psychology.

If one rejects a narrowly applied point of view, one can see that a considerable variety of research questions are relevant. Perhaps first in importance is the ubiquitous problem of how to measure complex human performances. The "criterion problem" has been with us for a long time. It has been solved in specific instances either by arbitrary determination or by selection on the basis of empirical data. But the need for a fundamental solution which seeks to account for the variances attributable to machine and man, and the relationships between behavioral processes and products, is pointed up clearly by the training device.

Bound up with the criterion problem is, of course, the perplexing methodological question of how to analyze job activities. For the kinds of performances to which most training devices relate, it is apparent that the methods of traditional job analysis are not particularly helpful. Yet the necessity of differentiating the critical activities from those which are routine and easy is one of the central problems actually faced by anyone who designs a training device, whether he is a psychologist or not.

Since in many situations a training device functions as a performance test, there is a need for the application of the principles and methods of test construction, analysis, and scoring. In fact, many problems of training device use point to the development of new principles and methodology, including perhaps a fresh examination of the concepts of reliability and validity.

Some of the most intriguing research problems in the design and use of training devices are fundamental problems of human learning and retention. When one inquires about the effectiveness of the characteristics of a device for training, one is really asking about the transfer of learning to some criterion performance. When one inquires how a training device may be designed to yield highest transfer, one is actually asking more intensively about the conditions of efficient learning. Since human beings are highly verbal, it is not too surprising to find many of the characteristics that can be manipulated in the design of a training device are those which may be presumed to change a self-instructional set, or a source of motivation. Other characteristics may determine the range of variety of responses practiced and the amount of information provided to the learner during practice. In other words, within the framework of training device research can be found some of the central problems of human learning.

SOME DEFINITIONS

It may be helpful at the outset to try to phrase some definitions and distinctions. Typically, there is a certain amount of disagreement over a terminology which comes partly from psychology and partly from engineering, with a little dash of administrative cliché thrown in for good measure. The following are working definitions and refer primarily to content of the present paper.

Training Device. As usually understood, *training device* refers to any piece of apparatus which is used for training individuals. A somewhat more precise and useful meaning, also common, is that a training device is used for the training of skills. Such a definition serves to distinguish the class training devices from the class training aids, which are objects and devices used to facilitate the presentation and teaching of informational knowledge. In this more restrictive sense, training devices are characterized by the possession of a *display* which presents information necessary to the correct operation of *controls,* which the student must learn to operate, in most cases during periods of practice on the device.

There are many occasions in which the actual equipment, rather than a substitute device, is used as a training device. A simple example is the rifle. In addition, no hard and fast distinction can be drawn between what piece of equipment is a training device and what is a piece of operational equipment. For example, an airplane used for basic flying training can certainly be considered a training device; yet an additional ground trainer may be used to establish skills used in the training airplane. What distinguishes a training device is not its appearance or construction, but rather *how* and *for what purpose* it is used.

Simulator. A simulator is generally understood to be a kind of training device which has a high degree of resemblance to operational equipment, particularly with respect to the display, the controls, and the way one affects the other when in operation. For example, a simulator used in training for a particular aircraft is expected to be like the aircraft in cockpit arrangement, furnishings, lighting, instrument panel, and controls, as well as in the effects which control movements have upon instrument readings or upon the attitude of a movable cockpit. It is, of course, well known that perfect simulation is never completely achieved, and that the more closely it is approached, the greater the expense is likely to be. It will be suggested later that the clearest usefulness of a simulator is that of proficiency measurement.

Evaluation. The phrase "evaluation of training devices" has a number of meanings which can be confusing indeed when used interchangeably. The following brief descriptions of evaluative activities may help to sort out these meanings.

(a) *Engineering evaluation* of a training device is usually done on the first manufactured model of a device when it is delivered to the agency which ordered it, for the purpose of determining whether the device meets required specifications, i.e., whether specified physical tolerance limits are met. This evaluation is usually done by engineers.

(b) *Field evaluation.* Just as is the case with a new weapon, it is reasonable that every new training device should have a tryout or evaluation "in the field." Many questions may be answered by such an evaluation, including those of acceptability of the device to instructors and students, ease of maintenance, facility of operation, and reliability of operation. Such questions are important, even though they can often be answered by relatively simple observation or by the collection of records over the interval of time when the device is in use. However, it would be a mistake to suppose that such data alone can yield any positive information about the effectiveness of the device for establishing desired skills.

The crucial question in a field evaluation is that of *training effectiveness.* The *training effectiveness* of any device for a particular job is the difference between measured performance on the job preceded by practice on the device, and measured performance on the job not preceded by practice on the device. This is the kind of evaluation for which there is a clear application of the technique of the controlled learning experiment and the measurement of transfer of training. It is often the most difficult experiment, either because of lack of a criterion performance measure or, frequently, for administrative reasons. It is reasonable to expect, though, that a psychologist's activities will always be based upon the knowledge that there is a clear-cut experimental method and a definite numerical answer to the question. "How effective is x hours of practice in device A in establishing the skills required in operation of equipment B?"

PERFORMANCE IMPROVEMENT VS. PERFORMANCE MEASUREMENT

When one looks at the situations in which training devices and simulators are used, one fact emerges with great forcefulness: these devices are used very frequently and extensively for the *measurement* of performance, as opposed to the improvement of performance through practice. A typical procedure in the use of instrument flying trainers, for example, is to require the student to "fly" standard missions with navigational and instrument "problems" which tax his knowledge and ingenuity. Another procedure consists in the systematic introduction of various failures in equipment which call upon the student to use emergency procedures. Now, it cannot be said that these procedures are designed to provide the kind of practice that is typical of actual aerial flight, because a great many more happenings of a critical nature are crowded into a ground trainer mission than would occur in the air. Instead, these procedures usually reflect a desire to "see what the student can do." In practice, this may be carried out by the instructor's observing how well the student performs in terms of some over-all judgment, by determining the limit of what the student can "take" in number of emergencies, or by some objective recording of his performance. However it is done, the aim is to measure proficiency.

To maintain that no training is accomplished when devices and simulators are used in this way would, of course, be incorrect. The important point, however, is that the procedures employed are usually not intended to give systematic practice, but primarily to provide hurdles or problems which sample the repertoire of knowledges and skills, as any good measure of proficiency is expected to do. There are many situations in which devices are employed in measuring proficiency, though this use is not always recognized or stated.

On the other hand, it is quite possible to separate the function of *performance improvement* from that of *performance measurement*. A device may be, and frequently is, designed for the sole purpose of providing practice in essential skills. Here it is obviously not essential for the device to possess a scoring system for the measurement of performance. There are many relatively simple devices in which the student is simply run through a series of exercises by an instructor, with perhaps some accompanying qualitative judgments about the student's improvement with practice. This type of device, though inelegant and perhaps inexpensive, can possess potentialities for performance improvement equal to those of more elaborately constructed devices which yield precise scores. Furthermore, its training effectiveness can as readily be evaluated by means of a transfer experiment.

Thus it is possible to conceive of a device which is primarily designed to measure performance, or one which is designed solely to provide opportunity for learning of critical skills. The two functions may or may not be present in the same piece of equipment. But the important point of the distinction is this: *Performance improvement may require different characteristics for effectiveness than does performance measurement.* For example, reliability of scores obtained from a device is relatively unimportant for improvement, but essential for measurement. Transfer of training from practice is the essential criterion for performance improvement, but obviously irrelevant to measurement.

This distinction also serves to put the question of simulation in its proper place —which is not conceived to be one of primary emphasis. According to this idea there are two essential questions: What characteristics of a device are essential for performance improvement? And what characteristics are essential for performance measurement? Within such a framework, simulation becomes merely one of many factors to be considered. It is important for training to the extent that it implies adequate representation of critical skills, but only to the extent that it does not impede the learning of these skills. It is important for performance measurement to the extent that it makes for validity, but only to the extent that it does not interfere with adequate reliability of measurement. A training device, in other words, cannot be justified on the grounds that it looks good (i.e., closely simulates an operational situation). It must do something, too, and that something is either training or measurement. Simulation becomes a secondary matter whenever it conflicts with these purposes.

Whether a device will be used for performance improvement or for performance measurement, or both, is an extremely important question which affects each and every stage in the life of a training device, from its initial planning through the various stages of development, evaluation, and use. It is also this question, we believe, which can bring the maximum of order to the planning and conduct of research programs.

RESEARCH ON TRAINING DEVICE DEVELOPMENT AND USE

What kinds of problems are generated when consideration is given to the design of a training device (or simulator) or to its use in a training program? These problems arise when decisions are made concerning what to build, as opposed to considerations of testing what is already built. It is reasonable that the first stage of this process is to determine what the job is like for which a training device is required. This involves some kind of *analysis of the job*, with the emphasis being placed upon the activities or behaviors present. It is also important that a device's characteristics necessary for adequate *performance measurement* be given separate consideration from those characteristics which make for *performance improvement*.

Analysis of the Task

Since the building of a training device or simulator is undertaken in the first place because of a need to represent the actual job (either more simply, less expensively, with smaller involvement of danger to the operator, etc.), it is obvious that some decisions must be made, at the very beginning, about what are the essential aspects of the job to represent. Although such decisions are always made when a training device is designed, they are not always made in a systematic manner. And sometimes they seem to be perverted, rather than clarified, by attempts to follow the principle: "Make the device as nearly like the actual equipment as possible."

In general terms, the way to go about describing and analyzing the job to gain knowledge of the requirements for training device design would seem to be to categorize the kinds of specific equipment-oriented behaviors and skills which are involved. To use a relatively simple example of an operator's job, flexible gunnery, it has been recognized for many years that some of the essential behaviors are tracking, ranging, and triggering (8, 9). The significance of this categorization of component behaviors may be very simply stated. It is impossible to design an adequate trainer for flexible gunnery unless one knows that these behaviors are, in fact, important in the job. The trainer is built, not to represent some abstraction called "flexible gunnery," but to create a situation in which these activities can be practiced on the ground under conditions leading to improvement in their performance in the air.

Some idea of the process of deciding what behaviors and skills are essential was recently gained by the writer when he participated in preliminary discussions of a projected helicopter trainer. The student helicopter pilot must, of course, learn to do many things. He must learn how to preflight the equipment; how to take off; how to coordinate rotor and engine speed; how to fly forward, backward, to the right, and to the left; how to correct for torque; how to hover, how to land; what to do in emergencies; and numerous other details. The first and most obvious point made in discussion was simply this: If one determined to build a trainer to represent all these behaviors, it would not be a trainer, but a helicopter. It became evident that what had to be decided was what were the critical skills which ground-trainer practice could be expected to facilitate. There was practically universal agreement among training experts that the most critical behavior was hovering, and the agreement on this human activity strikingly brought order and direction into the discussion. One could now speak of a "hovering-trainer," rather than a more abstract "helicopter-trainer"; the operating characteristics needed for such a trainer then became relatively easy to formulate.

What can the psychologist contribute at this stage of the process? First of all, he can insist that decisions about what aspects of the job a training device represents are arrived at by systematic consideration of operator activities, rather than by consideration of the equipment alone. Further, he can use his knowledge of human behavior to analyze and describe the categories of behavior which a job contains. But though he may do this, and perhaps succeed at it better than other specialists, it is this activity that makes him most keenly aware of the limitations of psychological knowledge.

What, in fact, are the categories of human behavior? Or even, of human equipment-operation behavior? This is one of the most challenging areas of research for the experimental psychologist. Few attempts have been made to bring scientific order into this field.[1] Although a kind of taxonomy must probably be involved, the important research problem appears to be the development of a theoretical system which will relate physical task variables to performance variables by means of conceptualized intervening processes. The lack of such a theory creates a void in this area of human behavioral knowledge. When he has developed such a theory, the psychologist will be able to say in what exact ways flying forward in a helicopter is like or unlike hovering, or taking off, or flying backwards, or landing; or, for that matter, is like or unlike flying forward in a conventional airplane, or driving forward in an automobile. At that point he will be able to state with considerable precision what behaviors should be represented in a training device.

The question of how closely a device should be made to simulate an operational situation can often be reduced to the question of critical skills. For example, if a flying trainer is designed to train the critical skill of landing, it is known that certain types of instruments are just not used in the practice of this skill, i.e., they provide no stimuli to which the learner must acquire appropriate responses. On the other hand, if flying an instrument navigation mission is the critical skill for which training is desired, then clearly some of these same instruments may be essential, although visual stimuli from the terrain are quite unnecessary. In many instances, the discrimination between what is essential or unessential for the representation of a critical skill requires no great amount of technical knowledge. Nevertheless, there may be considerable need for emphasis on the *purpose* to be served by a training device. When a need for a training device is developed, this in itself implies the belief that certain essential skills can be more simply represented than by means of a replication of the operational situation. Unless this purpose is constantly borne in mind throughout the development process, the demand for close simulation can readily lead one back to the operational situation itself. This is no solution.

Performance Measurement

If one accepts the purpose of performance measurement as a legitimate function of a simulator, desirable characteristics may be defined with considerable clarity. Any test or measure of performance should aim for high *validity*, i.e., some sort of assurance that the test measures what it is supposed to measure. When one mentions measurement, though, reliability as well as validity is implied, since it must also be determined that the test yields a performance score which differentiates between a superior and inferior individual with some degree of dependability. To the psychologist, these are well-known concepts. They appear to be applicable without change or reservation to the measurement of performance by means of a training device. Their application in this new field gives rise to research questions which have been with us for many years and serves to empha-

[1] Two important exceptions are articles by Brown and Jenkins (10) and by Craig and Ellson (12).

size these problems in particularly vivid fashion. Specifically, these are questions concerning the nature of criterion measurement.

Validity. It is apparent that close simulation has the aim of insuring high validity to the task presented by the trainer. It is not quite as easy as it sounds, though, to produce a highly valid measure of performance by exact simulation of the operational equipment. The chief reason is that there are some very difficult unsolved problems about performance criteria (cf. 29), all concerned with what aspect of performance shall be measured. For example, what is the ultimate criterion of an aerial combat gunner's performance? Is it number of planes shot down? Number of planes hit? Number of planes scared away? And should friendly and enemy planes be distinguished from one another in these frequency counts? It should be pointed out here that these questions may be legitimately asked independently of the question of reliability. They are questions concerning the meaning of criteria and the relationships between products of behavior and measures of the behavior itself. They represent an area of psychological knowledge which is not highly developed. And, obviously, they imply that the problem of simulation cannot be adequately solved without facing the more general problem of criterion performance measurement. Only when he knows some general rules about the relation of performance to product will the psychologist be able to predict with confidence what performances should be simulated to insure high validity.

In many industrial activities, the adoption of a product measure such as "number of units assembled" seems an entirely reasonable performance criterion. But there are many other human activities, both within and without the armed services, to which the application of such a measure appears impossible. There are others to which some form of product measure might be applied. For example, one can conceive of using "number of hits" as a criterion of performance in flexible gunnery during a specified number of aerial missions. What is usually employed is some more analytical behavior measure such as "miss distance." There is little evidence, and no theory, which enables a choice of the most desirable criterion measure, except on the basis of reliability.

Reliability. Even assuming that high degrees of simulation are desirable for validity, sacrifices at the expense of similarity must often be made for the sake of reliability of performance measurement. The position of the psychologist in this matter is, presumably, that regardless of the possibility of greater validity with high similarity, no measurement of performance is possible without some degree of reliability.

It is instructive to note here some studies which have investigated the measurement of student pilot performance in aerial missions.[2] In these studies the aerial task was designed to simulate exactly the maneuver taught; in fact, they were identical. Moreover, the observations of behavior were made as free from the likelihood of human error as possible, in order to remove this source of unrelia-

---

[2] Personal communication, Lt. Col. William V. Hagin, Commander, 6566th Research and Development Group (Pilot Training Research Laboratory), Goodfellow Air Force Base, Texas.

bility. Nevertheless, the results showed many of these performance measures to be quite unreliable—not because of observer error, but because the performance of the individual student varied from one time to the next. In this instance, regardless of their admitted validity, the degree of unreliability in these measures imposes a serious obstacle to their use. A similar conclusion would be drawn were these results obtained in a ground trainer which closely simulated aerial maneuvers.

There is at least one way in which a training device may have to be dissimilar from an operational task in order for reliability to be achieved. The device must be capable of providing an adequate sampling of the particular behavior to be measured. Stated in terms of analogy to a proficiency test, the reliability of a device will increase as the number of items measuring a given type of activity is increased. This principle is utilized in many present simulators, though in a somewhat unsystematic way. For example, aircraft simulators often provide for the running of "missions" which include a number of different instances of simulated equipment failure requiring the use of emergency procedures. Repetitions of these emergency situations are provided, not because they occur in an operational situation, but (at least partially) because of the need to obtain a large enough sample of the student pilot's behavior to permit a reliable assessment of his performance.

Actually, there is room for a great deal of improvement in simulators in the application of this principle of adequate behavior sampling. If the purpose of a training device is performance measurement, there must be increased acceptance of the fact that simulation has to be diluted to the extent that it is necessary to provide long enough, or large enough, or frequent enough activity sequences for performance to be measured with minimally acceptable reliability.

Research in this area should have the aim of formulating a set of principles relevant to the arrangement of conditions in a trainer situation so as to produce maximum reliability of performance measurement with the sacrifice of as little validity as possible. The general question is: How must the characteristics of an operational task be deliberately altered in order to make possible adequately reliable measurement of performance highly related to the operational task?

One problem in training device design which arises frequently is the degree of accuracy which must be specified for a system of scoring performance. Accuracy refers to repeatability of the limits of the scoring area; it should be clearly distinguished from accuracy in the sense of deviation of scoring area from some absolute physical value. Obviously, the greater the accuracy in the first sense, the better. Minimal acceptable accuracy must be estimated in terms of the relation between the amount of variance contributed by the device and the amount of variance contributed by the inherently variable human operator. Up to the present time, only by collecting empirical data on each particular performance which is being scored has it been possible to estimate the minimal acceptable apparatus accuracy. Is it possible that data on the variability of human motor activities could be cast in a systematic framework? Is it possible, for example, that one might eventually be able to predict the variability of motor response functions as well as is now done for verbal intelligence? The exploration of this area of human engineering appears a very worth-while research effort.

Investigation of how the length and arrangement of performance samples affect reliability is also needed. The construction of items for printed proficiency tests is an activity which can now be carried out on the basis of a number of verified principles pertaining to such matters as length of items, number of misleads, arrangement of items, and so on (cf. 1). To what extent can these principles be translated to a performance-measuring situation, in which the behavior to be measured is to a greater extent controlled by the apparatus, rather than being determined by the experimenter? How long should a single trial be for different types of responses? How can one arrange trials to yield a homogeneous proficiency measure in spite of changes in performance with practice? What is the relation between length of the behavior sample and reliability? Is it possible to arrive at generalizable principles of motor performance measurement which will apply to the great variety of human operator activities, so that the necessity for empirical determination of each case can be avoided?

Effectiveness for Training

If facilitation of performance improvement is the purpose being considered for a training device, a desirable aim of research is the determination of what characteristics of the trainer task will increase transfer of learning to the operational task. More specifically, research should tell us how the physical characteristics of a trainer may be designed to bring about the most rapid acquisition and the highest possible level of performance in the operational skills for which training is required. The question can even be expressed in this way: How must the characteristics of an operational task be deliberately altered in order to insure the most effective training by means of a training device?

This point of view may be exemplified in its broad aspects by considering the implications of an experiment by Lindahl (23). This investigator was concerned with the problem of effective training for the industrial job of disc cutting, in which discs must be cut from a tungsten rod by means of an abrasive wheel operated with a foot pedal. The standard training situation, with "exact simulation," would have been to provide new workers with regular periods of practice on the disc-cutting machine itself. This type of practice did not appear to Lindahl to be necessarily most effective. Instead, following an analysis of the foot movements involved, he arranged for new workers to practice a standard pattern of foot action in response to a visually presented model. Comparison of production records of new employees trained by this method with those of older employees who had simply practiced disc cutting itself showed that the former achieved a level of performance after 11 weeks of training equivalent to the level of those who had been on the job for 5 months.

Wolfle (32) considers that this experiment illustrates the effectiveness of a training method which emphasizes "process" rather than "product." Other possible conceptions may suggest themselves. The most important implication, however, is this: A training situation which was deliberately changed from a job situation in order to emphasize critical aspects showed transfer to the final task greater than that produced by practice in the job situation itself. More effective training resulted when "exact simulation" was sacrificed.

This example does not necessarily violate the hallowed principle that amount of transfer between tasks increases with the number of identical elements. Indeed, it may be said that transfer was high in this situation because the foot movements practiced were identical with those of the most highly skilled disc-cutter operators. The problem of effective training, however, was not one of making the tasks similar, but rather of arranging the conditions of practice in such a way that essential skills were most efficiently learned. One may generalize from this example that the answer to the problem of what makes a training device effective is to be sought, not in identity of all task elements, but rather in viewing a training device as a means of making conditions most effective for learning.

There are many ways of arranging the physical task or conditions of practice, each of which may be viewed as a deliberate alteration from the conditions of exact simulation, and any or all of which might conceivably have the effect of facilitating the learning of operational skill. Suggestions for these arrangements come from a variety of research sources. Since none of the results can at present provide definite answers to the problem of facilitation of skill acquisition, they have the status at present of questions, more or less adequately defined, which research can be designed to illuminate. Some of these possibilities are discussed briefly in the following paragraphs.

Repeated Practice. One essential condition for the learning of skill is, of course, practice. As has been pointed out, the provision of opportunities for repeated practice is one of the most obvious advantages a training device can have, as contrasted with an operational situation. The motor skills in an airplane maneuver can be practiced many times during a relatively short session in a training device, while an equivalent amount of practice in a training airplane requires a very much greater expenditure of time and effort. In a good trainer, the task is deliberately designed so as to make possible the repeated practice of critical skills.

There are some unknown things, however, about even this essential variable of practice. Is there a general answer to the question, how much practice is desirable for most effective transfer? Or, in terms of experimental variables, what is the relation between amount of initial practice and amount of transfer? At the present time, little more than a qualitative answer can be given to this question, to the effect that transfer increases with amount of initial training. For many problems of training device design, it will be necessary to provide answers in terms of how much. The problem has a number of facets, but perhaps the most obvious one is the need for a scale of measurement for transfer.

Although some attempts have been made to bring order to this problem of measurement (cf. 15), there is still a great need for research. Perhaps the most crucial question is that of deciding what should be meant by the upper limit of transfer, particularly when what is learned is increased response accuracy, rather than a specific set of "new" responses.

Motivation and Reinforcement. A training device may be designed to provide additional or different motivation from that of the operational task. Although such things as monetary rewards are not particularly relevant to the training device situation, a number of well-known forms of variation in competition and cooperation (cf. 26, pp. 470–499) appear to be. A recent study by Bilodeau (3) has explored one possible arrangement of the competitive situation in the learning of a motor skill. The negative findings of this study do not, of course, make additional research less desirable.

Knowledge of results, in its many forms, has often been considered a variable which has motivating properties in the sense of reinforcement. (Its information-giving properties will be considered in a subsequent paragraph.) In motor skill learning situations, studies (20, 25) have tended to show that telling the learner when he is on target or off target (comparable to saying "right" and "wrong") does not improve transfer, though it does have a fairly immediate effect on the level of performance. The results suggest that the added on-target knowledge becomes an extra cue for the establishment of certain responses which rapidly extinguish once the cue is removed. These results may have a decelerative effect on the tendency to make blanket recommendations for "knowledge of results" features in training devices. However, it can certainly not be maintained that the full range of reinforcement possibilities of this variable has yet been explored.

Another variation in reinforcement suggested by conditioned response studies is the frequency of application. Under certain conditions, results of these studies suggest a heightened resistance to extinction of the response established under less than 100 per cent frequency of reinforcement (cf. 17). Can such a finding be extended to the learning of human motor skills? Up to now, several studies (6, 20, 25) of reduced frequency of presentation of knowledge of results, at regular trial-intervals, have failed to give any clear indication of learning facilitation or heightened transfer. However, regularity, as opposed to randomness of reinforcement, may be a critical condition of these findings in view of the possibility that the learner may quickly acquire habits of reaction to the regular patterning of on-target information. Thus, a potentially fruitful research area is the effect of random presentation of knowledge of results on the learning and transfer of motor skills. There are many varieties of frequency patterning of reinforcement, as the work of Skinner (28) shows, each having different consequences in the performance of simple acts. Under what circumstances can these variations affect the acquisition of complex skills?

Set. It is conceivable that an important result of the use of a training device might be the establishment of a set which could not be reliably produced under conditions of exact simulation. It is possible, for example, that habits of attending to critical task elements might later facilitate performance of an operational task. Or, the division of attention which is required of the pilot in high speed aircraft might be established by suitable arrangements of the presentation of stimuli on an instrument trainer panel. Our present knowledge of the conditions of establishment of such sets, and the ways in which they influence the learning and performance of skills, is very inadequate. A recent study of flexible gunnery skill (7) shows that an alteration of the physical characteristics of the sight can produce significant performance differences presumably because of changes induced in the subject's attentive habits. At least the potential fruitfulness of this research approach is supported by these results.

The possibility of establishment of a learning set, described by Harlow (19), also appears to be a worthy field for investigation. It seems possible that systematic variations in the stimulus material, which one study (27) showed to be effective for transfer, may be one way of establishing such sets. There may be other ways, and the use of verbal processes available to human operators, but not to Harlow's monkeys, should not be overlooked (cf. the discussion by Wolfle, 32, pp. 1272–1275). The potentialities of "learning set" for training human beings who will probably be faced with the need to make frequent adjustments to rapidly changing equipment appear great.

Component Practice. It is doubtful that any complex skill is ever learned entirely by practice "on the total skill." Training in driving an automobile, for example, is very often conducted in sessions designed to give special emphasis, or additional practice, to such difficult part-skills as shifting gears or the maneuver of parallel parking. In this respect, practical skill-training methods have always departed more or less from the conditions of exact simulation.

If the complex skill for which training is desired may be analyzed logically into component activities which differ in difficulty, the research problem of the relation to total-skill performance of differential amounts of training can then be explored. For example, the task of B-29 gunnery sighting is usually described in terms of the components of tracking, framing, and triggering. Though no systematic study has been carried out, it is a reasonable hypothesis that differential treatment of these three components in terms of practice would yield differential amounts of transfer to the total task. The implication of such results might very well be that systematic arrangement of component practice produces greater transfer to complex skills than does a "natural" (i.e., highly simulative) arrangement. From a research point of view, this area may be expected to yield findings of considerable import for an understanding of how complex skills become integrated from simpler activities.[3]

Response Precision. Much effort and expense are sometimes devoted to making the degree of tolerance permitted by a control movement in a training device highly similar to that of the operating equipment; for example, the design specifications for an instrument flying trainer may state that the ratio of a given stick movement to a given instrument-needle displacement must simulate that of the aircraft exactly. Yet the fact is, we know little about the effects of variation in "target tolerance" on the learning and transfer of skills. Some recent studies on this subject (4, 5, 14) strongly suggest that the generalization "transfer increases with similarity" is of limited usefulness indeed when applied to the characteristic of on-target response tolerance. Instead, they support the use of a somewhat simpler principle, "transfer of skill is unaffected by wide variations in target tolerance." The factors which do determine these limits remain to be discovered by additional research; but a simple principle of "similarity" does not appear to be relevant.

The most obvious effect of increasing the size of on-target tolerance in a training task is a marked increase in the rate of improvement in performance scores during practice. One hypothesis would regard such an increase in success as a

[3] Research on component skill training bears only a slight resemblance to older studies of verbal part-whole learning, summarized by Hunter (21), or to the work of Batson (2) and Kao (22) on motor tasks. Generally, in this older work, interest centered on the learning of total *sequences* of part-activities, as opposed to the integration of component activities which must often be simultaneously exercised.

motivating or reinforcing factor. As already mentioned, however, the evidence obtained so far does not support such a hypothesis.

On the other hand, one might think that decreasing the target tolerance in a training task, despite making the task more difficult (in the sense that less success is achieved), would serve to encourage greater precision of responding during practice, i.e.. prevent the occurrence of responses outside the desired operational tolerance range. This, in turn, might have the effect of facilitating transfer to an operational task. Things do not seem to work this way; at least one study (5) indicates that the range of responses actually practiced is unaffected by target tolerance.

A somewhat different method of restricting the target tolerance permitted in a training task is the use of what may be called *enforced guidance* (cf. the studies of Carr, 11). One can think of designing a training device which would physically restrict the range of control responses permitted, in order to make it impossible for the learner to practice "over-controlling" responses or the making of gross errors. What would be the effect on transfer of a training situation which permitted the practice of only "correct" responses, as opposed to one which allowed the making and correction of errors? This question seems to have rather interesting roots in basic learning theory, including particularly that of Guthrie (18).

Surely the last word has not yet been said on this problem. There are many ways and degrees in which target tolerance can be varied throughout a practice session, and there are many different theoretical reasons leading to the prediction of changes in both rate of acquisition and transfer of motor skills, as a result of such changes.

Performance Feedback. It is not possible to distinguish this characteristic clearly from those discussed as knowledge of result and target tolerance, both of which include forms of performance information fed back to the learner. Nevertheless, there are some research hypotheses arising from a consideration of this characteristic as having only the function of information giving, as contrasted with the function of motivation or reward.

For example, what kind of information can best be utilized by the learner in the sense of producing greatest transfer? On a flexible gunnery trainer, one study found no transfer gains from an indicator that informed the learner when he was "on target" (6). On the other hand, a study conducted on a similar trainer (16) showed lasting (i.e., presumably transferable) skill changes when information was given about the direction and extent of errors. As yet, these results may be considered only suggestive of a generalizable conclusion. But they indicate that a general research problem may perhaps be stated as that of discovering principles of efficient coding of feedback information obtained by a learner during practice on a training device.

Besides the problem of frequency of feedback, which may be considered from the information-giving point of view as well as in terms of the previously mentioned concept of reinforcement, there is the matter of the size of "chunks" of information. The usual procedure is to restrict the performance information to a report of the last preceding trial. But it is conceivable that a reporting of performance on a greater number of preceding trials, whether in serial or summarized form, might have some differential effectiveness for learning and transfer.

In other words, this, or other methods of providing summarized feedback covering larger segments of performance, might be considered as another kind of variation in information coding. Systematic studies of the application of this concept of coding to feedback information have not yet appeared.

Summary. There are many ways in which the training device situation might be deliberately altered from the operational situation in order to serve the purpose of performance improvement. At the present time, we do not know that any of these changes would, in fact, produce transfer differences when compared with training conducted under conditions of close simulation. The problems offer a challenge to the research psychologist running somewhat as follows: "Can you discover principles which may be applied to training devices in such a way as to make them more efficient than haphazard on-the-job practice?" It should be emphasized, perhaps, that even if negative findings can be established for any or all of these factors, we shall be much farther along in our understanding of training device design requirements and shall also be able to make some money-saving recommendations.

EVALUATION OF TRAINING DEVICES

Once a training device has been constructed, it needs to be evaluated. The first type of evaluation should logically be an engineering effort to determine whether the physical specifications contracted for have been met by the manufacturer. When the training device is put into use, there are a number of evaluative questions which are the legitimate concern of the experimental psychologist.

The aims of field evaluation research on training devices are, generally speaking, quite different from research on design principles. In the field use situation, one is concerned with answering questions about a device which is being put to more or less specific use in a particular course of training. The research question may take the general form: What are the conditions which determine the most effective use of this device, either for training, for performance measurement, or both? Research of this sort may suggest a number of design principles, though this is not its specific aim; and such principles, of course, cannot be applied to the particular device in question, since it is already built.

The distinction between the aims of performance measurement and performance improvement can serve to clarify greatly the approaches of field evaluation research. Certain characteristics are plainly irrelevant to the value of a device for training, just as others are irrelevant to the value of a device for performance measurement. When the psychologist is given the job of evaluating a training device, the first question he should answer is, "What is the purpose for which the device will be used?"

Evaluation as a Performance-Measuring Device. The determination of reliability is of primary interest in a training device used for performance measurement. Studies are often required to determine more than the reliability of a single type of score alone. For example, the question of how long a behavior sample should be used or of how many trials of a particular type should be included in a measure are frequent research questions. The psychologist should be concerned with recommendations regarding the conditions under which performance measurement is to be accomplished, if it is to have the reliability essential to its usefulness in a training program.

In a field evaluation study, validity is determined in order to assess the degree of relationship between performance measures obtained in the device and performance involved in some criterion task. It should be borne in mind, on the one hand, that the purpose for which the validity coefficient is obtained is to evaluate the *device*, and not the criterion; though it is true, of course, that the size of the relationship may provide some information about the nature of the criterion. On the other hand, the limitations on the size of the validity coefficient imposed by the reliability of the criterion are of crucial importance for evaluating this relationship. One cannot determine the degree of resemblance of performance on a simulator to operational performance if the latter can only be measured with zero reliability. Since there are many instances of the latter situation at present, it would seem best frankly to face the fact that validity cannot be measured in these cases. If the training device yields reliable performance scores, it may be desirable to decide on rational grounds (cf. 29) to use the device itself as an intermediate criterion. But no amount of correlating and intercorrelating can by itself solve the problem of unreliability of the criterion.

It is not essential to the evaluation of a performance-measuring device to determine its learning and transfer characteristics. The device may measure performance in both a reliable and a valid manner without contributing to performance improvement. For example, relatively short single measurement sessions may yield adequately reliable scores which are highly related to criterion performance, without affecting the level of skill. Improvement in skill may require practice of considerably greater duration either outside or inside the training device situation. Conversely, indications of low transfer between the device and operational situations do not necessarily imply a low validity relationship, since the device may, and often does, *train* only a portion of the abilities exhibited in the criterion task.

Summarizing, the most important aspects of the evaluation of a device for performance measurement are the reliability of the measurement and the degree of its relationship to an operational performance criterion. Both are measured by a correlation coefficient. Measures of performance improvement, however, whether obtained in the trainer situation or in the operational situation, are irrelevant to this question.

Evaluation of Training Effectiveness. The primary concern of research designed to evaluate the training effectiveness of a device should be to obtain measures of the transfer of training on the device to an operational performance situation. Since a device seldom has one single effectiveness value, it is usually desirable to frame more analytical questions for research such as: At what stage in the course of training will practice on the device yield greatest transfer? Or, what amount or type of practice on the device produces maximal transfer to the operational task? Results of such studies may be expected to reveal, not a yes-no answer to the question of training effectiveness, but rather in what ways the device can be used so that it will be optimally effective.

Since reliable performance measures in operational tasks are frequently difficult to obtain, the problem of measuring transfer is also difficult. A partial answer to the question of training effectiveness may be obtained by measuring improvement in performance within the training device situation itself. In most situations one cannot expect transfer to occur unless there has been some improvement exhibited in the trainer itself. Such results are therefore useful, though they cannot actually replace transfer measures stated in terms of the criterion task

Training effectiveness cannot be measured by means of correlation coefficients. The concepts of reliability and validity of the measures obtained on the device therefore have no place in studies whose purpose it is to evaluate the effectiveness of a device for performance improvement. After all, the traditional psychometric approach deliberately tries to minimize practice effects, whereas in training they must be emphasized. A correlation coefficient indicating a high validity relationship between a device and an operational situation means simply that the same abilities are involved in both tasks. It does not show that practice on the first task can lead to improvement on the second; in fact, it tells nothing about this relationship. Obviously, too, practice on a device may bring about improvement whether behavior is measured or not. This implies, incidentally, that a scoring mechanism is not a necessity when the device is used for training, though it may be desirable for other reasons.

In summary, the training effectiveness of a device is determined in terms of the measurement of transfer from training on the device to performance in an operational situation. Research in this area is often concerned with more detailed questions such as the effects of changes in length, arrangement, and sequence of practice within a training course and their relations to amount of transfer. Coefficients of correlation measure the extent to which common abilities are involved in the device and operational tasks but do not provide measures of training improvement.

CONCLUDING REMARKS

This paper has attempted to describe and clarify some research issues which occur in connection with the development, use, and evaluation of training devices. The purpose has been to see whether such clarification can reveal a framework for psychological research in the training devices area.

It is maintained that the kinds of utilization of training devices are two: performance measurement and performance improvement. Although the two uses are frequently made of a single piece of equipment, they may be distinguished particularly in the characteristics of the device which are essential for each purpose. When the device is used for performance measurement, the important characteristics are reliability and validity. When the device is employed for improving performance, on the other hand, the characteristic of importance is the amount of transfer of learning to an operational task. In either case, degree of simulation becomes a secondary consideration.

When one considers these purposes, some of the opportunities for research on training devices become apparent. On the side of methodology, the range of these problems includes job analysis, training, proficiency measurement, and criterion development. On the side of theory, questions mentioned in the present discussion include those of the structure of skills, the determinants of human variability, relationships of set and motivation to learning, and the mechanisms of transfer of learning.

REFERENCES

1. Adkins, Dorothy C. *Construction and analysis of achievement tests.* Washington, D.C.: U.S. Government Printing Office, 1947.
2. Batson, W. H. Acquisition of skill. *Psychol. Monogr.*, 1916, **21**, No. 2 (Whole No. 91).
3. Bilodeau, E. A. Acquisition of skill on the Rudder Control Test with various forms of social competition. *USAF Hum. Resourc. Res. Cent., Res. Note P&MS*, 51-6, 1951.
4. Bilodeau, E. A. A preliminary study of the effects of reporting goals as a function of different degrees of response accuracy. *USAF Hum. Resourc. Res. Cent., Res. Bull.*, 52-4, 1952.
5. Bilodeau, E. A. A further study of the effects of target size and goal attainment upon the development of response accuracy. *USAF Hum. Resourc. Res. Cent., Res. Bull.*, 52-7, 1952.
6. Bilodeau, E. A. Some effects of various degrees of supplemental information given at two levels of practice upon the acquisition of a complex motor skill. *USAF Hum. Resourc. Res. Cent., Res. Bull.*, 52-15, 1952.
7. Bilodeau, E. A., & Morin, R. E. Proficiency on the Pedestal Sight Manipulation Test with and without the tracking pipper. *USAF Hum. Resourc. Res. Cent., Res. Bull.*, 51-27, 1951.
8. Bridgman, C. S., Gray, Florence, & Solomon, R. L. A graphical analysis of sighting performance on B-29 sights. OSRD Report No. 5221. Applied Psychology Panel, NDRC, Project AC-94, June 1945.
9. Brogden, W. J., & Smith, K. U. Summary of research and development on project AC-94; psychological factors in the operation of flexible gunnery equipment. OSRD Report No. 6212. Applied Psychology Panel, NDRC, Project AC-94, October 1945.
10. Brown, J. S., & Jenkins, W. O. An analysis of human motor abilities related to the design of equipment and a suggested program of research. In P. M. Fitts (Ed.), Psychological research on equipment design, *AAF Aviation Psychology Program Report* No. 19. Washington, D.C.: U.S. Government Printing Office, 1947. Pp. 35-63.
11. Carr, H. A. Teaching and learning. *J. genet. Psychol.*, 1930, **37**, 189-218.
12. Craig, D. R., & Ellson, D. G. The design of controls. In *National Research Council, Human factors in undersea warfare.* Washington, D.C.: Committee on Undersea Warfare, National Research Council, 1949. Pp. 133-151.
13. Edgerton, H. A., & Fryer, D. H. The development of an evaluation procedure of training aids and devices. Port Washington, L.I., N.Y.: U.S. Navy Special Devices Center, 1950. (Tech. Rep., SDC 383-2-1.)
14. Gagné, R. M. Learning and transfer of training in two forms of rudder control task. *USAF Hum. Resourc. Res. Cent., Res. Note P&MS*, 50-1, 1950.
15. Gagné, R. M., Foster, Harriet, & Crowley, Miriam E. The measurement of transfer of training. *Psychol. Bull.*, 1948, **45**, 97-130.
16. Goldstein, M., Rittenhouse, C. H., & Woods, J. P. Studies of performance on the E-26 Flexible Gunnery Trainer. *USAF Hum. Resourc. Res. Cent., Res. Bull.*, 52-17, 1952.
17. Grant, D. A., & Schipper, L. M. The acquisition and extinction of conditioned eyelid responses as a function of the percentage of fixed-ratio random reinforcement. *J. exp. Psychol.*, 1952, **43**, 313-320.

18. Guthrie, E. R. *The psychology of learning.* New York: Harper, 1935.
19. Harlow, H. The formation of learning sets. *Psychol. Rev.,* 1949, **56**, 51-65.
20. Houston, R. C. The function of knowledge of results in learning a complex motor skill. Unpublished master's thesis, Northwestern Univer., 1947.
21. Hunter, W. S. Learning: IV: Experimental studies of learning. In C. Murchison (Ed.), *Handbook of general experimental psychology.* Worcester: Clark Univer. Press, 1934. Pp. 497-570.
22. Kao, Dji-Lih. Plateaus and the curve of learning in motor skill. *Psychol. Monogr.,* 1937, **49**, No. 3 (Whole No. 219).
23. Lindahl, L. G. Movement analysis as an industrial training method. *J. appl. Psychol.,* 1945, **29**, 420-436.
24. Mahler, W. R., & Bennett, G. K. Psychological studies of advanced Naval Air Training: evaluation of operational flight trainers. Port Washington, L.I., N.Y.: U.S. Navy Special Devices Center, 1950. (Tech. Rep., SDC 999-1-1.)
25. Morin, R. E., & Gagné, R. M. Pedestal Sight Manipulation Test performance as influenced by variations in type and amount of psychological feedback. *USAF Hum. Resourc. Res. Cent., Res. Note P&MS,* 51-7, 1951.
26. Murphy, G., Murphy, Lois B., & Newcomb, T. M. *Experimental social psychology.* (Rev. Ed.) New York: Harper, 1937.
27. Seashore, H. G., Kurtz, A. K., Kendler, H., Stuntz, S. E., & Rappaport, C. Variation of activities in code classes: an experimental study of the problem of monotony in code learning. OSRD Report 4082. New York: The Psychological Corporation, 1944.
28. Skinner, B. F. Some contributions of an experimental analysis of behavior to psychology as a whole. *Amer. Psychologist,* 1953, **8**, 69-78.
29. Thorndike, R. L. *Personnel selection.* New York: Wiley, 1949.
30. Williams, A. C., Jr., & Flexman, R. E. An evaluation of the Link SNJ operational trainer as an aid in contact flight training. Port Washington, L.I., N.Y.: U.S. Navy Special Devices Center, 1949. (Tech. Rep., SDC 71-16-3.)
31. Wolfle, D. The use and design of synthetic trainers for military training. OSRD Report 5246. Washington, D.C., 1945.
32. Wolfle, D. Training. In S. S. Stevens (Ed.), *Handbook of experimental psychology.* New York: Wiley, 1951. Pp. 1267-1286.

39

Programed Instruction—Past, Present, Future

John S. Abma

INTRODUCTION

An ideal educational or training system would be one in which every student could be instructed by an expert tutor. This is the technique we actually use for

From Aerospace Medical Research Laboratories Technical Report No. 64-89, September 1964.

some difficult, specialized skills, as in pilot training, the training of astronauts, musical performance, sports and athletics, and advanced studies of some kinds at the college level. In all of these cases, an instructor spends time working with individual students. However, there are not enough instructors to go around for this kind of teaching when large numbers of students are involved. Here is where programed instruction can play a role. Programed instruction is one way to make the advantages of private tutoring available to large groups of students being trained in routine topics or skills. The key idea of programed instruction is that the material is taught in the same way a tutor would teach it, except that the material is presented in printed form, and sometimes by a "teaching machine."

As in tutoring, the major features of programed instruction are; first, each student can proceed at his own pace, not being held back by slower students nor rushed by the faster ones; second, each student is kept informed of how he is progressing at every step of the course; third, the system offers rewards or encouragement as the student tackles difficult subjects; and fourth, the emphasis is upon student involvement and activity—not passive reading or listening. In some types of programed instruction, the content of the course can be altered to meet the needs of each individual student.

Programed instruction is not new. It has been with us at least ten years, reckoning from B. F. Skinner's 1954 article in the Harvard Educational Review (11). Other calculations, based on the work of S. L. Pressey in the 1920's (7) or the invention of educational drill machines starting in the 1870's (6) would reveal an even longer history. Indeed, some writers compare programed instruction with the method employed by Socrates in his instructional dialogues (2). But in 1954, all the major current approaches to programed instruction had been enunciated. In this article we will examine these various approaches and assess the present status of programed instruction in the light of current research. We shall also attempt to predict what influence programed instruction will have on future educational and training methods.

PAST

Early terms used to describe this area were "Teaching Machines," "Automated Instruction," "Self-Tutoring," and "Programed Learning." These terms suggest the concern with automation, self-instruction and individual tutoring methods that should improve our educational and training activities. Today, there is a distinction between automation and programed instruction. Automation can be applied to any educational technique. It includes televised lectures, automatic test scoring and complex audio-visual aids, including language laboratories. Although programed instruction usually does reduce the role of the human instructor, it does not necessarily rely upon complex machinery and it may retain the human instructor for some of the tutorial functions. In this article we will deal with programed instruction, not automation.

Just as there are different ideas about what a good human tutor should do to teach effectively, there are different ideas about how programed instruction should be implemented. One approach we will consider is that of Sidney L. Pressey, starting in the 1920's. Dr. Pressey was originally concerned with automating the testing process and developed machines that provided automatic scoring of multiple-choice tests. Inasmuch as his machines gave feedback to the student indicating the correctness of each answer, they could also be used to teach drill material. Later, Pressey integrated the devices into courses of instruction that included the usual methods of lecture and text study. In this application, the devices provided "Adjunct Autoinstruction"—something new added to the usual methods of instruction. Pressey continues to recommend multiple-choice test items, never essay or short-answer tests. He believes more can be learned by distinguishing right answers from among wrong possibilities than from constructed answers.

Two distinct possibilities are presented by the self-testing idea. First, it is possible for the student to learn from the test itself. Second, it is possible for self-testing to make other sources of instruction more effective by showing the student where he is weak, for example. Figure 1 indicates both uses of self-testing.

Box A shows that a student or group of students would be exposed to classroom lectures and demonstrations, or text book assignments, or both. After an hour or more of such study, the students would leave that activity (arrow 1). Students would then take multiple-choice tests (box B), finding out immediately what items they missed. Such a test might indicate that they needed more study or another lecture on the topic, or the test might indicate that the student had learned the material. In either case, he would return (arrow 2) for more conventional instruction. The dashed line (3) around box B shows that some learning can take place when self-tests are used in isolation from conventional instruction. Students can repeat the self-tests often enough to discover correct answers by "trial and error." This is sufficient for the learning of some material.

In recent articles (8) (9) Pressey urges reconsideration of "Adjunct Autoinstruction." He cites as one of its advantages that it employs conventional materials and approaches. We can retain our courses and materials as they are, and simply add multiple-choice self-testing. This certainly is easier to implement than other systems of programed instruction, and experiments have shown that gains can be made from this procedure (5). However, some educators and psychologists are quite concerned about the teaching effectiveness of lectures and ordinary texts. Group lectures, for example, may leave some students behind and, at the same time, be too easy for others. When the progress of an individual is tied to that of the group, we are confronted with a "lock step" in education and training. Also, the ever-increasing size of lecture groups makes any personal interaction between student and instructor most unlikely. With regard to textbooks, it is possible that their logical organization, sometimes amounting to a catalogue of facts, is not the best organization for teaching purposes. Even though self-testing adds to the effectiveness of orthodox methods in some cases, many feel that greater improvements may be expected from the design of better teaching procedures.

Intrinsic programing, a development of Norman A. Crowder, also relies upon text study as the main source of information. However, the texts are specially written to provide for different student needs. After each page of text, a single multiple-choice test item appears. A different page number is printed by each alternative answer. After a student selects the answer he thinks is correct, he turns to the page number given with that alternative. On that page he finds out if his choice was correct. If it was not, he has reached a "wrong answer" page which tells him he was wrong and gives additional explanation. He is then usually told to return to the page he just left and try the same item again. When his answer is correct, he reaches a "right answer" page which says "You are correct," and continues with new material. Figure 2 represents an intrinsic programing system.

Each student starts with box A which contains one or two paragraphs of text instruction. As arrow 1 indicates, students then go to a multiple-choice test question (box B). (This question is usually printed on the same page as the text instruction, but, since their functions are different, we show them in separate boxes). At this point, students can branch out on their own, since they may choose different answers from the 2 to 5 alternatives usually given. (For this reason, intrinsic programs are also called branching programs). Students who choose a wrong answer reach a page with remedial or corrective instruction (arrow 2 to box C). They are then usually returned (arrow 3) to the same page they started with for re-study. However, they may simply be told to try the multiple-choice question again (arrow 4). In either case, they eventually pick the right answer to the multiple-choice question. When this happens, they reach another page (arrow 5) which gives new material. Obviously, the branching procedure can be made complex. For example, remedial instruction might be followed by a test question which would be the occasion for more branching, and, conceivably, some branches could even advise a student to take a different course as a needed prerequisite. What we have diagramed here is both the simplest and most common form of intrinsic programing.

The largest evaluation of intrinsic programing was conducted by the Air Force at Keesler Air Force Base from 1960 to 1962. The subject matter was electronics, and programs were presented both in booklets and on a special film viewer. Results of the study were equivocal, with no improvement in amount learned, but some indication that time was saved over ordinary methods.

The system requires clear writing in text portions and the very greatest skill in construction of multiple-choice questions. Possible student misunderstanding must be anticipated by the alternatives presented and no correct answers must be "give-aways." If they are easy, students will get right answers even though they might not comprehend the material they have just read.

The last approach we will consider is called "Linear Programing," and is a development of B. F. Skinner. This represents the most ambitious attempt to overhaul orthodox means of instruction. Whereas lectures and texts have in the past concentrated on mere presentation, Skinner's approach concentrates on what the student is doing as well as on clear presentation. Active student responding and participation, in contrast to a passive manner of listening or reading, may be the key to improved educational efficiency.

A linear program looks like a series of incomplete sentences. Each sentence, along with the blank usually found at the end of it, is called a "frame." The sentences are so written that students can usually figure out what word should be used to complete the sentence. After writing in the word the student thinks should go in the blank, he turns the page or moves a masking device to reveal the right answer. When he sees that he has written the correct answer, that answer is said to be "reinforced." In the future, he will be likely to give the same answer under similar questioning conditions. Figure 3 is a diagram of a single linear frame. Thousands of such frames may be used in a complete program.

The incomplete sentence is found in box A. After considering what word to put

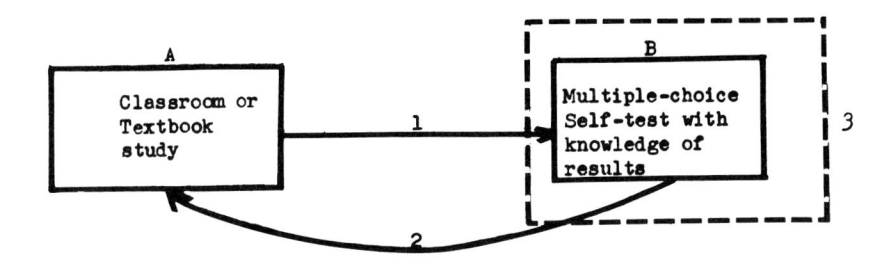

Figure 1. Adjunct autoinstruction.

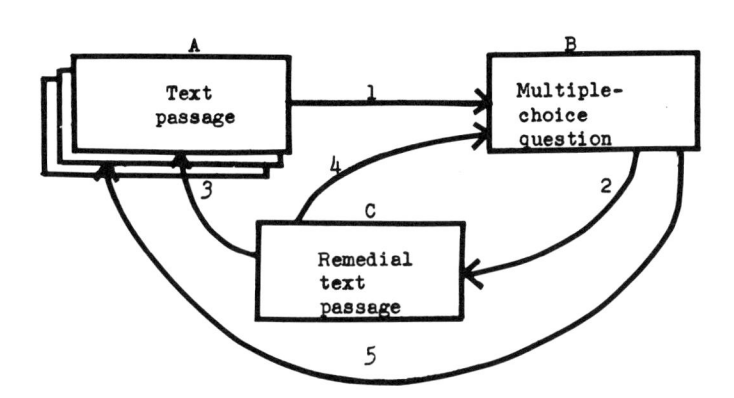

Figure 2. Intrinsic programing.

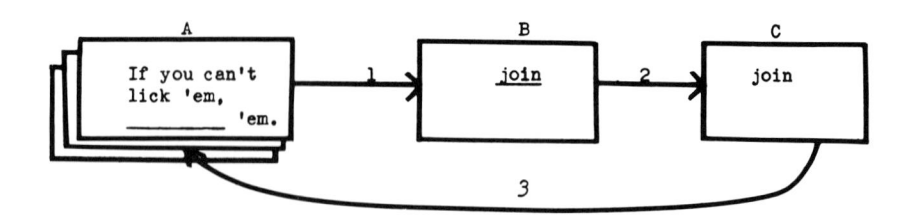

Figure 3. A linear program frame.

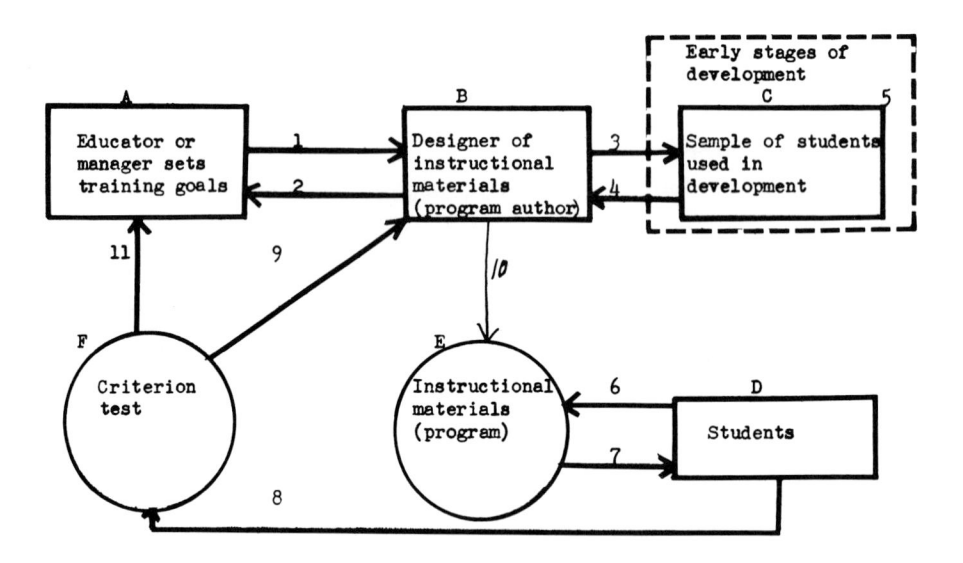

Figure 4. A systems view of instruction.

in the blank, the student follows arrow 1 to box B where he writes out the word. (He may write directly in the sentence portion, or, in other cases, on a separate answer sheet, as diagramed here). Then, as indicated by arrow 2, the student reaches box C where the correct answer is printed. He notes that he was correct and goes to the next frame (arrow 3). If his answer was wrong, he could correct it and go on. It is called a linear program because each student proceeds in a straight-line fashion; everyone takes the same frames in the same order. Students may differ, however, in the speed with which they work, and some may finish in half the time required by the slowest students. Although the first frames in a program would be easy for anyone to get right, later frames would be practically impossible unless the student had gone through the program. The frames gradually become more and more technical or "difficult," and students are able to take them because of what they have learned in earlier frames.

The linear programing approach is based on, or, at least, analogous to, effective animal training procedures worked out in the psychology laboratory. This has caused critics to be skeptical because people, of course, do differ from animals —most obviously in language skills. But this hardly constitutes proof that similar learning procedures will not work for different species, if properly adapted and applied.

In the next section we will examine some current research findings, especially for linear programing.

PRESENT

Research on learning and training methods presents special problems. The first requirement is to specify the goals of training. If these are not spelled out, then an advocate of one method or another can always say that some other goal was met by this method that was not sufficiently well measured in the experiment. Once all the goals have been defined, measurement techniques are necessary. It is not sufficient to specify "understanding" of a particular area because understanding cannot be measured directly. Instead, the exact performances and test items to be mastered are needed. Then the relative skill of students taught by different methods can be observed and measured. Developing that kind of test is time-consuming.

In addition to these difficulties, there are two unsolved problems in educational research. First, student motivation over a short period may make any method look good. Related to motivation is the novelty effect, which may also mask real differences among methods. The second unsolved problem is that of control when human instructions are involved. During an experiment, instructors may act differently than they do day in and day out.

Perhaps for these reasons, most of the experiments done to compare various types of programed instruction have shown "no significant differences." That is, even though the methods were varied systematically from group to group, the amount learned was so similar that statistical tests revealed no above-chance variations. Notwithstanding these problems, research does give valuable indications of training effectiveness, especially when pursued over extended periods. Brief experiments have their place, too. One of their purposes is to suggest the possible success of larger evaluations of the same approaches.

By far most of the programs that have been written have been linear programs and most of the research on programed instruction has centered on linear programing. Questions are raised about the efficiency of the approach. Current programs often take more time to complete than other means. To be balanced against this finding, however, are some of the inherent advantages of programed instruction. Students can work individually, they can schedule themselves, and they can study at remote sites where other training would be impossible.

Textbooks can also be used in this manner, of course. However, they provide little structure for the study activity, and some students simply cannot learn well from ordinary printed matter. In other cases the text itself may be too difficult or confusing. A program tells the student just how to proceed, and the program is revised until it is clear. It is usually possible to guarantee a given level of learning if a program is used as directed.

Research results call into question some features of linear programing. The requirement for students to write out their responses has been shown unnecessary time and again. Students apparently learn just as much if they only "think" their answers, and they learn in much less time. There is an interesting exception, however. Cummings and Goldstein (3) found that students actually had to draw complex diagrams rather than just think about how they would draw them. The point is that written responses are necessary whenever the response itself is something new to be learned, like making accurate technical drawings. In addition, if the response is complex, the student needs to have a written record to compare accurately with the given right answer. (During program development, there is no question that written responses are necessary because the programer needs these to assess the clarity of each frame. The question relates only to routine use of fully-developed programs.)

About 350 programs of all types are available. Costs vary from $2.00 to over $17.00, and lengths vary from under 50 to over 10,000 frames. Typical cost is $5.00 for a 1000-frame program. These figures correspond to study times of under one hour to over 100 hours. The most common length is 1000 frames, requiring 10 hours of study. Mathematics is the most frequently programed topic, with science subjects running second. Business Education, Economics, Grammar, and Social Studies are also programed. Catalogues of programs are available (4). It is difficult to estimate how wide-spread is the use of programed instruction. Most use, in public schools and colleges, is on an experimental basis so far. Industry uses many programed courses, and private individuals also use them for self-improvement. The Air Training Command of the U.S. Air Force has trained a large number of programers and is using programs in many locations.

Perhaps Pressey's adjunct autoinstruction has not attracted more research because it is so closely bound up with conventional lecture-text means. Investigators are content to believe that, even though adjunct autoinstruction has benefits in some situations, its value will always be affected by the conventional means employed. Improvement of the conventional means will be necessary to get the most out of Pressey's approach. In addition, the exclusive use of multiple-choice questioning seems arbitrary. Short-answer, constructed response items are valuable for some kinds of learning (12).

An experiment on Crowder's intrinsic programing has been completed at the

Behavioral Sciences Laboratory. For the particular program studied, the branching feature did not improve its effectiveness. College students did just as well when all the branches were removed from a program on computer arithmetic (10).

FUTURE

Programed instruction, especially of the linear type, is causing many educators to view learning as a closed-loop system. The system output (learning) is fed back into the instructional materials so that deficiencies can be corrected. This is in contrast to a lecture series, for example, where student learning is measured infrequently, and deficiencies may not bring about changes in the lectures. Figure 4 shows how the instructional process may be viewed as a system.

The educator or manager (box A) realizes that certain training goals must be met for the effective functioning of society, or business, or a military organization. Therefore, he contacts a designer of instructional materials and procedures, who is often an author of programed instruction (box B). The author considers the objectives presented by the manager, (arrow 1) and, interacting with the manager (arrow 2), transforms them into measurable specifications. Typically, a criterion test is developed at this time that will indicate the extent to which goals of training will be met by the materials yet to be developed. The author then selects one or two students (box C) to try out short portions of his materials. He administers the materials (arrow 3) and receives immediate knowledge of their effectiveness (arrow 4). Through this process, he revises materials until a potentially effective course has been designed. Typically, this process is used early in development, and then eliminated, as indicated by dashed line 5. Once the program or other material is written, it is used by students (box D) in a routine manner. They write, or perhaps just "think" their responses to the program (arrow 6). They find out from the program whether their responses are correct (arrow 7). After finishing the entire program (circle E) the students, following arrow 8, take the criterion test (circle F). Results of this test are fed (arrow 9) to the author. If results are unsatisfactory, he will revise the program (arrow 10) to make it more effective. In some cases the author may revert to the use of sample students, as in box C. Program revisions can be continued until the criterion test is passed by a required proportion of qualified students. At this stage we can refer to the program as "validated instruction." The educator or manager may receive criterion test results from the programer, or he may prefer to look for himself (arrow 11). The many feedback loops in this system make it possible for effective materials and procedures to be developed.

The fields of education and training are undergoing many changes. The most significant of these are the press of increasing student enrollments, and the increasing levels of skill required of graduates to cope with the problems of our technological age. Programed instruction will help alleviate these problems to some extent. The future will probably see greater use on a routine basis of well-designed programs in all subject-matter areas. Such use can reduce classroom burdens of instructors and make high-quality training available on an individual basis to students at home or during leisure hours. But training methods will also benefit from the side effects, or implications, of programed instruction. These are: better specifications of the goals of any course, with a corresponding drop in non-essential content; better assessment of the effectiveness of any educational procedure; and the system's concept of education.

Valuable though it is, programed instruction does not by itself represent a final solution to many educational problems. There is still a need for student contact with teachers, especially in discussions. There is still a need for student contacts with other students for formal and informal discussions and mutual help. There is still a need for excellent reference books, texts, films, recordings, libraries and laboratories.

One possibility being considered for the future is called "independent study," or "self-directed study." In this situation the teacher becomes a manager of instructional resources, and provides enough direction so that students can capitalize on the materials presented (including programed instruction). Class meetings are reduced, but small group discussions are increased. In cases where this has been explored, it has been found to work not only for the most intelligent students, but to be feasible also for the average or less qualified students (1).

Perhaps the future will reveal that a far-reaching consequence of programed instruction is the insistence upon proof that a given method or course really works. It makes good sense that procedures should be tried often enough and revised frequently enough to insure course effectiveness. Significantly, it is only through student responses that we can judge whether any given training procedure has been effective. These same responses are also important for the student's efficient progress through a course. The basic philosophy of programed instruction is a permanent contribution which will have an important effect upon future educational and training methods.

REFERENCES

1. Baskin, S., "Experiment in independent study (1956-1960)." *Antioch College Reports,* 1961, Antioch College, Yellow Springs, Ohio.
2. Cohen, Ira S., "Programed learning and the Socratic dialogue." *American Psychologist,* 1962, 17, 772-775.
3. Cummings, A., Goldstein, L. S., "The effect of overt and covert responding on two kinds of learning tasks." New York: The Center for Programed Instruction, 1962.
4. Hendershot, C. H., *Programed Learning—a bibliography of programs and presentation devices,* Carl H. Hendershot, 4114 Ridgewood Drive, Bay City, Michigan, 48707).
5. Little, J. K., "Results of use of machines for testing and for drill upon learning in educational psychology." *Journal of Experimental Education,* 1934, 3, 45-49.
6. Mellan, I., "Teaching and Educational Inventions." *Journal of Experimental Education,* 1936, 4, 291-300. Also in A. A. Lumsdaine and R. Glaser, eds., *Teaching machines and programed learning: a source book,* National Education Association, Washington, D.C., 1960.
7. Pressey, S. L., "A simple apparatus which gives tests and scores—and teaches." *School and Society,* 1926, 23, 373-76. Also in A. A. Lumsdaine and R. Glaser, eds., *Teaching machines and programed learning: a source book,* National Education Association, Washington, D.C., 1960.
8. Pressey, S. L., "Teaching machine (and learning theory) crisis." *J. Applied Psychol.,* 1963, 47, 1-6.
9. Pressey, S. L., "A puncture of the huge 'programing' boom." *Teachers College Record,* 1964, 65, 413-418.

10. Senter, R. J., Neibert, A., Abma, J. S., and Morgan, R. L., "*An evaluation of branching and motivational phrases in a scrambled book.*" Tech. Doc. Rep. No. AMRL-TDR-63-122, 6570th Aerospace Medical Research Laboratories, Wright-Patterson AFB Ohio, Nov. 1963. (AD 429 459)
11. Skinner, B. F., "The science of learning and the art of teaching." *Harvard Educational Review*, 1954, **24**, 86-97. Also in A. A. Lumsdaine and R. Glaser, eds., *Teaching machines and programed learning: a source book*, National Education Association, Washington, D.C., 1960.
12. Williams, Joanna P., "The effectiveness of constructed response and multiple-choice programing modes as a function of test mode." Presented at Eastern Psychological Association Convention, April 1964.

40

Computer-Assisted Instruction

R. C. Atkinson and H. A. Wilson

Ten years ago the use of computers as instructional devices was only an idea that was being considered by a handful of scientists and educators. Today that idea has become a reality. Computer-assisted instruction, like other aspects of electronic data processing, has undergone an amazingly rapid development. This rate of growth is partly attributable to the rich and intriguing potential of computer-assisted instruction for answering today's most pressing need in education—the individualization of instruction. Many useful ideas, however, have not achieved realization as quickly as computer-assisted instruction. The favored growth pattern of this method of instruction then must involve causes other than just a rich potential for meeting an educational need.

At least three other factors may be cited as contributing heavily to the growth of computer-assisted instruction. One of the most important was the development of programmed instruction. The surge of interest in programmed instruction during the 1950's, stemming primarily from the work of Skinner (1), focused the interest of educators on the problem of individualized instruction. Even though the actual results of programmed learning fell somewhat short of the glowing predictions of its early prophets, it left educators in a state of "rising expectations." The feeling remained that somehow through the use of science and technology the instructional process might eventually be tailored in a meaningful way to match the already known differences in motives and abilities among students.

The second factor contributing to the growth of computer-assisted instruction

From *Science*, **162**, October 4, 1968, 73-77. Copyright 1968 by the American Association for the Advancement of Science.

has been the mushrooming of electronic data processing in general. More specifically, however, the introduction of time-sharing systems and the design and production of third-generation computers has provided a major impetus to computer-assisted instruction. The early pioneering work at the University of Illinois on the Plato I system, which could handle only one student terminal at a time, furnished the foundation for further development. With the advent of time-sharing and the capability of the central processor to maintain more than one student terminal simultaneously, the wedding of programmed learning and electronic data processing got under way.

A third factor, and one of no less importance than those previously mentioned, has been the increasing aid to education by the federal government. In particular, the National Science Foundation and the various funding agencies which came into being under the Elementary and Secondary Education Act of 1965 have contributed substantially to the growth of computer-assisted instruction. Experimentation and development in the area of electronic data processing, particularly in the third-generation systems, has been an expensive process. Without supporting funds from the various government agencies and private philanthropic foundations (Carnegie, Ford, and others) the notion of applying electronic data processing capabilities to the problems of instruction might still be an idea discussed abstractly in a few technical journals.

Due to the interaction of the above factors, computer-assisted instruction has grown in less than 10 years to a point where during the school year of 1967–68 several thousand students ranging from elementary school to university level received a significant portion of their instruction in at least one subject area under computer control. In the Stanford projects alone approximately 3000 students were processed daily. Serious applications of computer-assisted instruction are now in progress in many universities throughout the United States: a list of those that have had major programs under way for two or more years includes Stanford University, University of California at Irvine, University of Texas, Florida State University, University of Illinois, Pennsylvania State University, University of Pittsburgh, State University of New York at Stony Brook, and Harvard University. The University of California at Irvine, which is a relatively new university, has made a serious attempt from its earliest planning stages to integrate computer-assisted instruction into its total instructional program (2).

Computer-assisted instruction has been used in university centers and is now moving into the public schools. Philadelphia's was the first major school system to implement computer-assisted instruction independent of university development or sponsorship. Philadelphia was followed closely by New York City where a significant project in computer-assisted instruction began its initial phase of operation during 1967–68 and will be in full operation during the school year 1968–69. Projects in several other school districts are in the planning stages and will be in an initial implementation phase during 1968–69.

Industry has also become deeply involved in the field of computer-assisted instruction, particularly in the design and production of totally integrated hardware-software systems. IBM was a pioneer in this area with the production of the 1500 Systems which will be in operation in over a dozen installations throughout the country during the 1968–69 school year. Philco-Ford was next, entering the market with the system currently in use in the Philadelphia public schools. More recently,

Instructional Systems was organized as a division of RCA. The RCA Instructional 70 System is now in its debugging phase in the New York public schools and will commence full-scale operation at the beginning of the 1968–69 school year. Applications of these commercial systems have covered a wide range of content and method, from relatively simple drill and practice in elementary arithmetic to sophisticated simulation exercises in college level science courses.

Of equal importance to the development of hardware and time-sharing systems is the development of instructional programs, the curriculums to be used with the system. Several major publishers are entering this vital area of computer-assisted instruction either alone or in collaboration with one of the hardware manufacturers. Harcourt, Brace and World, L. W. Singer, Harper and Row, and Science Research Associates all have programs in preparation. The heavy, long-range financial commitment of both publishers and hardware manufacturers is an index of the present reality and future development.

THE STANFORD PROJECT

The growth of the computer-assisted instruction project at the Institute for Mathematical Studies in the Social Sciences at Stanford University is illustrative of the development of the field over the past several years. Beginning in 1963 with a grant from the Carnegie Foundation, we set about to develop a small tutorial system. Since there were no integrated computer-assisted instruction systems available at that time, we assembled a system from components produced by several manufacturers. The central processor of that first Stanford system was a PDP-1 computer produced by Digital Equipment Corporation, working from a disk on an IBM 7090. The system used an IBM film-chip projector and a Philco cathode-ray tube, both equipped with light pens, as visual presentation and student response devices; also included in the system was a Westinghouse "random access" audio device. The technical difficulties of forging a unified system out of such diverse components were enormous. However, most of the difficulties were overcome, and the system went into operation. Six student stations functioned simultaneously, providing instruction mainly in elementary mathematics and language arts. Elementary school students were brought to the Stanford laboratory by bus and received instruction on a more or less regular daily basis.

Encouraged by our initial success on the first Stanford system a sizable grant was obtained from the U.S. Office of Education under Title IV of the Elementary and Secondary Education Act for the development and implementation of a computer-assisted instruction program in initial reading and mathematics for culturally disadvantaged children. At this point IBM, in collaboration with the Stanford group, undertook the design and development work on the IBM 1500 System and an author source language known as Coursewriter II. After major developmental efforts by both IBM and Stanford, the 1500 System was installed at the Brentwood Elementary School in the Ravenswood City school district in East Palo Alto and went into operation in the fall of 1967.

The 1500 System consists of an IBM 1800 Central Processing Unit with bulk storage maintained on tape and interchangeable disks, a station controller, and peripheral devices including a card reader and line printer. The student terminal interface consists of a cathode-ray tube, a typewriter keyboard, a light pen by means of which touch probe responses may be made on the face of the cathode-ray tube, an image projector with a capacity of 1000 frames which may be randomly accessed under computer control, and a set of earphones and a microphone (Figure 1). Audio messages may be played to the student from a bank of audio-tape playing and recording devices. One hundred and eighty minutes of audio messages may be stored on each of the three-track tapes and may be randomly accessed under computer control.

By the end of the second year of operation of this system (June 1968), approximately 400 students had received a major part of their daily instruction in either reading or mathematics under computer control. The 1500 System has been classified as a tutorial system in the sense that a very rich branching structure allows real-time instructional decisions to be made on what material is to be presented next based on the student's last response or upon an evaluation of some subset of his total response history. The Stanford-Brentwood laboratory was the first installation of its kind in an ongoing school environment, and it has therefore received considerable national attention from professional journals and the popular press and from television coverage. In addition, over 3000 visitors a year have observed students at work on the system. More importantly, significant gains in student achievement have been observed in each of the 2 years of operation (3).

Parallel to the development of the 1500 System, a second computer-assisted instruction system based on a considerably different design has been developed by the Stanford group (4). The system, known as the Stanford Drill and Practice System, uses a Digital Equipment Corporation PDP-1 central processing unit with a high-speed drum for bulk storage and model 33 teletype units at the student interface.

Although the hardware configuration on the drill and practice system is much simpler than that of the 1500 tutorial system, an even greater difference is found in the data management and branching structures. The drill and practice system does not have the real-time branching capability of the tutorial system. Individualization is accomplished through an off-line update where the performance of each student on day t is examined overnight and the appropriate lesson material is selected, based upon that performance record, for presentation to the student on day $t + 1$. The basic assumption in the drill and practice mode is that concepts are presented and developed by the teacher in the classroom, and the computer system furnishes intensified drill and practice on those previously developed concepts at a level of difficulty appropriate to each student.

During the first year of operation of the system (1965), 41 fourth-grade students received drill in elementary arithmetic computational skills at remote terminals in Grant School (Cupertino Union school district, near San Jose, California). In the 1967–68 school year approximately 300 students received daily lessons in arithmetic, spelling, logic, and elementary Russian in seven nearby schools and in locations as far distant as McComb, Mississippi, and Morehead, Kentucky, all under control of one central computer located at Stanford. With the addition of the logic and Russian programs, the distinction between drill-and-practice and tutorial programs becomes extremely blurred.

The Stanford project of computer-assisted instruction has expanded rapidly from its rather modest beginnings, and throughout its period of growth, the project

Figure 1. Student terminal used for tutorial instruction in initial reading.

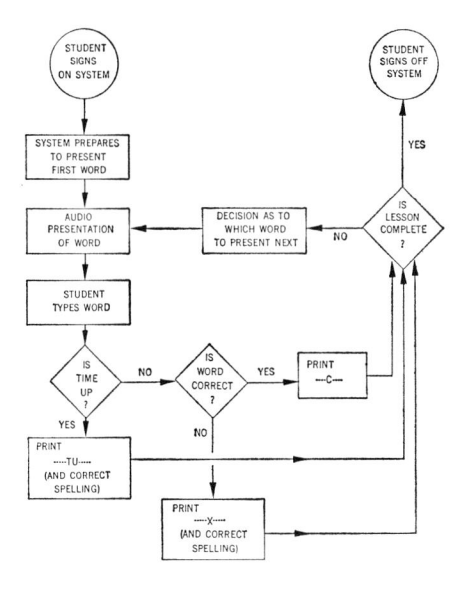

Figure 2. Flow chart for one type of spelling lesson used in the drill-and-practice program at Stanford.

has had an important influence on the development of computer-assisted instruction. Let us turn our attention now to other modes of development as exemplified by a few selected projects.

CURRENT MODES OF COMPUTER-ASSISTED INSTRUCTION

The tutorial and drill-and-practice procedures described above in the context of the Stanford project are by far the most prevalent modes of computer-assisted instruction. However, they are both essentially simulations of normal teacher-student interactions and home-work assignments. Their value lies in the degree of individualization of those activities and the increase in efficiency which can be brought about through the unique capabilities of electronic data management. Another mode of application of computers to the instructional process has been pursued by Systems Development Corporation (SDC). A college-level statistics course developed by SDC and implemented at University of California at Los Angeles uses a source language called *PLANIT* which provides the student with a powerful computational tool. By means of this system the student can manipulate large and complex data bases. This introduces an important element of realism, particularly in a course in statistics, and gives the student practice in handling realistic data. The SDC program is also illustrative of the general use of the computer as a laboratory tool in mathematics and science courses.

The use of games and simulations is being explored in a number of projects. An economics simulation has been developed by the Board of Cooperative Educational Services in Westchester County, New York, called the Sumerian Game, in which the student rules a mythical empire through his actions at critical decision points. The results of his decisions on the allocation of manpower and resources are extrapolated by the computer and the interactions of economic factors in complex situations are graphically illustrated to the student through his manipulation of the relevant parameters.

A computer simulation program involving laboratory experiments in chemistry has been developed by Bunderson (5) at the University of Texas. This program, which is an important component of a developing computer-assisted course in chemistry, frees the student from the time-consuming task of handling complex and sometimes dangerous equipment and allows him to concentrate on observation and the logical dynamics of analysis.

The ultimate computer-based instructional system is one in which the student could input free-form questions and statements which would be analyzed by the system and understood in the sense that the system would then compose and display appropriate replies (6). We are some distance from that goal at the present. However, the logic program developed by Suppes (4) at Stanford University is a step in that direction. In this program the student is required to carry out logical derivations and algebraic proofs. The system will accept any line in the proof or derivation that does not violate the rules of logic. Thus, the student and the system can achieve a kind of free interaction, at least within the confines of the very restricted language of elementary logic.

The above is but a brief sampling of the variety of applications of computers to education that are currently available. Let us turn our attention now to some of the problems that confront workers in the field of computer-assisted instruction.

CURRENT PROBLEMS

A variety of technical problems concerning both hardware and software design remain unsolved. The cathode-ray tube is the most flexible device for displaying graphic information, but at present, it has serious limitations. The resolution is not adequate for many purposes, and tubes must be placed usually at a distance of not more than 180 meters from the computer because of broadband transmission problems. By their very nature, cathode-ray tubes require continuous regeneration of the image. This requirement presents problems both of cost and limitations on the number of terminals that can be maintained on a given system. A plasma display tube is under development by Bitzer (7) at the University of Illinois, however, which may solve at least some of the problems encountered with the video display devices currently available. The plasma tube does not require image regeneration since the decay interval is extremely long. This will greatly decrease the cost of maintaining the image on the tube and increase the number of terminals that can be handled simultaneously.

Random-access audio tape units are plagued by a host of mechanical and physical problems, not the least of which is the trade-off between message capacity and search time. Work being carried on at Stanford and at other centers on audio problems points to the efficient use of digitized audio in the near future. The major problem in storing audio in digital form is the cost of both bulk and rapid access storage components. Partial relief on that problem is anticipated within the next 2 years in view of recent developments in the area of data storage.

Costs are a recurring problem in almost all aspects of computer-assisted instruction. Costs per terminal hour are relatively high even with the simplest systems available, and they increase with the addition of sophisticated audio and graphic display components. The major costs in this respect, however, are associated with the terminal hardware itself. Considerable reduction in these costs can be anticipated in the next few years as equipment design becomes more standardized and efficient production methods are brought into play. Telephone line charges also play an important role in the cost structure when maintaining terminals at remote locations.

In general, technical solutions to problems of hardware design can be expected to reduce the cost per terminal hour. Organizational solutions will also play a part in reducing costs by providing for maximum use of the system. Extension of the instructional day will have a desirable effect as will the prorating of the central processing unit's costs over tasks such as record keeping, budget planning, course scheduling, and others (8).

Of a much more serious nature than technical improvements or reduction of hardware costs is the problem of premature evaluation or of evaluation questions stated in the wrong terms. Attempts at a general evaluation of computer-assisted instruction in terms of cost and effectiveness are premature in two respects. The costs, as has been suggested above, are unrealistic in even the short-term sense. Hardware manufacturers are only beginning the transition from development to production. As the transition continues over the immediate future, the per unit costs will be reduced accordingly. Second, measurements of effectiveness are difficult to achieve given the current lack of a sound theoretical basis for describing levels of learning and chievement. What is needed is a definition of some standard unit, some "erg" of learning and forgetting. Definition of such a unit is far from realization.

At a more intuitive level it must be clearly understood that evaluation of a computer-assisted instruction program is only partially an evaluation of the system and equipment. Primarily it is an evaluation of the instructional program and as such is basically an evaluation of the program designer who is the real teacher in a computer-assisted instruction system.

The evaluation question then becomes, "To what extent did the curriculum designer provide the computer with an appropriate set of instructional materials and an adequate decision structure for branching among them?" Unfortunately, curriculum design is still more of an art than a science. However, computers are a unique instructional tool in that we can embody in their programs what scientific knowledge we currently possess about human learning; at the same time they hold the promise of increasing that knowledge at an astounding rate if proper use is made of the response data which they can collect. For example, Grubb (9) at IBM is developing a qualitatively new approach to computer-assisted instruction, and at the same time is investigating important differences in cognitive style through a learner-controlled statistics course. Similarly, analysis of data from the Stanford-Brentwood project will help us to better understand how young children acquire reading skills (3). As a further example, data from the drill-and-practice program in mathematics have been used to develop performance models that predict a variety of response statistics generated by arithmetic tasks (4).

One of the primary aims of computer-assisted instruction is to optimize the learning process. This is implicit in the concept of individualized instruction. A major focus of the research effort at Stanford is the development and testing of instructional strategies expressed as mathematical models. An important class of such models may be called optimization models since they prescribe the sequence of instructional events which will produce optimum learning within certain boundary conditions. Such optimization models are generally extremely difficult to investigate in a rigorous way for complex learning procedures. The problem can be attacked, however, at the level of fairly simple learning tasks; to be sure, these simple tasks do not encompass all of the instructional processes of interest even at the elementary-school level, but they include enough to warrant careful investigation. Analyses of these tasks will, it is hoped, provide guidelines for the investigation of the more cognitively oriented instructional procedures.

An example of an optimization procedure is provided by one type of spelling lesson used in the drill-and-practice program at Stanford (10). A list of N words are to be learned. The instruction essentially involves a series of discrete trials: on each trial the computer selects a word to be pronounced by the audio system, the student then responds by typing the word, and the computer evaluates the student's answer. If the response is correct the computer types –C–; if incorrect, –X– followed by the correct spelling. A flow chart summarizing this procedure is given in Figure 2. If n trials are allocated for teaching the list (where n is much larger than N), then the problem becomes one of finding a decision rule that will maximize the amount of learning. In general, such decision rules can be classified into two types: those that make use of the student's response history on a moment-to-moment basis to modify the flow of instructional materials, and those that do not. The resulting strategies have been termed response sensitive and response insensitive (11). The response-insensitive strategies are usually less complicated and can be specified completely in advance so that they do not require a system capable of branching during an instructional session. The programs developed by Skinner (1) and his associates are examples of response-insensitive strategies.

In order to illustrate a response-sensitive strategy, let us assume that the learning process for the spelling task described above is adequately described by the one-element model of stimulus sampling theory (12); in essence, this is a mathematical model which postulates that the learning of a given item occurs on an all-or-none basis. Under the assumptions of the model the optimum strategy is initiated by presenting the N items in any order on the first N trials, and a continuation of this strategy is optimal over the remaining $n - N$ trials if, and only if, it conforms to the following rules. (i) For each item set up two counters; one (designated the P-counter) to keep track of the number of times the item has been presented, and the other (the R-counter) to count the length of the most recent run of correct responses to the item. At the end of trial N set all the P-counters to 1, and all the R-counters to 0. (ii) On any trial, present an item if its R-count is least among the R-counts for all items. If several items are eligible, select from these the item that has the smallest P-count for presentation. If several items are still eligible under this condition, then select from this subset the item that had the slowest reaction time on its last presentation. (iii) Following a trial, increase the P-counter for the item presented by 1. Also, increase the R-counter for the presented item by 1 if the subject's response was correct, but reset it to 0 if his response was incorrect. Leave all other counters unchanged.

Even though these decision rules are fairly simple, they would be difficult to implement without the aid of a computer. Data from our experiments indicate that the above strategy is far better than one that presents the items equally often in a predetermined order. Another potentially more useful model may also be derived that fixes the achievement criterion at some specified level, and produces a set of decision rules which minimize the number of trials required to reach criterion.

These are examples of extremely simple optimization strategies. Others under investigation (3, 11, 13) make use of more realistic assumptions regarding the learning process and use more powerful mathematical techniques to derive optimum strategies. Of greater importance, they attempt to optimize performance not only within a given day's session, but from one unit of the curriculum to the next. The development and testing of viable models for optimizing instruction have just begun but show great promise for the future. These problems have received little attention in the past because optimization strategies that have been derived for even the simplest learning tasks are uually too complex to incorporate into an instructional setting without the data-managing capability of the computer.

SUMMARY

We have briefly reviewed the rapid growth of computer-assisted instruction from its beginning some 10 years ago to its present realization in many schools and universities. We have also characterized several different modes of application and discussed some current problems. The use of computers as educational tools is still extremely limited when one considers their potential for improving the instructional process. Many problems remain to be solved; the obvious problems of hardware and costs as well as the deeper problems of understanding the learning process more fully and applying that knowledge in both curriculum development and evaluation.

Because of the shortage of funds for research on learning, only a small segment of the scientific community is involved in work on computer-assisted instruction. However, the theoretical and practical problems to be solved in this area are exciting and engrossing for the scientist who wants to apply his skills to the pressing problems of society. There is every reason to expect that the area will be able to attract top-rank scientific talent and, in the not too distant future, make a direct impact on education.

REFERENCES AND NOTES

1. B. F. Skinner, *The Technology of Teaching* (Appleton-Century-Crofts, New York, 1968).
2. R. W. Gerard, in *Computers and Education*, R. W. Gerard, Ed. (McGraw-Hill, New York, 1967).
3. R. C. Atkinson, *Amer. Psychol.* **23**, 225 (1968).
4. P. Suppes, M. Jerman, D. Brian, *Computer-Assisted Instruction: Stanford's 1965-66 Arithmetic Program* (Academic Press, New York, 1968); P. Suppes, L. Hyman, M. Jerman, in *Minnesota Symposia on Child Psychology*, J. P. Hill, Ed. (University of Minnesota Press, Minneapolis, 1967).
5. V. Bunderson, "The role of computer-assisted instruction in university education," *Progress Report to the Coordination Board of the Texas College and University System* (University of Texas, Austin, 1967).
6. K. L. Zinn, *Rev. Educ. Res.* **37**, 618 (1967).
7. D. L. Bitzer, & H. G. Slottow, "Principles and application of the plasma display panel," *Proceedings of the OAR Research Applications Conference* (Institute for Defense Analyses, Washington, D.C., 1968), vol. 1, pp. a1-a43.
8. F. F. Kopstein, & R. J. Seidel, *Computer-Administered Instruction Versus Traditionally Administered Instruction: Economics*, Professional Paper 31-67 (Human Resources Research Office, Alexandria, Va., 1967).
9. R. Grubb, *Programmed Learning Educ. Tech.* **5**, 38 (1968).
10. E. Fishman, L. Keller, R. C. Atkinson. *J. Educ. Psychol.*, in press.
11. G. J. Groen, & R. C. Atkinson, *Psychol. Bull.* **66**, 309 (1966).
12. R. C. Atkinson & W. K. Estes, in *Handbook of Mathematical Psychology*, R. D. Luce, R. R. Bush, E. Galanter, Eds. (Wiley, New York, 1963), vol. 2; R. C. Atkinson & R. M. Shiffrin, in *The Psychology of Learning and Motivation: Advances in Research and Theory*, K. W. Spence and J. T. Spence, Eds. (Academic Press, New York, 1968), vol. 2.
13. R. D. Smallwood, *A Decision Structure for Teaching Machines* (MIT Press, Cambridge, Mass., 1962); G. Pask, in *Automaton Theory and Learning Systems*, D. Steward, Ed. (Academic Press, New York, 1966).
14. Supported by NASA grant NGR-05-020-244.

Chapter Twelve

Directions in Research

This chapter completes our survey of the training field. In the preceding portions of this part, we touched upon a host of research issues and areas in which research is sorely needed. By and large, however, we organized our comments and material around the major *problems* facing the training "establishment" today. Difficulties and shortcomings in both research and practice were freely intermingled in our discussion of each problem (e.g., *evaluation* of training). In this chapter, our emphasis shifts from particular *problems* to particular *strategies* or *methods*. We examine separately a few of the current techniques that seem most appropriate for dealing with these problems. The reader will find that some of the approaches developed here are not completely unfamiliar; some of them appeared earlier in connection with one or more of the problems.

The first article of this chapter describes the traditional but very critical *job analysis* approach. Although he addresses himself to a particular system, Morsh gives us a fairly broad look at the procedural details, the data, and the problems associated with this method. At the present time we must rely heavily upon job analysis for the specification of training objectives (i.e., criteria), and for translation of these objectives into a training program. We must remember, though, that until a program is evaluated (*if* it is evaluated), any deficiencies in the *analysis* procedure will remain hidden. Thus, Morsh's discussion of the *validation* of job analysis techniques merits particular attention.

It is often necessary to design a new training system, selection program, or environmental effects program without the benefit of clearly relevant prior data. Fleishman (1967), for example, complains that ". . . much of the accumulated data and experience of the past are largely inapplicable and . . . the problem of skill identification, training, and performance must be restudied almost from scratch" (p. 1). Are jobs and abilities really so different that we must continually start from the beginning? Hopefully not. What this problem highlights, however, is our lack of a good *task taxonomy*—a method for categorizing tasks according to a classification scheme which is related to the principles of learning and training. Such a scheme would allow us, with at least some degree of confidence, to generalize research or training data from one working environment to another—even if the "other" was an entirely new job situation. We noted earlier that one of the basic problems in training design is the shortage of useful theoretical formulations in the more complex areas of human learning, e.g., perceptual-motor skills. Fitts (1965) notes this difficulty in one of his many pleas for improved "taxonomizing." He says (p. 178):

References will be found in the General References at the end of the book.

The importance of an adequate taxonomy for skilled tasks is widely recognized in all areas of psychological theorizing today. A taxonomy should identify important correlates of learning rate, performance level, and individual differences. It should be equally applicable to laboratory tasks and to tasks encountered in industry and in military service.

A more recent analysis by Noble (1968) suggests that the problem is still with us. "Still," he says, "we are bedeviled with taxonomic difficulties and the lack of serviceable units of analysis . . ." (p. 210). In the second article of this chapter, Fleishman presents a methodology which strikes at the heart of the taxonomy issue. More important, he is able to support his case with data showing the ability of this approach to integrate at least *some* previously conflicting results. At the present time, it is impossible to predict how successful Fleishman's approach will ultimately be; it is certain, however, that our ability to generalize both training and research results in the future is going to depend heavily upon our success in establishing meaningful task dimensions.

The third article takes us back to the matter of individual differences. As we said earlier, learning theorists and training researchers have traditionally concentrated their attention on group performance. In this approach, individual differences are viewed as little more than an annoying "error" problem. During the past few years, however, there has been a growing interest in individual differences per se, and innovations in training apparatus (such as those discussed in Chapter Eleven) have increased our capability to deal with them effectively. Efforts are now being made to promote learning efficiency by designing the environment to complement the needs and capabilities of the individual learner. The importance of the individualized approach is underscored by a number of recent studies in which the effectiveness of particular training variables or techniques was shown to differ according to the subject's ability level. For instance, a study by Reed and Haymen (1962) showed that a programmed English text produced better performance than conventional instruction for *high* achieving students, but exactly the opposite effect for *low* achievers. In another study, Shay (1961) found that the size of "steps" used in presenting programmed material was important for low-intelligence students but not for more capable ones. Eckstrand's paper, then, is presented here as an elaboration upon the individual difference strategy in training. It affords an excellent discussion of methods used in this approach and of their implications for training.

The fourth paper in this chapter deals with a more specific methodological concern: the question of how best to train *teams*. Despite our technological achievements—and in some cases because of them—there are still many tasks that require the coordinated effort of several individuals working together as a team. We encountered examples of team tasks in our discussion of multiman-machine systems (Part Three). In that discussion we noted that there has been no lack of interest in such topics as interpersonal behavior and group performance (although the effort even in these areas has been somewhat diffuse). Where research has lagged, however, is in the area of team *training*; we know very little about how to establish effective coordination of effort among team members. What are the best training procedures? What basic learning or transfer principles are applicable to this situation, and how are they best implemented? And finally, how is team effectiveness most properly measured? We cannot, of course, offer any satisfactory answers to these three vital questions, although a start in this direction is afforded by Briggs and Johnston's work which we illustrate in our fourth selection (see also Briggs & Johnston, 1967).

The particular paper from this series which we have chosen to feature is noteworthy in several respects. First, it illustrates a promising research methodology. Second, it discusses some of the difficulties inherent in the study of coordinative training. Third, it presents some suggestive empirical findings. And finally, it shows indirectly the importance of adequate criterion specification. Training can affect a number of component skills and behaviors in a team situation; we must be certain which ones we wish to promote in designing our training or research vehicle. A more pointed illustration of this latter issue is provided by yet another of the Briggs and Johnston (1966) papers. There, transfer from a simple to a complex task criterion resulted in persistence of behavior appropriate to the *simple* criterion but inappropriate to the *complex* one. In research, then, as well as training practice, we must be ever mindful of the criterion problem. For further discussion of team training issues, the reader is referred to Boguslaw and Porter (1962), Glanzer (1965), and Johnston (1967).

The preceding chapters have emphasized the *interaction* of the various aspects of a training program, drawing together work on training requirements, development of training environments, and measurement of training evaluation. As our training programs increase in complexity, it will be necessary to consider the interactive effects of these components on the total training effort. The systems analysis approach as presented in Chapter Seven can serve as a model for the examination and evaluation of training programs.

41

Job Analysis in the United States Air Force

Joseph E. Morsh

In the search for a job analysis method that would have the greatest potentiality for the systematic collection, quantification, and organization of information about

From *Personnel Psychology*, 17, 1964, 7-17. By permission of Personnel Psychology, Inc.

The research reported in this paper was conducted both in service and by contract during the past three years under the sponsorship of the 6570th Personnel Research Laboratory, Aerospace Medical Division (AFSC), under Project 7734, Tasks 773401 and 773403. The results of this research have been incorporated in the January 1963 revision of Air Force Manual 35-2.

Air Force jobs, the literature was reviewed (Morsh, 1962a) and major government agencies who conduct job analyses were surveyed (Morsh and Ratliff, 1959). In order to show the background from which the Air Force method was developed, brief descriptions and summary evaluations of some of the more important job analysis methods are presented.

JOB ANALYSIS METHODS

The *questionnaire* is usually used to obtain occupational information by means of a mail survey. The respondent is asked to give identifying data and to describe his position in his own words. The method may prove useful for obtaining information from clerical and supervisory workers who write easily but may yield poor results from production and maintenance workers. Organizing the unstructured, handwritten information obtained from large numbers of incumbents presents a formidable problem. Furthermore, the data obtained tend to be incomplete because of the heavy demand placed upon recall.

In the *check list* method the incumbent checks the tasks he does from a list of task statements which describe a job. The check list is preceded by some kind of a job analysis and is usually followed by the development of work activity compilations or job descriptions. The scope of the task statements listed depends upon the judgment of the check list constructor. The sequencing of tasks and the relationships among tasks are not shown by the check list, and shared or two-man tasks are not identified. An over-all perspective of the job is virtually impossible from check list information. The check list, however, depends upon recognition rather than recall, and can be economically administered to large groups. Check list data are adaptable to machine tabulation.

The *individual interview* method of job analysis involves the selection of representative incumbents who are usually interviewed away from the job. The results of the interview are typically recorded in standardized form. Usually responses from several interviews are combined into a single job schedule. Like the questionnaire, the individual interview depends largely upon recall. The method is slow and costly and thus impractical for obtaining information from large samples.

The *observation interview* takes place at the work site while the incumbent performs activities being discussed. The analyst observes and questions the worker to obtain complete, specific, and accurate information. The disadvantages of the method are the requirement for an expert job analyst, its relative slowness, interference with operational activities, and greater cost than other methods.

For the usual application of the *group interview*, a large number of incumbents representative of the same job are selected. With the assistance of the analyst, the incumbents record identification data and recall and write down their work activities. They may also be asked for other information such as estimation of skills and knowledge, supervision, training, or physical requirements. At the conclusion of the meeting, the analyst combines the data derived from the incumbents into a single composite schedule for each job.

In the *technical conference* a group of experts, selected for their broad knowledge and experience, work together to record the activities that comprise the job under investigation. Frequently, however, experienced technical experts hold supervisory positions and have thus lost intimate contact with the work. They may also estimate job requirements in terms of their own extensive background rather than that of the average worker on the job.

The *daily diary* method requires incumbents to record their tasks day by day. The memory requirement is thus reduced and the frequency of task performance for any desired data collection period is readily determined. Tasks can be coded to facilitate statistical analysis and organization of the data into job descriptions.

The *work participation* method involves work performance by the job analyst. He may perform simple tasks with little or no instruction. In the case of more technical activities he must first learn the job and then work at it along with the regular incumbents. In a limited operation the method has certain advantages, but in a large and complex organization involving highly technical jobs the work participation method is obviously impractical as well as exorbitantly expensive.

The *critical incident* technique involves the collection of statements based on direct observation or recollection of behavior which typifies both excellent and unsatisfactory performance. The behavioral aspects of the job are emphasized and usually a great many incidents are collected. Such incidents may serve to identify the crucial requirements of the job but do not provide sufficient information for a complete job description.

THE AIR FORCE METHOD OF JOB ANALYSIS

The method developed for the Air Force combines features of the check list with those of the open-ended questionnaire and the observation interview. Using information derived from job descriptions, job training standards, and other available sources, a preliminary task inventory is constructed according to a standardized format. Subject matter experts act as consultants in the early stages of development to insure completeness of coverage and accepted use of terminology. The resulting instrument usually consists of from 200 to 300 task statements grouped under major functions which are called duties. A particular inventory covers all levels of one airman career ladder from apprentice through journeyman and supervisor to superintendent; or for officer jobs, all grades from junior officer through company grade to staff officer (Morsh, Madden and Christal, 1961).

The course outlines for each Air Force specialty published by the Air Training Command have proved the best sources of task and duty statements. Preliminary job inventories can be most readily constructed in those specialties which have long existing civilian counterparts, such as carpenter, cook, or dental technician. Tasks and duties in such areas have become standardized and published source materials are comprehensive and consistent. Inventory construction is most difficult for jobs in the new, complex, and rapidly changing fields, such as electronics (Archer and Fruchter, 1963).

A thorough field review is essential before using an inventory for an operational survey. The preliminary inventory is submitted to experienced incumbents at installations in several commands. Task statements judged to be inadequate are modified or deleted and new duties and tasks are added. The review information is then considered in the construction of a final operational form.

The revised inventory is administered to large samples of job incumbents who respond to the task statements. Since no incumbent performs all of the duties and tasks on the inventory, great care is taken in selecting representative samples. In specialties with large numbers of incumbents, a 10 to 50 per cent random sample may be sufficient. When the total number of incumbents in a specialty is small, the required sample may include all incumbents. For surveying airman jobs, group administration in base testing rooms is carried out by personnel trained to administer tests. Officer inventories are self-administered with all instructions contained in the booklets.

Incumbents are directed to add task statements that are not listed. In the past the compilation of such written-in information into consolidated job descriptions was a difficult and time-consuming process. Collation of added tasks is now simplified because statements tend to be written at the same specificity level as tasks already listed and are categorized by incumbents under the major functions of their jobs. The more complete the inventory, the fewer the tasks that are written.

Beginning each task statement with an action word and then arranging these alphabetically under major functional subdivisions facilitates recall of work done. The incumbent considers only one action word at a time. The inventory identifies tasks for the non-verbal incumbent who might have difficulty in writing down or telling a job analyst the many things he does. The fact that incumbents supply missing tasks which are incorporated in subsequent revisions, and delete or fail to respond to obsolete tasks makes the inventory a dynamic, self-correcting instrument for obtaining current job information.

One of the chief advantages of the Air Force method is its economy. Job inventories can be inexpensively constructed and administered by mail by Air Force personnel using existing facilities. The method permits sampling of individuals by name, by specialty, or by unit. Ten working days are allowed for completing the inventories after they have been delivered to a base. A 3000-case, world-wide survey of incumbents is possible within about thirty days. Thus changes in Air Force jobs may be detected early and routinely before they become problems to local commanders.

In addition to performance or non-performance of tasks, information about the incumbent and data descriptive of tasks performed are also obtained. In surveys thus far conducted, the following task characteristics are among those that have been examined: frequency of task performance, amount of supervision required or exercised, relative time needed to perform a task, complexity as estimated by the incumbent or by the supervisor, amount of training and knowledge required, difficulty as compared with other tasks, experience required for effective performance, and importance of the duty or task to unit mission. The rating of some of these attributes has presented a number of unexpected problems. For instance, when incumbents were asked to state why they found certain tasks difficult, the reasons given could be reliably classified into ten categories, nine of which represented different definitions of difficulty (Madden, 1962).

RELIABILITY OF JOB INVENTORY RESPONSES

Test-retest reliability in reporting frequency of task performance and length of task time has been fairly satisfactory, with mean coefficients of about .70. Frequency of task performance is more reliably reported than relative proportion of time spent, or task difficulty. A one-month recall period is more reliable than a six-month period for reporting time required and task difficulty, but a six-month period is more reliable for reporting task occurrence. The six-month period elicits more written-in tasks than a one-month period, but respondents prefer the one-month recall period. The mechanics of recording responses, whether by means of IBM mark-sense cards or checking in a booklet, apparently has no effect on the consistency of reporting task information. Consistency in reporting task occurrence, however, is not generally related to consistency in reporting other kinds of task information (McCormick and Ammerman, 1960). Supervisors and incumbents show no systematic differences in consistency in reporting information about incumbents' work (McCormick and Tombrink, 1960). It was found that incumbents who were required to report more (as opposed to fewer) kinds of information about their tasks tended to provide more reliable information. If incumbents must read each task statement closely in order to follow instructions they will give reliable information, but if they are required just to check the statements they may not read them carefully (McCormick, 1960).

While interest has centered on internal consistency and test-retest reliability, other measures of reliability have been investigated (McCormick and Ammerman, 1960). For example, certain tasks known not to be performed by the sample being surveyed were included in an inventory. In another approach, groups of tasks were identified which were performed as a unit by one person. Since the tasks were always done together, responses indicating frequency of performance should be identical. The same task statements worded in two different ways were included in inventories to see if responses to each member of the pair would be the same. Although these devices appeared to have potential value as indicators of reliability, responses obtained were unrelated to incumbents' difficulty or time estimates or to reports of task performance or nonperformance.

VALIDITY OF JOB INVENTORY RESPONSES

The ultimate measure of the validity of task inventory data is the determination of the correspondence between the reporting of task inventory information and actual performance on the job. Direct observation of incumbents in the work situation is not entirely satisfactory, since incumbents may perform tasks which cannot be properly observed. The presence of the observer, also, may have complex effects on task performance which are difficult to determine.

Another procedure for obtaining validity data is to compare supervisors' reports of incumbent performance with responses obtained from the incumbents themselves. While it is true that supervisors do not always know exactly what particular incumbents do, exploratory studies have revealed only minor discrepancies when interview data from supervisors were compared with incumbents' inventory responses.

If a large proportion of incumbents in the same job respond to a task statement in the same way, a kind of intrinsic item validation has been obtained. The alternative possibility, that these responses are due to a consistent and uniform error in reporting, seems highly unlikely. If, however, few incumbents respond to the statements, or if the responses show great range in variability, other measures of validity must be sought. The feasibility of using such criteria as time cards, job work orders, and maintenance records is being investigated. The problem of adequate validation is not peculiar to the task inventory method, but presents difficulties in all methods of job analysis (as in other psychometric areas). Although some progress has been made toward the validation of task inventory responses, much more study will be required to achieve satisfactory validity measures.

GROUPING WORK ACTIVITIES

Work activities are organized into groups at three levels in the Air Force. Tasks are grouped into jobs, jobs are clustered into specialties, and specialties are assembled into career or utilization fields. When tasks are grouped to form jobs, the aim is to include, in the job, tasks which are homogeneous with respect to the kind of talent called for. Such homogeneity of tasks within a job obviously facilitates the selection and training of personnel.

At the next level, jobs are grouped into specialties. The specialty is the basic occupational unit used to manage the selection, classification, training, assignment, and rotation of personnel. To lend flexibility to the Air Force personnel system, provision is made for men to move freely from one job to another within a specialty. The specialty, therefore, should be organized to minimize cross-training times among the jobs composing it.

At the third level of organization, the several hundred specialties are assembled into career fields. These groupings of specialties must allow for the orderly progression of an officer or airman up a career ladder.

The analysis of data derived from occupational surveys can make an appreciable contribution toward the solution of grouping problems (Bottenberg and Christal, 1961; Ward, 1961). When jobs are to be assembled into specialties, the first problem is to identify and describe the job types involved. A job type exists whenever essentially identical duties and tasks are performed by more than one man. By means of an iterative computer procedure, the tasks performed by each man are compared with the tasks performed by every other man in the survey sample. The two individuals performing the most similar jobs are located and a single description, based on tasks performed, time spent on tasks, or other criterion, is developed to replace the two separate job descriptions. The process continues by adding a third job to the first pair or by identifying a new pair of similar jobs. The composite descriptions each time are based on minimizing loss of descriptive accuracy associated with each of the many thousands of alternatives. During the later stages of the grouping process, if dissimilar jobs are combined there is a sharp decline in descriptive accuracy. This sudden decline helps to determine the number of identifiable job types.

The problems of grouping work activities are only partially solved, but it is anticipated that samples of 2000 cases can be grouped into job types with a reasonable expenditure of computer time. Problems have been encountered, also, in deciding what makes jobs similar. If similarity is the amount of overlapping work time for example, two jobs could involve identical tasks and still be classified as not alike. If, on the other hand, job similarity is identified as the number of overlapping tasks, the times spent on these tasks might be entirely different. Not only the numbers, but also the kinds of tasks that overlap, is an important consideration (Christal, 1962).

In a survey of 772 incumbents in the Airman Supply Career Field, the per cent of incumbents doing each task was found to be consistent with the relative time spent on the task. Both time spent and per cent performing indexes had appropriate skill-level validity for tasks and duties. This means that airmen at the higher skill levels tend to do more of the difficult tasks and spend more time on these tasks than do lower skill-level airmen. Negative skill-level validity, as shown by cases in which apprentice airmen reported doing supervisory tasks, was rarely found (Gragg, 1962).

Group analysis of job data obtained in a more extensive survey involving 1220 incumbents in the Airman Supply Career Field revealed ten relatively clear-cut job types. The work done in this career field was found to be so heterogeneous, however, that many individuals perform unique jobs. Their work is not characterized by any of the identifiable job types.

ANALYZING JOB INVENTORY DATA

At present, inventory data are key punched on cards and then placed on magnetic tape for analysis. With a view to the elimination of the key punching step, a pilot study was undertaken to test the feasibility of using IBM mark-sense cards for recording inventory information (Gragg, 1963). By using machine scanning, errors made in recording information on mark-sense cards can be reduced to an acceptable level. In the exploratory study, incumbents were asked to rate tasks they performed on time spent and on training and experience scales. Correlation coefficients between scales were .31, .41, and .50, respectively, for three conditions of administration: (1) first rating covered, (2) first rating in view, and (3) ratings made together. The differences were significant beyond the .01 level of confidence. The study showed that mark-sense cards could be used to gather job information in an operational setting. The mark-sense method, however, is only a stopgap measure. It appears that within the next two or three years optical scanning procedures will be routine. In consequence, format and item organization are being studied so that inventory response may be adapted to optical scanning devices.

With the advent of high-speed electronic data processing equipment, it is possible to analyze occupational information rapidly and accurately in a great variety of ways. It is now economically feasible to use samples of incumbents numbering in the hundreds with the assurance that the results will be available almost immediately. The forms and procedures have been devised in anticipation that analysis by means of electronic computers will often be practicable and expedient. At the same time the procedures are readily adaptable to less elaborate methods of data analysis.

A number of other problems associated with the job inventory method are under study or have been programmed for investigation. These problems have to do with writing adequate task statements; compiling written-in job information; determining the optimum recall period of inventory responses; collecting information about responsibilities; measuring the completeness of job information obtained; using importance as a task attribute; determining the frequency of specialty surveys, the desirable specificity of work activity statements, and the most useful inventory format.

From a purely pragmatic point of view, the feasibility of the new Air Force method, at least as applied to airman jobs, has been demonstrated in an operating setting. The whole area of job analysis, however, is comparatively unexplored by the use of modern scientific techniques. Partial answers have been obtained to some of the problems, but for the most part the stage has merely been reached when problems have been identified and defined.

REFERENCES

Archer, W., & Fruchter, Dorothy A. "Construction, Review, and Administration of Air Force Job Inventories." (In press)

Bottenberg, R. A., & Christal, R. E. "An Iterative Technique for Clustering Criteria which Retains Optimum Predictive Efficiency." *Technical Note WADD-TN-61-30*, Lackland Air Force Base, Texas, 1961.

Christal, R. E. "USAF Occupational Research Projects of Tri-Service Interest." In "Proceedings, Tri-Service Conference on New Approaches to Personnel Systems Research." Washington: Office of Naval Research, *ONR Symposium Report* ACR-76, 1962, 62-75.

Gragg, D. B. "An Occupational Survey of an Airman Career Ladder: Supply Warehousing-Inspection." *Technical Documentary Report* PRL-TDR 62-19, Lackland Air Force Base, Texas, 1962.

Gragg, D. B. "Mark-Sense Card Methodology for Collecting Occupational Information." (In press)

Madden, J. M. "What Makes Work Difficult?" *Personnel Journal*, XLI (1962), 341-344.

McCormick, E. J. "Effect of Amount of Job Information Required on Reliability of Incumbents' Check-List Reports." *Technical Note WADD-TN-60-142*, Lackland Air Force Base, Texas, 1960.

McCormick, E. J., & Ammerman, H. L. "Development of Worker Activity Check Lists for Use in Occupational Analysis." *Technical Report WADD-TR-60-77*, Lackland Air Force Base, Texas, 1960.

McCormick, E. J., & Tombrink, K. B. "A Comparison of Three Types of Work Activity Statements in Terms of the Consistency of Job Information Reported by Incumbents." *Technical Report WADD-TR-60-80*, Lackland Air Force Base, Texas, 1960.

Morsh, J. E. "Job Analysis Bibliography." *Technical Documentary Report* PRL-TDR-62-2, Lackland Air Force Base, Texas, 1962. (a)

Morsh, J. E. "Research Findings Obtained with a Task Inventory Method of Job Analysis." *American Psychologist*, XVII (1962), 395. (b)

Morsh, J. E., Madden, J. M., & Christal, R. E. "Job Analysis in the United States Air Force." *Technical Report WADD-TR-61-113*, Lackland Air Force Base, Texas, 1961.

Morsh, J. E., & Ratliff, F. R. "Occupational Classification in Some Major Government Agencies." *Technical Report WADC-TN-59-37*, Lackland Air Force Base, Texas, 1959.

42

Development of a Behavior Taxonomy for Describing Human Tasks: A Correlational-Experimental Approach

Edwin A. Fleishman

A fundamental prerequisite to the effective operation of any man-machine system is the acquisition, performance, and retention of skilled behaviors. For many years psychologists have studied human learning and performance, often in applied contexts, under numerous task and environmental conditions, and have accumulated vast quantities of data. And yet, as new systems are conceived for the exploration of space, for defense, for command and control, it appears that much of the accumulated data and experience of the past are largely inapplicable and that the problems of skill identification, training, and performance must be restudied almost from scratch. Why is this the case? Why such a waste of prior findings?

Superficially, each new system differs from other systems with respect to application, mission, and technology, and apparently in its task demands on its human operators. No two task analyses are ever quite the same. No two systems ever have identical job requirements. No training device or simulator ever quite seems to fit the requirements of any system except the one for which it was developed. Is it reasonable to conclude that the tasks of men in systems are so varied that there are no common dimensions with respect to the basic abilities required, the types of training needed for job proficiency, or the degradation of skilled behavior under given environmental conditions?

The present lack of a set of unifying dimensions underlying skilled behavior would appear to require one to answer the question in the affirmative. Why else have so many prominent psychologists (e.g., Melton, Fitts, Gagné, Miller) called so often for so long for a method for classifying human tasks—for a "task taxonomy"?

From *Journal of Applied Psychology*, **51**, 1967, 1-10. Copyright 1967 by the American Psychological Association, and reproduced by permission.

Contribution to symposium "Task Taxonomy and its Implications for Military Requirements," at the American Psychological Association Convention, Chicago, September 1965. Preparation of this paper was facilitated by Contract No. DA-49-193-MD-2632 with the Army Surgeon General's Office, and by National Science Foundation Grant No. GB-1742.

The problem is not only one of finding ways to generalize principles from one operational system to another. It also involves the generalization of findings from the laboratory to operational tasks. One reason why much of current learning research in the experimental laboratory appears so sterile to those who try to apply it to real-life training situations is the lack of concern by learning psychologists for the problem of task dimensions. This, of course, is often true for laboratory studies of the effects of environmental factors, the effects of motivational variables, the effects of drugs, etc., on human performance. It is not too long ago when a favorite distinction was between "motor" and "mental" tasks or between "cognitive" and "noncognitive" tasks. Such distinctions are clearly not very helpful in generalizing results to new situations, which involve a complex array of tasks and skills not adequately described by such all-inclusive terms. Tasks selected in laboratory research are not often based on any clear rationale about the class of task or skill represented. Most learning theory is devoid of any concern about task dimensions, and it is this deficiency which, many feel, makes it so difficult to apply these theories in the real world of tasks and people. What is needed is a learning and performance theory which ascribes task dimensions a central role.

At the other extreme, one finds task descriptions, often developed by job and systems analysts, which are highly detailed and highly specific. These are extremely useful and necessary for a variety of purposes, but their very specificity limits the kinds of generalizations which can be made to new classes of skills and tasks.

Categories which conceive of man-task interactions in terms of classes of functions certainly would seem to be steps in the right direction. Thus, Gagné (1964) tends to use categories like discrimination, identification, sequence learning, problem solving; and Robert Miller (1965) uses terms like scanning, identification of cues, interpretation, short-term and long-term memory, decision making, etc. Psychologists working with the Lockheed approach use dimensions like tracking, vigilance, arithmetic, and pattern comparison (Mangelsdorf, 1965).

These categories may turn out to be highly useful in both (a) organizing psychological data into categories of consistent principles, or (b) allowing more dependable predictions from laboratory to operations and from one operation to another. The demonstration, of course, still needs to be made. One needs to recognize, naturally, that these approaches are initially arm-chair, rational descriptive approaches. And there is nothing wrong with this, provided the necessary experimental predictive work is carried out to test the utility of these systems. However, I am skeptical that any small number of categories is going to be successful. For example, one of the most careful systematic approaches is that developed by Alluisi (1965) in his extension of the Lockheed approach. I am sure that this work represents a major step forward in task standardization and that we will learn much from this work. However, the basic battery will consist of from six to eight tests covering categories such as vigilance processes, memory functions, communication functions, intellectual functions, procedural functions, etc. Everything known about the correlations among human performances indicates a greater degree of specificity than this and considerable diversity of function within these categories. Since there are many types of human functions within each subarea, one probably cannot generalize too far within each area. It is my feeling that we will just have to admit that human performance is complex and consists of many components. The problem is to simplify the description as far as we can, seeking all the while to find the limits and generality of the categories developed. And the selection of measures, diagnostic and representative of these categories, is an empirical rather than an arm-chair question.

The major point of this presentation is that there are empirical-experimental approaches to developing task taxonomies and that these represent alternative ways of going about it. It is my feeling that we already know quite a bit about task dimensions from experimental-correlational studies already completed, and that these allow us to be much more specific about task dimensions than do the more general categorical terms previously described. And I believe that combinations of experimental and correlational methods can develop a taxonomy of human performance which is applicable to a large variety of tasks and situations. Furthermore, such an approach yields empirical indexes (e.g., factor loadings) for diagnostic measures of the categories developed. I would like to elaborate these issues.

CONCEPTUAL AND METHODOLOGICAL FRAMEWORK

First, I would like to define some concepts which have been developed. Second, I would like to present some illustrative data of past and current work, and finally point up some directions in which we might go.

I find it useful to distinguish between the concepts of ability and skill. As we use the term, *ability* refers to a more general trait of the individual which has been inferred from certain response consistencies (e.g., correlations) on certain kinds of tasks. These are fairly enduring traits which, in the adult, are more difficult to change. Many of these abilities are, of course, themselves a product of learning and develop at different rates, mainly during childhood and adolescence. Some abilities (e.g., color vision) depend more on genetic than learning factors, but most abilities depend on both to some degree. In any case, at a given stage of life, they represent traits or organismic factors which the individual brings with him when he begins to learn a new task. These abilities are related to performances in a variety of human tasks. For example, the fact that spatial visualization has been found related to performance on such diverse tasks as aerial navigation, blueprint reading, and dentistry, makes this ability somehow more basic.

The term *skill* refers to the level of proficiency on a specific task or limited group of tasks. As we use the term skill, it is task oriented. When we talk about proficiency in flying an airplane, in operating a turret lathe, or in playing basketball, we are talking about a specific skill. Thus, when we speak of acquiring the skill of operating a turret lathe, we mean that this person has acquired the sequence of responses required by this specific task. The assumption is that the skills involved in complex activities can be described in terms of the more basic abilities. For example, the level of performance a man can attain on a turret lathe may depend on his basic abilities of manual dexterity and motor coordination. However, these same basic abilities may be important to proficiency in other skills as well. Thus, manual dexterity is needed in assembling electrical components, and motor coordination is needed to fly an airplane.

Implicit in the previous analysis is the important relation between abilities and learning. Thus, individuals with high manual dexterity may more readily learn the specific skill of lathe operation. The mechanism of transfer of training probably operates here. Some abilities may transfer to the learning of a greater variety of specific tasks than others. In our culture, *verbal* abilities are more important in a greater variety of tasks than are some other types of abilities. The individual who has a great many highly developed basic abilities can become proficient at a great variety of specific tasks.

Elsewhere (Fleishman, 1964; Gagné & Fleishman, 1959) we have elaborated our analysis of the development of basic abilities. This included a discussion of their physiological bases, the role of learning, environmental and cultural factors, and evidence on the rate of ability development during the life span. With this much conceptualization in mind, we can say that in much of our previous work one objective has been to describe certain skills in terms of these more general ability requirements.

The original impetus for this program was a very applied problem. While I was with the Air Force Personnel and Training Research Center, one of our missions was to build better psychomotor tests for the prediction of pilot success. The wartime Air Force program had been highly successful in developing such tests. For example, the Complex Coordination Test had consistent validity for pilots. This seems not surprising, since the test seemed to be a "job sample" of aspects of the pilot's job. The pilot does manipulate stick and rudder controls. But there were many tests which had substantial validity but did not at all "resemble" the pilot's job. Cases in point are the Rotary Pursuit and Two Hand Coordination Tests. And there were other tests which seemed to resemble aspects of the pilot's job but had no validity. So it seemed to me that the first step was to discover the sources of validity in these tests. What ability factors were there in common to psychomotor tests which were common to pilot performance (see, e.g., Fleishman, 1953, 1956; Fleishman & Hempel, 1956)?

Perhaps a not too extreme statement is that most of the categorization of human skills, which is empirically based, comes from correlational and factor-analysis studies. Many of these studies in the literature are ill designed or not designed at all. This does not rule out the fact that properly designed, systematic, programmatic, correlational research can yield highly useful data about general skill dimensions. We can think of such categories as representing empirically derived patterns of *response consistencies* to task requirements varied in systematic ways. In a sense this approach describes tasks in terms of the common abilities required to perform them. As an example, let us take the term "tracking," a frequent behavioral category employed by laboratory and systems psychologists alike. But we can all think of a wide variety of different tasks in which some kinds of tracking are involved. Can we assume that the behavioral category of tracking is useful in helping us generalize results from one such situation to another? Is there a general tracking ability? Are individuals who are good at compensatory tracking also the ones who are good at pursuit tracking? Do people who are good at positional tracking also do well with velocity or acceleration controls? What happens to the correlations between performances as a function of such variations? It is to these kinds of questions that our program was directed.

SOME PREVIOUS RESEARCH

In subsequent years we have conducted a whole series of interlocking, experimental, factor-analytic studies, attempting to isolate and identify the common variance in a wide range of psychomotor performances. Essentially this is laboratory research in which tasks are specifically designed or selected to test certain hypotheses about the organization of abilities in a certain range of tasks (see, e.g., Fleishman, 1954). Subsequent studies tend to introduce task variations aimed at sharpening or limiting our ability-factor definitions. The purpose is to define the fewest independent ability categories which might be most useful and meaningful in describing performance in the widest variety of tasks.

Our studies generally start with some gross area of human performance. Thus, we have conducted studies of fine manipulative performances (Fleishman & Ellison, 1962; Fleishman & Hempel, 1954a), gross physical proficiency (Fleishman, 1963, 1964; Hempel & Fleishman, 1955), positioning movements and static reactions (Fleishman, 1958a), and movement reactions (Fleishman, 1958b; Fleishman & Hempel, 1956).

Thus far, we have investigated more than 200 different tasks administered to thousands of subjects (Ss) in a series of interlocking studies. From the patterns of correlations obtained, we have been able to account for performance on this wide range of tasks in terms of a relatively small number of abilities. In subsequent studies our definitions of these abilities and their distinctions from one another are becoming more clearly delineated. Furthermore, it is now possible to specify the tasks which should provide the best measure of each of the abilities identified.

There are about 11 psychomotor factors and 9 factors in the area of physical proficiency which consistently appear to account for the common variance in such tasks. Their labels are Control Precision, Multilimb Coordination, Response Orientation, Reaction Time, Speed of Arm Movement, Rate Control, Manual Dexterity, Arm-Hand Steadiness, Wrist-Finger Speed, Aiming. In the physical proficiency area, a program was recently completed and described in book form (Fleishman, 1964); the factors have names like Extent Flexibility, Dynamic Flexibility, Static Strength, Dynamic Strength, Explosive Strength, Trunk Strength, Gross Body Coordination, Gross Body Equilibrium, and Stamina.

Of course, there are detailed descriptions of the operations involved in each category (see Fleishman, 1960b, 1962, 1964, 1966, 1967); some of them are more general in scope than others. But it is important to know, for example, that it is not useful to talk about strength as a dimension, but that, in terms of what tasks the same people can do well, it is more useful to talk in terms of at least four general strength categories which may be differentially involved in a variety of physical tasks.

Perhaps it might be useful to provide some examples of how one examines the generality of an ability category and how one defines its limits. The definition of the Rate Control factor may provide an example. In early studies it was found that this factor was common to compensatory as well as following pursuit tasks. To test its generality, tasks were developed to emphasize rate control, which were not conventional tracking tasks (e.g., controlling a ball rolling through a series of alleyways). The factor was found to extend to such tasks. Later studies attempted to discover if emphasis on this ability is in judging the rate of the stimulus as distinguished from ability to respond at the appropriate rate. A task was developed involving only button pressing in response to judgments of moving stimuli. Performance on this task did *not* correlate with other rate control tasks. Finally, several motion picture tasks were adapted in which S was required to extrapolate the course of a plane moving across a screen. The only response required was on an IBM answer sheet. These tasks did not relate to the core of tasks previously found to measure "rate control." Thus, our definition of this ability was expanded to include measures beyond pursuit tasks, but restricted to tasks requiring the timing of a muscular adjustment to the stimulus change.

A similar history can be sketched for each ability variable identified. Thus, we know that S must have a feedback indicator of how well he is coordinating before the Multilimb Coordination factor is measured; we know that by complicating a simple reaction-time apparatus, by providing additional choice reactions, we measure a separate factor (Response Orientation), but that varying the stimulus modality in a simple reaction-time device does not result in measurement of a separate factor.

Some later studies using experimental-correlational approaches provide encouraging results which indicate that it is possible to build up a body of principles through systematic studies of ability-task interaction in the laboratory. The approach is to develop tasks which can be varied along specified physical dimensions, to administer these tasks, systematically varied along these dimensions, to groups of Ss who also receive a series of "reference" tasks, known to sample certain more generalized abilities (e.g., "Spatial Orientation," "Control Precision," certain "Cognitive Abilities"). Correlations between these reference tasks and scores on variations of the criterion task specify the ability requirements (and changes in these requirements) as a function of task variations. Thus far we have studied tasks varied along the following dimensions: degree of rotation of display panels relative to response panels; the predictability or nonpredictability of target course or response requirements; the extent to which the task allows S to assess the degree of coordination of multiple limb responses; the degree of stimulus-response compatibility in display-control relationships; whether there is a constant "set" or changing "set" from one stimulus presentation to the next; whether or not certain kinds of additional response requirements are imposed in a visual discrimination reaction task; whether or not certain kinds of feedback are provided. Hopefully, once such principles are established, it should be possible to look at new tasks, operational or otherwise, and specify the ability requirements.

Perhaps I can illustrate with one study in which we were able to show systematic changes in the abilities required to perform a task as the display-control relations in the task were systematically varied. The S was required to press a button within a circular arrangement of buttons on a response panel, in response to particular lights which appeared in a circular arrangement of lights on the display panel (see Figure 1). The display panel was rotated from 0° to 45°, 90°, 135°, 180°, 225°, and 270°. The same Ss performed under these task conditions and also performed on a series of spatial, perceptual, and psychomotor reference tasks. The results showed systematic changes in factors sampled by the criterion task as a function of degree of rotation of the display panel. Progressive rotation shifted the requirements from "Perceptual Speed" to two other factors "Response Orientation" and "Spatial Orientation" (see Table 1). Thus, individual differences along known dimensions are used to explore the relations between tasks and the characteristics of people who could perform the tasks most effectively. Of course, these are problems faced every day by personnel, training, engineering, and systems psychologists.

The results of other studies (e.g., the one in which we varied stimulus-response compatibility) showed that certain variations did make a difference but others did not. Such studies indicate that it should be possible to develop principles relating task dimensions to ability requirements, using laboratory experimental-correlational studies.

SOME APPLICATIONS OF THE TAXONOMY

I would like to mention a few areas in which the framework developed has proven useful in accounting for phenomena under investigation. For a number of years, I have been interested in the relations between reference ability measures, developed around our perceptual-motor taxonomy, and a variety of learning phenomena (see, e.g., Fleishman, 1957a, 1957b, 1960a; Fleishman & Hempel, 1954b, 1955; Fleishman & Rich, 1963). Our results have allowed us to show the differential role of these abilities at different stages of learning more complex tasks. These studies have as an additional goal the specification of abilities predictive of advanced levels of learning. The laboratory findings have been shown to hold with simulations of complex operational tasks such as piloting an air-intercept mission (Parker & Fleishman, 1960) and have been extended to real-life training situations involving skills other than perceptual-motor skills (Fleishman & Fruchter, 1960). We have recently summarized these findings elsewhere (Fleishman, 1966, 1967).

We have also used this paradigm to study part-whole task relationships (Fleishman, 1965; Fleishman & Fruchter, 1965), the prediction of associative interferences between two tasks, and the prediction of retention phenomena (Fleishman & Parker, 1962).

These categories have also proven useful in accounting for the interrelations among component proficiencies in fixed-wing (Fleishman & Ornstein, 1960) and helicopter piloting (Locke, Zavala, & Fleishman, 1965), and among driving proficiencies (Herbert, 1963). We have also been able to facilitate training in a simulated air-intercept tracking task using information developed on the role of these abilities at different stages of learning this complex task (Parker & Fleishman, 1961).

Are these categories useful as a means of task standardization and for generalizing research results to new tasks? We do not have any ready answer, but we are working on the problem. Let me at least mention a few studies in which tasks, representing these task dimensions, have been used. These studies have included research on the effects of stress—fear (Gorham & Orr, 1957), diet (Brozek, Fleishman, Harris, Lassman, & Vidal, 1955), and drugs (Baker, Elkin, Van Cott, & Fleishman, 1966; Elkin, Fleishman, Van Cott, Horowitz, & Freedle, 1965), and

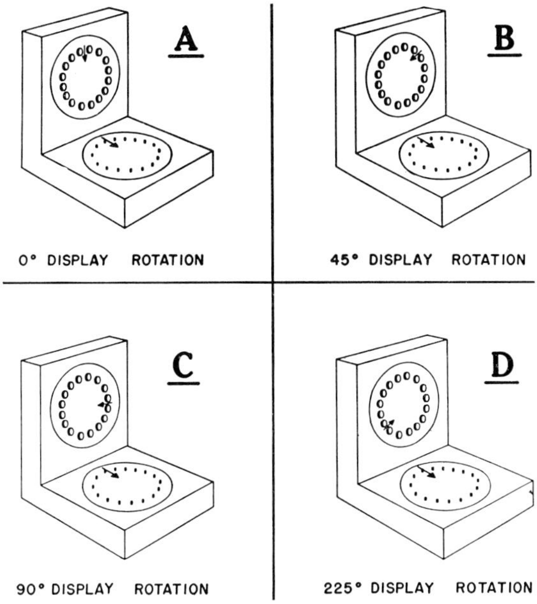

Figure 1. The Response Orientation Device, showing four conditions of display rotation (drawing has omitted certain details not relevant). (After Fleishman, 1957.)

TABLE 1
Factor Loadings of Response Measures for Different Conditions of Display Rotation

Display rotation		Factor	
	Perceptual speed	Response orientation	Spatial orientation
0° (↑)	.47	—	—
45° (↗)	.40	—	.34
90° (→)	—	—	.69
135° (↘)	—	.37	.48
180° (↓)	—	.40	.40
225° (↙)	—	.30	.35
270° (←)	—	—	.30
315° (↖)	.36	—	—

Note.—Loadings below .30 omitted. Adapted from E. A. Fleishman, 1957.

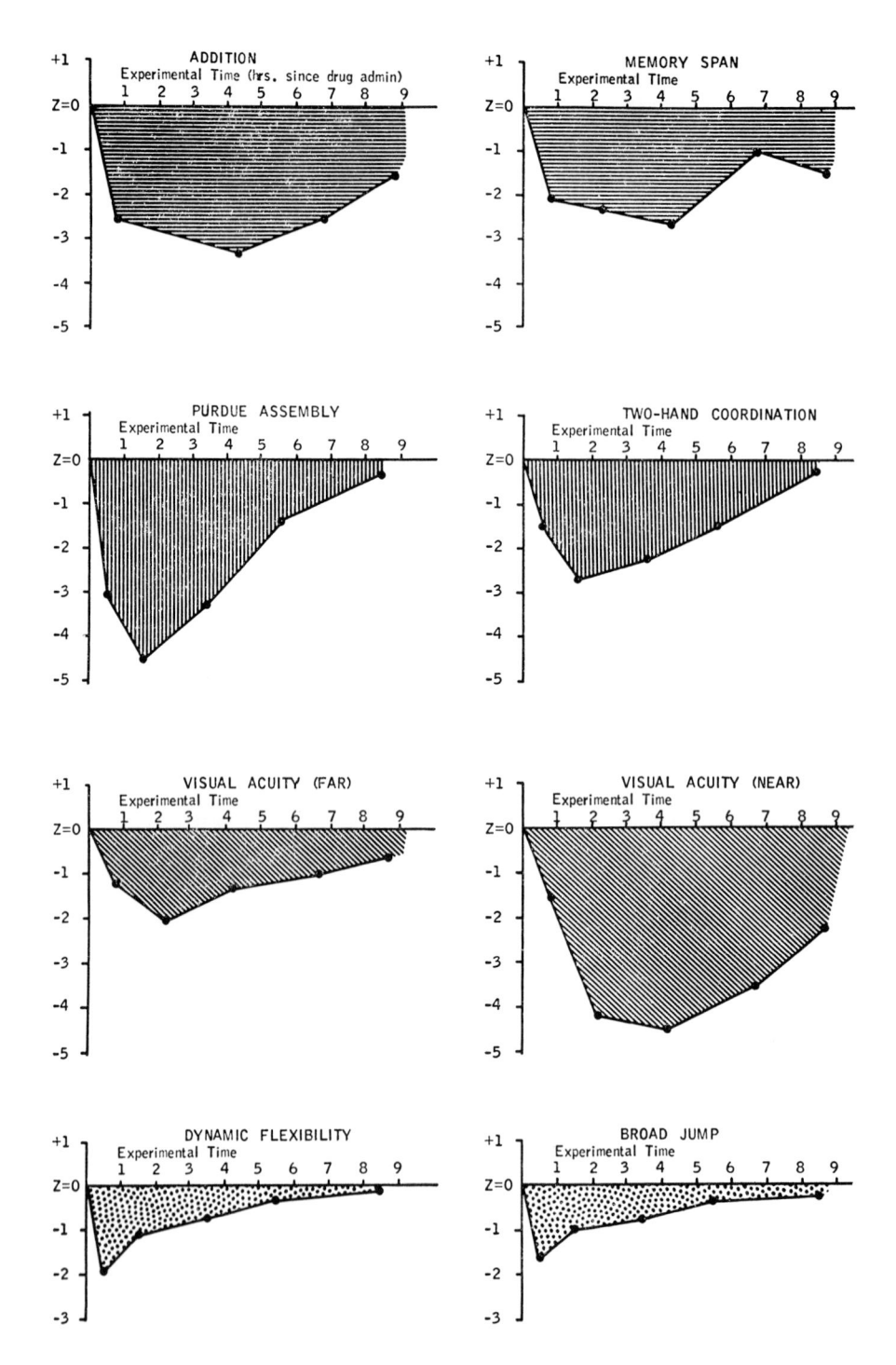

Figure 2. Performance decrement on different ability tasks as a function of time since a given dosage of drug was administered. (After Elkin et al., 1965.)

an ongoing study on the effects of high altitude (anoxia). For use in the space environment, in collaboration with James Parker, a console has been developed to measure in a compact unit 15 different factors for possible application in the manned orbital laboratory program (Parker, Reilly, Dillon, Andrews, & Fleishman, 1965). But it is in our drug work that we hope to make an approach to checking on the generalizations possible from standardized batteries of "basic ability" measures. We are observing the effects of a variety of drugs and dosages on measures of a variety of reference measures. These measures sample the perceptual, motor, sensory, and cognitive areas. We are getting differential effects, that is, some abilities within each area are more affected that others. Figure 2 gives a few illustrations of these events. The question is whether we get parallel results with other tasks representing the same factors. A further phase includes the development of complex tasks and the testing of these drug effects on such tasks. The question is whether our laboratory results, using component ability measures, could have predicted the drug effects on the complex tasks.

Finally, let me say that I am not at all sure about the ultimate utility of this kind of taxonomy. It may well be that the kind of taxonomy most useful to one set of applied problems (e.g., training) may be different from the one useful for another problem (e.g., selection, system design). But I am encouraged that the framework thus far developed has been useful in integrating some previously disparate data, that it has helped simplify some problems, and that it has led to some standardization of laboratory tasks used in a variety of research areas.

REFERENCES

Alluisi, E. A., & Thurmond, J. B. Behavioral effects of infectious diseases: Annual progress report. *University of Louisville, Performance Research Laboratory Report,* 1965, No. PR-65-3.

Baker, W. J., Elkin, E. H., Van Cott, H. P., & Fleishman, E. A. *Effects of drugs on human performance: The development of analytical methods and tests of basic human abilities.* (Report AIR-E-25-3/66-TR2) Washington, D.C.: American Institutes for Research, 1966.

Brozek, J., Fleishman, E. A., Harris, S., Lassman, F. M., & Vidal, J. H. Sensory functions and motor performance during maintenance of survival rations. *American Psychologist,* 1955, **10**, 502.

Elkin, E. H., Fleishman, E. A., Van Cott, H. P., Horowitz, H., & Freedle, R. O. *Effects of drugs on human performance—Phase I: Research concepts, test development, and preliminary studies.* First Annual Summary Report. Washington, D.C.: American Institutes for Research, October 1965.

Fleishman, E. A. Testing for psychomotor abilities by means of apparatus tests. *Psychological Bulletin,* 1953, **50**, 241-262.

Fleishman, E. A. Dimensional analysis of psychomotor abilities. *Journal of Experimental Psychology,* 1954, **48**, 437-454.

Fleishman, E. A. Psychomotor selection tests: Research and application in the United States Air Force. *Personnel Psychology,* 1956, **9**, 449-467.

Fleishman, E. A. A comparative study of aptitude patterns in unskilled and skilled psychomotor performances. *Journal of Applied Psychology,* 1957, **41**, 263-272. (a)

Fleishman, E. A. Factor structure in relation to task difficulty in psychomotor performance. *Educational and Psychological Measurement,* 1957, **17**, 522-532. (b)

Fleishman, E. A. An analysis of positioning movements and static reactions. *Journal of Experimental Psychology,* 1958, **55**, 13-14. (a)

Fleishman, E. A. Dimensional analysis of movement reactions. *Journal of Experimental Psychology*, 1958, **55**, 438-453. (b)

Fleishman, E. A. Abilities at different stages of practice in Rotary Pursuit performance. *Journal of Experimental Psychology*, 1960, **60**, 162-171. (a)

Fleishman, E. A. Psychomotor tests in drug research. In J. G. Miller & L. Uhr (Eds.), *Drugs and behavior*. New York: Wiley, 1960. Ch. 17. (b)

Fleishman, E. A. The description and prediction of perceptual-motor skill learning. In R. Glaser (Ed.), *Training research and education*. University of Pittsburgh Press, 1962. Ch. 5.

Fleishman, E. A. Factor analyses of physical fitness tests. *Educational and Psychological Measurement*, 1963, **23**, 647-661.

Fleishman, E. A. *The structure and measurement of physical fitness*. Englewood Cliffs, N.J.: Prentice-Hall, 1964.

Fleishman, E. A. The prediction of total task performance from prior practice on task components. *Human Factors*, 1965, **7**, 18-27.

Fleishman, E. A. Human abilities and the acquisition of skill. In E. A. Bilodeau (Ed.), *Acquisition of skill*. New York: Academic Press, 1966. Ch. 3.

Fleishman, E. A. Individual differences and motor learning. In R. M. Gagné (Ed.), *Learning and individual differences*. Columbus: Charles Merrill, 1967. Ch. 8.

Fleishman, E. A., & Ellison, G. D. A factor analysis of fine manipulative performance. *Journal of Applied Psychology*, 1962, **46**, 96-105.

Fleishman, E. A., & Fruchter, B. Factor structure and predictability of successive stages of learning Morse Code. *Journal of Applied Psychology*, 1960, **44**, 96-101.

Fleishman, E. A., & Fruchter, B. Component and total task relations at different stages of learning a complex tracking task. *Perceptual & Motor Skills*, 1965, **20**, 1305-1311.

Fleishman, E. A., & Hempel, W. E. A factor analysis of dexterity tests. *Personnel Psychology*, 1954, **7**, 15-32. (a)

Fleishman, E. A., & Hempel, W. E. Changes in factor structure of a complex psychomotor test as a function of practice. *Psychometrika*, 1954, **18**, 239-252. (b)

Fleishman, E. A., & Hempel, W. E. The relation between abilities and improvement with practice in a visual discrimination reaction task. *Journal of Experimental Psychology*, 1955, **49**, 301-312.

Fleishman, E. A., & Hempel, W. E. Factorial analysis of complex psychomotor performance and related skills. *Journal of Applied Psychology*, 1956, **40**, 96-104.

Fleishman, E. A., & Ornstein, G. N. An analysis of pilot flying performance in terms of component abilities. *Journal of Applied Psychology*, 1960, **44**, 146-155.

Fleishman, E. A., & Parker, J. F. Factors in the retention and relearning of perceptual-motor skill. *Journal of Experimental Psychology*, 1962, **64**, 215-226.

Fleishman, E. A., & Rich, S. Role of kinesthetic and spatial-visual abilities in perceptual-motor learning. *Journal of Experimental Psychology*, 1963, **66**, 6-11.

Gagné, R. M. *Conditions of learning*. New York: Holt, Rinehart & Winston, 1964.

Gagné, R. M., & Fleishman, E. A. *Psychology and human performance: An introduction to psychology*. New York: Holt, Rinehart & Winston, 1959.

Gorham, W. A., & Orr, D. B. *Research on behavior impairment due to stress*. Washington, D.C.: American Institutes for Research, 1957.

Hempel, W. E., & Fleishman, E. A. A factor analysis of physical proficiency and manipulative skill. *Journal of Applied Psychology*, 1955, **39**, 12-16.

Herbert, M. J. Anaysis of a complex skill: Vehicle driving. *Human Factors*, 1963, **5**, 363-372.

Locke, E. A., Zavala, A., & Fleishman, E. A. Studies of helicopter pilot performance: II. The analysis of task dimensions. *Human Factors*, 1965, **7**, 285-302.

Mangelsdorf, J. E. Empirical measurement: Rational approach. In H. P. Van Cott (Chm.), Task taxonomy and its implication for military requirements. Symposium presented at American Psychological Association, Chicago, September 1965.

Miller, R. B. Task analysis and task taxonomy: Inventive approach. In H. P. Van Cott (Chm.), Task taxonomy and its implication for military requirements. Symposium presented at American Psychological Association, Chicago, September 1965.

Parker, J. F., & Fleishman, E. A. Ability factors and component performance measures as

predictors of complex tracking behavior. *Psychological Monographs,* 1960, **74**, No. 503.
Parker, J. F., & Fleishman, E. A. Use of analytical information concerning task requirements to increase the effectiveness of skill training. *Journal of Applied Psychology,* 1961, **45**, 295-302.
Parker, J. F., Reilly, R. E., Dillon, R. F., Andrews, T. G., & Fleishman, E. A. Development of a battery of tests for measurement of primary perceptual-motor performance. Biotechnology, Inc. NASA Contract Report, 1965.

43

Individuality in the Learning Process: Some Issues and Implications

Gordon A. Eckstrand

Differential psychologists have studied psychological differences among individuals for many years, and applied psychologists have been quick to exploit these differences in such areas as selection, classification, etc. Experimental psychologists, however, have largely been content to admit and then ignore the fact that people differ. The invariance assumption which underlies this approach to behavior study, while parsimonious and attractive, may be invalid in many cases. This general problem has been discussed from different points of view by Cronbach (1957), Hirsh (1958), and Sidman (1960) and needs no further discussion here.

This article is devoted to a brief and preliminary consideration of individual differences in one of the most fundamental areas of behavior, that of learning. In no other area of behavior study has the topic of individual differences been more neglected. Perhaps this lack of interest in studying the learning behavior of individual subjects can be traced to the presumed irreversibility of the behavior phenomena involved. Whatever the reason, in the field of learning it is generally assumed not only that the process of learning is essentially the same among all individuals within a species, but among all species. Hull (1945) perhaps stated this assumption most explicitly when he wrote, "The natural science theory of behavior being developed by the present author and his associates assumes that all behavior of the individuals of a given species and that of all species of mammals, including man, occurs according to the same set of primary laws" (p. 56). In this article emphasis will be given to examining some of the implications

From *The Psychological Record,* **12**, 1962, 405-416. By permission of *The Psychological Record.*

of individuality in the learning process, particularly in the areas of research and application. It is important at this point to emphasize that the subject of interest in this paper is individual differences in the *processes* which together make up the functional changes we call learning, not in individual differences in *state* or in the effects which such differences in state have on learning when learning conditions are held constant. A great number of the latter type of studies have been conducted, but few, if any, studies of individual differences in learning processes have been carried out.

IMPLICATIONS OF INDIVIDUALITY FOR LEARNING RESEARCH

Most experiments in the field of human learning investigate the effect of a selected variable upon some measure of learning efficiency. In some experiments only two instances of the variable are considered, in which case some condition A is compared with some condition B in terms of their influence on learning. In other experiments, more than two instances of the variable are considered, and in these cases the aim is to establish some functional relationship between the variable and some measure of learning efficiency. In either case, a question of the following nature is asked, "What is the effect of conditions X, Y . . . N on learning efficiency?" In all such experiments it is tacitly assumed that a meaningful answer to this question can be arrived at by testing "random" groups of subjects under each of the conditions and then statistically assessing differences in the average performance of the groups. In other words, the assumption is made that, whatever the effect of the variables under consideration, it will be the same for all "normal" subjects, with the exception of experimental error. There are variations on this assumption to be sure; most investigators would admit of individual differences in magnitude of effect, insisting only that the general nature of the effect is invariant across subjects. Nevertheless, some variant of this uniformity assumption appears to underlie the vast majority of learning studies.

But what if this assumption is not justified; what if there are both quantitative and qualitative individual differences in the learning behavior of subjects in response to basic learning variables? Such a situation, which is certainly not altogether unlikely, could produce highly artifactual results in learning experiments which have not been designed to be sensitive to such differences. In order to illustrate this fact, consider the simple example outlined below.

Let us assume, first of all, that a study is conducted to determine the effects of three learning treatments, X, Y and Z on some measure of learning efficiency. Let us further assume that a degree of individuality does exist in the learning process and that for certain individuals one learning treatment is best while for other individuals a different treatment is best. For simplicity's sake, assume that there are three types of individuals A, B and C and that the learning efficiency of these types of individuals (as measured by trials to a criterion) under learning treatments X, Y and Z is as indicated in Table 1 below. Now if our hypothetical experiment is conducted using randomly selected groups of six subjects, each of which happens to include equal numbers of our three types of individuals, it is

	Individuals		
Treatments	A	B	C
X	10	20	30
Y	20	30	10
Z	30	10	20

apparent that no differences in mean performance would be obtained among the groups, each group requiring an average of 20 trials.

Obviously our conclusion from these data would be that there are no significant differences among the learning treatments in terms of learning efficiency. In a certain sense, of course, this conclusion has practical as well as statistical validity. If the practical problem is to select one learning treatment to be applied to a randomly selected group of individuals, then indeed the experiment has indicated that one treatment is as good as another. Learning, however, is an individual, not a group process, and if one were to conclude from these data that individual learning efficiency was independent of the learning treatment being used, then the conclusion would certainly be spurious. In this hypothetical study, the individual differences indeed represent the most significant data obtained. The example used here is, of course, an artificial one and the loading in an experimental sample would probably never be as uniform as that described. Nevertheless, the principle is clear, and it is easy to see the dangers inherent in learning research which treats individual differences as error without evidence to indicate that such an assumption is warranted.[1]

CURRENT TREATMENT OF INDIVIDUAL DIFFERENCES IN LEARNING

Thus far in this article it has been stated rather categorically that experimental psychologists in general have tended to neglect the concept of individual differences. To what extent has this been true in the field of learning theory and experimentation in particular? Hirsh (1962) has traced the concept of individual differences in behavior theory from Watson to the present time and concludes that while individual differences have sometimes been acknowledged and sometimes understood, they have, in general, been either minimized or neglected entirely.

In an effort to get some rough idea of the treatment currently given to individual differences by learning psychologists, three presumably representative textbooks in the field of learning were examined (Deese, 1958; Lawson, 1960; McGeoch and Irion, 1952). A brief analysis of the treatment given in these books to individual differences in the learning process is presented below.

The term "individual differences" does not appear either in the Table of Contents or in the Subject Index of the books by Deese and by Lawson. In both books there are a few brief references to individual differences scattered throughout the chapters, but in no case is the possibility of individuality in the learning process

[1] Williams (1960, pp. 178-179) has used a similar example to show how the concept of individuality has important implications for medical research on remedial agents for the prevention and treatment of disease.

seriously proposed or pursued. Each book carries a brief acknowledgment of the importance of individual differences in a complete learning theory (Lawson on p. 131 and Deese on p. 330). After a review of the books, however, one is left with the conclusion that attention has been given almost entirely to the treatment variables determining learning, and that the interactions with individual differences have been left for another time.

One complete chapter of approximately fifty pages is devoted to the topic of individual differences in the book by McGeoch and Irion. This chapter, however, deals almost exclusively with the measurement of individual differences in rate or amount of learning when conditions of learning are held constant. These data, while interesting and useful for both theory and practice, are not the most useful in establishing the nature and amount of individuality in the learning process. What are needed are data on the interactions between individual differences and conditions of learning and this book has little to say on this matter. In the Summary and Interpretation of this chapter, the following statement is made: "Nevertheless, variability of results represents an important obstacle to the accomplishment of research, particularly where the demonstration of small differences is important or where it is desired to corroborate quantitative theoretical predictions." This book, then, gives considerably more attention to individual differences than the other two but still treats them as factors to be controlled in good learning research rather than as a class of really important variables deserving study.

The above comments about lack of concern for individual differences are not necessarily criticisms of the textbooks reviewed. Their principal role as textbooks in learning is to present an accurate picture of this field and, consequently, any dearth of attention to individual differences is probably representative of the field of learning. A review of the chapters on Individual Differences and on Learning in recent volumes of the Annual Review of Psychology certainly lends confirmation to this viewpoint. In recent years, little attention has been given to the problem of individual differences in the learning process by researchers in the fields of differential or learning psychology.

In summary, it may be said that the present day psychology of learning presents a model of the organism which permits little in the way of individuality in the learning process. It is perhaps ironic that, whereas it provides a major mechanism for the creation of individuality, it has difficulty in handling what it has created. Again, this is not necessarily a criticism of current learning theory, for perhaps the model which it presents is the best that can be developed at the present time. Further work must correct this, however. As Deese (1958) has said, "The course of future work in learning and in the basic theory of behavior that goes along with it will be in the direction of introducing some flexibility and life into the model" (p. 332). To do this will mean, among other things, coming to grips with the problem of individual differences in learning behavior. There is some indication that movement in this direction is beginning to take place. Such concepts as anxiety, motivation and perception are playing a larger role in the psychology of learning and these will provide mechanisms for both qualitative and quantitative individual differences in learning. There is also an increased interest in relating physiology to learning and this too may focus interest on individual differences, since biological individuality is receiving increased attention (Williams 1956, 1960 a, 1960 b). Also, Zeaman and Kaufman (1955) in an interesting series of experiments have shown that it is possible, though cumbersome and unwieldy, to account for obtained individual differences within the framework of Hullian theory. But the real need now is for experimental data on individuality in the learning process, and it will be profitable for learning psychologists to devote some of their attention to this problem.

IMPLICATIONS OF INDIVIDUALITY FOR LEARNING APPLICATIONS

In a previous section, the implications of individuality for learning research were examined. Using a simple hypothetical example, it was demonstrated that if individual differences in the learning process do exist, they have important implications for the design and interpretation of learning research. If we make the additional assumptions that these individual differences are reasonably stable and are measurable, then it would appear that they also have important implications for education and training. By taking advantage of individual differences and tailoring the learning environment to each learner, it is at least theoretically possible to produce learning efficiency which is higher than can be achieved by ignoring individual differences and optimizing the learning environment for the "average learner." Lawrence (1954) has stated the issue about as succinctly as possible. In an article devoted to a general discussion of training efficiency, he has the following to say.

> If an individually tailored program is adopted, consideration of individual differences suggests other ways of reducing the cost by modifying the training procedures. As indicated previously, studies on training methods concerned with spaced versus massed practice, motivation level and the like are primarily directed at modifying the average rate of learning. Normally, however, a standardized procedure for the spacing of practice and like factors is adopted for all individuals of the group. It is doubtful if this produces the optimum effect for each individual. The optimum value for one individual might well differ considerably from the optimum value for other individuals in the group, depending upon the variability in levels of performance at any given time, the rates of learning, and other characteristics of the individuals involved. Consequently, the maximum possible change in rate of learning is not being achieved. If research were to show that individual differences in reactions to these training procedures were of considerable magnitude and could be predicted prior to training, allowance for such individual differences in the training program adopted might permit a worthwhile reduction in the average cost (p. 374).

In order to demonstrate this effect, we can return to the example used previously. In the hypothetical experiment described there, no difference in learning efficiency was found among the three learning treatments X, Y, and Z. Consequently, on the basis of this research, we would feel free to select any of these treatments for use in a particular training program involving tasks and trainees comparable to those used in the experiment. If we use the same figures presented there, assuming them to represent some measure like trials to criterion, it will take twenty trials on the average to accomplish the required training regardless of which learning treatment is used. If, on the other hand, we were able to predict in advance those individuals who learned best under treatment X, those who learned best under treatment Y, etc., considerable savings in learning trials would be possible. With individuals and learning treatments matched for learning efficiency, the training could be accomplished in only ten trials. Here again, of course, the example is artificially simple, but it does serve to illustrate a principle of considerable importance.

An equally valid case can be made concerning the importance of individual differences in transfer of training, but the argument will not be elaborated here. At least one aspect of the argument is convincingly presented by Lawrence (1954).

APPLICATIONS OF INDIVIDUALITY IN EDUCATION AND TRAINING

Individual differences are at present a nuisance for theory, and consequently, it is perhaps understandable that they have been largely neglected by learning theorists. But one would expect that they would be heavily exploited in education and training in much the same manner as they have been in personnel selection and classification. However, anyone familiar with the fields of education and training knows that such is not the case. This is not to imply that individual differences go unrecognized in these fields. Trow (1960), in his recent text, devotes the largest unit in the book to individual differences, and other recent books in the field of educational psychology also give heavy emphasis to the concept of individuality. Why, then, do we not find a greater responsiveness to this concept in the development and conduct of actual educational and training programs? Two reasons can be adduced to explain this situation. First of all, it appears administratively difficult and uneconomical to conduct educational and training programs which are individually tailored. Trow (1960) recognizes this fact when he writes, "Since students must be taught in groups, the adaptation of instruction to individual differences constitutes one of the major problems of teaching" (p. 11). Although many teachers and training experts recognize that so-called "lock-step" educational and training procedures are incompatible with the facts of individual differences, they have not been able to develop better procedures which also satisfy other criteria which are imposed by cultural and economic factors. Some attempts have been made, of course, such as the Winnetka plan and the Dalton plan, but they have proved cumbersome and have not been widely adopted. Many schools now employ a device called "ability grouping" with considerable apparent success, but such a plan is somewhat narrowly conceived in that it is based upon only one type of difference between individuals; differences in learning *rate*. In other words, fast, average and slow learners are grouped together. Little attention is given to possible qualitative differences in the learning process within and between such groups.

A second reason is the dearth of factual information relating to individual differences in the learning process. Therefore, whereas Trow (1960) can devote five chapters of his book to discussing individual differences and pointing out the necessity of adapting instruction to pupils who differ physically, mentally, socially and in their personality characteristics, he can find practically nothing to say about the manner in which this is to be accomplished. Such information is simply not available, and undoubtedly this has had a deterrent effect upon the development of programs which adapt instructional procedures to the individual student.

AUTOMATED TEACHING AND INDIVIDUAL DIFFERENCES

Historically, one instructional method has been available and has been employed which emphasizes and is responsive to the individual characteristics of the learner. This is the tutorial method. As implied above, however, it is expensive and has never been feasible for widespread use, being largely restricted to education of the wealthy and to providing special help for educational laggards. Possibly due to its restricted employment, the tutorial method has fostered little research and its employment has remained rather in the nature of an art.

Recently, however, the development of programmed learning materials and automated teaching devices or teaching machines has provided a class of teaching and training media which appear to offer almost unlimited possibilities for responsiveness to individual differences in the learning process. Automated teaching devices are devices which provide effective instruction to an individual student or trainee without the direct participation of a human instructor. In other words, these devices simulate, in one manner or another, the functions normally fulfilled by a human instructor in individualized teaching or training. In their simplest form they may amount to little more than question and answer machines, whereas in more complex forms they are theoretically able not only to present information and evaluate student mastery of the material but to select future instructional material and methods on the basis of the performance of individual students. No attempt will be made here to catalog the variety of devices available, since this is the subject of another report (Kopstein & Shillestad, 1961). Suffice it to say that they vary from specially written books to teaching stations controlled by large digital computers.

One might expect that the advent of automated instructional devices with their potential for adapting to individual differences would have awakened a greater interest in the characteristics of the learner as they affect learning efficiency. To a certain extent, at least, this has been true. Many writers in this field have at least implied that the concept of individual differences is important to an analysis and study of the variables which affect the efficiency of automated training and a number have explicitly mentioned or discussed the importance and implications of this concept (Carr, 1959; Coulson & Silberman, 1960, 1961; Eckstrand, Rockway, Kopstein & Morgan, 1960; Glaser, Homme & Evans, 1959). Some adaptive devices have also been built. Most of the discussions and devices to date, however, have involved a limited type of adaptability which is associated largely with differences in rate of learning. They are pre-programmed so that the individual student can "move at his own pace" and through a programming technique called "branching" be presented at each step in the instructional sequence with material which is appropriate to his present level of achievement. This type of machine reponsiveness, however, represents only a beginning in the direction of adapting instruction to individual differences. Automated training devices are theoretically capable of being programmed to use to advantage any measurable differences between individuals which affect the efficiency of learning. Furthermore, current work in the area of heuristics and self-organizing systems points to the possibility of automated instructional systems which will have the capability of selecting and altering the type and nature of their adaptability, once they have been given a definition of criterion performance to be achieved.

This is obviously an area which could profitably be treated in much greater detail. However, the intent here is merely to point to the possibilities offered by automated training for adapting instruction to individual differences as a means of emphasizing the importance and relevance of these developments to the study of individuality in the learning process.

NEED FOR RESEARCH

The assumption that the most individual learning situation is the most efficient is logically compelling but has never been demonstrated. Even if it is true, however, it will not be possible to make use of this principle, either in theory or in practice, until we know more about individual differences in the learning process. Therefore, the greatest need at the present time is for research data on the nature and magnitude of individual differences in a wide range of learning situations. Hovland (1951) seemed to have called for something like this when he wrote a decade ago, "Much of the literature on human learning is presented in terms of the broadest types of generalization possible. But every worker in the field is struck by the wide variability from individual to individual in even the simplest types of learning. Greatly needed are systematic investigations of the sources of variability" (p. 632). Unfortunately, the need is as great today as it was in 1951.

There are two reasons which probably contribute considerably to the dearth of experimental work on individual differences in the learning process. The first is methodological and the second philosophical.

It is not a difficult problem methodologically to establish that different individuals learn at different rates. One has merely to select a group of subjects, have them learn a task under standard conditions and measure the result by some accepted criterion of learning efficiency. Most of the research on individual differences in learning is of this nature. It also would not be too difficult methodologically to demonstrate that this group of subjects would order themselves in different ways with respect to learning efficiency as learning treatments are systematically varied. Although not nearly as well established as the interaction between learning treatments and learning tasks, few learning psychologists would deny the reality of an interaction between learning treatments and individuals. If, on the other hand, the objective is to demonstrate that the learning treatments which produce *optimal* learning performance are different for different individuals, then indeed, a difficult methodological problem exists. By its very nature, this problem requires the extensive study of the learning behavior of individuals, with comparisons being made between the individual's behavior under different conditions rather than between the behavior of two or more groups under different conditions. Repeated measurements on the same individual offer no particular problem when it can safely be assumed that the measurement process does not in some way produce an irreversible change in the individual with respect to the characteristic being measured. Where this condition does not exist, however, a methodological problem is posed. Learning, of course, by definition involves a relatively permanent modification of the organism under study.

Two general types of methodology suggest themselves and both present difficulties. The most direct approach would be to use different learning treatments as independent variables and attempt to demonstrate stable individual differences in learning behavior which are attributable to this variable. This approach, while desirable because of its directness, offers all of the problems of trying to use the individual as his own control in learning and transfer experiments. The second approach would be to state and test hypotheses about the relationship between certain measurable individual characteristics and learning efficiency under specified conditions. This approach avoids some of the difficulties inherent in the first approach but creates others. In the first place, it is a sort of "shot gun" approach and can proceed for a long time without producing conclusive results. This approach is also discouraging because of the many perils in research which seeks to establish individual "trait" correlates for anything. Nevertheless, this method is logically attractive and Cronbach (1957) has made a convincing case for the combination of experimental and correlational methods in attacking this particular problem. In addition, Sidman (1960) has recently discussed in a provocative way a number of techniques for research on individual behavior.

The second reason contributing to a lack of research on individual differences in the learning process is philosophical and is perhaps the more important of the two. It is generally regarded that the aim of science is the formulation of general laws and that the more general the law, the more useful it is. Although there are many ways of stating it, few would disagree with this general idea. The difficulty comes in a choice of ways to achieve the aim. Most experimental psychologists appear to feel that the formulation of general laws of learning and transfer is best approached by an intensive study of the average behavior of a large sample of subjects. They tend to regard any study of individual learning behavior as a "retreat into a psychology of individuals," which is inimical to the aims of science if not downright unscientific. But the aim of science can also be regarded as that of accounting for variance. Where there are no variations there is nothing to be explained. But how can one account for variance without first determining the amount and nature of the variance to be accounted for? It seems logical and defensible that the formulation of general laws must begin with the study of the unit for which the law is expected to provide prediction. If one wishes to predict the behavior of aggregates of cells, one studies these aggregates. If, on the other hand, one wishes to predict the behavior of individual cells, then one must spend at least some time in studying the behavior of individual cells. The same would appear to be true in human behavior theory. If we are interested in predicting the behavior of individuals, then some attention and study must be devoted to individual behavior. The issue is not whether an idiographic rather than nomothetic science of learning is desirable, but rather, whether an idiographic approach to the study of learning behavior can contribute significantly to the goal of formulating more general laws of learning. One can accept the conclusion that general laws and their interactions are potentially sufficient to account for individual learning behavior without rejecting the intensive study of individual learning behavior as a useful, and perhaps necessary, tool in the formulation of these general laws.

Idiographic methods in the study of learning behavior could serve at least two useful purposes. First, they could provide the data required to develop a better classification of learners such that assignment of an individual to a subclass would permit a more accurate prediction of his learning behavior than would otherwise be possible. Second, as Falk (1956) has pointed out, they could serve a valuable function in the delineation of new variables and working hypotheses in support of a more generalized conception of learning. Hull (1945) saw quite clearly the dilemma posed to a strictly natural science approach to behavior theory by individual and species differences. He felt that the dilemma was easily resolved, nevertheless, by changing the constants of his equations with each individual. Such an assumption could not be verified, however, much less have any practical validity, without intensive studies of individual learning behavior and Hull's plan has not yet been implemented in even a limited way.

Hirsh (1962) has pointed out that the scientific study of behavior involves the analysis of variation and covariation along four dimensions which may be labelled response, stimulus, time and individuals. So far, it is the last of these four which has been most neglected in the study of human learning. It would appear to be scientifically sound, theoretically valuable and practically profitable for more attention to be devoted to the study of individuality in the learning process. That there will be many difficulties in accomplishing such studies cannot be denied, but it appears that the time is ripe to combine the methodologies which study variance among organisms and those that study variance among treatments in an attack on this problem.[2] If such studies reveal little or no variability in the learning behavior of individuals, the concept of individuality in the learning process can be discarded. If, on the other hand, substantial variability is found in the nature of the learning process in different individuals, then it will be appropriate and necessary to attack other problems. One set of problems will concern theory, where a search must be made for conceptualizations which encompass the variability which is found among individuals. It seems not unlikely that a detailed knowledge of the nature and magnitude of individual variability in learning behavior will serve to aid in the discovery of the concepts appropriate to this more inclusive systematization. Another set of problems will involve the bringing of this variability under predictive control for exploitation in practical learning situations. Unless the characteristics of individual learning behavior can be measured, either before or during training, they cannot be used as a basis for adapting instructional materials and procedures to the individual learner.

REFERENCES

Carr, W. J. *Self-instructional devices: A review of current concepts.* WADD Technical Report 59-503, 1959.

Coulson, J. E., & Silberman, H. F. Designing an experimental teaching machine that

[2] It is quite likely that automated instructional devices, which stand to benefit from new knowledge about individuality in the learning process, also provide a very useful tool for conducting the research required to generate new knowledge. Such devices offer real potential for carrying out accurately controlled observation of individual learning behavior in real life learning situations.

responds to individual student differences: Present status and future plans. Proceedings of Teaching Machine Conference, University of Southern California, October 1960.

Coulson, J. E., & Silberman, H. F. Automated teaching and individual differences. *Audio-Visual Communications Review,* 1961, **9**, 5-15.

Cronbach, L. J. The two disciplines of scientific psychology. *Amer. Psychologist,* 1957, **12**, 671-684.

Deese, J. *The psychology of learning.* New York: McGraw-Hill, 1958.

Eckstrand, G. A., Rockway, M. R., Kopstein, F. F., & Morgan, R. L. *Teaching machines in the modern military organization.* WADD Technical Note 60-289, 1960.

Falk, J. L. Issues distinguishing idiographic from nomothetic approaches to personality theory. *Psych. Rev.,* 1956, **63**, 53-62.

Glaser, R., Homme, L. E., & Evans, J. An evaluation of textbooks in terms of learning principles. Paper read at Amer. Educ. Res. Assn., Feb. 1959.

Hirsh, J. Recent developments in behavior genetics and differential psychology. *Dis. nerv. Syst. (Monogr. Suppl.),* 1958, **19**, 17-24.

Hirsh, J. Individual differences in behavior and their genetic basis. In E. L. Bliss (Ed.), *Roots of behavior.* New York: Hoeber Medical Division of Harper & Brothers, 1962. Pp. 3-23.

Hovland, C. I. Human learning and retention. In S. S. Stevens (Ed.), *Handbook of experimental psychology.* New York: Wiley, 1951. Pp. 613-689.

Hull, C. L. The place of innate individual and species differences in a natural science theory of behavior. *Psychol. Rev.,* 1945, **52**, 55-60.

Kopstein, F. F., & Shillestad, Isabel, J. *A survey of auto-instructional devices.* ASD Technical Report 61-414, 1961.

Lawrence, D. H. The evaluation of training and transfer programs in terms of efficiency measures. *J. Psychol.,* 1954, **38**, 367-382.

Lawson, P. R. *Learning and behavior.* New York: Macmillan, 1960.

McGeoch, J. A., & Irion, A. I. *The psychology of human learning.* New York: Longmans, Green & Co., 1952.

Sidman, M. *Tactics of scientific research.* New York: Basic Books, Inc., 1960.

Trow, W. C. *Psychology in teaching and learning.* New York: Houghton Mifflin, 1960.

Williams, R. J. *Biochemical individuality.* New York: Wiley, 1956.

Williams, R. J. The biological approach to the study of personality. Paper read at the Berkeley Conference on Personality Development in Childhood, University of California, 1960. (a)

Williams, R. J. Etiological research in the light of the facts of individuality. *Texas Reports Biol. and Med.,* 1960, **18**, 168-185. (b)

Zeaman, D., & Kaufman, H. Individual differences and theory in a motor learning task. *Psychol. Monogr.,* 1955, **69** (6), No. 391.

44

Stimulus and Response Fidelity in Team Training

George E. Briggs and William A. Johnston

In an earlier experiment on team training, Briggs and Naylor (1965) manipulated fidelity of the response (R) mode in a training task and found during transfer that superior performance occurred following the higher R-fidelity condition than under a lower R-fidelity training situation. However, this superiority was short-lived and transfer performance was comparable after the first of four transfer sessions. The present study represents an extension of the earlier work to include systematic manipulations of both stimulus (S-) and R-fidelity during training. Further, the basic task utilized here in both training and transfer required a higher level of interaction among team members than did the task employed in the earlier research; thus, we can test the generality of the previous results on R-fidelity in a more demanding task.

METHOD

Transfer Apparatus and Procedure. The basic equipment used for transfer was the same as that described by Naylor and Briggs (1965): a special-purpose analog computer capable of generating and displaying simulated radar returns from target and interceptor aircraft on 14-inch CRTs. Experimenter assistants portrayed the interceptor pilots and executed heading and/or speed changes on target generators as directed to do so over simulated radio channels by the team members. A team consisted of two radar controllers (RCs) each of whom viewed his own display and was responsible for two interceptor aircraft. There was a constant load of two targets per RC and the targets followed non-evasive straight-line courses at

From *Journal of Applied Psychology*, 50, 1966, 114–117. Copyright © 1966 by the American Psychological Association, and reproduced by permission.

This research was carried out in the Laboratory of Aviation Psychology and was supported by the United States Navy under Contract No. N61339-1327, sponsored by the United States Naval Training Device Center, Port Washington, New York. Permission is granted for reproduction, translation, publication, use, and disposal in whole or in part for any purpose of the United States Government. James C. Naylor participated in the planning of this research.

airspeeds of from 350 to 600 knots. The interceptors were able to use speeds of from 200 to 1200 knots and could turn at 3 degrees/second.

The diameter of each display simulated 200 miles of airspace, and each display was marked into four quadrants as shown in Figure 1 which illustrates one of the initial conditions. Interceptors I_1 and I_3 were controlled by RC_1 while RC_2 directed I_2 and I_4. The goal was to have I_1 intercept (come within 1 mile) of target T_1 at the same time that I_2 intercepted T_2; thus, RC_1 and RC_2 were required to *coordinate* interceptions. Likewise, I_3 and I_4 were to be coordinated one with the other (but not necessarily with I_1 and I_2). In the earlier research (Briggs & Naylor, 1965) no coordination was required; thus, the present task was more demanding. Each target entered the airspace with one of 12 entry characteristics made up of four headings and three speeds. Pairs of targets entered symmetrically, as shown in Figure 1, and targets were reset following interception or after passing over a quadrant boundary.

The instructions to RC stressed the need to achieve coordinated interceptions and the teams received summary feedback on this aspect of performance following each 35-minute session on the transfer task. There were four such transfer sessions. Verbal communications were permitted between RCs over a channel other than those used to communicate with the pilots. An experimenter assistant observed the operation on a third simulated radar display and measured separations of pairs of targets and interceptors following a successful intercept by one of the two interceptors. This provided a performance index to be called degree of coordination which was used in all analyses. Perfect performance (maximum coordination) would yield a separation of 1 mile for T_1 and I_1 and 1 mile for T_2 and I_2 of Figure 1; thus, the average degree of coordination would be 1 mile for perfect performance. However, if I_2 was not perfectly coordinated with I_1, then when I_1 and T_1 were within 1 mile, I_2 and T_2 might be separated by, say, 5 miles; this would yield an average degree of coordination score of 3 miles.

Training Conditions. There were four groups of teams in the experimental design representing the four possible combinations of two training variables. Each training variable existing at two levels, high and low S-fidelity and high and low R-fidelity, defined as follows:

1. High S-fidelity: In this condition the RCs each viewed 5-inch CRT displays which were essentially identical in all respects, except size, to the transfer task displays and which were functionally related in the same way as the transfer displays to target generator consoles manipulated by pilots under direction from the RCs. Coordinated interceptions were required during training Sessions 2 through 4.

2. Low S-fidelity: A circular game board, identical to that employed by Briggs and Naylor (1965), with 3,690 squares was used to represent the radar surveillance area. An experimenter moved four target checkers one square every 20 seconds while an experimenter assistant (a pilot) moved the four interceptors either one or two squares at the same time interval as directed by the RCs. During training Sessions 2–4 the goal, of course, was to obtain coordinated interceptions by landing on the same square as that occupied by a target aircraft at the same time the other RC obtained a similar blockage of the comparable target. The RCs observed the board from an elevated platform.

3. High R-fidelity: The RCs issued commands to the pilots over a simulated

radio channel identical to those used in the transfer task. Thus, they received practice using verbal codes and phrases which would be appropriate in the transfer task.

4. Low R-fidelity: Manual communication devices, which were connected to pilot display panels, were employed in this condition. The RCs issued commands by positioning heading and speed switches to the desired values, and thus while the pilots received the necessary information, the RCs had no opportunity to acquire the efficient verbal communication procedures which would be required in the transfer task. Factors such as time to give and execute commands were held constant for the two levels of R-fidelity. Figure 2 indicates the training conditions for the four groups of teams.

Subjects and Training Procedures. There were seven two-man teams per group, and pairings of RCs in a team were made nonsystematically. The RCs were students enrolled at the University who had not served in the previous research in this program and who answered a newspaper ad to earn $10 for eight evening sessions.

The first training session involved a general explanation of the task requirements, practice in using the aircraft identification code to be experienced in the training and transfer tasks, and a 5-minute introductory period on the training task. The latter involved one interceptor-target pair per RC and did not require coordinated hits. Training Sessions 2–4 each lasted 35 minutes, and the RCs were required to obtain coordinated hits. Immediate feedback from the experimenter was provided following each interception. No verbal communication was permitted between RCs during training.

RESULTS

Training. Within each level of S-fidelity the two levels of R-fidelity were compared in terms of the degree of coordination measure: none of these comparisons indicated statistically significant differences in performance. Thus, the effects (see below) of R-fidelity on transfer performance cannot be explained in terms of difficulty differences between high and low R-fidelity training task conditions. It was not possible to obtain comparable measurement units for the two S-fidelity conditions and so no statistical tests were made of these treatments. However, the low S-fidelity task clearly was easier than the high S-fidelity condition.

Transfer. In the previous research (Briggs & Naylor, 1965; Naylor & Briggs, 1965), an efficiency score (fuel consumed per intercept) served as the major dependent variable. In the present study these data were recorded also but there were no statistically significant differences found except a significant practice effect, $F(3, 72) = 14.64, p < .001$.

An analysis of variance was made of the degree of coordination data and this indicated three statistically significant sources: S-fidelity, $F(1, 24 = 5.77, p < .01)$; a practice effect, $F(3, 72) = 10.61, p < .001$; and an interaction of Transfer Sessions × R-Fidelity, $F(3, 72) = 4.21, p < .01$. These results can be seen in Figure 2: on the average, Groups 1 and 2 (high S-fidelity) achieved lower scores (better coordination) than did Groups 3 and 4 (low S-fidelity); there is a general

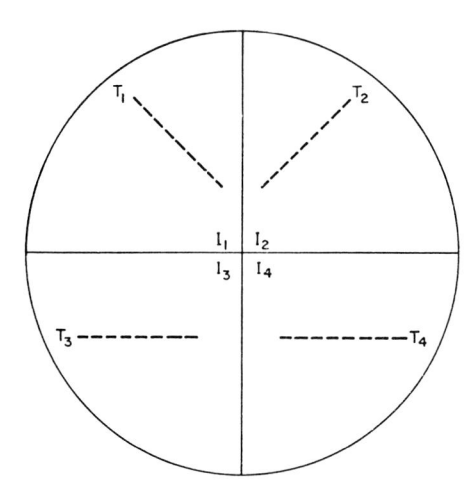

Figure 1. A schematic representation of one of the initial conditions used during transfer. (The Is and Ts represent interceptor and target positions, respectively. In actual practice clock codes appeared at these indicated positions and the dashed lines were not present; they indicate here the future positions of the targets.)

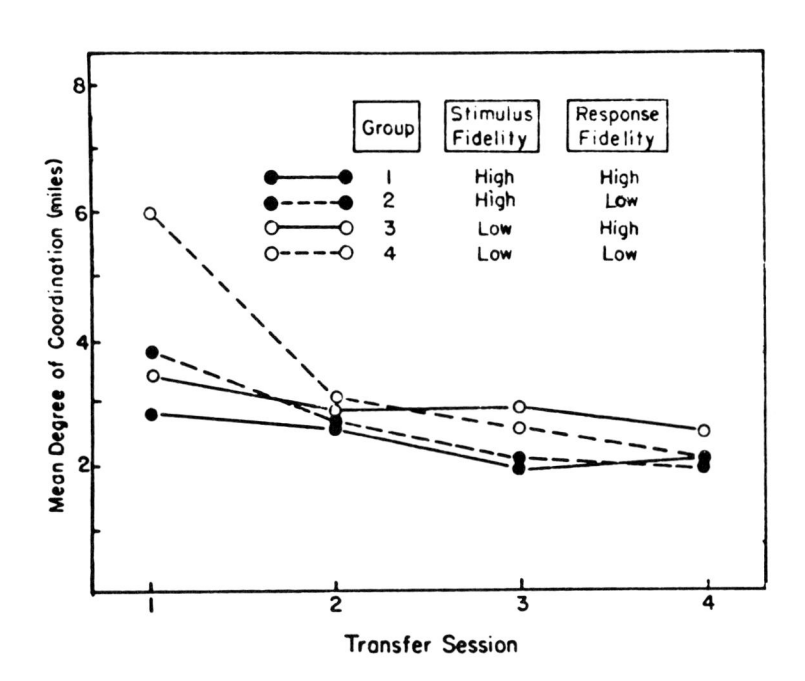

Figure 2. Degree of coordination trends during transfer as a function of stimulus and response fidelity during training.

TABLE 1

The Effect of R-Fidelity on Degree of Coordination and Number of Commands to Pilots as a Function of Transfer Sessions

Measure	R-fidelity	Transfer session 1	2	3	4
Degree of coordination	High	2.98	2.69	2.44	2.31
	Low	4.91	2.89	2.45	2.09
Number of commands	High	127.36	137.22	147.79	162.72
	Low	105.72	141.15	152.79	157.93

downward trend (improvement) for all groups which is particularly pronounced for Groups 2 and 4; and the average of Groups 1 and 3 (high R-fidelity) on the *first* transfer session is superior to that of Groups 2 and 4 (low R-fidelity) but no differences are apparent for the remaining three sessions, thus the interaction of R-fidelity with sessions.

DISCUSSION

It is clear that both high S- and high R-fidelity training conditions produced superior transfer task performance. However, the superiority of performance following high R-fidelity training was relatively short-lived, there being no statistically significant difference between high and low R-fidelity treatments after the first transfer session. On the other hand, the superior effects of high S-fidelity training were more extensive and longer lasting; therefore, one may conclude that while high fidelity training conditions are desirable for both input (S-fidelity) and output (R-fidelity) aspects of an operational (transfer) task, it is relatively more important that high input fidelity characteristics be utilized, the detrimental effects of low fidelity output conditions being easily overcome upon transfer to the operational task.

It must be cautioned that this conclusion should be restricted, at least tentatively, to those tasks in which verbal communication represents the primary output mode. Briggs and Wiener (1959), for example, show that in a complex tracking task, high fidelity of the response device during training was an important determinant of transfer performance. It is not surprising that there may be more justification for high R-fidelity in a motor skill task than in the present task: we spend far more time in verbal communication than in motor control operations; further, when necessary, the former can be utilized very effectively with little trouble adapting to environmental demands, while output utilizing the control device of a vehicle or other continuous processes requires rather precise human adjustment to the machine and other environmental dynamics. Thus, the relatively more extensive practice and the less precise environmental demands would permit more ready transfer of verbal-communication skills than one might expect for motor-control skills.

Table 1 indicates that the effect of R-fidelity on verbal communications between RC and pilot was of the same pattern as that noted above for the major dependent variable: degree of coordination. Apparently RCs trained on the manual communication device (low R-fidelity) found it more difficult than did the RCs trained in verbal communications to generate commands in the first transfer session, and it is logical, given the need for numerous rapid adjustments of interceptor speeds and headings, that this contributed to the significantly poorer performance of those teams at transfer in terms of coordination. An analysis of variance supports this observation by indicating a significant interaction of Transfer Sessions × R-Fidelity for the communications data, $F(3, 72) = 3.23$, $p < .05$.

The results on the significance of R-fidelity confirm the earlier findings by Briggs and Naylor (1965): high R-fidelity training does produce superior transfer performance but the superiority is of rather short duration. Since a more demanding task was employed in the present research, the earlier results are both confirmed and extended.

Finally, it is important to acknowledge that the efficiency index of performance was insensitive here to the two training variables even though in the earlier research (Briggs & Naylor, 1965; Naylor & Briggs, 1965) this same dependent variable was sensitive to differential treatments. It is felt that the following observation can account for this inconsistency: the efficiency score reflects the individual proficiency of team members more than it does the level of their "teamwork"; in the earlier research only a relatively low level of such teamwork was required, the task demanding primarily individual proficiency; thus, the measure was responsive to experimental treatments that affected individual proficiency. However, the present task requires a rather high level of teamwork, and while individual proficiency is necessary, it obviously is not a sufficient condition for successful coordination. Therefore, it is not too surprising that a score which is sensitive to individual proficiency would not necessarily be sensitive to team proficiency in a task which emphasizes teamwork so greatly.

REFERENCES

Briggs, G. E., & Naylor, J. C. Team versus individual training, training task fidelity, and task organization effects on transfer performance by three-man teams. *Journal of Applied Psychology,* 1965, 49, 387-392.

Briggs, G. E., & Wiener, E. L. Fidelity of simulation: I. Time sharing requirements and control loading as factors in transfer of training. *U.S. Nav. Tra. Dev. Cen. Tech. Rep.,* October 1959, No. 508-1.

Naylor, J. C., & Briggs, G. E. Team training effectiveness under various conditions. *Journal of Applied Psychology,* 1965, 49, 223-229.

PART FIVE

ENVIRONMENTAL AND ORGANISMIC CONDITIONS IN MAN-MACHINE SYSTEM PERFORMANCE

Scientific and technological developments of recent years have not only forced man to adjust to continually changing *task* requirements, they have made it necessary for him to perform these tasks under a variety of unusual circumstances. In some cases he must adapt to severely reduced sensory input levels; others tax his capabilities to the point of overload. Still others subject him to unusual atmospheric pressures, oxygen levels, temperatures, noise levels, or bodily vibrations. In combination with some of these newer complications even an ancient issue, such as what constitutes desirable lighting, takes on renewed urgency.

Scientific engineering psychology has always professed some interest in environmental factors as another class of variables which contribute to human performance in man-machine systems. By and large, however, it has been the industrial psychologist and the human factors engineer (by our definition) who have taken the lead in exploring their effects on the man-machine system, and most of the resulting research has had a distinctly pragmatic flavor. In very few instances has work in the area of environmental factors had any marked impact upon the broader conceptualizations of human performance; in other words, it seems to have contributed little to psychological theory.

In retrospect, this shortcoming—if one wishes to call it that—is not at all surprising. In the first place, many of the projects concerned with such variables have had only a peripheral interest in *behavioral* measures; more often, the emphasis has been upon *physiological* indices. Second, even when behavioral measures have been accorded some importance, it has usually been for their physiological or pathological, rather than psychological, implications. One does not typically invoke psychological concepts to explain why, for example, a person responds more slowly when he is cold. Third, environmental research has often been conducted under extremely adverse

References will be found in the General References at the end of the book.

experimental conditions, frequently in response to some immediate military need. Finally, some environmental effects having distinct behavioral implications (such as weightlessness) have only recently come to assume any *practical* significance. In essence, then, we might say that environmental research up to this time has fallen more directly within the province of *human factors engineering* than of *engineering psychology*, and perhaps rightly so. There are some areas, of course, (and that of reduced input is perhaps a prime example) in which the engineering psychologist has been deeply involved all along, and still other areas in which he seems to be developing a more fundamental interest. In Part Five we devote particular attention to these theoretically "promising" directions.

A word is in order concerning the fact that *organismic* as well as *environmental* conditions are mentioned in the part title. If the reader were to interpret these two terms with their usual broad connotations, he would be forced to regard this as a strictly miscellaneous section! Such, of course, is not our intention at all. We have in mind a very specific—albeit difficult to defend—distinction in using these terms. By now we should have made reasonably clear what we mean by *environmental effects:* these are performance functions which result from manipulation of specific ambient conditions in the environment (light, heat, etc.). What we mean by *organismic effects* is not so easily explained—especially since they, too, are usually manipulated through the environment. Our use of the term *organismic* is intended to accommodate those situations in which the experimental manipulations (environmental, task, or whatever) are carried out because they are presumed to control some general organismic "state" (such as *general activation* or *stress*) which is of principal interest. For example, Miller's article on information *overload* is chiefly concerned with hypothesized mechanisms whereby man adjusts to excessive input rates; Appley and Trumbull are interested in psychological *stress* (which presumably derives from many kinds of unusual environmental conditions and inhibits performance). We hold no particular brief for such "state" concepts or, for that matter, for the implication that environmental and organismic approaches will lead to fundamentally different conclusions when all the facts are in. We use both terms only to distinguish two *emphases* which seem to exist within what we consider to be an identifiable research area.

Part Five includes a sampling from a wide variety of studies concerned with unusual environments or their hypothesized psychological correlates and their effect upon human performance. Because of its relatively advanced state of theoretical and methodological development, the topic of *vigilance* is accorded an entire chapter. Another chapter is devoted to the effects of unusual input conditions: *reduced* input (sensory deprivation), *excessive* input (information rates), and the catch-all concept of *stress*. Finally, several illustrative studies from the massive ambient-environment literature are assembled into a chapter entitled *Environment Constraints*.

Chapter Thirteen

Vigilance

As we have seen, a great change has taken place in the man-machine relationship over the past several decades. No longer can we regard the machine simply as man's strong but witless servant; it has become his partner. Functions like vehicular control which once were considered uniquely "human" are now shared with machines or are completely automated. Others, such as medical diagnosis, which seem even further removed from the province of "nuts and bolts" are fast becoming cooperative ventures. Clearly, the age-old stereotype of the machine as a mere extension of human senses and limbs is woefully obsolete; to be fully accurate, the picture must now include the machine as an extension of man's perceptual and cognitive processes as well.

Today, then, virtually every "human" function is, in theory at least, a candidate for automation. Consequently, people who are responsible for charting the course of future systems continually face the problem of deciding how to divide up the work load between men and machines. Over the years, the cumulative effect of their decisions has been a lessening of man's active participation in the ongoing affairs of the system. Ultimate authority is still—and probably will always be—vested in man, but more and more of the subordinate activities are being carried out in complete absence of direct human control. It seems that man has reserved for himself the most *responsible*, yet the least *active*, positions in the entire system: those of monitoring and high-level decision making.

Undoubtedly mankind is much the richer—materially, at least—for this growing reliance upon machines. Even the noblest social advancements, however, produce *some* unpleasant side effects, and man's transition from moving force to monitor is certainly no exception. We have already considered the retraining problem created by rapidly changing job requirements. Leisure time, which people will have in abundance in the foreseeable future, presents another set of difficulties. Some visionaries go so far as to suggest that we must fortify ourselves psychologically and socially for the time when people will no longer have any work *at all* to do.

While less spectacular in their implications, problems of a more immediate nature have already appeared which could offset some of the advantages of automation. One of the first of these to attract the serious attention of engineering psychology was the matter of human vigilance. It has long been suspected that people are poorly suited for the passive role of monitoring. During World War II, for example, it was discovered that radar operators—even though apparently mo-

References will be found in the General References at the end of the book.

tivated by the importance of their assignment—became increasingly inept at detecting signals as their watch progressed. It seemed as though the enforced inactivity somehow dulled their senses or made them less alert.

If people are inherently incapable of maintaining vigilance over any reasonable period of time, one might argue that there is little sense in using men to monitor machines. Consider a sophisticated surveillance system with its elaborate sensing devices, data-processing equipment, and display techniques. Few pertinent changes in the environment are likely to escape its notice. Now insert the human monitor into the picture. As he sits before his console, it is his responsibility to scan incoming information for the occasional signal which *might* spell the difference between war and peace. If he is not vigilant at that moment when the critical information arrives, of what value are all the hardware components which precede him in the system? By divesting himself of the more active operations in the system, man would seem to have placed himself unwittingly in a very awkward position. He has taken on greater responsibility, but in so doing he may have created a work situation which is least conducive to his best efforts. Is it true, then, that man is not a proficient monitor? Did the declining proficiency reported for radar operators in World War II reflect a basic human shortcoming, or was it merely a unique occurrence specific to that situation?

As might be expected, a problem as critical as this was very quickly brought into the laboratory for further study. Here, under careful experimental control, the same decrement in vigilance was observed as had been reported in the field. The Mackworth article, which appears first in this chapter, serves to illustrate the earliest research in this area. Obviously, then, the problem was a real one. But what was the nature of the problem? To what psychological mechanism or mechanisms did it owe its existence, And in view of the seemingly unalterable trend of man-machine relations, what could be done about it?

Interest in all three of these questions developed almost simultaneously, with the result that early work on the problem was a bit disjointed. Explanations were borrowed principally from classical theories of conditioning, attention, and (later) physiological activation, but few of the experiments that were carried out actually favored any of these positions to the exclusion of the others (Frankmann & Adams, 1960). The mark of a good theory is that it generates unique hypotheses, and none of the classical theories seemed to qualify on this score. Techniques which seemed at least somewhat effective in *controlling* the human vigilance decrement —"dummy" signals, rest periods, rewards for detections—were likewise not clearly understood.

Reviewing the confusion which surrounded this earliest work on the vigilance problem, Holland proposed a fresh approach. He recommended that investigators forego the luxury of theorizing until a more reasonable quantity of empirical data became available upon which to base their theoretical constructs (e.g., attention, expectancy, activation). Moreover, he suggested that *observing behavior*, rather than vigilance per se, be the primary subject of investigation. Understanding which variables control man's observing responses, he argued, would necessarily lead to an explanation of *decrements* in observing proficiency. The second article in this chapter is therefore a reproduction of this very influential paper by Holland.

At the time this book is written, a considerable quantity of data has, in fact, been gathered on various monitoring tasks (see, for example, Adams, 1965; Buckner & McGrath, 1963; Jerison & Pickett, 1963). We know, for example, that decrements become less likely as signals occur more frequently, more predictably, more distinctly, and more persistently. Furthermore, we know that the observer can be induced to perform more adequately by incentives administered whenever he detects a signal, by specialized training, and by alteration of the values which he attaches to missed signals, false alarms, and detected signals. We know, too, that some individuals seem peculiarly resistant to decrements even under the worst of conditions (for the "average" observer).

Probably the most noteworthy of recent findings is the discovery by Adams (1963) that decrements are negligible or completely absent in certain types of *complex* monitoring situations. When asked to observe information displayed at reasonable levels of intensity and duration, experimental subjects were reported to have little difficulty detecting infrequent events, even after many hours of vigil. A degree of complexity added to the situation (such as requiring watch over six or eight information sources rather than one) only appeared to offset whatever decrement the passive monitoring task might have engendered.

If Adams's findings are taken completely at face value, the "viligance decrement" becomes more a phenomenon of academic interest than of practical importance; surely our technology will one day make it unnecessary for man to work with transient signals in cluttered, near-threshold displays. In systems of the future, we can expect man to have at his fingertips even greater quantities of information, but we can also anticipate that it will appear in some reasonably legible form. If decrements in human vigilance are truly obviated by this sort of display, then there would seem to be little cause for alarm over man's increasingly passive role in system affairs.

Unfortunately, however, we cannot rest assured that the vigilance problem will cease to exist when suitable displays become commonplace. Further investigation has revealed that serious decrements *can* occur under some circumstances even when complex, legible, persistent displays are used. Worse yet, the complex display decrements seem to be of a much more insidious nature than the rather straight-forward decline in performance found in the classical vigilance task; they may show up in the form of increasingly severe—yet intermittent—periods of *complete inattention*. Thus performance, especially time required to detect a signal, tends to exhibit increasing fluctuation over the period of vigil. This could be just as damaging to the operation of the man-machine partnership as the classical decline in sensitivity, and perhaps even more so. Obviously, we are a long way from understanding this new facet of the human monitoring problem. The topic and related problems are discussed in our fourth selection by Howell, Johnston, and Goldstein.

At the present time a new era of vigilance or monitoring research has arrived. Traditional theoretical constructs (such as expectancy and activation) and classical "threshold" tasks are exerting far less dominance over the experimentation than they did in the past. The premise that multiple factors are at work in real-life monitoring behavior has led to several new ways of approaching the problem. Two of these are illustrated in the final articles of this chapter. The Broadbent and Gregory paper, incidentally, is particularly noteworthy in that it represents still another application of the theory of signal detectability which we discussed in Chapter Three. Despite its currency, this study has already become something of a classic.

We see in the area of vigilance a rather vivid illustration of how the engineering psychologist seeks to accommodate both practical and scientific interests by engaging in research of value to both, and it is one of the areas in which he has achieved a measure of success in reconciling the objectives of both. He has decomposed a very immediate design problem, the question of how to offset human vigilance decrements, into a set of fundamental human performance issues: What variables control monitoring performance? How successful are existing psychological constructs in accounting for the effects of these variables? Can better models of human monitoring behavior be devised? Then, by exploring these basic issues, he has been able to offer progressively more informed recommendations concerning solutions to the original problem, while at the same time gaining some insight into neglected or poorly understood facets of human behavior.

45

The Breakdown of Vigilance During Prolonged Visual Search
N. H. Mackworth

I. INTRODUCTION

1. The General Problem

The deterioration in human performance resulting from adverse working conditions has naturally been one of the most widely studied of all psychological problems. Amongst other possibilities, the stress arising from an unusual environment may be due either to physico-chemical abnormalities in the surroundings or to an undue prolongation of the task itself. This paper is concerned with the latter form of stress, as it has been found to occur in one particular type of visual situation; a later publication will more fully discuss the implications of these and other visual and auditory experiments (Mackworth, 1948).

From *Quarterly Journal of Experimental Psychology*, 1, 1948, 6-21. By permission of the Experimental Psychology Society.

This paper has been based on one given to the first meeting of the Experimental Psychology Group.

The relevant literature on the deterioration in perceptual efficiency resulting from prolonged work makes depressing reading; not only is it scanty, but it is also rather contradictory. For example, in 1890, William James could confidently define the nature of attention—but nearly 50 years later both Woodworth (1938) and Bills (1934) were dubious about the whole concept of attention, the latter maintaining that the term had lost its meaning from an identification with the conscious results of the process rather than with the process itself. Head (1926) used the term vigilance to describe both a physiological and a psychological readiness to react, and the present writer also believes that vigilance is a useful word to adopt, particularly in describing a psychological readiness to perceive and respond, a process which, unlike attention, need not necessarily be consciously experienced.

Thorndike held that for tasks up to two hours in length "the work grows much less satisfying or much more unbearable, but not much less effective." Robinson (1934) criticised this view because it was based on experiments on accuracy of work at tasks which gave the subjects immediate and reliable information on any trend in their working efficiency. This was an interesting point because Wyatt and Langdon (1932) studied performance trends in work which had no such accuracy self-check and here they did find deterioration in accuracy of industrial inspection when the workers had been 30–45 minutes at the task; this decrease continued for about 90 minutes and was followed by an irregular recovery towards the end of the four-hour working spell. Ditchburn (1943), on the other hand, found that look-out duties in a practical form began to deteriorate almost as soon as the work was started and vigilance fell to a minimum level within only a few minutes. Anderson *et al.* (1944) undertook some interesting and detailed experiments with synthetic radar equipment which suggested that "in general, to ensure the most efficient performance, it does not appear wise to prolong daily operating periods more than 40 minutes, if such periods of operation are to be repeated daily without intervening days of rest. On the other hand, it is possible that occasional operating periods of as much as 4 hours in duration may be tolerated without marked reduction in efficiency."

2. The Specific Problem

The Second World War provided many situations which involved prolonged visual search. In 1943, for example, the Royal Air Force had to determine the optimum length of watch for airborne radar operators on anti-submarine patrol. It was suspected that working efficiency was deteriorating due to overlong spells at the radar screens. The situation was, in fact, thought to resemble one described by Shakespeare:

> For now they are oppress'd with travel, they
> Will not, nor cannot, use such vigilance
> As when they are fresh.—(*The Tempest*, Act 3, Scene 3.)

Common sense, however, had given no precise answer to the problem because discussion with operational personnel showed that in practice their working spells at the radar sets over the Atlantic were anything from one-half to 2 hours in length

(Craik and Mackworth, 1943). Laboratory studies were therefore started, initially for this wholly practical reason, to determine over what length of time accuracy could be maintained in work of this sort. It was arbitrarily decided that the synthetic laboratory situation need attempt to reproduce only the more general features of watch-keeping duties. It therefore seemed necessary to provide a series of visual signals which were all difficult to perceive because the subject had no more than a glimpse of each of these barely visible stimuli. It was also thought important to ensure that the signals were presented only occasionally, so that the lengthy searching nearly always drew a blank, although many rather similar signals of no importance at all were constantly being encountered and having to be disregarded. Isolation of the watcher seemed a significant factor in the situation, and care was taken to ensure that the subject had no reliable objective yardstick by which he could gauge his own performance.

II. METHOD, PROCEDURE AND SUBJECTS

1. *General Procedure.* The Clock test was devised to produce a visual situation into which these general psychological factors could be introduced. The subject sat in a wooden cabin in a room entirely by himself and looked at a black pointer 6 inches in length. This normally moved on in steps (like the seconds hand of a large clock) in front of a plain white vertical surface. It jerked on to a new position once every second, 100 of these movements making the full circle. There were no scale markings or reference points of any kind on this white background.

While the subject was looking at this ordinary movement, he was told that "every now and again, at long and irregular intervals, instead of the pointer moving like so, it will move through double the usual distance. This double movement gives the effect of one unusually large movement forward. Press the response key as soon as you notice one of these double-length movements. Though I want you always to press the key as quickly as possible, do not think it is ever too late to press it, if you suddenly remember having seen the signal."

The pointer was mounted in such a way that its tip traced out a circle 10 inches in diameter. Every ordinary movement of the pointer therefore moved the tip through 0.3 inch and every double movement pushed it on 0.6 inch. Since the subject was 7 feet away, this 0.3 inch difference between the two displacements subtended an angle of 12 minutes at the eye, and the subject was picking out 24-minute movements from the usual "background" of 12-minute movements. The width of the pointer itself subtended an angle of 8 minutes at the tip and 16 minutes at the broad end.

The subject had a preliminary run on the first five double-length stimuli of the series. During this practice spell the experimenter watched the clock with the subject and immediately after the first one or two double-length movements he said "Now!" or "That's one!" to make sure the subject had understood the task. In this preliminary period the experimenter told the subject whether his responses were right or wrong.

The subject then went on to do the actual experiment after being told that it would last for 2 hours. In this test the double-length movements of the hand occurred only 12 times in 20 minutes, at intervals of ¾ minutes, ¾, 1½, 2, 2, 1, 5, 1, 1,

2 and 3 minutes, in that order. Ten further minutes of the ordinary uneventful movement completed the first half-hour of the experiment. The second, third and fourth half-hours of the test were identical with the first, and followed straight on without any obvious break in the presentation. In case the subject realised this repetition he was asked to lend the experimenter his wrist watch for the duration of the test. As far as possible all extraneous noises were excluded from the room; the apparatus made a constant droning noise and also some other faint irregular sounds to mask all auditory cues to these double-length movements of the clock hand. The scoring part of the equipment was in an adjacent room and this recorded the time taken by the subject to press a morse key in response to each of the 48 stimuli given during the 2-hour experiment. If no response came from the subject within 8 seconds of the stimulus being applied, it was assumed that he had missed that particular signal. The experimenter gave the subject no knowledge of results at all during the actual test or at the end of the experiment.

2. *Specific Procedures.* The Clock test in this basic form has been given under a range of different conditions:

(a) *Two-hour watches.* 25 R.A.F. cadets were each tested once for a two-hour spell, and their results formed the scores shown in Figure 1.

(b) *One-hour watches.* 50 other R.A.F. cadets were also tested, one at a time, for one hour each. They reported in pairs; one man did the first hour of the test, and then the second man (who had been waiting outside the room) took over for the second hour. The 25 people who did the first hour were Group A in Figure 2; the 25 men who did the second hour were Group B. A comparison could therefore be made between the average performance of 50 men divided into two groups each working for one hour with that of the control group who worked for a two-hour spell.

(c) *Half-hour watches.* 50 further R.A.F. cadets were then tested in the following way. They reported in pairs—Subject X and Subject Y; Subject X did the first and third half-hours of the test and Subject Y did the second and fourth half-hours. This meant that results were obtained from the half-hour alternation of two groups (Group X and Group Y of Figure 3). During their half-hour break these subjects had no opportunity of determining their level of accuracy on their initial spell.

(d) *Two-hour watches with the telephone message.* 25 other R.A.F. cadets were then tested exactly as for the two-hour control group experiments, but one alteration was made to the conditions. The subject was given a hand microphone telephone and told he would receive various instructions over this from time to time during the test. In fact, only one standard message was given, and this always arrived in the middle of the ten minutes of negative stimuli ending the second half-hour of the test. The message therefore immediately preceded the beginning of the third half-hour, the results for which are plotted as point T_3 in Figure 4. The exact technique was that a warning buzzer sounded and as soon as the subject answered, he was asked over the telephone "to do even better for the rest of the test; to try to see even more of these double-length movements of the pointer."

(e) *The briefing experiment.* An investigation was also made of the effects of briefing the subjects. In this, the six subjects did a two-hour run exactly as the test for the two-hour control group, but again with one alteration in the conditions. These men were given an extra display board to watch as well as the Clock hand. The whole of their briefing was given before the test began and each subject was told that the board represented the position of his imaginary aircraft in relation to the area of sea being patrolled. During the two-hour experiment, a plotting arrow crept very slowly across the map, so that the subject had to glance at this display only occasionally to determine his supposed position. He was asked to concentrate on the Clock pointer and told to be ready at all times for the double-length movements of the hand, and warned to be especially on the alert when the indicator arrow was passing through the "danger area" marked on the map, as more signals were to be expected during that period. The plotting arrow was timed to enter this special area just before the third half-hour of the test, and left it at the end of the third half-hour.

III. RESULTS

1. *Results of Two-hour Watches (Control Group).* It is clear from Figure 1 that the men began to miss more signals after they had been working for half-an-hour at the task (see Table 1).

This reduction in working efficiency after half-an-hour was not likely to have been due to uncontrolled factors in the experiment, since the odds were at least 19 to 1 against the difference between C_1 and C_2 being due to chance; the difference between the two means was exactly three times its standard error, when twice would have shown a significant difference. The differences between C_1 and C_3 and between C_1 and C_4 were also statistically significant since the critical ratios here were 3.3 and 3.5 respectively. On the other hand, the slight fall in efficiency between C_2 and C_4 was not statistically definite since the critical ratio here was only 0·6.

2. *Results of One-hour Watches (Group A and Group B).* Here again there was a marked difference between the performance of the subjects during their first half-hour and their second half-hour (see Table 2).

The differences between A_1 and A_2 and between B_1 and B_2 were statistically sound since critical ratios of 4.4 and 3.7 were obtained. On the other hand there was no significant difference between A_1 and C_1 since here the critical ratio was 0.1—nor was there any real difference between A_2 and C_2, as the critical ratio was 1.2.

It will be noted that bringing in the fresh subjects (Group B) on the third half-hour of the test reduced the incidence of missed signals to one-half of the value obtained from the control group. This difference between C_3 (26.8% missed) and B_1 (13.0% missed) was statistically significant as the critical ratio was found to be be 4.2. But this advantage again lasted for only half-an-hour, because during the last half-hour of the test there was no real difference between C_4 and B_2 as the critical ratio here was only 1.1.

3. *Results of Half-hour Watches (Groups X and Y).* Statistical analysis showed that under the conditions of rapid, half-hour alternation there was no significant

EFFECTS OF A TWO-HOUR WATCH

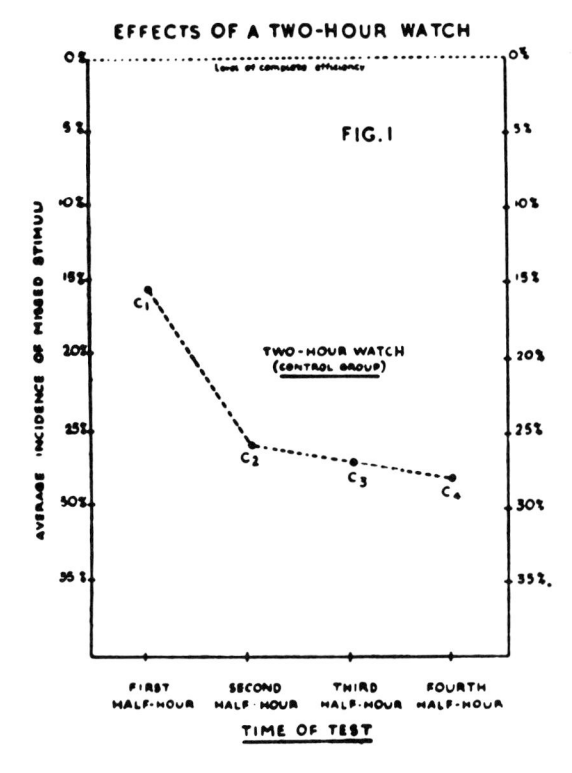

FIG. 1

Two-hour watch (control group)

Average incidence of missed stimuli vs. Time of test (First, Second, Third, Fourth Half-Hour)

TWO-HOUR WATCH COMPARED WITH ONE-HOUR SPELLS

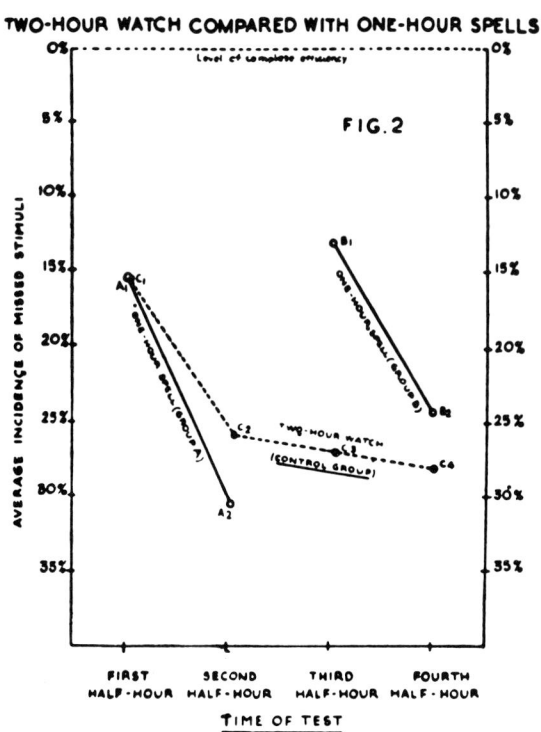

FIG. 2

One-hour spell (Group A), *One-hour spell (Group B)*, *Two-hour watch (control group)*

TWO-HOUR WATCH COMPARED WITH HALF-HOUR SPELLS

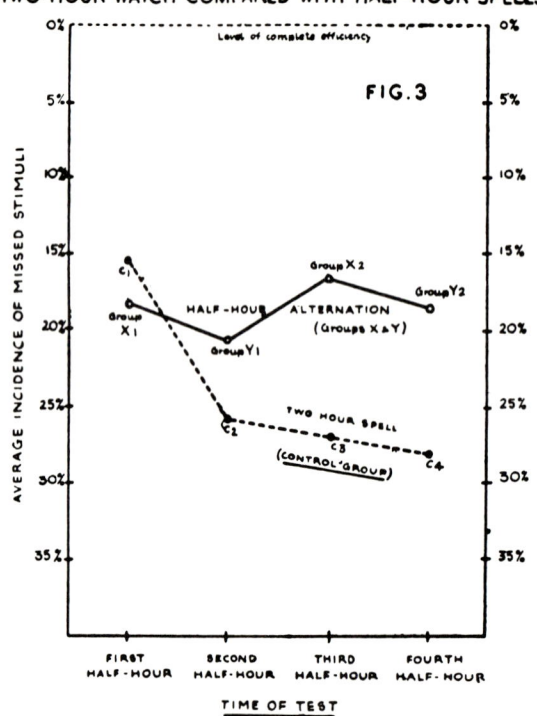

FIG. 3

EFFECT OF THE TELEPHONE MESSAGE

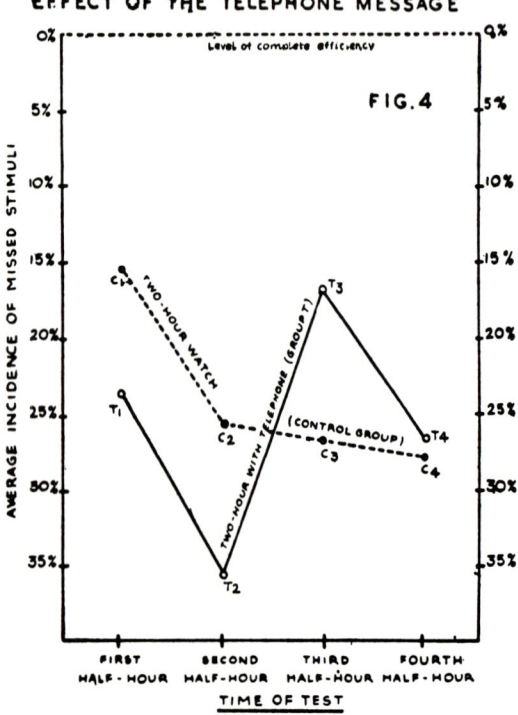

FIG. 4

TABLE 1

Mean Percentage Incidence of Missed Signals

First Half-hour (C_1)	Second Half-hour (C_2)	Third Half-hour (C_3)	Fourth Half-hour (C_4)
15·7% (46/293)	25·8% (75/291)	26·8% (79/295)	28·0% (82/293)

TABLE 2

Mean Percentage Incidence of Missed Signals

Nature of Group	First Half-hour (A_1)	Second Half-hour (A_2)	Third Half-hour (B_1)	Fourth Half-hour (B_2)
Group A	15·3% (45/294)	30·4% (91/296)	—	—
Group B	—	—	13·0% (38/292)	24·2% (72/297)

TABLE 3

Mean Percentage Incidence of Missed Signals

Nature of Group	First Half-hour (X_1)	Second Half-hour (Y_1)	Third Half-hour (X_2)	Fourth Half-hour (Y_2)
Group X	18·1% (53/293)	—	16·5% (51/289)	—
Group Y	—	20·7% (61/294)	—	18·5% (54/291)

TABLE 4

Total Incidence of Missed Signals

| | Time after Telephone Message ||||
Nature of Group	Within 13 minutes (Stimuli 1–7 of 3rd half-hour)	Within 18–25 minutes (Stimuli 8–12 of 3rd half-hour)	Within 35–43 minutes (Stimuli 1–7 of 4th half-hour)	Within 48–55 minutes (Stimuli 8–12 of 4th half-hour)
Telephone message group	28	22	45	33
Two-hour control group	49	30	48	34
Ratio	0.57	0.73	0.94	0.97

TABLE 5

Mean Percentage Incidence of Missed Signals

Nature of Group	First Half-hour	Second Half-hour	Third Half-hour	Fourth Half-hour
Briefed Group	18.1% (13/72)	26.4% (19/72)	27.8% (20/72)	27.8% (20/72)
Two-hour Control Group	15.7% (46/293)	25.8% (75/291)	26.8% (79/295)	28.0% (82/293)

TABLE 6

Effect of Length of Preceding Inter-Stimulus Time Interval on Mean Percentage Incidence of Missed Signals

Length of preceding blank spell..	¾–1 minute	1½–3 minutes	5–10 minutes
Average incidence of missed signals	25.9% (637/2460)	24.1% (594/2460)	20.4% (201/984)

FIG. 5

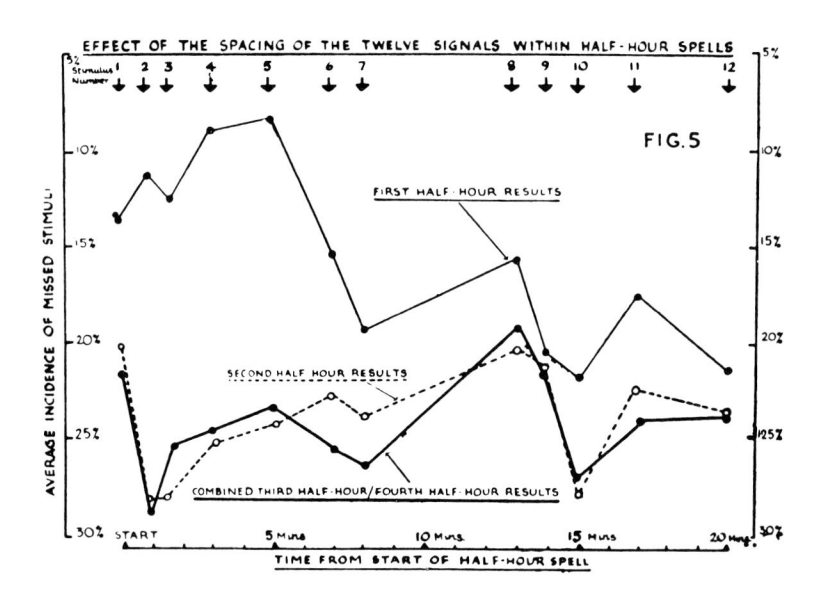

Effect of the spacing of the twelve signals within half-hour spells.

TABLE 7

Frequency Distribution Table for Missed Signals

Number of signals missed (out of 12)	Fresh Group %	Number	Tired Group %	Number
0	24·9	(56)	14·7	(33)
1	24·9	(56)	13·8	(31)
2	16·9	(38)	15·1	(34)
3	11·1	(25)	12·9	(29)
4	9·3	(21)	8·8	(20)
5	3·1	(7)	13·8	(31)
6	4·4	(10)	8·0	(18)
7	4·0	(9)	8·4	(19)
8	0·9	(2)	2·2	(5)
9	0·5	(1)	1·8	(4)
10	—	—	0·5	(1)
11	—	—	—	—
12	—	—	—	—
	Total ..	255 readings	Total ..	225 readings

TABLE 8

Conditioned Stimulus	Unconditioned Stimulus	Original Voluntary Response Conditioned Voluntary Response	Reinforcement
Long movement of the clock hand	Command of "Now!"	Pressing response key	Comment of "Yes, that was right," etc. (Knowledge of results).

The Breakdown of Vigilance During Prolonged Visual Search

difference between the average performance on the first half-hour of the test and that during any other period of the experiment. In Figure 3 the point X_1 was not statistically different from either Y_1 or X_2 or Y_2, the critical ratios being 0.8, 0.5 and 0.1 respectively (see Table 3).

Comparison with the two-hour control group in Figure 3 shows that on the first half-hour of the test the initial points X_1 and C_1 of the two curves were not significantly different (the critical ratio being 0.8) when both groups were fresh to the work. But during the last hour-and-a-half of the experiment there was a highly significant difference between the two-hour group and the half-hour alternating groups. With the points C_2, C_3 and C_4 taken together and the points Y_1, X_2, and Y_2 combined, the critical ratio was 4.1. Comparing these points separately there were significant differences between C_2 and X_2 and between C_4 and Y_2—critical ratios being 3.0 and 2.7, but the difference between C_2 and Y_1 was not significant, the critical ratio being 1.5. This was probably due to chance factors producing a slightly less efficient group of people for Group Y. (It might have been thought that because Group Y had to wait before doing the test, this would have had some deleterious effect on point Y_1 which would not have affected point X_1. This seems unlikely since in the previous experiment there was no significant difference between point A_1 and B_1 of Figure 2.)

Though there was no statistically sound difference in Figure 3 between X_1 and Y_1 or between C_1 and Y_1, the increased incidence of errors at C_2 over X_1 was quite reliable, the critical ratio being 2.2. The apparent improvement between points X_1 and X_2 and between Y_1 and Y_2 was not confirmed on statistical analysis as the critical ratios came to 0.5 and 0.7 respectively.

4. *Results of Two-hour Watches with the Telephone Message (Group T).* Analysis of the average incidence of missed signals (see Figure 4) suggested that listening for the telephone at the same time as looking for the double movements of the clock hand was a definite handicap; because during the first hour of the test (before the message was given) there was a significant difference between the two-hour control and the two-hour telephone groups. The critical ratio for the points C_1 and T_1 was 2.4, and for the points C_2 and T_2 was 2.5. But the telephone message dramatically reduced the number of missed signals; it raised the average level of efficiency for the third half-hour to a standard which was usually obtained by fresh subjects, not by men like these who had been working for more than an hour at the task (see Table 4).

The difference between C_3 and T_3 in Figure 4 showed a significant advantage in favour of the telephone group, as the critical ratio was 3.0. The point T_3 was not significantly different from C_1 (the critical ratio being 0.3), but T_3 was just significantly better than T_1 (the critical ratio being 2.0). This improvement did not last for more than half an hour since the difference between C_4 and T_4 was not statistically definite, and here the critical ratio was 0.4. A subjective improvement following the telephone message was reported by the majority of the subjects and this also lasted for about half an hour.

A further analysis of the data was made to ensure that the improvement in performance did in fact extend over all the stimuli in the third half-hour, because it was just possible that the effect had been confined to the signals given immediately afterwards. The total number of signals missed by the 25 men was calculated for the first seven signals given after the message and for the remaining five signals given in the third half-hour. The data from the fourth half-hour were subdivided in the same way—and the results from the two-hour control group provided a comparison. Since the telephone message was given five minutes before the beginning of the third half-hour, the evidence was obtained at the time intervals after the message. (See Table 4.)

In other words there was no doubt that the introduction of the telephone message produced an effect which lasted for 25 minutes, but this disappeared about 35 minutes after the application of the stimulus.

5. *Results of the Briefing Experiment.* Unlike the experiment with the telephone message, this briefing technique had no effect at all on efficiency during the third half-hour of the experiment, or at any other time. Table 5 compares the average incidence of missed signals obtained from this briefed group of subjects with that made by the two-hour control group. There was no significant difference between the two sets of scores.

6. *Results of a Stimulus-by-Stimulus Analysis of the Performance Trend.* An attempt was made to detect any fluctuations in working efficiency due to the spacing of the signals in the test series, i.e., the pattern made in time. Two difficulties arose which necessitated the use of average results taken from a large number of subjects in searching for these minute-by-minute variations in accuracy. The two disturbing factors were:—(1) each man had some entirely random fluctuations in his level of vigilance; and (2) all stimuli were not of entirely uniform difficulty because a double-length movement was probably easier to detect when the clock hand was pointing in certain directions than in others. This was important because a double-length movement could start at any one of the hundred possible positions of the clock hand; the spatial position of such test signals was therefore entirely randomised within the circle traced out by the tip of the clock hand.

For the *first and second half-hours* of the experiment, results were available from 214 subjects, i.e., the average incidence of missed signals was calculable on these subjects for each of the 24 stimuli taken in turn. Not enough data were available to do this for the *third and fourth half-hours* of the experiment, but there the two time patterns could be superimposed, and, for example, the evidence from the first, second, third and fourth, etc., signals in the third half-hour was taken together with that from the first, second, third and fourth, etc., signals respectively of the fourth half-hour. This meant there were 278 presentations of each stimulus in this combined third half-hour/fourth half-hour series.

The results are plotted in Figure 5; they confirmed the marked difference in overall accuracy between the first half-hour and the rest of the test. There was also some suggestion of a warming-up effect within the first five minutes of the experiment. This was followed by a downward trend in efficiency during the remainder of the first half-hour of the test. During the second and subsequent half-hours this falling-off was arrested, and performance oscillated around the same low level of accuracy. That the bulk of the experiment was being done by the subjects in an inaccurate manner is clear when one sees from Figure 5 that the subjects *usually* made twice or even three times as many mistakes compared with their optimum performance. Three to five minutes after the beginning of the experiment the men averaged 8% or 9% errors, but for the last three half-hours of the test, the same people gave average error scores which fluctuated between 19% and 28% missed signals.

Figure 5 emphasises that apart from the long-term trend, there were similarities between the stimulus-to-stimulus trends within the various half-hour spells. At first glance, there seems no good reason why, for example, the seventh stimulus in each of the half-hour series should, on the average, always have proved more difficult than the eighth. It is believed that the similarity between the stimulus-by-stimulus records for each half-hour is explicable only on the grounds that the spacing of the signals in time drew a definite pattern of performance by producing minute-by-minute changes in attitude which arose from the apparent occurrence or non-occurrence of the signals. If one compares the three curves shown in Figure 5, it seems clear that (apart from the initial warming-up effect during the first five minutes) the signals that were most difficult to see were those presented in relatively rapid succession at intervals of ¾–1 minute apart, i.e., the second and third stimuli, and the ninth and tenth stimuli in the series. This may have been due to the fact that the instructions and the test, as a whole, suggested that a good deal of time elapsed between signals. The subjects, therefore, lowered their vigilance after noticing the first or eighth signals—particularly as these followed, and were therefore in direct contrast with, blank spells lasting ten or five minutes, which were entirely free from meaningful activity. Once a signal has been seen, the subject may miss the next stimulus if he does not expect the next signal to appear for some little time, and it does in fact appear before he expected it.

An attempt has been made to analyse the data statistically to see whether any reliable evidence could be obtained about this effect of the length of the blank spell preceding the stimulus on accuracy of work. The data from the first half-hour of the experiment have been excluded owing to the marked general trend previously described and the study has, therefore, been confined to the last hour-and-a-half of the experiment. The results were classified into three main groups according to the length of the preceding inter-stimulus time interval (see Table 6).

The critical ratio between the 25.9% reading and the 20.4% reading came to 3.5, but the critical ratio between the 25.9% incidence and the 24.1% incidence was only 1.5; the critical ratio between the 24.1% score and the 20.4% score was also only 1.3. The subjects were, therefore, definitely worse on signals given after blank spells of one minute or less than they were with those presented after an inter-stimulus interval of 5 to 10 minutes.

7. *Results of a Study of the Individual Differences.* In considering the differences between people in their ability to do this kind of work, the index of performance selected was the number of missed signals per half-hour. In view of the sharp deterioration in efficiency found after half an hour at the work, it seemed reasonable to divide the results into two categories—fresh and tired subjects.

(i) *Fresh subjects:* Results from people who were doing their first half-hour on the test and from those starting again after half an hour's break.
(ii) *Tired subjects:* Results from men who had been working more than half an hour on the test.

A study of the frequency distribution (see Table 7) showed the wide range of ability amongst either fresh or tired subjects. Secondly, although the distributions obtained from the fresh and tired groups overlapped, the tired group tended to have a greater proportion of readings in the lower levels of efficiency. For example, it will be seen that about half the readings in the fresh group were in the categories where either all the signals were seen or only one stimulus was missed; only about a quarter of the records from the tired men reached this high pitch of efficiency. These marked individual differences are not related either to visual acuity grades or group intelligence test scores on verbal and spatial relationship material.

IV. DISCUSSION

1. Introductory Points

The above experiments suggest that lapses in visual perception become more frequent (i) when the Clock test has been proceeding for more than half-an-hour, (ii) when the subject has recently responded to one visual signal and is not expecting another since he imagines the stimuli are few and far between—and (iii) when the subject is expecting an auditory stimulus while doing the visual task. It is perhaps permissible to mention briefly here that other studies by the writer have also indicated that (iv) lack of sleep is liable to lead to more missed signals on this test and (v) so also to a lesser extent will an increase of the atmospheric temperature of the working environment to a reading about that of the normal body temperature.

Conversely, the conditions that are known to improve performance include (i) a short rest of half an hour and (ii) the receipt of a telephone message. Similarly, fewer signals pass unnoticed (iii) if while they do the test the subjects are allowed to have a progress record giving them immediate knowledge of their results, or (iv) if they take 10 mgms. of aphetamine sulphate by mouth one hour before the start of the test. Either of these last two procedures keeps performance at the initial high level of accuracy throughout the two-hour spell.

Various possible interpretations of these facts can be attempted but none of the explanations appears wholly satisfactory to the writer. It is at least clear from the various psychological influences that can be used to alter the achievement level that the phenomenon is central in origin and is not due to changes in the peripheral visual mechanisms. A relevant point is that Carpenter (1948) has shown that, although the average blink rate on the Clock test does give a rising trend similar to the half-hour to half-hour incidence of missed signals, this is a coincidental and not a causal relationship.

It might be thought that these researches were dealing only with alterations in the general attitude of the subjects towards the whole test situation—some fading of interest in external events, perhaps leading to a condition like the lowered standard phenomenon described by Bartlett (1941) where the subjects on a highly skilled task became less self-critical and allowed progressively larger and larger errors to occur before they regarded these as calling for action. This view of the Clock test results is very hard to maintain owing to the absence of effect with the briefing technique; presumably on this theory this ought to have given some performance improvement by modifying the general attitude of the objects during the part of the test spell that they had been told was particularly important. Similarly there was no "suggestion effect" improvement in performance when men did the test after taking some unspecified tablets which in fact were pharmacologically inert. If the subjects were simply becoming bored with the work, one feels that they ought also to have been *more*, and not less, alert immediately after the occurrence of some slightly unusual event in their environment such as one of these long signals. It would still be necessary to know what caused this fading of interest and why it was not usually progressively more marked with each half-hour at the work. On this last point Oldfield has suggested that the flattening of the work curve (which was also found by Anderson *et al.* (1944)), might be taken to indicate that the normal and relatively stable state was that represented by the large error score in the last hour-and-a-half of the test spell and that the first half-hour results were due to an abnormal and unstable level of alertness.

2. Discussion of the Main Downward Trend

The writer considers that the least unsatisfactory way of interpreting the findings on the Clock test is to regard the condition under consideration primarily as an example of a state of inhibition—the sense of the decrease or absence of a response which was the result of some form of positive stimulation. Since this decrease was apparently due to the repeated presentation of stimuli incorporated in the test situation itself, it would seem reasonable to regard the lowered efficiency as an example of internal inhibition. Other interpretations of the facts are possible and all are hard to bring to the experimental test.

In his paper on internal inhibition, Oldfield (1937) remarked that no attempt was being made to explain the intimate mechanics of all cortical activity; internal inhibition was regarded as a state which may descend on any one of a number of different cortical functions. The present writer has used the same approach; internal inhibition has been studied because it seems to have been found in circumstances which at first were being investigated entirely empirically to investigate a practical problem. The terminology of conditioned response theory has not been adopted with the unjustified assumption that the principles of the conditioned response provide a comprehensive solution of the mysteries of all cortical function.

In discussing the characteristics of situations leading to internal inhibition, Oldfield (1937) makes the uniformity of the total sensory field a first requirement; this uniformity is usually obtained in the form of simple, regular, discrete changes in the sensory environment, because after a time such regular changes tend to become accepted as uniformity. The Clock test satisfies this primary condition in view of the extreme regularity of the movements of the clock hand in *space*—round and round it goes, always taking the same short steps except for an occasional double-length signal. As regards regularity in *time*, the clock hand invariably moves with a regular rhythm which is not disturbed even by the double-length stimuli.

In his searching review of the relationship between conditioned response and conventional learning experiments, Hilgard (1937) made the point that the distinction between reinforcement and reward, sometimes permitted to separate conditioning from learning, is found to disappear when the conditioned response situation is carefully examined; Hilgard and Marquis (1940) have further developed this concept, including among their specific examples the learning experiments by Grindley (1932) and the studies on children by Ivanov-Smolensky (1927). Investigations of this sort have been grouped under the general heading of "Instrumental conditioning." The term instrumental conditioning is employed when the correctness or otherwise of the response determines whether a reward is given or withheld, and when the response can, therefore, be regarded as instrumental in obtaining the reward. The essential feature of instrumental conditioning is that the unconditioned stimulus leading to the original reaction is entirely different from the reinforcing stimulus which follows the conditioned response.

Food reward experiments satisfy a primary need of the organism, but it is clear that when the experimental subjects are adult human beings an effective reward need not necessarily satisfy a primary drive. Indeed a neutral stimulus may acquire the reinforcing properties of a reward if it has been repeatedly associated with a primary reward; Hilgard and Marquis (1940) term this "secondary reinforcement" or "derived reinforcement." In adult humans a very effective form of derived reinforcements is the token reward of knowledge of results, i.e., immediate information, supplied while the work is being done, about the success with which the prescribed task has been undertaken (Crawley, 1926; Grindley and Elwell, 1939; Mace, 1935).

When such investigations are borne in mind, it becomes clear that the Clock test situation readily lends itself to a description in conditioned response terms. During this account it has been found convenient to refer to the double-length movements of the Clock hand as the long signals, and to the ordinary single-length movements as the short signals.

The *original voluntary instructed response* was the movement of the arm and hand to press the reaction key.

The *unconditioned stimulus* was present only during the brief initial practice spell, and not at all during the main experiments. Even in the practice spell, it usually had to be given only once or twice in the first minute or so of the demonstration period. It took the form of the experimenter's remark, "Now!"—a shortened version of the implied instructions, "Press the key, now!" This comment was made immediately after a long signal, and it seems logical to regard it as the unconditioned stimulus, since by itself, without any other stimulus, it would have led to the original voluntary instructed response.

The *conditioned stimulus* was the long signal given by the Clock hand. This positive stimulus had to be differentiated from the mass of similar stimuli coming from the short signals made by the Clock hand.

The *conditioned voluntary response* occurred when the subject pressed the response key on detecting a long signal.

Reinforcement was normally given by the experimenter only during the practice period, and consisted of his comment "Yes, that was right" if the subject had responded correctly, *or* "No, that was wrong" if the key had been pressed after a short signal rather than a long one, *or* "You missed one there," if the subject had failed to notice a long signal. In other words the reinforcement was a derived reinforcement—the token reward of knowledge of successful results, and the token punishment of information on failures.

This description of the Clock test can be summarised as follows (see Table 8):—

Very rapidly, during the practice spell, this situation gave an entirely new significance to what was, at first, the quite neutral stimulus of a long movement of the clock hand. This result is in keeping with the experience of previous workers who have found that the conditioned voluntary response is established quickly if it is formed at all.

Since the term internal inhibition demands a more precise definition it is necessary to attempt to define the nature of the condition more fully. It is likely that the exact form it took was that of experimental extinction, since this can usually be considered as the absence or weakening of a conditioned response upon repeated application of the conditioned stimulus without any reinforcing agent. Under the particular circumstances in question, this was taken to be the repeated presentation of the long signal without any provision of knowledge of results. Further evidence in support of this view that the long-term downward trend in performance from half-hour to half-hour was a manifestation of experimental extinction was found in the prevention of the usual increase in error score by the supply of knowledge of results (Mackworth, 1948).

It was interesting also to find that the sudden presentation of a novel distracting stimulus in the form of the telephone message immediately restored the partially extinguished response because this strongly suggested a disinhibitory effect, particularly as the briefing experiment which introduced no sudden unexpected change in the experimental situation failed to improve performance. But the length of time during which this disinhibition was effective was much longer than has usually been reported in Pavlovian experiments (Hull, 1934).

3. Discussion of the Minor Fluctuations in Efficiency

Although the experimenter did not usually give progress information to the subjects, and although this was apparently the main factor in the performance decline obtained in the Clock test, it remained possible that the subjects had a vague and uncertain source of knowledge of results in their own experience of the test situation. Figure 5 suggests that in fact this was so—that the subjects were collecting *some* information about their performance from the task itself as they went along. If they had not done so, then there would not have been any resemblance at all between the shape of the stimulus-to-stimulus curve for one half-hour compared with that obtained for another. This is believed to have been why the experimental extinction was only partial and not complete, i.e., that to some extent the influence of expectancy and self-instructions replaced the initial reinforcing stimulus of knowledge of results. This resulted in second-order fluctuations being superimposed on the main trend of the working efficiency curve.

Gibson (1941) has emphasized in a valuable review the distinction between two forms of preparatory set—*expectancy* (expecting stimulus objects) and *intention* (intending to react). Mowrer, Rayman and Bliss (1940) have shown that expectancy can be varied independently from intention by demonstrating that subjects were definitely slower in reacting to a series of sounds when previous instructions had led them to expect either sounds or lights, than when they were told to expect only sounds. This divided attention effect may explain the deterioration in accuracy that was found during the first hour of the Clock test when the subjects had to listen for a telephone message while at the same time undertaking the ordinary visual task.

Similarly the finding by Mowrer (1940)—that a deterioration in auditory reaction time was more likely if the signal came in before it was expected, rather than after—would seem to be in keeping with the finding that errors in the Clock test were usually higher after the shorter inter-stimulus intervals.

V. SUMMARY

Laboratory researches on prolonged visual search led to the development of an experimental situation in which the trend of ability at synthetic look-out duties was studied throughout long watchkeeping spells by the automatic production of occasional brief and barely visible signals. These stimuli were given at irregular time intervals; sometimes one or two minutes elapsing between signals, sometimes as long as ten minutes passing without significant incident. The results have been expressed in terms of the average error score for groups of healthy young men, but each subject was tested in a room entirely by himself, the experimenter being in an adjoining room with the recording apparatus. These arrangements made it possible to keep the subjects largely unaware of their accuracies and inaccuracies at the task and so preserved an essential feature of vigilance testing.

Under these circumstances, efficiency at signal detection during a two-hour spell was definitely lower when the subjects had been watching the display for about half an hour. One-hour spells showed the same advantage in favour of the first half-hour compared with the second. But a half-hour rest following a half-hour watch was sufficient to allow a further half-hour spell to be undertaken straight away with no detrimental effect on accuracy. A sudden short telephone message in the middle of a two-hour watch produced a temporary improvement lasting only half an hour, but previous instructions that a particular period of the watch would be likely to contain rather more signals than usual had no demonstrable effect on performance.

Various possible interpretations of these changes in behaviour have been considered. The *main* downward trend in efficiency was eventually regarded as an example of the partial experimental extinction of a conditioned voluntary response, this extinction being the product of a lack of reinforcement arising from the absence of knowledge of results. The improvement following the distraction of the telephone message was interpreted as a form of disinhibition. There were also, however, *minor* stimulus-to-stimulus variations in efficiency which depended more on the general attitude of the subjects towards the test situation, particularly in regard to their opinion of the likelihood of a signal occurring at a given moment. For example, except in the first few minutes of the test the subjects were definitely less alert at detecting a signal when this was presented only one minute after the previous stimulus.

VI. ACKNOWLEDGMENTS

Professor Sir Frederic Bartlett has greatly helped these researches with extensive advice, and helpful discussions on the problems have also been held with the late Dr. K. J. W. Craik, with Mr. G. C. Grindley and with Mr. R. C. Oldfield, but the interpretation is that of the writer. The statistical analysis was assisted by Mr. E. G. Chambers, and the experimental apparatus was constructed by Major H. C. L. Holden and Dr. E. Schuster. Mrs. Nixon de Weber gave most of the tests and prepared the illustrative diagrams. The Royal Air Force arranged the supply of the experimental subjects, and the writer is indebted to the Medical Research Council for financial support.

VII. REFERENCES

1. Anderson, I. H. *et al.* (1944). Radar operator fatigue: the effects of length and repetition of operating periods on efficiency of performance. O.S.R.D. Report No. 3334, Research Report No. 6.
2. Bartlett, F. C. (1941). Fatigue following highly skilled work. 1941 Ferrier Lecture, published in 1943. *Proc. Roy. Soc. B.*, 131, 247-257.
3. Bills, A. G. (1934). *General Experimental Psychology*. New York, London and Toronto.
4. Carpenter, A. (1948). The rate of blinking during prolonged visual search. *J. exp. Psych.* (in the press).
5. Craik, K. J. W., & Mackworth, N. H. (1943). Preliminary notes on the optimum length of ASV Mark III watch. (Unpublished report to Coastal Command, Royal Air Force.)
6. Crawley, S. L. (1926). An experimental investigation of recovery from work. *Arch. Psychol.*, 13, No. 85.
7. Ditchburn, R. W. (1943). Some factors affecting efficiency of work of look-outs. (Unpublished Admiralty Research Laboratory Report ARL/RI/84.46/0).
8. Gibson, J. J. (1941). A critical review of the concept of set in contemporary experimental psychology. *Psychol. Bull.*, 38, 781-817.
9. Grindley, G. C. (1932). The formation of a simple habit in guinea pigs. *Brit. J. Psychol.*, 23, 127-147.
10. Grindley, G. C., & Elwell, J. L. (1939). The effect of knowledge of results on learning and performance. *Brit. J. Psychol.*, 29, 39-54.
11. Head, H. (1926). *Aphasia*, Vol. I. Cambridge.
12. Hilgard, E. R. (1937). The relationship between the conditioned response and conventional learning experiments. *Psychol. Bull.*, 34, 61-102.
13. Hilgard, E. R., & Marquis, D. G. (1940). *Conditioning and Learning*, pp. 429. New York.
14. Hull, C. L. (1934). *Handbook of General Experimental Psychology*, ed. by Murchison. Chapter 9. The factor of the conditioned reflex. Clark University Press.
15. Ivanov-Smolensky, A. G. (1927). On the methods of examining the conditioned food reflexes in children and in mental disorders. *Brain*, 50, 138-141.
16. James, W. (1890). *Principles of Psychology*, Vol. I, pp. 689. London.
17. Mace, C. A. (1935). Incentives. Medical Research Council, Industrial Health Research Board Report No. 72. H.M. Stationery Office, London.
18. Mackworth, N. H. (1944). Notes on the Clock test. Unpublished report to the Medical Research Council, A.P.U. 1, also Flying Personnel Research Committee Report No. 586.

19. Mackworth, N. H. (1948). Researches on the measurement of human performance. Medical Research Council Special Report Series. H.M. Stationery Office, London (in the press).
20. Mowrer, O. H. (1940). Preparatory set (expectancy)—some methods of measurement. *Psychol. Monogr.*, **52**, No. 233.
21. Mowrer, O. H., Rayman, N. N., & Bliss, E. L. (1940). Preparatory set (expectancy) —an experimental demonstration of its central locus. *J. exp. Psychol.*, **26**, 357-372.
22. Oldfield, R. C. (1937). Some recent experiments bearing on internal inhibition. *Brit. J. Psychol.*, **28**, 28-42.
23. Robinson, E. S. (1934). *Handbook of General Experimental Psychology*, ed. by Murchison. Chapter 12. Work of the integrated organism. Clark University Press.
24. Woodworth, R. S. (1938). *Experimental Psychology*. New York.
25. Wyatt, S., & Langdon, J. N. (1932). Inspection processes in industry. Medical Research Council, Industrial Health Research Board Report No. 63. H.M.S.O., London.

46

Human Vigilance

James G. Holland

Current interest in the classical problem of sustained efficiency in monotonous perceptual tasks has centered around situations in which human beings are required to monitor some display in search of critical, but infrequent, signals. Such tasks are numerous and of considerable practical importance. In air defense systems, operators must search radarscopes for extremely infrequent enemy targets. Increased automation requires human monitoring of equipment which seldom fails. In addition, cases involving assembly-line inspection of products represent another large group of monitoring tasks in which the critical signals may arise relatively infrequently.

Recent work on operators monitoring displays having infrequent signals indicates a drop in the percentage of signals detected as time on watch progresses. Mackworth (1) has shown a decrement in the subject's ability to detect signals as a two-hour watch progressed. The signals were double steps of a clock hand which normally stepped 0.3 inches every second but had 24 double steps per hour. Similar decrements have been demonstrated (2–6) when subjects were required to detect targets on simulated radar displays. Field studies also have

From *Science*, **128**, July 11, 1958, 61-67. By permission of the American Association for the Advancement of Science.

shown the decrement as time on watch passes. This has been found true for radar operators (6, 7) and for a variety of industrial inspectors (8). In addition, Bakan (9), using a modified threshold measurement technique, has demonstrated a decrement in a brightness discrimination task.

Not all investigators have found a decrement. One investigator (10), using the clock test, has shown an increased variance in the number of detections as the watch progresses, but no average decrement. Others (4, 11), using latency of detection of nontransient signals as a criterion, rather than the percentage of signals detected, have found an increase in variance but no increase in the average latency of detection.

Whether a decrement is found or not, the fact is clear that many signals well above absolute threshold are not detected either early or late in the session. Furthermore, if the frequency of signals increases, there is an increase in the percentage of signals detected (4, 5). For example, Deese found that with a display simulating a research-radar 'scope and using 10, 20, 30, or 40 targets per hour during a three-hour watch, 46, 64, 83, and 88 percent were detected, respectively.

In order to "account for" the decrement and the relation between signal frequency and detection probability, an abundance of theoretical constructs have been offered. The results obtained are said to reflect declines in, or waxing and waning of, attention, vigilance, or fatigue. Mackworth (1) tentatively postulated an excitatory state termed "vigilance" which is opposed by an inhibitory state that parallels the concept of external inhibition found in the literature on classical conditioning. More recently, Adams (2) has used Hull's I_R (reactive inhibition) in a similar manner. The performance decrement is supposed to be a partial extinction phenomenon reflecting the build-up of the inhibitory state. When a verbal message to the effect that the subject should "do even better for the rest of the test" was delivered, the percentage of signals detected returned to the initial level. This is explained as disinhibition and thus as evidence for the existence of an inhibitory state. When a 1-hour break was provided, again the performance returned to the initial level. This is said to reflect spontaneous recovery from the inhibitory state.

Several investigators have employed expectancy as an explanatory concept. Mackworth (6), Broadbent (11), and Deese (12) have used it to "explain" (i) the greater over-all percentage of detections when the number of signals per session increases (4, 5), and (ii) the increased probability of detection for the longer intersignal times that is observed when a signal-by-signal analysis is made (1, 5). [The latter finding has not always been confirmed (3, 5).] In addition, Broadbent has used the idea of stimulus selectivity (that is, attention or set) to explain not only the findings concerning monitoring behavior but classical conditioning as well.

In addition to these theories relative to psychic and conceptual states, a physiological theory has been advanced. Deese suggested that the waking center (12) of the hypothalamus may be involved and that the activity of the center depends on an influx of sensory stimulation. According to this theory, as it applies to problems of detecting infrequent signals, a varied sensory input is necessary to maintain the excitatory state in this center and thus to maintain a high level of detection.

NEED FOR AN ATHEORETIC APPROACH

The various theories have all been developed to account for a rather meager set of data. The parameters influencing the monitoring of low-signal-frequency displays are as yet poorly explored, with the result that inconsistencies are found among the findings. In view of this state of affairs it might be well to forego the luxury of developing explanatory concepts until the empirical relations are better established.

Indeed, the necessity for theories has been sensibly challenged by Skinner (13) with regard to theories of learning. Theories, in Skinner's sense, are explanations of data that make use of events at another level of observation and are not to be equated with empirically defined concepts which refer to the behavioral level of observation. The latter permit generalization of empirical principles. His arguments seem at least as relevant to theories of "vigilance."

Such entities as vigilance, attention, inhibition, expectancy, and waking-center activity all fall into the category of theories, as defined above. They, like concepts in learning theories, are "at some other level of observation, described in different terms, and measured, if at all, in different dimensions" (13, p. 193). These concepts give the appearance of explaining the data because of the syntax of the statements. The subject is said to make a detection because he is, at that moment, vigilant or attentive or expecting a signal. But the concepts are no less mysterious than the phenomena they purport to explain. There remains the task of discovering the events which influence vigilance, attention, or expectancy. Once having done this, we may be little better off than if we had simply searched for the conditions controlling the probability of detection, since this is assumed to be directly related to the intervening explanatory concept.

The argument that theories generate research does not seem to apply to theories in this area. With one exception (5), the theories seem to have been offered as explanations of data already collected. But even should they generate research in the future, it is by no means obvious that this research would be of greater significance than research directed toward an empirical and behavioral systematization of the field.

However, the use of theories is by no means surprising in view of the types of measure used. In practice, only the percentage of signals detected, latency of detection, or change in threshold intensity is measured. The investigator is then faced with the problem of saying what it is that changes during the monitoring task. It is unsatisfactory to say that the percentage of signals detected *is* the vigilance rather than a *result* of vigilance, attention, or expectancy, just as the learning theorist is unsatisfied in saying that decrease in errors *is* learning. Instead of proceeding to search for a satisfactory datum on the behavioral level of observation, the investigator postulates events for other levels of observation. Signal detection is said to *reflect* states of vigilance, attention, or expectancy. The result of this is that the search for an appropriate behavioral datum is impeded, and in its place assumed causes are used which are mental, physiological, or conceptual events not desirable in behavioral terms.

One approach to discovering a satisfactory datum is to consider the behavior which may be involved in monitoring and then to determine the variables which control that behavior. Success in detecting signals may depend on the emission of responses which will make the detection possible. These could be responses of orienting toward the correct portion of the display and fixating or scanning the display. Such responses can be termed observing responses in that they bring about the observation of signals (14). Furthermore, these observing responses might follow the same principles as instrumental responses and thus be subject to control by the same type of environmental variables. It is suggested that the observing responses which make detections possible follow the principles of operant behavior. The reinforcement for these observing responses could be the detection of the signals. That is to say, the detection itself could exert control over the rate or probability of emission of observing responses in exactly the same manner as food reinforcement controls the rate of operant responses in animals.

SIGNAL DETECTION AS REINFORCEMENT

In order to evaluate this formulation of "vigilance" it was first necessary to determine whether signal detection really could serve to reinforce an observing response. To do this, subjects (Navy enlisted men), working in the dark, were required to report deflections of a pointer on a dial; but the pointer could be seen only when the subject pressed a key which provided a brief flash of light that illuminated the face of the dial. When the key was pressed, the light flashed for a period of only 0.07 second, even if the subject held the key down. Thus he had to release and redepress the key to obtain another look at the dial. When the subject observed a pointer deflection he reported it by pressing another key, which reset the pointer. The pointer remained deflected until this key was pressed. The deflections of the pointer were programmed so as to make possible various schedules of detections (or reinforcements). Each subject was advised that his only aim should be to make as many detections as he could and to reset the pointer as rapidly as possible. At the end of each session he was informed of the number of detections made and the average time per detection. He was not informed that the experimenter was in any way concerned with the frequency with which he flashed the light. Cumulative response records were made of his responses on the light-flashing key. This type of recording, commonly used in operant conditioning (see 15, 16), consists of a pen which moves in small discrete steps across the recorder paper as responses are made, while the paper moves slowly in a direction perpendicular to the direction of pen movement. The result is a tracing in which the slope of the line reflects the rate of responding.

In order to determine whether signal detection can serve as reinforcement for observing behavior, various schedules of signal presentation were used, analogous to the scheduling of more conventional reinforcers, such as food and water, employed in operant conditioning with animals. Throughout all of the various schedules to be discussed below the subjects were *never* told anything about the nature of the schedule.

Fixed Interval. The first schedules used were of the fixed-interval type. Five subjects began with a ½-minute fixed-interval schedule. That is to say, the needle was deflected for ½ minute after each detection and remained deflected until it was reset by the subject. After eight 40-minute sessions, the interval was increased to 1, 2, 3, and finally 4 minutes, with eight successive sessions on each.

Figure 1 presents data from comparable portions of records for a typical subject on several schedules, all of different fixed intervals. Each curve is a segment of record from the last session which the subject had on the indicated fixed interval. The individual curves are displaced along the horizontal axis. The lines cutting across the records indicate signal detections. Shortly after each detection there is a period in which no observing responses are emitted, as indicated by the flat portions of the curves. Then responding (observing) resumes in an accelerated fashion and reaches a high rate before the next signal. These "scallops" are analogous to those obtained with animals working for food reinforcement on a fixed-interval schedule (15–17). In either case the data represent a temporal discrimination. Responses immediately after reinforcement are not reinforced, so a discrimination is formed for "no responding following reinforcement." Responding resumes after time passes and the conditions become appropriate for reinforcement.

Examining the records, one could, if so inclined, speculate that they reflect "fluctuations of attention" or the course of "subjective expectancies." However, the temptation to do so should not be great since the dependency of observing rate on detection, or reinforcement, is clear. To postulate states accompanying the changes in observing rate adds nothing of use in controlling or predicting the observing rate.

Additional insight into the role of signal detection is provided when no further signals occur (that is, during extinction). Extinction data are provided in Figure 2 for the same subject for whom data were given in Figure 1. This is a complete record for a 1-hour session. Three signals were first provided on the 4-minute fixed-interval schedule which had maintained the observing behavior for six previous sessions; then no further signals were provided. Following each signal detection in the early portion of the record, the characteristic fixed-interval scallops are found. After the third and final detection there is again a scallop, with the high rate continuing for a time and then gradually declining to a very low value. This decline in rate of observing response is dependent upon the absence of signal detection. It cannot be interpreted as physiological fatigue, since on other schedules higher rates have been maintained, without decrement, for more than three hours.

Fixed Ratio. To pursue further the analogy between signal detection and reinforcement as found in typical operant conditioning situations, fixed-ratio schedules were employed. These schedules make reinforcement contingent on the number of responses emitted rather than on the passage of time. To begin with, seven subjects were tested on a fixed-ratio schedule of 36 responses per detection. That is, a needle deflection occurred only after 36 observing responses were made following the immediately preceding detection. After six 40-minute sessions on this schedule the ratio was increased, in blocks of six sessions, to 60, 84, 108, 150, and finally 200 responses per detection. Presented in Figure 3 is a family of curves for the various ratios for a typical subject. These curves are equivalent segments of the subject's final sessions on the indicated schedules. Tests with

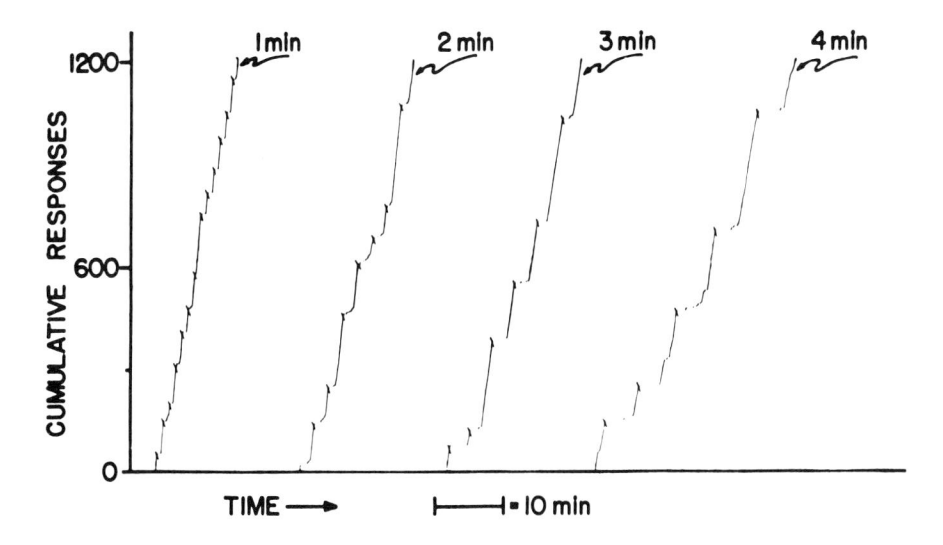

Figure 1. Cumulative response records for 1-, 2-, 3-, and 4-minute fixed-interval schedules of pointer deflections. Detections are indicated by lines cutting across the records.

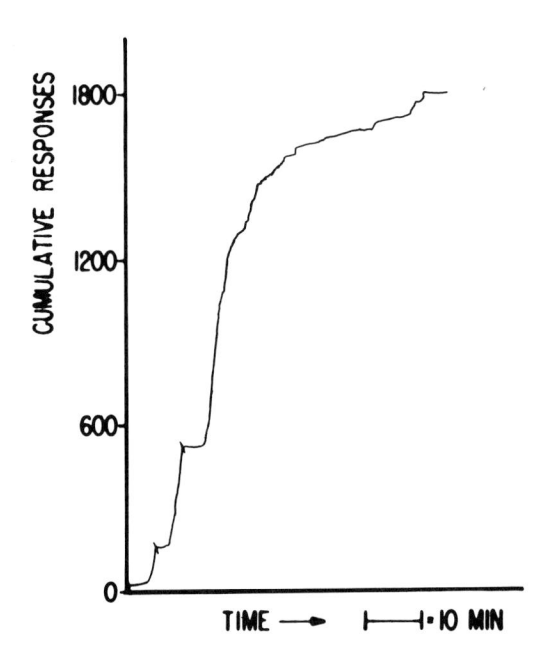

Figure 2. Cumulative response record showing effect of withholding pointer deflections following a fixed-interval schedule. After three detections (indicated by lines cutting across the record) no further pointer deflections occurred.

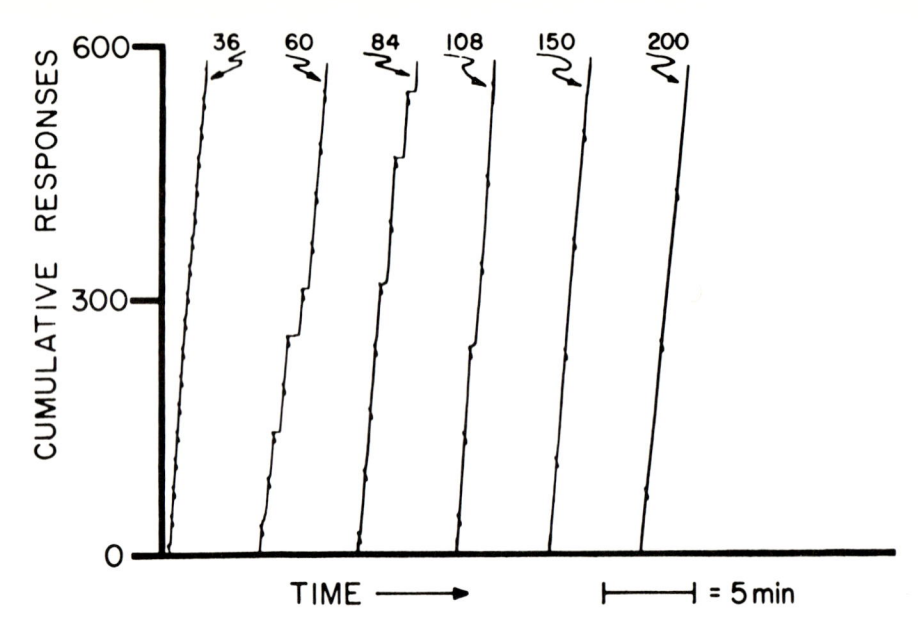

Figure 3. Cumulative response records for 36-, 60-, 84-, 108-, 150-, and 200-response fixed-ratio schedules of pointer deflections. Detections are indicated by lines cutting across records.

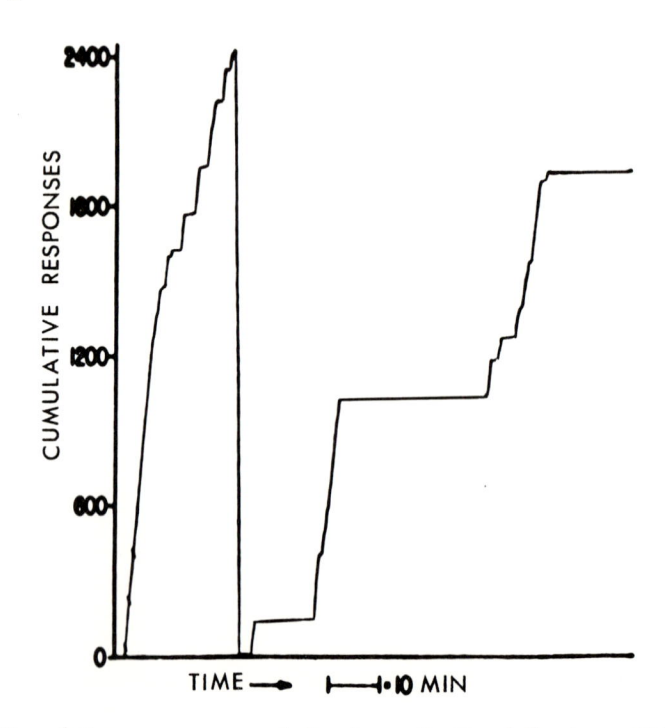

Figure 4. Cumulative response record showing extinction following a 200-response fixed-interval schedule of pointer deflections. After three detections (indicated by lines cutting across the record) no further pointer deflections occurred.

these schedules, unlike most monitoring tasks, permit the subject to minimize the number of signals by not responding. Instead, however, he tends to maximize the number of signals by emitting responses at a high rate. Occasionally short breaks or periods of no responding occur, but only immediately following a detection. These results are also characteristic of those obtained with conventional reinforcement on fixed-ratio schedules (15–17).

An additional demonstration of the control exerted by the schedule of signal detection or reinforcement is seen in the extinction following fixed-ratio schedules. A 1½-hour extinction record is presented in Figure 4 for the same subject for whom data were given in Figure 3. Three needle deflections were provided on the 200-response ratio schedule which he had experienced for the preceding six sessions. After that, no more signals were given. The second portion of the record, following resetting of the pen at the vertical line, is continuous with the first. This record resembles extinction following fixed-ratio reinforcement with animals (15, 17) but is decidedly unlike typical extinction following fixed-interval reinforcement (see Figure 2). Instead of the gradual decline seen for extinction following fixed-interval schedules, the rate, when the subject responds at all, is high. Immediately after the last reinforcement the subject continues at his normal rate for more than 800 responses. He then begins showing occasional periods of no responding, but in each case responding resumes at the original high rate. As extinction progresses the periods of no responding increase, but, throughout the session, when there is a single response there is a run of responding at the high rate that prevailed during reinforcement.

Previous analysis (13, 17) has indicated that the form of the extinction curves for various schedules depends on the presence or absence of conditions which were present at the moment of reinforcement in the past. In the case of fixed-ratio schedules there tends to be reinforcement for groups of closely spaced responses (see 15). Thus, high rates are reinforced, and these high rates come to characterize ratio schedules. As a result, when a response is made during extinction, conditions are like those that prevailed at the time of reinforcement. During extinction, therefore, intermediate rates are lacking. The subject either responds rapidly or not at all.

Multiple Schedule. It has also proved possible in operant conditioning to generate behavior appropriate to more than one schedule in a single organism during the same session (17). To do so, stimuli are provided to indicate which schedule is in effect at a given moment. The stimuli used have been alternation of schedules (called mixed schedules), different colored stimulus lights, or both. I have successfully combined a 40-response fixed-ratio and a 3-minute fixed-interval schedule, using four subjects. These tests began with six 40-minute sessions in which a small red light indicated a fixed ratio of 23 responses to be in effect and a small green light indicated a ½-minute fixed interval to be in effect. The order of appearance for these two schedules was randomly determined. Then for sessions 6 through 11, the schedules were changed to a 40-response fixed ratio and a 3-minute fixed interval. These two schedules were alternated regularly. Then for the twelfth and final 40-minute session the two schedules appeared randomly, with only the stimulus light providing the basis for discrimination. A typical record for this session is presented in Figure 5. The 3-minute fixed-interval portions of the record are labeled *I*, and the 40-response fixed-ratio portions are labeled *R*. It can be seen that when the interval was in effect (green light on) the subject's observing rate provided the fixed-interval scallop. (There is a rougher grain to the scallop than to that found in Figure 1. This is probably due to the experience on fixed-ratio schedules.) When the fixed-ratio schedule was in effect (red light on), the subject's observing rate was that typical for fixed-ratio reinforcement. Thus, like other operant behavior, the observing response can be brought under stimulus control. There remains no need to appeal to another level of analysis by speaking of "attention" being dependent on "context" or "meaning." Such proposed constructs are unnecessary when the control exerted by the schedule of detection under correlated stimuli can be directly demonstrated.

Differential Reinforcement of Low Rates. One further schedule which attests to the control of observing rate by detections is one which makes detections (reinforcements) contingent on low rates of responding. Two subjects were placed on such a schedule. The needle was deflected only after they had failed to emit an observing response for 30 seconds. A record of one subject's fourth 1-hour session is presented in Figure 6. It can be seen that this schedule provides a very low rate of responding, like that found in other operant conditioning experiments (18, 19) for which similar schedules were used. The few short bursts of higher rates tend to occur after the subject responded just a little sooner than the required 30 seconds. Even this detail parallels results with animals working for food on this schedule (18).

Conclusion. The results reported thus far demonstrate that signal detections can control the rate or probability of emission of observing responses. Furthermore, this control is of the same nature as that exerted by conventional reinforcers, thereby permitting the conclusion that signal detections serve as reinforcements for observing responses.

OBSERVING RATES AND "VIGILANCE"

There remains the problem of determining whether the schedules used in classical vigilance studies will generate observing rates which parallel the probability-of-detection data found in those studies. A decrement in probability of detection during the course of a session has been shown (5) for 20 signals per hour when the signals were arranged randomly through the session with the intersignal times drawn from a rectangular distribution. Such a schedule, in operant conditioning terms, would be a variable-interval schedule having an average interval of 3 minutes. Four subjects were placed on this schedule. Figure 7 shows the records for two of these subjects during their first session. (Vigilance studies frequently have only one session.) These records were chosen by way of illustration because these two subjects were the two extremes in terms of decrement of response rate as the session progressed. All four subjects showed periods of lower observing rates in the latter portions of the session. The drop in rate as the session progresses is brought about by the fact that reinforcement frequently is insufficient to maintain the higher initial rate, which results in part from the sub-

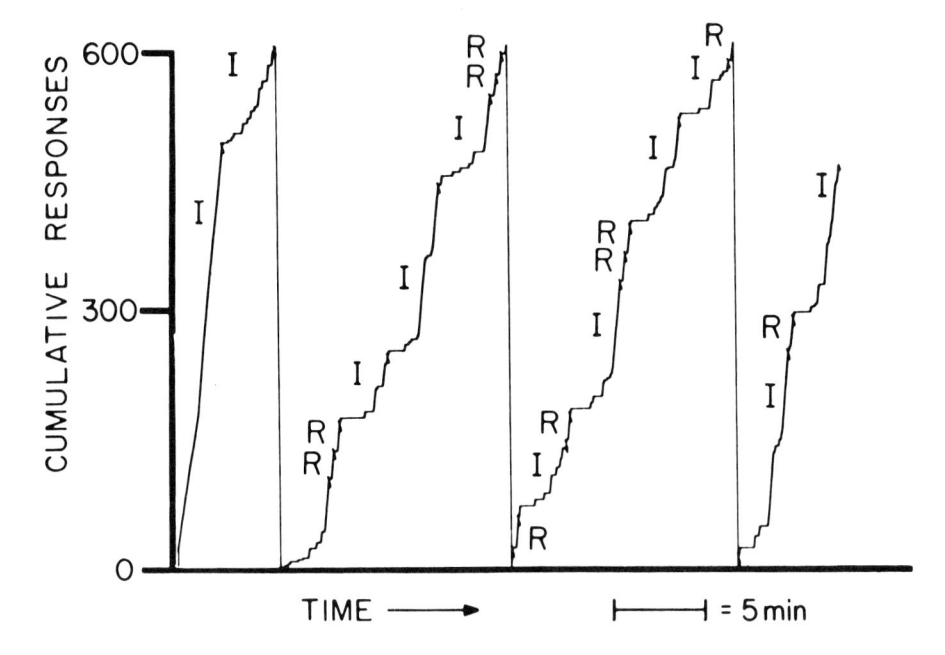

Figure 5. Cumulative response record for a multiple schedule consisting of a 3-minute fixed interval (*I*) and a 40-response fixed ratio (*R*). Lines cutting across the record indicate detections.

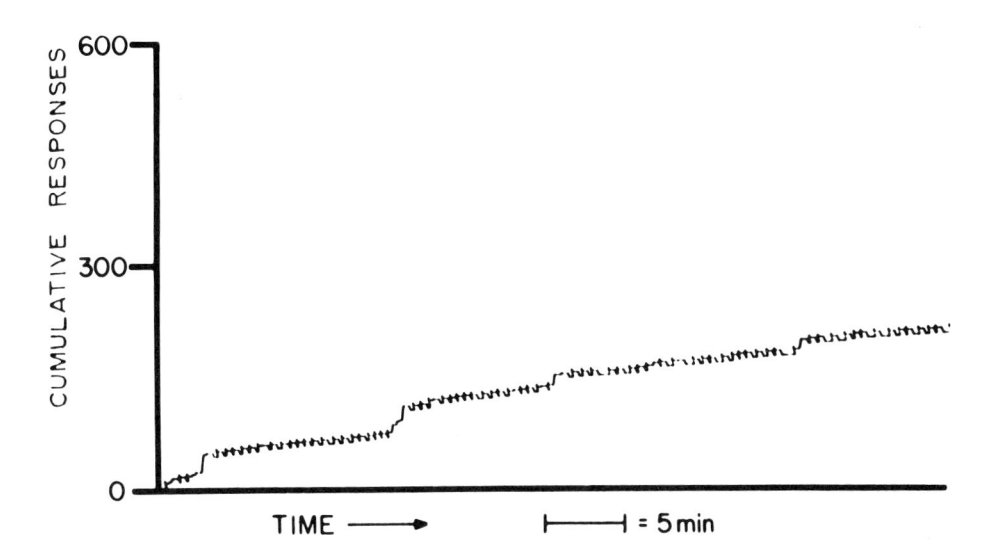

Figure 6. Cumulative response record for differential reinforcement of a low rate. Downward deflection of the pen indicates pointer deflection, and upward deflection indicates detection. Pointer deflections occurred only after no observing response was emitted for 30 seconds.

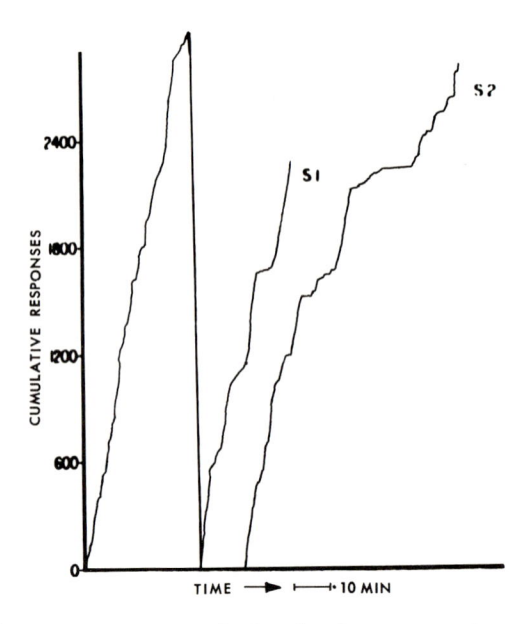

Figure 7. Cumulative response records for the first session for two subjects (S1 and S2) on a variable-interval schedule with average interval of 3 minutes (rectangular distribution ranging from 5 seconds to 6 minutes).

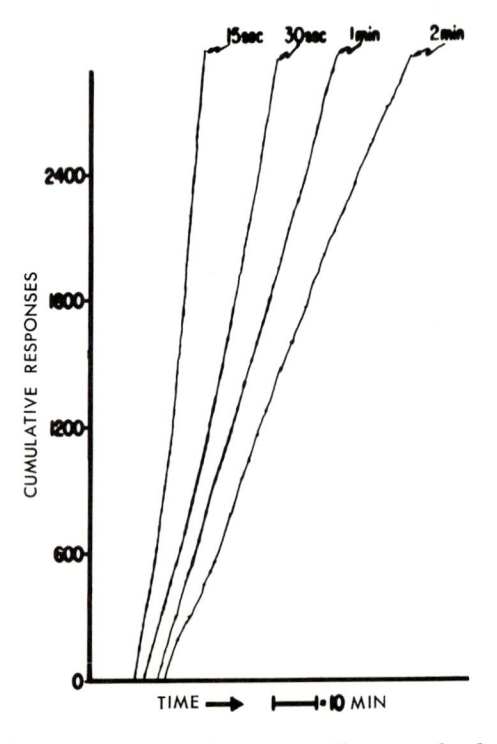

Figure 8. Cumulative response records for variable-interval schedules with average intervals of 15 seconds, 30 seconds, 1 minute, and 2 minutes, respectively (rectangular distributions ranging from 5 seconds to double the average interval). All records are from the same subject. In each case the record was made after three previous sessions on the schedule.

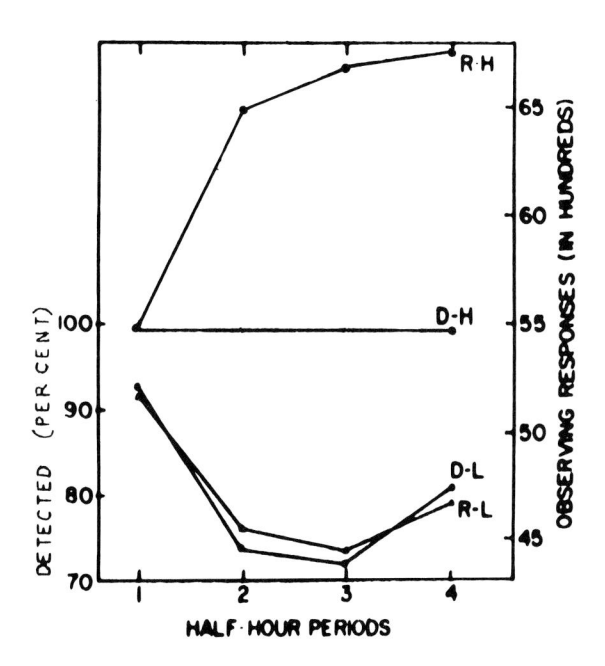

Figure 9. Mean percentage of signals detected and mean number of observing responses per half-hour period for a two-hour session on the Mackworth schedule. Curve *R-H*, observing response data for the high-detection group; curve *D-H*, detection data for the high-vigilance group; curve *R-L*, observing response data for the low-detection group; curve *D-L*, detection data for the low-vigilance group.

ject's past experience. However, some decline does continue to appear within each session for as many as 18 additional 1-hour sessions. Similarly data on pigeons (17) show a within-session decline in rate on a variable-interval schedule when the average interval is long. Furthermore, the drop in observing rate parallels the frequent finding, in vigilance studies, of a decline in the percentage of signals detected.

It has also been demonstrated in vigilance studies that the percentage of signals detected increases as the signal frequency increases. To determine whether rate of observing responses also increases, two subjects were tested on various variable-interval schedules; first there were three 1-hour sessions in which the average interval was 15 seconds (240 per hour), then the interval was increased, in blocks of three sessions, to 30 seconds (120 per hour), 1 minute (60 per hour), and finally 2 minutes (30 per hour). In each case the distribution of intervals was rectangular, varying from 5 seconds to double the average interval. In Figure 8 there is shown a family of curves for one subject for these various average intervals. These records are for the first 3000 responses of the final session on each schedule. It can clearly be seen that the rate of observing is highest for the high signal rate and decreases as the signal rate decreases. Again this finding parallels the results of classical vigilance studies in the higher percentage of detection for higher signal rates (5), and at the same time it parallels other operant conditioning research with variable-interval reinforcement which also shows high response rates to be associated with schedules having a low average interval (15, 17).

The curvature seen in the records in Figure 8 is also of some interest. For the average interval of 2 minutes there is a decline in observing rate as the interval progresses, while for the 15- and 30-second average intervals there is an increase in observing rate. The decrease shown in the case where the smallest number of signals is used is another illustration of a decrement in "vigilance." When larger numbers of signals are used, the "vigilance" literature reports and the present study shows that the decline during the session disappears. Actually, most studies are incapable of showing a rise in probability of detection because the signal is set so that initial detection is nearly always made.

OBSERVING BEHAVIOR WITH THE MACKWORTH SCHEDULE

Additional evidence for the adequacy of the observing-behavior analysis of "vigilance" is seen when the schedule used by Mackworth (1) is employed in the present study. The aims of this study were (i) to determine whether the schedule of signals actually used by Mackworth would confirm the data on decrement in percentage of detections found by him and at the same time provide data on decrement in observing rate, and (ii) to determine whether the data on observing rate would parallel the data on percentage of signals detected. In all of the experiments reported above, signals which remained until detected (nontransient signals) were used in order that the schedule of detections would be under the experimenter's control. The result was that signals would never be missed. But in order to determine whether the typical vigilance measure of percentage of signals detected is paralleled by the observing rate, it was necessary to make the signal automatically disappear if it was not detected within a short time (these are called transient signals). The general procedure was identical with that previously used except for the fact that when the needle was deflected it returned to its original position after 1¼ seconds unless the subject previously detected and reset it in the usual fashion by pressing the key which indicated a detection. The schedule of pointer deflections was identical with the schedule of double jumps used by Mackworth (1) in his clock tests, which stand as the classics in the area of vigilance. This sequence of intervals between needle deflections was ¾, ¾, 1½, 2, 2, 1, 5, 1, 1, 2, 3, and 10 minutes, in that order, and the sequence was repeated four times during the 2-hour sessions. Thus there were twelve signals each half-hour, the shortest interval between signals being ¾ minute and the longest, 10 minutes. Sixteen subjects served in two 2-hour sessions. Cumulative records were made of their observing responses. In addition, a record was kept of their successes and failures in making detections.

In Mackworth's studies, as well as in the present study, there were some important individual differences. Mackworth found that 29 percent of his subjects missed not more than one signal in the last three half-hour periods. In the present study 39 percent of the subjects missed not more than one signal in the entire two hours. The vigilance decrement is thus due to the performance of the other subjects. It turns out that the high-detection subjects show rather different observing response rates than the others. Therefore, in treating the data the subjects were divided into two groups—a high-detection group, made up of those who missed not more than one signal in a 2-hour session, and a low-detection group, made up of those who missed more than one signal per session.

The results for both the percentage of signals detected and for observing responses are summarized in Figure 9. The data for the two 2-hour sessions are combined and show the means for each half-hour period for both measures. The curve labeled $D-H$ (open circles) represents the percentage-detection data for the high-vigilance group. It shows, of course, nearly perfect detection throughout, since this was the basis for assignment to this group. The curve labeled $R-H$ (open triangles) shows the mean number of observing responses for this high-vigilance group. Interestingly, these subjects actually show a rise in response rate as the session progressed. Their percentage-detection data cannot reflect this rise because these subjects are already detecting nearly all the signals. It is probable that this group has an increased detection efficiency which cannot be revealed by the detection measure. Classical vigilance studies have had no measure of observing rate and therefore have been unable to show such a phenomenon.

The low-vigilance group's detection results are shown in the curve labeled $D-L$ (solid circles) and their observing response results, in the curve labeled $R-L$ (solid triangles). By the second half-hour there is a drop both in the percentage of signals detected and in the rate of observing responses. In the first half-hour members of this group detected 93 percent of the signals and emitted an average of approximately 5100 observing responses, while in the second half-hour they detected 74 percent of the signals and emitted an average of about 4550 observing responses. The drop from the first to the second half-hour is significant at the 1 percent level for both measures. The slight decline from the second to the third half-hour is not significant for either measure. But the rise in the fourth half-hour is significant at the 5 percent level for both measures. This end-spurt is probably due to the fact that the subjects knew that the session was 2 hours long. Mackworth found no such end-spurt, but other studies (5) have shown that knowledge of the length of the session can produce such an effect.

In general, then, the vigilance decrement found by Mackworth was confirmed in this study, and a parallel decrement in observing rate was shown as well. It should be recalled that detections (that is, reinforcements) on variable-interval schedules show that the lower the rate of signals, the lower the rate of responding (see Figure 8). Thus, when signals are missed, this might have the effect of lowering the rate, since the subject is then on a different variable-interval schedule with a higher average interval.

One further factor may have an influence on the response rate in this study. When transient signals are used, the situation is analogous to work with animals in which a variable-interval schedule is used, with the added contingency that when the program is set up for reinforcement, the animals have only a brief time in which to respond before reinforcement is no longer available. Such a schedule (17) used with animals (called "variable interval with limited hold") has shown that the use of a limited hold considerably increases the rate of response over that for the same variable-interval schedule with unlimited hold. This presumably results from the differential reinforcement of high rates, since high rates of responding are more likely to be reinforced in the case of limited hold. There may well be an analogous effect in this study. Those subjects who detect almost all signals are very probably being reinforced for high response rates, with the result that their rate increases and thus maintains maximum detection proficiency.

ADDITIONAL PARALLELS BETWEEN RESPONSE RATE AND DETECTION DATA

The similarity in the shapes of the curves in Figure 9 for the observing rate data and the detection data for the low-detection group offers support to the position that the finding of classical vigilance studies could reflect observing behavior. Additional evidence may be adduced for this in parallels between vigilance data and work on operant behavior from animal laboratories. For example, (i) Mackworth (6) finds that giving subjects 10 milligrams of benzedrine raises the level of detection. Similarly Brady (20) has shown that doses of benzedrine administered to rats provide high response rates when the rats are on a variable-interval schedule. (ii) In addition, Mackworth (6) has shown that high room temperatures result in lower levels of detection; and, similarly animals on a variable-interval schedule show lower response rates when the room temperature is high (21). (iii) Nicely and Miller (22) have investigated the effect of unequal spatial distribution of signals on a radar display. The strobe line rotated at 6 revolutions per minute. One quadrant had signals on an average of one every five rotations, while the remainder of the display had signals on an average of one every 30 rotations. Nicely and Miller found that the percentage of signals detected increased for the high signal-frequency area and declined for the low signal-frequency area. After 30 minutes the detection-data curve for the high signal-frequency area had approached a higher asymptote than had that for the low signal-frequency area. This situation is analogous to a multiple schedule having a 40-second average variable-interval schedule with one stimulus (one area) and a 5-minute average variable interval with another stimulus (the other area). Ferster and Skinner (17) have shown that animals on such a multiple schedule show a lower response rate in the presence of the stimulus correlated with the long variable interval than in the presence of the stimulus correlated with the short variable interval. (iv) It has been demonstrated that rest periods restore the detection efficiency to nearly what it was at the beginning of the session (1, 2). Similarly, Ferster and Skinner (17) have found that response rates on variable-interval schedules are increased by interspersing rest periods.

CONCLUSIONS AND IMPLICATIONS

This analysis (23) has demonstrated that detections of signals can serve as reinforcements for observing responses and, further, that the detection data of vigilance studies may reflect the observing response rates generated by the particular schedules employed. Thus a means of analysis is provided which does not appeal to a nonbehavioral level.

In other vigilance studies the observing behavior has probably been fixation and scanning with the head and eyes as well as perhaps more subtle responses. It would be of interest to extend the present technique to some of these responses, although for many problems the topography of the response may be unimportant and the present methods entirely sufficient.

So far as application is concerned, the striking fact is the rather precise control exerted by the environment over the human operator's observing behavior. Thus, in a man-machine system it should be possible for the machine to maintain control over the operator's monitoring behavior. The ideal manner for exerting such control remains to be worked out. It is hoped that this will be the goal of much additional research in this area. But one obvious way is to provide a high rate of realistic artificial signals on a schedule which would provide the desired observing rate. The most promising schedule for many situations would be a variable-interval schedule of signals having a short duration, like the limited hold in animal work. Other do's and don'ts of the engineering of monitoring tasks must be worked out. To this end it is clear that the abundant amount of systematic research on operant behavior that has been done with animals should be a fruitful source of ideas for developmental research as well as for educated guesses in designing man-machine systems requiring monitoring by human beings.

REFERENCES AND NOTES

1. N. H. Mackworth, *Quart. J. Exptl. Psychol.* **1**, 6 (1948).
2. J. A. Adams, *J. Exptl. Psychol.* **52**, 204 (1956).
3. S. C. Bartlett, R. L. Beinert, J. R. Graham, "Study of visual fatigue and efficiency in radar observation," Rome Air Development Center, Tech. Rept. RADC 55-100 (1955).
4. H. M. Bowen and M. M. Woodhead, Royal Air Force Research Unit Interim Rept. FPRC 955 (Applied Psychology Research Unit, Cambridge, England).

5. J. Deese and E. Ormond, "Studies of detectability during continuous visual search." Wright Air Development Center, Tech. Rept. WADC 53-8 (1953).
6. N. H. Mackworth, "Researches on the measurement of human performance," *Med. Research Council (Brit.) Spec. Rept. ser. No. 268* (1950).
7. D. B. Lindsley et al., "Radar operator 'fatigue': The effects of length and repetition of operating periods on efficiency of performance," Office of Scientific Research and Development, Rept. OSRD 3334 (1944).
8. R. M. Belbin, "Compensating rest allowances: Some findings and implications for management arising from recent research," College of Aeronautics, Cranfield, England, Note CoA 54 (1956); S. Wyatt and J. N. Langdon, "Inspection processes in industry." *Ind. Health Research Board Rept., No. 63* (1932).
9. P. Bakan, *J. Exptl. Psychol.* **50**, 387 (1955).
10. D. C. Fraser, *Quart. J. Exptl. Psychol.* **2**, 176 (1950).
11. D. E. Broadbent, *Psychol. Rev.* **60**, 331 (1953).
12. J. Deese, *ibid.* **62**, 359 (1955).
13. B. F. Skinner, *ibid.* **57**, 193 (1950).
14. For a use of a similar type of response applied to discrimination learning problems, see L. S. Reid, *J. Exptl. Psychol.* **46**, 107 (1953); L. B. Wykoff, Jr., *Psychol. Rev.* **59**, 431 (1952).
15. B. F. Skinner, *The Behavior of Organisms* (Appleton-Century-Crofts, New York, 1938).
16. ———, *Am. Psychologist* **8**, 69 (1953).
17. C. B. Ferster and B. F. Skinner, *Schedules of Reinforcement* (Appleton-Century-Crofts, New York, 1957).
18. M. Sidman, *J. Comp. and Physiol. Psychol.* **49**, 459 (1956).
19. M. P. Wilson and F. S. Keller, *ibid.* **46**, 190 (1953).
20. J. V. Brady, *Science* **123**, 1033 (1956).
21. R. J. Herrnstein, personal communication.
22. P. E. Nicely and G. A. Miller, *J. Exptl. Psychol.* **53**, 195 (1957).
23. I am grateful to Dr. Richard J. Herrnstein and Dr. Murray Sidman of the Walter Reed Army Research Center and to Dr. William D. Garvey and Dr. Franklin V. Taylor of the U.S. Naval Research Laboratory for their encouragement and advice in connection with this study. This article is based on research carried out when I was at the Naval Research Laboratory.

47

Effects of Noise and of Signal Rate upon Vigilance Analysed by Means of Decision Theory

Donald E. Broadbent and Margaret Gregory

INTRODUCTION

Early studies of vigilance considered only the proportion of signals which were correctly reported by an operator, and thus made a sharp line between detection and failure of detection. A decision theoretic approach can however be taken to vigilance, as to more abstract laboratory psychophysics (Egan, Greenberg and Schulman 1961; Jerison and Pickett 1963). In simplified terms, such an approach suggests that a signal often produces in a man a doubtful state in which he is less than completely certain that a signal has in fact occurred. The extent to which he is prepared to report such doubtful cases may have a very large effect upon the objectively measured incidence of correct detections. Any increase in the operator's willingness to report in doubtful situations will produce an increase in the number of "false alarms," if the number of doubtful situations itself remains constant. This change in false alarms, however, may be very slight compared with the change in detections, and it is this feature which distinguishes the theoretical approach based on decision theory from the traditional psychophysics which uses the conventional guessing correction for false alarms. (See Swets 1964). In fact it has been demonstrated by Broadbent and Gregory (1963) that, in vigilance situations, the acceptance of responses made with lower degrees of confidence does give a relationship supporting the decision theoretic rather than the traditional point of view. Similar results have been reported by Loeb and Binford (1965).

Since this is so, we must enquire about the various well known changes in vigilance performance which in some cases occur as a function of signal rate or of prolonged watch sessions. Are these changes due to a change in the actual evidence on which the operator bases his decisions to report, or are they due to a change in the criteria which he sets himself before reporting in doubtful cases? Rather different remedial measures might be appropriate in one case from those which one might use in the other. In fact, if the false alarm rate as well as the

From *Human Factors*, 7, 1965, 155-162. By permission of the Human Factors Society.

detection rate is recorded, it is possible to calculate two parameters from the performance of any operator. One of these, d', is a measure of the strength of the evidence on which the operator is deciding, while the other, beta, is a measure of his degree of caution or riskiness in accepting doubtful situations. The method of calculation depends on the properties of the normal distribution, and is given in outline by Broadbent and Gregory (1963) or in fuller detail in Swets (1964).

Unfortunately, there is an immediate conflict of evidence on this point, since Mackworth and Taylor (1963) in a long and valuable series of experiments, found that d' shows a decrease during a period of continuous watch. This suggests that the fall in detections reported in some earlier experiments in vigilance is to be interpreted as an actual shift in the sensitivity of the operator, that is, in the evidence on which he decides whether or not to report. Broadbent and Gregory (1963) on the other hand, found no indication of a drop in d' as the watch period proceeded. They found rather a rise in Beta, which would suggest that an operator who has been working for some time demands more adequate evidence of the presence of a signal before he is prepared to report it. The results of Loeb and Binford (1965) agree with those of Broadbent and Gregory, so that the point clearly demands further examination. Experiments reported in this paper are intended therefore to take up the following three points.

The Nature of Changes during the Run

Some of the experiments performed by Mackworth and Taylor were superior to those of the other authors mentioned, in the following way. They informed the subjects that in each ten second period of the task there would certainly be a signal either in the first five seconds or in the second: and the subject had to indicate which period contained the signal. The percentage of correct responses in such a forced-choice procedure should be directly related to d', at least if subjects have no bias in favour of one alternative or the other. Each of the other studies, however, required the subject merely to watch the display and to respond when he had some suspicion of the presence of a signal. With these instructions it is unclear whether, for example, the occurrence of 36 false alarms in an hour is to be regarded as a probability of one in a hundred of making a false alarm, or a probability of one in ten. The first figure would be correct if the subject uses an observation interval of one second, but the latter figure would be appropriate if the observation interval were ten seconds. Other observation intervals would give other values for the probability. The value of d' thus depends quite considerably upon the assumption made at this point. Accordingly, in the present study, essentially the same equipment and procedure as in the earlier study of Broadbent and Gregory were employed, except that the display consisted of a slowly flashing light, and a response was demanded to every flash, non-signal flashes receiving a response of "no".

Effects of Noise

The earlier study of Broadbent and Gregory showed that, for visual signals high intensity acoustic noise produced a drop in the number of responses made with intermediate degrees of confidence. This means that if one considered an operator working cautiously with few false alarms, there would be no drop in detections in noise: but that a man who was performing with a higher false positive rate might show a drop in detections in noise. This was held to explain the disparity in results on noise between Jerison (1957) and Jerison (1959). Nevertheless, Jerison himself had attributed the difference between his two experiments to the fact that one task involved several different sources of signals, while the other involved only one. He suggested that noise affects primarily the division of attention, and this view is supported by results of Sanders (1961). The earlier results of Broadbent and Gregory concerned merely performance on one particular test and they drew their conclusion from the analysis of responses made with different degrees of confidence: it therefore seemed essential to examine directly the effects of changing the signal rate and the number of sources, upon the size of any noise effects.

The Effects of Signal Rate

Although the trend of performance during a continuous watch is one of the major interests of research in vigilance, there is a separate and very important feature of such situations. This is the change in overall level rather than trend of performance, which can be produced by a change in signal rate. According to decision theory, the optimal position of a criterion should be less cautious when the probability of a signal is high (see Swets 1964), and it is therefore to be expected that Beta should be different when signal rate is different. This point also therefore was examined: however, as this was a subsidiary purpose, the experiment was not designed in such a way as to make it possible to check whether changes in d' also occurred as signal rate varied. In a number of preliminary experiments, it was found that the general level of performance for the same signal strength was greatly inferior with three channels than with one. A different strength of signal was therefore employed in order to keep the number of detections approximately the same. The two primary purposes of the experiment could have been held to be vitiated by any major change in detection rate in some of the subgroups of the experiment.

PROCEDURE

Three groups of subjects, all Royal Navy enlisted men between the ages of 18 and 25, were tested. Twelve subjects were assigned to each group, and each group worked only with a single frequency to avoid transfer effects. All subjects were screened by pure tone audiometry to eliminate any man with more than a 30 dB hearing loss.

Group A: Three Channel High Signal Frequency. The subject was seated 8½-ft. away from the display which consisted of three fluorescent tubes arrayed vertically on a black panel each on a level with the others and each separated by 12 ins. from its neighbor. The tubes flashed rhythmically and simultaneously once every 3.5 secs for 0.30 sec. A signal was a single slightly brighter flash which might

Effects of Noise and of Signal Rate upon Vigilance Analysed by Decision Theory 545

occur on any of the three tubes. The brightness of the background was 0.16 ft. lamberts, that of the flashing tubes 100 ft. lamberts and that of the signal 180 ft. lamberts.

The subjects was provided with a panel which he rested on his knees, on which were mounted three buttons, the one on the left being marked "Sure Yes" the centre one "Unsure" and the right hand one "Sure No". He was instructed to press one of the buttons after every flash: he was to use the "Sure Yes" if he was certain that he had seen a signal, the "Sure No" if he was sure that no signal was present and the "Unsure" if he was at all uncertain either way.

Four signals were presented in every 20 flashes (i.e., there were 3.43 signals per minute) the shortest interval between signals containing one flash and the longest eight: all intervals occurred with equal frequency. During the test period the subjects received a total of 1200 flashes containing 240 signals and lasting 70 minutes. The subjects were told that on the average there would be one signal in every five flashes.

Each subject was tested for two periods of 70 minutes, one on each of two successive days. On one day he was in an environment of 100dB noise while on the other day the same noise was present at only 75dB (which will be called "Quiet"). The order of conditions was counterbalanced between subjects. The noise as before consisted of amplified valve noise at approximately equal level per octave 100-5000 cyc/sec. It was delivered through a pair of loud speakers mounted in the wall of the experimental room which was windowless and insulated from outside sounds although internally reverberant. The experimenter was outside the room.

On the day before the experiment each subject was given a practice session containing 200 flashes (40 signals). After each signal a bell was rung after the subject had made his response. At the end of the period his score of detections and false positives was given to him and if he had not made use of the centre "Unsure" button but had not shown perfect performance he was asked to remember to use it when he was at all unsure. Before the practice session and each of the test sessions the signal was demonstrated for each of the three lights; then the subject had to detect one signal correctly from each light. Following the demonstration period each of the main sessions was preceded by a practice period of 100 flashes (20 signals) during which knowledge of results was given. Instructions about the use of the Sure and Unsure buttons were repeated if necessary. It should be especially noted that the signal rate in practice sessions corresponded to that in experimental sessions.

Group B: One Channel High Signal Frequency. The procedure was as described above except that all the signals were presented on the centre tube and the brightness of the signal was slightly reduced being in this case 150 ft. lamberts.

Group C: One Channel Low Signal Frequency. Again the signal was presented only on the centre tube, the brightness being 150 ft. lamberts. In this condition the signal rate was ⅓ of that in the other two conditions, being 20 signals every 300 flashes, i.e., 80 signals in a 70 minute session.

The shortest interval between signals contained three flashes and the longest

27. The subjects were told that a signal would occur on the average once in 15 flashes and in the practice sessions the signal rate corresponded to the rate used in the test sessions.

RESULTS

The proportion of detections and of false positives made with each of the two degrees of caution in each section of the test are shown in Figure 1. It will be noted that in general the total number of detections does not decrease during a watch in quiet, but that less of them seem to be made with high confidence. False positives do seem to decrease. Groups A and B show rather fewer doubtful responses in noise than they do in quiet, especially at the end of the run.

The value of d' and of Beta was worked out for each subject individually for the first quarter and last quarter of the test, and an analysis of variance was carried out for the first and last periods of the quiet run for each subject. In the case of Beta, the more cautious of the two criteria was used for this analysis, and logarithmic transformation was necessary: the mean values of d' and of log Beta are given in Table 1, and the analyses of variance in Table 2.

Changes during a Watch

It will be seen that there is no indication whatever of a fall in d' as a watch in quiet proceeds. On the contrary, there is an almost significant increase, presumably due to practice. Once again therefore this experiment shows no indication that the end of a watch is inferior to the beginning when we take a score that eliminates the effect of the subject's own criterion. The analysis of Beta, however, shows that there is a significant interaction between the trend during the watch and the day of the experiment. This is due to an increase in Beta taking place during the watch on the first day of the experiment, but not on the second day: the increase on the first day is significant when analysed alone $P = .025$. This result therefore confirms the earlier findings of an increase in caution of criterion setting during the watch, which would of course in many cases result in a drop in the number of recorded detections. The similar results of Broadbent and Gregory (1963) and of Loeb and Binford (1965) may also well be due solely to the first day's performance: and we may therefore add the caution that this phenomenon may not appear in experienced operators.

As in our earlier study, there was no indication of a similar shift of the more risky criterion. Thus in terms of Figure 1, the number of detections remain much the same, but they shifted from being reported as confident to being reported as doubtful as the watch proceeded.

Effects of Noise

In the study of Broadbent and Gregory (1963), the effect of noise was to produce a reduction in the number of responses of intermediate confidence in the last part of a watch in noise. That result was obtained with a fairly high signal rate, divided

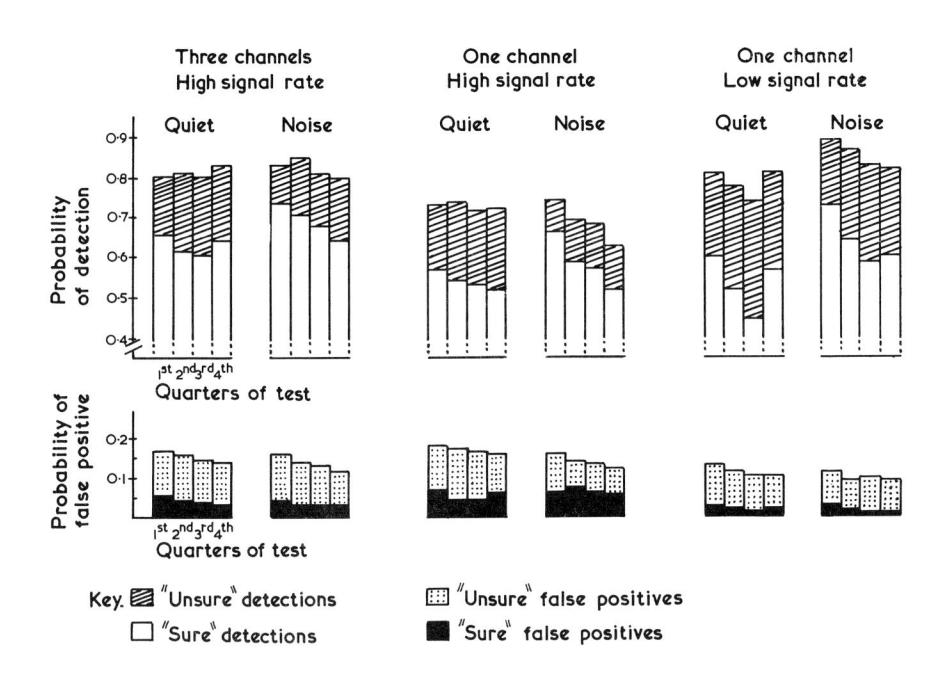

Figure 1. Directions and false positives under various conditions. Note that a different signal strength was used in the three channel high signal rate case.

TABLE 1
Means of d' and Log β for Quiet Runs Under All Conditions

		Condition A Three channels High Signal Rate		Condition B One channel High Signal Rate		Condition C One channel Low Signal Rate	
		d'	log β	d'	log β	d'	log β
Day 1	First Quarter	1.99	0.38	1.78	0.41	2.22	0.73
	Last Quarter	2.78	0.68	1.66	0.59	2.40	1.00
Day 2	First Quarter	2.80	1.05	1.92	0.55	2.54	1.20
	Last Quarter	2.93	1.13	1.90	0.58	2.54	1.01

TABLE 2
Analysis of Variance for d' and log β

Source	df	d' Mean Square	F	Log β Mean Square	F
Between subjects	35				
Conditions	2	4.209	3.49*	1.2375	3.59*
Days	1	1.598		1.5168	4.40*
Conditions × Days	2	0.472		0.3732	
Error (b)	30	1.206		0.3444	
Within subjects	36				
First and last	1	0.472	3.04	0.2193	2.65
First and last × Days	1	0.276		0.3496	4.23*
First and last × Conditions	2	0.471		0.0331	
First and last × Days × Conditions	2	0.198		0.0287	
Error (w)	30	0.155		0.0826	

* $p < 0.05$.

between three display sources. The corresponding group (A) in this study shows a similar effect, the drop in the number of "doubtful" responses being significant $P = .01$. (The test used in this case was a comparison of the Run 2 − Run 1 difference in the QN and NQ subgroups, by U-test.) The same effect appears in the one channel high signal frequency group, for which $P < .03$: but it is quite untrue for the one channel low frequency group. ($P > .40$). Thus the harmful effects of noise seem likely to be less serious with a low signal frequency, not simply because that will encourage the adoption of a cautious criterion, but also because responses of intermediate confidence do not decrease in such a situation.

Special interest attaches to the one channel high frequency group, because of the evidence provided by Jerison and by Sanders that a separation of sources of information helps effects of noise to appear. Nevertheless, in the last period of the watch the number of confident "no" responses in the presence of a signal is significantly greater in noise than in quiet by sign test, $P = .038$. Thus it is not essential to have several sources of signals to get effects of noise: a high signal rate from one source will do as well.

Effects of Signal Rate

As already explained, the significant difference in d' between the three groups is meaningless, since the signal strength was known to be different. It will be noted however that there is also a significant difference in Beta. The low signal frequency produces a very cautious criterion, and the high signal frequency a risky one, when the signals come from one source. When the high signal rate is divided between three sources, the criterion is intermediate, as is perhaps reasonable since the probability of a signal on any one channel is then low although the overall probability is high. It is thus clear that some of the effects of signal rate are due to the criterion set by the subject.

Attention should be drawn however to the fact that there is no interaction between the trend in Beta during the run, and the signal rate. This suggests firstly that the trend is not due to a change in subjective probability of a signal, which would in any case not be expected in view of the nature of the practice given to the subjects. In addition however it does not seem altogether consistent with the fact that a trend during the run is only found within any one group when the more cautious criterion is considered, and not when the more risky one is used.

DISCUSSION

To take the simplest points first, the fact that the effects of a low signal rate are at least in part due to reduced willingness to report uncertain signals is theoretically gratifying. It agrees with the theoretically rational optimal decision (Swets (1964). From a practical point of view, it suggests that methods of persuading operators to report doubtful signals, such as manipulation of payoff, will be effective in conditions of low signal rate.

It is also satisfactory to find that effects of noise seem to be more serious at high signal rates than low: as has been pointed out by Broadbent (1963) there is evidence that noise produces too high a level of arousal, so that it has more serious effects under high incentives than low (Wilkinson 1963) and tends to cancel out its ill-effects with those of sleeplessness (Corcoran 1962; Wilkinson 1963). Since it is known that sleeplessness shows a bigger effect with a low signal rate (Corcoran 1963) it might be expected that noise would be more serious with a high signal frequency. This is not of course to deny the importance of division of attention in making tasks susceptible to noise, since the presence of many input channels may be an important factor in arousing the operator. Division of attention is not however essential for deterioration due to noise.

The most serious remaining problem, however, is the discrepancy between the type of experiment performed here and those of Mackworth and Taylor, in terms of trend during a run. In view of the number of experiments which they have performed, and the repeatability of the results given in the present study, it seems fairly clear that the discrepancy is not due to some minor accident in running the experiment, but rather to some systematic difference of procedure. Certain possibilities suggest themselves, although as will be seen only one of them seems very plausible.

Preliminary Practice

As will have been seen, our own subjects were very familiar with the type of signal employed before the experimental trials: while at least in some cases this seems to have been less true for the subjects of Mackworth and Taylor. Their results might conceivably therefore be due to a form of short term memory decay for the nature of the signal, which might be expected to affect d'. On the other hand, some of Mackworth and Taylor's groups were tested more than once, and gave very similar results on later testing. This does not therefore seem a very satisfactory explanation, although it cannot be ruled out completely.

Individual Differences

The nature of the subjects used in the conflicting studies is of course very different, and in addition Mackworth and Taylor normally presented values of d' calculated from the scores of the entire group, rather than working out d' for each subject individually. Thus it is not known how far the results of any one of their groups were statistically significant, but merely that a satisfactory number of groups behaved in a similar way. Thus if the population sampled in Canada were to include some people of a type not met in the U.K. and U.S.A. samples (for example, because they were female) the discrepancy might be explained. As against this, the similarity of Mackworth and Taylor's various groups makes such an explanation seem very improbable.

The Precise Model Used

The statistics d' and Beta are of course calculated on certain assumptions which may be unjustified either in one case or in the other. Indeed, Mackworth and Taylor refrained from calculating Beta on the grounds that it depends crucially upon the particular model employed; although it could well be argued that this is even more true of d'. Thus for example in our own technique the subject might in fact make two successive observations within a single flash: and in Mackworth and Taylor's forced-choice technique, he might similarly make several observations within a single five-second period. The true value of d' would then be somewhat different from the calculated one. Data obtained with confidence ratings such as our own or those of Loeb and Binford allow tests of the applicability of the model used, and we can therefore be moderately sure of its validity. The techniques of Mackworth and Taylor, on the other hand, do not allow testing of the appropriateness of the model.

No rational ground appears, however, for supposing that the model is adequate for one task and not for the other. Furthermore, it seems likely that failure of the assumption will only displace the value of d' up or down and not introduce a spurious trend; unless the failure occurs to a greater extent at one end of the watch than at the other. Again, no rational ground can be suggested for a failure of the model limited to the beginning or end of a work period. Thus this factor seems capable only of an ad hoc explanation of the discrepancy.

Effects of Signal Strength

Mackworth and Taylor, in plotting their results, use a scale of log d', on which results of all groups show a parallel decline. This of course means that the effect on d' is much greater when the initial value is large than when it is small. Indeed, those groups tested by Mackworth and Taylor with a forced-choice method and with a relatively faint signal might not by themselves be regarded as giving very adequate evidence for a drop in d' during a run. While it is difficult to compare discriminability of two quite different types of signal, it may seem plausible that the drop in d' during a watch is important only with a rather conspicuous one. This would be consistent with the conclusions of Broadbent (1965) who, from a mathematical model devised for other purposes, concludes that the value of d' should change with overall changes in level of motivation or arousal, but that the change will be relatively unimportant when d' is small. The major difficulty in this view is that the actual values of d' in the present study and in that of Mackworth and Taylor are not very different: and any appeal to this explanation must therefore also involve some extra assumptions. For example, one might suppose that Mackworth and Taylor's subjects did not in fact use a five-second observation interval, but rather made five separate one second observations within that interval. The value of d' calculated on the assumption of a 5-second interval would then under-estimate the true value of d', even in the forced-choice situation.

The Importance of Coupling

Perhaps the best suggestion for the discrepancy was made by Loeb and Binford (1965), who suggested that it was due to purely peripheral sensory adjustment or "coupling" to the display being an important feature of the work of Mackworth and Taylor but not of the other studies. Loeb and Binford used an auditory signal, and one of the studies by Broadbent and Gregory (1963) was also auditory: so that it could be argued that Mackworth and Taylor's fall in d' was due to their subjects looking away from the display with increasing frequency as the watch proceeded. In simple form, this interpretation cannot be maintained in view of the visual results of the present paper and indeed of our previous one. These visual results are however obtained with an intermittent stimulus, which leaves regular intervals of time sufficiency long for the subject to look away from the display and back again without missing a stimulus. The situation of Mackworth and Taylor used as a signal a brief pause in a continuously moving pointer, and within any ten second observation period there was therefore no opportunity for looking away. At the present time therefore the least improbable interpretation of the difference between the two sets of results is that situations involving difficult sensory adjustments may actually show a drop in the sensitivity of the operator during the watch, while others will merely show an increased reluctance to report signals of which he is doubtful.

REFERENCES

Broadbent, D. E. Differences and interactions between stresses. *Quart. J. exp. Psychol.*, 1963, **15**, 205-211.

Broadbent, D. E. A reformulation of the Yerkes-Dodson Law. *Brit. J. math. stat. Psychol.*, 1965, in press.

Broadbent, D. E., & Gregory, M. Vigilance considered as a statistical decision. *Brit. J. Psychol.*, 1963, **54**, 309-323.

Corcoran, D. W. Noise and loss of sleep. *Quart. J. exp. Psychol.*, 1962, **14**, 178-182.

Corcoran, D. W. J. Doubling the rate of signal presentation in a vigilance task during sleep deprivation. *J. Appl. Psychol.*, 1963, **47**, 412-415.

Egan, J. P., Greenberg, G. Z., & Schulman, A. I. Operating characteristics, signal detection, and the method of free response. *J. acoust. Soc. Amer.*, 1961, **33**, 993-1007.

Jerison, H. J. Performance on a simple vigilance task in noise and quiet. *J. acoust. Soc. Amer.*, 1957, **29**, 1163-1165.

Jerison, H. J. Effects of noise on human performance. *J. Appl. Psychol.*, 1959, **43**, 96-101.

Jerison, H. J., & Pickett, R. M. Vigilance: a review and re-evaluation. *Human Factors*, 1963, **5**, 211-238.

Loeb, M., & Binford, J. R. Vigilance for auditory intensity changes as a function of preliminary feedback and confidence level. *Human Factors*, 1964, **6**(5), 445-458.

Mackworth, J. F., and Taylor, M. M. The d' measure of signal detectability in vigilance-like situations. *Canad. J. Psychol.*, 1963, **17**, 302-325.

Sanders, A. F. The influence of noise on two discrimination tasks. *Ergonomics*, 1961, **4**, 253-258.

Swets, J. A. (Ed.). *Signal detection and recognition by human observers.* 1964, Wiley.

Wilkinson, R. T. Interaction of noise with knowledge of results and sleep deprivation. *J. exp. Psychol.*, 1963, **66**, 332-337.

48

Complex Monitoring and Its Relation to the Classical Problem of Vigilance

William C. Howell, William A. Johnston, and Irwin L. Goldstein

Psychology's interest in human vigilance developed under rather trying circumstances. It originated during World War II with the alarming discovery that man is poorly suited to function in the role of a monitor (McGrath, 1963a). Viewed in the context of an automation-oriented society, this discovery had obvious and serious implications: since man seemed destined to become monitor of increasingly broad and important operations—whether suited for it or not—it seemed imperative that psychologists find ways to render him more proficient.

Considering the urgency of the problem, it is scarcely surprising that neither a systematic accumulation of facts nor an orderly approach to theory development ensued. The generality of a vigilance *decrement* was assumed, and all efforts were bent upon eliminating it. There being few data from which to construct a coherent vigilance theory, those who were inclined toward generalizing from their data borrowed directly from existing models of conditioning, motivation, or attention. Foremost among the classical mechanisms adapted to this purpose were *inhibition* (Mackworth, 1948), *expectancy* (Baker, 1963; Deese, 1955), *arousal* (Hebb, 1955; Zuercher, 1965), and selective attention or *filtering* (Broadbent, 1954). In addition, Holland (1958) argued for a strict empirical approach based upon the principles of operant conditioning; while not a formal theory, this approach evolved into what might be termed the *reinforcement* hypothesis (Jerison and Pickett, 1964). Since these positions have been described fully, and discussed critically, in a number of comprehensive reviews (Baker, 1963; Deese, 1955; Frankmann and Adams, 1960; Jerison and Pickett, 1963), we may forego elaboration of them here. Suffice it to say that none has achieved general acceptance because, even

From *Organizational Behavior and Human Performance*, 1, 1966, 129-150. By permission of Academic Press, Inc.

This study was carried out in the Human Performance Center and was supported by the Air Force Systems Command, Research and Technology Division, Rome Air Development Center, Griffith Air Force Base, N. Y. 13442, under Contract No. AF 30(602)-3622 with the Ohio State University Research Foundation.

at a very gross level of explanation, none has been able to account successfully for all of the reported phenomena. Even more disturbing, none has been particularly fruitful in generating testable hypotheses (Frankmann and Adams, 1960; McGrath, 1963b).

Added to these theoretical difficulties has been a growing suspicion that even the primary vigilance phenomenon—the decrement—is not as universal a one as had originally been assumed. Absence of decrements has been particularly notable in monitoring tasks characterized by a degree of task complexity (Jerison, 1963). Adams (1963), for example, summarizes a long series of complex monitoring studies with the conclusion that man is a very acceptable monitor in tasks other than the near-threshold, transient-signal variety. Considering the fact that modern technology has been redirecting the human monitoring function away from just such simple tasks toward greater display and response complexity (Kibler, 1965), it has become increasingly apparent that a restatement of the vigilance problem is in order.

A BROADENED VIEW OF VIGILANCE

In proposing restatement of the vigilance problem, it is not implied that all classical work has been misdirected. Rather, it is the present view that the classical problem represents but a small and arbitrary part of the much broader issue of how man functions as a *monitor*. Such a broadened view of the problem area has several basic implications. First, it encourages the evaluation of absolute levels of performance as well as time-dependent functions (i.e., decrements); many variables, such as those related to display complexity, may have more of a bearing upon overall monitoring proficiency than they do upon maintenance of proficiency over time. Second, it invites consideration of more extensive response functions than the mere *detection* of transient signals so characteristic of the classical vigilance task. As pointed out by Kibler (1965), the possibility exists that monitoring inadequacy may appear in the *use* of detected information, even though detection per se is accomplished perfectly. Third, it promotes the integration of explanatory concepts, classical or otherwise, by recognizing the possibility that monitoring performance may be controlled by a number of interacting mechanisms. Certainly, the classical work has illustrated that at least *some* task can be contrived which will produce consistent evidence for the arousal, the expectancy, or the reinforcement position. If any task generality is to be achieved for these notions, some effort must be made to establish the conditions under which each is expected to play a prominent role, and this can only be accomplished within a context which is sufficiently broad to enable all of them to operate.

Finally, and perhaps most important, regarding vigilance as part of a more comprehensive problem area promotes consideration of new directions or strategies for research. Classical theories have, until only recently, dominated all aspects of vigilance research: the questions asked, variables manipulated, tasks employed, and measures taken. Since these theories have been of limited heuristic value, as mentioned earlier, the development of new strategies would seem particularly advantageous. Evidence in support of this contention is available in two recent departures from tradition.

Holland (1957, 1958) adopted an empirical strategy based upon the notion that observing behavior is controlled by the reinforcing efficacy of detected signals. This approach introduced all the variables of operant conditioning into the domain of vigilance, providing a wealth of new relationships to be explored.

The second new strategy was an outgrowth of the current wave of enthusiasm surrounding the theory of signal detection (Swets, Tanner, and Birdsall, 1961). According to this view, monitoring is regarded as a complex function involving the same two processes which underlie all other perceptual judgment tasks: discrimination and decision. In this context, the discrimination process comes into play in connection with the distinctiveness of signal and nonsignal distributions, while the decision process serves to establish a criterion level for attributing any event to one or the other of these distributions. Analyses of the relative frequency with which missed signals, false alarms, and correct detections occur offers a means of estimating the contribution of each process to monitoring performance in any given situation. It is quite apparent that this approach has also broadened the scope of vigilance research to include many variables which would never have arisen within the classical framework (Broadbent and Gregory, 1963; Jerison and Pickett, 1963; Jerison, Pickett, and Stenson, 1965; Mackworth and Taylor, 1963).

It should be apparent from the foregoing discussion that restatement of the vigilance problem in broader terms is necessary both to describe the diversity of current research and to represent projected real-life monitoring requirements. What, then, should define the broadened area of monitoring? While any definition is somewhat arbitrary since each dimension of the problem extends well into other arbitrarily defined segments of human behavior, it is convenient to adopt at least a working definition. One which appears most consistent with current trends includes all situations in which man is required to produce an act or sequence of actions in response to the occurrence of relatively infrequent and unpredictable (in time, space, or both) stimulus changes or signals over relatively long periods of time. The term *relatively* is used to indicate that only the directtions, not the absolute limits, can be specified with regard to these characteristics (McGrath, 1963b). For example, signal frequency is itself an important variable in monitoring research; as frequency is raised to the point of time-stress or perceptual-motor skill involvement, however, the task becomes less applicable to this definition of monitoring.

While the present definition is not as extensive as that proposed by Kibler (1965) to encompass all present-day monitoring applications, it is sufficient to include most of the situations encountered in the laboratory. It also is broad enough to permit development of more complex task situations and more comprehensive research strategies. The remainder of this paper is devoted to a discussion of the various aspects of one such approach.

OBJECTIVES AND STRATEGY

The present work was stimulated initially by the marked differences which seem to characterize performance in simple vigilance and complex monitoring situations (Adams, 1963; Jerison, 1963). Most obvious, of course, is the common absence or triviality of a classical decrement in even the most elementary of complex tasks. A series of investigations by Adams and his co-workers (Adams, 1963; Adams and Boulter, 1962; Adams and Humes, 1963; Adams, Humes, and Sieveking, 1963; Adams, Humes, and Stenson, 1962; Adams, Stenson, and Humes, 1961; Montague, Webber, and Adams, 1965) is even more notable in that none of the variables derived from classical theories (and hence based directly upon simple vigilance data) produced any substantial influence upon performance in the complex situation.

Apparently, then, complex monitoring is either controlled by entirely *different* mechanisms than are involved in simple monitoring (vigilance), or it is a function of some *combination* or *interaction* of such mechanisms. In either case, it is doubtful that any traditional approach is broad enough to expose the most important relationships involved in complex monitoring. By its very nature, in fact, the complex situation seems to work directly against isolation of mechanisms such as expectancy, alertness, or criterion level by the usual procedures, i.e., manipulation of variables appropriate to only one of them in a given experiment. This is because the complex situation affords the subject a considerable measure of control over the amount, kind, and even the distribution of stimulation which he receives and the activity whereby he receives it. It would be surprising indeed if, faced with increasing boredom or worsening performance, a subject were to ignore the possibility of adjusting his behavior with respect to the stimuli. By way of contrast, the simple task places most of this control in the hands of the experimenter; the subject has little choice, within the task structure, but to await whatever stimulation is bestowed upon him.

Considered in this fashion, the complex monitoring problem would appear to demand a somewhat different research strategy than those applicable to the simple task situation. One such strategy would have as its ultimate objective the description of general programs of behavior engaged in by the subject over time, and the identification of variables contributing to these programs at various stages in their development. In time, of course, it would be hoped that at least certain aspects of such programs of behavior could be interpreted in terms of basic mechanisms of the sort postulated for vigilance.

A significant step toward a new strategy for complex monitoring was taken by Jerison (1963) in his proposal that all monitoring situations involve two basic factors: *scanning* and *alertness*. Presumably, alertness would be the sole factor involved in the simple situation while both would contribute in some unspecified way to complex monitoring. Assuming this dichotomy, one might proceed to construct a situation in which variables likely to relate to these processes could be manipulated within a common framework. Measures could be taken, not only of behavioral consequences, but of physiological processes associated with scanning and alertness.

The present conceptualization agrees with Jerison to the extent that complex monitoring is viewed as involving a complex interplay of more fundamental processes. It diverges, however, when the attempt is made to specify these processes. There would appear to be very little evidence at present to justify postulation of a two-factor or any other specific model for complex monitoring. The classical data base seems inappropriate for generating such a model, and the complex situation has produced largely negative results. The present position, then, is that if any real progress is to be made toward understanding the mechanisms responsible for complex monitoring, the initial orientation must be in the direction of establishing an appropriate data base.

The present research strategy was to establish a number of purely empirical relationships between rather gross characteristics of complex displays and a variety of monitoring indices. After identifying the molar variables of greatest consequence, the intention was to subject these variables to further refinement; to determine, if possible, the components of which such molar variables might be composed. Progressive refinement of variables was expected to result in a data base from which truly meaningful hypotheses could be drawn concerning the manner in which a monitor carries out his task. Briefly, then, the present strategy may be described as one of progressive refinement of molar variables studied within a complex task situation. The studies reported below are intended to illustrate the strategy. It will be noted that variables were selected for investigation because of their apparent contribution to complexity, not because of their probable effect upon alertness, expectancy, decision criterion, or any other organismic state.

RESEARCH

The proposed strategy called for a task in which a wide variety of variables, related in one way or another to complexity, could be manipulated without changing the basic structure of the situation. Such a task was devised, and a number of molar variables were explored in at least a cursory fashion in the initial stages of the program. Table 1 provides a summary of these variables. Since much of this work has been reported in detail elsewhere (Howell, Johnston, and Goldstein, 1966; Johnston, Howell, and Goldstein, 1966; Johnston, Howell, and Zajkowski, 1967), only those portions which illustrate one or another particular phase of the current approach will be repeated here. Three aspects of the program will be described: the basic task paradigm, two studies illustrating the molar-empirical phase, and one study illustrating the refinement phase.

Task

Apparatus and Stimulus Material. All programming of input and scoring of responses was performed on a combined IBM 1401/7094 digital computer system. Four CRT display consoles were linked to this system in such a way that four subjects could serve simultaneously under independent experimental conditions. Information appeared on the face of each 19-inch CRT in the form of stimuli located within an 8 × 8 matrix. While the format and amount of information presented were not identical in all studies, the basic design always included alpha-numeric symbols as stimuli and 64 spatial locations. A sample display is illustrated in Figure 1.

A detection button and a light pen served as response devices for each subject. With the former he indicated his observation of any change in the display, and with the latter he attempted to describe it: its *location* by aiming at the appropriate cell, and its *identification* by aiming at the appropriate alpha-numerics in the list provided beneath the display (see Figure 1). Cells in the matrix which did not contain stimuli were always provided with an asterisk to permit location responses for those stimuli removed from the display. Three other symbols appeared on the CRT outside the display area. These were all included for particular response contingencies: *EXC* to instruct the computer to record the response, *DNO* to plead ignorance in the location or identification of a detected signal, and *ERS* to erase any unintended response.

A computer program was developed which permitted updating of the display in any way specified at 10-second intervals for a period of 50 minutes. When longer sessions (watch periods) were employed, as was usually the case, additional 50-minute cycles were added, each having the same statistical characteristics as the first. All variables listed in Table 1 were handled conveniently by this program. Records of all signal and response occurrences were also taken automatically, but the matching of each signal with its response and the calculation of latencies was, of necessity, carried out manually.

Procedure

Each subject was seated before his console and instructed to report, as rapidly and accurately as possible, any relevant changes which occurred in the display. He was informed that stimuli might either appear or disappear and that in either case he was to respond in a prescribed sequence: (*a*) press detect button with left hand as rapidly as possible after detecting a change,[1] (*b*) indicate location with light pen held in right hand, and (*c*) identify stimulus using light pen and alpha-numerics at the bottom of the CRT. He was not encouraged to guess about locations and identifications, and was provided with a "don't know" response for these occasions. He was, however, instructed to give any part of the information that he had available. Following each location or identification response, the event at which he had aimed began blinking to indicate that the response was accepted for recording; it continued to blink until the EXC symbol was activated to execute the recording. If, prior to EXC, the subject wished to change his response, or if proper recording was not signaled, he designated ERS with his light pen and corrected the response. Completion of the entire response sequence was required before reporting a new signal.[2]

Four subjects served at once in the same room, but interaction among them was minimized by arrangement of consoles, by ambient noise generators, by instructions, and by the computer program (all displays were driven by independent programs). In addition, all means of keeping time were removed from the room at the start of each session. Subjects were permitted to report machine malfunction to the experimenter by use of a telephone provided.

All characteristics of signal occurrence were determined in advance and incoporated into the computer program. Stimuli were added (*adds*) and omitted (*omits*) with equal frequency over a session; specific stimuli to be added or omitted were selected at random; order of occurrence was randomized; and within

[1] The left hand rested at all times on this button.
[2] Since practiced subjects could complete the entire sequence in several seconds, it was not anticipated that overlap with new signals would be a problem. In some of the studies, however, overlapping did occur.

TABLE 1

Number and Kind of Investigations Devoted to Each of the Major Categories of Independent Variables Explored

Independent variable category	Number of studies Exploratory	Number of studies Formal	Range of values explored
Stimulus density	2	2	4–46
Signal frequency	2	2	10–100/hour
Intersignal interval			
Average duration	1	3	(30–300 seconds, as per frequency)
Variability	2	—	2 levels
Display format	1	1	4 types
Length of service			
Within session	1	1	1–4 hours
Over sessions	1	—	2–20 days
Response requirements	2	—	2 kinds
Irrelevant signals			
Rel/irrel ratio	—	2	1/9–2/1
Total frequency	—	2	10–270
Heterogeneity	—	2	3 levels
Rel-irrel similarity	—	1	3 levels
Signal kind (add vs. omit)		3	2 kinds

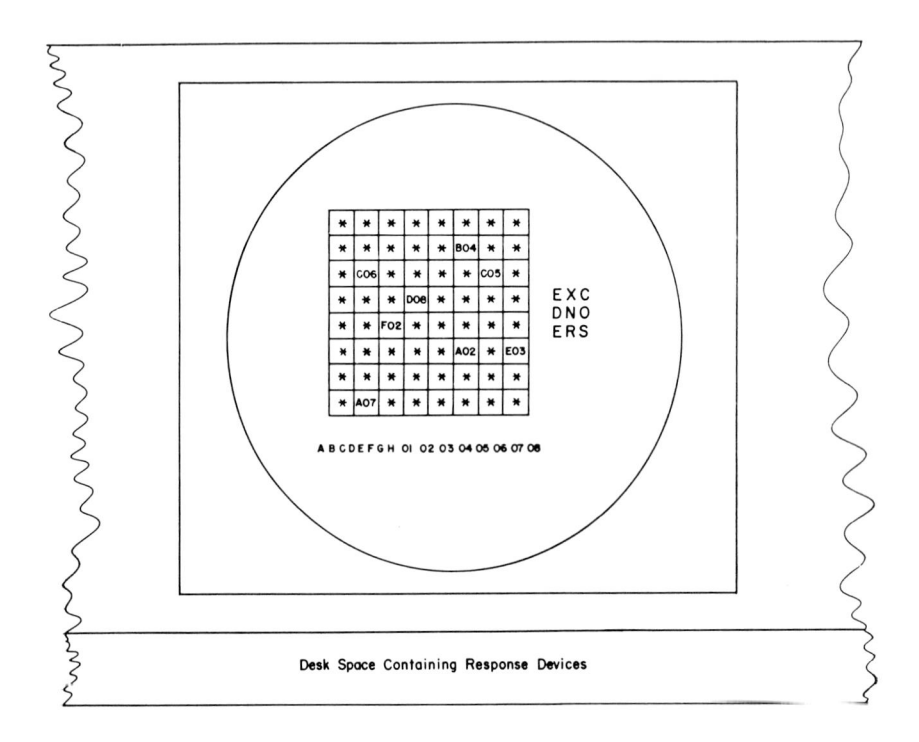

Figure 1. Front view of pertinent features of subject's display. Asterisks denote empty cells and signals occurred either as additions (adds) or deletions (omits) of the alpha-numeric stimuli.

the restrictions of intersignal interval (ISI) distribution parameters called for by experimental conditions, temporal occurrence of particular ISI's was randomized. In short, the subject had little basis for predicting what, when, or how a signal would occur other than those introduced experimentally.

One very important characteristic of this task was the relative persistence of signal states. When a signal occurred in any cell, the new state of that cell (i.e., signal *present* or *absent*) remained in effect indefinitely subject only to the rules governing selection of *any* cell for change. Thus, add signals could only occur in vacant cells, and omits only in occupied cells. Furthermore, since signals did not disappear after a specified time interval, latency measures were regarded as the most appropriate indices of performance.

There were two reasons for designing the task in the above fashion. First, it was believed that the distribution of latency scores would provide some insight into the manner in which the subject approached his task, particularly if such scores were not subject to an arbitrary upper limit. Very *long* latencies in a given segment, for example, would suggest a period of complete inattention, while a gradual increase in the *shortest* latencies might reflect something in the nature of a threshold shift. Such distinctions would not be as apparent using the more traditional procedure in which transient signals occur and the major index of performance is missed signals. A second reason for using the persistent signal approach was to permit a meaningful comparison of performance relative to added and omitted signals. Since in the present framework the response sequence (detect, locate, identify) for adds requires less reliance upon memory than it does for omits, the present procedure furnished a built-in means of estimating the role of information storage in monitoring functions.

Although latency was the measure of greatest concern, several other indices were also recorded. These included frequency of false alarms, missed signals, location errors, and identification errors. In addition, a separate category of latency was defined for occasions upon which one or more signal-response sequences intervened between a given signal and its response. This was termed *delayed latency*.

Molar Empirical Studies

Stimulus Density and Signal Frequency. Certainly one distinguishing feature of a complex monitoring task is the presence of multiple events to be observed. Within the context of the present task, this feature was directly manipulable in terms of the average number of alpha-numeric stimuli appearing on the display at any given time. If the total area of the display is held constant, as was the case in the present task, the most appropriate designation of this variable would be *average stimulus density*. Density, therefore, was one of the first variables selected for study.

A second variable was selected on somewhat different grounds. *Signal frequency* has always occupied a prominent position in classical vigilance research: typically it is under relatively low input frequencies that the most serious performance decrements occur (Deese, 1955). Since a very low density level would approximate the classical vigilance situation, factorial exploration of density and frequency might provide a basis for ultimate resolution of the discrepancy between simple and complex monitoring performance. On the strength of this reasoning, stimulus density (at levels of 4, 8, 16, and 32 stimuli per display) and signal frequency (30 and 75 per hour) were incorporated into one of the earliest studies in the series (Johnston et al., 1966). Consistent with the nature of the task, of course, two other variables were evaluated concurrently: signal *kind* (adds vs. omits) and signal *blocks* (reflecting duration of watch).

Two groups of eight subjects each served under one of the two frequency conditions and all four of the density levels. Since all subjects were highly trained on the task, having served in six preparatory sessions under a variety of conditions, a single 2-hour session was considered sufficient for each condition; thus, each subject served a total of four sessions on 4 consecutive days. The order in which they experienced the four density levels, of course, was balanced. For each frequency level, the distribution of ISI's was normal, about means of 40 and 100 seconds for high and low levels, respectively.

Before discussing the outcome of this study, a word of explanation is necessary regarding the method of combining data. To summarize performance over time, it is necessary to obtain average scores for successive portions of each session. There are, however, two ways to fractionate a session: by successive blocks of *time* and by successive blocks of *signals*. Each has a disadvantage: the former describes a varying number of signals; the latter, a varying interval of time. The alternative used consistently in this work was that of *signal blocks* because of its greater stability. Furthermore, time is at least represented ordinally for a given subject (i.e., all Block 1 signals precede all those in Block 2), and when scores for a given block are averaged over subjects, the *average* time of occurrence approximates a constant fraction of the session's length.[3]

The major findings are illustrated in Figures 2–4 for the most sensitive of the performance measures, nondelayed detection latency. Identification accuracy also was sensitive to the experimental variables, but this measure is not plotted since it merely duplicated the findings for latency.

Each point in Figure 2 represents a mean block score for eight subjects; each subject's block score, in turn, represents mean performance on five signals under both high- and low-frequency conditions. Since twice as many signals occurred under the high-frequency conditions, only half of these scores were used to obtain block averages; those high-frequency signals selected were the ones which occurred closest in time to the five low-frequency signals in each block.

An analysis of variance applied to these data revealed two significant effects: density, $F(3, 42) = 5.01$, $p < .01$, and the blocks × frequency interaction, $F(5, 210) = 2.42$, $p < .05$. Clearly, the poorest performance, in terms of both absolute latencies and decrements over time, occurred under conditions of low frequency (30 signals per hour) and highest density (32 stimuli per display). Duncan range tests supported this conclusion. It is interesting to note that the functions for adds and omits are quite similar; apparently, whatever the process or processes are that underlie complex monitoring, they are not at all affected by a storage requirement.

It should also be noted that decrements do not occur under the simplest conditions, even for low input frequencies. Since decrements have usually been observed

---

[3] For example, a 5-signal block in a 30-signal session of 100-minute duration would represent an average duration of $.17 \times 100 = 15$ minutes.

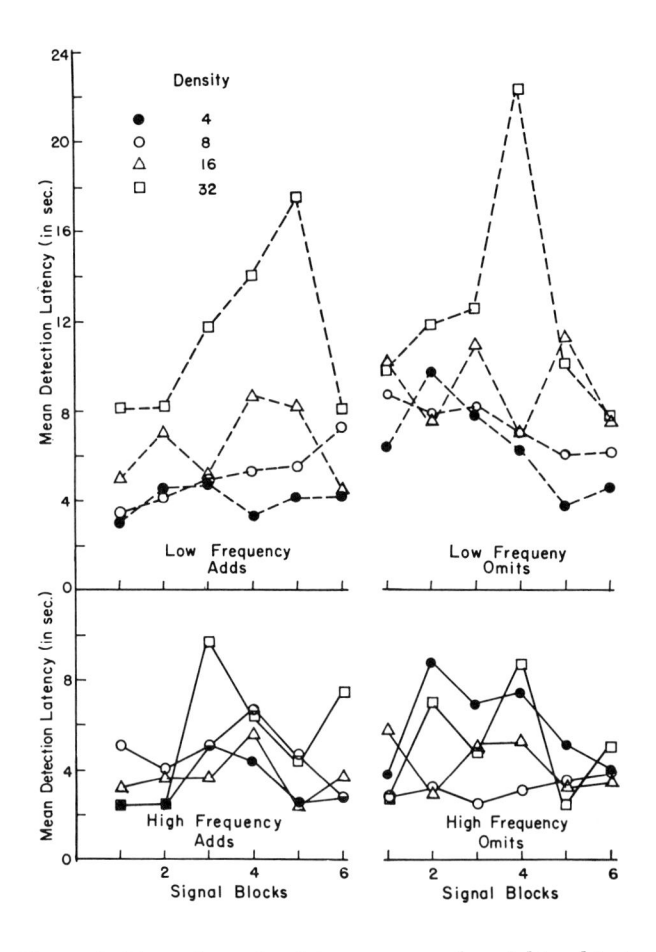

Figure 2. Mean detection latency scores for eight subjects.

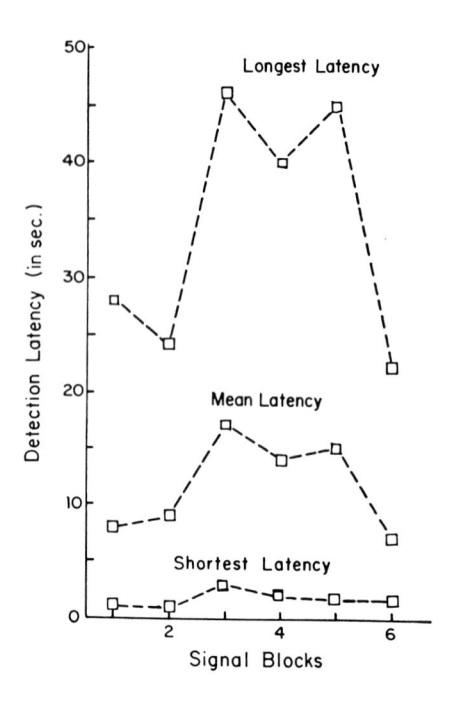

Figure 3. Mean detection latency scores for the low-frequency, high-density condition using the mean, shortest, or longest latencies per signal block for each subject (adds and omits pooled).

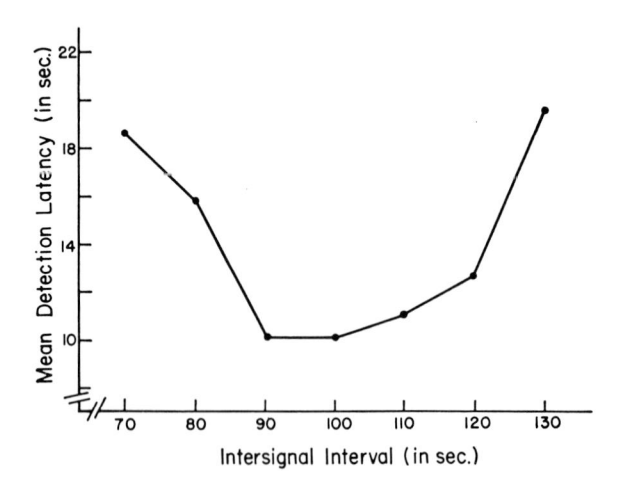

Figure 4. Detection latency as a function of ISI in the low-frequency, 32-density condition (adds and omits pooled).

in simple monitoring tasks, this suggests that even four events is enough complexity to block or obscure the operation of whatever mechanisms control such decrements. It also suggests that the high-density decrements obtained here may be attributable to entirely different mechanisms.

In view of the significant decrement incurred under the 32–30 condition, these specific data were examined in several other ways. First, in order to eliminate possible artifacts incurred by separating omits and adds, the signals of both types were put back in their original serial arrangement and then reorganized into six blocks of 10 signals each. In order to gain some insight into the reasons for the decrement, the mean, longest, and shortest latencies for each block were determined and viewed separately. As shown in Figure 3, the longest latencies show a marked decrement, while the shortest latencies remain constant over blocks. Clearly, then, the decrement in mean latencies is attributable to the increased occurrence of *very long* latencies rather than to a general increase of *all* latencies. This suggests that decrements occur, not because of reduced sensitivity, but because of increased inattention to the display for short periods of time.

A second summary of the 32–30 condition data was aimed at evaluating the effect of signal predictability on latency scores. It will be recalled that the duration of intersignal intervals (ISI's) was normally distributed about a mean value of 100 seconds for the low-frequency condition. If a subject were to attempt to guess when the next signal would occur, therefore, his best guess would be the mean (most frequent) ISI, or 100 seconds. As Figure 4 clearly shows, latencies were indeed shortest for the most probable interval and increased systematically with less likely intervals of both greater and lesser magnitude.

Finally, an analysis was made of the missed-signal and false-alarm errors obtained under the 32–30 condition. While neither type occurred with sufficient frequency to warrant statistical treatment, it was found that 82% of those which did occur were missed signals, and that the frequency of occurrence seemed to agree very closely with the latency functions of Figures 2–4; that is, missed signals became more common as the watch progressed and as the ISI deviated from the mean value.

Taken together, these findings indicate that sheer simplicity is not a necessary condition for monitoring decrements; that, in fact, excessively complex as well as excessively simple display circumstances may well result in poor performance. Also suggested by these data is a molar explanation of why high-density decrements occur. Both missed signal and latency indices support the notion that short periods of inattention are responsible rather than an overall restriction in level of sensitivity.

Irrelevant Information. If monitoring decrements for highly complex displays occur only when signal frequency is low, it becomes meaningful to ask whether the important aspect of frequency is the number of *overt responses* required or the amount of *stimulus change* occurring per unit time. To explore this distinction further, an experiment was designed in which some of the signals which occurred required no overt response; such signals were termed *irrelevant*.

Two variables were manipulated in this study: total signal frequency (30 or 60 per hour), and ratio of relevant to irrelevant signal frequency (1/2, 1/1, and 2/1). The resulting conditions may thus be described in terms of specific releevant/irrelevant signal frequency ratios as follows: 20/40, 30/30, and 40/20 for the 60-signal-per-hour conditions; 10/20, 15/15, and 20/10 for the 30-signal-per-hour conditions. A seventh (control) condition of a 20/20 ratio was administered before and after all others to evaluate any overall changes (i.e., learning or intersession decrement) and to permit comparisons with certain of the other conditions (especially 15/15 and 30/30). Eight subjects served under all eight experimental conditions in a balanced order.

Throughout the experiment, density was constant at 30 stimuli per display and the occurrence of both relevant and irrelevant signals was randomized over time. All stimuli were letter trigrams and their composition determined relevancy: identical letters (e.g., AAA) denoted relevant stimuli; dissimilar letters (e.g., ABC or ABB), irrelevant ones. A normal distribution of ISI's was used and all response measures were recorded.

Since very few errors and delayed latencies occurred under any condition, analysis was again restricted to the nondelayed latency index. Figure 5 presents the mean scores for all conditions summarized over 10-signal blocks. A separate portion of this figure is devoted to the three 1/1 ratio conditions (the two 40-per-hour conditions yielded practically identical scores so were combined for presentation here).

An analysis of variance revealed that none of the differences was sufficiently great to achieve statistical significance ($p > .05$). The closest approximation was the frequency effect, $F(1, 7) = 3.64$, $p < .10$, which suggests that poorer performance is likely under low-frequency conditions.

One result of this study is, however, noteworthy. For both the high- and low-frequency conditions latencies were considerably higher in this than in the previous study: overall means were nearly twice as great in both cases. While it would be inappropriate to subject this difference to statistical test, so great a difference would strongly suggest that irrelevant information per se may actually be *detrimental* to performance. At least it would appear highly unlikely that monitoring efficiency in the complex situation is furthered by mere stimulus change. Involvement of an overt response would seem to be a necessary condition for any beneficial influence of frequency to be felt.

While totally unexpected, the failure of either independent variable to produce a significant effect is not entirely without explanation. Probably the foremost reason is that a critical aspect of density may not have been adequately controlled. It will be recalled that *total* density was held constant at 30 stimuli; consequently, density of *relevant* stimuli increased with the proportion of relevant to irrelevant signals. If it is actually the relevant stimulus density that is most detrimental to performance, then the wrong kind of density was selected for control. This variable may well have worked in direct opposition to both ratio and frequency, obscuring their effects. Later work is planned to resolve this issue.

Refinement of Variables

In accordance with the adopted strategy, molar variables showing some influence upon complex monitoring behavior were to be subjected to refinement in later studies. The following study serves as an example of one such attempt to reduce molar variables to a more fundamental level.

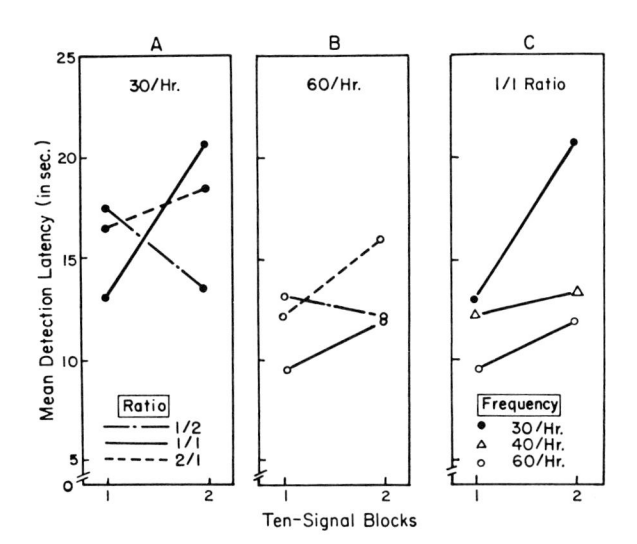

Figure 5. Mean detection latency scores for 10-signal blocks obtained under the various frequency and relevant-irrelevant ratio conditions. The A and B figures include all major experimental conditions; C illustrates specifically the frequency effect for the two 1/1 ratio conditions (in A and B) and for the 40/hour control condition which was also a 1/1 ratio.

Figure 6. An illustration of the three display formats compared. Slashes denote permanently unused cells and asterisks, cells in which an add signal may appear. The numbers 3(03)–6(06) were designated as relevant.

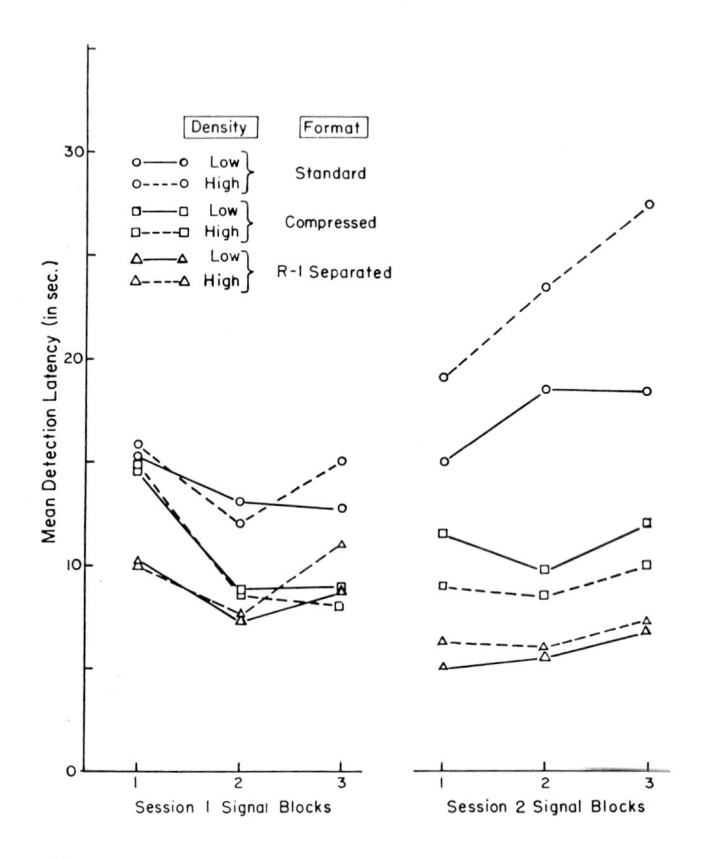

Figure 7. Mean detection latency scores for the six combinations of format and density (adds and omits combined). Session I was devoted to training.

The Density Effect and Display Format. It was quite apparent in the earlier studies that high stimulus density and the occurrence of irrelevant signals both serve to degrade monitoring performance. Furthermore, when combined with low signal frequency, the density effect occurs as a decrement over time. Considering these effects, it was reasoned that both increased density and the presence of irrelevant information might serve primarily to complicate scanning behavior: in each case more scanning activity would be required to detect the occurrence of the same number of signals. In a sense, then, both variables can be viewed as contributing to the uncertainty in the task. This uncertainty, however, also seems divisible into two components: that attributable to the sheer number of events to be monitored, and that attributable to the spatial extent over which such events are spread. The objective of the present study was to isolate the density effect with respect to these two characteristics.

The basic method adopted was one in which the same information was displayed using one of three formats (see Figure 6), with each format representing an aspect of density. The first format (*standard*) had relevant and irrelevant information interspersed over the entire area of the matrix; it was intended to maximize spatial extent and number of events scanned. The second (*compressed*) included the same information in only four rows of the matrix, reducing the extent by one-half but retaining the number of events. The third (*R-I separated*) allocated all relevant information to four rows and all irrelevant information to the other four rows; thus, the number of events to be scanned as well as the spatial extent covered by them was effectively halved. In all three formats, however, the *total* number of events displayed and changed over a session remained constant. Stimulus density was introduced at two levels (10 and 20 events per display), but the relevant/irrelevant stimulus and signal ratio was constant at 1/1. The lowest frequency used in other studies (30 signals per hour with 15 relevant and 15 irrelevant) was also used here, and detection latency supported by the location response again constituted the primary measure.

The 48 student subjects were divided into six groups of eight each with each group assigned to a different one of the six format-density conditions. Each group served for two 100-minute sessions under the same condition and the first session was considered training. In all other respects, this study was conducted just as the earlier ones had been.

The results for both training and experimental sessions are summarized in Figure 7. Only the latter, however, are discussed here since the present interest is not in acquisition processes. Clearly, it is the spatial factor which contributes most heavily to the density effect. Latencies are considerably longer for the standard format than for either of the spatially reduced ones, and only the standard format shows a serious density effect. These conclusions were supported by an analysis of variance which indicated a significant format effect, $F(2, 42) = 6.49$, $p < .01$. Duncan's test revealed that the significance was attributable solely to the inferiority of the standard format; none of the differences involving the compressed or R-I separated formats was significant.

Several of the findings in Figure 7 are of interest in light of the other two studies. The only tendency toward a decrement appears in connection with the standard format; this is basically the same format used in both of the other studies. Since this tendency was also apparent under comparable conditions involving both irrelevant information (Figure 5c at 30 per hour) and no irrelevant information (Figure 2 for a density of 32 and low frequency), it would seem likely that decrements are also related to area scanned. It is interesting to note too that the absolute level of performance obtained under the standard format is very similar to that appearing in Figure 5c for a very comparable set of conditions; those obtained under the two reduced formats are much more comparable to the low density levels of Figure 2. This again suggests that reducing area scanned is effectively the same as reducing density insofar as overall latency is concerned.

One final result is worthy of mention even though it is based upon very tenuous data. A tabulation of all responses made to irrelevant stimuli was carried out for each of the six conditions. While none of the frequencies were particularly great (never more than 10% of the total irrelevant event occurrences), such errors were considerably more frequent under the compressed and standard formats than under the R-I separated format. Furthermore, most of this difference was attributable to the high density condition. This would seem to indicate that compressing a great deal of information into a small spatial area may reduce discrimination among elements even though it may enhance detection by reducing scanning requirements.

CONCLUSIONS

The present research does not offer anything approximating a complete explanation or even description of complex monitoring behavior. What it does is to illustrate an approach through which some understanding of such behavior may be gained. The findings with regard to stimulus density, signal frequency, and, to some extent, signal relevancy, suggest that it may be profitable to proceed from a molar level of description toward an ultimate explanation in terms of fundamental mechanisms. Using this strategy, it was possible to show, for example, that sizable decrements can occur in the complex situation under conditions of high stimulus density and low signal frequency, that overall performance is adversely affected by the presence of irrelevant signals, and that both latencies and missed signal frequencies are greater for signals which occur at improbable times. Pursuing the density effect a step further, it was also possible to show that the spatial extent over which information is displayed is critical to the influence of density on behavior. In short, a large number of events can be monitored adequately only if they occur within a restricted area; a smaller number of events can be handled effectively even if they are spread over a wide area.

From the data obtained in these studies, it is possible to formulate at least a general hypothesis regarding some of the internal processes which operate in the complex monitoring situation. As Jerison (1963) has suggested, it would appear that scanning behavior is of critical importance: the monitor seems to engage in a periodic search, the pattern of which is adjusted to the spatio-temporal characteristics of signal occurrence. Thus, in accordance with both expectancy (Baker, 1963) and reinforcement principles (Holland, 1958), high signal frequencies would promote frequent scanning and a high, stable level of overt performance. Low frequencies, especially when coupled with the effort required to scan a large number of elements spread over a wide area, would promote infrequent scanning and poor performance. Furthermore, the effort involved in scanning such displays would seem to foster less frequent scanning as the watch progresses; presumably, scanning behavior would become increasingly restricted to high-probability time intervals. This would account for both the obtained decrement and the *nature* of the decrement, i.e., the increased occurrence of very long latencies, rather than a threshold shift, over time.

It should be noted in closing that the current program is still in its formative stages. Much remains to be learned regarding even the molar characteristics of complex monitoring behavior. At this point, however, it would appear that reinforcement principles and perhaps the expectancy notion (if, in fact, this can be distinguished from reinforcement) have the best chance of ultimately explaining such behavior. Certainly, the present data provide no evidence in support of differential arousal effects. It is likely that the stimulation afforded by complex displays is sufficient to maintain arousal, particularly since the subject has considerable control over stimulation received. The basic question from the standpoint of arousal theory, then, would concern the manner in which he exerts this control; from an operational standpoint, this question is indistinguishable from that of how he scans the display.

REFERENCES

Adams, J. A. Experimental studies of human vigilance. *United States Air Force ESD Technical Documentary Report,* 1963, No. 63-320.

Adams, J. A., & Boulter, L. R. An evaluation of the activationist hypotheses of human vigilance. *Journal of Experimental Psychology,* 1962, **64,** 495-504.

Adams, J. A., & Humes, J. M. Monitoring of complex displays: IV. Training for vigilance. *Human Factors,* 1963, **5,** 147-154.

Adams, J. A., Humes, J. M., & Sieveking, N. A. Monitoring of complex visual displays: V. Effects of repeated sessions and heavy visual load on human viligance. *Human Factors,* 1963, **5,** 385-390.

Adams, J. A., Humes, J. M., & Stenson, H. H. Monitoring of complex visual displays: III. Effects of repeated sessions on human vigilance. *Human Factors,* 1962, **4,** 149-158.

Adams, J. A., Stenson, H. H., & Humes, J. M. Monitoring of complex visual displays: II. Effects of visual load and response complexity on human vigilance. *Human Factors,* 1961, **3,** 213-221.

Baker, C. H. Further toward a theory of vigilance. In Buckner, D. N., and McGrath, J. J. (Eds.), *Vigilance: A symposium.* New York: McGraw-Hill, 1963. Pp. 127-154.

Broadbent, D. E. Some effects of noise on visual performance. *Quarterly Journal of Experimental Psychology,* 1954, **6,** 1-5.

Broadbent, D. E., & Gregory, M. Vigilance considered as a statistical decision. *British Journal of Psychology,* 1963, **54,** 309-323.

Deese, J. Some problems in the theory of vigilance. *Psychological Review,* 1955, **62,** 359-368.

Frankmann, J. P., & Adams, J. A. Theories of vigilance. *United States Air Force AFCCDD Technical Note,* 1960, No. 60-25.

Hebb, D. O. Drives and the C.N.S. (conceptual nervous system). *Psychological Review,* 1955, **62,** 243-254.

Holland, J. G. Technique for behavioral analyses of human observing. *Science,* 1957, **125,** 348-350.

Holland, J. G. Human vigilance—the rate of observing an instrument is controlled by the schedule of signal detections. *Science,* 1958, **128,** 61-67.

Howell, W. C., Johnston, W. A., & Goldstein, I. L. Influence of stress variables on display design. *United States Air Force RADC Technical Report,* 1966, No. 66-42.

Jerison, H. J. On the decrement function in human vigilance. In Buckner, D. N., and McGrath, J. J. (Eds.), *Vigilance: A symposium.* New York: McGraw-Hill, 1963. Pp. 199-212.

Jerison, H. J., & Pickett, R. M. Vigilance: A review and re-evaluation. *Human Factors,* 1963, **5,** 211-238.

Jerison, H. J., & Pickett, R. M. Vigilance: The importance of the elicited observing rate. *Science,* 1964, **143,** 970-971.

Jerison, H. J., Pickett, R. M., & Stenson, H. H. The elicited observing rate and decision processes in vigilance. *Human Factors,* 1965, **7,** 107-128.

Johnston, W. A., Howell, W. C., & Goldstein, I. L. Human vigilance as a function of signal frequency and stimulus density. *Journal of Experimental Psychology,* 1966 (in press).

Johnston, W. A., Howell, W. C., & Zajkowski, M. M. The regulation of attention to complex displays. *Journal of Experimental Psychology,* 1967, in press.

Kibler, A. W. The relevance of vigilance research to aerospace monitoring tasks. *Human Factors,* 1965, **7,** 91-99.

McGrath, J. J. Irrelevant stimulation and vigilance performance. In Buckner, D. N., and McGrath, J. J. (Eds.), *Vigilance: A symposium.* New York: McGraw-Hill, 1963. Pp. 3-19. (a)

McGrath, J. J. Some problems of definition and criteria in the study of vigilance performance. In Buckner, D. N., and McGrath, J. J. (Eds.), *Vigilance: A symposium.* New York: McGraw-Hill, 1963. Pp. 227-237. (b)

Mackworth, J. F., & Taylor, M. M. The d' measure of signal detectability in vigilance-like situations. *Canadian Journal of Psychology,* 1963, **17,** 303-325.

Mackworth, N. H. The breakdown of vigilance during prolonged visual search. *Quarterly Journal of Experimental Psychology,* 1948, **1,** 6-21.

Montague, W. E., Webber, C. E., & Adams, J. A. The effects of signal and response complexity on eighteen hours of visual monitoring. *Human Factors,* 1965, **7,** 163-172.

Swets, J. A., Tanner, W. D., Jr., & Birdsall, T. G. Decision processes in perception. *Psychological Review,* 1961, **68,** 301-340.

Zuercher, J. D. The effects of extraneous stimulation on vigilance. *Human Factors,* 1965, **7,** 101-105.

Chapter Fourteen

Other Extreme Input Effects

Man is remarkably adept at adjusting to changes in his environment. He is endowed with an assortment of mechanisms that enable him—within certain limits—to amplify or attenuate, simplify or embellish, speed up or slow down, the flow of information from a particular milieu in which he finds himself through his "processing" systems. There are, however, limits to his adaptive capabilities and some of them are better understood than others. We know a great deal, for example, about sensory limits, and at least something about storage, reaction-time, and "psychological refractoriness." We have just examined the phenomenon of vigilance decrements, which again would seem to imply some sort of limitation—in this case on attention—as a function of infrequent stimulus input. We now proceed to consider some apparent limitations that are much less clearly understood. In this chapter and in Chapter Fifteen, the limits of primary interest are those on skilled performance (other than monitoring). This chapter considers what happens when stimulation from external sources is severely curtailed and prescribed processing requirements are reduced to zero; it also examines the opposite case in which input information exceeds man's processing capabilities. How is he affected by these input extremes?

First, we take up the matter of reduced input or *sensory deprivation* (SD). As Corso (1967) points out, there is nothing new about the phenomena associated with isolation and confinement; anecdotal reports of peculiar experiences and behaviors observed under these conditions can be traced as far back as the history of man. What *is* new is the scientific investigation of these phenomena; a rather vigorous effort has been made in the last fifteen years or so to sort fact from fiction, to quantify verifiable effects, and to postulate mechanisms responsible for these effects.

Numerous experimental techniques have been devised to limit stimulation experimentally and a host of variables have been explored (see, for example, Solomon, 1961; Burns and Kimura, 1963; Wheaton, 1959). Findings, while not always consistent, have at least produced *some* generalizations. The main ones may be classified under three broad headings: experiential or introspective effects, physiological correlates, and behavioral effects. The first two categories are not of particular interest here. Suffice it to say that hallucinations, images, and a host of other experiences (such as anxieties and fantasies) are commonly reported after long periods of SD, and the kind and degree of these experiences seems to

References will be found in the General References at the end of the book.

depend a great deal upon personality characteristics of individual subjects. Interest in the physiological aspect of the problem has centered around similarities between SD effects and the reactions associated with certain drugs, together with the postulated role of nonspecific pathways of the lower brain stem in regulation of central and peripheral "activation levels." Both lines of inquiry, of course, bear upon the issue of exactly how manipulation of input level might lead to such diverse behavioral and experiential effects as have been observed.

The third set of generalizations concerns overt behavior—in particular how acquisition, performance, and retention of various tasks is affected by prolonged periods of SD. Except for an occasional report of transient improvement in learning, retention, or visual efficiency, most studies have shown either no effect or a decrease as a result of SD. However, the pattern of these effects does not present a particularly coherent picture. If there are general task dimensions which bear a systematic relationship to SD parameters, they have so far escaped notice. To be sure, specific *tasks* can be described which will consistently suffer as the result of SD—especially after several days of confinement—and the first paper in this chapter presents several examples falling within this category (together with some that consistently show *no effect*). But predicting what will happen in a completely *new* task on the basis of current knowledge is still a risky proposition. Unquestionably, to use the old cliché, more research is needed in this intriguing and potentially significant area. It should be noted in passing, however, that SD research is far easier said than done. Few areas present methodological difficulties of the magnitude that this one does.

The second paper in this chapter deals with the opposite extreme: information overload. The issue here is, what does a person do when he is presented with more information than he can hope to handle? We have touched upon this area earlier in connection with human information processing (especially Chapters Four and Five). There, however, our concern was chiefly with performance functions below the point of overload. Here we wish to explore both the strategies that a person uses to cope with overload and the general limitations that such "stressful" conditions impose upon his behavior. Miller's article is addressed primarily to the first of these objectives: discovering the ways in which man adjusts to overload. Few investigators have given much attention to this as an experimental problem, and certainly none has pursued it more vigorously over the years than Miller. The second paper, therefore, summarizes conclusions drawn from a substantial research base.

One of the concepts that has arisen in conjunction with excessive input or task demands is the ubiquitous notion of stress. In its most general form, the idea seems to be that taxing conditions are capable of reducing man's performance, not only to the level that one would expect knowing his limited processing capacity, but beyond—even to the point of response disorganization. It is as though another mechanism were at work, actively disrupting performance in proportion to its level of output. This hypothetical mechanism is called *stress*. It is easy to see how appealing such a construct might be on purely common sense and introspective grounds: we are all familiar with the athlete who is said to "choke" in difficult situations, the underachiever who claims that he "does poorly because he tries too hard," the "pressure" we feel when speaking or performing before a large audience, and the utter hopelessness we experience when work piles up on our desk. How eminently reasonable it seems to attribute all these undesirable effects to a common organismic state or process! When we have gone this far, it is but a short conceptual step to a lot of other disruptive circumstances. Why not include extreme temperature effects (they tend to be disruptive), and instructions to subjects (they certainly can be disruptive), or any of a thousand other things? Very soon we reach a point at which our explanatory concept has lost any explanatory power it might otherwise have had; *any* operation resulting in performance decrement and/or unpleasant or "tense" feelings is said to be producing "stress," and thus the behavior to be accounted for becomes the means of defining the "explanatory" concept (a clear case of circularity).

It may be overstating the case somewhat, but we believe that a fate much like that just described has befallen the stress concept. There have been attempts to rescue the concept, such as that by Fitts and Posner (1967) in which the term is restricted to a description of task demands, but these, too, encounter difficulties. By the Fitts and Posner definition, for example, the term becomes little more than a compendium of all possible input variables, and one is led to such curious concepts as "optimal stress" (i.e., input levels producing the best performance) and "stress tolerance" (beyond which performance deteriorates). It is difficult to see what purpose the concept serves when defined in this way.

In any event, the term *stress* appears frequently in the literature of engineering psychology and, as the above discussion implies, in a variety of contexts. One of the most common places it occurs is that of extreme input conditions. Hence the historical-analytic paper by Appley and Trumbull (1967) with which we close the chapter. This paper, incidentally, originally appeared as the initial chapter in a book edited by Appley and Trumbull and devoted entirely to research and theory in this troubled area. The reader who wishes to pursue the topic further would do well to read the remainder of this comprehensive and insightful volume.

49

Effect of Sensory Deprivation
on Some Perceptual and Motor Skills

Jack A. Vernon, Thomas E. McGill,
Walter L. Gulick, and Douglas K. Candland

Interest in the effects of isolation upon human behavior (1, 2, 3) has prompted at Princeton a series of investigations (4, 5, 6) one of which the present paper reports. The conditions of isolation employed in our studies so reduced the sensory stimulation available to the individual as to cause us to refer to the confinement as Sensory Deprivation (S.D.). Obviously, S.D. did not produce a total deprivation of sensory stimulation but, as shall be seen, it did effect a drastic reduction in the sensory input.

METHOD

Subjects. Volunter Ss were male graduate students between the ages of 20 and 28 yr. Each passed a physical examination administered by the University Medical Staff. For each day of confinement there was a payment of $20.00. There were nine experimental Ss and nine controls. Control Ss were paid $5.00 for each of the three test sessions. Control Ss were not unwilling to enter the S.D. condition but were unable to do so because of the amount of time required for confinement. We requested all confined Ss to make available to us a block of six days of which we would take whatever amount we needed. We adopted this procedure because we considered it desirable that a confined S not know exactly how long he was to be in S.D. Some volunteers could not take six days from their normal school schedule so we used them in one of the control subgroups. It seems important to make this point so that the reader will not misunderstand and believe that the control Ss were merely individuals who were unwilling to enter the S.D. condition due to fear of confinement or the like.

Reprinted with permission of author and publisher: Vernon, J. A., McGill, T. E., Gulick, W. L., and Candland, D. K. Effect of sensory deprivation on some perceptual and motor skills. *Perceptual and Motor Skills*, 1959, 9, 91-97.

The Princeton Project on Sensory Deprivation is supported by a grant-in-aid for research from the Office of the Surgeon General of the Army, Contract Number DA-49-007-MD-671.

Procedure. Ss were confined in a dark, light-proof, sound-proof (80 db loss) "floating chamber" which measured 15' × 9'. The confinement cell (4' × 9') at one end of the floating chamber contained a bed, a portable ice box, two relief bottles, a panic button, and a large tray of sodium hydroxide placed beneath the bed to absorb S's accumulated CO_2. Ss were allowed access to a chemical toilet located outside the confinement cell but within the floating chamber. The confined individual wore a gauntlet-type glove which was intended to restrict movements and to remind Ss that they were to make as little noise as comfortably possible. The noise of their own breathing and bodily movements was not shut out by ear plugs for the simple reason that the "seashell" effect produced by the plugs provided greater auditory stimulation.

Each individual was monitored throughout his confinement by a concealed inter-communication system which revealed almost perfect adherence to the instructions against making noise.

Each of the five tests, to be described later, was administered three times to all 18 Ss. The first testing (A) was prior to confinement, the second testing (B) was upon release from confinement, and the third testing (C) was 24 hr. after release from confinement. Test A provided a "normal" measure, Test B provided an opportunity for the effects of S.D., if any, to be demonstrated, and Test C provided a measure of the relative persistence of any S.D. effects. Since each test was given three times, a change in results at B or C could reflect either the effects of S.D. or a practice effect or both. This dilemma was resolved by comparing the data for the confined group with those of the control group. The comparison was made more meaningful by matching the control Ss and confined Ss. The basis for the matching was the performance at Test A for all five tests. Some 16 individuals were tested before a matching control group of 9 Ss could be selected. The significance of any difference at Test B between confined and control groups was evaluated by a Wilcoxon test (7), and $p < .05$ was accepted as significant.

The periods of confinement were 24, 48, and 72 hr., hence the nine experimental Ss were assigned to three subgroups of three Ss each. This arrangement made the time between Test A and Test B 24 hr. for one subgroup, 48 hr. for another subgroup, and 72 hr. for the remaining subgroup. The same time lapses between Tests A and B were arranged for the comparable control subgroups.

Tests. The five tests which were administered are fairly well known so only a brief description of them will be given. They were (a) color perception, (b) depth perception, (c) pursuit rotor, (d) mirror drawing, and (e) rail walking.

(a) *Color perception* was tested with the Dvorine Color Test, Plates I through XX. Each plate was presented for 3 sec. under normal illumination. The order of presentation of the plates was randomized.

(b) *Depth perception* was tested with the Howard-Dohlman apparatus. Each test of depth perception was the mean of six binocular trials where half of the trials started with the variable stimulus behind the standard and the other half with it in front. Ss were 20' distant from the apparatus. A head holder prevented the utilization of head movements. Between trials Ss were required to drop the strings to the floor while E, blocking S's view, set the stimuli for the next trial.

(c) *Pursuit rotor* performance was measured by the amount of time a stylus (21 cm. long) was held in contact with a rotating disc (1 cm. in diameter) when the disc rotated at 44 rpm in a circle 13 cm. in diameter. The test lasted 60 sec., and the performance measure was the total time on target which was measured automatically. The stylus and the disc were electrically connected so that, when they contacted, the circuit to an electrical clock was completed and the clock was activated. The clock ran only when contact was maintained thus cumulating time on target. The scoring procedure was explained to S.

(d) *Mirror drawing* performance was tested utilizing the standard concentric six-pointed star where the separation between lines was ¼" and the total length of path to be traced was 22½". Ss were required to trace a path between the pattern lines with a pencil with reference to nothing but the mirror image of the star pattern. The performance measures were the amount of time required to complete the task and the number of errors committed. An error was considered to be any crossing of the lines of the pattern outline. This procedure was carefully explained to each S before testing.

(e) *Rail walking* performance was an attempt to get at a more gross motor ability, a type of activity which S.D. Ss had little opportunity to exercise. The test was of S's ability to walk upon the narrow edge of a 2" × 4" U-shaped rail which was 18' in length. The performance measure was the total time required by S to successfully negotiate the railing without falling off. If S fell or stepped off the rail he was required to return to the starting point again. The time recorded was the total time spent by S on the rail; time required to return to the start after a fall was not counted. The scoring procedure was carefully explained to S.

RESULTS

In evaluating the results the reader is urged to be cautious because the Ns are small and because of the variability of performance measures for some tasks.

Table 1 presents the results of the color perception tests. The data presented are the changes in the number of correct identifications of the Dvorine Color Plates from Test A to Test B expressed as percent. It is obvious that the control subgroups remained unchanged from Test A to Test B in their ability to perceive color regardless of whether the tests were separated by 24, 48, or 72 hr. The S.D. subgroups, on the other hand, showed a loss in color perception from Test A to Test B. When the two tests were separated by 48 or 72 hr. of S.D., the change in color perception was significantly greater than that of the comparable sub-

TABLE 1
Percent Change in Color Perception from Test A to Test B

Group	Hours between Tests A and B		
	24	48	72
Control	0.0	0.0*	1.0
Confined	−3.3	−13.3	−10.0

*Significant statistical comparisons shown in italics are between control and confined groups within a given period between Tests A and B.

group. A statistically significant comparison between subgroups is indicated by italic numerals for the two subgroups being compared. Thus in Table 1 the performance of the 24-hr. S.D. subgroup is not significantly different from that of the 24-hr. control subgroup. The 48-hr. S.D. subgroup is significantly poorer than the 48-hr. control subgroup as is true also for the two 72-hr. subgroups.

The color perception test was not conducted until S had been out of confinement for a half hour so that he was normally light adapted. Examination of the color plates missed at Test B showed that no particular color was consistently missed and that the failures occurred primarily for the desaturated hues.

The data for Test C are not presented; however, they reveal that the losses showed by the S.D. subgroups no longer obtain 24 hr. after release from confinement. Test C data are not presented in tabular form for any of the tasks for the simple reason that in all cases they can be easily verbalized and because their inclusion would greatly complicate the tables.

The results of the depth perception tests are presented below in Table 2. The data of Table 2 are the changes in mm. from Test A to Test B. They can be explained by considering the case of the 24-hr. control group. The entry, + 7.9

TABLE 2

Change in Depth Perception Performance (mm.) from Test A to Test B

Group	Hours between Tests A and B		
	24	48	72
Control	+ 7.9	+1.5	−3.1
Confined	−12.9	+1.4	−1.6

mm., means that at Test B the average error was 7.9 mm. less than that at Test A; therefore, in Table 2 a plus sign indicates improvement at Test B over Test A. Note, in the case of the control subgroups, as the time increased between Test A and Test B, that the practice effect steadily decreased, becoming negative at 72 hr.

Performance of the confined subgroups shows no consistent trend except that the greatest loss occurs for the shortest period of confinement. These data *are not* statistically significant due to the large individual variation. The data for depth perception are presented primarily to suggest one kind of performance upon which S.D. appears to have no significant effect. In fact the data would not have been worth including here were it not for the large loss in performance for the 24-hr. S.D. subgroup.

The results for the rotary pursuit tests presented in Table 3 are increases in the amount of time on target at Test B in comparison with Test A. For example, the 24-hr. control subgroup was on target at Test B on the average 1.5 sec. longer than at Test A. Note the rather unusual practice effect displayed by the control subgroups, that is, as the interval between Tests A and B increased, improvement is observed. The S.D. subgroups confined for either 24 or 72 hr. also show an improvement similar to that shown by the comparable control subgroup. The 48-hr. confinement subgroup, however, not only fails to improve at Test B

TABLE 3
Rotary Pursuit Improvement (sec.) from Test A to Test B

Group	Hours between Tests A and B		
	24	48	72
Control	1.5	4.3	7.4
Confined	2.4	−0.3	6.0

but actually loses slightly in performance level. The difference between the performance of two 48-hr. subgroups is statistically significant.

The data for Test C reveal that the losses suffered by the 48-hr. confinement subgroups were substantially recovered and the other subgroups benefited very little by the additional practice provided by Test C.

With regard to the performance of the 48-hr. confinement subgroup, it is of interest that most S.D. Ss report that subjectively the "worst" part of a long confinement is somewhere around 48 hr. This duration of confinement also seems critical for mirror tracing performance.

Performance on the mirror tracing task is presented in Tables 4 and 5. Table 4 gives the time data in terms of the percentage of improvement at Test B over

TABLE 4
Percent Improvement in Time (sec.) Required for Mirror Tracing from Test A to Test B

Group	Hours between Tests A and B		
	24	48	72
Control	55	49	50
Confined	51	17	58

TABLE 5
Improvement in Errors (%) in Mirror Tracing from Test A to Test B

Group	Hours between Tests A and B		
	24	48	72
Control	78	57	88
Confined	43	76	30

Test A. For example, the 24-hr. control subgroup performed at Test B 55% faster on the average than at Test A. Note that *all* subgroups improve in time at Test B over Test A; however, the improvement of the 48-hr. confinement subgroup is so slight as to be significantly less than that of the comparable control subgroup, or, for that matter, for any of the other subgroups.

Why the 48-hr. confinement subgroup fails to improve much as indicated by the time scores is partially explained by the error data in Table 5. These data are presented in the same manner as in Table 4 so that, for example, the 24-hr.

control subgroup shows 78% fewer errors on Test B than on Test A. Once again note that all subgroups show a reduction in errors for Test B over Test A. The interesting point is that the subgroups confined for 24 and 72 hr. significantly fail to improve as much as their comparable control subgroups. The 48-hr. confinement subgroup reduces its errors as much as the control groups. In fact, it seems to improve more than its comparable control subgroup, which for some unexplained reason performs worse, but not statistically significantly worse, than the other two control subgroups. Thus, considering Tables 4 and 5 together, it appears that the 48-hr confinement subgroup has sacrificed improvement in time in order to improve accuracy and eliminate errors.

Once again the results at Test C reveal the disappearance of previous differences between the various subgroups.

Performance for the rail walking tests was measured in time required to complete the task, and in Table 6 the performance at Test B is compared with that

TABLE 6
Change in Time (%) Required for Rail Walking from Test A to Test B

Group	Hours between Tests A and B		
	24	48	72
Control	−36	13	46
Confined	−21	−56	−50

at Test A in terms of percent. For example, the 24-hr. control subgroup required 36% more time for Test B than for Test A, or in other words a loss of −36%. The 24-hr. confined subgroup shows about the same result, but with that, similarities in Table 6 cease. Note that the 48- and 72-hr. control subgroups show progressive improvement while the 48- and 72-hr. S.D. subgroups show losses. The difference between the two 72-hr. subgroups is statistically significant. The loss, however, was not permanent as indicated by the recoveries made at Test C.

SUMMARY

The effect of Sensory Deprivation (S.D.) lasting 24, 48, or 72 hr. was measured for several perceptual and motor skills. It was found that color perception was adversely affected especially when confinement extended to 48 and 72 hr. There was no significant effect of S.D. upon depth perception, probably because of the large individual variations in performance. Rotary pursuit ability was significantly adversely affected by S.D. only when confinement lasted 48 hr. A similar finding resulted for the time required to perform a standard mirror tracing task. However, the errors made on the mirror tracing task were fewer after 48 hr. of confinement and significantly increased after 24 and 72 hr. of confinement. Gross motor behavior as measured by a rail walking task was adversely affected by S.D., especially after 72 hr. of confinement. All S.D. effects were somewhat temporary in nature since tests given 24 hr. after release from confinement revealed a tendency toward elimination of effects produced by S.D. Although results are based upon small Ns and single brief measures at each test period, the effects of S.D. upon performance are fairly consistent.

REFERENCES

1. Bexton, W. H., Heron, W., & Scott, T. H. Effects of decreased variation in the sensory environment. *Canad. J. Psychol.*, 1948, **8**, 70-76.
2. Heron, W. The pathology of boredom. *Sci. Amer.*, 1957, 196(1), 52-56.
3. Heron, W., Doane, B. K., & Scott, T. H. Visual disturbances after prolonged perceptual isolation. *Canad. J. Psychol.*, 1956, **10**, 13-18.
4. Vernon, J. A., & Hoffman, J. Effect of sensory deprivation on learning rate in humans. *Sci.*, 1956, **123**, 1074-1075.
5. Vernon, J. A., & McGill, T. E. The effect of sensory deprivation upon rote learning. *Amer. J. Psychol.*, 1957, **70**, 637-639.
6. Vernon, J. A., McGill, T. E., & Schiffman, H. Visual hallucinations during perceptual isolation. *Canad. J. Psychol.*, 1958, **12**, 31-34.
7. Wilcoxon, F. *Some rapid approximate statistical procedures.* Stamford, Conn.: American Cyanamid Co., 1949.

50

Adjusting to Overloads of Information

James G. Miller

The response of living systems to overloads of information has interesting similarities whether the system in question is a neuron or a human group. With all their obvious differences, both are organizations of genetically linked protoplasm. They exist in the earthly environment, which requires similar, very specific characteristics of everything that survives in it. And they find their places in a hierarchy of systems which, composed of nonliving molecules, increase in complexity from cells through organs, organisms, groups, social organizations and societies, to supranational institutions. The units of each are systems of the next lower order.

From Waggoner, R. W., & Carek, D. J. (Eds.) *Disorders of Communication.* Research Publications, Association for Research in Nervous and Mental Disease, **42**, 1964, 87-100. Reprinted by permission of the author and the Association for Research in Nervous and Mental Disease.

Supported in part by grants from Project MICHIGAN of the United States Army and from the Carnegie Foundation. The author wishes to thank Dr. William Horvath, Dr. Stanley Moss, Mr. Kent Marquis, Mr. Bertram Peretz and Dr. Paul Halick for their assistance in this research.

Each of these may be regarded as a level which has its own special characteristics. At each level there are differences among the species and among the individual examples of each species. Over the whole range of living systems, however, there are some characteristics which are common to all.

They are all, for instance, open systems which maintain themselves in a changing environment by regulating inputs and outputs of matter or energy and information and by preserving internal steady states of critical variables by governance of subsystems.

A representative of the level upon which interest is focused at a given time is referred to as a "system." Those at lower levels are subsystems or subsubsystems and those at higher levels, suprasystems or suprasuprasystems. Each subsystem carries out a specific function. It may be localized in a single component or member of the system or dispersed to several of them. Sometimes a single component may carry out a number of subsystem functions.

Unless a living system exists in parasitism or symbiosis with another system, it must carry out for itself certain critical subsystem functions. Some of these involve information, some matter-energy processing. A free-living cell, like an ameba, has subsystems which perform all these essential processes, while a cell which is part of an organ may depend for some of them upon other specialized cells. An isolated group, like a roving gypsy family, will be found to have all the critical subsystems, while groups which are parts of social organizations often do not.

THE CRITICAL SUBSYSTEMS

Table VII.1 lists the subsystems which a living system must have in order to survive and provide for the continuance of its species.

A system possessing all of these subsystems can be called totipotent. Of course, matter-energy and information never flow separately, but it is the energy in an intravenous injection of glucose and the information in his doctor's conversation which affect the patient. It is on this basis (which aspect of them affects the receiver) that matter-energy flows are distinguished from information flows. Three subsystems process both matter-energy and information, and the others transmit only one or the other. Because all living systems must be open to both matter-energy and information, the *boundary* provides for the passage of both. The *decider*, as the central controlling subsystem, must regulate both matter-energy and information transmissions, though the former are usually governed by information feedbacks. The *reproducer* is largely concerned with the information transmitted in the genes, but this must travel and divide and grow in a matter-energy substrate.

Within a system a subsystem function may be dispersed to subsubsystems or even lower levels. The reproductive function, for example, is carried on in animal species by a dyad, a mating pair of organisms, or by an individual organism. Levels of systems above the organism or group arrange that the reproductive function is carried out by enough of their units (the "household sector" in human societies) to assure the continuation of the larger system, but they do not themselves reproduce.

Table VII.1 shows that there are a number of rough parallels in the subsystem handling of matter-energy and information. The way a given function is carried out may be very different at different levels. The *ingestor,* for example, which takes required matter-energy into the system, may be an opening in the boundary of a cell, a mouth in an organism, a group of supply personnel in an organization or a set of agricultural, mining and importing organizations in a society. The input *transducer* has a parallel function to the ingestor, bringing information into the system and passing on these patterns to other subsystems. A sense organ does this for an organism, a scout does it for an army and an embassy does it for a nation. The *internal* transducer receives and passes on information which arises within the system. This the proprioceptors do in an animal and the organizations which discover attitudes of individuals and groups do within a society.

The *distributor* carries matter and energy about the system to the places where they are needed. Information is comparably transmitted among the parts of a living system by a *channel and net* subsystem which may be a nerve net in an organ, an entire nervous system or an elaborate set of communication devices used by individuals in a community and operated by an organization of communication specialists. The word "channel" is used two ways in electronics without confusion. It may refer only to the flow route for information or it may include along that route as well components of various sorts which alter the character of that flow. We follow the second usage because the noise, distortions and transformations of these components are factors of great importance in the communications of living systems.

The parallel is not so close between the *decomposer,* which breaks down input energy and raw materials into simpler forms that meet the specific internal needs of the system, and the *decoder.* Every system, however, has an internal language or languages which differ from the languages of its suprasystem, and the decoder alters messages from the external form to a code which can be used internally.

The *producer* unites decomposed inputs to fabricate particular substances required for the internal operation of that particular system. These may (like an organism's protein molecules) differ somewhat from products synthesized by any other system. Perhaps slightly parallel is the function of the *learner,* which establishes reliable and enduring associations (by conditioning, imprinting or learning) between certain elementary information inputs and other elements of information from outside or inside the system, to "synthesize" items of knowledge that will be especially useful to that particular system.

Both energy and information are stored in living systems. *Matter-energy storage* may be in the form of fat, glucose or adenosine triphosphate in a cell, in body fat depots in an animal and in reservoirs and warehouses in a society. The information storage or memory of an animal is in the brain; of a group, perhaps in books kept by a secretary; and of a nation, in libraries and filing cabinets maintained by multitudes of librarians and file clerks.

Beyond the central decider of the system, matter-energy and information are prepared for output. Unneeded materials and energy are outputted by an *excretor,* so that the system will not become glutted as more inputs arrive. The information output from a living system is changed by the *encoder* from the system's internal language into a language used in the suprasystem.

The *motor,* by which all or part of a system moves in space, may in a nation be a different set of organizations (armies, navies, merchant marines) from the *output transducer* (radio broadcasters, writers, diplomats) which sends information out across system boundaries, although the matter-energy available to the society is used by both. In organisms, however, very comparable nervous and muscular apparatuses are both motor (hands, feet) and output transducer (larynx, face, hands, feet).

It is usually possible in organisms to show by anatomical and physiological studies what structures perform the subsystem functions relating to the use of energy and materials. This is not true, at least at present, for all the information-processing subsystems. Some of these, like the input transducer (all the external sense organs taken together) have been identified and studied with care. It is possible to specify such characteristics of sense organs as transducers, as their bandwidths, channel capacities, signal-to-noise ratios, distortions and response lags. Many, but not all, channels over which input and output information flows are also known. If, however, an experimenter wanted to use electrodes to study the functional characteristics of the decider, he would not know where in the nervous system to place them. The literature also makes it clear that the structures and physiological processes related to learning and memory cannot yet be definitely identified. We do have some evidence that the various "speech areas" of the cortex mediate language encoding and output transducing in human beings.

THE INDIVIDUAL ORGANISM AS A CHANNEL

Without precise knowledge about the relationship of structure to information processing in the brain, it becomes necessary to conceptualize the human organism in a communication situation as a "black box," whose inputs and outputs can be observed and measured, but not the different flows through specific components with their particular delays, distortions and breakdowns. Experimental controls can sometimes be used in a psychological research with human subjects when instructions are given so that it seems possible to study an aspect of function of a single information-processing subsystem holding constant all other subsystems. For instance, in complex reaction-time experiments one can calculate accurately the amount of time which is added to the response by a choice. Among alternate behaviors, in a situation where the muscular response time remains constant, this decision-time falls to zero as the task is better practiced and the choice becomes automatic (1). Again, Brown and Lenneberg (2) have demonstrated that encoding time is longer when subjects are naming unusual colors than when they are naming common ones.

One can look upon an individual as a single channel, as Broadbent has done (3). His information-flow diagram of the organism, which is based upon research in perception, learning, memory and other aspects of information processing, includes the sense modalities, a short-term memory store, a selective filter, a channel of limited capacity, a "store of conditional probabilities of past events," a "system for varying output until some input is secured" (an example of which is the continued appetitive behavior of an animal until it is able to reduce its drive) and effectors. The orderly progress of information processing in such a system can be delayed or disrupted by any of the components or by functional difficulties in any of their connecting links. The capacity of the short-term memory is limited. There may be retrieval problems from long-term stores. It may be necessary to establish new, learned input-output relationships. The motor task required of the subject may be difficult and so it may slow his responses. It is also possible to present the system, whether it is a cell, an organism, or any other living system, with underload (sensory deprivation) or overload of input information, both of which may lead to pathological function.

INFORMATION INPUT OVERLOAD

In our information input overload research (4), the living system is regarded as a single channel whose capacity can be exceeded. The responses to various rates of information input up to and beyond the maximum rate which the system can process, and the ways of adjusting to overloads have been studied for different levels of living systems.

Quastler and Wulff (5) have experimented upon the upper limits of a human being's capacity to process information. Their tasks were designed so that neither peripheral input nor peripheral output handling ability were limiting factors. Rates were calculated for information transmission by reading, typing, playing the piano, doing mental arithmetic, glancing at displays of letters and playing cards, scales or dials. They found that people are able to make up to five or six successful associations per second, assimilate about 15 bits at a glance and process 22 bits per second in piano playing, 24 in reading aloud and 24 in mental arithmetic.

OUR IOTA RESEARCH

We have measured the channel capacity and studied the responses of individuals to overloads by the use of especially built equipment, known as the IOTA (Information Overload Testing Aid). This apparatus consists of a ground-glass screen about 3 by 4 feet in size, upon which stimuli are presented. The subject responds by pushing buttons arrayed before him. The stimuli are thrown on the back of the screen by a projector, known as a Perceptoscope, which is capable of showing moving picture film at rates of from 1 to 24 frames per second. The program for this experiment presents black arrows in one of eight angular positions, like clock hands, on a white background. These appear in from one to four of four 2-inch wide vertical slots, made up of 12 squares superimposed on one another, which run down the screen. There is a set of eight buttons, corresponding to the eight possible positions of the arrows, for each of the slots being used. A subject can see stimuli in a maximum of four slots at once, and therefore has 32 buttons. If an arrow in position A appears in slot 2, the correct response is to push button A for slot 2, and any other response is an error. A lack of response is an omission. The subject can control the number of squares with slots which are open by using a foot pedal. At the beginning of each test, only the top square in each slot is open. He can open all 12, or close as many as he wishes. As the film is run, an arrow in position B which first appears in slot 3 in frame 1 goes to the next lower square and then to the next lower, until it has gone through

TABLE VII.1
The Critical Subsystems

Matter-energy Processing Subsystems	Subsystems Which Process Both Matter-energy and Information	Information Processing Subsystems
	Boundary	
Ingestor		Input transducer
		Internal transducer
Distributor		Channel and net
Decomposer		Decoder
Producer		Learner
Matter-energy storage		Memory
	Decider	
Excretor		Encoder
Motor		Output transducer
	Reproducer	

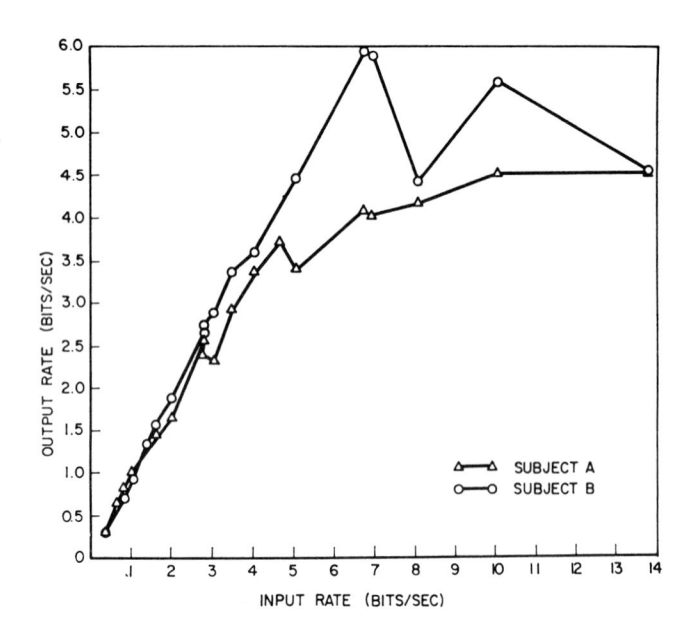

Figure VII.1. Performance curves for *Subject A* and *Subject B*.

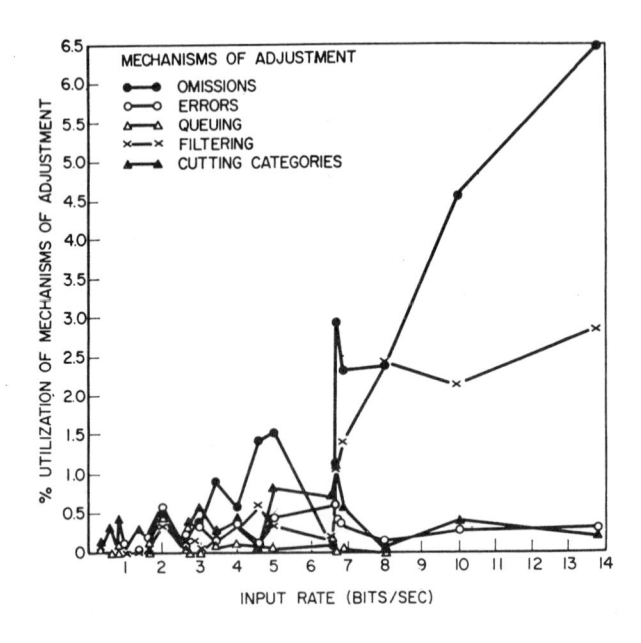

Figure VII.2. Mean utilization of adjustment processes by both subjects at various input rates.

Adjusting to Overloads of Information

all 12 positions. While this stimulus is passing through the lower squares, others appear at the top of the same slot or in other slots. By pushing the pedal, the subject gives himself more time to respond before an arrow disappears.

With this apparatus the amount of information presented can be controlled in several different ways: first, it is possible to change the number of alternate positions of the arrows from two (1 bit) to eight (3 bits); second, the movie can be run more or less rapidly; third, more or less slots can be used simultaneously; and fourth, the regularity or randomness of the presentations can be changed.

In the original study, two male college students were our subjects. A kymograph record of their button pushes was made, and this was compared with the program of arrows which had been presented to them. These data were fed into a computer which calculated the input and output rates in bits per second, using the Shannon information statistic.

The channel capacity of our subjects was about six bits per second (Figure VII.1). This is lower than maximum capacities which have been found by others in other situations, but the subjects had less training; their task was more difficult in this problem, and it has relatively high stimulus-response incompatibility.

A review of the literature had yielded a curve which we believed might be demonstrated at all levels of living systems as characteristic of response to information input overload. We also were able to identify several adjustment processes which appeared at various levels and which, while perhaps not exhaustive, seemed to be the fundamental adjustments possible to this type of stress. These included: (1) omission, which is not processing information if there is an overload; (2) error, processing incorrectly and failing to correct for it; (3) queuing, delaying responses during heavy load periods and catching up during any lulls that occur; (4) filtering, systematic omission of certain types of information, usually according to a priority scheme; (5) approximation, a less precise response given because there is no time for details; (6) multiple channels, making use of parallel subsystems if the system has them at its disposal; (7) decentralization, a special case of multiple channels; (8) escape, either leaving the situation or taking other steps which cut off the input of information.

For the purpose of our experiment, we trained the subjects to know how to make all these adjustments on the IOTA. Omissions and errors appeared without training and this fact was pointed out to the subjects. Queuing is possible because a subject can keep more squares open in a slot and see the arrows longer as they move down the openings, giving himself more time to make the required response. Of course, new stimuli keep appearing at the same time. A subject could filter with this apparatus by disregarding arrows in certain positions or paying attention to only certain of the slots. He could approximate by pushing all the buttons corresponding to right-hand positions of the arrow, if he had observed only that the arrow pointed to the right and not the exact position, or pushing all the buttons for a slot if he did not know the direction or position of the arrow but did know the slot. An ambidexterous subject could use multiple channels by working with both hands at the same time. Escape would occur if the subject refused to continue.

At slow rates of transmission our subjects used few adjustment processes. At medium rates they attempted them all. At higher rates filtering was preferred, but as the maximum channel capacity was reached, both subjects used chiefly omission (Figure VII.2).

The curve which had been hypothesized after study of the literature showed the output rising as a more-or-less linear function of input until channel capacity is reached, then leveling off and finally decreasing as breakdown or confusion occurs. Data from these subjects produced curves of the general shape we had expected when information input is plotted against information output in bits per second. However, within the range that was used there was no breakdown or confusion. We probably had not raised the input rate sufficiently. Klemmer and Muller (6), in a comparable study, did find overload (Figure VII.3).

We also tested two three-man groups with the IOTA apparatus and found they had similar performance curves, with lower channel capacities of about 3 to 4 bits per second. The groups used adjustment processes in generally similar ways. A social organization made up of three echelons of three-man groups (an air raid warning system simulation at Systems Development Corporation) gave similar results.

OUR PULSE-INTERVAL CODE RESEARCH

The IOTA apparatus conveys its information by a code based upon presentation of different symbols from an ensemble or alphabet of arrow positions rather than upon different temporal intervals between its signals. It is a code founded on the nature of its signal pulses rather than its pulse intervals. Evidence is increasing, however, that pulse-interval codes are common, if not universal, in the nervous system. If we wish to study true formal identities in information channel capacities across several levels of living systems, from cells to social institutions, therefore, we must employ pulse-interval codes rather than the type used in the IOTA apparatus. We have done this by constructing equipment which delivers pulses or signals at different intervals to systems at various levels (7).

At the level of the cell, we built an electronic stimulator which delivered two or more pulses at various brief intervals through electrodes attached to single fibers of the sciatic nerve of the frog. The interval between the input and output pulses, obtained through microelectrodes on the axons of the same fiber, was measured with an accuracy of one microsecond. This was repeated enough times to determine the variance in the latency (time between stimulus and response). This, together with the absolute refractory period or dead time, enabled us to calculate the theoretical maximum possible channel capacity of the neuron by a formula developed by Rapoport and Horvath (8). The curve obtained is the top one in Figure VII.4. The output curve rises rapidly to peak at a maximum slightly over 4000 bits per second and then declines gradually as pulse input rate increases.

A similar procedure was used at the organ level. Pulses at different intervals were delivered to the optic nerve of a rat, and the resultant evoked potentials were registered through a microelectrode on the optic cortex. The Rapoport-Horvath formula was applied to the data and, consequently, a curve of similar shape was obtained, peaking at about 55 bits per second (Figure VII.4).

Ten individual human subjects were presented with light flashes at differing intervals and were instructed to push a lever to respond to each flash. The intervals of the responses were compared with those of the flashes, and an average curve obtained from the 10 subjects' data was calculated, using the Rapoport-

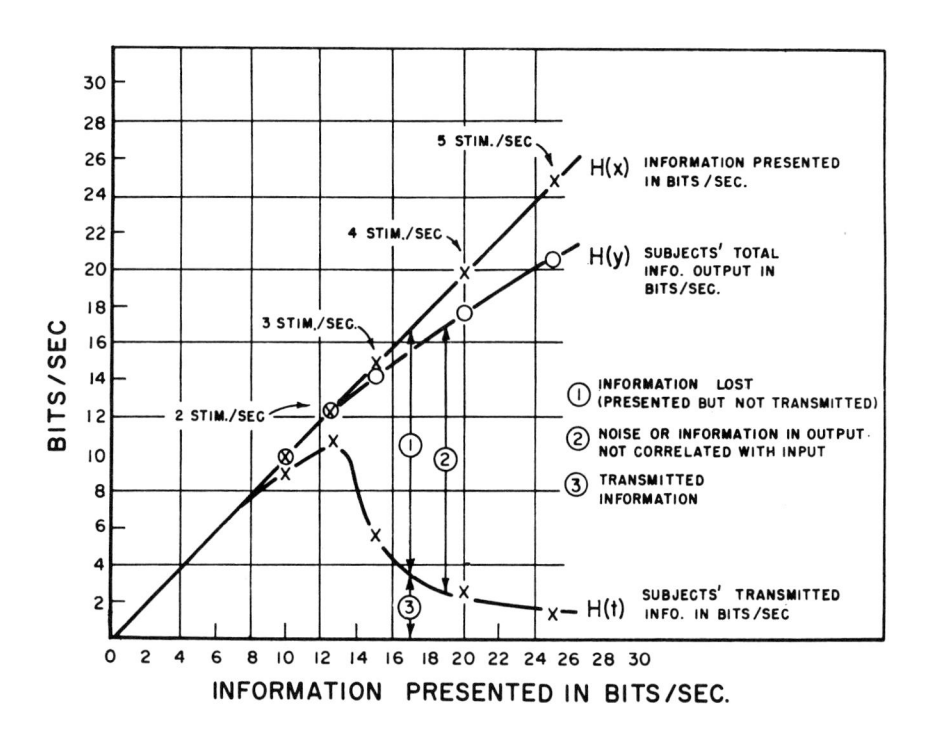

Figure VII.3. Subjects' total informational output and transmitted information as functions of input rate (from Klemmer and Muller (6)).

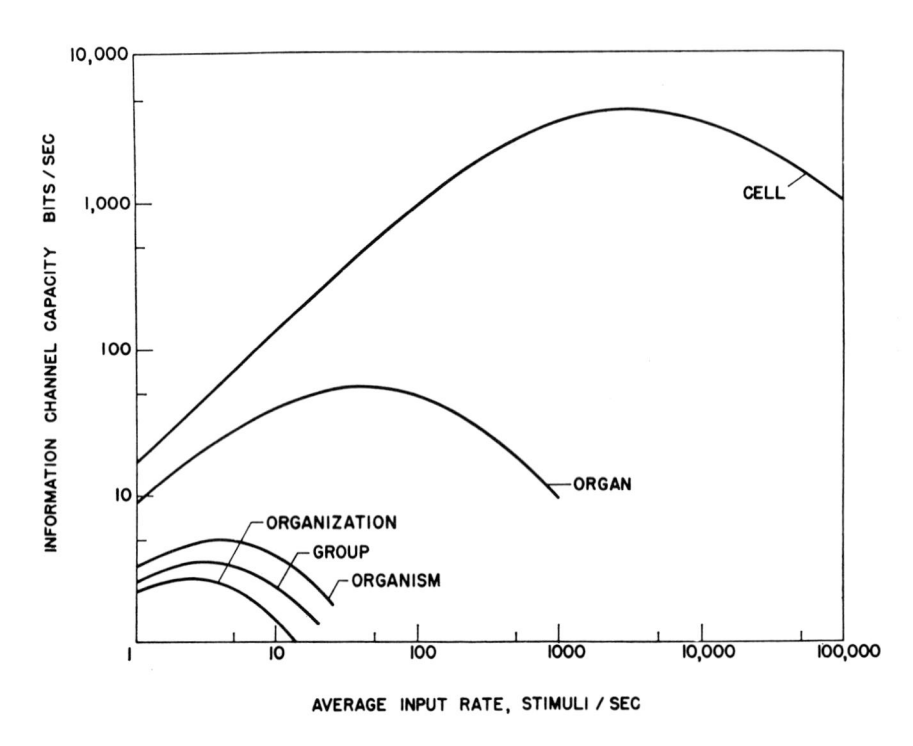

Figure VII.4. Performance curves for five levels of systems calculated from pulse-interval data.

Horvath formula. It, of course, had the same general form and peaked at a maximum theoretical channel capacity of about 5.75 bits per second (Figure VII.4).

The face-to-face group level was studied using three separate groups of three members each. In each group subject A saw the first light flash and subject B saw the second. Each pushed a lever when he saw his light. Each of these lever pushes flashed a light before subject C, who responded to each flash by pushing a similar lever, creating the system output signals. Calculations using the same formula produced a comparable output curve peaking at about 3.75 bits per second (Figure VII.4).

A small multiechelon social organization has also been subjected to a similar experiment. It was made up of a primary input echelon of two groups of two subjects each, a secondary echelon of one group of two subjects each and a tertiary, output echelon of one subject—a total of seven subjects. Each subject in the first group saw the first of a pair of two flashes with differing intervals between them and responded as rapidly as possible with his lever. Each subject in the second group of this echelon (which was physically separate from the first group) responded similarly to the second flash when he saw it. One of the two subjects in the second echelon saw the two flashes caused by the lever pushes of the subjects in the first group, and responded when he saw both. The other subject in this echelon—physically separate from the first—did the same thing for the two flashes from the second group. (This echelon responded only after seeing *two* flashes from the input echelon, and so corrected for any erroneous extra pushes of a member of the input echelon—but not for erroneous omissions.) The one subject in the third echelon pushed his lever each time he saw a flash from either member of the second echelon. His signals were the output from the organization. The same formula applied to these data yielded a similar curve peaking at about 2.5 bits per second.

Thus, a formal identity of theoretical maximum channel capacity and over-all performance curves to increasing information input rates were found from experiments on five levels of living systems—cell, organ, organism, groups and organizations. These reached maxima descending hierachically from about 4000 bits per second for the cell to about 2.5 for the multiechelon organization (Figure VII.4). Apparently the more components there are in a channel, the lower is its channel capacity. This probably is because no channel is faster than its slowest component, and the more components a channel has, the more likely it is to have a slow one. Also, some information is lost in the recoding process which must occur at the boundary of each component. These are probably explanations for findings by Roby and Lanzetta (9) that work groups which deployed their personnel so that a minimum number of communication links were required, performed their tasks best.

PSYCHOPATHOLOGY OF INFORMATION OVERLOAD

Information input underload (sensory deprivation) is believed by many to be capable of producing psychopathology (10). Overload probably does also. We have received personal communications from clinicians concerning patients who appeared to break down in their performances on the job and in psychotherapy from information input which reached them faster than they could handle it.

More objective evidence is available. Incorrect associative responses have been emphasized by clinicians as characteristic of the thought deviations of schizophrenics (11–14). Stressful conditions which can produce such thinking in normals have been reported to include brain damage, centrally active drugs, oxygen deprivation, distraction, fatigue, interruption of ongoing activity, environmental changes and emotional excitement (15). Flavell et al. (16) used a word-association test to compare differences between schizophrenic and normal responses, with differences between normals who set their own rates of response and normals under time pressure. They found the responses of the last group to be more like those of schizophrenics than the responses of normals under no time pressure.

A well-controlled study by Usdansky and Chapman (17) also is highly relevant. To 28 apparently normal college student subjects, they presented a task of associating given words with one of three alternate words. The given word was on a card which the subject was instructed to put in one of three trays, each of which had a different alternate word displayed behind it. For each new card to be sorted, a new set of three alternate words appeared behind the trays. One of these three alternatives was always a "correct" conceptual response, one was an "associative distractor" and one an irrelevant response. For instance, to the word "gold" the conceptual response was "steel," the "associative distractor" was "fish," and the irrelevant response was "typewriter." "Fish" was considered an associative distractor because of the commonly experienced contiguity of the words "gold" and "fish." Other distractors were based on commonly experienced contiguity of objects (e.g., "hat"–"head") or on rhymes (e.g., "dragon"–"wagon"). Each subject sorted 45 cards under self-paced conditions (average sorting time 220 seconds) and 45 cards under forced-paced conditions, seeing how fast they could get through them while still trying to be correct (average sorting time 114 seconds). Half the subjects did the forced-pace task first, and half did the self-paced task first. Both the irrelevant and the associative errors increased significantly under the fast, forced-paced conditions, just as errors increased in our research at rapid input rates. Associative errors, moreover, increased significantly more than irrelevant errors. Because, in a previous study, Chapman (18) had found the excess of associative errors over irrelevant errors to be much greater for schizophrenics than for normals, Usdansky and Chapman (17) concluded that increasing the speed of response (in our terms, the rate of information processing) in normals increased a kind of error characteristic of schizophrenic thought. One might speculate from such findings, and from ours reported above, that schizophrenia (by some as-yet-unknown process, perhaps a metabolic fault which increases neural "noise") lowers the capacities of channels involved in cognitive information processing. Schizophrenics consequently have difficulties in coping with information inputs at standard rates like the difficulties experienced by normals at rapid rates. As a result, schizophrenics make errors at standard rates like those made by normals under fast, forced-input rates.

SUMMARY

Living systems at several levels process energy and information through a number of critical subsystems essential for their survival. Characteristic functions of these critical subsystems are discussed. The total human being may be viewed as an information-processing channel with a maximum channel capacity. As information input rates to such channels increase, output rates increase, then level off at the channel capacity and then decrease. Adjustment processes are employed at the higher input rates which serve to maintain output as information arrives even faster. Two sorts of research are reported, measuring individual human information processing performances under various input rates. In one sort the information is coded in discrete symbols; in the other it is coded in terms of the durations of interpulse intervals. These studies demonstrate how performance peaks and then declines and how adjustment processes are used as inputs to single human subjects get faster. Similarities in information processing characteristics of cells, organs, organisms, groups and multiechelon organizations are investigated in these researches. Their possible relevance to psychopathology, including schizophrenia, is discussed.

REFERENCES

1. Mowbray, G. H., & Rhoades, M. V. On the reduction of choice reaction times with practice. *Quart. J. Exper. Psychol.*, 11, 16-22, 1959.
2. Brown, R., & Lenneberg, E. A study in language and cognition. *J. Abnorm. & Social. Psychol.*, 49, 454-462, 1954.
3. Broadbent, D. E. Perception and Communication, pp. 297-301. Pergamon Press, Inc., New York, 1958.
4. Miller, J. G. Information input overload and psychopathology. *Amer. J. Psychiat.*, 116, 695-704, 1960.
   Miller, J. G. The individual as an information processing system. In Information Storage and Neural Control, by Fields, W. S. and Abbott, W., pp. 1-28. Charles C Thomas, Publisher, Springfield, Ill., 1963.
5. Quastler, H., & Wulff, V. J. Human Performance in Information Transmission. Report R-62, Control Systems Laboratory, University of Illinois, Urbana, 1955.
6. Klemmer, E. T., & Muller, P. T., Jr. The rate of handling information: key pressing responses to light patterns. H.F.O.R.L. Memo Report No. 34, March, 1953.
7. Miller, J. G. Living systems and mechanical systems. Preprint 112. Mental Health Research Institute, University of Michigan, Ann Arbor, Michigan, 1963.
8. Rapoport, A., & Horvath, W. J. The theoretical channel capacity of a single neuron as determined by various coding systems. Information and control, 3, 335, 1960.
9. Roby, T. B., & Lanzetta, J. T. Work group structure, communication, and group performance. *Sociometry*, 19, 105, 1956.
10. Lilly, J. In Illustrative Strategies for Research on Psychopathology in Mental Health, pp. 13-20. Group for the Advancement of Psychiatry, Symposium No. 2, June, 1956.
11. Kraepelin, A. Dementia Praecox and Paraphrenia. E. & S. Livingstone, Ltd., Edinburgh, 1919.
12. Jung, C. G. The psychology of dementia praecox. Nerv. & Ment. Dis. Monogr. Series, No. 3, 1936.
13. Bleuler, E. Dementia Praecox or the Group of Schizophrenias. International Universities Press, Inc., New York, 1950.
14. Arieti, S. Interpretation of Schizophrenia. Brunner, New York, 1955.
15. Cameron, N. A., & Margaret A. Behavior Pathology. Riverside Editions, Cambridge, Mass., 1951.
16. Flavell, J. H., Draguns, J., Feinberg, L. D., & Budin, W. A microgenetic approach to word association. *J. Abnorm. & Social. Psychol.*, 57, 1-8, 1958.
17. Usdansky, G., & Chapman, L. J., Schizophrenic-like responses in normal subjects under time pressure. *J. Abnorm. & Social. Psychol.*, 60, 143-146, 1960.
18. Chapman, L. J. Intrusion of associative responses into schizophrenic conceptual performance. *J. Abnorm. & Social. Psychol.*, 56, 374-379, 1958.

51

On the Concept of Psychological Stress

Mortimer H. Appley and Richard Trumbull

The concept of *stress* was first introduced into the life sciences by endocrinologist Hans Selye in 1936 and elaborated in successive papers, leading to a full theoretical statement in book form in 1950. The wide appeal of the concept was evidenced by the fact that the literature on its primarily physiological aspects alone was close to six thousand publications per annum by the early 1950's, when Selye began his *Annual report of stress series* (see Selye, 1951 on). The use of the term in psychological research had an accelerated growth curve following Selye's invited address to the American Psychological Association in 1955. Initially, as Harris et al. (1956) have suggested, this interest developed "because of the importance of physiological variables as independent measurable indicators of a stressed organism" (p. 3). However, the use of the concept has spread through many facets of psychology (cf. Appley, 1957a, b; Cofer & Appley, 1964, pp. 441–465) and has been applied even where no physiological or endocrine factors were subject to study.

There are at least three reasons for the apparent popularity of stress as a psychological concept. The first might be called a bandwagon effect. Since the term gained some attention, and apparently some status, as a research topic, it has been used as a substitute for what might otherwise have been called anxiety, conflict, emotional distress, extreme environmental conditions, ego-threat, frustration, threat to security, tension, arousal, or by some other previously respectable terms. Secondly, because of its wide use in the biological field, the use of the term suggested both apparent and real possibility of correlating psychological events with physiological substrata, a prestigious and hopeful pre-occupation of psychologists these days. Thirdly, of course, is the genuine interest in stress phenomena, stimulated in part by concern with the effects of the unusual environments in which men are being placed these days in military and space operations, and in part by the exciting possibilities of real links being established between areas of clinical, psychosomatic, and various types of traditional experimental research.

From: *Psychological Stress*, Edited by Mortimer H. Appley and Richard Trumbull. Copyright © 1967 by Meredith Publishing Company. Reprinted by permission of Appleton-Century-Crofts.

This chapter combines and extends the introductory remarks made by the two authors to the Opening Session of the Conference on Psychological Stress, May 10, 1965.

That the use of common language might lead to the establishment of relationships is surely an advantage. That the use of common terms in different ways may lead to confusion, however, is quite apparent to anyone studying or confronted by the burgeoning stress literature.

The present volume provides an opportunity to examine the concept of stress across a wide spectrum of situations, levels, and views. Its ultimate usefulness as a psychological concept will depend upon the adequacy with which differentiations can be made among stimulus-, organismic-, and response-elements of stress situations, and whether stress researchers can avoid the inviting trap into which many personologists, for example, have fallen, namely that of treating their subject as though it were a unitary, all-or-none phenomenon. The papers and discussions which compose this volume will make clear some of the difficulties which are unique to the field of stress research as well as show how some of these problems are common to psychological investigation and scientific research generally.

SYSTEMIC STRESS

To begin at the beginning one should look at Selye's original conception of systemic stress. In arriving at this concept, Selye made much of the point that although different disease syndromes have unique attributes and symptoms, they have features in common, and it is those features that are common among them that constitute stress. He wrote:

> . . . if we abstract from these specific reactions, there remains a common residual response that is non-specific as regards its cause and can be elicited with such diverse agents as cold, heat, X-rays, adrenalin, insulin, tubercle bacilli, or muscular exercise. This is so despite the essentially different nature of the evocative agents themselves and despite the coexistence of highly specific adaptive reactions to any one of these agents. (1959, pp. 406–407).
>
> . . . the stereotypical response, which is superimposed upon all specific effects, represents the somatic manifestations of non-specific "stress" itself. (1953, p. 18).

And in further elaboration in another context, Selye explains:

> Among other things . . . stress is not necessarily the result of damage but can be caused by physiologic function and . . . it is not merely the result of a non-specific action but also comprises the defenses against it. (1955, p. 626)

Selye found antecedents for his work in the concept of Hippocrates that disease not only includes suffering (pathos) but a reaction of the body seeking restoration through toil (panos) as well. Claude Bernard's (1859) description of and evidence for the necessity for maintenance of the "constancy of the internal milieux" was an acknowledged base for Selye (1956), as well as for Cannon (1932) in the development of his concept of homeostasis. This general notion of systemic equilibration had still another early advocate in Herrick, who wrote:

The material and energy of the living body are in constant flux, yet the pattern of their manifestation persists. When this pattern is deformed by external violence or by changes of internal state it is said in current biological descriptions that there is a tedency to return to the typical condition. This restoration of the original pattern after deformation is termed regulation. . . . In view of the fact that the ordinary conditions of life involve constant changes in the relations of the body to its environment, regulation in the broad sense means simply the continuous readjustment of the organism to the flux of surrounding conditions. (1924, pp. 279–280)

It would be impossible to do justice to Selye's concept of stress (or to its precursors) in a few paragraphs here. His own work on this subject has extended over thirty years and resulted in a thousand publications (Institut de Médecine et de Chirurgie Expérimentales, 1964). But as background for the papers which follow, we shall try to indicate a few of the salient emphases in his development of the stress concept.

Systemic stress is manifested by a *General Adaptation Syndrome* (GAS). The first stage of this syndrome, or the *alarm reaction,* includes an initial *shock phase* (in which resistance is lowered) and a *countershock phase* (in which defensive mechanisms become active). A second *stage of resistance* follows, during which maximum adaptation occurs. Should the stressor persist, however—or the defensive reaction prove ineffective—a *stage of exhaustion* is reached in which adaptive mechanisms collapse.

The *alarm reaction* is typically characterized by autonomic excitability, adrenaline discharge, heart rate, muscle tone and blood content changes, and gastrointestinal ulceration. Adrenocortical enlargement and hyperactivity are ordinarily observed in the countershock phase.[1] Depending on the nature and intensity of the stressor and the condition of the organism at the time of exposure, the periods of resistance may be foreshortened or prolonged and the severity of symptoms may vary from mild invigoration to what Selye has called the "diseases of adaptation."[2]

Throughout his extensive writings, Selye emphasizes the fact that a variety of circumstances gives rise to a highly stereotyped bodily reaction, as well as reactions peculiar or specific to the nature of the insulting agent. It is the general (or common) rather than the specific reaction of the organism which constitutes stress as we have earlier noted. Both the events in the environment which induce stress and certain concomitant and resultant responses may be quite varied. The *systemic stress* response, however, is invariant.

This pattern might be described in terms of an hourglass model, feeding from a wide source of stimuli through a narrow *common* element to a spectrum of re-

---

[1] For a more detailed description of characteristic symptoms see Selye (1950, 1952, 1959). General schemata of the stress syndrome have been offered elsewhere by Appley (1961); Cofer and Appley (1964).

[2] "*Diseases of Adaptation* are those in which imperfections of the GAS play the major role. Many diseases are actually not the direct result of some external agent but rather the consequences of the body's inability to meet these agents by adequate adaptive reactions. Maladaptation plays a major role in diseases of the heart and blood vessels, diseases of the kidney, eclampsia, rheumatism and rheumatoid arthritis; inflammatory diseases of the skin and eyes, infections, allergy and hypersensitivity, nervous and mental diseases, sexual derangements, digestive and metabolic diseases, and cancer" (Institut de Médecine et de Chirurgie Expérimentales, 1964, p. 23).

sponses. Although psychologists do not usually measure changes in terms of adrenal weight or stomach ulcers, preferring such indices as the galvanic skin response (GSR) or changes in heart rate, they nevertheless have assumed and sought to identify *common* elements of body change to index the stress response. In other words, something like this hourglass model underlies much of the prolific research which is carried on by psychologists on the physiological changes accompanying emotion. This belief in a *common* state in aroused organisms follows from a conviction of psychosomatic unity and from a desire to have at least one stable anchor in their descriptive system. It seems most likely, however, as research findings accumulate, that such a conception will turn out to be an oversimplification of the facts. Individual differences, styles, patterns of response, and prepotent tendencies appear to be the rule rather than the exception in studies of psychological stress. As laboratories in universities become more affluent and can afford to add channels to their psychophysiological recording equipment, as measurement techniques become more sophisticated, and as computers permit more exacting and exhaustive data analysis, we may find a common response pattern. It seems more likely, however, that the study of idiosyncratic psychobiological patterns will emerge as a major area of stress research.

Turning now to psychological stress more specifically, and looking at how people have dealt with the concept, we see that it is in fact an inference from either manipulations of the environment (external and/or internal) or from measurements of change in response (internal and/or external).

PSYCHOLOGICAL STRESS

1. On the *stimulus* side, the term has been used to describe situations characterized as new, intense, rapidly changing, sudden or unexpected, including (but not requiring) approach to the upper thresholds of tolerability. At the same time, stimulus deficit, absence of expected stimulation, highly persistent stimulation, and fatigue-producing and boredom-producing settings, among others, have also been described as stressful, as have stimuli leading to cognitive misperception, stimuli susceptible to hallucination, and stimuli calling for conflicting responses. Any one of these procedures has at some time actually been used as an operational means for defining and producing stress.

2. On the *response* side, the presence of emotional activity has been used *post facto* to define the existence of stress. This usually refers to any bodily response in excess of "normal or usual"—states of anxiety, tension, and upset—or for that matter any behavior which deviates momentarily or over time from normative value for the individual in question or for an appropriate reference group. Indices used include such overt emotional responses as tremors, stuttering, exaggerated speech characteristics, and loss of sphincter control—or such performance shifts as perseverative behaviors, increased reaction time, erratic performance rates, malcoordination, error increase, and fatigue.

3. The existence of a stress state *within* the organism has alternatively been inferred from one or more of a number of partially correlated indices, such as a change in blood eosinophils, an increase in 17-ketosteroids in the urine, an increase in ACTH-content or gluco-corticoid concentration in the blood, or changes in any number of psychophysiological variables, such as heart rate, galvanic skin

response (GSR), change in critical flicker fusion (CFF) threshold, inspiration: expiration $\left(\frac{I}{E}\right)$ ratio, and so on. (These are response measures, of course, but they can be usefully distinguished from responses which are of the order of overt performance changes or observable symptoms of emotionality, such as those noted above.)

The Problem of Definition

Investigators have usually sought both a condition which produces stress and a measure which indicates its presence as the most frequent combination of circumstances in which to study the phenomenon. Unfortunately, the choices have been selective—as often governed by the convenience or tradition of a given laboratory as by rational considerations. There are, then, clearly wide variations in specific uses, specific definitions, and specific purposes with which the term *stress* has been associated. However, one is reminded of Whitehorn's comments:

> We may be able to get some use out of the term stress, even if it is left vague and not very clearly defined, provided we succeed in specifying fairly sharply some of the aspects of the biological reactions to stress.
> If we were dealing with inanimate objects, the conceptual and terminological problem would be greatly simplified, because in physics action and reaction are equal, and stress can be expressed in dynes per square centimeter; but in biology this is not so. Living organisms are specially organized to accumulate and spend energy on their own discriminately, and not in exact equality to the forces acting upon them. We take this one step further and recognize the psychological factors which further influence this discriminating function and appreciate that our difficulties of description and evaluation have been geometrically expanded. (1953, p. 3)

Without a doubt, this geometrical expansion has served as a deterrent to exploring the psychological complex which produces the inequality of response to the forces acting.

Let us look next at the pattern of stress experiments. Typically, the experimenter manipulates the environment in a manner intended to produce a response, and then measures the extent and/or direction of the behavior change produced. (This is, of course, the pattern of all psychological experiments. Stress studies are usually distinguishable primarily in the selection of stimulating conditions).

Experimenters in these studies choose environmental manipulations which would, in their consideration, serve to produce not just change in the direction of ongoing behavior—which after all is what any response must be, and would not distinguish stress studies—but a disruption of behavior, or its disorganization. Accompanying such disorganization one expects to find certain physiological changes, and it is here that much of this research is concentrated. In fact, the most widely accepted types of operational definitions of the existence of stress are changes in physiological indices. Unfortunately, one investigator relies on the GSR, a second on blood volume changes, a third on pulse rate or heart rate, a fourth on muscle action potential, and so on. In animals, the presence of stress is often inferred from feces counts, trembling, "freezing" or washing behavior, and so on. The use of these measures rests on the simple assumption that certain environmental conditions induce not only overt behavioral effects but common autonomic and other internal effects as well.

We would not so much disagree with the logic of this argument as with the facts it has produced. In the studies with which we have acquaintance there are marked individual differences, as has already been noted. All subjects apparently do not respond to given environmental conditions as a given experimenter expected or intended. We may either conclude that the conditions were not stress-producing or, if we insist that they were, we must face the problem of explaining why some subjects were thereby *not* put into a stress state. We surely needn't labor this point which could be made in connection with most psychological studies. It is of particular importance here, however, because of what we have described as the assumption of the commonality of intervening responses in stress situations. Further, the dependence on the stimulus to define a situation as stressful is obviously too limiting.

If we accept that a physiological index is a proper monitor of the presence of stress, which may be reasonable in some respects, we must nevertheless parallel our studies of these responses with investigations of those conditions which precipitate stress in the individuals under study. It is clear that if we cannot rely upon the stimulus we must look for some pattern of stimulus-organism interaction to understand why stress occurs in some exposed organisms and not in others.

With the exception of extreme and sudden life-threatening situations, it is reasonable to say that no stimulus is a stressor to all individuals exposed to it.[3] The earlier assumption of a common all-or-none psychophysiological stress state is untenable in the face of evidence to the contrary. James Miller and his associates (1953), after surveying the available stress-sensitive tests more than a dozen years ago, concluded that "in a specific situation it becomes necessary to recognize the many different kinds of stress" (p. A-4). Lazarus, Deese, and Osler (1952), after reviewing the literature on the effects of stress on performance, concluded that these effects are not general, but "will depend upon what the individual expects or demands of himself" (p. 296). In a more recent paper, Lazarus (1964) suggested what he called "cognitive appraisal" as a mediating condition for such determination, and Appley (1962) has placed emphasis on the importance of "threat perception" as a mediator of stressfulness. A similar point was made earlier by Pascal (1951) in defining stress 'in terms of a perceived environmental situation which threatens the gratification of needs . . .' (p. 177).

Basowitz, Persky, Korchin, and Grinker (1955), in an elaborate study of anxiety and stress in paratroopers, started with a situational definition but concluded that "in future research . . . we should not consider stress as *imposed* upon the organism, but as its *response* to internal or external processes which reach those threshold levels that strain its physiological and psychological integrative capacities close to or beyond their limits" (pp. 288–289).

Cofer and Appley (1964) defined stress as "the state of an organism where he perceives that his well-being (or integrity) is endangered and that he must divert all his energies to its protection" (p. 453). In all of these instances we see an emphasis on individual determination of when stress will or will not occur.

---

[3] An interesting contrast can be made within the category of life-endangering situations. Attempted strangulation, for example, may be a psychological stressor of great import, whereas significant levels of irradiation (which may be near fatal but are not "detected" by the organism) may have no discernible effect on behavior at all.

It is further evident from the definitions cited that another area in which separation of psychological from physical aspects is required is that of "threat." In reports of studies involving physical "threat" we sometimes do not know and cannot determine if the "threat" is really "perceived" by the subject as such. Often the experimenter merely assumes that the situation *should have been* threatening or would have been threatening had he been the subject. The point to be made is that the first necessary step in such studies is to determine how the subject perceives the stimulus or situation presented. We know many ways to change deliberately his perception of the situation, but seem often to ignore other subtle forces which may be present and acting to change the value of an "objective" stressor. The extreme stimuli which some experimenters have used—such as electric shock or pistols fired close to the ear—indicate their awareness of this point. However, the very use of such strong stimuli probably obscures the influence of intervening perceptual factors, the understanding of which is so important. Obviously, it would be unreasonable to insist that the experimenter know what constitutes a stressor for his subject before beginning an investigation, when often that is to be its end point. However, the implications of this are found in Haggard's (1949) discussion of emotional stress:

> An individual experiences emotional stress when his over-all adjustment is threatened, when his adaptive mechanisms are severely taxed and tend to collapse. Some of the factors which influence an individual's ability to tolerate and master stress include: the nature of his early identifications and his present character structure, and their relation to the demands and gratifications of the present stress-producing situation; the nature of his reactions to the situation; his ability to master strong and disturbing emotional tensions; the extent to which he knows about all aspects of the situation, so that he is not helplessly unaware of the nature and source of threat; his available skills and other means of dealing effectively with it; and the strength and pattern of his motivations to do so. (p. 458)

It would take months if not years to know a subject well enough to meet these demands. However, this does delineate those interests which are primarily psychological in nature, and we will briefly note a few studies which emphasize them. Lazarus et al. (1952) summarize their extensive review in the following terms:

> Very little information has been obtained about the relationship between various measures of personality and reaction to stress. The problem has theoretical as well as practical importance. On the one hand, while great individual differences in response to stress have been recognized, few fruitful attempts have been made to discover their nature. On the other hand, it would be most useful to be able to predict which people will be adversely affected by a stressful situation. (p. 307)

> Essentially, this boils down to a consideration of interactions between persons and types of stress. It would be interesting to know what kind of individual develops anxiety reactions to task-induced stress. We might guess that such people are highly motivated to perform well. The successful understanding of any individual's performance under stress depends upon some way of measuring the kinds and strength of his motivations and relating them to the characteristics of the situation in which he must perform. The fulfillment of his aim is, indeed, no simple affair. (p. 314)

If measurement of performance is the criterion, there would have to be decrement but one would never be certain as to why. The teasing out of truly psychological factors would appear to be the major contribution to come from psychologists.

Once again, literature reviews show us how pursuit of these many psychological factors without system has produced little or nothing.

> Some investigators have studied the effects of subject variables upon performance under short-term stress situations. Stopol (1954) tested twelve hypotheses involving Rorschach responses . . . and found no relationship between such responses and performance under stress. Lofchie (1955), however, found that subjects who scored high on a Rorschach index of perceptual maturity performed better on a psychomotor task under distraction stress than did subjects who scored low on the index. Katchmar (1953) selected anxious and non-anxious subjects by the Taylor Manifest Anxiety Scale. Under failure stress conditions, the anxious subjects did worse than the non-anxious subjects on a form-naming and substitution task, although the performance of both groups showed a decrement over control conditions. Hutt (1947) found that maladjusted children did better than well-adjusted children on the Stanford-Binet Test under failure stress conditions. Further studies on the personality correlates of behavior under stress are cited by Lazarus et al. (1952). (Harris, Mackie, and Wilson, 1956, p. 34)

These definitions and comments generally suggest that stress is a response state and that its induction depends on the mediation of some appraising, perceiving, or interpreting mechanism. As was suggested earlier, certain universally adequate stimuli may be expected to lead to stress more rapidly than others—as, for example, cutting off the air supply. This should lead to a stress state in *all* persons, with little variation in the rate of its development. However, any less severe stimulation—and particularly where the effectiveness of the stimulation is dependent on prior conditioning (as in the case of social stimuli)—will give rise to response patterns that vary greatly from person to person and may induce anxiety or stress much more rapidly in one person than another.

What must be taken ito account is not only the objective reality of any given situation, as perceived by an independent observer, but the series of subtle, subjective equations comprising the individual's own assessments of possible success or failure in motive satisfaction. Simultaneous equations must be solved for the multiple motives, multiple response modes (as these are evaluated in terms of situational feasibility), and for motive-mode interactions. We must also recognize *time* as an important additional factor, altering both absolute and relative strengths of motives and the efficacy of different response modalities. We have not even mentioned the to-be-expected effects of stress, as these feed back into the subjective equations which modify the thresholds for threat perception from moment to moment.

Stress and Individual Vulnerability

We have emphasized the personal equation in assessing reactions to stress. It is consistently found that these reactions vary in intensity from person to person under exposure to a same environmental event. (This has been shown when the conditions studied were combat, oppressive leadership, internment, threat to life, threat to status, threat to livelihood, and others.) It has also been noted that, with few extreme exceptions, the *kind* of situation which arouses a stress response in a particular individual must be related to significant events in that person's life. Many people have used the terms "ego-strength," "stress-tolerance," and "frustration-tolerance." It is perhaps doubtful that there is such a thing as a general stress-tolerance in people. There is more likely to be a greater or lesser insulation from the effects of certain kinds of stress-producers rather than others. The common idea of a threshold of tolerance for stress implies that stress-producing agents must reach a given strength in order to arouse this response. It seems more likely that there are differing thresholds, depending on the kinds of threats that are encountered and that individuals would be differentially vulnerable to different types of stressors. In other words, *not only must a situation be of a given intensity to lead to stress, it must also be of a given kind for a particular person.* To know what conditions of the environment are likely to be effective for the particular person, the motivational structure and prior history of that individual would have to be taken into account. Where the particular motives are known—where it is known what a person holds important and not important, what kinds of goals he will work for and why, what kinds of situations have for him been likely to increase anxiety or lead to aversive or defensive behavior,—a reasonable prediction of stress proneness might be made. It is clear that what we would then have is a *vulnerability profile*, perhaps analogous to the industrial psychologist's job-profile, but based on strengths of motives and of motive-satisfying possibilities in situations rather than on strengths of skills and of skill requirements in particular jobs.[4]

If one tries to gain some overall perspective on what stress studies have so far revealed, and especially on their relation to studies of frustration, conflict, and anxiety, one is led to these kinds of general observations.

1. Stress is probably best conceived as a state of the total organism under extenuating circumstances rather than as an event in the environment.

2. A great variety of different environmental conditions is capable of producing a stress state.

3. Different individuals respond to the same conditions in different ways. Some enter rapidly into a stress state, others show increased alertness and apparently improved performance, and still others appear to be "immune" to the stress-producing qualities of the environmental conditions.

4. The same individual may enter into a stress state in response to one presumably stressful condition and not to another.

5. Consistent *intra*-individual but varied *inter*-individual psychobiological response patterns occur in stress situations. The notion of a *common* stress reaction needs to be reassessed.

6. The behaviors resulting from operations intended to induce stress may be the same or different, depending on the context of the situation of its induction.

7. The intensity and the extent of the stress state, and the associated behaviors, may not be readily predicted from a knowledge of the stimulus conditions alone, but require an analysis of underlying motivational patterns and of the context in which the stressor is applied.

[4] See Cofer and Appley (1964, pp. 449-465) for related discussion for this and subsequent section.

8. Temporal factors may determine the significance of a given stressor and thus the intensity and extent of the stress state and the optimum measurement of effect.

Extra-Individual Factors

In addition to emphasizing the role of the individual, which we have so far done, attention must also be given to social factors—the influence of other individuals or of the social and cultural milieu in producing or reducing stress. The pace setter, the previously habituated, acclimated, or trained companion, or the equally naive sharer of a stressful situation may significantly influence stress reaction. The role of social facilitation, leadership, and that nebulous influence called "social motivation," as related to perception and actual experiencing of physical stressors, further extends the continuum of variables demanding attention. The social anthropologist will remind us, too, of cultural norms in such stressors as pain—as experienced in puberty and other rituals—and of role expectations and the effects of discrepancies between prescribed and attained roles as factors in stress proneness.

It would be an error to conclude that the concern which has been shown for psychological, social, and cultural variables precludes consideration of the impact of physical and physiological factors in stress research. On the contrary, considerabe interest has existed in the study of effects of extreme environmental factors on the organism and on performance effectiveness in particular (cf. Trumbull, 1965; Burns, Chambers, & Hendler, 1963; and Weybrew, 1967). One of the particular purposes of the symposium on which this volume is based was to provide an opportunity to examine side-by-side some of the physical, physiological, social, and cultural aspects of stress. It was hoped thereby to open the way to further and more sophisticated studies of stress in which interactions of variables across levels as well as within levels would be involved. The focus of this volume has been at the psychological level by design, but the range of concern has been kept deliberately broad.

As the above discussion suggests, stress is a concept which can have interpretation and relevance at many levels of human organization, from cellular to cultural. The starting point here has been with Selye's concept of systemic stress. It could have been with influences on cell division, cancer development, or nucleic acid production. All of these levels have environments which are compatible with their normal function, and all are susceptible to environmental factors which disrupt normal functioning. Such disrupting influences are usually referred to as stressors, but the effects of incompatibility need not always be assumed to be destructive. Environmental changes can lead to extensions of the range of tolerance so that adaptation (or natural selection) takes place. Thus the absolute levels of stressors should not be expected to be constant over time (or at different time periods) within systems.

Further, environmental change seldom produces uni-dimensional stressors, and the interaction of one stressor with others, sequentially or concomitantly, must be understood (cf. Trumbull, 1965, on cross-adaptation). In some instances effects will be additive, in other instances cancelling. While the study of situations involving single stressors is undoubtedly simpler, extrapolation in anticipation of complex settings is tenuous.

Periodicity and Stress

Finally, note should be made of the fact that environmental context not only influences stressor levels but exerts a more subtle influence upon periodicity of function. Recent interest in the influence of temporal factors on psychological stress derives in part, at least, from the pioneering physiological studies of Halberg (cf. 1961, 1962a).

As is well known, but perhaps not appropriately appreciated, there are definite cycles in the various functions of the human system. The more general sleep-waking cycle has enjoyed much recognition in North America since the work of Kleitman (1939, 1963). Unfortunately, not enough attention has been paid to this variable as it relates to studies of vigilance, theories of expectancy, sleep-loss influence on functions, and other attempts at assaying human performance. Indeed, one finds little or no reference in such research reports as to time of day, whether after meals, etc. Maybe one can argue that the impact of these factors might be so small as to merit no further consideration. We submit that we do not know but that unless future research in these areas reports such factors we cannot assess their significance. The time of day one is tested after a period of sleep loss should be correlated with the position it had in his pretest cycle, for example. This is of particular importance in stress research. Recognizing such cycles in breathing rate, basal metabolism (BMR), alpha rhythms, blood pressure, and urine volume, should lead those using such stressors as air mixtures, atmospheric pressure, drugs, and related variables to concern themselves with the coincidence of stressor initiation and the present or anticipated position or level of function of the organ of primary influence.

Attention has here been directed to the probably significant influence of such cycles because we so often find in research that things which seemingly enjoy simple relationships do so only because of our own simple conceptions of them. Growing sophistication in technique then parallels our own growing appreciation of the multiple factors involved. This is the geometric expansion to which Whitehorn referred. The chapters which follow will serve to extend all of our "natural" or "usual" worlds, and we will come to appreciate the scope and complexity that our experimental models will have to encompass for a fuller understanding of psychological stress.

REFERENCES

Appley, M. H. (Ed.), 1957a. *Psychological stress and related concepts: A bibliography.* Proj. NR 172-288, Contr. Nonr 996(02), New London, Conn.: Connecticut College.
Appley, M. H., 1957b. Psychological stress. (Appendix II of *A study of operational safety requirements in the submarine polaris missile system*). Groton, Conn.: Electric Boat Div., General Dynamics Corp. (February), dittoed.
Appley, M. H., 1961. Neuroendocrine aspects of stress. In B. E. Flaherty (Ed.), *Psychophysiological aspects of space flight.* New York: Columbia, pp. 139-157.
Appley, M. H., 1962. Motivation, threat perception and the induction of psychological stress. *Proc. Sixteenth Internat. Congr. Psychol., Bonn, 1960.* Amsterdam: North Holland Publ., pp. 880-881 (Abstract).
Basowitz, H., Persky, H., Korchin, S. J., & Grinker, R. R., 1955. *Anxiety and stress.* New York: McGraw-Hill.

Bernard, C., 1859. *Leçons sur les proprietes physiologiques et les alterations pathologiques des liquides de l'organisme.* Vols. I and II. Paris: Balliere.
Burns, N. M., Chambers, R. M., & Hendler, E. (Eds.), 1963. *Unusual environments and human behavior.* New York: Collier-Macmillan.
Cannon, W. B., 1932. *The wisdom of the body.* (2nd ed.) New York: Norton.
Cofer, C. N., & Appley, M. H., 1964. *Motivation: Theory and research.* New York: Wiley.
Haggard, E. A., 1949. Psychological causes and results of stress. In D. B. Lindsley et al., *Human factors in undersea warfare.* Washington, D.C.: National Research Council, pp. 441-461.
Halberg, F., 1961. Circadian rhythms: A basis of human engineering for aerospace. In B. E. Flaherty (Ed.), *Psychophysiological aspects of space flight.* New York: Columbia, pp. 166-194.
Halberg, F., 1962a. Physiological 24-hour rhythms: A determinant of response to environmental agents. In K. E. Schaefer (Ed.), *International symposium on submarine and space medicine, Vol. 1: Man's dependence on the earthly atmosphere.* New York: Macmillan, pp. 48-96.
Harris, W., Mackie, R. R., & Wilson, C. L., 1956. *Performance under stress: A review and critique of recent studies.* Tech. Rep. VI, Los Angeles, Calif.: Human Factors Research Corp. (July), (ASTIA AD No. 103779).
Herrick, C. G., 1924. *Neurological foundations of animal behavior.* New York: Holt.
Hutt, M. L., 1947. A clinical study of "consecutive" and "adaptive" testing with the revised Stanford-Binet. *J. consult. Psychol.,* 11, 93-103.
Institut de Médecine et de Chirurgie Expérimentales, 1964. Montreal: Université de Montréal, Faculté de Médecine.
Katchmar, L., 1953. *Indicators of behavior decrement: 22. The effects of stress, anxiety, and ego involvement on "shift" task performance.* Proj. DA-49-007-MD-222, Tech. Rep. 22. College Park, Md.: University of Maryland, Army Med. Res. & Devel. Bd.
Kleitman, N., 1939; rev. and enlarged, 1963. *Sleep and wakefulness.* Chicago: University of Chicago.
Lazarus, R. S. A laboratory approach to the dynamics of psychological stress. *Amer. Psychol.,* 1964, 19, 400-411.
Lazarus, R. S., Deese, J., & Osler, Sonia F., 1952. The effects of psychological stress upon performance. *Psychol. Bull.,* 49, 293-317.
Lofchie, S. H., 1955. The performance of adults under distraction stress: A developmental approach. *J. Psychol.,* 39, 109-116.
Miller, J. G., Bouthilet, Lorraine, & Eldridge, Carmen, 1953. *A bibliography for the development of experimental stress-sensitive tests for predicting performance in military tasks.* PRB Tech. Rep. 1079, Res. Note 22, Washington, D.C.: Psychological Research Associates.
Pascal, G. R., 1951. Psychological deficit as a function of stress and constitution. *J. Personal.,* 20, 175-187.
Selye, H., 1951-1956. *Annual report on stress.* Montreal: Acta, Inc., 1951, Selye, H., & Horava, A., 1952, 1953, Selye, H., & Heuser, G., 1954. M. D. Publicat. (New York), 1955-1956.
Selye, H., 1952. *The story of the adaptation syndrome.* Montreal: Acta, Inc.
Selye, H., 1955. Stress and disease. *Science,* 122, 625-631.
Selye, H., 1956. *The stress of life.* New York: McGraw-Hill.
Selye, H., 1959. Perspectives in stress research. *Perspect. Biol. Med.,* 2, 403-416.
Stopol, M. S., 1954. The consistency of stress tolerance. *J. Personal.,* 23, 13-29.
Trumbull, R., 1965. *Environment modification for human performance.* Washington, D.C.: Office of Naval Research, Rep. ACR-105.
Weybrew, B. B., 1967. Patterns of psychophysiological response to military stress. In M. H. Appley and R. Trumbull (Eds.), *Psychological Stress.* New York: Appleton-Century-Crofts, pp. 324-354.
Whitehorn, J. C., 1953. Introduction and survey of the problems of stress. In Army Medical Service Graduate School (Walter Reed Army Medical Center), *Symposium on stress.* Washington, D.C.: Army Med. Serv. Gr. Sch., pp. 2-7.

Chapter Fifteen

Environment Constraints

In this chapter we take a very hasty look at performance limitations imposed by ambient conditions in the environment. The cursory treatment accorded this general topic has been explained earlier and should not be taken as any indication of the amount of effort that has gone into this line of research. Actually, each of the conditions illustrated here (body vibration, noise, temperature, and lighting) has been the subject of intensive study, and a number of others (e.g., acceleration forces, weightlessness, intermittent photic and acoustic stimulation, radiation, anoxia) have received some attention. In many of these areas, however, the psychological research has been—often of necessity—poorly controlled, unsystematic, and idiosyncratic, and dependent variables of major interest have been physiological or introspective rather than behavioral (in the sense of performance functions). The most compelling consequences of whole body vibration, for example, are nausea and tissue damage; it is only natural that such traumatic effects as these should command priority as research topics over mere performance functions. Similarly, human *comfort* is certainly as important a consideration in the design of space or of military clothing as is measured human *proficiency*. In fact, severe discomfort would seem almost certain to produce performance degeneration in the long run, even if the two measures are uncorrelated in short-term laboratory investigations (as has been reported, at times).

For all of these reasons the results of environmental research have had little impact on scientific psychology as a whole, although they have played a very import role in design applications. There are exceptions to this rule, of course, the most notable of which are to be found in the study of acceleration (or "G" force) and noise effects. Recently, too, there has been some systematic research on performance under vibration and temperature conditions at levels well below those producing severe trauma (see, for example, Poulton 1966a; Buckhout, 1964). All of these areas save the "G" force work are represented in the present chapter. The latter research was omitted only because the most representative illustrations were too extensive to be included in this book. Rather than do the research an injustice by severely abridging it, we refer the reader to some excellent papers (Loftus & Hammer, 1963; White & Monty, 1963). Before introducing particular papers, we shall pause briefly to consider a few more general characteristics of environmental research.

The first point to be made is that work in this area has usually been prompted by

References will be found in the General References at the end of the book.

a particular design problem. The manned space program, for example, provided the major impetus for studying visual, intellectual, and particularly psychomotor performance under unusual "G" conditions. Development of high-speed jet aircraft has been the occasion for intensified research on a variety of noise effects. Warfare in arctic and tropical climates has demanded that we learn more about temperature effects, and so on. Because these are real and urgent problems, there has been some reluctance to depend upon answers derived from controlled laboratory investigation. Also, the cost, time, and facilities required to carry out adequate laboratory simulation have further discouraged this line of attack. And finally, some conditions are simply *impossible* to reproduce with any degree of fidelity in the laboratory setting. For these reasons, much of our existing knowledge comes from "the field," where it is extremely difficult to implement any semblance of experimental control.

A second point, not completely unrelated to the first, concerns research objectives. Since the stakes in most of these problem areas are high, the researcher cannot afford to become deeply involved in trivial consequences. His main concern must be whether, within the limits that can be expected to occur in the field, an environmental condition will constitute a handicap of *practically* significant proportions. Often the research issue seems to be framed in somewhat the following terms: (a) Which human states, activities, or functions is it most critical to preserve in the field situation? Vigilance? Visual performance? Perceptual-motor skill? (b) What are the ranges of environmental conditions that characterize the field situation? (c) What are the ranges of acceptable performance? And finally, (d) Does performance on the critical activities over the significant range of conditions fall within the tolerable limits? Or, alternatively, can we drive performance *out* of these limits experimentally? Whether or not it is phrased in exactly these terms, therefore, the question seems usually to focus upon *decrements*, in *relevant* activities, and in significant *amounts*. If we recall our earlier discussion of the concept of stress, it is not surprising that the term *stress* is often invoked in this general context. It is also not surprising to find that results receiving the most attention involve either conditions that produce measured decrements, or conditions that would seem, on intuitive grounds, to be detrimental but *fail* to produce measured decrements.

While this sort of negative emphasis may be understandable, it serves to play down several equally promising directions. First, there is the possibility that ambient environmental conditions might sometimes serve to *enhance* performance. Second, there is the possibility that *interactive* effects might be important: e.g., several "tolerable" conditions may summate to produce an intolerable one, or one might partially offset another. Very little work has been done on such combined effects (Broadbent, 1963). Finally, there is the point that trivial effects—whether good or bad—can contribute to our understanding of human performance even if they are of no consequence for the issue immediately at hand.

The first of the selected readings deals with the noise problem. Interest in industrial and military noise has had a long and fruitful history. Furthermore, information gathered in connection with the problem of noise has usually proven valuable to the hearing theorist as well as to the acoustic engineer. No effort will be made to prove the point here, since this work has been the subject of a number of good reviews (e.g., Broadbent, 1955; Jerison, 1959; Kryter, 1950, 1966).

We have chosen, instead, to represent the area with another paper by Kryter, largely because of its currency and the unique nature of the problem it poses.

The issue of concern in the Kryter paper is the *annoyance* to the general public of noise associated with high-speed jet aircraft. Few persons realize the extent to which engineering psychologists have been caught up in this problem, and probably even fewer appreciate the progress they have made toward quantifying such difficult attributes as actual and potential annoyance value. It is indeed interesting to follow, with Kryter, the ramifications of this very imminent and serious problem bestowed upon us by our unrelenting technology. Strictly apart from the noise problem, it is interesting to compare Kryter's views on the objectives of engineering psychology with the position adopted throughout this book. The discrepancy is very clear. Lest the reader take Kryter's position too lightly, we must call particular attention to the fact that this paper was delivered to the Society of Engineering Psychologists as the 1967 Presidential Address!

Moving on to the subject of vibration, we chose the Harris and Shoenberger paper mainly to show that despite the aforementioned difficulties inherent in this particular kind of stimulation, well-conceived studies of human performance under whole-body vibration can and have been carried out. In our opinion, however, they are in the distinct minority. On the other hand, systematic *physiological* research—especially that conducted using non-human subjects—has flourished, and the results have proven of considerable practical value.

The paper by Poulton and Kerslake is particularly noteworthy in terms of a comment made earlier. It is one of the few studies in *any* of the "ambient" areas that considers possible enhancement effects. Moreover, the transient increments in performance are not merely reported as empirical facts (as is usually the case in environmental research), they are interpreted in terms of a recognized—albeit controversial—theoretical construct, *arousal*. The problem with *arousal* is closely akin to that discussed in connection with *stress*: i.e., it has often been used in a completely circular fashion. Nevertheless, it has some promise as an explanatory concept, even if the promise is as yet largely unfulfilled, because independent measurement operations do exist (in particular, physiological indicants as discussed in Berlyne, 1960). In any event, Poulton and Kerslake, in a fashion seemingly much more characteristic of British than of American investigators working in this area, seek at least to develop *some* theoretical explanation for the phenomenon. Whether or not we accept their interpretation, there is no disputing the fact that more efforts in this direction are desperately needed.

Finally, we come to one of the oldest problem areas of all—the question of how ambient lighting influences performance functions. In contrast to the noise research, work on lighting has had a negligible impact on visual theory. A probable reason is the fact that the scientific aspects of the illumination issue have been all but obscured by its commercial and political ramifications. Light is a notably salable commodity; so, too, are lighting systems. Imagine, for example, the profit that could be reaped by power companies and lighting system manufacturers if tomorrow every town and city in the country were to increase its street lighting by as little as one percent. Or think of the financial losses that would result if it were proven to everyone's satisfaction that illuminated signs are hazardous to night driving and steps were taken to enforce their immediate removal!

Undoubtedly there are many other reasons for the theoretical sterility of much that has happened in the area of lighting. For example, our understanding of human visual processes is probably more advanced than that in any other area of psychology. Why, then, should we theorize about lighting at a *gross molar* level when we already have a reasonably sound *molecular* base on which to build? Whatever the reasons, lighting research has tended to center around strictly practical questions such as: What are the most *efficient* kinds and levels of general lighting? The most *comfortable?* The most *restful?* The best for particular purposes (e.g., for viewing cathode-ray-tube displays, for reading, for driving at night)? Obviously, the key issue underlying all of these questions is the matter of criterion specification. Until recently, there was no generally accepted basis upon which to compare various levels or systems of illumination. Now, thanks largely to the work of Blackwell (1959), some agreement has been reached (see McCormick, 1964, for a good review of this and related work). Even this, however, has not solved all dimensions of the lighting problem, as the final paper in this chapter clearly illustrates.[1]

Guth and Eastman discuss the importance of considering *individual differences* in the specification of illumination standards. More typically, illumination criteria are set according to the *average* performance of large groups of people rather than their *variability*. As we noted earlier, this problem has an interesting parallel in the field of training, where individual differences set a limit on the efficiency with which one can use any single training technique. One way of handling this problem in training is to match the learning procedures to the individual characteristics of the trainees. While it would be unrealistic to specify lighting requirements for every single individual, it could well prove feasible (and highly desirable in terms of speed and accuracy of performance, accident prevention, etc.) to match the standards to particular *groups* of individuals employed (e.g., younger vs. older workers). As an illustration of the importance of such considerations, Guth and Eastman report that people between 46 and 65 years of age need two to three times the illumination required by those in the 16 to 25-year range. Without adequate compensation for age, then, we can expect older workers to have difficulty on perceptual-motor tasks under "optimal" lighting conditions simply because they are unable to see what they are doing!

We should be remiss if we were to end this section without mentioning, for the reader's further study, a few of the more recent books devoted to the review and discussion of general ambient effects. Most of the areas of current interest are reviewed thoroughly in Burns, Chambers, and Hendler (1963), *Unusual Environments and Human Behavior*. McCormick (1964) also devotes considerable attention to environmental factors in his widely-used text, *Human Factors Engineering*. And finally, in a still more practical vein, there are Morgan, Cook, Chapanis, and Lund (1963), *Human Engineering Guide to Equipment Design*, and Bennett, Degan and Spiegel (1963), *Human Factors in Technology*.

---

[1] Despite the fact that the Guth and Eastman paper antedates much of Blackwell's work, the issue it raises is still a valid one. Moreover, this paper is particularly useful in emphasizing the multidimensional nature of the criterion problem.

52

An Example of "Engineering Psychology": The Aircraft Noise Problem

K. D. Kryter

Engineering psychology is usually defined as the application of psychological research information to the design and/or operation of man-machine systems. It also, of course, involves the doing of the research when that is required, as it often is; but without the application intent, the activity, I believe, does not qualify as "engineering" psychology.

Traditionally, a man-machine system has been taken to be a human operator or operators plus a simple or complex set of electronic or mechanical devices performing some useful function. The purpose of the engineering psychology in this context is, of course, to increase the efficiency and efficacy of a particular man-machine system.

A second, perhaps sometimes secondary, role of engineering psychology has been to provide design criteria or information relative to the protection of the well-being of the operator or user against psychophysiological harm from the machine, even though such harm may not interfere with the ostensible performance of the man-machine system in question. Here we would include design criteria for excessive noise, vibration, light, etc., that could bring annoyance or eventual physiological damage to the operator. This aspect of engineering psychology represents a broadening of responsibility of the engineering psychologist; for example, partial permanent deafness resulting from exposure to the excessive noise made by a piece of machinery may be a handicap to the operator of the machine only when he is *not* operating the machine itself—when he is in the relative quiet of his home or in an office, etc.

The subject of this paper is concerned with a third, even more remote, type of engineering psychology. It has to do with the fact that machines sometimes have

From *American Psychologist*, 23, 1968, 240-244. Copyright © 1968 by the American Psychological Association, and reproduced by permission.

Presidential Address presented to the Society of Engineering Psychologists at the meeting of the American Psychological Association, Washington, D.C., September 1967.

a way of reaching out and affecting people other than the direct operators or users of them; in particular, we will be concerned with the effects upon people of the external sounds from aircraft. The neighborhood noise from ground-based transportation vehicles and heavy industry is, of course, another similar example.

Engineering psychology qualifies for involvement in this problem area merely by broadening, I trust justifiably, our definition of a system to include all the people affected directly through their senses as the result of the operation of the machine part of the system. Extending the definition of man-machine systems this way probably seems reasonable to most of us, but the kinds of research information required for this somewhat "global" man-machine system and the avenues of application of this information are sometimes a bit startling, as I will attempt to show.

THE AIRCRAFT NOISE PROBLEM

Fundamentally, the aircraft noise problem requires two kinds of psychological research information for the man-machine system problem I wish to discuss:

1. Basic behavior or characteristics of the auditory system as a receptor of acoustic energy and

2. The reactions of people to aircraft noise in the environment of, primarily, their homes.

The latter is obviously the true criterion against which we must work and evaluate the results of the basic laboratory-generated information and the results of any human engineering system design recommendations that might be made. These engineering design recommendations will be:

1. For the design of aircraft engines and the operation of the aircraft to produce the least objectionable kind and amount of sound and/or

2. For the design of the airport-community system so that the sounds and the communities are compatible, i.e., placing the airport away from residential areas or zoning the areas near airports for industrial use only.

But the design recommendations to be made with respect to either, and especially the second, of these parts of the system run head-on into extremely complicated economic, social, legal, and political matters on both a national and international basis. The engineering psychologists who wish or have the opportunity to work on the aircraft noise problem need to consider and, to some extent, understand these practical, real-life parts of the problem if they are to behave and interact sensibly with the people who are responsible for creating and solving the problem.

So before presenting some of the research facts and data that might be used for the "best" engineering of the aircraft noise problem, I will burden you with a few brief comments on the more political-legal aspects of the problem. Most of my comments to follow in this regard apply strictly only to the United States, although the arguments can usually be applied to others countries.

In some countries, such as the United States of America, aviation is a private enterprise and has the right, if not the obligation, to promote its own interests first. If making noise results from these activities, restraints on making noise will be self-imposed by the aviation industry only if the noise hurts aviation business; such restraints may also be government-imposed if it creates a public nuisance, damages health, or destroys the value of property.

Some parts of the aviation industry are making valiant efforts to self-impose noise limits for the benefit of persons on the ground near airports as a matter of good public relations and public responsibility. However, the aviation noise problem has become so acute, and promises to become even worse (Greatrex, 1963), that some government participation in setting and enforcing limitations on aviation noise seems unavoidable at national, if not international, levels. But, regardless of who sets tolerable limits for aircraft noise in a community, a rational reason for setting these limits must be developed.

Three bases for such action have been argued from time to time: that noise (a) is a public nuisance, (b) damages health, and (c) destroys property. Let me remove from consideration the question of "damage to health." I think, although some may disagree with me, that aircraft noise as we know it is not demonstrably dangerous to the health of people in a community near an airport—and I am including not only direct physiological effects but possible indirect effects from loss of sleep, startle, etc. There is no convincing evidence, in my opinion, that significant adverse effects of this sort occur in real life as the result of exposure to aircraft noise per se. Fortunately, man, at least physiologically, seems to be able to adapt more or less completely to most noises.

The question of "public nuisance" is also a slippery basis for predicting the need for the establishment of aircraft noise limits. In the first place, what bothers some people is acceptable to others; but more importantly, a nuisance can be made legal if it is in the general interest of the public to have the nuisance. Aircraft noise, to a considerable extent, qualifies as legalizable nuisance, inasmuch as aviation has become such an important part of our economy and way of life. Ultimately, this balance between different and conflicting "values" can probably only be settled by application of some form of governmental judgment.

It would seem, however, that damages to property values may provide legal grounds for limiting aircraft noise in communities. (I do not mean to say that in some courts of law and in some legislatures aircraft noise above certain limits will not be considered as hazardous to health and well-being and, therefore, an illegal nuisance. This is certainly a possibility.) In the United States of America and elsewhere it is maintained that neither the government nor any private party can take or destroy property without adequately compensating the owner of the property. Property can, of course, be partly taken or destroyed, and if the presence of aircraft noise at a person's house makes that house less desirable as a house, its value is reduced and the property has been partly "taken" by the presence of the noise, be the noise in the public interest or not. In short, noise may damage or cause a relative decline in the value of a property because it is not acceptable to people trying to live on the property.

THREE CRITERIA FOR ACCEPTABLE AIRCRAFT NOISE

Let me now turn to a discussion of possible criteria of acceptability of aircraft noise in a community. The term "criterion" needs to be defined because it is often misused. By "criterion" I here mean the behavior or response to sound, such as airplane noise, that is deemed to be on the borderline between acceptable and unacceptable. It is not the noise level that produces the behavior that is the criterion, although it is common practice to refer to these just tolerable (according to the criterion) noise conditions as "noise criteria."

In any event, I have recently (Kryter, 1966) had the temerity to describe how, on the basis of existing acoustical, psychological, and sociological data, one could: (*a*) specify criteria of acceptability of aircraft noise in a community, and (*b*) specify the noise conditions that would result in behavior that just on the average meets these criteria. Most of the next few paragraphs are taken from the above-referenced article and also were presented at the Inaugural Meeting of the British Acoustical Society on Aircraft Noise (Kryter, 1967).

Criterion 1

A new or novel noise environment that is comparable in basic noisiness to a noise environment known and considered by the average person to be sigificantly unacceptable at a residence will likewise be considered significantly unacceptable at a residence. Obviously, the expressions "average person" and "significantly unacceptable" render this criterion open to interpretation and adjudication. But the approach may have some merit in that it allows persons to evaluate a noise environment that is relatively unknown to them with another with which they are more familiar. Many of the people making decisions about the possible effects of aircraft noise upon people in communities near airports have not been repeatedly exposed to such a noise environment.

Figure 1 suggests that aircraft noise having a perceived noise level (Johnson & Robinson, 1967) in excess of 100 PNdB [1] might be considered by a significant number of people to be unacceptable in their homes, inasmuch as that is the approximate noise level 50 feet (15 m) from trucks or motorcycles at maximum highway speed or in the course of acceleration, or 200 feet from a diesel train going 30 to 50 miles per hour.

These comparisons, to be most meaningful, should include not only peak PNdB levels, but also the number and duration of occurrences. In these respects the exposures to aircraft, truck, motorcycle, and train noise differ greatly, not always in favor of the aircraft noise. Two very similar methods have been developed whereby

---

[1] So-called perceived noise level in PNdB is presently being used for a basic unit for measuring the sound from aircraft and other sources in terms of its most probable "annoyance" effect on people (Kryter, 1963). A PNdB is found by making certain calculations on octave band or one-third octave band sound pressure level measurements of a sound; the effects on annoyance or the "noisiness" of a sound in terms of pure-tone content (which is an important contributor to the annoyance value of a sound) and duration of a sound can also be evaluated by "corrected" PNdB units.

It might be noted that the information developed for and contained in the PNdB values for a given sound is at least potentially of direct use by the engineer designing aircraft engines, in that the engineer can control the spectrum and frequency loci of pure-tone components and thereby make the engine noise as compatible as possible to the person on the ground under the aircraft. Likewise, engine power settings and other landing and take-off procedures on the aircraft can be specified on the basis of perceived noise levels to reduce community noise in presumably the most effective way possible.

PNdB values and numbers of daily occurrences of intense sounds are used to depict the total daily noise environment present in a community, as will be described below.

Criterion 2

A noise environment in which vigorous complaints and concerted group action against the noise are made is considered to be an unacceptable noise environment. These are the expected responses from a community when a composite noise rating (CNR) of 100 to 115 is present, see Figure 2. A CNR is calculated, incidentally, according to the following formula: $CNR = PNdB - 12 + 10 \log_{10} N$, where N is number of aircraft flyover events.

Criterion 3

It has been found that in a noise environment having a noise and number index (NNI) of 45 about 50% of the people will report that they are disturbed by the noise in various ways, and that it tends to be rated the worst aspect of a residential environment. Figure 3 illustrates the type of sociological data that substantiates the NNI method of measuring daily exposure to aircraft noise. NNI is calculated as follows: $NNI = PNdB - 80 + 15 \log_{10} N$, where N is number of aircraft flyover events.

In short, it is deduced that a noise, repeated fairly often during each day, having a peak level of 100 PNdB would probably be considered as unacceptable; thus 30 to 40 daily repetitions of an aircraft noise at 100 PNdB would be rated unacceptable by each of three rating methods described above.

SONIC BOOM

Finally, let me make a few remarks about a noise from a proposed commercial aircraft of the future—the so-called sonic boom. This new noise will be a significant problem, it appears, not because it will have any worse effects upon people than the noise from present-day subsonic aircraft near airports—as a matter of fact, research in the United States (Kryter, Johnson, & Young, 1967; Pearsons & Kryter, 1964) and Great Britain (Broadbent & Robinson, 1964; Johnson & Robinson, 1967) indicates that the effects of sonic booms and noise from subsonic jets near airports may actually be roughly comparable—but because the sonic boom will be heard by so many more people and because it may cause some slight amount of structural damage, the overall noise problem could become much worse. For example, it is estimated that transcontinental SST operations over the United States could expose 50,000,000 or so people to 15 or so booms per day. I think that the "absolute" number of bothered people becomes important for two reasons:

1. Practically speaking, there probably is a "critical mass" of people required to exert significant political and social action against a nuisance, and the number

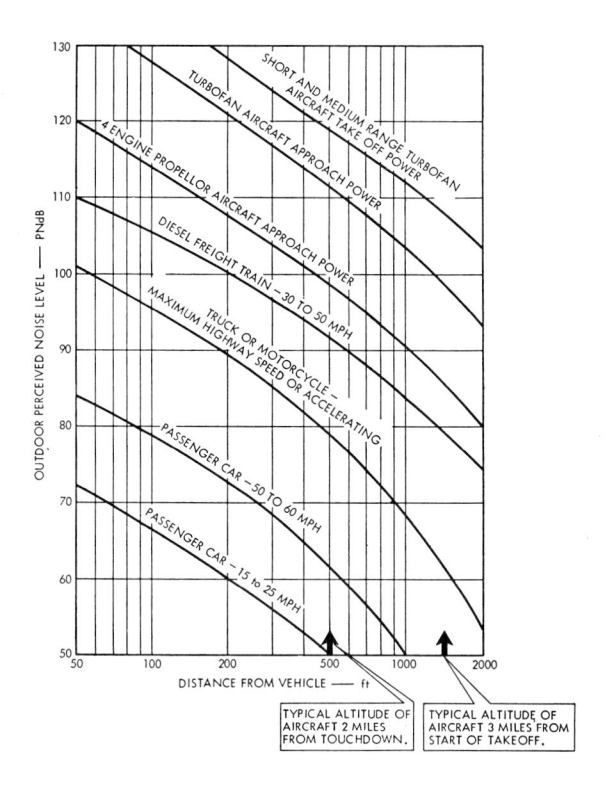

Figure 1. Typical levels of intermittent noise produced by vehicles. (An increase of 10 PNdB is usually equivalent to a 100% increase in judged noisiness. See Kryter, 1966.)

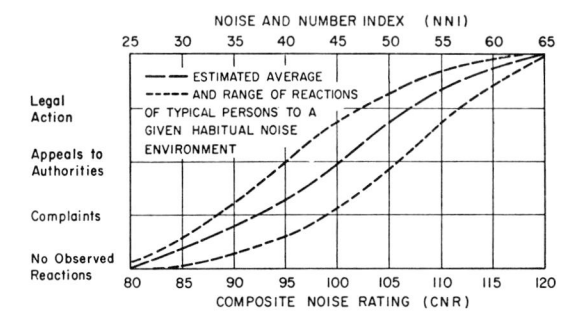

Figure 2. General relation between community response to aircraft or other noises and composite noise rating or noise and number index. (See Bolt, Beranek and Newman, 1964; Wilson, 1963.)

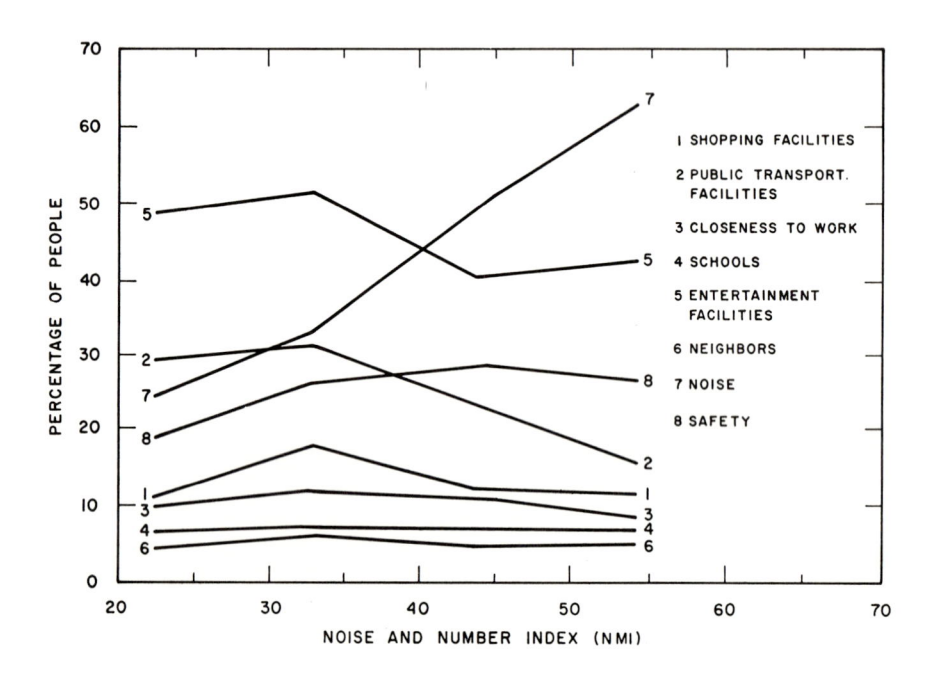

Figure 3. Results of interviews in communities within a 10-mile radius of Heathrow Airport, London, showing percentages of people rating their area as a poor, or very poor, place to live for various reasons. (See Wilson, 1963.)

of people near present airports appears in many cases to be fewer than this critical size or number.

2. Also practically speaking, whereas it is conceivable that compensation for taking property around airports might be economically feasible, compensation for taking of property in the United States by sonic booms (the property of 50,000,000 people) is hardly conceivable.

The problem of setting maximum tolerable exposures to sonic booms for communities would perhaps be amenable to solution a priori if some resolution of the question of the acceptability of the noise from subsonic aircraft were forthcoming and if some realistic and convincing estimate could be given as to the political response to complaints of possibly millions of persons, in comparison to the present-day complaints about aircraft noise from but a few tens of thousands near major airports. Indeed, whether the SST will be permitted to operate supersonically when over populated land areas may be largely decided on the basis of research information bearing on these specific points—information which I like to think belongs to the field of engineering psychology, broadly defined.

REFERENCES

Broadbent, D. E., & Robinson, D. W. Subjective measurements of the relative annoyance of simulated sonic bangs and aircraft noise. *Journal of Sound and Vibration*, 1964, 1(2), 162.

Greatrex, F. B. Take-off and landing of the supersonic transport. *Aircraft Engineering*, 1963, August, 1-5.

Johnson, D. R., & Robinson, D. W. On the subjective evaluation of sonic bangs. *Acustica*, 1967, 18, 241-258.

Kryter, K. D. Psychological reactions to aircraft noise. *Science*, 1966, 151, 1346-1355.

Kryter, K. D. Acceptability of aircraft noise. *Journal of Sound and Vibration*, 1967, 5(2), 364-369.

Kryter, K. D., Johnson, P. J., & Young, J. R. Psychological experiments on sonic booms. Annex B of Sonic Boom Experiments at Edwards Air Force Base, 1967, United States Department of Commerce, Springfield, Va.

Kryter, K. D., & Pearsons, K. S. Some effects of spectral content and duration on perceived noise level. *Journal of the Acoustical Society of America*, 1963, 35, 866.

Wilson, A. (Chmn.) *Noise*. (Report of the Committee on the Problem of Noise.) London: Her Majesty's Stationery Office, 1963.

Pearsons, K. S., & Kryter, K. D. Laboratory tests of subjective reactions to sonic boom. NASA Report No. CR-187, 1964.

Bolt, Beranek and Newman Inc. Land use planning relating to aircraft noise. Washington, D.C.: Federal Aviation Agency, 1964.

53

Effects of Frequency of Vibration on Human Performance

Charles S. Harris and Richard W. Shoenberger

Vibration has been studied with reference to its effects on truck drivers, helicopter pilots, airline passengers, operators of farm equipment, operators of heavy construction equipment, operators of space craft, and pilots of low altitude high speed aircraft (Buckhout, Sherman, Goldsmith & Vitale, 1963; Coermann, 1961; Dean, McGlothlen, & Monroe, 1964; Hornick, 1961; Lippert, 1955; Rawson, 1963; Simons & Schmitz, 1958). One point of commonality among many of these studies, regardless of the context in which vibration was investigated or the particular vehicle under consideration, is that at least some of the same frequencies of vibration are involved. In studying subjective human tolerance to vibration, Magid et al (1960) have restricted their investigations to the frequency range of 1-20 cps. Some of the reasons they give for this restriction are: (a) these are the frequencies to be expected in low altitude high speed flight; (b) pilots can easily be protected against vibration above 20 cps by mechanical damping systems; and (c) important bodily resonances occur within these frequencies. There is also reason to think that these low frequencies are likely to be the ones having the most dire consequences for human performance.

In addition to the wide variety of situations in which information concerning the effects of vibration has been sought, an almost equally wide variety of different types of vibration effects have been investigated. Some investigators have been interested in the mechanical response of the body, particularly with respect to frequency of vibration (Coermann, 1960; Edwards & Lange, 1964). Other investigators have been concerned with the physiological effects of vibration, such as changes in respiration rate, heart rate, rate of oxygen consumption, blood pressure, and blood chemistry (Ashe, 1960; Gaeuman, Hoover, & Ashe, 1962; Guignard, 1964; Guillemin & Wechsberg, 1953; Hoover, 1962; Hoover & Ashe, 1962; Nadel, 1963). Still other researchers, as mentioned previously, have been interested

From *Journal of Engineering Psychology*, 5, 1966, 1-15. By permission of Elias Press, Ltd., Atlantic City, N.J. 08404.

The research reported in this paper was sponsored by the Aerospace Medical Research Laboratories, Aerospace Medical Division, Air Force Systems Command, Wright-Patterson Air Force Base, Ohio. Further reproduction is authorized to satisfy needs of the U.S. Government. This paper has been identified by Aerospace Medical Laboratories as AMRL-TR-66-152.

in subjective tolerance to vibration, usually emphasizing the point at which individuals refuse to accept an increase in the intensity of vibration at various frequencies (Magid, Coermann, & Ziegenruecker, 1960; Mandel & Lowry, 1962; Parks & Snyder, 1961). Finally, the effects of vibration environments on man's ability to perform various tasks has been studied (Catterson, Hoover, & Ashe, 1962; Guignard, 1964; Harris, Chiles, & Touchstone, 1964; Hornick, 1962; Loeb, 1954; Mozell & White, 1958; Schmitz, 1959; Simons & Schmitz, 1958).

Many different experimental approaches have been taken to the study of vibration by these investigators. Even limiting ourselves to only those studies in which some type of performance was assessed, we find that the diversity of methodologies used and the plethora of different performance measuring instruments employed make it difficult to come up with any generalizable statements about the performance effects of vibration.

The unique feature of vibration that distinguishes it from other motion seems to be its oscillatory nature. There is reason to believe that this oscillatory characteristic may also be an important variable in determining performance decrement under vibration. The reason that frequency of vibration is suspect has come from studies on the nature of the bodily response to mechanical vibration. In describing the body's response to vibration, Cope (1959) has stated:

> In general, at a frequency of 0-1 cps, the head vibrates at somewhere near the amplitude of the seat, but, as frequency increases, the amplitude of the head increases and reaches a peak somewhere between 3-6 cps. At this frequency—the head vibrates at an amplitude equal to 150-300% of the seat amplitude.—Seat-head transmission decreases progressively at the higher frequencies, so that when one has reached 70 cps, only about 10% of the amplitude of seat vibration may be expected to reach the head.

Some investigators have been active in research attempting to relate bodily resonance to subjective tolerance levels (Magid, Coermann, & Ziegenruecker, 1960; Mandel & Lowry, 1962), to visual acuity (Dennis, 1960; Lange & Coermann, 1962), and to performance (Buckhout, 1964; Coermann, Magid, & Lange, 1962). It might be expected from studies on the mechanical characteristics of the human body that the point of least subjective tolerance to vibration would correspond to the major resonant frequency of the body, which is approximately 5 cps (Coermann, 1961). This seems to be the case, at least where short time tolerance to vibration is being considered. Visual acuity also seems to be related to frequencies of vibration (Harris & Shoenberger, 1964). The evidence from tracking studies, although less clearcut and sometimes contradictory, does seem to indicate that tracking performance is affected most adversely at or near the frequency of maximum bodily resonance (Harris and Shoenberger, 1964). The idea that tracking performance error is directly determined by the ratio of input G as measured at the seat of the vibrating chair to output G as measured at the chest or head is probably the most obvious hypothesis. However, this approach seems to consider man as a hydraulic and mechanical system without taking into account such factors as fatigue and psychological stress, which are also logical candidates for producing performance decrements. Although mechanical interference cannot be considered to be the only causative factor operating in a vibration environment, it seems logical that those frequencies where the most G is transmitted to the head are not only likely to produce the most mechanical interference with a man's ability to perform a task but are also likely to produce the most fatigue and the greatest psychological stress.

Buckhout (1964) studied tracking performance during vibration at 25, 30, and 35 percent of the one-minute subjective tolerance level (as determined by Magid et al (1960)) at frequencies of 5, 7, and 11 cps. All of these vibration levels produced large decrements in tracking performance. This finding suggests that performance will be impaired at magnitudes well below vibration intensities that would be expected to create medical problems. We extended this study to find the minimum G level necessary to produce a significant increase in tracking error over a control condition. The purpose of our investigations was to provide tracking performance data which we felt would fill important gaps in the experimental literature on the effects of vibration. First, we selected four G values at each of the three vibration frequencies of 5, 7, and 11 cps that, on the basis of the above mentioned experiment and other literature, appeared likely to bracket the G value at which performance decrement would just reach significance. These values were approximately 10, 15, 20, and 25 percent of the one-minute subjective tolerance level. Second, performance was measured in 5 sequential segments during a 20-minute exposure. And third, each subject received all experimental conditions within a given study (one frequency), which allowed us to control for between subject effects. With this design, we hoped not only to be able to isolate the critical G level for tracking performance decrement at each of the frequencies, but also to examine possible sensitization or adaptation effects. In addition, we would be able to examine the occurrence of such effects over repeated exposures to the vibration environment. Finally, through a comparison of the results of all three experiments, we hoped to provide additional evidence in support of frequency as an important variable in determining tracking performance decrement.

The frequencies of 5, 7, and 11 cps were chosen because of the belief that these are ones most likely to produce performance decrements at moderate G levels. Five cps vibration is nominally the frequency of maximum bodily resonance, 7 cps is near the range of maximum body resonance and is also thought to be the resonant frequency of the heart, while 11 cps is near the point of secondary body resonance and evidence exists which suggests that this frequency is particularly troublesome for visual acuity.

METHOD

Subjects. Thirty male military personnel, members of a "vibration panel," were used as subjects. Ten subjects were used in each experiment. A separate experiment was conducted at each of the three frequencies of 5, 7, and 11 cps. The subjects in becoming vibration panel members had undergone a physical examination and had received indoctrination rides under vibration.

Procedure. An electromagnetically driven shake table (C-25H) was used to produce the vibration. The subject sat in a simulated cockpit on a wooden seat mounted on top of the shake table. His legs were extended at approximately a 30-degree angle with his feet resting on foot rests. The subject was firmly restrained by a shoulder harness and seat belt during vibration and control runs. A 5-inch cathode ray tube (CRT) was mounted in front of the subject at appromixately eye level, and at a distance of 3.5 feet from the back of the subject's chair. A white dot was displayed on the CRT and was driven in both the vertical and horizontal axes by a random noise signal recorded on two channels of an Ampex tape recorder. This forcing function was fed through ultra-low-frequency bandpass filters (Krohn-Hite Model 330A). It was hoped at the beginning of experimentation that exactly the same task could be used in all three experiments. However, equipment failure and the requirement to return borrowed equipment necessitated a change in the upper cut-off frequency for the 11 cps experiment. For the first two experiments (5 and 7 cps), the bandpass filter was set at .02 cps lower cut-off frequency and .27 cps upper cut-off frequency for the vertical dimension of the task, and at .02 cps lower cut-off frequency and .50 cps upper cut-off frequency for the horizontal dimension. In the last experiment (11 cps) both the horizontal and vertical portions of the task were set at .02 to .50 cps. The subject, in all experiments, attempted to keep the dot centered on the CRT through the use of a high gain, rate-controlled stick, located on the right side of the chair directly at the end of the right arm-rest. Control dynamics of the sidearm controller were programmed on a Donner Model 30 Analog computer. Subjects received x and y scores which consisted of the integral of the absolute error over a 3-minute period for the horizontal and vertical axes, respectively. These scores were represented by voltages at the outputs of two Philbrick scoring integrators which automatically rest when they accumulated 100 volts. Each reset advanced a digital counter, and the score was read from the counter plus the remaining voltage on the meter. An Eagle Cyclic Timer was used to program both trial time and intertrial intervals. During the intertrial intervals the computer was in the control mode.

Table 1 shows the different amplitudes and peak G administered to the subjects in the three experiments. Each subject was administered all amplitudes within an experiment in a 5 x 5 Latin square design. Two subjects were assigned to each of the five orders of presentation. Three weeks of preliminary training were given in the static mode. Each subject received twenty-one, 3-minute trials presented in blocks of seven trials during three different sessions. Then, a preliminary vibration run was given for indoctrination purposes. Scores obtained up to this point were not included in the data analysis. Thereafter, each subject was tested during one session per week for five weeks according to one of the orders of the Latin square. Regardless of the particular treatment a subject was to receive, he was given one 3-minute static trial as a warmup. He was then presented the five 3-minute test trials during 20 minutes of continuous vibration followed by a final 3-minute static trial. In all instances there was a 1-minute interval between trials. Subjects were urged to do their best and were told that their performance during vibration would be compared with their performance during the control condition. They were also informed of their scores on both the dimensions of the tracking task after each trial.

RESULTS AND DISCUSSION

Tables 2, 3, and 4 present the analyses of variance for the vertical tracking scores for all three experiments. It should be noted that a different type of analysis was performed for the 7 cps data. This was due to the withdrawal of two subjects in that study. In all three analyses, vertical error showed a significant effect due to G levels. Table 5 gives the mean vertical error for each vibration condition in the three experiments, along with the significance of the differences between the means. It can be seen from this table that the G values chosen for testing at each of the three frequencies did indeed bracket the levels necessary to produce a significant decrement in tracking performance. Figure 1 graphically presents the G levels at which subjects were tested at the three different frequencies, and indicates which levels produced a significant decrement. Tracking performance was significantly impaired at .20 G at 5 cps, at .25 G at 7 cps, and at .37 G at 11 cps. This finding agrees with the thinking of Von Gierke and Hiatt in 1961. They stated at that time that the capacity to perform certain tasks and make visual observations are adversely affected by sinusoidal vibration above .20 to .50 G. The curves for vertical tracking error show systematic increases with increasing G levels (see Figure 2). Figure 3 presents these data in terms of percentage increase in mean error over the control condition.

It is rather striking that the scores across experiments are as close together as they are, since different frequencies and for the most part different G levels were used. One common factor which could account for the similarity was that the G levels represented approximately the same percentage value of the one-minute subjective tolerance level. The curves are, in fact, plotted on the abscissa as the percentage of the one minute tolerance. The G level at which a decrement first occurred at 5 cps was 20%, the level at 7 cps was 21%, and the point at 11 cps was 17% of the one-minute subjective tolerance level. Therefore, it would seem that the physical response of the body to vibration is a factor in determining tracking performance as well as subjective tolerance. The generality of these results can only be determined by future experimentation. Huddleston (1964) has presented some evidence which suggests that fine manual skill is affected according to body resonance phenomena, whereas, degradation on tasks requiring more "mental" type of performance does not appear to be frequency dependent. However, the G level used in his experiment was higher than those found to produce a significant decrement in tracking performance in the present study. Although it may be true that the major limiting factor in performing in a vibration environment is difficulty of motor control, further experimentation will be necessary, using tasks not affected directly by vibration, to determine the minimum levels which produce central nervous system effects (non-specific stress).

A significant trials effect was obtained at 5 cps, and 11 cps, and closely approached significance at 7 cps (10% level of confidence). The performance of the subjects improved through each five-trial testing period in the 5 and 7 cps studies. There is some evidence that subjects were adapting to the vibration. This is particularly true at 7 cps. At this frequency, subjects showed greater improvement in vertical error scores at the higher G levels than at the lower G levels. This can

TABLE 1
Levels of Vibration Administered in Experiments

Frequency (cps)	Double Amplitude	Peak G
Control	0	0
5	.08 in.	.10
5	.12 in.	.15
5	.16 in.	.20
5	.20 in.	.26
Control	0	0
7	.06 in.	.15
7	.08 in.	.20
7	.10 in.	.25
7	.12 in.	.30
Control	0	0
11	.04 in.	.25
11	.06 in.	.37
11	.08 in.	.49
11	.10 in.	.62

TABLE 2
Summary Analysis of Variance for Vertical Tracking Scores

(5 cps)

Source of Variance	df	SS	MS	Error Term	F	
Between-Subjects	9	1362.85				
AB (b) (Order)	4	185.42	46.36	(b)	.20	NS*
error (b)	5	1177.43	235.49			
Within-Subjects	240	765.85				
A (G Levels)	4	127.37	31.84	1 (w)	4.16	p<.025
B (Sessions)	4	59.83	14.96	1 (w)	1.95	NS
C (Trials)	4	98.70	24.67	2 (w)	24.65	p<.001
AB (w)	12	86.35	7.20	1 (w)	.94	NS
AC	16	13.90	.87	3 (w)	.74	NS
BC	16	18.11	1.13	3 (w)	.96	NS
AB (b) C	16	25.86	1.16	2 (w)	1.16	NS
AB (w) C	48	68.35	1.42	3 (w)	1.21	NS
error (w)	120	267.38	2.23			
error 1 (w)	20	153.13	7.66			
error 2 (w)	20	20.02	1.00			
error 3 (w)	80	94.23	1.18			
TOTAL	249	2128.70				

* Not Significant.

TABLE 3

Summary Analysis of Variance for Vertical Tracking Scores

(7 cps)

Source of Variance	df	SS	MS	F	
S (Subjects)	7	700.27			
A (G Levels)	4	78.44	19.61	3.84	p<.025
B (Trials)	4	14.33	3.58	2.65	p<.10
AB	16	18.95	1.18	3.93	p<.001
AS	28	142.92	5.10		
BS	28	37.87	1.35		
ABS	112	33.82	.30		
TOTAL	199	1026.60			

TABLE 4

Summary Analysis of Variance for Vertical Tracking Scores

(11 cps)

Source of Variance	df	SS	MS	Error Term	F	P
Between-Subjects	9	261.12				
AB (b) (Order)	4	117.90	29.64	(b)	1.03	
error (b)	5	143.22	28.64			
Within-Subjects	240	203.03				
A (G Levels)	4	43.10	10.77	1 (w)	5.45	<.005
B (Sessions)	4	23.70	5.92	1 (w)	3.00	<.05
C (Trials)	4	12.12	3.03*	2 (w)	13.80	<.001
AB (w)	12	29.67	2.47	1 (w)	1.25	
AC	16	3.54	.22	3 (w)	.81	
BC	16	3.75	.23	3 (w)	.86	
AB (b) C	16	10.95	.68	2 (w)	3.12	<.01
AB (w) C	48	10.49	.22	3 (w)	.80	
error (w)	120	65.76	.55			
error 1 (w)	20	39.51	1.97			
error 2 (w)	20	4.39	.22			
error 3 (w)	80	21.86	.27			
TOTAL	249	464.15				

* Not Significant.

TABLE 5

Means and Mean Differences for G Levels at Each Frequency for Vertical Tracking Scores

Treatments (G Levels)	A	5 cps B	C	D	E	$\overline{X}$
A (control)		104	94	178**	204**	526
B (.10 G)			10	74	100	630
C (.15 G)				84	110	620
D (.20 G)					26	704
E (.26 G)						730

Treatments (G Levels)	A	7 cps B	C	D	E	$\overline{X}$
A (control)		75	46	141**	172**	494
B (.15 G)			29	66	97	569
C (.20 G)				95	126*	540
D (.25 G)					31	635
E (.30 G)						666

Treatments (G Levels)	A	11 cps B	C	D	E	$\overline{X}$
A (control)		21	74*	75*	116**	554
B (.25 G)			53	54	95**	575
C (.37 G)				1	42	628
D (.49 G)					41	629
E (.62 G)						670

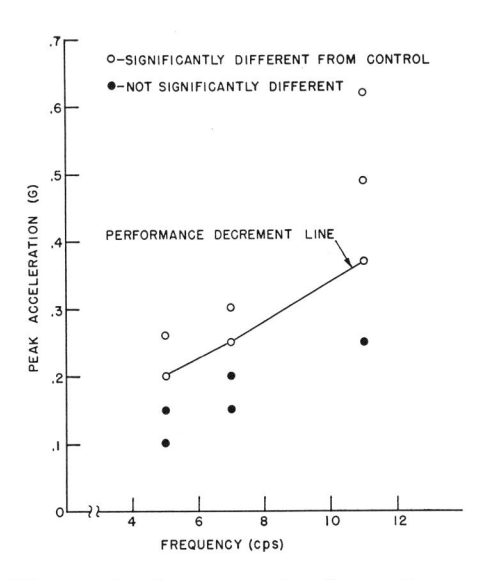

* $p < .05$ level
** $p < .01$ level

Figure 1. Vibration levels presented at 5 cps, 7 cps, and 11 cps.

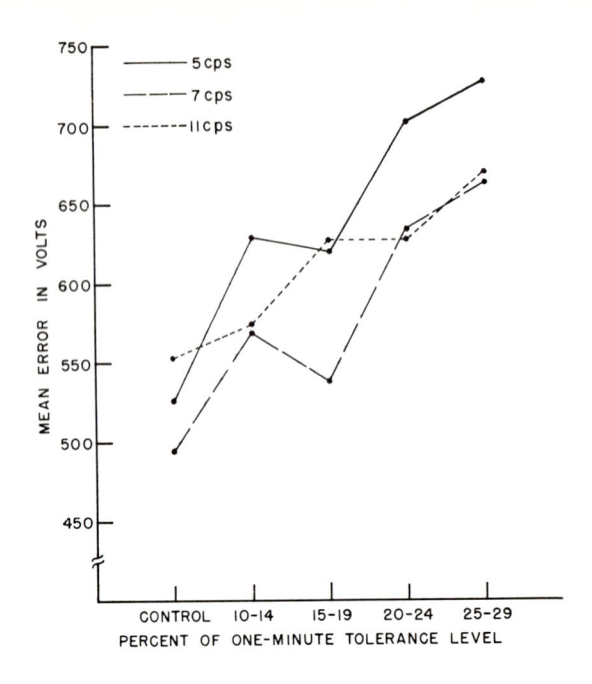

Figure 2. Mean vertical tracking error as a function of percent of one-minute subjective tolerance level at each frequency.

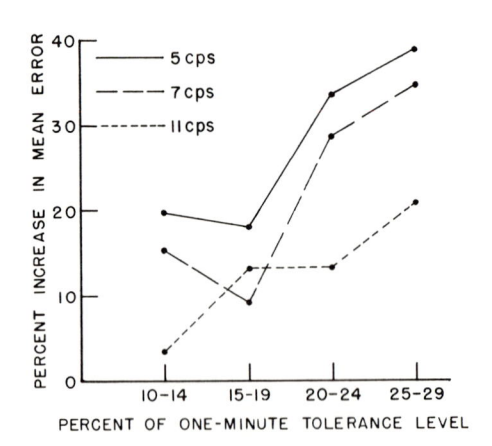

Figure 3. Percent increase in mean vertical tracking error as a function of percent of one-minute subjective tolerance level at each frequency.

be seen in the significant interaction of G levels and trials obtained in the analysis of variance for that group. Somewhat the same trend was seen for the 5 cps group, i.e., more adaptation occurred at the higher G levels. However, no significant interaction was obtained between G levels and trials in the 5 cps study. The failure to find a significant interaction was probably due to the fact that there was a large improvement over trials for the control group. At 11 cps, although a significant trials effect was obtained, the same pattern as described for 5 and 7 cps could not be seen. In fact, there was no consistent improvement in scores across trials. There was, however, a significant sessions effect at 11 cps. Thus it appears that at this frequency there was improvement in vertical tracking from one session to another, but no evidence that any adaptation occurred across trials within a session. It would be tempting to postulate that individuals adapt to vibration within a session at 5 and 7 cps but do not at 11 cps. However, this must remain a question for further study. We feel that a more clear-cut answer to this question can be obtained in future research where longer than 20 minutes of continuous vibration is used.

Relatively large differences in horizontal error scores were observed between the control conditions of the three experiments. Apparently, this portion of the task was more difficult in the 7 and 11 cps experiments than in the 5 cps experiment. Nevertheless, the amount of increase in horizontal error as a result of vibration was negligible for the G levels used for all three frequencies. Because of the obvious lack of differences, analyses of variance were not performed on the horizontal scores. There are a number of possible explanations for this difference in the effect of vibration on the horizontal and vertical scores, however, because of the change in the difficulty of the horizontal task from experiment to experiment, none will be offered.

Two subjects complained of pain produced by the vibration. Both of these individuals experienced pain during the 5 cps experiment. One complained of pain at .20 G and the other at .25 G. Both of these complaints occurred toward the end of a 20-minute vibration session. The subjects reported that the pain was slight and they were able to complete the experimental session. None of the remaining subjects reported any pain at any of the conditions in any of the experiments.

Many problems remain concerning the effects of vibration on human performance. Of particular importance for future research would be the study of the effects on tasks not directly affected by the vibration, the study of the relative effects of sinusoidal and random vibration, and the study of long term vibration exposure.

REFERENCES

Ashe, W. F. *Physiological and Pathological Effects of Mechanical Vibration.* NIH RF Project 862, Progress Report No. 3, National Institutes of Health, Washington, D.C., September 1960.

Buckhout, R. Effect of Whole Body Vibration on Human Performance. *Human Factors,* 1964, 6, 157-163.

Buckhout, R., Sherman, H., Goldsmith, C. T., & Vitale, P. A. *The Effects of Variations in Motion Fidelity During Training on Simulated Low-Altitude Flight.* AMRL Technical Report 63-108, Aerospace Medical Research Laboratories, Wright-Patterson AFB, Ohio, December 1963.

Catterson, A. D., Hoover, G. N., & Ashe, W. F. Human Psychomotor Performance During Prolonged Vertical Vibration. *Aerospace Medicine*, 1962, 33, 598-602.

Clark, C. C. *Human Control Performance and Tolerance under Severe Complex Waveform Vibration with a Preliminary Historical Review of Flight Simulation*. Martin-Baltimore Engineering Report 12406, The Martin Company, Baltimore, Maryland, April 1962.

Coermann, Rolf R. *The Mechanical Impedance of the Human Body in Sitting and Standing Position at Low Frequencies*. ASD Technical Report 61-492, Wright Air Development Center, Wright-Patterson AFB, Ohio, September 1961.

Coermann, R. R., Magid, E. B., & Lange, K. O. Human Performance Under Vibrational Stress. *Human Factors*, 1962, 4, 315-324.

Cope, F. W. *Problems in Human Vibration Engineering*, Project NM 180112.4, Report No. 2, Aviation Medical Acceleration Laboratory, U.S. Naval Air Development Center, Johnsville, Pa., 1959.

Dean, R. D., McGlothlen, C. L., & Monroe, J. L. *Performance and Physiological Effects of CH-46A Noise and Vibration*. Technical Report D2-90583, The Boeing Company, Seattle, Washington, May 1964.

Dennis, J. P. *The Effect of Whole Body Vibration on a Visual Performance Task*, Report No. 104, Directorate of Physiological and Biological Research, C.E.P.R.E. (United Kingdom), AD 247, 249, August 1960.

Edwards, R. G., & Lange, K. O. *A Mechanical Impedance Investigation of Human Response to Vibration*. AMRL Technical Report 64-91, Aerospace Medical Research Laboratories, Wright-Patterson AFB, Ohio, October 1964.

Gaeuman, J. V., Hoover, G. N., & Ashe, W. F. Oxygen Consumption During Human Vibration Exposure. *Aerospace Medicine*, 1962, 33, 469-474.

Goldman, D. F., & Von Gierke, H. E. *The Effects of Shock and Vibration on Man*. NMRI Lecture and Review Series No. 60-3, Naval Medical Research Institute, Bethesda, Maryland, January 1960.

Guignard, J. C. *Test of the Type 7 Anti-g Suit as an Anti-Vibration Device*. IAM Technical Memorandum No. 236, Royal Air Force Institute of Aviation Medicine, Farnborough, December 1964.

Guignard, J. C. *The Physiological Effects of Mechanical Vibration: A Selected Bibliography*. IAM Report No. 124, Royal Air Force, Institute of Aviation Medicine, Farnborough, August 1959.

Guillemin, V., & Wechsberg, P. *Physiological Effects of Mechanical Vibration*. Project Number 21-1203-0002, USAF School of Aviation Medicine, Randolph Field, Texas, January 1953.

Harris, C. S., Chiles, W. D., & Touchstone, R. M. *Human Performance as a Function of Intensity of Vibration at 5 cps*. AMRL Technical Report 64-83, Aerospace Medical Research Laboratories, Wright-Patterson AFB, Ohio, September 1964.

Harris, C. S., & Shoenberger, R. W. *Human Performance During Vibration*. AMRL-TR-65-204 (AD 624 196), Aerospace Medical Research Laboratories, Wright-Patterson AFB, Ohio, November 1965.

Hoover, G. N. *The Biology of Whole Body Mechanical Vibration: An Annotated Bibliography*. Report 862, National Institutes of Health Grants RG-5348 and OH-6, Research Foundation, Columbus, Ohio, June 1962.

Hoover, G. N., & Ashe, W. F. "Respiratory Response to Whole Body Vertical Vibration." *Aerospace Medicine*, 1962, 33, 980-984.

Hornick, R. J. Effects of Whole-Body Vibration in Three Directions Upon Human Performance. *Journal of Engineering Psychology*, 1962, 1, 93-101.

Hornick, R. J. Problems in Vibration Research. *Human Factors*, 1962, 4, 325-330.

Hornick, R. J. *The Effects of Tractor Vibration on Work Performance*. BRL Report No. 131, Bostrom Research Laboratories, Milwaukee, Wisconsin, June 1961.

Huddleston, H. F. *Human Performance and Behaviour in Vertical Sinusoidal Vibration*. IAM Report No. 303, Institute of Aviation Medicine, Farnborough, July 1964.

Lange, K. O., & Coermann, R. R. Visual Acuity under Vibration. *Human Factors*, 1962, 4, 291-300.

Lippert, S. *Human Response to Vertical Vibration*. Report No. SM-20021, Douglas Aircraft Company, Santa Monica, California, April 1955.

Loeb, M. *Further Investigation of the Influence of Whole Body Vibration and Noise on Tremor and Visual Acuity.* AMRL Report No. 165, Project 6-95 20-001, U.S. Army Medical Research Laboratory, Fort Knox, Kentucky, October 1954.

Magid, E. B., Coermann, R. R., & Ziegenruecker, G. H. Human Tolerance to Whole Body Sinusoidal Vibration. *Aerospace Medicine,* 1960, 31, 915-924.

Mandel, M. J., & Lowry, R. D. *One-Minute Tolerance in Man to Vertical Sinusoidal Vibration in the Sitting Position.* AMRL Technical Report 62-121, Aerospace Medical Research Laboratories, Wright-Patterson AFB, Ohio, October 1962.

Mozell, M. M., & White, D. C. *Behavioral Effects of Whole Body Vibration.* NADC-MA-5802, Project NM 180112.4, Report No. 1, U.S. Naval Air Development Center, Johnsville, Pennsylvania, January 1958.

Nadel, Aaron B. "Vibration" in N. M. Burns, M. Chambers, and E. Hendler (eds.). *Unusual Environments and Human Behavior,* Toronto: The Free Press of Glencoe, 1963, pp. 379-394.

Notess, Charles B. *A Triangle-Flexible Airplanes, Gusts, Crew.* Cornell Aeronautical Laboratory, Inc. Full-Scale Division Memorandum No. 343, May 1963.

Parks, D. L. Defining Human Reaction to Whole-Body Vibration. *Human Factors,* 1962, 4, 305-314.

Parks, D. L. *A Comparison of Sinusoidal and Random Vibration Effects on Human Performance,* Technical Report D3-3512-2, The Boeing Co., Wichita, Kansas, 28 July 1961.

Parks, D. L., & Snyder, F. W. *Human Reaction to Low Frequency Vibration.* Technical Report D3-3512-1, The Boeing Company, Wichita, Kansas, July 1961.

Rawson, H. E. (Principal Investigator). *Flight Simulation Study of Human Performance During Low-Altitude, High-Speed Flight,* ATRECOM technical report 63-52 (AD431 739), US Army Transportation Research Command, Fort Eustis, Virginia, November 1963.

Schmitz, M. A. *The Effect of Low Frequency, High Amplitude Vibration on Human Performance,* Bostrom Research Laboratories Progress Report No. 2a, for Office of Surgeon General, Dept. of the Army, Wash. 25, D.C., January 1959.

Simons, A. K., & Schmitz, M. A. *The Effect of Low Frequency, High Amplitude Whole Body Vibration on Human Performance.* Progress Report No. 1, Research and Development Division, Office of the Surgeon General, January 1958.

Von Gierke, H. E., & Hiatt, Edwin P. "Biodynamics of Space Flight," in S. F. Singer (ed.), *Progress in the Astronautical Sciences,* Vol. I, Amsterdam: North-Holland Publishing Company, 1962, pp. 344-401.

54

Initial Stimulating Effect of Warmth upon Perceptual Efficiency

E. C. Poulton, and D. McK. Kerslake

The cockpit of a large commercial aircraft standing in the mid-day sun on an aerodrome in the Middle East may reach a temperature of up to about 54° C (113° F). If special air-conditioning equipment is not available, the pilot has to work at this temperature until the aircraft becomes airborne. He cannot even have a fan on until he has checked the electrical equipment. The present experiment was designed to measure relatively short-lived changes in performance within a few minutes of exposure to a temperature of 45° C, as compared with a more acceptable temperature of 25° C (77° F).

The effects of hot environments upon human performance have been summarized by Pepler (8). In general, experimenters do not appear to have been on the lookout for transient changes in performance on first entering a warm environment. Yet the exposed skin surfaces are stimulated immediately on entering, and such stimulation could affect efficiency.

In order to measure transient effects produced by a relatively small increase in environmental temperature, it was decided (a) to introduce the man into the warm climate while he was at work, so that a continuous measure of performance would be available from the time he entered; (b) to use tasks in which each 2-minute period was equivalent; and (c) to use tasks which were likely to be sensitive to relatively small changes in conditions (9). The tasks were designed to extend the man's perceptual capacities to the full, so that he would have no reserve capacity to use to compensate for the less favorable condition, and were commenced after the minimum of practice.

METHODS

Subjects. The 12 men were all on the staff of the R.A.F. Institute of Aviation Medicine. Seven were medical officers and the rest technicians. Most had had

From *Aerospace Medicine*, 36, 1965, 29-32. By permission of the Aerospace Medical Association.

From the British Medical Research Council, Applied Psychology Research Unit, Cambridge, England, and the Royal Air Force Institute of Aviation Medicine, Farnborough, Hants., England.

some previous experience in the Hot Room, and 3 were very experienced. The average age was 36, range 29-44 years.

Tasks. Two tasks had to be performed simultaneously. The *dials* task involved searching for signals which could appear on any one of 5 ammeters, diameter 3 inches, mounted at head height at a distance of about 18 inches. One ammeter was directly in front of the man, the others were 45° and 90° to his right and left. A signal moved the needle smartly to the vertical from its resting position 45° to the left. He had to cancel the signal as quickly as possible by pressing a key 2 x 1.5 inches mounted below the ammeter. The signal was cancelled automatically after 3.5 seconds. In 2 minutes there were 20 signals, 4 on each of the 5 dials, arranged in the quasirandom sequence described by Leonard (7). The intervals between signals ranged from 4 to 8 seconds, and were also quasirandom. The *listening* task involved listening through headphones to 9 of the letters of the alphabet which Conrad (5) has found to be the least easily confused. One letter was presented each second, and in a random sequence of 10 consecutive letters a single letter was repeated. The paired letters never appeared in the first or last position of the sequence, and were always separated by at least 2 other letters. After the 10 letters the man heard the question "which?" Before the start of the next sequence there was then an interval of a second during which he had to call out the paired letter into a microphone. In each period of 2 minutes he had to make 10 responses.

Design and Procedure. The 12 men were divided into two comparable groups by training and experience in heat. The 6 men in Group Iw performed first in the warm at 45° C (113° F), and then in the cool at 25° C (77° F). The 6 men in Group IIw had the two temperatures in the reverse order. The absolute humidity was held constant at 10 millimeters of mercury, and the air movement at 500 feet per minute. This gave effective temperatures (E.T.) on the normal scale of 30° C (86° F) and 19° C (65° F) respectively. Each part of the experiment consisted of 10 comparable 2-minute periods. Normally the two parts were performed on consecutive days, but one man in Group IIw performed them on the morning and afternoon of the same day.

Immediately before the experiment the men had been performing sedentary duties in an environment with a temperature of about 21° C (70° F). Each man wore his normal indoor clothing, but removed his jacket and pullover (if worn), and rolled up his sleeves. He sat on a chair mounted on a low trolley which carried the experimental apparatus. Before each part of the experiment he received minimal practice at the two tasks in an antechamber. He was instructed to select an efficient method of visual scanning which gave as much emphasis to the 2 peripheral ammeters as to the 3 more central ones. Special emphasis was placed upon the importance of listening, since otherwise the man tended not to notice what he was hearing while he was responding to the dials. Knowledge of results on the two tasks was available after the practice and after each part of the experiment.

Just before the experiment proper two copper-constantan thermocouples forming part of an elastic headband were placed in position on the forehead to register skin temperature, which was recorded automatically. The man started the experiment seated on the trolley in the antechamber. After about 45 seconds an attendant entered, opened the door of the Hot Room, pushed the trolley inside, and closed the door again. The maneuver was completed about 15 seconds before the end of the first 2-minute period. The man was told to pay as little attention to it as possible. The trolley was left so that the 5 dials were well lit by overhead strip lighting, and the wind blew into the man's face.

Scoring and Calculations. On the listening task errors and omissions were summed. On the dials task all responses with times of 2 seconds or longer were counted as "delayed" ("errors" in Figure 1). Responses with times of 3 seconds or longer were counted as "very slow" (Figure 2). Measures of this kind have been found by Poulton *et al* (10) to be more sensitive than either the mean response time or the mean variability in response time. This is believed to be because they are affected both by increases in mean response time and by increases in variability.

In the analysis of variance the sum of squares for Tukey's "additivity" (12, 13) which has one degree of freedom, was taken out of the residual sum of squares. This checked the assumption underlying the analysis that the subject, temperature and order effects were additive. Apart from the analysis of variance, nonparametric tests (11) were used to assess statistical reliability, even when 5 per cent fiducial limits are given in the text. Except where stated, all p values took account of both tails of the theoretical distribution of scores.

RESULTS

These are shown in Figure 1 by comparable periods of 2 minutes. The first period of Day 1 provides the baseline, since most of it was spent either in the antechamber or on the way into the Hot Room. On all three measures the two groups were comparable at this time. The top two functions in the figure show that on both days in the warm climate (filled points) the *temperature of the skin* of the forehead rose by an average of 3.8° C, 5 per cent fiducial limits ± .6, and then gradually fell a little. In the cool climate (unfilled points) the temperature fell by an average of 1.3 ± .2° C, and then remained approximately constant. These changes were shown by all 12 men, and were highly reliable ($p < .001$).

The middle two functions in Figure 1 show the sum of errors and omissions on the listening task. Performance in the warm remained approximately constant on both days, the mean for the two days being 21.7 ± 11.4 per cent in the first 2-minute period and 23.3 ± 11.6 per cent in the tenth ($p > .05$). In contrast, performance deteriorated in the cool to a statistically reliable extent from a mean for the two days of 14.2 ± 7.4 per cent to a mean of 27.5 ± 9.8 per cent ($p < .02$); the first reliable difference in the cool occurred in the third 2-minute period ($p < .01$).

If anything Group IIw (broken line) started Day 1 slightly better than Group Iw (unbroken line) but the difference was not statistically reliable ($p > .05$). The first reliable difference between the two groups occurred in the seventh 2-minute period, when Group IIw performed reliably worse in the cool than Group Iw in the warm ($p < .01$). Over the middle three 2-minute periods of Day 1 Group IIw had an average error of 38.9 ± 19.0 per cent in the cool while Group

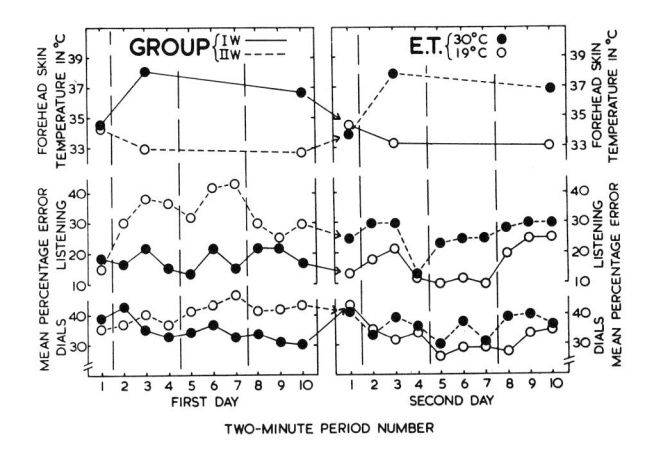

Figure 1. Results for each equivalent period of two minutes. Every point represents the average of 6 men.

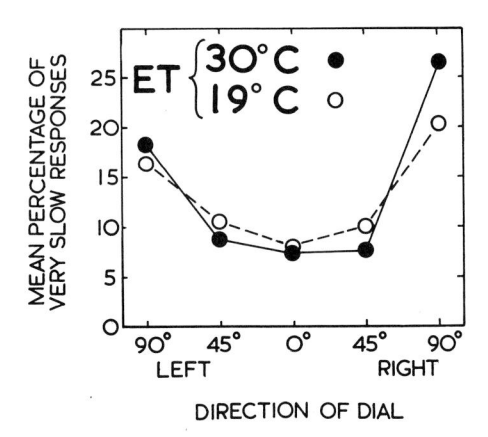

Figure 2. Proportion of response times of 3 seconds or longer on the dials task according to the direction of the dial. Every point represents 40 responses made by each of the same 12 men.

Iw had a reliably smaller average of 16.7 ± 11.5 per cent in the warm (p = .025). The results on Day 2 were complicated by the carryover from Day 1. Both groups started the second day at about the same level of performance as that at which they had finished the first day. This meant that Group Iw started off rather better in the cool than Group IIw in the warm, and was in fact just reliably better over the middle three 2-minute periods (p < .05). However by the end of the second day the performance of Group Iw in the cool had deteriorated almost to the level of the performance of Group IIw in the warm, which had hardly changed.

Delays of 2 seconds or more on the *dials* task, which are given at the bottom of Figure 1, produced a rather similar picture. Group IIw (broken line) started the first day with a mean of 35.0 ± 8.1 per cent of delayed responses, which was slightly but not reliably less than the 38.4 ± 8.5 per cent of Group Iw (unbroken line). Whereas over the last three 2-minute periods Group IIw's mean of 41.9 ±7.6 per cent in the cool was reliably greater than the mean of 31.4 ± 9.3 per cent of Group Iw in the warm (p = .025). On this measure there were no statistically reliable differences between the two groups when conditions were reversed on the second day.

Figure 2 shows how the proportion of delays of 3 seconds or more varied according to the position of the dial. There were reliably more very slow responses to the two peripheral dials than to the two next most peripheral dials in both the warm (p < .001) and the cool (p < .01). On the two peripheral dials there were reliably more very slow responses in the warm than in the cool (p < .02), whereas the difference in temperature had no statistically reliable effect on responses to the three central dials. These results are in line with those of Bursill (4). Delays of 2 seconds or more showed a similar U relationship with position of dial, but there were no statistically reliable effects related to the difference in temperature. There were no reliable differences attributable to the duration of the interval between signals, and the reported method of scanning the dials was also unrelated to performance.

At the end of the 20 minutes in the warm environment none of the men reported being uncomfortably warm, and there was no unevaporated sweat on their skin. At the end of the 20 minutes in the cool no one reported being uncomfortably cool. One man developed pink conjunctivae which he attributed to the wind in his face.

DISCUSSION

Performance on both tasks was maintained better over the brief period of less than 20 minutes after entering the stimulating environment of 45° C (113° F) with an absolute humidity of 10 millimeters of mercury and a wind movement of 500 feet per min., than after entering the less stimulating environment of 25° C (77° F) with a similar absolute humidity and wind movement. This suggests that a man may be able to work efficiently for a short period on first entering an enclosed space such as a cockpit with a temperature of 45° C, provided that he starts reasonably cool, and that humidity is kept low and the air movement adequate.

Performance in Heat as a Function of Time

The experiment was designed to measure relatively early changes in performance in the warm, such as could be produced by changes in cutaneous stimulation. It revealed a statistically reliable difference related to temperature within 12 minutes of entering the new environment. At this time the rectal temperature in the warm environment would in all probability have fallen (1). The experiment is thus complementary to that of Wilkinson *et al.* (15) in which performance was measured with the body temperature raised and then held constant by an appropriate manipulation of the environment. In a rather simpler perceptual task than the combination used here, these authors reported some evidence of deterioration at the slightly-raised body temperature of 37.3° C (99.1° F), but improvement at their highest body temperature of 38.5° C (101.3° F). They attributed these changes to a fall followed by an increase in the level of arousal.

If the results of the present experiment are considered together with those of Wilkinson *et al.*, entering a hot environment appears to produce the following sequence of changes:

1. First, the initial stimulation from warmth on the skin, with the corresponding fall in rectal temperature, produces an increase in the level of arousal. This is illustrated by Day 1 of the present experiment, on which perceptual efficiency was maintained better in the warm than in the cool.

2. As the temperature receptors of the skin adapt to the thermal stimulation and the rectal temperature rises, the level of arousal falls below normal. This gives the typical picture summarized by Pepler (8) of the man who is inefficient in heat.

3. As the body temperature rises still further, the man becomes uncomfortably hot. This raises his level of arousal above normal again, as described by Wilkinson *et al.* (15).

4. And finally, all performances deteriorate as the man begins to collapse from heat (8).

The performance of tasks which are affected by the man's level of arousal may thus improve or deteriorate in a hot environment, depending upon when performance is measured in relation to the sequence of changes. Experiments in which the duration of exposure to a hot environment is not controlled, or in which performance is averaged over a period of several hours, may show only the resultant of two or more parts of the sequence of changes; the results may thus be difficult to interpret in detail.

Transfer Effects in Latin-Square Designs

The results of the listening task in Figure 1 are an example of the transfer effects to which latin-square experimental designs are prone, and which complicate the analysis of the results. Although the two groups were comparable at the start of Day 1, they were no longer comparable at the start of Day 2. Group IIw which performed badly in the cool on Day 1 (broken line) started off badly on Day 2 in the warm and stayed bad. While Group Iw which maintained a steady level of performance in the warm on Day 1 (unbroken line) started off well on Day 2 in the cool and only gradually got worse. Its average on Day 2 was somewhat the better, although it was experiencing the less favorable of the two conditions.

On a straightforward analysis of variance using the performance of each man averaged over a whole day, the effect of temperature was statistically reliable ($p < .05$) only on a one-tailed test. Without the analysis of the trends within days, this level of reliability would not have been acceptable, since performance was better in the warm whereas a worse performance was predicted from previous work (8). Thus the latin-square design followed by a conventional analysis of variance using performance averaged over each day would have given a negative result for the effect of temperature. However the sum of squares for additivity (12, 13) was found to be reliable ($p < .02$). This shows that the data did not fit the usual model which assumes additive subject, temperature and order effects, and if calculated should suggest to the experimenter that he analyse the data in the way described in the results section.

A somewhat similar transfer effect was reported by Welford *et al.* (14) when they measured the performance of civilian aircrew before and after a fatiguing journey. Transfer effects have also been reported with continuous loud noise by Jerison (6) and by Broadbent (2, 3) and with compressed air by Poulton *et al.* (10). But this appears to be the first time that the experimental subjects were shown to be matched initially.

Acknowledgments

The research was carried out for the British Flying Personnel Research Committee under the direction of Mr. D. E. Broadbent and Air Commodore W. K. Stewart, C.B.E., A.F.C. We are grateful to Dr. J. A. Leonard for help with the experimental apparatus, to Squadron Leader J. M. Clifford for arranging for the supply of experimental subjects, and to Mr. P. Freeman for statistical help.

REFERENCES

1. Bazett, H. C. The regulation of body temperatures. *In* Newburgh, L. M. (Ed.), Physiology of Heat Regulation and the Science of Clothing, Philadelphia, Saunders, 1949, Chap. 4, pp. 109-192.
2. Broadbent, D. E. Effects of noises of high and low frequency on behaviour. *Ergonomics*, 1, 21-29, 1957.
3. Broadbent, D. E. Effect of noise on an "intellectual" task. *J. Acoust. Soc. Amer.*, 30, 824-827, 1958.
4. Bursill, A. E. The restriction of peripheral vision during exposure to hot and humid conditions. *Quart. J. exp. Psychol.*, 10, 113-129, 1958.
5. Conrad, R. Acoustic confusions in immediate memory. *Brit. J. Psychol.*, 55, 75-84, 1964.
6. Jerison, H. J. Differential effects of noise and fatigue on a complex counting task. USAF WADC Tech. Rep. 55-359, 1956.
7. Leonard, J. A. 5-choice serial reaction apparatus. Applied Psychology Unit Report No. 326, Cambridge, England, 1959.
8. Pepler, R. D. Performance and well-being in heat. *Temperature—Its Measurement and Control in Science and Industry*, 3, 319-336, 1963.
9. Poulton, E. C. On increasing the sensitivity of measures of performance. *Ergonomics*, 8, 69-76, 1965.

10. Poulton, E. C., Catton, M. J., & Carpenter, A. Efficiency at sorting cards in compressed air. *Brit. J. Indust. Med.*, **21**, 242-245, 1964.
11. Siegel, S. Nonparametric Statistics for the Behavioral Sciences, New York, McGraw-Hill, 1956.
12. Tukey, J. W. One degree of freedom for non-additivity. *Biometrics*, **5**, 232-242, 1949.
13. Tukey, J. W. Answer to query 113. *Biometrics*, **11**, 111-113, 1955.
14. Welford, A. T., Brown, R. A., & Gabb, J. E. Two experiments on fatigue as affecting skilled performance in civilian aircrew. *Brit. J. Psychol.*, **40**, 195-211, 1950.
15. Wilkinson, R. T., Fox, R. H., Goldsmith, R., Hampton, I. F. G., & Lewis, H. E. Psychological and physiological responses to raised body temperature. *J. Appl. Physiol.*, **19**, 287-291, 1964.

55

Lighting for the Forgotten Man

Sylvester K. Guth and A. A. Eastman

A fundamental objective of lighting practice is to provide illumination upon a visual task so that it can be seen quickly, accurately and easily. During the past several decades countless researches involving many criteria have been devised and conducted to obtain fundamental data upon which the specification of light and lighting could be based. Nearly all these visual data have been presented in terms of the averages obtained by a group of observers with a given level of illumination, such as the average speed and accuracy of performance, visibility, acuity, or contrast sensitivity. The data also have been expressed in terms of the average footcandles required by the group to achieve a given level of performance, visibility, etc. The variations among the individual observers generally have been used to indicate the statistical reliability of the data.

Averaged data serve a very useful purpose by providing information regarding the relationships among the various factors involved in light, vision and seeing. However, in studying the effects of illumination upon visual performance, the responses of the individual observers would seem to be at least of equal, and perhaps of greater importance than the average response. When the data are expressed as average footcandles to obtain a given performance level, it should be obvious that only half of the individuals can achieve the specific level of performance. The question arises whether or not the differences among the group average and the individual averages are so significant as to produce a considerably lower than average visual performance by many observers when they are provided with

From *American Journal of Optometry*, 32, 1955, 413-421. By permission of the American Journal of Optometry.

the average footcandle level. In other words, data need to be presented in such a way that the responses of the *forgotten half* of the observers can be used to predict the levels of illumination necessary to enable them to achieve the desired performance level. Only in this way can those individuals who need them most be given the seeing conditions they require.

Unfortunately, much of the experimental visual data is obtained with only a few observers. Therefore, it is practically impossible to analyze the data in terms of individual performance. While data obtained with highly trained and experienced observers are invaluable for developing the relationships among fundamental factors, they do not provide a basis for determining the constants which must be used for establishing absolute values of the factors for typical seeing situations. Furthermore, threshold data usually are difficult to relate to the suprathreshold conditions involved in everyday seeing.

During the past two decades we have obtained extensive data on the relationships between visibility and illumination for a large variety of visual tasks. These studies have emphasized the usefulness of the Luckiesh-Moss Visibility Meter (1) for determining the average footcandles necessary to provide a given visibility level. A method for determining a personal correction factor has been devised (2) so that data obtained by observers who deviate from the average can be brought into line. That is, every effort was made to minimize individual differences. This has been extremely useful for enabling the comparing of data on a relative basis. It also has been shown that evaluations of visibility are related to the absolute threshold footcandle level (3) and to the brightness difference criterion (4).

Recently in a study of brightness and comfort (5), it was emphasized that differences among individual observers are important. A brightness that is comfortable for one person may be extremely uncomfortable for another. It seems axiomatic that the effectiveness of illumination—or task brightness—should be appraised on a similar basis in order to give all individuals an equal opportunity for seeing. Such factors as economics and ability to provide these more adequate footcandle levels are ultimately important, but are not necessary to the development of this broader concept.

In the present study, visibility measurements were made by 40 observers. Of this group, 32 were inexperienced observers. Most of them had never used the L-M visibility meter. All observers possessed so-called normal vision and wore corrective lenses if necessary. Their ages ranged from 18 to 61 years. The task was a well printed sample of 8-point type on white paper and was viewed at a distance of 16 inches. Measurements were made with 10, 20, 40, and 80, and 100 footcandles. Care was taken to insure that the surroundings had minimal effect upon the observations.

Since the L-M visibility meter and the technique of its use have been described in detail elsewhere (1, 2), only a brief summary is given here to illustrate the significance of measurements of visibility. The instrument basically is a "reference" meter. That is, the scale has been calibrated in terms of the sizes of test objects that can be just barely discriminated through various portions of a circular gradient filter when viewed under standardized conditions of illumination. In essence, it is determined how dense a filter placed before the eyes must be to reduce the visual task to a threshold condition. The test objects were parallel bars, which consisted of black squares with the center one-third removed. Thus, the two bars and the space between them were identical in size. The observers were required to discriminate the width of the space which is calibrated in minutes visual angle. That portion of the gradient through which a 2-minute parallel-bar test object can just be seen by the standard observer is given a scale value of 2. Scale points of other values are similarly determined for a range of test object sizes. A visibility level, which can be assigned to any visual task, means that the task can be just barely seen through the same portion of the gradient filters as a certain size parallel-bar test object. It is simply a means for referring unlike visual tasks to a series of standard tasks and assigning a meaningful value to their visibilities. It must be emphasized that a scale value of 4, which indicates a visibility level twice that of a scale value of 2 does not necessarily mean that the former is twice as easy to see as the latter. In this respect it is similar to a thermometer which indicates degrees Fahrenheit; doubling the temperature from 20° to 40° does not mean that the latter is twice as warm. Nevertheless, both are rational and usable scales.

The experimental data are summarized in Table I, in which the observers have been arranged in order of increasing visibility as obtained with the 10-footcandle level. It can be seen that the range in visibility for a specific footcandle level is relatively large. For example, for 10 footcandles it ranges from about 2 to 5. These values indicate that to one observer the printed matter was equal in visibility to that of a 2-minute parallel-bar test object. On the other hand, another observer indicated that the visibility corresponded to a 5-minute test object. In other words, the perceptual demands of the two observers are considerably different in order for the task to be at threshold.

As the level of illumination is increased, the visibility level of the task improves. This occurs for all observers. An interesting comparison is the footcandle level with which each of several observers will obtain a given visibility level. Using a visibility level of 4.95 as a basis, observer No. 40 required 10 footcandles, while observers No. 25, 10 and 2 needed 20, 40 and 80 footcandles, respectively, in order to just barely discriminate the test object through the same portion of the gradient. Such differences are neither unusual nor unexpected and have been encountered by most researchers.

A more significant evaluation of the data is presented in Table II. The footcandle levels required by each observer to obtain specific visibility levels have been obtained from curves of their respective observed data. Thus, for a visibility level of 5, the illumination requirements range from 10 to 88, with an average value of 32 footcandles. It is obvious that observer No. 1 can obtain only a relatively low visibility level with this average illumination. In fact, his data indicate that the visibility of the specific task used in this investigation with 30 footcandles will be 3. This means that for the average observer the printed matter corresponds to the visibility of a 5-minute parallel-bar test object viewed with certain standard conditions, but to the low observer it corresponds to a 3-minute test object. On the other hand, if only 10 footcandles are specified for this particular visual task, it is seen from Table I that for observer No. 40 it will have a visibility level of about 5, but for observer No. 1 it will be only 2. Even though there are considerable differences among the observers and in the visibility level each obtains with a given level of illumination, the relative effectiveness of higher footcandles is about the same. Starting with a visibility level of 4, about a 50 per cent increase in illumination is required for a visibility level of 5 and a three-fold increase will

TABLE I

Visibility Levels Obtained by 40 Observers with 5 Levels of Illumination When Viewing a Sample of 8-point Type at a Distance of 16 Inches

Observer Number	\multicolumn{5}{c}{Footcandles}				
	10	20	40	80	100
1	2.00	2.55	3.4	4.8	5.4
2	2.20	2.75	3.6	5.0	5.8
3	2.25	2.85	3.8	5.3	6.1
4	2.30	3.00	4.1	6.0	7.0
5	2.30	3.00	4.4	6.6	7.5
6	2.33	2.80	3.6	4.8	5.4
7	2.35	3.15	4.4	6.8	7.9
8	2.55	3.15	4.1	5.7	6.4
9	2.60	3.50	4.9	6.9	7.7
10	2.80	3.70	5.0	6.6	7.3
11	2.90	4.25	6.3	9.7	11.2
12	3.00	3.95	5.5	7.9	8.9
13	3.10	4.00	5.3	7.2	8.0
14	3.10	4.05	5.8	8.9	10.5
15	3.15	4.05	5.3	7.1	7.9
16	3.15	4.20	5.8	8.5	9.6
17	3.20	4.05	5.4	7.6	8.5
18	3.25	4.30	6.1	9.5	11.2
19	3.35	4.35	5.8	8.0	8.9
20	3.40	4.50	6.4	9.9	11.7
21	3.50	4.65	6.3	8.8	9.8
22	3.54	4.80	6.8	10.0	11.5
23	3.55	4.70	6.5	9.2	10.5
24	3.55	4.90	7.1	10.5	12.0
25	3.60	4.95	7.4	12.0	14.3
26	3.60	5.35	8.3	13.2	15.5
27	3.65	5.05	7.3	11.0	12.5
28	3.70	4.85	6.5	8.7	9.5
29	3.80	5.10	7.7	13.8	18.0
30	3.90	5.35	7.5	11.3	13.0
31	3.95	5.50	8.2	14.5	19.0
32	4.00	5.45	7.7	11.2	12.8
33	4.00	5.45	8.0	13.5	15.3
34	4.40	6.00	8.8	14.5	17.5
35	4.55	6.70	9.9	15.5	18.0
36	4.75	6.90	10.7	18.0	21.5
37	4.90	6.60	9.0	12.5	14.0
38	4.90	6.80	9.9	14.7	16.8
39	4.95	7.30	11.3	18.8	22.5
40	4.95	7.60	12.5	21.0	25.0
Average	3.48	4.72	6.7	10.2	11.9

TABLE II

Footcandles Required by Each of 40 Observers to Obtain Various Visibility Levels When Viewing a Sample of 8-point Type at a Distance of 16 Inches.

Observer Number	\multicolumn{5}{c}{Visibility Level}
	3	4	5	7	10
1	30.0	56	88	155	275
2	25.0	49	77	145	260
3	23.0	46	71	125	210
4	20.0	38	58	100	170
5	20.0	33	50	89	160
6	25.0	53	86	160	275
7	18.0	33	48	83	140
8	17.5	38	62	120	215
9	14.0	26	42	83	165
10	12.0	24	41	92	195
11	10.5	18	27	47	83
12	10.0	20	37	64	120
13	8.3	20	35	75	165
14	9.0	19	30	55	94
15	8.6	20	35	77	170
16	9.0	18	30	57	110
17	8.0	20	34	69	130
18	8.0	17	28	50	86
19	7.4	16	28	60	130
20	7.2	16	25	46	80
21	6.8	14	23	49	105
22	6.6	14	22	42	79
23	6.3	14	23	47	93
24	6.8	13	21	39	73
25	6.4	13	20	37	63
26	7.1	12	18	31	53
27	6.1	12	20	37	69
28	5.8	12	22	48	115
29	4.8	11	20	35	57
30	4.5	10	18	35	66
31	5.3	10	17	31	53
32	4.8	10	17	33	65
33	4.3	10	17	32	56
34	3.1	8	14	27	48
35	4.2	8	12	22	40
36	3.6	7	11	20	36
37	2.7	6	11	23	50
38	3.0	6	10	21	41
39	2.9	6	10	19	33
40	4.1	7	10	18	30
Average	9.8	20	32	60	111

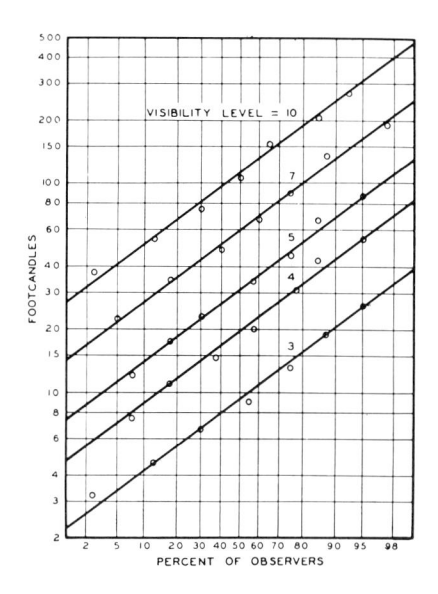

Figure 1. The relationships between the footcandles necessary for various percentages of observers to obtain given visibility levels.

TABLE III

Visibility Levels Obtained by a Representative Observer on Ten Different Days

Day	10	20	40	80	100
1	3.10	4.05	5.8	8.9	10.5
2	3.21	4.85	6.1	8.7	10.5
3	3.47	4.72	6.2	8.6	10.4
4	3.33	4.90	6.6	10.0	11.8
5	3.38	4.63	6.2	9.9	12.3
6	3.41	4.63	6.4	9.4	11.7
7	3.30	4.51	5.9	9.7	11.6
8	3.08	3.89	5.2	7.4	8.9
9	3.11	4.18	5.7	8.6	10.3
10	3.31	4.47	5.6	8.7	9.8
Average	3.27	4.48	6.0	9.0	10.8

TABLE IV

Footcandles Required by a Representative Observer to Obtain Various Visibility Levels. Measurements Were Made on Ten Different Days.

		Visibility Levels			
Day	3	4	5	7	10
1	9.0	19	30	55	94
2	8.2	17	27	51	93
3	6.8	15	26	50	94
4	8.0	15	23	43	79
5	7.3	15	25	48	80
6	7.3	15	25	48	83
7	8.0	16	25	48	83
8	9.3	22	38	71	115
9	8.9	19	30	57	97
10	7.4	17	29	58	103
Average	8.0	17	28	53	92

TABLE V

Visibility Levels Obtained by Various Age Groups of Observers with 5 Levels of Illumination.

Age Group	No. of Observers	\multicolumn{5}{c}{Footcandle Levels}
		10	20	40	80	100
16 - 25	11	3.95	5.6	8.0	12.4	14.6
26 - 35	16	3.43	4.7	6.8	10.3	11.9
36 - 45	6	3.58	4.7	6.5	9.3	10.5
46 - 55	6	2.48	3.2	4.5	7.0	8.4
56 - 65	1	2.30	3.0	4.4	6.6	7.5

raise the visibility level to 7. With relatively minor variations, this holds true for all 40 observers.

These visibility levels can be converted into equivalent type sizes (6). If 30 footcandles and a visibility level of 5 are taken as the basis for comparison, visibility levels of 4, 3 and 2 correspond to equivalent type sizes of 6, 4 and 2 points, respectively. In other words, on this basis, a visibility level of 2 indicates that even with 30 footcandles it is as though observer No. 1 were reading 2-point type. Certainly, it is a much more difficult visual task for him than for the other observers. These reduced visibility levels also can be converted into effective decreased contrasts (7). Thus, if a visibility level of 5 is obtained with a 95-per cent contrast test object, with a specific footcandle level, then visibility levels of 4, 3, and 2, with the same footcandle level, correspond to contrasts of 80, 55 and 30 per cent, respectively.

The footcandle levels required for various percentages of observers to obtain various visibility levels are plotted in Figure 1. These values were derived from Table II. For calculating purposes, the footcandle levels in each column were arranged into seven groups. The percentages have been plotted on a normal probability integral scale. Since the plotted points may be represented by a series of straight lines, normal or Gaussian distributions of the individual footcandle levels are indicated. Thus, the group of 40 observers can be considered representative of a much larger group.

The relationships of Figure 1 represent the percentage of observers who will obtain at least a given visibility level with various footcandle levels. Thus, a visibility level of 5 is obtained by 95, 50 and 5 per cent of the observers with 85, 30, and 11 footcandles, respectively. These relationships emphasize the importance of considering individual subjective variations when determining required levels of illumination. About three times the average footcandle level is required so that 95 per cent of the observers will be able to achieve the visibility level indicated by the average.

Another interpretation can be made from Figure 1. For example, 20 footcandles may be considered for a specific visual task for an average visibility level of 4. However, only 50 per cent of the workers will be expected to obtain the average or expected visibility level. About 25 per cent will have a visibility level of 5 or more, but 10 per cent will have a visibility level of 3 or less. If this visual task is a particularly critical operation in a factory and errors may be costly, it is evident that a higher footcandle level will enable more of the workers to achieve a greater certainty of seeing.

It has been shown that differences among individual observers are great. In Tables III and IV are presented data for a representative observer. Visibility measurements were made by observer No. 14 on ten different days. Each datum represents the average of a series of ten individual observations. This was one of the inexperienced observers and the observations of the first day only were included in Tables I and II. The day-to-day variability of a single observer is considerably less than was found among the entire group of 40 observers. Furthermore, there is no indication of a learning trend. The average deviation of the ten-day series of observations is about 10 per cent. This compares with an average deviation of about 50 per cent for all the observers. In other words, the average deviation for the latter would indicate that a relatively large increase in illumination would be required before there would be a statistically significant improvement in visibility. On the other hand, the individual data indicate that smaller increases in illumination will be significant. It certainly seems worthwhile to consider individual consistency rather than to emphasize interobserver variability.

While the 40 observers are not uniformly distributed in various age groups, an evaluation of the data does indicate a definite trend towards lower visibility levels for the older observer. This is illustrated in Table V. It is evident that these older observers require a higher footcandle level to obtain a given visibility level than do the younger observers. For example, the 46-55 age group needs about 100 footcandles for a visibility level of 8 as compared with 40 footcandles required by the youngest group. Other comparisons indicate that the two oldest groups require from two to three times more illumination than the 16-25 age group for a given visibility level. This is in general agreement with Weston's results for visual performance which involved speed and accuracy (8).

This brief summary illustrates the importance of considering individual ability and performance when developing footcandle recommendations for a specific task. If all visual responses were identical, it would be a very simple thing to do. The *apparent* lack of statistical reliability of data should be considered as positive rather than negative information. Averages are useful, but not all informative.

While this approach is relatively new to evaluating the effectiveness of illumination, it is not new in other fields. In optometry, for example, because the patient can read only 20/40 Snellen, does anyone specify a certain spherical correction without determining causes for this deficiency? Is a 55-year-old patient arbitrarily given an added correction of plus 2.25 diopters for near vision without consideration of many other factors?

It may seem idealistic to specify lighting requirements in terms of individual visual performance. Nevertheless, it is not idealistic in developing a broad concept. To be sure, while it may be practical to test everyone's eyes on this basis, it generally will be uneconomical to provide individualized lighting in typical workworld situations. But, would it not be worthwhile to consider that everyone will benefit from these more adequate footcandle levels? Those who are unfortunate and possess a relatively low visual ability can be raised to a reasonable average level. At the same time the others will be given even better seeing conditions. The mere fact that these levels may be difficult or uneconomical to obtain today should not prevent us from setting them as our goal for the future. Too often the goal of yesterday has become the millstone of today, holding us back from accomplishing what we know should be done. Our ultimate and invariable goal should be the ideal.

REFERENCES

1. Visibility—Its Measurement and Significance in Seeing. Matthew Luckiesh and F. K. Moss, Journal of the Franklin Institute, **220**, 1935, 431.
2. Technique of Using the Luckiesh-Moss Visibility Meter. Matthew Luckiesh, A. A. Eastman and S. K. Guth, Illuminating Engineering, **43**, 1948, 223.
3. A New Scale of Relative Footcandles for the Luckiesh-Moss Visibility Meter. Matthew Luckiesh, S. K. Guth and A. A. Eastman, Illuminating Engineering, **45**, 1950, 211.
4. Brightness Difference—A Basic Factor in Suprathreshold Seeing. Sylvester K. Guth,

A. A. Eastman and R. C. Rodgers, Illuminating Engineering, **48**, 1953, 233.
5. BCD Brightness Ratings in Lighting Practice, Sylvester K. Guth, Illuminating Engineering, **47**, 1952, 184.
6. Reading as a Visual Task. Matthew Luckiesh and Frank K. Moss. D. Van Nostrand Company, Inc., New York, 1942.
7. Footcandles for Critical Seeing. Matthew Luckiesh and A. A. Eastman, Illuminating Engineering, **41**, 1946, 828.
8. Age and Illumination in Relation to Visual Performance, H. C. Weston, Transactions of the Illuminating Engineering Society (London), **14**, 1949, 281.

General References

Adams, J. A. Some considerations in the design and use of dynamic flight simulators. In H. W. Sinaiko (Ed.), *Selected papers on human factors in the design and use of control systems.* New York: Dover Publications, 1961. Pp. 88–114.

Adams, J. A. Experimental studies of human vigilance. *Electronic Systems Division Technical Documentary Report* No. 63-320, 1963, United States Air Force, L. G. Hanscom Field, Bedford, Mass.

Adams, J. A. (Ed.), Vigilance. Special issue *Human Factors,* 1965, 7 (Whole No. 2).

Adams, J. A. *Human memory.* New York: McGraw-Hill, 1967.

Appley, M. H., & Trumbull, R. *Psychological stress.* New York: Appleton-Century-Crofts, 1967.

Atkinson, R. C. Computerized instruction and the learning process. *American Psychologist,* 1968, 23, 225–239.

Attneave, F. *Applications of information theory to psychology.* New York: Holt-Dryden, 1959.

Bahrick, H. P., Fitts, P. M., & Briggs, G. E. Learning curves—facts or artifacts? *Psychological Bulletin,* 1957, 54, 256–268.

Baker, C. A., & Grether, W. F. Visual presentation of information. WADC Technical Report 54-160, August 1954, Wright Air Development Center, United States Air Force, Wright-Patterson Air Force Base, Ohio.

Bartlett, F. C. *Thinking: An experimental and social study.* London: George Allen & Unwin, 1958.

Becker, G. M., & McClintock, G. G. Value: Behavioral decision theory. *Annual Review of Psychology,* 1967, 18, 239–286.

Bennett, E., Degan, J., & Spiegel, J. *Human factors in technology.* New York: McGraw-Hill, 1963.

Berlyne, D. E. *Conflict, arousal, and curiosity.* New York: McGraw-Hill, 1960.

Bilodeau, E. A. (Ed.) *Acquisition of skill.* New York: Academic Press, 1966.

Blackwell, H. R. Development and use of a quantitative method for specification of interior illumination levels on the basis of performance data. *Illuminating Engineering,* 1959, 54, 317–353.

Boguslaw, R., & Porter, E. H. Team functioning and training. In R. M. Gagné (Ed.), *Psychological principles in system development.* New York: Holt, Rinehart & Winston, 1962. Pp. 387–416.

Bourne, L. E., & Battig, W. F. Complex processes. In J. B. Sidowski (Ed.), *Experimental methods and instrumentation in psychology.* New York: McGraw-Hill, 1966. Pp. 541–576.

Briggs, G. E. The generality of research on transfer functions. In A. W. Melton (Ed.), *Categories of human learning.* New York: Academic Press, 1964. Pp. 286–292.

Briggs, G. E. Comments on Dr. Poulton's Paper. In E. A. Bilodeau (Ed.), *Acquisition of skill.* New York: Academic Press, 1966. Pp. 411–424.

Briggs, G. E., Fitts, P. M., & Bahrick, H. P. Transfer from a single to a double integral tracking system. *Journal of Experimental Psychology,* 1958, 55, 135–142.

Briggs, G. E., & Johnston, W. A. Influence of a change in systems criteria on team performance. *Journal of Applied Psychology*, 1966, **50**, 467–472.

Briggs, G. E., & Johnston, W. A. Team training, *Technical Report* No. 1327-4, June, 1967, Naval Training Device Center, Orlando, Fla.

Briggs, G. E., & Waters, L. K. Training and transfer as a function of component interaction. *Journal of Experimental Psychology*, 1958, **56**, 492–500.

Broadbent, D. E. Noise: Its effect on behavior. *Royal Society Health Journal*, 1955, **75**, 541–545.

Broadbent, D. E. *Perception and communication*. Oxford: Pergamon Press, 1958.

Broadbent, D. E. Differences and interactions between stresses. *Quarterly Journal of Experimental Psychology*, 1963, **15**, 205–211.

Brooks, H. Applied science and technological progress. *Science*, 1967, **156**, 1706–1712.

Buckhout, R. Effects of whole body vibration on human performance. *Human Factors*, 1964, **6**, 157–163.

Buckner, D. N., & McGrath, J. J. (Eds.) *Vigilance: A symposium*. New York: McGraw-Hill, 1963.

Bugelski, B. R., & Cadwallader, T. C. A reappraisal of the transfer and retroaction surface. *Journal of Experimental Psychology*, 1956, **52**, 360–366.

Burck, G. *The computer age*. New York: Harper & Row, 1965.

Burns, N. M., Chambers, R. M., & Hendler, E. (Eds.) *Unusual environments and human behavior*. New York: Free Press, 1963.

Burns, N. M., & Kimura, D. Isolation and sensory deprivation. In N. M. Burns, R. M. Chambers, & E. Hendler (Eds.), *Unusual environments and human behavior*. New York: Free Press, 1963. Pp. 167–192.

Campbell, D., & Stanley, J. *Experimental and quasi-experimental designs for research*. Chicago: Rand McNally, 1966.

Castle, P. F. C. Evaluation of human relations training for supervisors. *Occupational Psychology*, 1952, **26**, 191–205.

Chambers, R. M. Operator performance in acceleration environments. In N. M. Burns, R. M. Chambers, & E. Hendler (Eds.), *Unusual environments and human behavior*. New York: Free Press, 1963. Pp. 193–319.

Chapanis, A. *Man-machine engineering*. Belmont, Calif.: Wadsworth, 1965.

Chapanis, A., Garner, W. R., & Morgan, C. T. *Applied experimental psychology: Human factors in engineering design*. New York: Wiley, 1949.

Christensen, J. M. The emerging role of engineering psychology. Aerospace Medical Research Laboratories Technical Report No. 64-88, September, 1964, Wright-Patterson Air Force Base, Ohio.

Churchman, C. W., Ackoff, R. L., & Arnoff, E. L. *Introduction to operations research*. New York: Wiley, 1957.

Corso, J. F. *The experimental psychology of sensory behavior*. New York: Holt, Rinehart & Winston, 1967. Pp. 550–589.

Dunlap, J. W. Men and machines. *Journal of Applied Psychology*, 1947, **31**, 565–579.

Edwards, W. A perspective on automation and decision making. In D. Willner (Ed.), *Decisions, values, and groups*. New York: Pergamon Press, 1960. Pp. 3–8.

Edwards, W. Costs and payoffs are instructions. *Psychological Review*, 1961, **68**, 275–284.

Edwards, W., & Tversky, A. (Eds.) *Decision making*. Baltimore: Penguin, 1967.

Ellis, H. *The transfer of learning*. New York: Macmillan, 1965.

Estes, W. K. The problem of inference from curves based on group data. *Psychological Bulletin*, 1956, **53**, 134–140.

Fitts, P. M. Engineering psychology and equipment design. In S. S. Stevens (Ed.), *Handbook of Experimental Psychology*. New York: Wiley, 1951. Pp. 1287–1340.

Fitts, P. M. Engineering psychology. *Annual Review of Psychology*, 1958, **9**, 267–294.

Fitts, P. M. Engineering psychology. In S. Koch (Ed.), *Psychology: A study of a science*. New York: McGraw-Hill, 1963. Pp. 908–933.

Fitts, P. M. Perceptual-motor skill learning. In A. E. Melton (Ed.), *Categories of human learning*. New York: Academic Press, 1964. Pp. 243–285.

Fitts, P. M. Factors in complex skill learning. In R. Glaser (Ed.), *Training research and education*. New York: Wiley, 1965. Pp. 177–198.

Fitts, P. M., & Posner, M. I. *Human performance*. Belmont, Calif.: Wadsworth, 1967.

Fitts, P. M., & Seeger, C. M. S-R compatibility: Spatial characteristics of stimulus and response codes. *Journal of Experimental Psychology*, 1953, **46**, 199–210.

Fleishman, E. A. Development of a behavior taxonomy for describing human tasks: A correlational-experimental approach. *Journal of Applied Psychology*, 1967, **51**, 1–10.

Fleishman, E. A., Harris, E. F., & Burtt, H. E. Leadership and supervision in industry. Bureau of Educational Research, Report No. 33, The Ohio State University, 1955.

Frankmann, J. P., & Adams, J. A. Theories of vigilance. United States Air Force AFCCDD Technical Note No. 60-25, 1960.

French, S. H. Measuring progress toward industrial relations objectives. *Personnel*, 1953, **30**, 338–347.

Gagné, R. M. Problem solving and thinking. *Annual Review of Psychology*, 1959, **10**, 147–152.

Gagné, R. M. Military training and principles of learning. *American Psychologist*, 1962, **17**, 83–91. (a)

Gagné, R. M. *Psychological principles in system development*. New York: Holt, Rinehart & Winston, 1962. (b)

Galanter, E. H. (Ed.) *Automatic teaching: The state of the art*. New York: Wiley, 1959.

Garner, W. R. *Uncertainty and structure as psychological concepts*. New York: Wiley, 1962.

Geldard, F. A. Military psychology: Science or technology? *American Journal of Psychology*, 1953, **66**, 335–348.

Glanzer, M. Experimental study of team training and team functioning. In R. Glaser (Ed.), *Training research and education*. New York: Wiley, 1965. Pp. 379–407.

Glaser, R. Psychology and instructional technology. In R. Glaser (Ed.), *Training research and education*. New York: Wiley, 1965. Pp. 1–30.

Goodyear Aircraft Corp. Investigation of control "feel" effects on the dynamics of a piloted aircraft system. Goodyear Aircraft Corp. Report No. GER-6726, April, 1955.

Grether, W. F. Engineering psychology in the United States. *American Psychologist*, 1968, **23**, 743–751.

Guion, R. M. *Personnel testing*. New York: McGraw-Hill, 1965.

Hilgard, E. R., & Bower, G. H. *Theories of learning*. (3rd ed.) New York: Appleton-Century-Crofts, 1966.

Holding, D. H. *Principles of training*. London: Pergamon, 1965.

Holt, H. O. Programmed instruction. *Bell Telephone Magazine*, Spring, 1963.

Howell, W. C. Some principles for the design of decision systems: A review of six years of research on command-control system simulation. Aerospace Medical Research Laboratories Technical Documentary Report No. 67-136, September, 1967, Wright-Patterson Air Force Base, Ohio.

Hunt, D. P., and Zink, D. L. (Eds.) Predecisional processes in decision making: Proceedings of a symposium. Aerospace Medical Research Laboratories Technical Documentary Report No. 64-77, 1964, Wright-Patterson Air Force Base, Ohio.

Jerison, H. J. Effects of noise on human performance. *Journal of Applied Psychology*, 1959, **43**, 96–101.

Jerison, H. J., & Pickett, R. M. Vigilance: A review and re-evaluation. *Human Factors*, 1963, **5**, 211–238.

Johnston, W. A. Individual performance and self-evaluation in a simulated team. *Journal of Organizational Behavior and Human Performance*, 1967, **2**, 309–328.

Kennedy, J. L. The systems approach. *Human Factors*, 1962, **4**, 25–52.

Keppel, G. Verbal learning and memory. *Annual Review of Psychology*, 1968. **19**, 169–202.

Kidd, J. S. A summary of research methods, operator characteristics, and system design specifications based on the study of a simulated radar air traffic control system. WADC Technical Report 59-236, July, 1959, Wright Air Development Center, United States Air Force, Wright-Patterson Air Force Base, Ohio.

Kleinmuntz, B. (Ed.) *Formal representation of human judgment*. New York: Wiley, 1968.

Kryter, K. D. The effects of noise on man. *Journal of Speech and Hearing Disorders*, 1950 (Monogr. Suppl. 1).

Kryter, K. D. Psychological reactions to aircraft noise. *Science*, 1966, **151**, 1346–1355.

Laming, D. R. J. *Information theory of choice-reaction time*. New York: Academic Press, 1968.

Lindahl, L. G. How to build a training program. *Personnel Journal*, 1949, **27**, 417–419.

Loftus, J. P., Jr., & Hammer, L. R. Weightlessness. In N. M. Burns, R. M. Chambers, & E. Hendler (Eds.), *Unusual Environments and Human Behavior*. New York: Free Press, 1963, pp. 353–377.

Lumsdaine, A. A., & Glaser, R. *Teaching machines and programmed learning: A source book*. Washington, D.C.: National Education Association, 1960.

McCormick, E. J. *Human factors engineering*. New York: McGraw-Hill, 1964.

McGehee, W., & Livingstone, D. W. Training reduces material waste. *Personnel Psychology*, 1952, **5**, 115–123.

McGehee, W., & Livingstone, D. W. Persistence of the effects of training employees to reduce waste. *Personnel Psychology*, 1954, **7**, 33–39.

McGehee, W., & Thayer, P. W. *Training in business and industry*. New York: Wiley, 1964.

McGrath, J. E., & Altman, J. *Small group research*. New York: Holt, Rinehart & Winston, 1966.

Meister, D., & Rabideau, G. F. *Human factors evaluation in system development*. New York: Wiley, 1965.

Melton, A. W. Implication of short-term memory for a general theory of memory. *Journal of Verbal Learning and Verbal Behavior*, 1963, **2**, 1–21.

Melton, A. W., & Briggs, G. E. Engineering psychology. *Annual Review of Psychology*, 1960, **11**, 71–98.

Miller, J. W., Malecki, G. S., & Farr, M. J. ONR's role in human factors engineering. *Naval Research Reviews*, 1967, **20**, 1–10.

Morgan, C. T., Cook, J. S., III, Chapanis, A., & Lund, M. W. (Eds.) *Human engineering guide to equipment design.* New York: McGraw-Hill, 1963.

Murdock, B. B. Transfer designs and formulas. *Psychological Bulletin,* 1957, **54,** 313–326.

Naylor, J. C., & Briggs, G. E. Long-term retention of learned skills: A review of the literature. Aerospace Medical Laboratory, Aerospace System Division Technical Report 61-390, August, 1961, Wright-Patterson Air Force Base, Ohio.

Neisser, U. *Cognitive psychology.* New York: Appleton-Century-Crofts, 1967.

Noble, C. E. The learning of psychomotor skills. *Annual Review of Psychology,* 1968, **19,** 203–250.

Osgood, C. E. The similarity paradox in human learning: A resolution. *Psychological Review,* 1949, **56,** 132–143.

Pickrell, E. W., & McDonald, T. A. Quantification of human performance in large, complex systems. *Human Factors,* 1964, **6,** 647–662.

Pierce, J. F. (Ed.) *Operations research and the design of management information systems.* New York: Technical Association of the Pulp and Paper Industry, 1967.

Poulton, E. C. Engineering psychology. *Annual Review of Psychology,* 1966, **17,** 177–200. (a)

Poulton, E. C. Tracking behavior. In E. A. Bilodeau (Ed.), *Acquisition of skill.* New York: Academic Press, 1966. (b)

Reagan, M. D. Basic and applied research: A meaningful distinction? *Science,* 1967, **155,** 1383–1386.

Reed, J. E., & Haymen, J. L., Jr. An experiment involving use of English 2600, an automated instruction text. *Journal of Educational Research,* 1962, **55,** 476–484.

Shay, C. B. Relationship of intelligence to step size on a teaching machine program. *Journal of Educational Psychology,* 1961, **52,** 98–103.

Shelley, M. W., III, & Bryan, G. L. (Eds.) *Human judgment and optimality.* New York: Wiley, 1964.

Sidman, M. A note on functional relations obtained from group data. *Psychological Bulletin,* 1952, **49,** 263–269.

Smith, E. E. Choice reaction time: An analysis of the major theoretical positions. *Psychological Bulletin,* 1968, **69,** 77–110.

Smith, W. I., & Moore, J. W. (Eds.) *Programmed learning.* Princeton: Van Nostrand, 1962.

Solomon, P. (Ed.) *Sensory deprivation.* Cambridge, Mass.: Harvard University Press, 1961.

Swets, J. A., Tanner, W. P., Jr., & Birdsall, R. G. Decision processes in perception. *Psychological Review,* 1961, **68,** 301–340.

Taylor, D. W. Thinking. In M. H. Marx (Ed.), *Theories in contemporary psychology.* New York: Macmillan, 1963. Pp. 475–493.

Taylor, F. V. Psychology and the design of machines. *American Psychologist,* 1957, **21,** 120–125.

Taylor, F. V. Human engineering and psychology. In S. Koch (Ed.), *Psychology: A study of a science.* New York: McGraw-Hill, 1963. Pp. 831–907.

Walker, D. E. (Ed.) *Information system science and technology.* Washington, D.C.: Thompson, 1967.

Weinberg, E. Old worker's performance in industrial retraining programs. *Missouri Labor Review,* 1963, **86,** 935–939.

Welford, A. T. The measurement of sensory-motor performance: Survey and reappraisal of twelve years' progress. *Ergonomics,* 1960, **3,** 189–230.

Wheaton, J. L. *Fact and fancy in sensory deprivation studies.* Report No. 5-59, 1959, School of Aviation Medicine, Brooks Air Force Base, Texas.

White, W. J., & Monty, R. A. Vision and unusual gravitational forces. *Human Factors*, 1963, **5**, 239–264.

Woodson, W. E. *Human engineering guide for equipment designers.* Berkeley: University of California Press, 1954.

Author Index

Abma, J. S., 440, 459, 465, 467
Aborn, M., 140
Ackoff, R. L., 303
Adams, D. K., 392, 397
Adams, J. A., 77, 131, 168, 169, 174, 176, 177, 178, 189, 190, 194, 195, 197, 204, 207, 209, 211, 439, 514, 515, 530, 531, 540, 551, 552, 553, 554, 563, 564, 626, 628
Adkins, D. C., 450, 458
Alberoni, F., 239, 241
Alluisi, E. A., 45, 55, 56, 348, 351, 353, 354, 487, 493
Altman, J., 300
Ammerman, H. L., 482, 485
Anderson, D. M., 174, 193
Anderson, I. H., 517, 525, 529
Anderson, N. H., 247, 249
Anderson, N. S., 110, 115
Andrews, T. G., 493, 495
Appley, M. H., 512, 567, 584, 586, 589, 592, 594, 595, 626
Archer, W., 218, 221, 222, 223, 225, 227, 228, 229, 480, 485
Arginteanu, J., 281, 286
Arieti, S., 582, 583
Arnoff, E. L., 302
Arnoult, M. D., 180, 190
Aseltine, J. A., 330
Ashby, W. R., 308, 314
Ashe, W. F., 606, 607, 611, 612
Atkinson, R. C., 250, 441, 467, 470, 473, 474, 475, 626
Attneave, F., 130, 217, 228
Attridge, B. F., 410, 417
Autor, S. M., 373, 379
Averbach, E., 104, 111, 115, 125, 128
Azuma, H., 244, 245, 249

Baddeley, A. D., 124, 128
Baganoff, F., 213
Bahrick, H. P., 44, 69, 170, 190, 194, 195, 196, 207, 209, 211, 212, 276, 288, 290, 291, 292, 387, 439, 626
Bakan, P., 531, 541
Baker, C. A., 43, 63
Baker, C. H., 551, 562, 563
Baker, W. J., 492, 493
Baratta, P., 254, 255, 256, 266

Barnes, C. B., 245, 250
Bartlett, F. C., 25, 51, 168, 174, 178, 190, 194, 212, 215, 218, 228, 524, 529
Bartlett, S. C., 530, 531, 540
Baskin, S., 466
Basowitz, H., 589, 594
Bass, B. M., 426, 427
Bateson, G., 22, 26, 142
Batson, W. H., 453, 458
Battig, W. F., 215, 216, 227, 228, 229
Baxt, N., 106, 107, 108, 109, 111, 115
Bazett, H. C., 618, 619
Beach, L. R., 216, 230, 233, 234, 236, 237, 238, 239, 240, 243, 244, 249, 251
Beardslee, D. L., 255, 266
Becker, G. M., 216, 231, 241, 242, 249, 321, 330
Beebe-Center, J. G., 91, 93, 101
Beinert, J. R., 530, 531, 540
Belbin, R. M., 531, 541
Bellman, R., 324, 325, 330, 331
Bennett, E., 276, 599, 626
Bennett, G. K., 441, 459
Bennett, W. F., 170, 190, 290, 291
Bentley, A. F., 8, 15
Berger, C., 46, 55
Berlyne, D. E., 598, 626
Bernard, C., 585, 595
Bernoulli, D., 254, 263, 266
Besnard, G. G., 406, 408
Bexton, W. H., 568, 574
Biederman, I., 159, 162, 222, 225, 229
Biel, W. C., 286
Bills, A. G., 517, 529
Bilodeau, E. A., 211, 213, 386, 451, 452, 453, 454, 458, 626
Binder, A., 244, 249
Binford, J. R., 542, 543, 546, 549, 550
Birdsall, T. G., 35, 75, 76, 78, 553, 564
Birmingham, H. P., 172, 174, 175, 182, 190, 193, 308, 314
Bitzer, D. L., 472, 475
Bjorkman, M., 245, 249
Blackwell, H. R., 78, 599, 626
Blank, G., 143, 157
Bleuler, E., 582, 583
Bliss, E. L., 527, 530
Boguslaw, R., 310, 314, 478, 626
Boring, E. G., 17, 21, 26
Bottenberg, R. A., 483, 485

633

Boulter, L. R., 554, 563
Bourne, L. E., 215, 216, 219, 227, 228, 229, 230, 240, 251
Bousfield, W. A., 100, 101
Bouthilet, L., 589, 595
Bowen, H. M., 65, 68, 530, 531, 540
Bower, G. H., 385, 387, 628
Bower, J. L., 171, 173, 190
Bradley, J. V., 274, 275, 276, 279, 281, 286
Brady, J. V., 539, 541
Bramson, A. D., 333, 341
Brandalise, B. B., 184, 190
Brandt, R., 331
Brayfield, A. H., 425, 427
Brennan, T. N. N., 286
Brian, D., 470, 471, 473, 475
Bricker, P. D., 217, 218, 221, 222, 223, 224, 225, 229
Bridgman, C. S., 446, 458
Briggs, G. E., 27, 35, 40, 131, 170, 189, 190, 194, 195, 196, 205, 207, 211, 212, 356, 360, 361, 371, 387, 389, 439, 478, 506, 507, 508, 509, 510, 626, 627, 630
Briggs, L. J., 398, 403, 404, 406, 407, 408
Broadbent, D. E., 31, 35, 77, 109, 110, 111, 113, 115, 122, 125, 128, 131, 160, 162, 194, 212, 221, 229, 515, 531, 541, 542, 543, 546, 547, 549, 550, 551, 553, 563, 577, 583, 597, 604, 605, 619, 627
Brogden W. J., 188, 191, 446, 458
Brooks, H., 30, 35
Brown, D. L., 243, 249
Brown, F. G., 227, 228
Brown, F. R., 46, 55
Brown, G. S., 171, 173, 190
Brown, J., 111, 115, 117, 119, 122
Brown, J. S., 447, 458
Brown, R., 577, 583
Brown, R. A., 619, 620
Brown, W. L., 243, 251
Brozek, J., 492, 493
Bruner, J., 217, 229
Brunswik, E., 230, 249
Bryan, G. L., 215, 216
Buchanan, P. C., 432
Buckhout, R., 596, 606, 607, 608, 611, 627
Buckner, D. N., 515, 627
Budin, W. A., 582, 583
Bugelski, B. R., 389, 627
Bunderson, V., 471, 475
Burck, G., 381, 627
Burke, C. J., 247, 250
Burnett, A., 414, 417
Burns, N. M., 565, 593, 595, 599, 627
Bursill, A. E., 617, 619
Burtt, H. E., 421, 628

Busgang, G., 331
Bush, R. R., 274, 317, 323

Cadwallader, T. C., 389, 627
Cameron, N. A., 582, 583
Campbell, D., 421, 627
Campbell, D. P., 171, 173, 190
Campbell, D. T., 218, 229
Candland, D. K., 568, 569, 570, 571, 572, 573, 574
Canges, L., 237, 253
Cannon, W. D., 585, 595
Carmichael, L., 100, 101
Carpenter, A., 524, 529, 616, 619, 620
Carr, H. A., 454, 458
Carr, W. J., 501, 504
Castle, P. F. C., 419, 627
Cattell, J. McK., 104, 115
Catterson, A. D., 607, 612
Catton, M. J., 616, 619, 620
Chambers, A. N., 286
Chambers, R .M., 593, 595, 599, 627
Chambers, R. W., 207, 211
Chang, J. J., 161, 162, 225, 229
Chapanis, A., 21, 25, 26, 28, 35, 43, 93, 101, 276, 292, 295, 296, 298, 599
Chapman D. W., 95, 101
Chapman, L. J., 582, 583
Chernikoff, R., 170, 174, 182, 184, 190
Cherry, E. C., 115, 128, 140
Child, I. L., 188, 190
Chiles, W. D., 607, 612
Christal, R. E., 480, 483, 484, 485
Christensen, J. M., 1, 16, 19, 21, 26, 27, 28, 33, 35, 45
Christie, L. S., 161, 162, 268, 273
Christner, L. A., 56
Churchill, E., 27
Churchman, C. W., 302
Clark, C. C., 612
Coermann, R. R., 606, 607, 608, 612, 613
Cofer, C. N., 584, 586, 589, 592, 595
Coffey, L. J., 56
Cohen, B. H., 100, 101
Cohen, I. S., 460, 466
Cohen, J., 239, 249
Cole, M., 247, 250
Conklin, J. E., 170, 190
Conrad, R., 123, 125, 128, 177, 185, 190, 212, 615, 619
Cook, J. S., 43, 276, 298, 599, 630
Cook, T. W., 410, 411, 417
Coombs, C. H., 255, 266
Cope, F. W., 607, 612
Corcoran, D. W. J., 548, 550
Corso, J. F., 565, 627
Coulson, J. E., 501, 504

Author Index

Cowan, J. D., 344, 347, 354
Craig, D. R., 175, 190, 199, 287, 458, 477
Craik, K. J. W., 174, 182, 183, 190, 518, 529
Crawford, M. P., 400, 408
Crawley, S. L., 526, 529
Creamer, L. R., 195, 209, 211
Crockett, W. H., 425, 427
Cronbach, L. J., 244, 245, 249, 393, 397, 495, 503, 505
Crook, M. N., 46, 52, 53, 55
Cross, K., 196, 197, 198, 199, 201, 202, 209, 210, 211, 212, 213
Crossman, E. R. F. W., 220, 221, 222, 223, 225, 229
Crowder, N. A., 406, 407, 408
Crowley, M. E., 451, 458
Cummings, A., 464, 466

Dale, H. C. A., 74
Damon, A., 27
Daniel, D. R., 333, 341
Dart, R. R., 18, 26
Dashiell, J. F., 392, 393
David, H. A., 161, 162
Davis, R. L., 174, 182, 188, 191, 266
de Groot, S., 65, 68
Dean, R. D., 606, 612
Dearnaley, E. J., 239, 249
Deese, J., 205, 206, 212, 497, 498, 505, 530, 531, 532, 536, 537, 539, 541, 551, 557, 563, 589, 590, 591, 595
Degan, J., 276, 599, 626
Dennis, J. P., 607, 612
Derwort, A., 290, 292
Dillon, R. F., 493, 495
Ditchburn, R. W., 517, 529
Doan, B. K., 568, 574
Dodge, R., 103, 104, 115
Dodson, J. D., 246, 249, 359, 360, 371
Dolanskey, L., 140
Dolanskey, M. P., 140
Dorfman, D. D., 237, 249
Dougherty, D. J., 44, 64
Dowda, F., 237, 253
Draguns, J., 582, 583
DuCharme, W. M., 235, 240, 242, 251
Dudycha, L. W., 244, 249
Duggar, B. C., 300, 342, 372, 373, 379
Duncan, D. B., 51, 55
Dunlap, J. W., 388, 627
Dupertuis, C. W., 27

Eastman, A. A., 599, 620, 621, 623, 624, 625
Eckstrand, G. A., 286, 477, 495, 501, 505

Edgerton, H. A., 441, 458
Edwards, R. G., 606, 612
Edwards, W., 69, 214, 215, 216, 231, 234, 235, 236, 237, 239, 240, 241, 242, 243, 246, 249, 250, 251, 252, 253, 255, 257, 258, 262, 265, 266, 267, 269, 273, 301, 304, 315, 316, 317, 321, 322, 331, 342, 355, 360, 364, 371
Egan, J. P., 542, 550
Ekman, G., 190, 212
Eldridge, C., 589, 595
Elithirn, A., 174, 182, 190
Elkin, E. H., 492, 493
Ellis, D. S., 402, 403, 408
Ellis, H., 389, 627
Ellison, G. D., 490, 494
Ellson, D. G., 172, 175, 191, 207, 212, 447, 458
Elwell, J. L., 526, 529
Emmons, W. H., 106, 115
Enoch, J. M., 47, 55, 373, 379
Erdmann, B., 103, 104, 115
Eriksen, C. W., 91, 93, 94, 101, 224, 229
Erlick, D. E., 232, 238, 250
Eslinger, R., 196, 197, 213
Estes, W. K., 118, 119, 122, 244, 247, 250, 317, 387, 474, 475, 627
Evans, J., 501, 505
Exner, S., 106, 115

Falk, J. L., 504, 505
Fano, R. M., 140
Fant, C. G. M., 102
Farr, M. J., 28, 35
Feinberg, L. D., 582, 583
Feldman, S. E., 244, 249
Fenn, W. O., 289, 292
Ferguson, G. A., 409, 417
Ferster, C. B., 534, 535, 537, 539, 540, 541
Fick, R., 291, 292
Ficks, L., 93, 94, 102
Fischer, R. A., 138
Fishman, M., 473, 475
Fiske, D. W., 422, 426, 427
Fitch, F. B., 122
Fitts, P. M., 16, 17, 18, 25, 26, 27, 28, 29, 30, 31, 32, 34, 35, 40, 45, 50, 56, 131, 159, 162, 167, 168, 170, 172, 177, 182, 183, 184, 190, 191, 192, 193, 194, 195, 196, 207, 209, 211, 212, 216, 217, 219, 220, 222, 225, 229, 267, 273, 274, 286, 289, 290, 291, 292, 296, 298, 308, 314, 343, 344, 348, 350, 351, 353, 354, 386, 387, 439, 476, 567, 626, 628
Flanagan, J. C., 415, 417
Flavell, J. H., 582, 583

Fleishman, E. A., 168, 169, 191, 414, 417, 421, 476, 477, 486, 489, 490, 492, 493, 494, 495, 628
Flexman, R. E., 441, 459
Floyd, W. F., 174, 191
Folley, J. D., Jr., 406, 408
Forbes, T. W., 169, 191
Forrin, B., 218, 222, 223, 225, 229
Foster, H., 451, 458
Fowler, F., 200, 201, 202, 204, 206, 207, 212, 213
Fox, R. H., 618, 620
Frankmann, J. P., 514, 551, 552, 563, 628
Fraser, D. C., 531, 541
Freedle, R. O., 492, 493
French, R. S., 226, 229, 406, 407, 408
French, S. H., 419, 628
Frick, F. C., 93, 96, 102, 140, 141
Friedman, M. P., 247, 250
Fruchter, D. A., 480, 485, 492, 494
Fry, G. A., 47, 55
Fryer, D. H., 441, 458

Gabb, J. E., 619, 620
Gaeuman, J. V., 606, 612
Gagné, R. M., 215, 300, 305, 387, 388, 397, 408, 439, 440, 441, 451, 452, 453, 459, 487, 489, 494
Galanter, E. H., 232, 252, 274, 322, 331, 440, 628
Gardner, J. E., 395, 397
Garner, W. R., 28, 35, 90, 91, 101, 130, 140, 141, 159, 162, 196, 212, 218, 226, 229
Garret, A. G., 272, 273
Garvey, W. D., 171, 173, 175, 193
Geldard, F. A., 28, 30, 35, 91, 288, 292
Gerard, R. W., 217, 229, 468, 475
Gerstman, L. S., 142
Ghiselli, E. E., 424, 425, 428
Gibbs, C. B., 168, 183, 191
Gibson, J. J., 527, 529
Gilbert, G. M., 188, 191
Gilbert, L. C., 106, 115
Glanzer, M., 478, 628
Glaser, R., 385, 440, 501, 505, 628, 629
Goldbeck, R. A., 404, 408
Goldman, D. F., 612
Goldscheider, A., 288, 292
Goldsmith, C. T., 606, 611
Goldsmith, R., 618, 620
Goldstein, I. L., 27, 301, 355, 515, 551, 555, 558, 564
Goldstein, L. S., 464, 466
Goldstein, M., 402, 403, 408, 454, 548
Goodacre, D. M., 432, 433
Goode, H. H., 171, 173, 191

Goodnow, R., 244, 250
Goodyear Aircraft Corp., 131
Gordon, N. B., 168
Gorham, W. A., 492, 494
Goss, A. E., 180, 191
Gottsdanker, R. M., 180, 183, 184, 189, 190, 191
Gragg, D. B., 484, 485
Graham, J. R., 530, 531, 540
Grant, D. A., 200, 212, 317, 452, 458
Gray, C. W., 245, 250
Gray, F., 446, 458
Gray, F. E., 207, 212
Greatrex, F. B., 602, 605
Grebstein, L., 245, 250
Green, D. M., 237, 238, 241, 250, 253
Green, P. E., 241, 242, 250
Greenberd, G. Z., 542, 550
Gregg, L. W., 188, 191, 221, 222, 229
Gregory, M., 126, 161, 162, 515, 542, 543, 546, 549, 550, 553, 563
Grether, W. F., 1, 28, 35, 43, 63
Grindley, G. C., 526, 529
Grinker, R. R., 589, 594
Groen, G. J., 474, 475
Grossberg, J. M., 237, 249
Grubb, R., 473, 475
Guignard, J. C., 606, 607, 612
Guilford, J. P., 415, 417
Guillemin, V., 606, 612
Guion, R. M., 420, 422, 428, 628
Gulick, W. L., 568, 574
Guth, S. K., 599, 620, 621, 624, 625
Guthrie, E. R., 186, 191, 454, 459

Haggard, E. A., 590, 595
Hake, H. W., 69, 91, 101, 140, 141, 200, 212, 224, 229
Halberg, F., 594, 595
Halbert, M. H., 241, 242, 250
Hall, M., 122
Halle, M., 102
Halsey, R. M., 93, 101
Hammer, L. R., 596, 629
Hammond, K. R., 238, 244, 245, 247, 250, 251, 252
Hampton, I. F., 618, 620
Hansel, L. E. M., 239, 249
Hanson, J. A., 46, 52, 55
Harker, G. S., 46, 52, 53, 55
Harlow, H., 452, 459
Harris, C. S., 598, 606, 607, 612, 613
Harris, E. F., 421, 628
Harris, S., 492, 493
Harris, W., 584, 591, 595
Harter, G. A., 50, 56, 344, 347, 354
Harter, H. L., 287

Hartley, H. O., 161, 162
Hartley, R. V., 141
Hartman, B. O., 170, 184, 188, 191
Hartmann, G. W., 24, 191
Houty, G. T., 174, 192
Hayes, J. R. M., 97, 98, 102
Haygood, R. L., 227, 229
Haymen, J. L., Jr., 477, 630
Hays, W. L., 235, 252, 323
Head, H., 517, 529
Hebb, D. O., 417, 551, 563
Heise, G. A., 141
Helson, H., 169, 174, 178, 192, 194, 212, 290, 292
Hempel, W. E., 169, 191, 414, 417, 489, 490, 492, 494
Hendershot, C. H., 464, 466
Hendler, E., 593, 595, 599, 627
Henneman, R. H., 187, 192
Herbert, M. J., 492, 494
Herman, L. M., 44, 69
Heron, W., 568, 574
Herrera, L., 287
Herrick, C. G., 585, 586, 595
Herrnstein, R. J., 539, 541
Hertzberg, H. T., 21, 27
Hiatt, E. P., 610, 613
Hick, W. E., 130, 141, 143, 146, 157, 174, 182, 184, 192, 272, 273
Highland, R. W., 406, 408
Hilgard, E. R., 117, 122, 385, 387, 393, 397, 525, 526, 529, 628
Hill, H., 175, 190
Hille, B., 125, 128
Hirsh, J., 495, 497, 504, 505
Hitt, W. D., 44, 56
Hixson, W. C., 344, 347, 354
Hoehn, A. J., 407, 408
Hoffman, A. C., 46, 52, 53
Hoffman, J., 568, 574
Hofstatter, P. R., 233, 250
Hogan, H. P., 100, 101
Holding, D. H., 385, 388, 389, 628
Holland, J. G., 530, 541, 551, 553, 562, 563
Holt, H. O., 418, 628
Homme, L. E., 501, 505
Hoover, G. N., 606, 607, 612
Horenstein, B. R., 174, 192
Hornick, R. J., 606, 607, 612
Hornseth, J. P., 200, 212
Horowitz, H., 492, 493
Horvath, W. J., 580, 583
Houston, R. C., 452, 459
Hovland, C. I., 122, 228, 229, 502, 505
Howe, W. H., 290, 292
Howell, W. C., 27, 44, 45, 53, 216, 272, 273, 342, 515, 551, 555, 558, 564

Howes, D. H., 78
Howland, D., 170, 192, 289, 290, 291, 292
Huddleston, H. F., 610, 612
Hull, C. L., 116, 121, 122, 174, 190, 192, 392, 397, 495, 504, 505, 527, 529
Humes, J. M., 553, 554
Humphrey, C. E., 169, 177, 192
Humphreys, L. G., 200, 212
Hunt, D. P., 215, 287
Hunt, E. B., 218, 228, 229
Hunter, W. S., 453, 459
Hursch, C. J., 244, 250
Hursch, J. L., 244, 250
Husén, T., 190, 212
Hutt, M. L., 591, 595
Hyman, L., 470, 471, 473, 475
Hyman, R., 69, 200, 212, 272, 274

Inhelder, B., 238, 250
Irion, A. I., 497, 505
Irwin, F. W., 236, 241, 242, 250, 321, 331
Ivanov-Smolensky, A. G., 526, 529

Jack, O., 141
Jacobson, H., 141
Jakobson, R., 102
James, W., 106, 115, 517, 529
Janke, M., 237, 251
Jarvik, M. E., 243, 250
Javal, L. E., 103, 115
Jeantheau, G., 351, 354
Jeffress, L. A., 192, 212
Jenkins, H. M., 228, 229, 238, 250, 253
Jenkins, W. L., 288, 291, 292
Jenkins, W. O., 287, 447, 458
Jerison, H. J., 178, 192, 515, 542, 544, 550, 551, 552, 553, 554, 562, 564, 597, 619, 629
Jerman, M., 470, 471, 473, 475
Johansson, G., 190, 212
Johnson, D. R., 603, 604, 605
Johnson, P. J., 604, 605
Johnson, S., 331
Johnston, W. A., 478, 506, 510, 515, 551, 555, 558, 564, 627, 629
Jonckheere, A. R., 72, 73
Jones, L. V., 225, 230
Jones, R. E., 287
Jung, C. G., 582, 583

Kao, Dji-Lih, 453, 459
Kalaba, R., 324, 325, 331
Kanareff, Z. T., 241, 250
Kaplan, R. J., 360, 371
Karlin, J. E., 220, 222, 223, 229, 331

Katchmar, L., 591, 595
Katzell, R. A., 425, 428, 429
Kaufman, E. L., 95, 102
Kaufman, H., 498, 505
Keller, L., 247, 250, 473, 475
Kelley, C. R., 65, 68
Kelley, G. A., 390, 397
Keller, F. S., 536, 541
Kelly, J., 324, 325, 331
Kendler, H., 452, 459
Kennedy, J. L., 46, 52, 53, 300
Keppel, G., 77
Kerslake, D. McK., 598, 614, 620
Kibler, A. W., 552, 553, 564
Kidd, J. S., 303, 306, 309, 314, 342, 343, 351, 352, 354
Kimble, G. A., 174, 192
Kimura, D., 565, 627
Kincaid, W. M., 78
Kinkade, R. G., 314
Kleinmuntz, B., 215
Kleitman, N., 594, 595
Klemmer, E. T., 91, 93, 96, 102, 273, 274, 580, 583
Koch, S., 35, 228, 250, 331
Kochen, M., 322, 331
Kopstein, F. F., 472, 475, 501, 505
Korchin, S. J., 589, 594
Kraepelin, A., 582, 583
Kraepelin, E., 146, 157
Kraft, C. L., 44, 45, 53, 343, 348, 350, 351, 352, 354
Kramer, E. E., 241, 242, 250
Kreidler, D. L., 272, 273
Kroeker, L., 237, 249
Kryter, K. D., 597, 598, 600, 603, 604, 605, 629
Külpe, O., 95, 102, 103, 115
Kurtz, A. K., 452, 459

Lacey, J. I., 415, 417
Ladd, G. T., 106, 115
Laemmel, A. E., 141
Laming, D. R. J., 130
Landauer, T. K., 160, 162
Landsdell, H., 46, 56
Langdon, J. N., 517, 530, 531, 541
Lange, K. O., 606, 607, 612
Lanzetta, J. T., 160, 162, 320, 581, 583
Laplace, P. S., 230, 251
Lashley, K. S., 182, 192, 194, 212
Lassman, F. M., 492, 493
Lathrop, R. G., 233, 251
Lawrence, C., 174, 182, 192
Lawrence, D. H., 499, 500, 505
Lawson, P. R., 497, 505
Lazar, R., 163

Lazarus, R. S., 589, 590, 591, 595
Learner, D. B., 348, 354
Lee, W., 237, 251
Lehrer, R. N., 23, 24, 25, 27
Leibermann, B., 239, 251
Lenneberg, E., 577, 583
Leonard, J. A., 168, 174, 180, 192, 194, 212, 615, 619
Lev, J., 118, 122
Lewis, D., 174, 192
Lewis, H. E., 618, 620
Lichten, W., 141
Licklider, J. C. R., 323, 331
Lilly, J., 581, 583
Lindahl, L. G., 394, 397, 420, 450, 459, 629
Lindbom, T. R., 428, 430, 431, 432, 433
Lindenbaum, L. E., 296, 298
Lindman, H., 234, 235, 236, 239, 243, 246, 251, 267, 273
Lindsley, D. B., 106, 115, 531, 541
Lintz, L. M., 236, 251
Lippert, S., 606, 612
Little, J. K., 461, 466
Little, K. B., 236, 251
Livingstone, D. H., 393, 394, 397, 420, 629
Locke, E. A., 492, 494
Lockhead, G. R., 295, 296, 298
Loeb, M., 542, 543, 546, 549, 550, 607, 613
Loftus, J. P., Jr., 596, 629
London, I. D., 188, 192
Long, E. R., 187, 192
Long, G. E., 344, 354
Lord, M. W., 95, 102
Loschie, S. H., 591, 595
Lowry, R. D., 607, 613
Luce, R. D., 161, 162, 268, 273, 274
Luckiesh, M., 621, 623, 624, 625
Lumsdaine, A. A., 440, 629
Lund, M. W., 43, 276, 298, 599, 630

McClintock, C. G., 231, 249
McConnell, T. R., 392, 397
McCormick, E. J., 276, 425, 428, 482, 485, 599, 629
McDonald, T. A., 300
Mace, C. A., 526, 529
McGehee, W., 385, 386, 387, 390, 393, 394, 395, 397, 420, 629
McGeoch, J. A., 497, 505
McGill, T. E., 568, 574
McGill, W. J., 267, 274
McGlothlin, C. L., 606, 612
McGrath, J. E., 300

McGrath, J. J., 515, 551, 552, 553, 564, 627
McGuire, J. C., 351, 352, 354
McHale, T. J., 245, 251
Machol, R. E., 171, 173, 191
MacKay, D. M., 141
McKey, M. J., 238, 253
Mackie, R. R., 584, 591, 595
MacKinney, A. C., 421, 432, 437
Mackworth, J. F., 178, 186, 192
Mackworth, N. H., 48, 55, 56, 178, 186, 192, 516, 518, 527, 529, 530, 531, 537, 538, 539, 540, 541, 543, 550, 551, 553, 564
McMillan, B., 273
Madden, J. M., 480, 481, 485
Magid, E. B., 606, 607, 608, 613
Mahler, W. R., 441, 459
Mainland, D., 287
Malecki, G. S., 28, 35
Mancini, A. R., 330
Mandel, M. J., 607, 613
Mangelsdorf, J. E., 487, 494
Mankin, D. A., 276, 292
Marcus, M., 325, 331
Margaret, A., 582, 583
Markowitz, H., 257, 266
Marks, M. R., 141
Marquis, B. G., 526, 529
Martin, H. B., 45, 55
Marx, M. H., 211
Matheny, W. G., 44, 64, 65, 68
Matthews, D. H. C., 183, 192, 288, 292
Mayfield, J. F., 236, 250, 321, 331
Meister, D., 300, 305
Mellan, I., 460, 466
Melton, A. W., 27, 35, 77, 158, 161, 169, 192, 389, 402, 408, 629
Merkel, J., 143, 144, 146, 157
Messick, D. N., 241, 242, 251
Metzger, R., 226, 227, 228, 229
Meyer, D. R., 187, 192
Middleton, D., 331
Miller, A. J., 235, 237, 240, 242, 251
Miller, G. A., 76, 77, 87, 102, 104, 115, 129, 130, 132, 140, 141, 158, 161, 217, 219, 225, 229, 539, 541
Miller, J. G., 512, 566, 574, 578, 580, 583, 589, 595
Miller, J. W., 28, 29, 35, 39
Miller, R. B., 406, 408, 487, 494
Milward, R. B., 247, 250
Minas, J. S., 74, 241, 242, 250
Monroe, J. L., 606, 612
Montague, W. E., 554, 564
Monty, R. A., 594, 631
Moore, D. E., 373, 379
Moore, J. W., 440, 630

Morant, G. M., 286
Morgan, C. T., 28, 35, 43, 276, 298, 599, 640
Morgan, R. L., 465, 467, 501, 505
Morganstern, O., 254, 263, 366
Morin, R. E., 218, 222, 223, 225, 229, 452, 458, 459
Morrison, E. J., 398, 403, 408
Morsh, J. E., 476, 478, 479, 480, 485
Moss, F. K., 621, 623, 624, 625
Mosteller, F., 254, 255, 256, 257, 266, 317, 323
Mowbray, G. H., 577, 583
Mowrer, O. H., 527, 528, 530
Mozell, M. M., 607, 613
Muller, P. F., Jr., 45, 50, 56, 273, 274, 348, 351, 353, 354
Muller, P. T., Jr., 580, 583
Murdock, B. B., 111, 115, 389, 630
Murphy, G., 451, 459
Murphy, L. B., 451, 459

Nadel, A. B., 606, 613
Nagle, B. F., 423, 425, 428
Nash, H., 232, 251
Naylor, J. C., 204, 212, 244, 245, 249, 389, 506, 507, 508, 509, 510, 630
Neibert, A., 465, 467
Neimark, E., 320, 331
Neisser, U., 130, 161, 162, 163, 164, 166, 168, 215
Neumann, J. R., 251
Newcomb, T. H., 451, 459
Newell, A., 114, 115
Newman, E. B., 141, 142
Newman, J. R., 360, 371
Newman, S. E., 406, 407, 408
Nicely, T. E., 102, 539, 541
Noble, C. E., 386, 477, 630
Noble, M. E., 13, 132, 170, 177, 183, 192, 193, 194, 195, 196, 197, 198, 199, 200, 201, 202, 204, 205, 206, 207, 209, 210, 212, 213, 214, 289, 290, 291, 292
Nogee, P., 254, 255, 256, 257, 266
Notess, C. B., 613
Novick, R., 163

O'Connell, D. N., 91, 93, 101
Oldfield, R. C., 525, 530
Ormond, E., 530, 531, 532, 536, 537, 539, 541
Ornstein, G. N., 44, 69, 492, 494
Orr, D. B., 492, 494
Osgood, C. E., 180, 192, 389, 630
Osler, S. F., 589, 590, 591, 595
Osterberg, W., 428, 430, 432, 433

Overall, J. E., 243, 251

Parke, N. G., 142
Parker, J. F., 492, 493, 494, 495
Parks, D. L., 607, 613
Parks, T. E., 237, 251
Pascal, G. R., 589, 595
Payne, R. B., 174, 192
Pearsons, K. S., 604, 605
Pepler, R. D., 614, 618, 619
Perkins, D. T., 122
Persky, H., 589, 594
Peterson, C. R., 216, 230, 235, 237, 238, 239, 240, 242, 243, 244, 245, 247, 249, 251
Peterson, J. R., 219, 229, 267, 273
Peterson, L., 77, 111, 116
Peterson, M. J., 77, 111, 116
Pew, R. W., 217
Phillips, L. D., 235, 239, 246, 250, 251, 252, 323
Phillips, L. W., 126, 128
Piaget, J., 238, 248, 250
Pickett, R. M., 515, 542, 550, 551, 553, 564, 629
Pickrel, E. W., 300
Pierce, J. R., 220, 222, 223, 229, 302
Pillsbury, W. B., 117, 122
Pitz, G. F., 232, 252
Pollack, I., 89, 90, 92, 93, 94, 97, 102, 104, 116, 161, 162, 223, 224, 229
Porter, E., 310, 314
Porter, E. H., 478, 626
Porterfield, J., 200, 201, 206, 213
Posner, M. I., 31, 35, 39, 159, 162, 195, 207, 212, 216, 217, 222, 223, 224, 226, 229, 567, 628
Postman, L., 126, 128
Poulton, E. C., 27, 28, 29, 35, 40, 131, 170, 174, 178, 179, 180, 182, 184, 192, 193, 194, 195, 196, 207, 209, 212, 213, 596, 598, 614, 616, 619, 620, 630
Pressey, S. L., 460, 461, 466
Preston, M. G., 254, 255, 256, 266
Primoff, E., 405, 408

Quastler, H., 229, 578, 583
Quigley, J., 199, 204, 213

Rabbitt, P. M., 221, 229
Rabideau, G. F., 300, 305
Radford, B. K., 273
Raiffa, H., 241, 252, 325, 326, 331
Rapoport, A., 161, 162, 247, 252, 268, 274, 580, 583

Rappaport, C., 452, 459
Ratliff, F. R., 479, 485
Ratoosh, P., 74
Rawson, H. E., 606, 613
Ray, H. W., 56
Rayman, N. N., 527, 530
Reagan, M. D., 30, 35
Reed, J. E., 477, 630
Reese, T. W., 95, 102
Reid, L. S., 533, 541
Reilly, R. E., 493, 495
Reynolds, P., 174, 190
Rhine, R. J., 305, 332
Rhoades, M. V., 577, 583
Rich, S., 492, 494
Richardson, J., 224, 225, 229
Rittenhouse, C. H., 402, 408, 454, 458
Robbins, H., 331
Robbins, J., 331
Robinson, D. W., 603, 604, 605
Robinson, E. S., 517, 530
Robinson, G. H., 232, 247, 252
Roby, T. B., 581, 583
Rock, M. L., 46, 56
Rockway, M. R., 501, 505
Rodgers, R. C., 621, 625
Rogers, M. S., 91, 93, 101
Ross, R. T., 122, 261, 266
Rubenstein, H., 140
Ruesch, J., 142
Rund, P. A., 175, 193
Ryan, T. A., 188, 193

Saltz, E., 407
Sampson, H. A., 410, 417
Samson, E. W., 142
Sanders, A. F., 126, 128, 544, 550
Sandstrom, C. I., 190, 212
Sarture, C. W., 330
Savage, L. J., 234, 236, 250, 267, 273, 325, 327, 331
Scales, E. M., 296, 298
Schapiro, H. B., 46, 48, 56
Schenck, A., 244, 245, 252
Schiffman, H., 568, 574
Schipper, L., 343, 350, 351, 352, 354, 452, 458
Schlaifer, R., 241, 252, 325, 326, 327, 331
Schmitz, M. A., 606, 607, 613
Schneider, A., 290, 291
Schneider, R. J., 170, 190, 235, 251
Schulman, A. I., 542, 550
Schultheis, P. M., 171, 173, 190
Schultz, H. G., 56
Schum, D. A., 301, 355, 356, 360, 361, 371
Scodel, J., 74

Author Index

Scopp, T. S., 234, 237, 238, 240, 249
Scott, T. H., 568, 574
Searle, L. V., 172, 174, 175, 182, 193
Seashore, H. G., 452, 459
Seeger, C. M., 274
Seidel, R. J., 472, 475
Selfridge, J. A., 141
Selfridge, O. G., 164, 168
Selye, H., 584, 585, 586, 595
Senders, V. L., 142
Senter, R. J., 465, 467
Shackel, B., 296, 298
Shannon, C. E., 50, 56, 83, 87, 142, 144, 145, 158, 177, 193, 325, 326
Shay, C. B., 477, 630
Shelly, C. H., 195, 207, 209, 212
Shelly, M., 215, 216, 343, 351, 352, 354
Shepard, R. N., 161, 162, 225, 228, 229
Sherman, H., 606, 611
Shiffrin, R. M., 474, 475
Shillestad, I. J., 501, 505
Shipley, E. F., 238, 253
Shoenberger, R. W., 598, 606, 607, 612, 613
Shuford, E. H., 69, 232, 239, 252
Siddall, G. J., 174, 193
Sidman, M. A., 387, 495, 503, 505, 536, 541, 630
Sidorsky, R. C., 45, 56, 348, 351, 354
Siegel, S., 73, 616, 620
Sieveking, N. A., 554, 563
Silberman, H. F., 501, 504
Simon, H. A., 114, 115
Simons, A. K., 606, 607, 613
Simpson, W., 232, 252
Skinner, D. F., 452, 459, 460, 467, 474, 475, 532, 533, 534, 535, 537, 539, 540, 541
Slivinske, A. J., 45, 56, 348, 351, 354
Slottow, H. G., 472, 475
Smallwood, R. D., 474, 475
Smedslund, J., 238, 244, 248, 252
Smith, E. E., 130
Smith, K. U., 446, 458
Smith, P. R., 406, 408
Smith, S. L., 99, 342, 372, 374, 379
Smith, T. A., 225, 230
Smith, W. A. S., 236, 241, 242, 250, 302, 321, 331
Smith, W. I., 440, 630
Smode, A. F., 350, 351, 352, 354
Snyder, F. W., 607, 613
Solomon, P., 565, 630
Solomon, R. L., 446, 458
Southard, J. F., 301, 355, 356, 360, 361, 371
Sowards, A., 142
Spencer, J., 233, 252

Sperling, G., 77, 102, 104, 106, 107, 111, 112, 115, 116, 125, 128, 219, 230
Spiegel, J., 276, 599, 626
Spragg, S. D. S., 46, 56
Stanley, J., 421, 627
Stenson, H. H., 553, 554, 563, 564
Sternberg, S., 130, 158, 161, 162
Stevens, S. S., 35, 122, 141, 191, 232, 252, 292
Stilson, D. W., 240, 251
Stolurow, L. M., 245, 251
Stone, M., 161, 162, 237, 252, 267, 268, 273, 274
Stopol, M. S., 591, 595
Stuntz, S. E., 452, 459
Summers, D. A., 238, 244, 245, 247, 251, 252
Summers, S. A., 244, 252
Suppes, P., 470, 471, 473, 475
Sutcliffe, M., 287
Swain, A. D., 286
Swensson, R. G., 233, 243, 249
Swets, J. A., 31, 35, 75, 76, 78, 237, 238, 241, 252, 253, 268, 274, 542, 543, 544, 547, 550, 553, 564
Swink, J., 205, 207, 213
Switzer, G., 167, 168
Sylvester, A., 117, 122

Tanner, W. D., Jr., 553, 564
Tanner, W. P., Jr., 31, 35, 75, 76, 78, 319, 320, 322
Taylor, C. D., 46, 56
Taylor, F. V., 1, 3, 16, 17, 25, 27, 28, 29, 34, 35, 38, 170, 171, 172, 173, 174, 175, 182, 184, 190, 193, 215, 308, 314
Taylor, M. M., 543, 550, 553, 564
Telford, C. W., 181, 193
Thayer, P. W., 329, 385, 386
Thomas, S., 225, 230
Thompson, J. E., 169, 177, 192
Thorndike, L. R., 422, 428, 434, 435, 436, 448, 456, 459
Thrall, R. M., 266
Thurmond, J. B., 487, 493
Thurstone, L. L., 206, 211, 213, 410, 417
Tinker, M. A., 46, 56
Tipton, L. L., 175, 193
Toda, M., 230, 243, 253
Todd, F. J., 244, 245, 250, 253
Tombrink, K. B., 482, 485
Toops, H. A., 425, 428
Touchstone, R. M., 607, 612
Trow, W. C., 500, 505
Trumbo, D., 131, 132, 194, 196, 197, 198, 199, 200, 201, 202, 204, 205, 206, 207, 210, 212, 213, 214

Trumbull, R., 512, 567, 584, 593, 595, 626
Tucker, J. A., Jr., 406, 408
Tukey, J. W., 616, 619, 620
Tune, G. S., 243, 253
Tversky, A. N., 216, 239, 253

Uhl, C. N., 244, 245, 253
Ulehla, Z. J., 237, 240, 244, 251, 253
Ulrich, L., 198, 205, 207, 210, 212, 213
Underwood, B. J., 116, 119, 122, 401, 404, 408
Usdansky, G., 582, 583

Van Cott, H. P., 492, 493
Vernon, J. A., 568, 574
Versace, J., 351, 352, 354
Vidal, J. H., 492, 493
Vince, M. A., 174, 175, 180, 181, 182, 193
Vitale, P. A., 606, 611
Volkmann, A. W., 103, 116
Volkmann, J., 95, 102
Von Gierke, H. E., 610, 612, 613
von Neumann, J., 254, 263, 266
Voss, J. F., 232, 252

Wagner, R. C., 177, 193, 194, 213
Wald, A., 241, 253, 268, 274, 325, 331
Walker, C. N., 227, 230, 300
Walker, H., 118, 122
Wallace, S. R., 422, 428
Waller, H. S., 406, 408
Wallis, R. A., 178, 192
Walter, A. A., 100, 101
Ward, W. C., 238, 250, 253, 483
Warren, C. E., 170, 183, 192, 207, 212, 344, 347, 354
Warrick, M. J., 170, 193
Waters, L. K., 439, 627
Weaver, W., 50, 56, 142, 158, 177
Webber, C. E., 554, 564
Wechsberg, P., 606, 612
Weinberg, E., 419, 630
Weiss, B., 170, 193
Weisz, A., 46, 52
Weitz, J., 422, 428
Welford, A. T., 130, 161, 162, 168, 174, 182, 191, 193, 619, 620
Wendt, G. R., 188, 190
Wertheimer, M., 219, 230
Weston, A., 17, 22, 23, 27
Weston, H. C., 624, 625
Weybrew, B. B., 593, 595

Wheaton, J. L., 565, 631
Wherry, R. J., 422, 425, 428
White, D. C., 607, 613
White, H. E., 291, 292
White, R. M., 27
White, W. J., 596
Whitehorn, J. C., 588, 595
Whiteley, B. R., 287
Whitfield, D., 296, 298
Wiener, E. L., 509, 510
Wiener, N., 142, 144, 158, 326
Wiesen, R. A., 232, 252
Wilcoxon, F., 120, 122, 287, 569, 574
Wilkerson, L. E., 65, 68
Wilkinson, E. F., 245, 250
Wilkinson, R. T., 548, 550, 618, 620
Williams, A. C., Jr., 441, 459
Williams, E. J., 25, 279, 287
Williams, J. P., 464, 467
Williams, R. J., 497, 498, 505
Willis, J. M., 44, 64
Wilson, A., 605
Wilson, C. L., 584, 591, 595
Wilson, H. A., 467, 475
Wilson, M. P., 536, 541
Wolfle, D., 394, 397, 441, 450, 453, 459
Wolpe, G., 219, 229, 267, 273
Woodbury, M. A., 142, 256
Woodhead, M. M., 530, 531, 540
Woods, J. P., 454, 458
Woodson, W. E., 43, 276
Woodworth, R. S., 97, 102, 143, 158, 194, 213, 517, 530
Wulff, J. J., 407, 408
Wulff, V. J., 578, 583
Wundt, W., 112, 116
Wyatt, S., 517, 530, 531, 541
Wyckoff, L. B., 178, 193, 533, 541

Xhignesse, L. V., 195, 197, 209, 211

Young, J. R., 604, 605
Young, R. K., 405, 408

Zajkowski, M. M., 555, 564
Zavala, A., 492, 494
Zeaman, D., 498, 505
Ziegenruecker, G. H., 606, 607, 608, 613
Zink, D. L., 215
Zinn, K. L., 471, 475
Zuercher, J. D., 551, 564

Subject Index

Acquisition
abilities, 414–416
assumptions, 400–404
classification learning, 224–228
concept learning, 226–228
generalizations for training, 392–396, 404–407
goal seeking, 394–395
individual differences, 495–505. See also Individual differences
job analysis. See Job analysis
learning curves, 386–387
learning process, 392–393
learning theory and training, 382, 384–389, 390–408
plateau, 387
principles, 388, 401–404
probability learning, 317–318
relation to transfer, 388, 416–417. See also Transfer
rote learning, 224–226
skill, 180, 197–203
variables
component practice, 453
distinctiveness of elements, 401, 403
distribution of practice, 394, 401, 404
feedback (KOR), 394, 396, 402, 454, 460
goal setting, 392–393
reinforcement, 386, 401–402, 404, 451–452, 532–540, 551, 553, 562–563
repeated practice, 402–403, 451
response availability, 401
response precision, 453–454
set, 452
Aggregation. See Probabilistic information processing
Anticipation
receptor vs. perceptual, 194
in tracking, 178–181, 194–211
Attention. See also Memory, Scanning
span, 96–98, 158
Arousal theory, 598–599. See also Vigilance theories
Automation
in system design, 41, 327–331, 355–358, 370–371, 513–515
in training, 381–382
in vigilance, 513–516

Aviation psychology laboratory, 344, 353, 355

Bayes theorem
choice reaction time, 267–269
defined, 84, 267, 358–359
dynamic decision theory, 325–327
intuitive statistical inference, 234–238
modified, 359–360
signal detection theory, 84–85
system design, 358–360
Bionics, 24
Bit, 50, 89, 133–134
Blur, 49

Cam hypothesis, 182–184
Channel capacity, 89, 137, 577–583
Choice reaction time
decision making, 267–273
defined and discussed, 128–129, 143–155
feature-testing model, 153–155
memory, 151–155, 158–164
template-matching model, 153–155
Chunk, 98
Cognitive processes. See also Acquisition
choice reaction time, 267–273
defined, 40, 214–217
in skill acquisition, 207–211
Coherence
stimulus, 195
task, 197–211
Command as system function, 336–337
"Common sense" approach, 42–43
Confusion matrix, 52
Computer-assisted instruction (CAI), 383, 440–441, 467–478
costs, 472
evaluation, 472–475. See also Training evaluation
historical foundations, 467–469
individualized instruction, 470–471, 473. See also Individual differences
tutorial systems, 469–471
Conservatism, human inference, 234–236
Contrast, 49
Control
as system function, 333–336

643

Control (cont.)
 center, 57
Controls
 coding, 276–286
 physical properties, 288–291
 vis-a-vis displays, 292–298

Decision making
 definitions, 70, 78–86, 214–217, 315–318
 diagnostic. See Probabilistic information processing
 dynamic vs. static, 315–319
 military, 327–330, 355–371
 multiman-machine systems, 302–305, 315–330, 355–371
 perception. See Signal detectability
 theories and models, 78–86, 230–248, 253–266, 315–327
 vigilance, 516, 542–550, 551, 553
Display
 criteria, 56–63, 66–68
 variables (system), 58–60, 64–66, 347–348, 372–379
Displays
 alpha-numeric, 44–45
 cathode ray tube (CRT), 45, 348, 555–556
 "inside-out" vs. "outside-in," 44, 64–68
 intelligence, 56–63
 probabilistic, 44, 69–74
 shared (group), 56–63, 372–373, 506–508
Distribution of practice, 394, 401, 404. See also Acquisition
Dynamic programming, 324

Encoding, 576–577
Engineering
 definition, 17
 history, 18–20
Engineering properties of man, 11–13
Engineering Psychologists, Society of, 4, 25
Engineering psychology
 curriculum, 23–26, 32–34
 history, 3–5, 18–22, 28–30
 science and, 13–15, 27–35, 598, 600
 technology and, 5–13, 16–35
Entropy, 50, 144–145, 159
Environmental conditions. See also Lighting; Noise; Temperature; Vibration
 defined and discussed, 511–512, 596–599
Equivocation, 50, 137, 272
Evaluation of training. See Training evaluation

Expected utility, 254–255
Expected value, 84, 254

False alarm, 82, 546, 557
Fatigue. See Overload; Stress
Feature testing, 153–155
Feedback (KOR), 394, 396, 402, 454, 460. See also Acquisition
Forgetting. See Memory

"G" forces, 596–597, 608–611
Gating, 219–228. See also Human information processing
General-principle research, 43, 274

Hick's law, 130, 143–157. See also Human information processing
Human factors engineering
 defined, 6, 29
 technology, 5
Human information processing. See also Load; Display; Speed-accuracy trade-off; Choice reaction time; Memory
 cognitive processes, 214–217
 conservation, 39, 215–228
 creation, 39, 215–228
 definitions, 38–41, 75–77, 128–132, 214–217
 discrete vs. continuous, 39, 50, 131
 gating, 219–228
 Hick's law, 130, 143–157
 overload, 278–282
 probabilistic, 215–217, 230–248, 253–266, 315–330, 355–371
 proprioceptive, 288–291
 Rapoport-Horvath formula, 580–581
 receiving, 38, 75–76, 78–96
 reduction, 39, 215–228
 sequential, 138–140
 Shannon information statistic, 579
 storing, 38, 76–77, 96–128
 transmitting, 39, 50, 129–131, 217–219
Human performance theory, 31–32

Individual differences, 383, 477, 495–505
 computer-assisted instruction, 470–471, 473
 learning
  automated teaching, 501–502
  current methods, 497–499
  implications, 496–497
  training applications, 499–500
 lighting, 599, 620–624

Individual differences (*cont.*)
  stress, 591–592
  teaching machines, 460, 465
  vigilance, 515, 523–524, 538, 548
Inference. *See also* Probabilistic information processing
  intuitive statistical, 234–248
Information. *See also* Human information processing
  average amount, 156
  decision making, 319
  measurement, 39, 50, 88–92, 132–140, 144–145
  rate of gain, 143–157
  systems model, 303
  theory (concepts), 89, 132–140, 143–157
Information processing. *See* Human information processing
Input, excessive. *See* Overload
Intellectual processes. *See* Cognitive processes; Human information processing
Interference, memory, 111, 114, 124
Intermittency hypothesis, 181–184

Job analysis, 383, 388, 405–407, 476, 478–485
  job groupings, 483–485
  methods, 479–481
  reliability and validity, 481–483
  simulators, 446–447
  task taxonomy, 476–477. *See also* Task taxonomy
  teaching machines, 463
  training, 425–427
Judgment. *See also* Inference
  absolute
    unidimensional, 89–92
    multidimensional, 92–95
  confidence, 236
  mean, 233–234
  population parameter, 234–242
  probabilistic, 230–248
  proportion, 232
  variance, 233–234
Kinesthesis, 67, 183–184
Knowledge of results (KOR). *See* Feedback; Acquisition

Layout, control panel, 292–298
Learning. *See* Acquisition
Legibility, 44–55, 378
Lighting, 620–625, 698–699
  age of observer, 624
  individual differences, 620–625
Likelihood ratio (LR), 81–84, 267–269

Linearity. *See* Transfer function
Load. *See* Overload
Luckiesh-Moss visibility meter, 621–622

Man-machine system
  air traffic control, 44–45, 342
  closed loop, 5, 171
  command and control, 332–341, 355–371
  defined, 4, 27, 37, 300
  intelligence, 327–330, 355–358
  management, 332–341
  multi-, 299–301
  open loop, 5
  PIP. *See* Probabilistic information processing
  sage, 333–337
  taxonomy, 309–312
  threat evaluation, 327–330, 355–371
  vigilance, 513–516
Memory. *See also* Choice reaction time; Perceptual-motor skill
  auditory information storage, 104–106
  basic concepts, 95, 109, 111, 114, 123, 130, 158–161
  immediate, 77, 102, 158–161
  long-term, 77
  organization in, 98–101
  short-term, 76–77, 96–101, 116–121, 123–127
  span, 96–98
  visual information storage, 104–106
Monitoring as system function, 338–339
Motor response models, 181–184

Noise, 597–598
  aircraft noise problem, 600–605
    criteria, 603–605
    sonic boom, 604
  information theory, 88, 137
  memory, 109
  signal detectability, 80
  vigilance, 543–547

Observing response
  tracking, 176–177, 185–186
  vigilance, 514, 532–540
Operations as system function, 339–340
Operations research, 301–302
Optional stopping, 241–242, 320–321, 325–327
Organismic conditions. *See also* Overload; Sensory deprivation; Stress
  defined and discussed, 511–512
Organization
  memory, 98–101
  response, 194–211

Overload, 512, 566, 574–583
  communication channel, 577–578
  information, 347, 578–583
  information flow, 577–578
  psychopathology, 581–582
  subsystems, 575–577
  tracking, fatigue, 174

Paced task, 176
Pandemonium model, 164
Parallel search (processing), 130, 164
Partial report technique, 105
Perceptual-motor skills
  defined, 40, 131, 169
  organization, 40, 194–211
  proprioceptive cues, 288–291
Plateau, 387. See also Acquisition
Practice, repeated, 402–403, 451. See also Acquisition
Prediction
  tracking, 178–181, 186
  dynamic decision making, 322
  sequential, 322
Probability
  a priori, 83
  conditional, 83
  density function, 70
  subjective, 230–248, 257–266, 315–330
Probabilistic information processing, 215–217, 230–249, 253–266, 315–327
  system design, 304, 327–330, 355–371
Problem solving, 215
Proprioception, 288–291

Random walk model, 268–273
Reaction time. See Choice reaction time
Receiver operating characteristic (ROC), 85–86
Redundancy, 138–140
Rehearsal. See Memory
Refractoriness, psychological, 181–184
Rejection
  correct (in TSD), 82
  incorrect (miss), 82
Retention. See Memory

Scanning (search), 106–109, 130
  exhaustive, 160–161
  self terminating, 160–161
  serial, 130
  shared displays, 372–379
  vigilance, 554–555, 562–563
Search. See Scanning

Self-paced task, 176
Sensory deprivation, 512, 565–566, 568–574
Sensory processes
  kinesthetic, 67, 183–184
  proprioceptive, 276, 288–291
  visual, 44–55, 104–106
  tactual, 267–286
Servo theory, 131, 171–173
Signal detectability
  decision making, 319–320
  theory (TSD), 75, 78–86, 237–238
  vigilance, 516, 542–550, 551, 553
Simulation, 20, 303–304, 312–314, 383, 439–459
  air traffic control, 343–348
  definition, 443
  evaluation, 443–444, 455–457. See also Training evaluation
  job analysis, 446–447. See also Job analysis
  learning variables, 450–455. See also Acquisition, variables
    component practice, 453
    feedback, 454
    fidelity, 304, 363–371, 439, 448–455
    reinforcement, 451–452
    repeated practice, 451
    response precision, 453–454
    set, 452
  performance improvement, 444–445
  performance measurement, 447–450
  research issues, 438–440
  threat diagnosis, 357–360
Single (limited) channel hypothesis, 123, 187–188, 211, 577
Skill. See Perceptual-motor skill; Tracking
Sonic boom, 604. See also Noise
Specific-comparison approach, 44, 274
Speed-accuracy trade-off, 216, 267–273, 378–379
S-R compatibility, 220, 274, 292–298
Stationarity, 245–248, 315–318
Storage. See Memory
Stress, 512, 566–567
  definition, 588–591
  extra-individual factors, 593
  general adaptation syndrome (GAS), 586–587
  individual differences, 591–592
  man-machine systems, 368–371
  periodicity, 594
  psychological, 587–588
  systemic, 585–587
Subitizing, 95–96, 102
Subjectively expected utility, 254–255, 315
Systems. See Man-machine systems; Systems research

Subject Index

Systems research, 299–302, 306–314. See also Displays; Controls; Human information processing
  air traffic control, 342–353
  criteria, 349–350
  methodology, 312–314, 348–351, 358–360
  PIP, 315–330, 350–371
  taxonomy, 309–312
  threat evaluation, 327–330, 355–371

Tactual coding, 276–286
Task analysis. See Job analysis
Task taxonomy, 383, 476–477, 486–495
  applications, 492–493
  methodology, 488
  need for, 486–488
  principles for training, 405–407
  task analysis, 476–477, 486–487
Teaching machines, 383, 440, 459–467
  adjunct autoinstruction, 460–461
  closed loop system, 465–466
  costs, 464
  evaluation, 463–466. See also Training evaluation
  independent study, 466, 500–502
  intrinsic programming, 461–462, 464–465
  linear programming, 462–463
Team training, 383–384, 477–478, 506–510
Temperature, 598, 614–620
Template matching, 151–153
Threat evaluation, 329–330, 357–358
Threshold, 76–87
Time sharing, 187–188
Tracking
  acceleration, 183–184
  bisensory, 187–188
  compensatory, 184
  described and defined, 39, 131, 168, 175–176
  one-dimensional, 175–184
  pursuit, 176–184
  two-dimensional, 185–186
Training, 381–510. See also Acquisition; Individual differences; Job analysis; Task taxonomy; Team training; Training apparatus; Training evaluation; Transfer of training
  purpose of, 381–383
Training apparatus, 383, 438–475. See also Computer-assisted instruction; Simulation; Teaching machines
  definition, 442–443

Training evaluation, 383
  criteria, 420–421, 422–427
    job analysis, 425–427
    measurement, 422–424
    multi-dimensionality, 424–425
    relevance of, 435–437
      objective scores, 436
      subjective scores, 436–437
  designs, 431–441
    controlled experimental study, 434
    lower level designs, 435
  levels of, 421–430
    employee on-the-job behavior, 431
    supervisor
      classroom behavior, 429–430
      on-the-job behavior, 430–431
  neglect of, 418–420
  training devices
    computer-assisted, 473–475
    simulators, 443–444, 455–457
    teaching machines, 463–466
Transfer effect, asymmetrical, 618–619
Transfer function, 171–173
Transfer of training, 382–383, 409–417. See also Acquisition
  abilities, 410–411
  concept of, 410–414
  measurement difficulties, 389
  relation to acquisition, 388
  relevance to learning theory, 416–417
  stimuli and response similarity, 388–389
Transmission, information, 50–51, 128, 136–138, 156–157

Uncertainty. See also Information
  tracking, 197–211
Utility, 254–261

Vibration, 598, 606–613
Vigilance, 513–564
  complex monitoring, 515, 553–555
  experimental variables
    briefing, 519–520, 522
    display format, 561–562
    irrelevant information, 559–560
    monitoring duration, 519–521, 543, 546
    noise, 543–547
    preliminary practice, 548
    schedules of reinforcement, 533–536
    signal rate, 544–547, 557–559
    stimulus density, 557–559, 561–562
    telephone interruptions, 519–521
  historical foundations, 514, 517–519
  individual differences, 515, 523–524, 538, 548
  man-machine relationship, 513–516

Vigilance (cont.)
  observing response, 514, 532–540
  tasks, 518–520, 533, 544–546, 555–557
  theoretical considerations
    arousal, 531, 551
    decision, 516, 542–550, 551, 553
    expectancy, 531, 534, 551, 562–563
    filtering, 551

Vigilance (cont.)
  inhibition, 518–519, 524–528, 531, 555
  reinforcement, 532–540, 551, 553, 562–563
Vision, 44–45, 104–106

Weber fraction, 233